A Survey of Computational Physics

A Survey of Computational Physics

Introductory Computational Science

Rubin H. Landau

Manuel José Páez

Cristian C. Bordeianu

PRINCETON UNIVERSITY PRESS • PRINCETON AND OXFORD

Published by Princeton University Press,
41 William Street, Princeton, New Jersey 08540
In the United Kingdom: Princeton University Press,
6 Oxford Street, Woodstock, Oxfordshire OX20 1TW

ISBN: 978-0-691-13137-5
Library of Congress Control Number: 2007061029
British Library Cataloging-in-Publication Data is available

Printed on acid-free paper. ∞
pup.princeton.edu

Printed in the United States of America

1 3 5 7 9 10 8 6 4 2

In memory of our parents

Bertha Israel Landau, Philip Landau, and Sinclitica Bordeianu

CONTENTS

PREFACE

In the decade since two of us wrote the book *Computational Physics* (CP), we have seen a good number of computational physics texts and courses come into existence. This is good. Multiple texts help define a still-developing field and provide a choice of viewpoints and focus. After perusing the existing texts, we decided that the most worthwhile contribution we could make was to extend our CP text so that it surveys many of the topics we hear about at computational science conferences. Doing that, while still keeping the level of presentation appropriate for upper-division undergraduates or beginning graduate students, was a challenge.

As we look at what we have assembled here, we see more than enough material for a full year's course (details in Chapter 1, "Computational Science Basics"). When overlapped with our new lower-division text, *A First Course in Scientific Computing*, and when combined with studies in applied mathematics and computer science, we hope to have created a path for undergraduate computational physics/science education to follow.

The ensuing decade has also strengthened our view that the physics community is well served by having CP as a prominent member of the broader *computational science and engineering* (CSE) community. This view affects our book in two ways. First, we present CP as a multidisciplinary field of study that contains elements from applied mathematics and computer science, as well as physics. Accordingly, we do not view the pages we spend on these subjects as space wasted not studying physics but rather as essential components of a multidisciplinary education. Second, we try to organize and present our materials according to the steps in the scientific problem-solving paradigm that lie at the core of CSE:

$$\text{Problem} \leftrightarrow \text{theory} \leftrightarrow \text{model} \leftrightarrow \text{method} \leftrightarrow \text{implementation} \leftrightarrow \text{assessment}.$$

This format places the subject matter in its broader context and indicates how the steps are applicable to a wider class of problems. Most importantly, educational assessments and surveys have indicated that some students learn science, mathematics, and technology better when they are presented together in context rather than as separate subjects. (To some extent, the loss of "physics time" learning math and CS is made up for by this more efficient learning approach.) Likewise, some students who may not profess interest in math or CS are motivated to learn these subjects by experiencing their practical value in physics problem solving.

Though often elegant, we view some of the new CP texts as providing more of the theory behind CP than its full and practical multidisciplinary scope. While this may be appropriate for graduate study, when we teach from our texts we advocate a learn-by-doing approach that requires students to undertake a large number of projects in which they are encouraged to make discoveries on their own. We attempt to convey that it is the students' job to solve each problem professionally, which includes understanding the computed results. We believe that this "blue-collar" approach engages and motivates the students, encompasses the fun and excitement of CP, and stimulates the students to take pride in their work.

As computers have become more powerful, it has become easier to use complete problem-solving environments like Mathematica, Maple, Matlab, and Femlab to solve scientific problems. Although these environments are often used for serious work, the algorithms and numerics are kept hidden from the user, as if in a black box. Although this may be a good environment for an experienced computational scientist, we think that if you are trying to *learn* scientific computation, then you need to look inside the black box and get your hands dirty. This is probably best done through the use of a compiled language that forces you to deal directly with the algorithm and requires you to understand the computer's storage of numbers and inner workings.

Notwithstanding our viewpoint that being able to write your own codes is important for CP, we also know how time-consuming and frustrating debugging programs can be, especially for beginners. Accordingly, rather than make the reader write all their codes from scratch, we include basic programs to modify and extend. This not only leaves time for exploration and analysis but also more realistically resembles a working environment in which one must incorporate new developments with preexisting developments of others.

The choice of Java as our prime programming language may surprise some readers who know it mainly for its prowess in Web computing (we do provide Fortran77, Fortran95, and C versions of the programs on the CD). Actually, Java is quite good for CP education since it demands proper syntax, produces useful error messages, and is consistent and intelligent in its handling of precision (which C is not). And when used as we use it, without a strong emphasis on object orientation, the syntax is not overly heavy. Furthermore, Java now runs on essentially all computer systems in an identical manner, is the most used of all programming languages, and has a universal program development environment available free from Sun [SunJ], where the square brackets refer to references in the bibliography. (Although we recommend using shells and jEdit [jEdit] for developing Java programs, many serious programmers prefer a development platform such as Eclipse [Eclipse].) This means that practitioners can work just as well at home or in the developing world. Finally, Java's speed does not appear to be an issue for educational projects with present-day fast computers. If more speed is needed, then conversion to C is straightforward, as is using the C and Fortran programs on the CD.

In addition to multilanguage codes, the CD also contains animations, visualizations, color figures, interactive Java applets, MPI and PVM codes and tutorials, and OpenDX codes. More complete versions of the programs, as well as programs left for exercises, are available to instructors from RHL.There is also a digital library version of the text containing streaming video lectures and interactive equations under development.

Specific additions to this book, not found in our earlier CP text, include chapters and appendices on visualization tools, wavelet analysis, molecular dynamics, computational fluid dynamics, MPI, and PVM. Specific subjects added to this text include shock waves, solitons, IEEE floating-point arithmetic, trial-and-error searching, matrix computing with libraries, object-oriented programming, chaotic scattering, Lyapunov coefficients, Shannon entropy, coupled predator–prey systems, advanced PDE techniques (successive overrelaxation, finite elements, Crank–Nicholson and Lax–Wendroff methods), adaptive-step size integrators, projectile motion with drag, short-time Fourier transforms, FFT, Fourier filtering, Wang–Landau simulations of thermal systems, Perlin noise, cellular automata, and waves on catenaries.

Acknowledgments

This book and the courses it is based upon could not have been created without financial support from the National Science Foundation's CCLI, EPIC, and NPACI programs, as well as from the Oregon State University Physics Department and the College of Science. Thank you all and we hope you are proud.

Immature poets imitate;
mature poets steal.
— T. S. Elliot

Our CP developments have followed the pioneering path paved with the books of Thompson, Koonin, Gould and Tobochnik, and Press *et al.*; indubitably, we have borrowed material from them and made it our own. We wish to acknowledge the many contributions provided by Hans Kowallik, who started as a student in our CP course, continued as a researcher in early Web tutorials, and has continued as an international computer journeyman. Other people have contributed in various places: Henri Jansen (early class notes), Juan Vanegas (OpenDX), Connelly Barnes (OOP and PtPlot), Phil Carter and Donna Hertel (MPI), Zlatko Dimcovic (improved codes and I/O), Joel Wetzel (improved figures and visualizations), Oscar A. Restrepo (QMCbouncer), and Justin Elser (system and software support).

It is our pleasure to acknowledge the invaluable friendship, encouragement, helpful discussions, and experiences we have had with our colleagues and students over the years. We are particularly indebted to Guillermo Avendaño-Franco, Saturo S. Kano, Bob Panoff, Guenter Schneider, Paul Fink, Melanie Johnson, Al Stetz, Jon Maestri, David McIntyre, Shashikant Phatak, Viktor Podolskiy, and Cherri Pancake. Our gratitude also goes to the reviewers Ali Eskanarian, Franz J. Vesely, and John Mintmire for their thoughtful and valuable suggestions, and to Ellen Foos of Princeton University Press for her excellent and understanding production work. Heartfelt thanks goes to Vickie Kearn, our editor at Princeton University Press, for her encouragement, insight, and efforts to keep this project alive in various ways.

In spite of everyone's best efforts, there are still errors and confusing statements for which we are responsible.

Finally, we extend our gratitude to the wives, Jan and Lucia, whose reliable support and encouragement are lovingly accepted, as always. ▌

1

Computational Science Basics

Some people spend their entire lives reading but never get beyond reading the words on the page; they don't understand that the words are merely stepping stones placed across a fast-flowing river, and the reason they're there is so that we can reach the farther shore; it's the other side that matters.

—*José Saramago*

As an introduction to the book to follow, we start this chapter with a description of how computational physics (CP) fits into the broader field of computational science, and what topics we will present as the contents of CP. We then get down to basics and examine computing languages, number representations, and programming. Related topics dealing with hardware basics are found in Chapter 14, "High-Performance Computing Hardware, Tuning, and Parallel Computing."

1.1 Computational Physics and Computational Science

This book adopts the view that CP is a subfield of computational science. This means that CP is a multidisciplinary subject combining aspects of physics, applied mathematics, and computer science (CS) (Figure 1.1), with the aim of solving realistic physics problems. Other computational sciences replace the physics with biology, chemistry, engineering, and so on, and together face grand challenge problems such as

Climate prediction	Materials science	Structural biology
Superconductivity	Semiconductor design	Drug design
Human genome	Quantum chromodynamics	Turbulence
Speech and vision	Relativistic astrophysics	Vehicle dynamics
Nuclear fusion	Combustion systems	Oil and gas recovery
Ocean science	Vehicle signature	Undersea surveillance

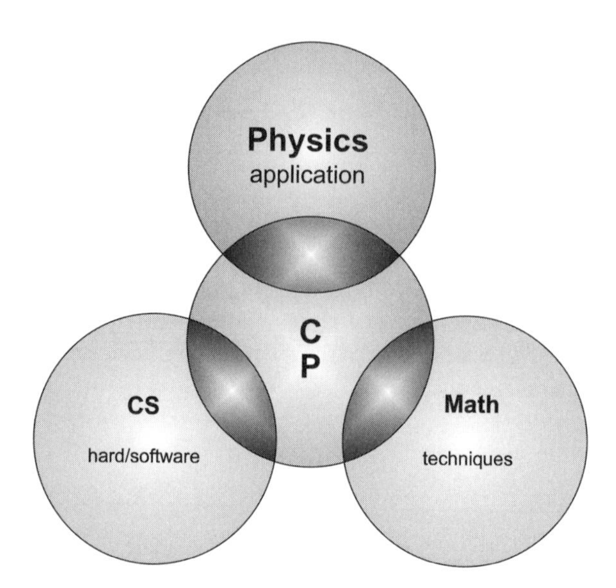

Figure 1.1 A representation of the multidisciplinary nature of computational physics both as an overlap of physics, applied mathematics, and computer science and as a bridge among the disciplines.

Although related, computational science is not computer science. Computer science studies computing for its own intrinsic interest and develops the hardware and software tools that computational scientists use. Likewise, applied mathematics develops and studies the algorithms that computational scientists use. As much as we too find math and computer science interesting for their own sakes, our focus is on solving physical problems; we need to understand the CS and math tools well enough to be able to solve our problems correctly.

As CP has matured, we have come to realize that it is more than the overlap of physics, computer science, and mathematics (Figure 1.1). It is also a bridge among them (the central region in Figure 1.1) containing core elements of it own, such as computational tools and methods. To us, CP's commonality of tools and a problem-solving mindset draws it toward the other computational sciences and away from the subspecialization found in so much of physics.

In order to emphasize our computational science focus, to the extent possible, we present the subjects in this book in the form of a problem to solve, with the components that constitute the solution separated according to the scientific problem-solving paradigm (Figure 1.2 left). Traditionally, physics employs both experimental and theoretical approaches to discover scientific truth (Figure 1.2 right). Being able to transform a theory into an algorithm requires significant theoretical insight, detailed physical and mathematical understanding, and a mastery of the art of programming. The actual debugging, testing, and organization of scientific programs is analogous to experimentation, with the numerical simulations of nature being essentially virtual experiments. The synthesis of

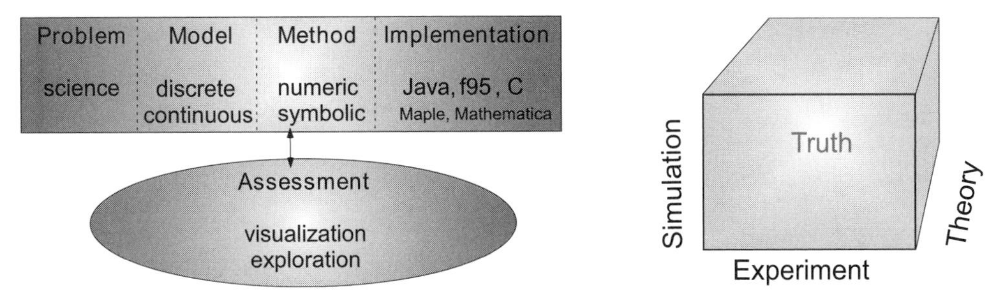

Figure 1.2 *Left:* The problem-solving paradigm followed in this book. *Right:* Simulation has been added to experiment and theory as a basic approach of science and its search for underlying truths.

numbers into generalizations, predictions, and conclusions requires the insight and intuition common to both experimental and theoretical science. In fact, the use of computation and simulation has now become so prevalent and essential a part of the scientific process that many people believe that the scientific paradigm has been extended to include simulation as an additional dimension (Figure 1.2 right).

1.2 How to Read and Use This Book

Figure 1.3 maps out the CP concepts we cover in this book and the relations among them. You may think of this concept map as the details left out of Figure 1.1. On the left are the hardware and software components from computer science; in the middle are the algorithms of applied mathematics; on the right are the physics applications. Yet because CP is multidisciplinary, it is easy to argue that certain concepts should be moved someplace else.

A more traditional way to view the materials in this text is in terms of its use in courses. In our classes [CPUG] we use approximately the first third of the text, with its emphasis on computing tools, for a course in scientific computing (after students have acquired familiarity with a compiled language). Typical topics covered in the 10 weeks of such a course are given in Table 1.1. Some options are indicated in the caption, and, depending upon the background of the students, other topics may be included or substituted. The latter two-thirds of the text includes more physics, and, indeed, we use it for a two-quarter (20-week) course in computational physics. Typical topics covered for each term are given in Table 1.2. What with many of the latter topics being research level, we suspect that these materials can easily be used for a full year's course as well.

For these materials to contribute to a successful learning experience, we assume that the reader will work through the problem at the beginning of each chapter or unit. This entails studying the text, writing, debugging and running programs, visualizing the results, and then expressing in words what has been done and what

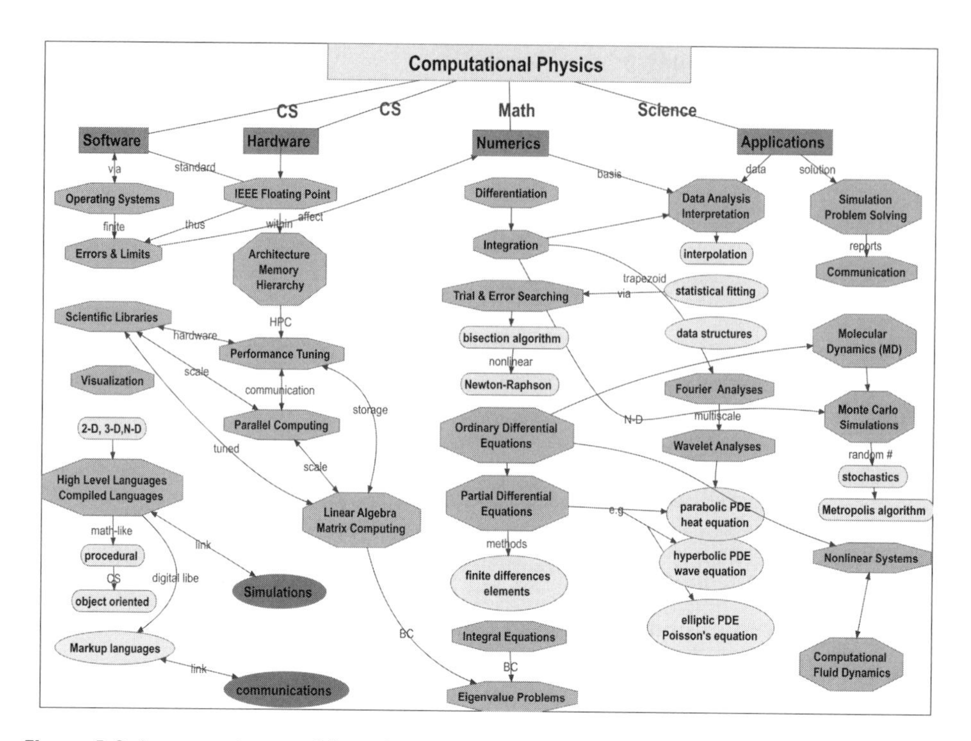

Figure 1.3 A concept map of the subjects covered in this book. The rectangular boxes indicate major areas, the angular boxes indicate subareas, and the ovals give specifics.

can be concluded. Further exploring is encouraged. Although we recognize that programming is a valuable skill for scientists, we also know that it is incredibly exacting and time-consuming. In order to lighten the workload somewhat, we provide "bare bones" programs in the text and on the CD. We recommend that these be used as guides for the reader's own programs or tested and extended to

TABLE 1.1
Topics for One Quarter (10 Weeks) of a scientific computing Course.*

Week	*Topics*	*Chapter*	*Week*	*Topics*	*Chapter*
1	OS tools, limits	1, (4)	6	Matrices, *N*-D search	8I
2	Errors, visualization	2, 3	7	Data fitting	8II
3	Monte Carlo, visualization	5, 3	8	ODE oscillations	9I
4	Integration, visualization	6, (3)	9	ODE eigenvalues	9II
5	Derivatives, searching	7I, II	10	Hardware basics	14I, III⊙

* Units are indicated by I, II, and III, and the visualization, here spread out into several laboratory periods, can be completed in one. Options: week 3 on visualization; postpone matrix computing; postpone hardware basics; devote a week to OOP; include hardware basics in week 2.

TABLE 1.2
Topics for Two Quarters (20 Weeks) of a computational Physics Course.*

Computational Physics I			*Computational Physics II*		
Week	*Topics*	*Chapter*	*Week*	*Topics*	*Chapter*
1	Nonlinear ODEs	9I, II	1	Ising model, Metropolis algorithm	15I
2	Chaotic scattering	9III	2	Molecular dynamics	16
3	Fourier analysis, filters	10I, II	3	Project completions	—
4	Wavelet analysis	11I	4	Laplace and Poisson PDEs	17I
5	Nonlinear maps	12I	5	Heat PDE	17III
6	Chaotic/double pendulum	12II	6	Waves, catenary, friction	18I
7	Project completion	12I, II	7	Shocks and solitons	19I
8	Fractals, growth	13	8	Fluid dynamics	19 II
9	Parallel computing, MPI	14II	9	Quantum integral equations	20I (II)
10	More parallel computing	14III	10	Feynman path integration	15III

*Units are indicated by I, II, and III. Options: include OpenDX visualization (§3.5, Appendix C); include multiresolution analysis (11II); include FFT (10III) in place of wavelets; include FFT (10III) in place of parallel computing; substitute Feynman path integrals (15III) for integral equations (20); add several weeks on CFD (hard); substitute coupled predator-prey (12III) for heat PDE (17III); include quantum wave packets (18II) in place of CFD; include finite element method (17II) in place of heat PDE.

solve the problem at hand. As part of this approach we suggest that the learner write up a mini lab report for each problem containing

Equations solved	Numerical method	Code listing
Visualization	Discussion	Critique

The report should be an executive summary of the type given to a boss or manager; make it clear that you understand the materials but do not waste everyone's time.

One of the most rewarding uses of computers is *visualizing* and analyzing the results of calculations with 2-D and 3-D plots, with color, and with animation. This assists in the debugging process, hastens the development of physical and mathematical intuition, and increases the enjoyment of the work. It is essential that you learn to use visualization tools as soon as possible, and so in Chapter 3, "Visualization Tools," and Appendix C we describe a number of free visualization tools that we use and recommend. We include many figures showing visualizations (unfortunately just in gray scale), with color versions on the CD.

We have tried to make the multifaceted contents of this book clearer by use of the following symbols and fonts:

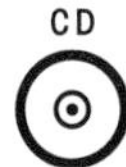

in the margin	Material on the CD
⊙	Optional material
▌ at line's end	End of exercise or problem
`Monospace font`	Words as they would appear on a computer screen
Italic font	Note at beginning of chapter to the reader about what's to follow
Sans serif font	Program commands from drop-down menus

We also indicate a user–computer dialog via three different fonts on a line:

`Monospace computer's output >` **`Bold monospace user's command`** Comments

Code listings are formatted within a shaded box, with *italic* key words and **bold** comments (usually on the right):

```
for ( i = 0; i <= Nxmax; i++ ) {                    // Comment: Fluid surface
  u[i][Nymax] = u[i][Nymax-1] + V0*h;
  w[i][Nymax-1] = 0. ;
}

   public double getI() { return (2./5.)* m * r* r; }           // Method getI
```

Note that we have tried to structure the codes so that a line is skipped before each method, so that each logical structure is indented by two spaces, and so that the ending brace } of a logical element is on a separate line aligned with the beginning of the logic element. However, in order to conserve space, sometimes we do not insert blank lines even though it may add clarity, sometimes the commands for short methods or logical structures are placed on a single line, and usually we combine multiple ending braces on the last line.

Although we try to be careful to define each term the first time it is used, we also have included a glossary in Appendix A for reference. Further, Appendix B describes the steps needed to install some of the software packages we recommend, and Appendix F lists the names and functions of the various items on the CD.

1.3 Making Computers Obey; Languages (Theory)

> Computers are incredibly fast, accurate, and stupid; humans are incredibly slow, inaccurate, and brilliant; together they are powerful beyond imagination.
>
> — *Albert Einstein*

As anthropomorphic as your view of your computer may be, keep in mind that computers always do exactly as they are told. This means that you must tell them

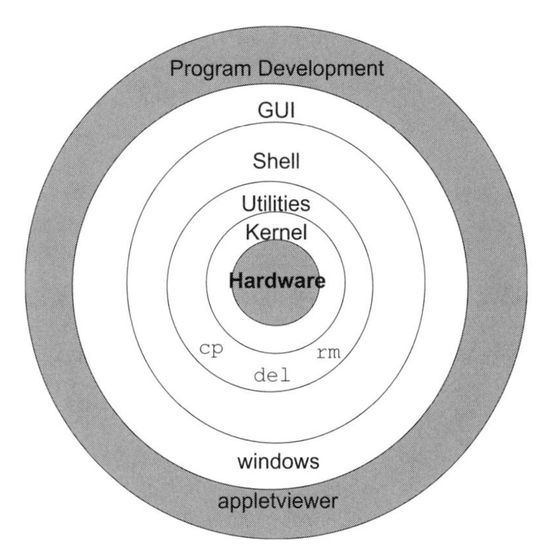

Figure 1.4 A schematic view of a computer's kernel and shells.

exactly everything they have to do. Of course the programs you run may have such convoluted logic that you may not have the endurance to figure out the details of what you have told the computer to do, but it is always possible in principle. So your first **problem** is to obtain enough understanding so that you feel well enough in control, no matter how illusionary, to figure out what the computer is doing.

Before you tell the computer to obey your orders, you need to understand that life is not simple for computers. The instructions they understand are in a *basic machine language*[1] that tells the hardware to do things like move a number stored in one memory location to another location or to do some simple binary arithmetic. Very few computational scientists talk to computers in a language computers can understand. When writing and running programs, we usually communicate to the computer through *shells*, in *high-level languages* (Java, Fortran, C), or through *problem-solving environments* (Maple, Mathematica, and Matlab). Eventually these commands or programs are translated into the basic machine language that the hardware understands.

A *shell* is a *command-line interpreter*, that is, a set of small programs run by a computer that respond to the commands (the names of the programs) that you key in. Usually you open a special window to access the shell, and this window is called a shell as well. It is helpful to think of these shells as the outer layers of the computer's operating system (OS) (Figure 1.4), within which lies a *kernel* of elementary operations. (The user seldom interacts directly with the kernel, except

[1] The Beginner's All-Purpose Symbolic Instruction Code (BASIC) programming language of the original PCs should not be confused with basic machine language.

possibly when installing programs or when building an operating system from scratch.) It is the job of the shell to run programs, compilers, and utilities that do things like copying files. There can be different types of shells on a single computer or multiple copies of the same shell running at the same time.

Operating systems have names such as *Unix, Linux, DOS, MacOS,* and *MS Windows*. The *operating system* is a group of programs used by the computer to communicate with users and devices, to store and read data, and to execute programs. Under Unix and Linux, the OS tells the computer what to do in an elementary way, while Windows includes various graphical elements as part of the operating system (this increases speed at the cost of complexity). The OS views you, other devices, and programs as input data for it to process; in many ways, it is the indispensable office manager. While all this may seem complicated, the purpose of the OS is to let the computer do the nitty-gritty work so that you can think higher-level thoughts and communicate with the computer in something closer to your normal everyday language.

When you submit a program to your computer in a *high-level language*, the computer uses a compiler to process it. The *compiler* is another program that treats your program as a foreign language and uses a built-in dictionary and set of rules to translate it into basic machine language. As you can probably imagine, the final set of instructions is quite detailed and long and the compiler may make several passes through your program to decipher your logic and translate it into a fast code. The translated statements form an *object* or compiled code, and when *linked* together with other needed subprograms, form a load module. A *load module* is a complete set of machine language instructions that can be *loaded* into the computer's memory and read, understood, and followed by the computer.

Languages such as *Fortran* and *C* use compilers to read your entire program and then translate it into basic machine instructions. Languages such as *BASIC* and *Maple* translate each line of your program as it is entered. Compiled languages usually lead to more efficient programs and permit the use of vast subprogram libraries. Interpreted languages give a more immediate response to the user and thereby appear "friendlier." The Java language is a mix of the two. When you first compile your program, it interprets it into an intermediate, universal *byte code*, but then when you run your program, it recompiles the byte code into a machine-specific compiled code.

1.4 Programming Warmup

Before we go on to serious work, we want to ensure that your local computer is working right for you. Assume that calculators have not yet been invented and that you need a program to calculate the area of a circle. Rather than use any specific language, write that program in pseudocode that can be converted to your favorite language later. The first program tells the computer:[2]

[2] Comments placed in the field to the right are for your information and *not* for the computer to act upon.

```
Calculate area of circle                                    // Do this computer!
```

This program cannot really work because it does not tell the computer which circle to consider and what to do with the area. A better program would be

```
read radius                                                 // Input
calculate area of circle                                    // Numerics
print area                                                  // Output
```

The instruction calculate area of circle has no meaning in most computer languages, so we need to specify an *algorithm*, that is, a set of rules for the computer to follow:

```
read radius                                                 // Input
PI = 3.141593                                               // Set constant
area = PI * r * r                                           // Algorithm
print area                                                  // Output
```

This is a better program, and so let's see how to implement it in Java (other language versions are on the CD). In Listing 1.1 we give a Java version of our area program. This is a simple program that outputs to the screen and has its input entered via statements.

```
//  Area.java: Area of a circle, sample program

public class Area
{
  public static void main(String[] args) {                  // Begin main method

    double radius, circum, area, PI = 3.141593;             // Declaration
    int modelN = 1;                                         // Declare, assign integer

    radius = 1.;                                            // Assign radius
    circum = 2.* PI* radius;                                // Calculate circumference
    area = radius * radius * PI;                            // Calculate area
    System.out.println("Program number = " + modelN);       // number
    System.out.println("Radius = " + radius);               // radius
    System.out.println("Circumference = " + circum);        // circum
    System.out.println("Area  = " + area);                  // area
  }                                                         // End main method
}                                                           // End Area class
/*
To Run:
>javac Area.java
>java Area
OUTPUT:
Program number = 1
Radius = 1.0
Circumference = 6.283186
Area  = 3.141593
*/
```

Listing 1.1 The program **Area.java** outputs to the screen and has its input entered via statements.

1.4.1 Structured Program Design

Programming is a written art that blends elements of science, mathematics, and computer science into a set of instructions that permit a computer to accomplish a desired task. Now that we are getting into the program-writing business, you will benefit from understanding the overall structures that you should be building into your programs, in addition to the grammar of a computer language. As with other arts, we suggest that until you know better, you follow some simple rules. A good program should

- Give the correct answers.
- Be clear and easy to read, with the action of each part easy to analyze.
- Document itself for the sake of readers and the programmer.
- Be easy to use.
- Be easy to modify and robust enough to keep giving correct answers after modifications are made.
- Be passed on to others to use and develop further.

One attraction of object-oriented programming (Chapter 4; "Object-Oriented Programs: Impedance & Batons") is that it enforces these rules automatically. An elementary way to make any program clearer is to *structure* it with indentation, skipped lines, and braces placed strategically. This is done to provide visual clues to the function of the different program parts (the "structures" in structured programming). Regardless of the fact that compilers ignore these visual clues, human readers are aided by having a program that not only looks good but also has its different logical parts visually evident. Even though the space limitations of a printed page keep us from inserting as many blank lines as we would prefer, we recommend that you do as we say and not as we do!

In Figure 1.5 we present basic and detailed *flowcharts* that illustrate a possible program for computing projectile motion. A flowchart is not meant to be a detailed description of a program but instead is a graphical aid to help visualize its logical flow. As such, it is independent of a specific computer language and is useful for developing and understanding the basic structure of a program. We recommend that you draw a flowchart or (second best) write a pseudocode before you write a program. *Pseudocode* is like a text version of a flowchart that leaves out details and instead focuses on the logic and structures:

```
Store g, Vo, and theta
Calculate R and T
Begin time loop
    Print out "not yet fired" if t < 0
    Print out "grounded" if t > T
    Calculate, print x(t) and y(t)
    Print out error message if x > R, y > H
End time loop    End program
```

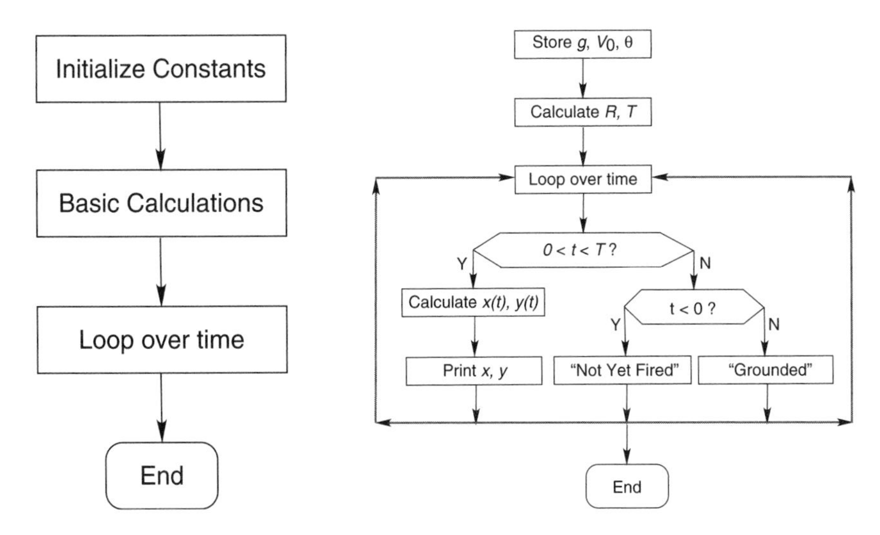

Figure 1.5 A flowchart illustrating a program to compute projectile motion. On the left are the basic components of the program, and on the right are some of its details. When writing a program, first map out the basic components, then decide upon the structures, and finally fill in the details. This is called *top-down programming.*

1.4.2 Shells, Editors, and Execution

1. To gain some experience with your computer system, use an editor to enter the program `Area.java` that computes the area of a circle (yes, we know you can copy it from the CD, but you may need some exercise before getting down to work). Then write your file to disk by saving it in your home (personal) directory (we advise having a separate subdirectory for each week). *Note:* For those who are familiar with Java, you may want to enter the program `AreaScanner.java` instead (described in a later section) that uses some recent advances in Java input/output (I/O).
2. Compile and execute the appropriate version of `Area.java`.
3. Change the program so that it computes the volume $\frac{4}{3}\pi r^3$ of a sphere. Then write your file to disk by saving it in your home (personal) directory and giving it the name `AreaMod.java`.
4. Compile and execute `AreaMod` (remember that the file name and class name must agree).
5. Check that your changes are correct by running a number of trial cases. A good input datum is $r = 1$ because then $A = \pi$. Then try $r = 10$.
6. Experiment with your program. For example, see what happens if you leave out decimal points in the assignment statement for r, if you assign r equal to a blank, or if you assign a letter to r. Remember, it is unlikely that you will "break" anything by making a mistake, and it is good to see how the computer responds when under stress.

7. Revise Area.java so that it takes input from a file name that you have made up, then writes in a different format to another file you have created, and then reads from the latter file.
8. See what happens when the data type given to output does not match the type of data in the file (e.g., data are doubles, but read in as ints).
9. Revise AreaMod so that it uses a main method (which does the input and output) and a separate method for the calculation. Check that you obtain the same answers as before.

1.4.3 Java I/O, Scanner Class with printf

In Java 1.5 and later, there is a new Scanner class that provides similar functionality as the popular scanf and printf methods in the C language. In Listing 1.2 we give a version of our area program incorporating this class. When using printf, you specify how many decimal places are desired, remembering to leave one place for the decimal point, another for the sign, and another to have some space before the next output item. As in C, there is an f for fixed-point formatting and a d for integers (digits):

```
System.out.printf("x = %6.3f, Pi = %9.6f, Age = %d %n", x, Math.PI, 39)
System.out.printf("x = %6.3f, "+" Pi = %9.6f, "+" Age = %d %n", x, Math.PI, 39)
x = 12.345, Pi = 3.142, Age = 39                                    Output from either
```

Here the %6.3f formats a double or a float to be printed in fixed-point notation using 6 places overall, with 3 places after the decimal point (this leaves 1 place for the decimal point, 1 place for the sign, and 1 space before the decimal point). The directive %9.6f has 6 digits after the decimal place and 9 overall, while %d is for integers (digits), which are written out in their entirety. The %n directive is used to indicate a new line. Other directives are

\"	double quote	\0NNN	octal value NNN	\\	backslash
\a	alert (bell)	\b	backspace	\c	no further output
\f	form feed	\n	new line	\r	carriage return
\t	horizontal tab	\v	vertical tab	%%	a single %

Notice in Listing 1.2 how we read from the keyboard, as well as from a file, and output to both screen and file. Beware that unless you first create the file Name.dat, the program will take exception because it cannot find the file.

1.4.4 I/O Redirection

Most programming environments assume that the standard (default) for input is *from* the keyboard, and for output *to* the computer screen. However, you can

```
// AreaScanner: examples of use of Scanner and printf (JDK 1.5)
import java.io.*;                                           // Standard I/O classes
import java.util.*;                                         // and scanner class

public class AreaScanner {
  public static final double PI = 3.141593;                              // Constants
  public static void main(String[] argv) throws IOException, FileNotFoundException {
  double r, A;                                                   // Declare variables
  Scanner sc1 = new Scanner(System.in);                  // Connect to standard input
  System.out.println("Key in your name and r on 1 or more lines");
  String name = sc1.next();                                            // Read String
  r = sc1.nextDouble();                                                // Read double
  System.out.printf("Hi " + name);
  System.out.printf("\n radius = " + r);
  System.out.printf("\n\n Enter new name and r in Name.dat\n");
  Scanner sc2 = new Scanner(new File("Name.dat"));                       // Open file
  System.out.printf("Hi %s\n", sc2.next());                     // Read, print line 1
  r = sc2.nextDouble();                                                // Read line 2
  System.out.printf("r = %5.1f\n", r);                                // Print line 2
  A = PI * r * r;                                                      // Computation
  System.out.printf("Done, look in A.dat\n");                         // Screen print
  PrintWriter q = new PrintWriter(new FileOutputStream("A.dat"),true);
  q.printf("r = %5.1f\n" , r);                                         // File output
  q.printf("A = %8.3f\n", A);
  System.out.printf("r = %5.1f\n" , r);                              // Screen output
  System.out.printf("A = %8.3f\n" , A);
  System.out.printf("\n\n Now key in your age as an integer\n");         // int input
  int age = sc1.nextInt();                                                // Read int
  System.out.printf(age + "years old, you don't look it!" );
  sc1.close(); sc2.close();                                           // Close inputs
 }                                                                        // End main
}                                                                        // End class
```

Listing 1.2 The program **AreaScanner.java** uses Java 1.5's **Scanner** class for input and the **printf** method for formatted output. Note how we input first from the keyboard and then from a file and how different methods are used to convert the input string to a double or an integer.

easily change that. A simple way to read from or write to a file is via *command-line redirection* from within the shell in which you are running your program:

% **java A < infile.dat** Redirect standard input

redirects standard input from the keyboard to the file infile.dat. Likewise,

% **java A > outfile.dat** Redirect standard output

redirects standard output from the screen to the file outfile.dat. Or you can put them both together for complete redirection:

% **java A < infile.dat > outfile.dat** Redirect standard I/O

1.4.5 Command-Line Input

Although we do not use it often in our sample programs, you can also input data to your program from the command line via the argument of your main method. Remember how the main method is declared with the statement void main(String[ ]

argv). Because main methods are methods, they take arguments (a *parameter list*) and return values. The word void preceding main means that no argument is returned to the command that calls main, while String[ ] argv means that the argument argv is an array (indicated by the [ ]) of the data type String. As an example, the program CommandLineArgs.java in Listing 1.3 accepts and then uses arguments from the command line

```
> java CommandLineArgs 2 1.0 TempFile
```

Here the main method is given an integer 2, a double 1.0, and a string TempFile, with the latter to be used as a file name. Note that this program is not shy about telling you what you should have done if you have forgotten to give it arguments. Further details are given as part of the documentation within the program.

```
/* CommandLineArgs.java: Accepts 2 or 3 arguments from command line, e.g.:
                java CommandLineArgs anInt aDouble [aString].
  [aString] is optional filename. See CmdLineArgsDemo on CD for full documentation
  Written by Zlatko Dimcovic */

public class CommandLineArgs {

  public static void main(String[] args) {
    int intParam = 0;                                              // Other values OK
    double doubleParam = 0.0;                                  // Defaults, args optional
    String filename = "baseName";                               // Will form/read in rest
    if (args.length == 2 || args.length == 3) {                    // Demand 2 or 3 args
      intParam     = Integer.parseInt  ( args[0] );
      doubleParam = Double.parseDouble( args[1] );
      if ( args.length == 3 ) filename = args[2];               // 3rd arg = filename
      else filename += "_i" + intParam + "_d" + doubleParam + ".dat";
    }
    else {                                            // No else, exit with instruction
      System.err.println("\n\t Usage: java CmdLineArgs intParam doubleParam [file]");
                                                   // "\n" not portable; use println()
      System.err.println("\t 1st arg must be int, 2nd double (or int),"
                         + "\n\t (optional) 3rd arg = string.\n");
      System.exit(1);
    }                                      // System.err, used to avoid accidental redirect
    System.out.println("Input arguments: intParam (1st) = " + intParam
                         + ", doubleParam (2nd) = " + doubleParam);
    if (args.length == 3) System.out.println("String input: " +filename);
    else if (args.length == 2) System.out.println("No file, use" + filename);
    else {
      System.err.println("\n\tERROR ! args.length must be 2 or 3.\n");
      System.exit(1);
    }
} }
```

Listing 1.3 The program **CommandLineArgs.java** (courtesy of Zlatko Dimcovic) demonstrates how arguments can be transferred to a main program via the command line.

1.4.6 I/O Exceptions: FileCatchThrow.java

You may have noted that the programs containing file I/O have their main methods declared with a statement of the form

```
main( String[] argv ) throws IOException
```

This is required by Java when programs deal with files. *Exceptions* occur when something goes wrong during the I/O process, such as not finding a file, trying to read past the end of a file, or interrupting the I/O process. In fact, you may get more information of this sort reported back to you by including any of these phrases:

FileNotFoundException   EOFException   InterruptedException

after the words throws IOException. As an instance, AreaScanner in Listing 1.2 contains

```
public static void main(String[] argv) throws IOException, FileNotFoundException
```

where the intermediate comma is to be noted. In this case we have added in the class (subclass) FileNotFoundException.

Dealing with I/O exceptions is important because it prevents the computer from freezing up if a file cannot be read or written. If, for example, a file is not found, then Java will create an Exception object and pass it along ("throw exception") to the program that called this main method. You will have the error message delivered to you after you issue the java command to run the main method.

```
// FileCatchThrow.java:  throw, catch IO exception

import java.io.*;

public class FileCatchThrow {

  public static void main(String[] argv)     {                              // Begin main
    double r, circum, A, PI = 3.141593;                                  // Declare, assign
    r = 2;
    circum = 2.* PI* r;                                                  // Calculate circum
    A = Math.pow(r,2) * PI;                                                  // Calculate A
    try {
      PrintWriter q = new PrintWriter(new FileOutputStream("ThrowCatch.out"), true);
      q.println("r = " + r + ", length, A = " + circum + ", " +A);}
      catch(IOException ex){ex.printStackTrace(); }                                // Catch
} }
```

Listing 1.4 FileCatchThrow.java reads from the file and handles the I/O exception.

Just how a program deals with (*catches*) the thrown exception object is beyond the level of this book, although Listing 1.4 does give an example of the try-catch construct. We see that while the declaration of the main method does not contain any statement about an exception being thrown, in its place we have a try-catch construct. The statements within the try block are executed, and if they throw an exception, it is caught by the catch statement, which prints a statement to that effect. In summary, if you use files, an appropriate throws IOException statement is required for successful compilation.

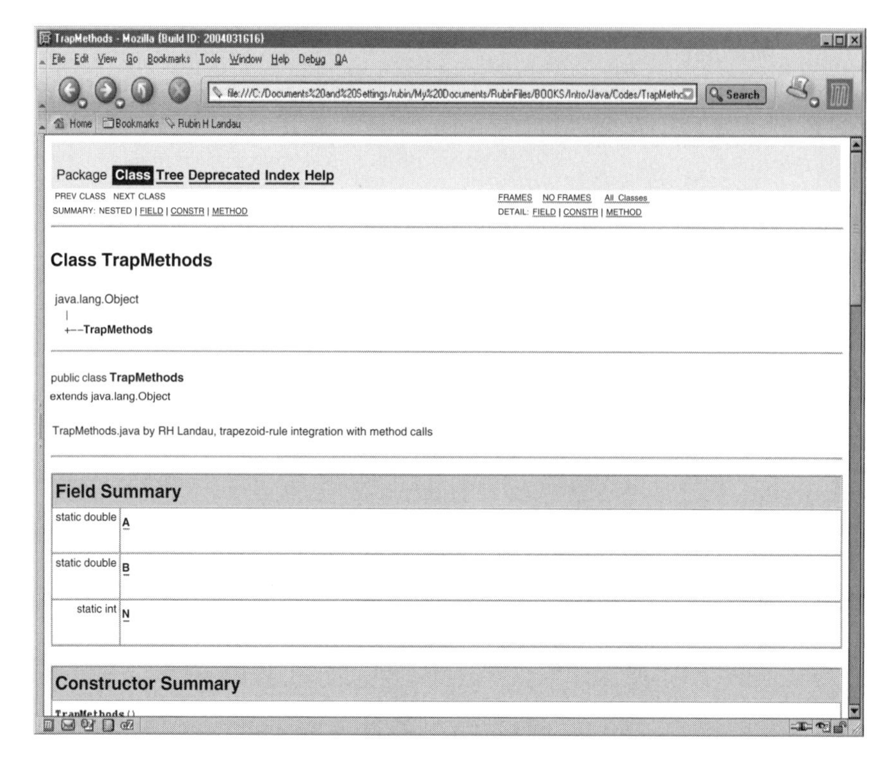

Figure 1.6 A sample of the automatic code documentation produced by running javadoc.

1.4.7 Automatic Code Documentation ⊙

A nice feature of Java is that you can place comments in your code (always a good idea) and then have Java use these comments to create a professional-looking Hypertext Markup Language (HTML) documentation page (Figure 1.6). The comments in your code must have the form

```
/**  DocDemo.java: with javadoc comments */
public class TrapMethods {
    public static final double A = 0., B = 3.;
    /** main method sums over points
    calls wTrap for trapezoid weight
    calls f(y) for integrand
    @param N number of data points
    @param A first endpoint
    @param B second endpoint     */
```

Here the comments begin with a /**, rather than the standard /*, and end with the standard /*. As usual, Java ignores the text within a comment field. For this to work, the comments must appear before the class or method that they describe and

TABLE 1.3
Tag Forms for javadoc

@author Loren Rose	Before class
@version 12.3	Before class
@parameter sum	Before method
@exception <exception name>	Before method
@return weight for trap integration	Before method
@see <class or method name>	Before method

must contain key words, such as @param. The documentation page in Figure 1.6 is named TrapMethods.html and is produced by operating on the TrapMethods.java file with the javadoc command

```
% javadoc DocDemo.java                                    Create documentation
```

Not visible in the figure are the specific definition fields produced by the @param tags. Other useful tags are given in Table 1.3.

1.5 Computer Number Representations (Theory)

Computers may be powerful, but they are finite. A problem in computer design is how to represent an arbitrary number using a finite amount of memory space and then how to deal with the limitations arising from this representation. As a consequence of computer memories being based on the magnetic or electronic realization of a spin pointing up or down, the most elementary units of computer memory are the two binary integers (*bits*) 0 and 1. This means that all numbers are stored in memory in *binary* form, that is, as long strings of zeros and ones. As a consequence, N bits can store integers in the range $[0, 2^N]$, yet because the sign of the integer is represented by the first bit (a zero bit for positive numbers), the actual range decreases to $[0, 2^{N-1}]$.

Long strings of zeros and ones are fine for computers but are awkward for users. Consequently, binary strings are converted to *octal*, *decimal*, or *hexadecimal* numbers before the results are communicated to people. Octal and hexadecimal numbers are nice because the conversion loses no precision, but not all that nice because our decimal rules of arithmetic do not work for them. Converting to decimal numbers makes the numbers easier for us to work with, but unless the number is a power of 2, the process leads to a decrease in precision.

A description of a particular computer system normally states the *word length*, that is, the number of bits used to store a number. The length is often expressed in *bytes*, with

$$1\,\text{byte} \equiv 1\,\text{B} \stackrel{\text{def}}{=} 8\,\text{bits}.$$

Memory and storage sizes are measured in bytes, kilobytes, megabytes, gigabytes, terabytes, and petabytes (10^{15}). Some care should be taken here by those who chose to compute sizes in detail because K does not always mean 1000:

$$1\,\mathrm{K} \stackrel{\text{def}}{=} 1\,\mathrm{kB} = 2^{10}\,\text{bytes} = 1024\,\text{bytes}.$$

This is often (and confusingly) compensated for when memory size is stated in K, for example,

$$512\,\mathrm{K} = 2^9\,\text{bytes} = 524,288\,\text{bytes} \times \frac{1\,\mathrm{K}}{1024\,\text{bytes}}.$$

Conveniently, 1 byte is also the amount of memory needed to store a single letter like "a", which adds up to a typical printed page requiring $\sim$3 kB.

The memory chips in some older personal computers used 8-bit words. This meant that the maximum integer was $2^7 = 128$ (7 because 1 bit is used for the sign). Trying to store a number larger than the hardware or software was designed for (*overflow*) was common on these machines; it was sometimes accompanied by an informative error message and sometimes not. Using 64 bits permits integers in the range 1–$2^{63} \simeq 10^{19}$. While at first this may seem like a large range, it really is not when compared to the range of sizes encountered in the physical world. As a case in point, the ratio of the size of the universe to the size of a proton is approximately 10^{41}.

1.5.1 IEEE Floating-Point Numbers

Real numbers are represented on computers in either *fixed-point* or *floating-point* notation. *Fixed-point notation* can be used for numbers with a fixed number of places beyond the decimal point (radix) or for integers. It has the advantages of being able to use *two's complement* arithmetic and being able to store integers exactly.[3] In the fixed-point representation with N bits and with a two's complement format, a number is represented as

$$N_{\text{fix}} = \text{sign} \times (\alpha_n 2^n + \alpha_{n-1} 2^{n-1} + \cdots + \alpha_0 2^0 + \cdots + \alpha_{-m} 2^{-m}), \tag{1.1}$$

where $n + m = N - 2$. That is, 1 bit is used to store the sign, with the remaining $(N - 1)$ bits used to store the α_i values (the powers of 2 are understood). The particular values for $N, m,$ and n are machine-dependent. Integers are typically 4 bytes (32 bits) in length and in the range

$$-2147483648 \le \text{4-B integer} \le 2147483647.$$

[3] The *two's complement* of a binary number is the value obtained by subtracting the number from 2^N for an N-bit representation. Because this system represents negative numbers by the two's complement of the absolute value of the number, additions and subtractions can be made without the need to work with the sign of the number.

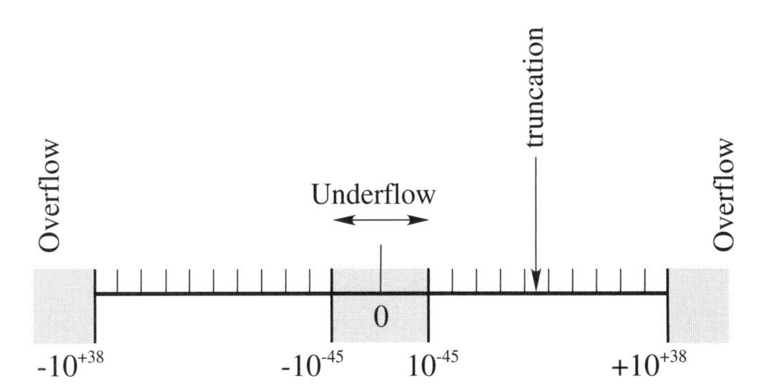

Figure 1.7 The limits of single-precision floating-point numbers and the consequences of exceeding these limits. The hash marks represent the values of numbers that can be stored; storing a number in between these values leads to truncation error. The shaded areas correspond to over- and underflow.

An advantage of the representation (1.1) is that you can count on all fixed-point numbers to have the same absolute error of 2^{-m-1} [the term left off the right-hand end of (1.1)]. The corresponding disadvantage is that *small* numbers (those for which the first string of α values are zeros) have large *relative* errors. Because in the real world relative errors tend to be more important than absolute ones, integers are used mainly for counting purposes and in special applications (like banking).

Most scientific computations use double-precision floating-point numbers ($64\,\text{b} = 8\,\text{B}$). The *floating-point representation* of numbers on computers is a binary version of what is commonly known as *scientific* or *engineering notation*. For example, the speed of light $c = +2.99792458 \times 10^{+8}$ m/s in scientific notation and $+0.299792458 \times 10^{+9}$ or 0.299795498 E09 m/s in engineering notation. In each of these cases, the number in front is called the *mantissa* and contains nine *significant figures*. The power to which 10 is raised is called the *exponent*, with the plus sign included as a reminder that these numbers may be negative.

Floating-point numbers are stored on the computer as a concatenation (juxtaposition) of the sign bit, the exponent, and the mantissa. Because only a finite number of bits are stored, the set of floating-point numbers that the computer can store exactly, *machine numbers* (the hash marks in Figure 1.7), is much smaller than the set of real numbers. In particular, machine numbers have a maximum and a minimum (the shading in Figure 1.7). If you exceed the maximum, an error condition known as *overflow* occurs; if you fall below the minimum, an error condition known as *underflow* occurs. In the latter case, the software and hardware may be set up so that underflows are set to zero without your even being told. In contrast, overflows usually halt execution.

The actual relation between what is stored in memory and the value of a floating-point number is somewhat indirect, with there being a number of special cases and relations used over the years. In fact, in the past each computer operating system and each computer language contained its own standards

TABLE 1.4
The IEEE 754 Standard for Java's Primitive Data Types

Name	*Type*	*Bits*	*Bytes*	*Range*
boolean	Logical	1	$\frac{1}{8}$	true or false
char	String	16	2	'\u0000' $\leftrightarrow$ '\uFFFF' (ISO Unicode characters)
byte	Integer	8	1	$-128 \leftrightarrow +127$
short	Integer	16	2	$-32,768 \leftrightarrow +32,767$
int	Integer	32	4	$-2,147,483,648 \leftrightarrow +2,147,483,647$
long	Integer	64	8	$-9,223,372,036,854,775,808 \leftrightarrow 9,223,372,036,854,775,807$
float	Floating	32	4	$\pm 1.401298 \times 10^{-45} \leftrightarrow \pm 3.402923 \times 10^{+38}$
double	Floating	64	8	$\pm 4.94065645841246544 \times 10^{-324} \leftrightarrow \pm 1.7976931348623157 \times 10^{+308}$

for floating-point numbers. Different standards meant that the same program running correctly on different computers could give different results. Even though the results usually were only slightly different, the user could never be sure if the lack of reproducibility of a test case was due to the particular computer being used or to an error in the program's implementation.

In 1987, the Institute of Electrical and Electronics Engineers (IEEE) and the American National Standards Institute (ANSI) adopted the IEEE 754 standard for floating-point arithmetic. When the standard is followed, you can expect the primitive data types to have the precision and ranges given in Table 1.4. In addition, when computers and software adhere to this standard, and most do now, you are guaranteed that your program will produce identical results on different computers. However, because the IEEE standard may not produce the most efficient code or the highest accuracy for a particular computer, sometimes you may have to invoke compiler options to demand that the IEEE standard be strictly followed for your test cases. After you know that the code is okay, you may want to run with whatever gives the greatest speed and precision.

There are actually a number of components in the IEEE standard, and different computer or chip manufacturers may adhere to only some of them. Normally a floating-point number x is stored as

$$\boxed{x_{\text{float}} = (-1)^s \times 1.f \times 2^{e-\text{bias}},} \tag{1.2}$$

that is, with separate entities for the sign s, the fractional part of the mantissa f, and the exponential field e. All parts are stored in binary form and occupy adjacent segments of a single 32-bit word for singles or two adjacent 32-bit words for doubles. The sign s is stored as a single bit, with $s = 0$ or 1 for a positive or a negative sign.

TABLE 1.5
Representation Scheme for Normal and Abnormal IEEE Singles

Number Name	*Values of s, e, and f*	*Value of Single*
Normal	$0 < e < 255$	$(-1)^s \times 2^{e-127} \times 1.f$
Subnormal	$e = 0, \ \ f \neq 0$	$(-1)^s \times 2^{-126} \times 0.f$
Signed zero (± 0)	$e = 0, \ \ f = 0$	$(-1)^s \times 0.0$
$+\infty$	$s = 0, \ \ e = 255, \ \ f = 0$	+INF
$-\infty$	$s = 1, \ \ e = 255, \ \ f = 0$	-INF
Not a number	$s = u, \ \ e = 255, \ \ f \neq 0$	NaN

Eight bits are used to stored the exponent e, which means that e can be in the range $0 \leq e \leq 255$. The endpoints, $e = 0$ and $e = 255$, are special cases (Table 1.5). *Normal numbers* have $0 < e < 255$, and with them the convention is to assume that the mantissa's first bit is a 1, so only the fractional part f after the *binary point* is stored. The representations for *subnormal numbers* and the special cases are given in Table 1.5.

Note that the values $\pm$INF and NaN are not numbers in the mathematical sense, that is, objects that can be manipulated or used in calculations to take limits and such. Rather, they are signals to the computer and to you that something has gone awry and that the calculation should probably stop until you straighten things out. In contrast, the value -0 can be used in a calculation with no harm. Some languages may set unassigned variables to -0 as a hint that they have yet to be assigned, though it is best not to count on that!

The IEEE representations ensure that all normal floating-point numbers have the same relative precision. Because the first bit is assumed to be 1, it does not have to be stored, and computer designers need only recall that there is a *phantom bit* there to obtain an extra bit of precision. During the processing of numbers in a calculation, the first bit of an intermediate result may become zero, but this is changed before the final number is stored. To repeat, for normal cases, the actual mantissa ($1.f$ in binary notation) contains an implied 1 preceding the binary point.

Finally, in order to guarantee that the stored biased exponent e is always positive, a fixed number called the *bias* is added to the actual exponent p before it is stored as the biased exponent e. The actual exponent, which may be negative, is

$$p = e - \text{bias}. \tag{1.3}$$

1.5.1.1 EXAMPLE: IEEE SINGLES REPRESENTATIONS

There are two basic, IEEE floating-point formats, singles and doubles. *Singles* or *floats* is shorthand for *single-precision floating-point numbers*, and *doubles* is shorthand for *double-precision floating-point numbers*. Singles occupy 32 bits overall, with 1 bit for the sign, 8 bits for the exponent, and 23 bits for the fractional mantissa (which gives 24-bit precision when the phantom bit is included). Doubles occupy 64 bits overall, with 1 bit for the sign, 10 bits for the exponent, and 53 bits for the fractional mantissa (for 54-bit precision). This means that the exponents and mantissas for doubles are not simply double those of floats, as we see in Table 1.4. (In addition, the IEEE standard also permits *extended precision* that goes beyond doubles, but this is all complicated enough without going into that right now.)

To see this scheme in action, look at the 32-bit float representing (1.2):

	s	e		f	
Bit position	31	30	23	22	0

The sign bit s is in bit position 31, the biased exponent e is in bits 30–23, and the fractional part of the mantissa f is in bits 22–0. Since 8 bits are used to store the exponent e and since $2^8 = 256$, e has the range

$$0 \le e \le 255.$$

The values $e = 0$ and 255 are special cases. With $\text{bias} = 127_{10}$, the full exponent

$$p = e_{10} - 127,$$

and, as indicated in Table 1.4, for singles has the range

$$-126 \le p \le 127.$$

The mantissa f for singles is stored as the 23 bits in positions 22–0. For *normal numbers*, that is, numbers with $0 < e < 255$, f is the fractional part of the mantissa, and therefore the actual number represented by the 32 bits is

$$\text{Normal floating-point number} = (-1)^s \times 1.f \times 2^{e-127}.$$

Subnormal numbers have $e = 0,\ f \neq 0$. For these, f is the entire mantissa, so the actual number represented by these 32 bit is

$$\text{Subnormal numbers} = (-1)^s \times 0.f \times 2^{e-126}. \tag{1.4}$$

The 23 bits m_{22}–m_0, which are used to store the mantissa of normal singles, correspond to the representation

$$\text{Mantissa} = 1.f = 1 + m_{22} \times 2^{-1} + m_{21} \times 2^{-2} + \cdots + m_0 \times 2^{-23}, \tag{1.5}$$

with $0.f$ used for subnormal numbers. The special $e = 0$ representations used to store ± 0 and $\pm\infty$ are given in Table 1.5.

To see how this works in practice (Figure 1.7), the largest positive normal floating-point number possible for a 32-bit machine has the maximum value for e (254) and the maximum value for f:

$$X_{\rm max} = 01111\ 1111\ 1111\ 1111\ 1111\ 1111\ 1111\ 111$$
$$= (0)(1111\ 1111)(1111\ 1111\ 1111\ 1111\ 1111\ 111), \tag{1.6}$$

where we have grouped the bits for clarity. After putting all the pieces together, we obtain the value shown in Table 1.4:

$$s = 0, \qquad e = 1111\ 1110 = 254, \qquad p = e - 127 = 127,$$
$$f = 1.1111\ 1111\ 1111\ 1111\ 1111\ 111 = 1 + 0.5 + 0.25 + \cdots \simeq 2,$$
$$\Rightarrow (-1)^s \times 1.f \times 2^{p=e-127} \simeq 2 \times 2^{127} \simeq 3.4 \times 10^{38}. \tag{1.7}$$

Likewise, the smallest positive floating-point number possible is subnormal ($e = 0$) with a single significant bit in the mantissa:

$$0\ 0000\ 0000\ 0000\ 0000\ 0000\ 0000\ 0000\ 001.$$

This corresponds to

$$s = 0, \qquad e = 0, \qquad p = e - 126 = -126$$
$$f = 0.0000\ 0000\ 0000\ 0000\ 0000\ 001 = 2^{-23}$$
$$\Rightarrow (-1)^s \times 0.f \times 2^{p=e-126} = 2^{-149} \simeq 1.4 \times 10^{-45} \tag{1.8}$$

In summary, single-precision (32-bit or 4-byte) numbers have six or seven decimal places of significance and magnitudes in the range

$$\boxed{1.4 \times 10^{-45} \leq \text{single precision} \leq 3.4 \times 10^{38}}$$

Doubles are stored as two 32-bit words, for a total of 64 bits (8 B). The sign occupies 1 bit, the exponent e, 11 bits, and the fractional mantissa, 52 bits:

	s		e		f		f (cont.)	
Bit position	63	62		52	51	32	31	0

As we see here, the fields are stored contiguously, with part of the mantissa f stored in separate 32-bit words. The order of these words, and whether the second word with f is the most or least significant part of the mantissa, is machine- dependent. For doubles, the bias is quite a bit larger than for singles,

$$\text{Bias} = 1111111111_2 = 1023_{10},$$

so the actual exponent $p = e - 1023$.

The bit patterns for doubles are given in Table 1.6, with the range and precision given in Table 1.4. To repeat, if you write a program with doubles, then

TABLE 1.6
Representation Scheme for IEEE Doubles

Number Name	*Values of s, e, and f*	*Value of Double*
Normal	$0 < e < 2047$	$(-1)^s \times 2^{e-1023} \times 1.f$
Subnormal	$e = 0, \;\; f \neq 0$	$(-1)^s \times 2^{-1022} \times 0.f$
Signed zero	$e = 0, \;\; f = 0$	$(-1)^s \times 0.0$
$+\infty$	$s = 0, \;\; e = 2047, \;\; f = 0$	+INF
$-\infty$	$s = 1, \;\; e = 2047, \;\; f = 0$	−INF
Not a number	$s = u, \;\; e = 2047, \;\; f \neq 0$	NaN

64 bits (8 bytes) will be used to store your floating-point numbers. Doubles have approximately 16 decimal places of precision (1 part in 2^{52}) and magnitudes in the range

$$4.9 \times 10^{-324} \leq \text{double precision} \leq 1.8 \times 10^{308}. \tag{1.9}$$

If a single-precision number x is larger than 2^{128}, a fault condition known as an *overflow* occurs (Figure 1.7). If x is smaller than 2^{-128}, an *underflow* occurs. For overflows, the resulting number x_c may end up being a machine-dependent pattern, not a number (NAN), or unpredictable. For underflows, the resulting number x_c is usually set to zero, although this can usually be changed via a compiler option. (Having the computer automatically convert underflows to zero is usually a good path to follow; converting overflows to zero may be the path to disaster.) Because the only difference between the representations of positive and negative numbers on the computer is the sign bit of one for negative numbers, the same considerations hold for negative numbers.

In our experience, *serious scientific calculations almost always require at least 64-bit (double-precision) floats*. And if you need double precision in one part of your calculation, you probably need it all over, which means double-precision library routines for methods and functions.

1.5.2 Over/Underflows Exercises

1. Consider the 32-bit single-precision floating-point number

	s	*e*		*f*	
Bit position	31	30	23	22	0
Value	0	0000 1110		1010 0000 0000 0000 0000 000	

a. What are the (binary) values for the sign s, the exponent e, and the fractional mantissa f. (*Hint:* $e_{10} = 14$.)
b. Determine decimal values for the biased exponent e and the true exponent p.
c. Show that the mantissa of A equals 1.625000.
d. Determine the full value of A.

2. Write a program to test for the **underflow** and **overflow** limits (within a factor of 2) of your computer system and of your computer language. A sample pseudocode is

```
under = 1.
over = 1.
begin do N times
          under = under/2.
          over = over * 2.
          write out: loop number, under, over
end do
```

You may need to increase N if your initial choice does not lead to underflow and overflow. (Notice that if you want to be more precise regarding the limits of your computer, you may want to multiply and divide by a number smaller than 2.)

a. Check where under- and overflow occur for single-precision floating-point numbers (floats). Give answers as decimals.
b. Check where under- and overflow occur for double-precision floating-point numbers (doubles).
c. Check where under- and overflow occur for integers. *Note:* There is no exponent stored for integers, so the smallest integer corresponds to the most negative one. To determine the largest and smallest integers, you must observe your program's output as you explicitly pass through the limits. You accomplish this by continually adding and subtracting 1. (Because integer arithmetic uses *two's complement* arithmetic, you should expect some surprises.) ▌

1.5.3 Machine Precision (Model)

A major concern of computational scientists is that the floating-point representation used to store numbers is of limited precision. In general for a 32-bit-word machine, *single-precision numbers are good to 6–7 decimal places, while doubles are good to 15–16 places*. To see how limited precision affects calculations, consider the simple computer addition of two single-precision words:

$$7 + 1.0 \times 10^{-7} = ?$$

The computer fetches these numbers from memory and stores the bit patterns

$$7 = 0\ \ 10000010\ \ 1110\ 0000\ 0000\ 0000\ 0000\ 000, \tag{1.10}$$

$$10^{-7} = 0\ \ 01100000\ \ 1101\ 0110\ 1011\ 1111\ 1001\ 010, \tag{1.11}$$

in *working registers* (pieces of fast-responding memory). Because the exponents are different, it would be incorrect to add the mantissas, and so the exponent of the smaller number is made larger while progressively decreasing the mantissa by *shifting bits* to the right (inserting zeros) until both numbers have the same exponent:

$$\begin{aligned} 10^{-7} &= 0\ \ 01100001\ \ 0110\ 1011\ 0101\ 1111\ 1100101\ (0) \\ &= 0\ \ 01100010\ \ 0011\ 0101\ 1010\ 1111\ 1110010\ (10) \qquad (1.12) \\ &\dots \\ &= 0\ \ 10000010\ \ 0000\ 0000\ 0000\ 0000\ 0000\ 000\ (0001101 \cdots 0 \\ \Rightarrow & \qquad 7 + 1.0 \times 10^{-7} = 7. \qquad (1.13) \end{aligned}$$

Because there is no room left to store the last digits, they are lost, and after all this hard work the addition just gives 7 as the answer (truncation error in Figure 1.7). In other words, because a 32-bit computer stores only 6 or 7 decimal places, it effectively ignores any changes beyond the sixth decimal place.

The preceding loss of precision is categorized by defining the *machine precision* ϵ_m as the maximum positive number that, on the computer, can be added to the number stored as 1 without changing that stored 1:

$$\boxed{1_c + \epsilon_m \stackrel{\text{def}}{=} 1_c,} \tag{1.14}$$

where the subscript c is a reminder that this is a computer representation of 1. Consequently, an arbitrary number x can be thought of as related to its floating-point representation x_c by

$$x_c = x(1 \pm \epsilon), \quad |\epsilon| \le \epsilon_m,$$

where the actual value for ϵ is not known. In other words, except for powers of 2 that are represented exactly, we should assume that all single-precision numbers contain an error in the sixth decimal place and that all doubles have an error in the fifteenth place. And as is always the case with errors, we must assume that we do not know what the error is, for if we knew, then we would eliminate it! Consequently, the arguments we put forth regarding errors are always approximate, and that is the best we can do.

1.5.4 Determine Your Machine Precision

Write a program to determine the machine precision ϵ_m of your computer system (within a factor of 2 or better). A sample pseudocode is

```
eps = 1.
begin do N times
    eps = eps/2.                                          // Make smaller
    one = 1. + eps                        // Write loop number, one, eps
 end do
```

A Java implementation is given in Listing 1.5, while a more precise one is ByteLimit.java on the instructor's CD.

```
// Limits.java: Determines machine precision

public class Limits  {

  public static void main(String[] args)  {
    final int N = 60;
    int i;
    double eps = 1., onePlusEps;
    for ( i = 0;  i < N;  i=i + 1)  {
      eps = eps/2.;
      onePlusEps = 1. + eps;
      System.out.println("onePlusEps = " +onePlusEps+", eps = "+eps);
} } }
```

Listing 1.5 The code **Limits.java** determines machine precision within a factor of 2. Note how we skip a line at the beginning of each class or method and how we align the closing brace vertically with its appropriate key word (in *italics*).

1. Determine experimentally the precision of single-precision floats.
2. Determine experimentally the precision of double-precision floats.

To print out a number in decimal format, the computer must make a conversion from its internal binary format. This not only takes time, but unless the number is a power of 2, there is a concordant loss of precision. So if you want a truly precise indication of the stored numbers, you should avoid conversion to decimals and instead print them out in octal or hexadecimal format (printf with \0NNN).

1.6 Problem: Summing Series

A classic numerical problem is the summation of a series to evaluate a function. As an example, consider the infinite series for $\sin x$:

$$\sin x = x - \frac{x^3}{3!} + \frac{x^5}{5!} - \frac{x^7}{7!} + \cdots \qquad \text{(exact)}.$$

Your **problem** is to use this series to calculate $\sin x$ for $x < 2\pi$ and $x > 2\pi$, with an absolute error in each case of less than 1 part in 10^8. While an infinite series is exact in a mathematical sense, it is not a good algorithm because we must stop summing at some point. An algorithm would be the finite sum

$$\sin x \simeq \sum_{n=1}^{N} \frac{(-1)^{n-1} x^{2n-1}}{(2n-1)!} \quad \text{(algorithm)}. \tag{1.15}$$

But how do we decide when to stop summing? (Do not even think of saying, "When the answer agrees with a table or with the built-in library function.")

1.6.1 Numerical Summation (Method)

Never mind that the algorithm (1.15) indicates that we should calculate $(-1)^{n-1}x^{2n-1}$ and then divide it by $(2n-1)!$ This is not a good way to compute. On the one hand, both $(2n-1)!$ and x^{2n-1} can get very large and cause overflows, even though their quotient may not. On the other hand, powers and factorials are very expensive (time-consuming) to evaluate on the computer. Consequently, a better approach is to use a single multiplication to relate the next term in the series to the previous one:

$$\frac{(-1)^{n-1}x^{2n-1}}{(2n-1)!} = \frac{-x^2}{(2n-1)(2n-2)} \frac{(-1)^{n-2}x^{2n-3}}{(2n-3)!}$$

$$\Rightarrow \quad n\text{th term} = \frac{-x^2}{(2n-1)(2n-2)} \times (n-1)\text{th term}. \tag{1.16}$$

While we want to ensure definite accuracy for $\sin x$, that is not so easy to do. What is easy to do is to assume that the error in the summation is approximately the last term summed (this assumes no round-off error, a subject we talk about in Chapter 2, "Errors & Uncertainties in Computations"). To obtain an absolute error of 1 part in 10^8, we then stop the calculation when

$$\left| \frac{n\text{th term}}{\text{sum}} \right| < 10^{-8}, \tag{1.17}$$

where "term" is the last term kept in the series (1.15) and "sum" is the accumulated sum of all the terms. In general, you are free to pick any tolerance level you desire, although if it is too close to, or smaller than, machine precision, your calculation may not be able to attain it. A pseudocode for performing the summation is

```
term = x, sum = x, eps = 10^(-8)                        // Initialize do
    do term = -term*x*x/((2n-1)/(2n-2));                // New wrt old
    sum = sum + term                                    // Add term
    while abs(term/sum) > eps                           // Break iteration
end do
```

1.6.2 Implementation and Assessment

1. Write a program that implements this pseudocode for the indicated x values. Present the results as a table with the headings

   x imax sum $|\text{sum} - \sin(x)|/\sin(x)$

   where `sin(x)` is the value obtained from the built-in function. The last column here is the relative error in your computation. Modify the code that sums the series in a "good way" (no factorials) to one that calculates the sum in a "bad way" (explicit factorials).
2. Produce a table as above.
3. Start with a tolerance of 10^{-8} as in (1.17).
4. Show that for sufficiently small values of x, your algorithm converges (the changes are smaller than your tolerance level) and that it converges to the correct answer.
5. Compare the number of decimal places of precision obtained with that expected from (1.17).
6. Without using the identity $\sin(x + 2n\pi) = \sin(x)$, show that there is a range of somewhat large values of x for which the algorithm converges, but that it converges to the wrong answer.
7. Show that as you keep increasing x, you will reach a regime where the algorithm does not even converge.
8. Now make use of the identity $\sin(x + 2n\pi) = \sin(x)$ to compute $\sin x$ for large x values where the series otherwise would diverge.
9. Repeat the calculation using the "bad" version of the algorithm (the one that calculates factorials) and compare the answers.
10. Set your tolerance level to a number smaller than machine precision and see how this affects your conclusions.

Beginnings are hard.

—*Chaim Potok*

2

Errors & Uncertainties in Computations

To err is human, to forgive divine.

—*Alexander Pope*

Whether you are careful or not, errors and uncertainties are a part of computation. Some errors are the ones that humans inevitably make, but some are introduced by the computer. Computer errors arise because of the limited precision with which computers store numbers or because algorithms or models can fail. Although it stifles creativity to keep thinking "error" when approaching a computation, it certainly is a waste of time, and may lead to harm, to work with results that are meaningless ("garbage") because of errors. In this chapter we examine some of the errors and uncertainties that may occur in computations. Even though we do not dwell on it, the lessons of this chapter apply to all other chapters as well.

2.1 Types of Errors (Theory)

Let us say that you have a program of high complexity. To gauge why errors should be of concern, imagine a program with the logical flow

$$\text{start} \rightarrow U_1 \rightarrow U_2 \rightarrow \cdots \rightarrow U_n \rightarrow \text{end}, \tag{2.1}$$

where each unit U might be a statement or a step. If each unit has probability p of being correct, then the joint probability P of the whole program being correct is $P = p^n$. Let us say we have a medium-sized program with $n = 1000$ steps and that the probability of each step being correct is almost one, $p \simeq 0.9993$. This means that you end up with $P \simeq \frac{1}{2}$, that is, a final answer that is as likely wrong as right (not a good way to build a bridge). The problem is that, as a scientist, you want a result that is correct—or at least in which the uncertainty is small and of known size.

Four general types of errors exist to plague your computations:

Blunders or bad theory: typographical errors entered with your program or data, running the wrong program or having a fault in your reasoning (theory), using the wrong data file, and so on. (If your blunder count starts increasing, it may be time to go home or take a break.)

Random errors: imprecision caused by events such as fluctuations in electronics, cosmic rays, or someone pulling a plug. These may be rare, but you have no control over them and their likelihood increases with running time; while you may have confidence in a 20-s calculation, a week-long calculation may have to be run several times to check reproducibility.

Approximation errors: imprecision arising from simplifying the mathematics so that a problem can be solved on the computer. They include the replacement of infinite series by finite sums, infinitesimal intervals by finite ones, and variable functions by constants. For example,

$$\sin(x) = \sum_{n=1}^{\infty} \frac{(-1)^{n-1} x^{2n-1}}{(2n-1)!} \quad \text{(exact)}$$

$$\simeq \sum_{n=1}^{N} \frac{(-1)^{n-1} x^{2n-1}}{(2n-1)!} = \sin(x) + \mathcal{E}(x, N), \quad \text{(algorithm)} \tag{2.2}$$

where $\mathcal{E}(x, N)$ is the approximation error and where in this case $\mathcal{E}$ is the series from $N+1$ to ∞. Because approximation error arises from the algorithm we use to approximate the mathematics, it is also called *algorithmic error*. For every reasonable approximation, the approximation error should decrease as N increases and vanish in the $N \to \infty$ limit. Specifically for (2.2), because the scale for N is set by the value of x, a small approximation error requires $N \gg x$. So if x and N are close in value, the approximation error will be large.

Round-off errors: imprecision arising from the finite number of digits used to store floating-point numbers. These "errors" are analogous to the uncertainty in the measurement of a physical quantity encountered in an elementary physics laboratory. The overall round-off error accumulates as the computer handles more numbers, that is, as the number of steps in a computation increases, and may cause some algorithms to become *unstable* with a rapid increase in error. In some cases, round-off error may become the major component in your answer, leading to what computer experts call *garbage*. For example, if your computer kept four decimal places, then it will store $\frac{1}{3}$ as 0.3333 and $\frac{2}{3}$ as 0.6667, where the computer has "rounded off" the last digit in $\frac{2}{3}$. Accordingly, if we ask the computer to do as simple a calculation as $2(\frac{1}{3}) - \frac{2}{3}$, it produces

$$2\left(\frac{1}{3}\right) - \frac{2}{3} = 0.6666 - 0.6667 = -0.0001 \neq 0. \tag{2.3}$$

So even though the result is small, it is not 0, and if we repeat this type of calculation millions of times, the final answer might not even be small (garbage begets garbage).

When considering the precision of calculations, it is good to recall our discussion in Chapter 1, "Computational Science Basics," of *significant figures* and of scientific notation given in your early physics or engineering classes. For computational purposes, let us consider how the computer may store the floating-point number

$$a = 11223344556677889900 = 1.12233445566778899 \times 10^{19}. \tag{2.4}$$

Because the exponent is stored separately and is a small number, we can assume that it will be stored in full precision. In contrast, some of the digits of the mantissa may be truncated. In double precision the mantissa of a will be stored in two words, the *most significant part* representing the decimal 1.12233, and the *least significant part* 44556677. The digits beyond 7 are lost. As we see below, when we perform calculations with words of fixed length, it is inevitable that errors will be introduced (at least) into the least significant parts of the words.

2.1.1 Model for Disaster: Subtractive Cancellation

A calculation employing numbers that are stored only approximately on the computer can be expected to yield only an approximate answer. To demonstrate the effect of this type of uncertainty, we model the computer representation x_c of the exact number x as

$$x_c \simeq x(1+\epsilon_x). \tag{2.5}$$

Here ϵ_x is the relative error in x_c, which we expect to be of a similar magnitude to the machine precision ϵ_m. If we apply this notation to the simple subtraction $a = b - c$, we obtain

$$\begin{aligned} a = b - c \quad &\Rightarrow \quad a_c \simeq b_c - c_c \simeq b(1+\epsilon_b) - c(1+\epsilon_c) \\ &\Rightarrow \quad \frac{a_c}{a} \simeq 1 + \epsilon_b \frac{b}{a} - \frac{c}{a}\epsilon_c. \end{aligned} \tag{2.6}$$

We see from (2.6) that the resulting error in a is essentially a weighted average of the errors in b and c, with no assurance that the last two terms will cancel. Of special importance here is to observe that the error in the answer a_c increases when we subtract two nearly equal numbers ($b \simeq c$) because then we are subtracting off the most significant parts of both numbers and leaving the error-prone least-significant parts:

$$\frac{a_c}{a} \stackrel{\text{def}}{=} 1 + \epsilon_a \simeq 1 + \frac{b}{a}(\epsilon_b - \epsilon_c) \simeq 1 + \frac{b}{a}\max(|\epsilon_b|, |\epsilon_c|). \tag{2.7}$$

This shows that even if the relative errors in b and c may cancel somewhat, they are multiplied by the large number b/a, which can significantly magnify the error. Because we cannot assume any sign for the errors, we must assume the worst [the "max" in (2.7)].

If you subtract two large numbers and end up with a small one, there will be less significance, and possibly a lot less significance, in the small one.

We have already seen an example of subtractive cancellation in the power series summation for $\sin x \simeq x - x^3/3! + \cdots$ for large x. A similar effect occurs for $e^{-x} \simeq 1 - x + x^2/2! - x^3/3! + \cdots$ for large x, where the first few terms are large but of alternating sign, leading to an almost total cancellation in order to yield the final small result. (Subtractive cancellation can be eliminated by using the identity $e^{-x} = 1/e^x$, although round-off error will still remain.)

2.1.2 Subtractive Cancellation Exercises

1. Remember back in high school when you learned that the quadratic equation

$$ax^2 + bx + c = 0 \tag{2.8}$$

has an analytic solution that can be written as either

$$x_{1,2} = \frac{-b \pm \sqrt{b^2 - 4ac}}{2a} \quad \text{or} \quad x'_{1,2} = \frac{-2c}{b \pm \sqrt{b^2 - 4ac}}. \tag{2.9}$$

Inspection of (2.9) indicates that subtractive cancellation (and consequently an increase in error) arises when $b^2 \gg 4ac$ because then the square root and its preceding term nearly cancel for one of the roots.

a. Write a program that calculates all four solutions for arbitrary values of a, b, and c.

b. Investigate how errors in your computed answers become large as the subtractive cancellation increases and relate this to the known machine precision. (*Hint*: A good test case employs $a = 1, b = 1, c = 10^{-n}$, $n = 1, 2, 3, \ldots$.)

c. Extend your program so that it indicates the most precise solutions.

2. As we have seen, subtractive cancellation occurs when summing a series with alternating signs. As another example, consider the finite sum

$$S_N^{(1)} = \sum_{n=1}^{2N} (-1)^n \frac{n}{n+1}. \tag{2.10}$$

If you sum the even and odd values of n separately, you get two sums:

$$S_N^{(2)} = -\sum_{n=1}^{N} \frac{2n-1}{2n} + \sum_{n=1}^{N} \frac{2n}{2n+1}. \tag{2.11}$$

All terms are positive in this form with just a single subtraction at the end of the calculation. Yet even this one subtraction and its resulting cancellation can be avoided by combining the series analytically to obtain

$$S_N^{(3)} = \sum_{n=1}^{N} \frac{1}{2n(2n+1)}. \tag{2.12}$$

Even though all three summations $S^{(1)}$, $S^{(2)}$, and $S^{(3)}$ are mathematically equal, they may give different numerical results.

a. Write a single-precision program that calculates $S^{(1)}$, $S^{(2)}$, and $S^{(3)}$.
b. Assume $S^{(3)}$ to be the exact answer. Make a log-log plot of the relative error *versus* the number of terms, that is, of $\log_{10} |(S_N^{(1)} - S_N^{(3)})/S_N^{(3)}|$ *versus* $\log_{10}(N)$. Start with $N = 1$ and work up to $N = 1,000,000$. (Recollect that $\log_{10} x = \ln x / \ln 10$.) The negative of the ordinate in this plot gives an approximate value for the number of significant figures.
c. See whether straight-line behavior for the error occurs in some region of your plot. This indicates that the error is proportional to a power of N.

3. In spite of the power of your trusty computer, calculating the sum of even a simple series may require some thought and care. Consider the two series

$$S^{(\text{up})} = \sum_{n=1}^{N} \frac{1}{n}, \qquad S^{(\text{down})} = \sum_{n=N}^{1} \frac{1}{n}.$$

Both series are finite as long as N is finite, and when summed analytically both give the same answer. Nonetheless, because of round-off error, the numerical value of $S^{(\text{up})}$ will not be precisely that of $S^{(\text{down})}$.

a. Write a program to calculate $S^{(\text{up})}$ and $S^{(\text{down})}$ as functions of N.
b. Make a log-log plot of $(S^{(\text{up})} - S^{(\text{down})})/(|S^{(\text{up})}| + |S^{(\text{down})}|)$ *versus* N.
c. Observe the linear regime on your graph and explain why the downward sum is generally more precise. ▮

2.1.3 Round-off Error in a Single Step

Let's start by seeing how error arises from a single division of the computer representations of two numbers:

$$a = \frac{b}{c} \Rightarrow a_c = \frac{b_c}{c_c} = \frac{b(1+\epsilon_b)}{c(1+\epsilon_c)},$$

$$\Rightarrow \frac{a_c}{a} = \frac{1+\epsilon_b}{1+\epsilon_c} \simeq (1+\epsilon_b)(1-\epsilon_c) \simeq 1+\epsilon_b-\epsilon_c,$$

$$\Rightarrow \frac{a_c}{a} \simeq 1 + |\epsilon_b| + |\epsilon_c|. \tag{2.13}$$

Here we ignore the very small ϵ^2 terms and add errors in absolute value since we cannot assume that we are fortunate enough to have unknown errors cancel each other. Because we add the errors in absolute value, this same rule holds for multiplication. Equation (2.13) is just the basic rule of error propagation from elementary laboratory work: You add the uncertainties in each quantity involved in an analysis to arrive at the overall uncertainty.

We can even generalize this model to estimate the error in the evaluation of a general function $f(x)$, that is, the difference in the value of the function evaluated at x and at x_c:

$$\mathcal{E} = \frac{f(x) - f(x_c)}{f(x)} \simeq \frac{df(x)/dx}{f(x)}(x - x_c). \tag{2.14}$$

So, for

$$f(x) = \sqrt{1+x}, \qquad \frac{df}{dx} = \frac{1}{2}\frac{1}{\sqrt{1+x}} \tag{2.15}$$

$$\Rightarrow \quad \mathcal{E} \simeq \frac{1}{2}\frac{1}{\sqrt{1+x}}(x - x_c). \tag{2.16}$$

If we evaluate this expression for $x = \pi/4$ and assume an error in the fourth place of x, we obtain a similar relative error of 1.5×10^{-4} in $\sqrt{1+x}$.

2.1.4 Round-off Error Accumulation After Many Steps

There is a useful model for approximating how round-off error accumulates in a calculation involving a large number of steps. We view the error in each step as a literal "step" in a *random walk*, that is, a walk for which each step is in a random direction. As we derive and simulate in Chapter 5, "Monte Carlo Simulations," the total distance covered in N steps of length r, is, on the average,

$$R \simeq \sqrt{N}\, r. \tag{2.17}$$

By analogy, the total relative error ϵ_{ro} arising after N calculational steps each with machine precision error ϵ_m is, on the average,

$$\epsilon_{\text{ro}} \simeq \sqrt{N}\, \epsilon_m. \tag{2.18}$$

If the round-off errors in a particular algorithm do not accumulate in a random manner, then a detailed analysis is needed to predict the dependence of the error on the number of steps N. In some cases there may be no cancellation, and the error may increase as $N\epsilon_m$. Even worse, in some recursive algorithms, where the error generation is coherent, such as the upward recursion for Bessel functions, the error increases as $N!$.

Our discussion of errors has an important implication for a student to keep in mind before being impressed by a calculation requiring hours of supercomputer time. A fast computer may complete 10^{10} floating-point operations per second. This means that a program running for 3 h performs approximately 10^{14} operations. Therefore, if round-off error accumulates randomly, after 3 h we expect a relative error of $10^7\epsilon_m$. For the error to be smaller than the answer, we need $\epsilon_m < 10^{-7}$, which requires double precision and a good algorithm. If we want a higher-precision answer, then we will need a very good algorithm.

2.2 Errors in Spherical Bessel Functions (Problem)

Accumulating round-off errors often limits the ability of a program to calculate accurately. Your **problem** is to compute the spherical Bessel and Neumann functions $j_l(x)$ and $n_l(x)$. These function are, respectively, the regular/irregular (nonsingular/singular at the origin) solutions of the differential equation

$$x^2 f''(x) + 2x f'(x) + \left[x^2 - l(l+1)\right] f(x) = 0. \tag{2.19}$$

The spherical Bessel functions are related to the Bessel function of the first kind by $j_l(x) = \sqrt{\pi/2x}J_{n+1/2}(x)$. They occur in many physical problems, such as the expansion of a plane wave into spherical partial waves,

$$e^{i\mathbf{k}\cdot\mathbf{r}} = \sum_{l=0}^{\infty} i^l \,(2l+1) j_l(kr)\, P_l(\cos\,\theta). \tag{2.20}$$

Figure 2.1 shows what the first few j_l look like, and Table 2.1 gives some explicit values. For the first two l values, explicit forms are

$$j_0(x) = +\frac{\sin\,x}{x}, \qquad j_1(x) = +\frac{\sin\,x}{x^2} - \frac{\cos\,x}{x} \tag{2.21}$$

$$n_0(x) = -\frac{\cos\,x}{x}. \qquad n_1(x) = -\frac{\cos\,x}{x^2} - \frac{\sin\,x}{x}. \tag{2.22}$$

2.2.1 Numerical Recursion Relations (Method)

The classic way to calculate $j_l(x)$ would be by summing its power series for small values of x/l and summing its asymptotic expansion for large x values. The approach we adopt is based on the *recursion relations*

$$j_{l+1}(x) = \frac{2l+1}{x} j_l(x) - j_{l-1}(x), \quad \text{(up)}, \tag{2.23}$$

$$j_{l-1}(x) = \frac{2l+1}{x} j_l(x) - j_{l+1}(x), \quad \text{(down)}. \tag{2.24}$$

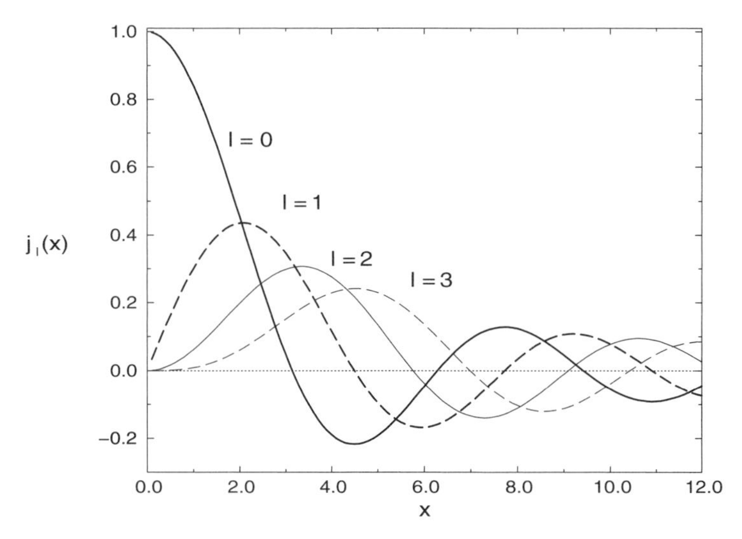

Figure 2.1 The first four spherical Bessel functions $j_l(x)$ as functions of x. Notice that for small x, the values for increasing l become progressively smaller.

TABLE 2.1
Approximate Values for Spherical Bessel Functions of Orders 3, 5, and 8 (from Maple)

x	$j_3(x)$	$j_5(x)$	$j_8(x)$
0.1	$+9.518519719\,10^{-6}$	$+9.616310231\,10^{-10}$	$+2.901200102\,10^{-16}$
1	$+9.006581118\,10^{-3}$	$+9.256115862\,10^{-05}$	$+2.826498802\,10^{-08}$
10	$-3.949584498\,10^{-2}$	$-5.553451162\,10^{-01}$	$+1.255780236\,10^{+00}$

Equations (2.23) and (2.24) are the same relation, one written for upward recurrence from small to large l values, and the other for downward recurrence to small l values. With just a few additions and multiplications, recurrence relations permit rapid, simple computation of the entire set of j_l values for fixed x and all l.

To recur upward in l for fixed x, we start with the known forms for j_0 and j_1 (2.21) and use (2.23). As you will prove for yourself, this upward recurrence usually seems to work at first but then fails. The reason for the failure can be seen from the plots of $j_l(x)$ and $n_l(x)$ *versus* x (Figure 2.1). If we start at $x \simeq 2$ and $l = 0$, we will see that as we recur j_l up to larger l values with (2.23), we are essentially taking the difference of two "large" functions to produce a "small" value for j_l. This process suffers from subtractive cancellation and always reduces the precision. As we continue recurring, we take the difference of two small functions with large errors and produce a smaller function with yet a larger error. After a while, we are left with only round-off error (garbage).

To be more specific, let us call $j_l^{(c)}$ the numerical value we compute as an approximation for $j_l(x)$. Even if we start with pure j_l, after a short while the computer's lack of precision effectively mixes in a bit of $n_l(x)$:

$$j_l^{(c)} = j_l(x) + \epsilon n_l(x). \tag{2.25}$$

This is inevitable because both j_l and n_l satisfy the same differential equation and, on that account, the same recurrence relation. The admixture of n_l becomes a problem when the numerical value of $n_l(x)$ is much larger than that of $j_l(x)$ because even a minuscule amount of a very large number may be large. In contrast, if we use the upward recurrence relation (2.23) to produce the spherical Neumann function n_l, there will be no problem because we are combining small functions to produce larger ones (Figure 2.1), a process that does not contain subtractive cancellation.

The simple solution to this problem (*Miller's device*) is to use (2.24) for downward recursion of the j_l values starting at a large value $l = L$. This avoids subtractive cancellation by taking small values of $j_{l+1}(x)$ and $j_l(x)$ and producing a larger $j_{l-1}(x)$ by addition. While the error may still behave like a Neumann function, the actual magnitude of the error will *decrease* quickly as we move downward to smaller l values. In fact, if we start iterating downward with arbitrary values for $j_{L+1}^{(c)}$ and $j_L^{(c)}$, after a short while we will arrive at the correct l dependence for this value of x. Although the numerical value of $j_0^{(c)}$ so obtained will not be correct because it depends upon the arbitrary values assumed for $j_{L+1}^{(c)}$ and $j_L^{(c)}$, the relative values will be accurate. The absolute values are fixed from the know value (2.21), $j_0(x) = \sin x/x$. Because the recurrence relation is a linear relation between the j_l values, we need only normalize all the computed values via

$$j_l^{\text{normalized}}(x) = j_l^{\text{compute}}(x) \times \frac{j_0^{\text{analytic}}(x)}{j_0^{\text{compute}}(x)}. \tag{2.26}$$

Accordingly, after you have finished the downward recurrence, you obtain the final answer by normalizing all $j_l^{(c)}$ values based on the known value for j_0.

2.2.2 Implementation and Assessment: Recursion Relations

A program implementing recurrence relations is most easily written using subscripts. If you need to polish up on your skills with subscripts, you may want to study our program `Bessel.java` in Listing 2.1 before writing your own.

1. Write a program that uses both upward and downward recursion to calculate $j_l(x)$ for the first 25 l values for $x = 0.1, 1, 10$.
2. Tune your program so that at least one method gives "good" values (meaning a relative error $\simeq 10^{-10}$). See Table 2.1 for some sample values.
3. Show the convergence and stability of your results.

4. Compare the upward and downward recursion methods, printing out l, $j_l^{(\mathrm{up})}$, $j_l^{(\mathrm{down})}$, and the relative difference $|j_l^{(\mathrm{up})} - j_l^{(\mathrm{down})}|/|j_l^{(\mathrm{up})}| + |j_l^{(\mathrm{down})}|$.
5. The errors in computation depend on x, and for certain values of x, both up and down recursions give similar answers. Explain the reason for this.

```
// Bessel.java: Spherical Bessels via up and down recursion

import java.io.*;

public class Bessel {                                          // Global class variables
  public static double xmax = 40., xmin = 0.25, step = 0.1;
  public static int order = 10, start = 50;

  public static void main(String[] argv) throws IOException, FileNotFoundException {
    double x;
    PrintWriter w = new PrintWriter(new FileOutputStream("Bessel.dat"), true);
    for (x = xmin; x <= xmax; x += step) w.println(" " +x+" "+down(x,order,start));
    System.out.println("data stored in Bessel.dat");
  }                                                                    // End main

  public static double down (double x, int n, int m) {                 // Recur down
   double scale, j[] = new double[start + 2];
   int k;
   j[m + 1] = j[m] = 1.;                                       // Start with anything
   for ( k = m; k>0 ; k--)   j[k-1] = ((2.*k+1.)/x)*j[k] - j[k+1];
   scale = (Math.sin(x)/x)/j[0];                      // Scale solution to known j[0]
   return j[n] * scale;
  }
}
```

Listing 2.1 Bessel.java determines spherical Bessel functions by downward recursion (you should modify this to also work by upward recursion).

2.3 Experimental Error Investigation (Problem)

Numerical algorithms play a vital role in computational physics. Your **problem** is to take a general algorithm and decide

1. Does it converge?
2. How precise are the converged results?
3. How expensive (time-consuming) is it?

On first thought you may think, "What a dumb problem! All algorithms converge if enough terms are used, and if you want more precision, then use more terms." Well, some algorithms may be asymptotic expansions that just approximate a function in certain regions of parameter space and converge only up to a point. Yet even if a uniformly convergent power series is used as the algorithm, including more terms will decrease the algorithmic error but increase the round-off errors. And because round-off errors eventually diverge to infinity, the best we can hope for is a "best" approximation. Good algorithms are good not only because they are fast but also because being fast means that round-off error does not have much time to grow.

Let us assume that an algorithm takes N steps to find a good answer. As a rule of thumb, the approximation (algorithmic) error decreases rapidly, often as the inverse power of the number of terms used:

$$\epsilon_{\text{approx}} \simeq \frac{\alpha}{N^{\beta}}. \tag{2.27}$$

Here α and β are empirical constants that change for different algorithms and may be only approximately constant, and even then only as $N \to \infty$. The fact that the error must fall off for large N is just a statement that the algorithm converges.

In contrast to this algorithmic error, round-off error tends to grow slowly and somewhat randomly with N. If the round-off errors in each step of the algorithm are not correlated, then we know from previous discussion that we can model the accumulation of error as a random walk with step size equal to the machine precision ϵ_m:

$$\epsilon_{\text{ro}} \simeq \sqrt{N}\epsilon_m. \tag{2.28}$$

This is the slow growth with N that we expect from round-off error. The total error in a computation is the sum of the two types of errors:

$$\boxed{\epsilon_{\text{tot}} = \epsilon_{\text{approx}} + \epsilon_{\text{ro}}} \tag{2.29}$$

$$\epsilon_{\text{tot}} \simeq \frac{\alpha}{N^{\beta}} + \sqrt{N}\epsilon_m. \tag{2.30}$$

For small N we expect the first term to be the larger of the two but ultimately to be overcome by the slowly growing round-off error.

As an example, in Figure 2.2 we present a log-log plot of the relative error in numerical integration using the Simpson integration rule (Chapter 6, "Integration"). We use the $\log_{10}$ of the relative error because its negative tells us the number of decimal places of precision obtained.[1] As a case in point, let us assume $\mathcal{A}$ is the exact answer and $A(N)$ the computed answer. If

$$\frac{\mathcal{A} - A(N)}{\mathcal{A}} \simeq 10^{-9}, \quad \text{then} \quad \log_{10}\left|\frac{\mathcal{A} - A(N)}{\mathcal{A}}\right| \simeq -9. \tag{2.31}$$

We see in Figure 2.2 that the error does show a rapid decrease for small N, consistent with an inverse power law (2.27). In this region the algorithm is converging. As N is increased, the error starts to look somewhat erratic, with a slow increase on the average. In accordance with (2.29), in this region round-off error has grown larger than the approximation error and will continue to grow for increasing N. Clearly

[1] Most computer languages use $\ln x = \log_e x$. Yet since $x = a^{\log_a x}$, we have $\log_{10} x = \ln x / \ln 10$.

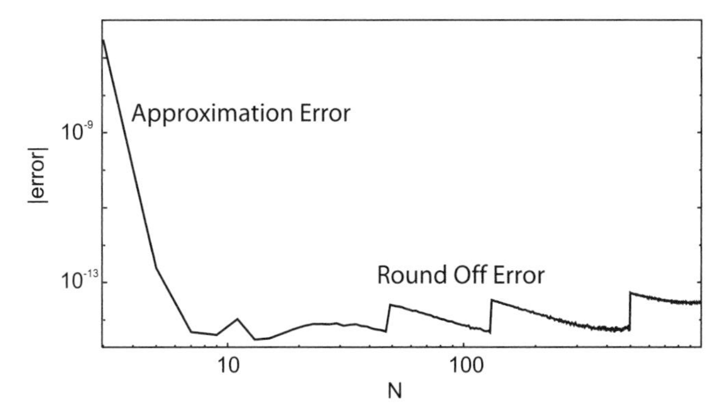

Figure 2.2 A log-log plot of relative error *versus* the number of points used for a numerical integration. The ordinate value of $\simeq 10^{-14}$ at the minimum indicates that ~ 14 decimal places of precision are obtained before round-off error begins to build up. Notice that while the round-off error does fluctuate, on the average it increases slowly.

then, the smallest total error will be obtained if we can stop the calculation at the minimum near 10^{-14}, that is, when $\epsilon_{\text{approx}} \simeq \epsilon_{ro}$.

In realistic calculations you would not know the exact answer; after all, if you did, then why would you bother with the computation? However, you may know the exact answer for a similar calculation, and you can use that similar calculation to perfect your numerical technique. Alternatively, now that you understand how the total error in a computation behaves, you should be able to look at a table or, better yet, a graph (Figure 2.2) of your answer and deduce the manner in which your algorithm is converging. Specifically, at some point you should see that the mantissa of the answer changes only in the less significant digits, with that place moving further to the right of the decimal point as the calculation executes more steps. Eventually, however, as the number of steps becomes even larger, round-off error leads to a fluctuation in the less significant digits, with a gradual increase on the average. It is best to quit the calculation before this occurs.

Based upon this understanding, an approach to obtaining the best approximation is to deduce when your answer behaves like (2.29). To do that, we call $\mathcal{A}$ the exact answer and $A(N)$ the computed answer after N steps. We assume that for large enough values of N, the approximation converges as

$$A(N) \simeq \mathcal{A} + \frac{\alpha}{N^{\beta}}, \tag{2.32}$$

that is, that the round-off error term in (2.29) is still small. We then run our computer program with $2N$ steps, which should give a better answer, and use that answer to eliminate the unknown $\mathcal{A}$:

$$A(N) - A(2N) \simeq \frac{\alpha}{N^{\beta}}. \tag{2.33}$$

To see if these assumptions are correct and determine what level of precision is possible for the best choice of N, plot $\log_{10}|[A(N) - A(2N)]/A(2N)|$ *versus* $\log_{10} N$, similar to what we have done in Figure 2.2. If you obtain a rapid straight-line drop off, then you know you are in the region of convergence and can deduce a value for β from the slope. As N gets larger, you should see the graph change from a straight-line decrease to a slow increase as round-off error begins to dominate. A good place to quit is before this. In any case, now you understand the error in your computation and therefore have a chance to control it.

As an example of how different kinds of errors enter into a computation, we assume we know the analytic form for the approximation and round-off errors:

$$\epsilon_{\text{approx}} \simeq \frac{1}{N^2}, \qquad \epsilon_{\text{ro}} \simeq \sqrt{N}\epsilon_m, \tag{2.34}$$

$$\Rightarrow \quad \epsilon_{\text{tot}} = \epsilon_{\text{approx}} + \epsilon_{\text{ro}} \simeq \frac{1}{N^2} + \sqrt{N}\epsilon_m. \tag{2.35}$$

The total error is then a minimum when

$$\frac{d\epsilon_{\text{tot}}}{dN} = \frac{-2}{N^3} + \frac{1}{2}\frac{\epsilon_m}{\sqrt{N}} = 0, \tag{2.36}$$

$$\Rightarrow \quad N^{5/2} = \frac{4}{\epsilon_m}. \tag{2.37}$$

For a single-precision calculation ($\epsilon_m \simeq 10^{-7}$), the minimum total error occurs when

$$N^{5/2} \simeq \frac{4}{10^{-7}} \quad \Rightarrow \quad N \simeq 1099, \quad \Rightarrow \quad \epsilon_{\text{tot}} \simeq 4 \times 10^{-6}. \tag{2.38}$$

In this case most of the error is due to round-off and is not approximation error. Observe, too, that even though this is the minimum total error, the best we can do is about 40 times machine precision (in double precision the results are better).

Seeing that the total error is mainly round-off error $\propto\sqrt{N}$, an obvious way to decrease the error is to use a smaller number of steps N. Let us assume we do this by finding another algorithm that converges more rapidly with N, for example, one with approximation error behaving like

$$\epsilon_{\text{approx}} \simeq \frac{2}{N^4}. \tag{2.39}$$

The total error is now

$$\epsilon_{\text{tot}} = \epsilon_{\text{ro}} + \epsilon_{\text{approx}} \simeq \frac{2}{N^4} + \sqrt{N}\epsilon_m. \tag{2.40}$$

The number of points for minimum error is found as before:

$$\frac{d\epsilon_{\text{tot}}}{dN} = 0 \quad \Rightarrow \quad N^{9/2} \quad \Rightarrow \quad N \simeq 67 \quad \Rightarrow \quad \epsilon_{\text{tot}} \simeq 9 \times 10^{-7}. \tag{2.41}$$

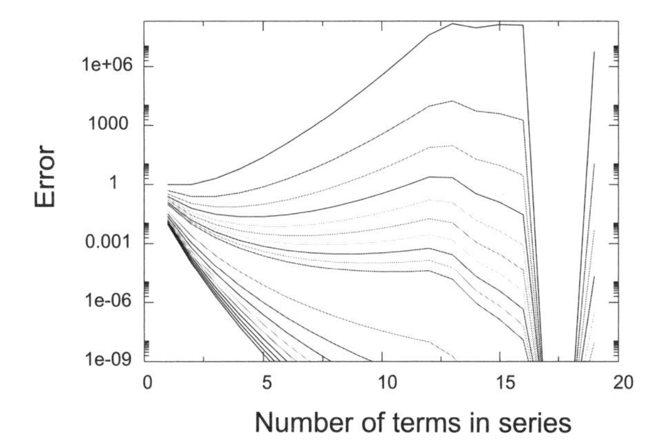

Figure 2.3 The error in the summation of the series for e^{-x} *versus* N. The values of x increase vertically for each curve. Note that a negative initial slope corresponds to decreasing error with N, and that the dip corresponds to a rapid convergence followed by a rapid increase in error. (Courtesy of J. Wiren.)

The error is now smaller by a factor of 4, with only $\frac{1}{16}$ as many steps needed. Subtle are the ways of the computer. In this case the better algorithm is quicker and, by using fewer steps, it produces less round-off error.

Exercise: Repeat the error estimates for a double-precision calculation. ▮

2.3.1 Error Assessment

We have already discussed the Taylor expansion of $\sin x$:

$$\sin(x) = x - \frac{x^3}{3!} + \frac{x^5}{5!} - \frac{x^7}{7!} + \cdots = \sum_{n=1}^{\infty} \frac{(-1)^{n-1} x^{2n-1}}{(2n-1)!}. \tag{2.42}$$

This series converges in the mathematical sense for all values of x. Accordingly, a reasonable algorithm to compute the $\sin(x)$ might be

$$\sin(x) \simeq \sum_{n=1}^{N} \frac{(-1)^{n-1} x^{2n-1}}{(2n-1)!}. \tag{2.43}$$

While in principle it should be faster to see the effects of error accumulation in this algorithm by using single-precision numbers, C and Java tend to use double-precision mathematical libraries, and so it is hard to do a pure single-precision computation. Accordingly, do these exercises in double precision, as you should for all scientific calculations involving floating-point numbers.

1. Write a program that calculates $\sin(x)$ as the finite sum (2.43). (If you already did this in Chapter 1, "Computational Science Basics," then you may reuse that program and its results here.)
2. Calculate your series for $x \le 1$ and compare it to the built-in function `Math.sin(x)` (you may assume that the built-in function is exact). Stop your summation at an N value for which the next term in the series will be no more than 10^{-7} of the sum up to that point,

$$\frac{|(-1)^N x^{2N+1}|}{(2N-1)!} \le 10^{-7} \left| \sum_{n=1}^{N} \frac{(-1)^{n-1} x^{2n-1}}{(2n-1)!} \right|. \tag{2.44}$$

3. Examine the terms in the series for $x \simeq 3\pi$ and observe the significant subtractive cancellations that occur when large terms add together to give small answers. [Do not use the identity $\sin(x+2\pi) = \sin x$ to reduce the value of x in the series.] In particular, print out the near-perfect cancellation around $n \simeq x/2$.
4. See if better precision is obtained by using trigonometric identities to keep $0 \le x \le \pi$.
5. By progressively increasing x from 1 to 10, and then from 10 to 100, use your program to determine experimentally when the series starts to lose accuracy and when it no longer converges.
6. Make a series of graphs of the error *versus* N for different values of x. (See Chapter 3, "Visualization Tools.") You should get curves similar to those in Figure 2.3. ▌

Because this series summation is such a simple, correlated process, the round-off error does not accumulate randomly as it might for a more complicated computation, and we do not obtain the error behavior (2.32). We will see the predicted error behavior when we examine integration rules in Chapter 6, "Integration."

3

Visualization Tools

If I can't picture it, I can't understand it.
— *Albert Einstein*

In this chapter we discuss the tools needed to visualize data produced by simulations and measurements. Whereas other books may choose to relegate this discussion to an appendix, or not to include it at all, we believe that visualization is such an integral part of computational science, and so useful for your work in the rest of this book, that we have placed it right here, up front. (We do, however, place our OpenDx tutorial in Appendix C since it may be a bit much for beginners.)

ALL THE VISUALIZATION TOOLS WE discuss are powerful enough for professional scientific work and are free or open source. Commercial packages such as Matlab, AVS, Amira, and Noesys produce excellent scientific visualization but are less widely available. Mathematica and Maple have excellent visualization packages as well, but we have not found them convenient when dealing with large numerical data sets.[1] The tools we discuss, and have used in preparing the visualizations for the text, are

PtPlot: Simple 2-D plotting callable from within a Java program, or as stand-alone application; part of the Ptolemy package. Being Java, it is universal for all operating systems.

Gnuplot: 2-D and 3-D plotting, predominantly stand-alone. Originally for Unix operating systems, with an excellent windows port also available.

Ace/gr (Grace): Stand-alone, menu-driven, publication-quality 2-D plotting for Unix systems; can run under MS Windows with Cygwin.

OPenDX: Formerly IBM DataExplorer. Multidimensional data tool for Unix or for Windows under Cygwin (tutorial in Appendix C).

3.1 Data Visualization

One of the most rewarding uses of computers is visualizing the results of calculations. While in the past this was done with 2-D plots, in modern times it

[1] Visualization with Maple and Mathematica is discussed in [L 05].

is regular practice to use 3-D (surface) plots, volume rendering (dicing and slicing), and animation, as well as virtual reality (gaming) tools.These types of visualizations are often breathtakingly beautiful and may provide deep insights into problems by letting us see and "handle" the functions with which we are working. Visualization also assists in the debugging process, the development of physical and mathematical intuition, and the all-around enjoyment of your work. One of the reasons for visualization's effectiveness may arise from the large fraction ($\sim$10% direct and $\sim$50% with coupling) of our brain involved in visual processing and the advantage gained by being able to use this brainpower as a supplement to our logical powers.

In thinking about ways to present your results, keep in mind that the point of visualization is to make the science clearer and to communicate your work to others. It follows then that you should make all figures as clear, informative, and self-explanatory as possible, especially if you will be using them in presentations without captions. This means labels for all curves and data points, a title, and labels on the axes.[2] After this, you should look at your visualization and ask whether there are better choices of units, ranges of axes, colors, style, and so on, that might get the message across better and provide better insight. Considering the complexity of human perception and cognition, there may not be a single best way to visualize a particular data set, and so some trial and error may be necessary to see what looks best. Although all this may seem like a lot of work, the more often you do it the quicker and better you get at it and the better you will be able to communicate your work to others.

3.2 PtPlot: 2-D Graphs Within Java

PtPlot is an excellent plotting package that lets you plot directly from Java programs.[3] PtPlot is free, written in Java (and thus runs under Unix, Linux, Mac OS, and MS Windows), is easy to use, and is actually part of Ptolemy, an entire computing environment supported by the University of California Computer Science Department. Figure 3.1 is an example of a PtPlot graph. Because PtPlot is not built into Java, a Java program needs to import the PtPlot package and work with its classes. You can download the most recent version over the Web.

The program `EasyPtPlot.java` in Listing 3.1 is an example of a how to construct a simple graph of $\sin^2(x)$ *versus* x with PtPlot (Figure 3.1 top). On line 2 we see the statement `import ptolemy.plot.*;` that imports the PtPlot classes from the `ptolemy` directory. (In order for this to work, you may have to modify your CLASSPATH environmental variable or place the `ptolemy` directory in your working directory.) PtPlot represents your plot as a `Plot` *object*, which we name `plotObj` and create on line 9 (objects are discussed in Chapter 4, "Object-Oriented Programs: Impedance & Batons"). We then add various features, step by step, to `plotObj` to make it just the plot we want. As is standard with objects in Java, we first give the name of the object and then modify it with "dot modifiers." Rather than tell PtPlot what ranges

[2] Although this may not need saying, place the independent variable x along the abscissa (horizontal), and the dependent variable $y = f(x)$ along the ordinate.

[3] Connelly Barnes assisted in the preparation of this section.

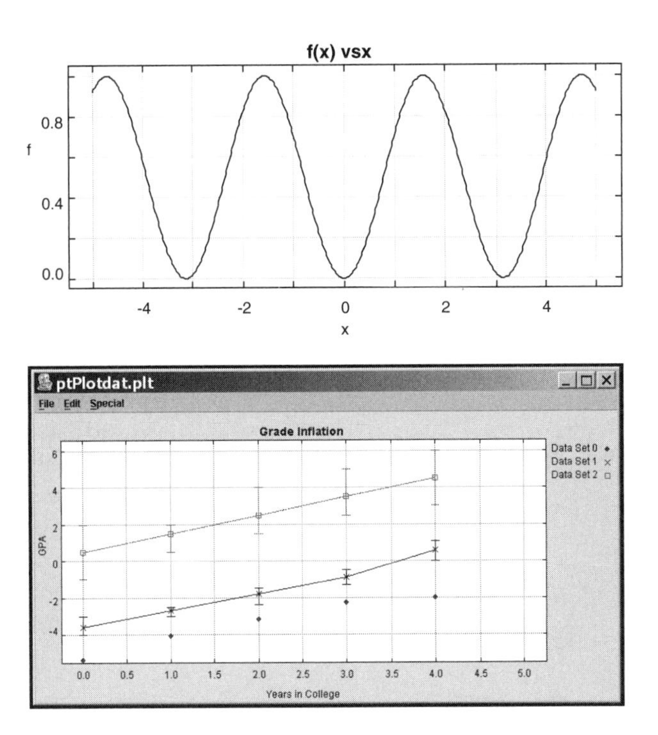

Figure 3.1 *Top:* Output graph from the program EasyPtPlot.java. *Bottom:* Output PtPlot window in which three data sets (set number = the first argument in addPoint) are placed in one plot. Observe the error bars on two of the sets.

```
// EasyPtPlot.java: Simple PtPlot application
import ptolemy.plot.*;

public class EasyPtPlot {
  public static final double Xmin = -5., Xmax = 5.;                    // Graph domain
  public static final int Npoint = 500;

  public static void main(String[] args) {
    Plot plotObj = new Plot();                                     // Create Plot object
    plotObj.setTitle("f(x) vs x");
    plotObj.setXLabel("x");
    plotObj.setYLabel("f(x)");
 // plotObj.setSize(400, 300);
 // plotObj.setXRange(Xmin, Xmax);
 // plotObj.addPoint(int Set, double x, double y, boolean connect)
    double xStep = (Xmax - Xmin) / Npoint;
    for ( double x = Xmin; x <= Xmax; x += xStep) {
      double y = Math.sin(x)*Math.sin(x);
      plotObj.addPoint(0, x, y, true);
    }
    PlotApplication app = new PlotApplication(plotObj);               // Display Plot
  }
}
```

Listing 3.1 EasyPtPlot.java plots a function using the package PtPlot. Note that the PlotApplication object must be created to see the plot on the screen.

to plot for x and y, we let it set the x and y ranges based on the data it is given. The first argument, `0` here, is the data set number. By having different values for the data set number, you can plot different curves on the same graph (there are two in Figure 3.1 right). By having `true` as the fourth argument in `myPlot.addPoint(0, x, y, true)`, we are telling PtPlot to connect the plotted points. For the plot to appear on the screen, line 21 creates a `PlotApplication` with the plot object as input.

We encourage you to make your plot more informative by including further options in the commands or by using the pull-down menus in the PtPlot window displaying your plot. The options are found in the description of the methods at the PtPlot Web site [PtPlot] and include the following.

Calling PtPlot from Your Program	
Plot myPlot = new Plot();	Name and create plot object `myPlot`
PlotApplication app = new PlotApplication(myPlot);	Display
myPlot.setTitle("f(x) vs x");	Add title to plot
myPlot.setXLabel("x");	Label x axis
myPlot.setYLabel("f(x)");	Label y axis
myPlot.addPoint(0, x, y, true);	Add (x, y) to set 0, connect points
myPlot.addPoint(1, x, y, false);	Add (x, y) to set 1, no connect points
myPlot.addLegend(0, "Set 0");	Label data set 0 in legend
myPlot.addPointWithErrorBars(0, x, y, yLo, yHi, true);	Plot $(x, y - y\text{Lo})$, $(x, y + y\text{Hi})$ + error bars
myPlot.clear(0);	Remove all points from data set 0
myPlot.clear(false);	Remove data from all sets
myPlot.clear(true);	Remove all points, default options
myPlot.setSize(500, 400);	Set plot size in pixels (optional)
myPlot.setXRange(–10., 10.);	Set an x range (default fit to data)
myPlot.setYRange(–8., 8.);	Set a y range (default fit to data)
myPlot.setXLog(true);	Use log scale for x axis
myPlot.setYLog(true);	Use log scale for y axis
myPlot.setGrid(false);	Turn off the grid
myPlot.setColor(false);	Color in black and white
myPlot.setButtons(true);	Display zoom-to-fit button on plot
myPlot.fillPlot();	Adjust x, y ranges to fit data
myPlot.setImpulses(true, 0);	Lines from points to x axis, set 0
myPlot.setMarksStyle("none," 0);	Draw `none`, `points`, `dots`, `various`, `pixels`
myPlot.setBars(true);	Display data as bar charts
String s = myPlot.getTitle();	Extract title (or other properties)

Once you have a PtPlot application on your screen, explore some of the ways to modify your plot from within the application window:

1. Examine the Edit pull-down menu (underlined letters are shortcuts). Select Edit and pull it down.

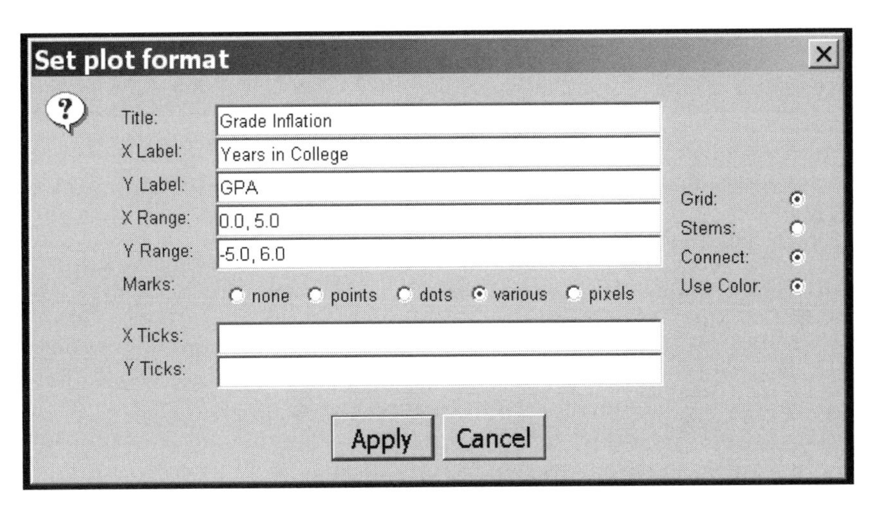

Figure 3.2 The Format submenu located under the Edit menu in a PtPlot application. This submenu controls the plot's basic features.

2. From the Edit pull-down menu, select Format. You should see a window like that in Figure 3.2, which lets you control most of the options in your graph.
3. Experiment with the Format menu. In particular, change the graph so that only points are plotted, with the points being pixels and black, and with your name in the title.
4. Select a central portion of your plot and **zoom in** on it by drawing a box (with the mouse button depressed) starting from the upper-left corner and then moving down before you release the mouse button. You **zoom out** by drawing a box from the lower-right corner and moving up. You may also resize your graph by selecting Special/Reset Axes or by resetting the x and y ranges. And of course, if things get really messed up, you always have the option of starting over by closing the Java window and running the `java` command again.
5. Scrutinize the File menu and its options for printing your graphs, as well as exporting them to files in Postscript (.ps) and other formats.

It is also possible to have PtPlot make graphs by reading data from a file in which the x and y values are separated by spaces, tabs, or commas. There is even the option of including PtPlot formatting commands in the file with data. The program `TwoPlotExample.java` on the CD and its data file `data.plt` show how to place two plots side by side and how to read in a data file containing error bars with various symbols for the points. In its simplest form, *PtPlot Data Format* is just a text file with a single (x, y) point per line. For example, Figure 3.1 was produced from the data file `PtPlotdat.plt`:

Sample PtPlot Data file PtPlotdat.plt

```
# This is a comment: Sample data for PtPlot TitleText: Grade
Inflation XRange: 0,5 YRange: -5, 6 Grid: on XLabel: Years in College
YLabel: GPA Marks: various NumSets:3 Color: on DataSet: Data Set 0
Lines:off 0,-5.4 1,-4.1 2,-3.2 3,-2.3 4, -2 DataSet: Data Set 1
Lines:on 0,-3.6, -4,-3 1,-2.7, -3, -2.5 2,-1.8, -2.4,-1.5 3,-0.9,
-1.3, -0.5 4, 0.6, 0,1.1 DataSet: Data Set 2 0,0.5, -1,2 1, 1.5, 0.5,
2 2, 2.5, 1.5, 4 3, 3.5, 2.5, 5 4, 4.5, 3, 6
```

To plot your data files directly from the command line, enter

> **java ptolemy.plot.PlotApplication dataFile**	Plot data in dataFile

This causes the standard PtPlot window to open and display your data. If this does not work, then your CLASSPATH variable may not be defined properly or PtPlot may not be installed. See "Installing PtPlot" in Appendix B.

Reading in your data from the PtPlot window itself is an alternative. Either use an already open window or issue Java's run command:

> **java ptolemy.plot.PlotApplication**	Open PtPlot window

To look at your data from the PtPlot window, choose File → Open → FileName. By default, PtPlot will look for files with the suffix .plt or .xml. However, you may enter any name you want or pull down the Filter menu and select * to see all your files. The same holds for the File → SaveAs option. In addition, you may Export your plot as an Encapsulated PostScript (.eps) file, a format useful for inserting in printed documents. You may also use drawing programs to convert PostScript files to other formats or to edit the output from PtPlot, but you should *not* change the shapes or values of output to alter scientific conclusions.

As with any good plot, you should label your axes, add a title, and add what is needed for it to be informative and clear. To do this, incorporate PtPlot commands with your data or work in the PtPlot window with the pull-down menus under Edit and Special. The options are essentially the same as the ones you would call from your program:

TitleText: f(x) vs. x	Add title to plot
XLabel: x	Label x axis
YLabel: y	Label y axis
XRange: 0, 12	Set x range (default: fit to data)
YRange: –3, 62	Set y range (default: fit to data)
Marks: none	(Default) No marks at points, lines connects points
Marks: points	or: dots, various, pixels
Lines: on/off	Do not connect points with lines; default: on
Impulses: on/off	Lines down from points to x axis; default: off
Bars: on/off	Bar graph (turn off lines) default: off
Bars: width (, offset)	Bar graph; bars of width and (optional) offset
DataSet: *string*	Specify data set to plot; *string* appears in legend

x, y	Specify a data point; comma, space, tab separators
move: *x, y*	Do not connect this point to previous
x, y, yLo, yHi	Plot $(x, y - y\text{Lo})$, $(x, y + y\text{Hi})$ with error bars

If a command appears before DataSet directives, then it applies to all the data sets. If a command appears after DataSet directives, then it applies to that data set only.

3.3 Grace/ACE: Superb 2-D Graphs for Unix/Linux

Our favorite package for producing publication-quality 2-D graphs from numerical data is *Grace/xmgrace*. It is free, easy to work with, incredibly powerful, and has been used for many of the (nicer) figures in our research papers and books. Grace is a WYSIWYG tool that also contains powerful scripting languages and data analysis tools, although we will not get into that. We will illustrate multiple-line plots, color plots, placing error bars on data, multiple-axis options and such. Grace is derived from *Xmgr*, aka *ACE/gr*, originally developed by the Oregon Graduate Institute and now developed and maintained by the Weizmann Institute [Grace]. Grace is designed to run under the Unix/Linux operating systems, although we have had success using in on an MS Windows system from within the Cygwin [CYG] Linux emulation.[4]

3.3.1 Grace Basics

To learn about Grace properly, we recommend that you work though some of the tutorials available under its Help menu, study the user's guide, and use the Web tutorial [Grace]. We present enough here to get you started and to provide a quick reference. The first step in creating a graph is to have the data you wish to plot in a text (ASCII) file. The data should be broken up into columns, with the first column the abscissa (x values) and the second column the ordinate (y values). The columns may be separated by spaces or tabs but *not* by commas. For example, the file Grace.dat on the CD and in the left column of Table 3.1 contains one abscissa and one ordinate per line, while the file Grace2.dat in the right column of Table 3.1 contains one abscissa and two ordinates (y values) per line.

CD

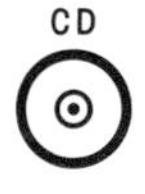

1. **Open Grace** by issuing the grace or xmgrace command from a Unix/Linux/Cygwin command line (prompt):

   > **grace**   Start Grace from Unix shell

[4] If you do this, make sure to have the Cygwin download include the xorg-x11-base package in the X11 category (or a later version), as well as xmgrace.

TABLE 3.1
Text Files Grace.dat and Grace2.dat*

Grace.dat		Grace2.dat		
x	y	x	y	z
1	2	1	2	50
2	4	2	4	29
3	5	3	5	23
4	7	4	7	20
5	10	5	10	11
6	11	6	11	10
7	20	7	20	7
8	23	8	23	5
9	29	9	29	4
10	50	10	50	2

*The text file Grace.dat (on the CD under Codes/JavaCodes/Data) contains one x value and one y value per line. The file Grace2.dat contains one x value and two y values per line.

In any case, make sure that your command brings up the user-friendly graphical interface shown on the left in Figure 3.3 and not the pure-text command one.

2. **Plot a single data set** by starting from the menu bar on top. Then
   a. Select progressively from the submenus that appear, Data/Import/ASCII.
   b. The Read sets window shown on the right in Figure 3.3 appears.
   c. Select the directory (folder) and file in which you have the data; in the present case select Grace.dat.
   d. Select Load as/Single set, Set type/XY and Autoscale on read/XY.
   e. Select OK (to create the plot) and then Cancel to close the window.
3. **Plot multiple data sets** (Figure 3.3) is possible in step 2, only now
   a. Select Grace2.dat, which contains two y values as shown in Table 3.1.
   b. Change Load as to NXY to indicate multiple data sets and then plot.
   Note: We suggest that you start your graphs off with Autoscale on read in order to see all the data sets plotted. You may then change the scale if you want, or eliminate some points and replot.
4. **Label and modify the axis properties** by going back to the main window. Most of the basic utilities are under the Plot menu.
   a. Select Plot/Axis properties.
   b. Within the Axis Property window that appears, select Edit/X axis or Edit/Y axis as appropriate.
   c. Select Scale/Linear for a linear plot, or Scale/Logarithmic for a logarithmic or semilog plot.
   d. Enter your choice for Axis label in the window.
   e. Customize the ticking and the numbers' format to your heart's desire.
   f. Choose Apply to see your changes and then Accept to close the window.

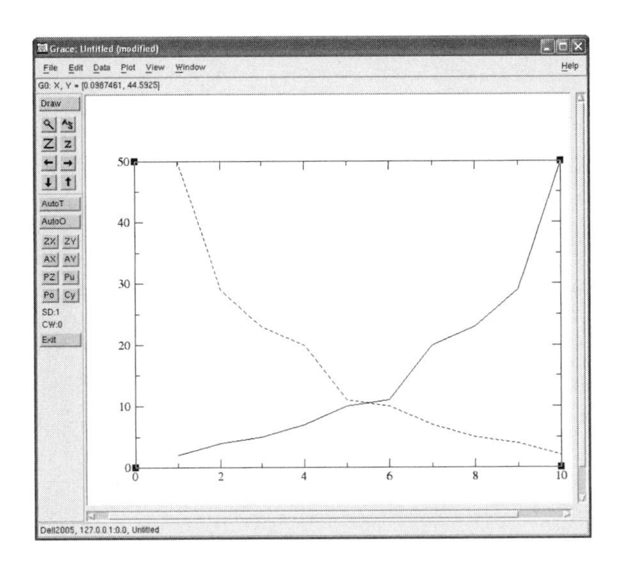

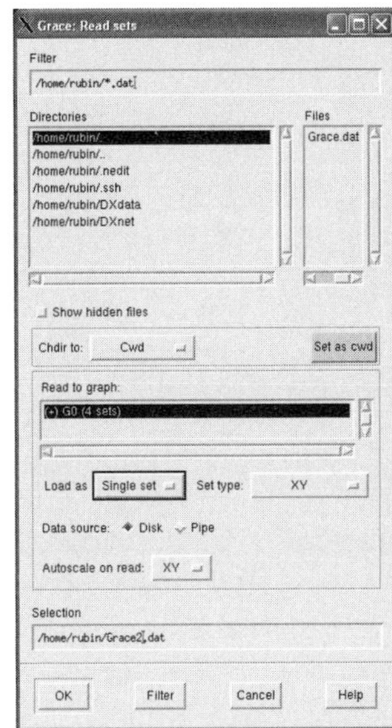

Figure 3.3 *Left:* The main Grace window, with the file `Grace2.dat` plotted with the title, subtitle, and labels. *Right:* The data input window.

5. **Title the graph,** as shown on the left in Figure 3.3, by starting from the main window and again going to the Plot menu.
   a. Select Plot/Graph appearance.
   b. From the Graph Appearance window that appears, select the Main tab and from there enter the title and subtitle.
6. **Label the data sets,** as shown by the box on the left in Figure 3.3, by starting from the main window and again going to the Plot menu.
   a. Select Plot/Set Appearance.
   b. From the Set Appearance window that appears, highlight each set from the Select set window.
   c. Enter the desired text in the Legend box.
   d. Choose Apply for each set and then Accept.
   e. You can adjust the location, and other properties, of the legend box from Plot/Graph Appearance/Leg. box.
7. **Plotting points as symbols** (Figure 3.4 left) is accomplished by starting at the main menu. Then
   a. Select Plot/Set Appearance.
   b. Select one of the data sets being plotted.
   c. Under the Main/Symbol properties, select the symbol Type and Color.
   d. Choose Apply, and if you are done with all the data sets, Accept.

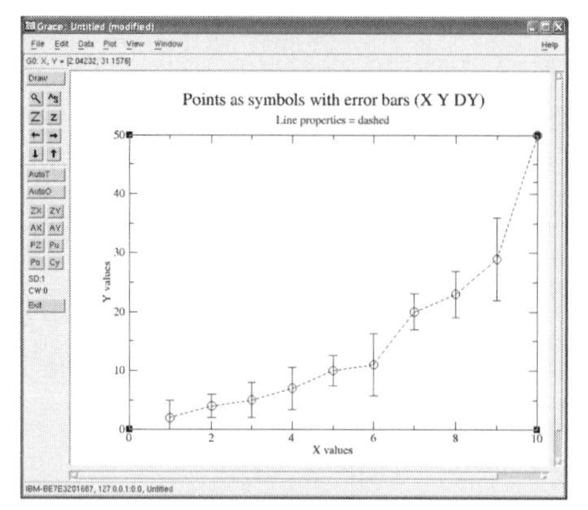

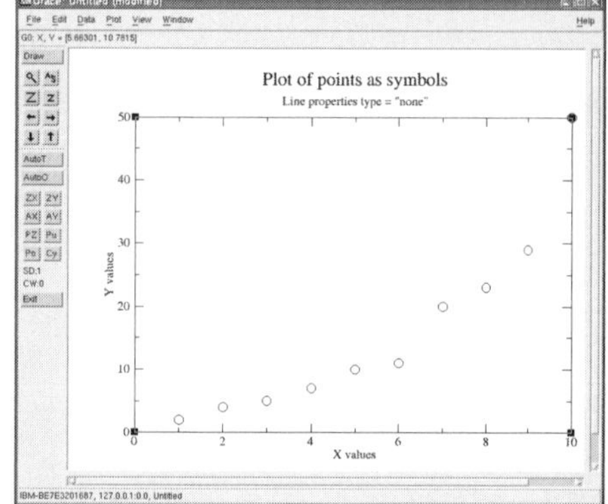

Figure 3.4 *Left:* A plot of points as symbols with no line. *Right:* A plot of points as symbols with lines connecting the points and with error bars read from the file.

8. **Including error bars** (Figure 3.4 left) is accomplished by placing them in the data file read into Grace along with the data points. Under Data/Read Sets are, among others, these possible formats for Set type:

   (X Y DX), (X Y DY), (X Y DX DX), (X Y DY DY), (X Y DX DX DY DY)

   Here DX is the error in the x value, DY is the error in the y value, and repeated values for DX or DY are used when the upper and lower error bars differ; for instance, if there is only one DY, then the data point is Y $\pm$ DY, but if there are two error bars given, then the data point is Y + DY_1, $-DY_2$. As a case in point, here is the data file for (X Y DY):

X	1	2	3	4	5	6	7	8	9	10
Y	2	4	5	7	10	11	20	23	29	50
DY	3	2	3	3.6	2.6	5.3	3.1	3.9	7	8

9. **Multiple plots on one page** (Figure 3.5 left) are created by starting at the main window. Then
   a. Select Edit/Arrange Graphs.
   b. An Arrange Graphs window (Figure 3.5 right) opens and provides the options for setting up a matrix into which your graphs are placed.
   c. Once the matrix is set up (Figure 3.5 left), select each space in sequence and then create the graph in the usual way.
   d. To prevent the graphs from getting too close to each other, go back to the Arrange Graphs window and adjust the spacing between graphs.

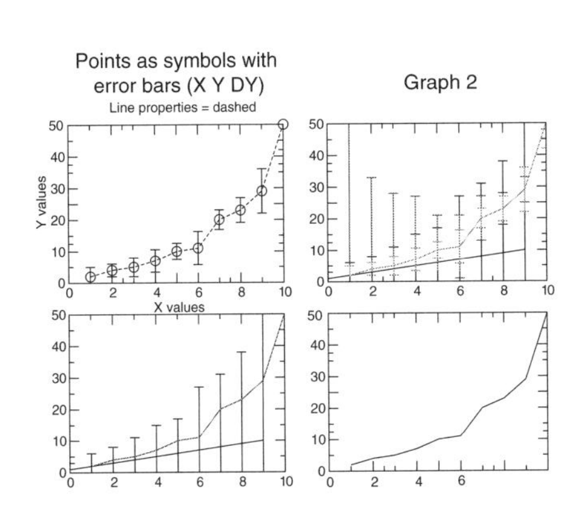

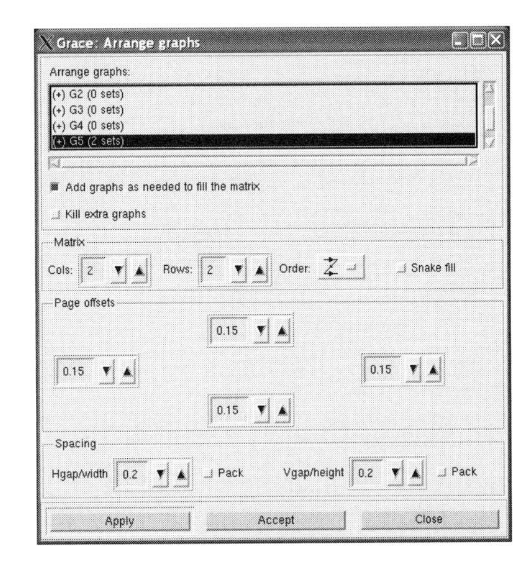

Figure 3.5 *Left:* Four graphs placed in a 2 × 2 matrix. *Right:* The window that opens under Edit/Arrange Graphs and is used to set up the matrix into which multiple graphs are placed.

10. **Printing and saving plots**
    a. To save your plot as a complete Grace project that can be opened again and edited, from the main menu select File/Save As and enter a `filename.agr` as the file name. It is a good idea to do this as a backup before printing your plot (communication with a piece of external hardware is subject to a number of difficulties, some of which may cause a program to "freeze up" and for you to lose your work).
    b. To print the plot, select File/Print Setup from the main window
    c. If you want to save your plot to a file, select Print to file and then enter the file name. If you want to print directly to a printer, make sure that Print to file is not selected and that the selected printer is the one to which you want your output to go (some people may not take kindly to your stuff appearing on their desk, and you may not want some of your stuff to appear on someone else's desk).
    d. From Device, select the file format that will be sent to the printer or saved.
    e. Apply your settings when done and then close the Print/Device Setup window by selecting Accept.
    f. If now, from the main window, you select File/Print, the plot will be sent to the printer *or* to the file. Yes, this means that you must "Print" the plot in order to send it to a file.

If you have worked through the steps above, you should have a good idea of how Grace works. Basically, you just need to find the command you desire under a menu item. To help you in your search, in Table 3.2 we list the Grace menu and submenu items.

TABLE 3.2
Grace Menu and Submenu Items

Edit	Data
Data sets	Data set operations
Set operations	sort, reverse, join, split, drop points
copy, move, swap	Transformations
Arrange graphs	expressions, histograms, transforms,
matrix, offset, spacing	convolutions, statistical ops, interpolations
Overlay graphs	Feature extraction
Autoscale graphs	min/max, average, deviations, frequency,
Regions	COM, rise/fall time, zeros
Hot links	Import
Set/Clear local/fixed point	Export
Preferences	
Plot	**View**
Plot appearance	Show locator bar (default)
background, time stamp, font, color	Show status bar (default)
Graph appearance	Show tool bar (default)
style, title, labels, frame, legends	Page setup
Set appearance	Redraw
style, symbol properties, error bars	Update all
Axis properties	
labels, ticks, placement	
Load/Save parameters	
Window	
Command	Font tool
Point explorer	Console
Drawing objects	

3.4 Gnuplot: Reliable 2-D and 3-D Plots

Gnuplot is a versatile 2-D and 3-D graphing package that makes Cartesian, polar, surface, and contour plots. Although PtPlot is good for 2-D plotting with Java, only Gnuplot can create surface plots of numerical data. Gnuplot is classic open software, available free on the Web, and supports many output formats.

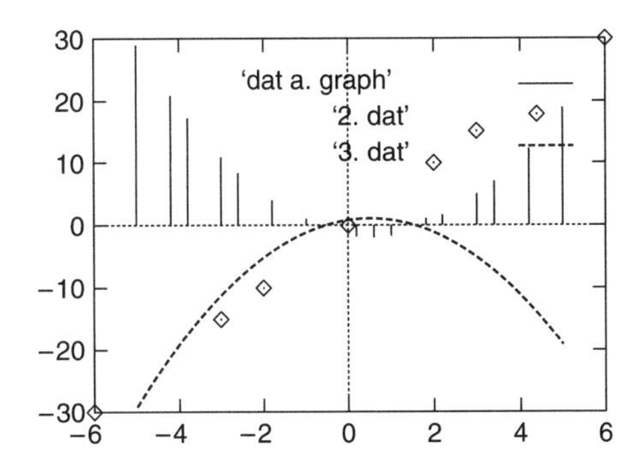

Figure 3.6 A Gnuplot graph for three data sets with impulses and lines.

Begin Gnuplot with a file of $(x,\ y)$ data points, say, graph.dat. Next issue the **gnuplot** command from a shell or from the Start menu. A new window with the Gnuplot prompt gnuplot> should appear. Construct your graph by entering Gnuplot commands at the Gnuplot prompt or by using the pull-down menus in the Gnuplot window:

> **gnuplot**	Start Gnuplot program
Terminal type set to 'x11'	Type of terminal for Unix
gnuplot>	The Gnuplot prompt
gnuplot> **plot "graph.dat"**	Plot data file graph.dat

Plot a number of graphs on the same plot using several data files (Figure 3.6):

gnuplot> **plot 'graph.dat' with impulses, '2.dat', '3.dat' with lines**

The general form of the 2-D plotting command and its options are

plot {ranges} function {title} {style} {, function . . . } Command

with points	Default. Plot a symbol at each point.
with lines	Plot lines connecting the points.
with linespoint	Plot lines and symbols.
with impulses	Plot vertical lines from the x axis to points.
with dots	Plot a small dot at each point (scatterplots).

For Gnuplot to accept the name of an external file, that name must be placed in 'single' or "double" quotes. If there are multiple file names, the names must be separated by commas. Explicit values for the x and y ranges are set with options:

gnuplot> **plot [xmin:xmax] [ymin:ymax] "file"**	Generic
gnuplot> **plot [−10:10] [−5:30] "graph.dat"**	Explicit

3.4.1 Gnuplot Input Data Format ⊙

The format of the data file that Gnuplot can read is not confined to (x, y) values. You may also read in a data file as a C language *scanf* format string xy by invoking the using option in the plot command. (Seeing that it is common for Linux/ Unix programs to use this format for reading files, you may want to read more about it.)

plot 'datafile' { using { xy | yx | y } {"scanf string"} }

This format explicitly reads selected rows into x or y values while skipping past text or unwanted numbers:

gnuplot>	**plot "data" using "%f%f"**	Default, 1st x, 2nd y.
gnuplot>	**plot "data" using yx "%f %f"**	Reverse, 1st y, 2nd x.
gnuplot>	**plot "data" xy using "%*f %f %*f %f"**	Use row 2,4 for x, y.
gnuplot>	**plot "data" using xy "%*6c %f%*7c%f"**	

This last command skips past the first six characters, reads one x, skips the next seven characters, and then reads one y. It works for reading in x and y from files such as

```
theta: -20.000000 Energy: -3.041676 theta: -19.000000 Energy:
-3.036427 theta: -18.000000 Energy: -3.030596 theta: -17.000000
Energy: -3.024081 theta: -16.000000 Energy: -3.016755
```

Observe that because the data read by Gnuplot are converted to floating-point numbers, you use %f to read in the values you want.

Besides reading data from files, Gnuplot can also generate data from user-defined and library functions. In these cases the default independent variable is x for 2-D plots and (x, y) for 3-D ones. Here we plot the acceleration of a nonharmonic oscillator:

gnuplot>	**k = 10**	Set value for k
gnuplot>	a(x) = .5*k*x**2	Analytic expression
gnuplot>	**plot [-10:10] a(x)**	Plot analytic function

A useful feature of Gnuplot is its ability to plot analytic functions along with numerical data. For example, Figure 3.7 compares the theoretical expression for the period of a simple pendulum to experimental data of the form

```
# length(cm)    period (sec)
10                   0.8
20                   0.9
30                   1.2
40                   1.3
50                   1.5
```

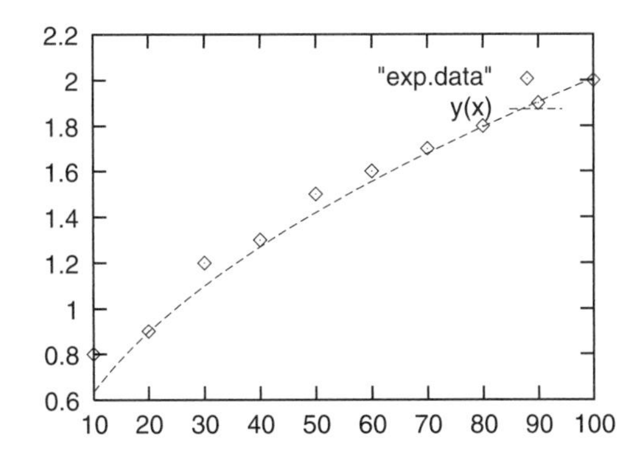

Figure 3.7 A Gnuplot plot of data from a file plus an analytic function.

Note that the first line of text is ignored since it begins with a # . We plot with

gnuplot> **g = 980**	Set value for g
gnuplot> **y(x) = 2*3.1416*sqrt(x/g)**	Period $T = y$, length $L = x$
gnuplot> **plot "exp.data", y(x)**	Plot both data and function

3.4.2 Printing Plots

Gnuplot supports a number of printer types including PostScript. The safest way to print a plot is to save it to a file and then print the file:

1. Set the "terminal type" for your printer.
2. Send the plot output to a file.
3. Replot the figure for a new output device.
4. Quit Gnuplot (or get out of the Gnuplot window).
5. Print the file with a standard print command.

For a more finished product, you can import Gnuplot's output .ps file into a drawing program such as CorelDraw or Illustrator and fix it up just right. To see what types of printers and other output devices are supported by Gnuplot, enter the `set terminal` command without any options into a `gnuplot` window. Here is an example of creating a PostScript figure and printing it:

gnuplot> **set terminal postscript**	Choose local printer type
Terminal type set to 'postscript'	Gnuplot response
gnuplot> **set term postscript eps**	Another option

gnuplot> **set output "plt.ps"**	Send figure to file
gnuplot> **replot**	Plot again so file is sent
gnuplot> **quit**	Or get out of gnu window
% **lp plt.ps**	Unix print command

3.4.3 Gnuplot Surface (3-D) Plots

A 2-D plot is fine for visualizing the potential field $V(r) = 1/r$ surrounding a single charge. However, when the same potential is expressed as a function of Cartesian coordinates, $V(x, y) = 1/\sqrt{x^2 + y^2}$, we need a 3-D visualization. We get that by creating a world in which the z dimension (mountain height) is the value of the potential and x and y lie on a flat plane below the mountain. Because the surface we are creating is a 3-D object, it is not possible to draw it on a flat screen, and so different techniques are used to give the impression of three dimensions to our brains. We do that by rotating the object, shading it, employing parallax, and other tricks.

The surface (3-D) plot command is splot, and it is used in the same manner as plot—with the obvious extension to (x, y, z). A surface (Figure 3.8) is specified by placing the $z(x, y)$ values in a matrix but without ever giving the x and y values explicitly. The x values are the row numbers in the matrix and the y values are the column values (Figure 3.8). This means that only the z values are read in and that they are read in row by row, with different rows separated by blank lines:

row 1 (blank line) row 2 (blank line) row 3 ... row N.

Here each row is input as a column of numbers, with just one number per line. For example, 13 columns each with 25 z values would be input as a sequence of 25 data elements, followed by a blank line, and then another sequence followed by a blank line, and so on:

```
0.0
0.695084369397148
1.355305208363503
1.9461146066793003
...
-1.0605832625347442
-0.380140746321537
[blank line]
0.0
0.6403868757235301
1.2556172093991282
...
2.3059977070286473
2.685151549102467
   [blank line]
2.9987593603912095
...
```

Although there are no explicit x and y values given, Gnuplot plots the data with the x and y assumed to be the row and column numbers.

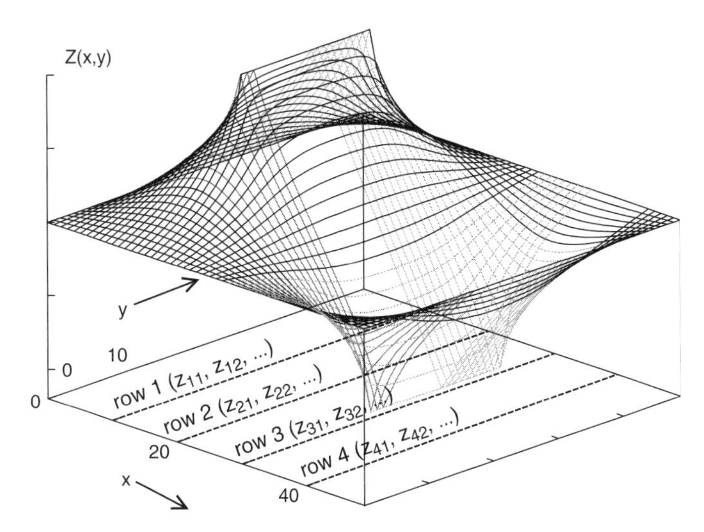

Figure 3.8 A surface plot $z(x, y) = z$ (row, column) showing the input data format used for creating it. Only z values are stored, with successive rows separated by blank lines and the column values repeating for each row.

Versions 4.0 and above of Gnuplot have the ability to rotate 3-D plots interactively. You may also adjust your plot with the command

```
gnuplot> set view rotx, rotz, scale, scalez
```

where $0 \leq \text{rotx} \leq 180°$ and $0 \leq \text{rotz} \leq 360°$ are angles in degrees and the scale factors control the size. Any changes made to a plot are made when you redraw the plot using the replot command.

To see how this all works, here we give a sample Gnuplot session that we will use in Chapter 17, "PDEs for Electrostatics & Heat Flow," to plot a 3-D surface from numerical data. The program Laplace.java contains the actual code used to output data in the form for a Gnuplot surface plot.[5]

```
> gnuplot                                      Start Gnuplot system from a shell
gnuplot> set hidden3d                          Hide surface whose view is blocked
gnuplot> set nohidden3d                        Show surface though hidden from view
gnuplot> splot 'Laplace.dat' with lines        Surface plot of Laplace.dat with lines
gnuplot> set view 65,45                        Set x and y rotation viewing angles
gnuplot> replot                                See effect of your change
gnuplot> set contour                           Project contours onto xy plane
gnuplot> set cntrparam levels 10               10 contour levels
```

[5] Under Windows, there is a graphical interface that is friendlier than the Gnuplot subcommands. The subcommand approach we indicate here is reliable and universal.

gnuplot> **set terminal epslatex**	Output in Encapsulated PostScript for LaTeX
gnuplot> **set terminal PostScript**	Output in PostScript format for printing
gnuplot> **set output "Laplace.ps"**	Plot output to be sent to file Laplace.ps
gnuplot> **splot 'Laplace.dat' w l**	Plot again, output to file
gnuplot> **set terminal x11**	To see output on screen again
gnuplot> **set title 'Potential V(x,y) vs x,y'**	Title graph
gnuplot> **set xlabel 'x Position'**	Label x axis
gnuplot> **set ylabel 'y Position'**	Label y axis
gnuplot> **set zlabel 'V(x,y)'; replot**	Label z axis and replot
gnuplot> **help**	Tell me more
gnuplot> **set nosurface**	Do not draw surface; leave contours
gnuplot> **set view 0, 0, 1**	Look down directly onto base
gnuplot> **replot**	Draw plot again; may want to write to file
gnuplot> **quit**	Get out of Gnuplot

3.4.4 Gnuplot Vector Fields

Even though it is simpler to compute a scalar potential than a vector field, vector fields often occur in nature. In Chapter 17, "PDEs for Electrostatics & Heat Flow," we show how to compute the electrostatic potential $U(x, y)$ on an $x + y$ grid of spacing Δ. Since the field is the negative gradient of the potential, $\mathbf{E} = -\vec{\nabla} U(x, y)$, and since we solve for the potential on a grid, it is simple to use the central-difference approximation for the derivative (Chapter 7 "Differentiation & Searching") to determine $\mathbf{E}$:

$$E_x \simeq \frac{U(x+\Delta, y) - U(x-\Delta, y)}{2\Delta} = \frac{U_{i+1,j} - U_{i-1,j}}{2\Delta}, \tag{3.1}$$

$$E_y \simeq \frac{U(x, y+\Delta) - U(x, y-\Delta)}{2\Delta} = \frac{U_{i,j+1} - U_{i,j-1}}{2\Delta}. \tag{3.2}$$

Gnuplot contains the vectors style for plotting vector fields as arrows of varying lengths and directions (Figure 3.9).

```
> plot 'Laplace_field.dat' using 1:2:3:4 with vectors        Vector plot
```

Here Laplace_field.data is the data file of (x, y, Ex, Ey) values, the explicit columns to plot are indicated, and additional information can be provided to control arrow types. What Gnuplot actually plots are vectors from (x, y) to $(x + \Delta x, y + \Delta y)$, where you input a data file with each line containing the $(x, y, \Delta x, \Delta y)$ values. Thousands of tiny arrows are not very illuminating (Figure 3.9 left), nor are overlapping arrows. The solution is to plot fewer points and larger arrows. On the right in Figure 3.9 we plot every fifth point normalized to unit length via

$$\Delta x = \frac{E_x}{N}, \quad \Delta y = \frac{E_y}{N}, \quad N = \sqrt{E_x^2 + E_y^2}. \tag{3.3}$$

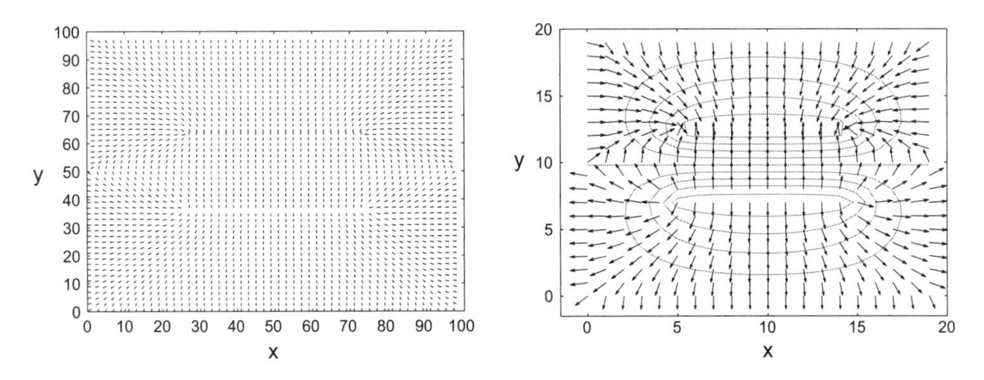

Figure 3.9 Two visualizations created by Gnuplot of the same electric field within and around a parallel plate capacitor. The figure on the right includes equipotential surfaces and uses one-fifth as many field points, and longer vectors, but of constant length. The *x* and *y* values are the column and row indices.

The data file was produced with our Laplace.java program with the added lines

```
ex = -( U[i+1][j] - U[i-1][j] ) ;                          // Compute field components
ey = -( U[i][j+1] - U[i][j-1] );
enorm = Math.sqrt( ex*ex + ey*ey );                              // Normalization factor
w.println(" "+i/5+" "+j/5+" "+ex/enorm +" "+ey/enorm +" ");                  // Output
```

We have also been able to include contour lines on the field plots by adding more commands:

```
gnuplot> unset key
gnuplot> set nosurface
gnuplot> set contour base
gnuplot> set cntrparam levels 10
gnuplot> set view 0,0,1,1
gnuplot> splot 'Laplace_pot1.dat' with lines
gnuplot> set terminal push
gnuplot> set terminal table
gnuplot> set out 'equipot.dat'
gnuplot> replot
gnuplot> set out
gnuplot> set terminal pop
gnuplot> reset
gnuplot> plot 'equipot.dat' with lines
gnuplot> unset key
gnuplot> plot 'equipot.dat' with lines, 'Laplace_field1.dat' with vectors
```

By setting terminal to table and setting out to equipot.dat, the numerical data for equipotential lines are saved in the file equipot.dat. This file can then be plotted together with the vector field lines.

3.4.5 Animations from a Plotting Program (Gnuplot) ⊙

An *animation* is a collection of images called *frames* that when viewed in sequence convey the sensation of continuous motion. It is an excellent way to visualize the time behavior of a simulation or a function $f(\mathbf{x}, t)$, as may occur in wave motion or heat flow. In the Codes section of the CD, we give several sample animations of the figures in this book and we recommend that you view them.

Gnuplot itself does not create animations. However, you can use it to create a sequence of frames and then concatenate the frames into a movie. Here we create an animated *gif* that, when opened with a Web browser, automatically displays the frames in a rapid enough sequence for your mind's eye to see them as a continuous event. Although Gnuplot does not output .gif files, we outputted *pixmap* files and converted them to gif's. Because a number of commands are needed for each frame and because hundreds or thousands of frames may be needed for a single movie, we wrote the script MakeGifs.script to automate the process. (A *script* is a file containing a series of commands that might normally be entered by hand to execute within a shell. When placed in a script, they are executed as a single command.) The script in Listing 3.2, along with the file samp_color (both on the CD under Codes/Animations_ColorImages/Utilities), should be placed in the directory containing the data files (run.lmn in this case).

CD

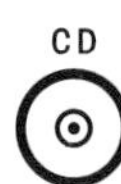

The #! line at the beginning tells the computer that the subsequent commands are in the korn shell. The symbol $i indicates that i is an argument to the script. In the present case, the script is run by giving the name of the script with three arguments, (1) the beginning time, (2) the maximum number of times, and (3) the name of the file where you wish your gifs to be stored:

```
% MakeGifs.script 1 101 OutFile          Make gif from times 1 to 101 in OutFile
```

The > symbol in the script indicates that the output is directed to the file following the >. The ppmquant command takes a pixmap and maps an existing set of colors to new ones. For this to work, you must have the map file samp_colormap in the working directory. Note that the increments for the counter, such as i=i+99, should be adjusted by the user to coincide with the number of files to be read in, as well as their names. Depending on the number of frames, it may take some time to run this script. Upon completion, there should be a new set of files of the form OutfileTime.gif, where Outfile is your chosen name and Time is a multiple of the time step used. Note that you can examine any or all of these gif files with a Web browser.

The final act of movie making is to merge the individual gif files into an animated gif with a program such as gifmerge:

```
> gifmerge -10 *.gif > movie          Merge all .gif files into movie
```

```
#! /bin/ksh

unalias rm
integer i=$1
while test i -lt $2
do

 if test i -lt 10
  then
  print "set terminal pbm small color; set output\"$3t=0000$i.ppm\"; set noxtics; set
       noytics;
  set size 1.0, 1.0; set yrange [0:.1];
  plot 'run.0000$i' using 1:2 w lines, 'run.0000$i' using 1:3 w lines;
  " >data0000$i.gnu
  gnuplot data0000$i.gnu
  ppmquant -map samp_colormap $3t=0000$i.ppm>$3at=0000$i.ppm
  ppmtogif -map samp_colormap $3at=0000$i.ppm > $30000$i.gif
  rm $3t=0000$i.ppm $3at=0000$i.ppm data0000$i.gnu
  i=i+99
  fi

if test i -gt 9 -a i -lt 1000
then
print "set terminal pbm small color; set output\"$3t=00$i.ppm\"; set noxtics; set
     noytics;
set size 1.0, 1.0; set yrange [0:.1];
plot 'run.00$i' using 1:2 w lines, 'run.00$i' using 1:3 w lines;
" >data00$i.gnu
gnuplot data00$i.gnu
ppmquant -map samp_colormap $3t=00$i.ppm>$3at=00$i.ppm
ppmtogif -map samp_colormap $3at=00$i.ppm > $300$i.gif
rm $3t=00$i.ppm $3at=00$i.ppm data00$i.gnu
i=i+100
fi

if test i -gt 999 -a i -lt 10000
then
print "set terminal pbm small color; set output\"$3t=0$i.ppm\"; set noxtics; set
     noytics;
set size 1.0, 1.0; set yrange [0:.1];
plot 'run.0$i' using 1:2 w lines, 'run.0$i' using 1:3 w lines;
" >data0$i.gnu
gnuplot data0$i.gnu
ppmquant -map samp_colormap $3t=0$i.ppm>$3at=0$i.ppm
ppmtogif -map samp_colormap $3at=0$i.ppm > $30$i.gif
rm $3t=0$i.ppm $3at=0$i.ppm data0$i.gnu
i=i+100
fi

done
```

Listing 3.2 MakeGifs.script, a script for creating animated gifs.

Here the -10 separates the frames by 0.1 s in real time, and the * is a wildcard that will include all .gif files in the present directory. Because we constructed the .gif files with sequential numbering, gifmerge pastes them together in the proper sequence and places them in the file movie, which can be viewed with a browser.

3.5 OpenDX for Dicing and Slicing

See Appendix C and the CD.

3.6 Texturing and 3-D Imaging

In §13.10 we give a brief explanation of how the inclusion of *textures* (Perlin noise) in a visualization can add an enhanced degree of realism. While it is a useful graphical technique, it incorporates the type of correlations and coherence also present in fractals and thus is in Chapter 13, "Fractals & Statistical Growth." In a related vein, in §13.10.1 we discuss the graphical technique of ray tracing and how it, especially when combined with Perlin noise, can produce strikingly realistic visualizations.

Stereographic imaging creates a virtual reality in which your brain and eye see objects as if they actually existed in our 3-D world. There are a number of techniques for doing this, such as virtual reality caves in which the viewer is immersed in an environment with images all around, and projection systems that project multiple images, slightly displaced, such that the binocular vision system in your brain (possibly aided by appropriate glasses) creates a 3-D image in your mind's eye.

Stereographics is often an effective way to let the viewer see structures that might otherwise be lost in the visualization of complex geometries, such as in molecular studies or star creation. But as effective as it may be, stereo viewing is not widely used in visualization because of the difficulty and expense of creating and viewing images. Here we indicate how the low-end, inexpensive viewing technique known as *ChromaDepth* [Chrom] can produce many of the same effects as high-end stereo vision without the use of special equipment. Not only is the technique easy to view and easy to publish, it is also easy to create [Bai 05]. Indeed, the OpenDX color images (visible on the CD) work well with ChromaDepth.

ChromaDepth consists of two pieces: a simple pair of glasses and a display methodology. The glasses contain very thin, diffractive gratings. One grating is blazed so that it shifts colors on the red end of the spectrum more than on the blue end, and this makes the red elements in the 3-D scene appear to be closer to the viewer. This often works fine with the same color scheme used for coding topographic maps or for scientific visualizations and so requires little or no extra work. You just write your computer program so that it color-codes the output in a linear rainbow spectrum based on depth. If you do not wear the glasses, you still see the visualization, yet with the glasses on, the image appears to jump out at you.

4

Object-Oriented Programs: Impedance & Batons

This chapter contains two units dealing with object-oriented programming (OOP) at increasing levels of sophistication. In most of the codes in this book we try to keep our programming transparent to a wide class of users and to keep our Java examples similar to the ones in C and Fortran. Accordingly, we have deliberately avoided the use of advanced OOP techniques. Nevertheless, OOP is a key element in modern programming, and so it is essential that all readers have at least an introductory understanding of it. We recommend that you review Unit I so that you are comfortable declaring, creating, and manipulating both static and dynamic objects. Unit II deals with more advanced aspects of OOP and, while recommended, may be put off for later reading, especially for those who are object-challenged at this stage in their computing careers. (Connelly Barnes helped to prepare Unit II.)

4.1 Unit I. Basic Objects: Complex Impedance

Problem: We are given a circuit containing a resistor of resistance R, an inductor of inductance L, and a capacitor of capacitance C (Figure 4.1 left). All three elements are connected in series to an alternating voltage source $V(t) = V_0 \cos \omega t$. Determine the magnitude and time dependence of the current in this circuit as a function of the frequency ω.

We solve this RLC circuit for you and assign as *your particular problem* that you repeat the calculation for a circuit in which there are two RLC circuits in parallel (Figure 4.1 right). Assume a single value for inductance and capacitance, and three values for resistance:

$$L = 1000\,\text{H}, \qquad C = \frac{1}{1000}\,\text{F}, \qquad R = \frac{1000}{1.5}, \ \frac{1000}{2.1}, \ \frac{1000}{5.2}\,\Omega. \tag{4.1}$$

Consider frequencies of applied voltage in the range $0 < \omega < 2/\sqrt{LC} = 2/\text{s}$.

4.2 Complex Numbers (Math)

Complex numbers are useful because they let us double our work output with only the slightest increase in effort. This is accomplished by manipulating them as

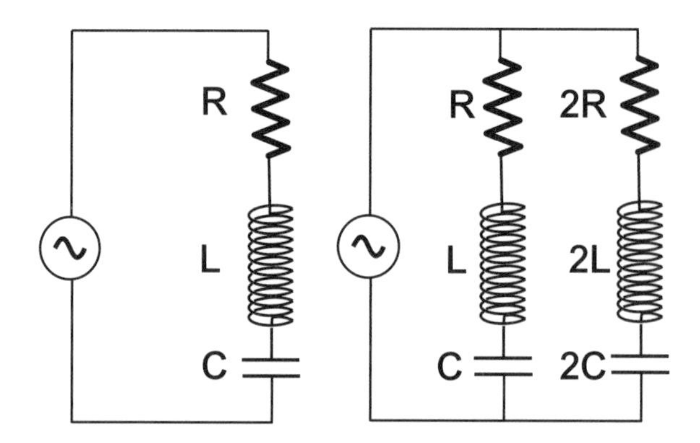

Figure 4.1 *Left:* An *RLC* circuit connected to an alternating voltage source. *Right:* Two *RLC* circuits connected in parallel to an alternating voltage. Observe that one of the parallel circuits has double the values of *R*, *L*, and *C* as does the other.

if they were real numbers and then separating the real and imaginary parts at the end of the calculation. We define the symbol z to represent a number with both real and imaginary parts (Figure 4.2 left):

$$z = x + iy, \quad \text{Re } z = x, \quad \text{Im } z = y. \tag{4.2}$$

Here $i \stackrel{\text{def}}{=} \sqrt{-1}$ is the imaginary number, and the combination of real plus imaginary numbers is called a *complex* number. In analogy to a *vector* in an imaginary 2-D space, we also use *polar coordinates* to represent the same complex number:

$$r = \sqrt{x^2 + y^2}, \qquad \theta = \tan^{-1}(y/x), \tag{4.3}$$

$$x = r\cos\theta, \qquad y = r\sin\theta. \tag{4.4}$$

The essence of the computing aspect of our problem is the programming of the rules of arithmetic for complex numbers. This is an interesting chore because while most computer languages contain all the rules for real numbers, you must educate them as to the rules for complex numbers (Fortran being the well-educated exception). Indeed, since complex numbers are not *primitive data types* like doubles and floats, we will construct complex numbers as *objects*. We start with two complex numbers, which we distinguish with subscripts:

$$z_1 = x_1 + i\,y_1, \quad z_2 = x_2 + i\,y_2. \tag{4.5}$$

Complex arithmetic rules derive from applying algebra to z's Re and Im parts:

Addition: $$z_1 + z_2 = (x_1 + x_2) + i(y_1 + y_2), \tag{4.6}$$

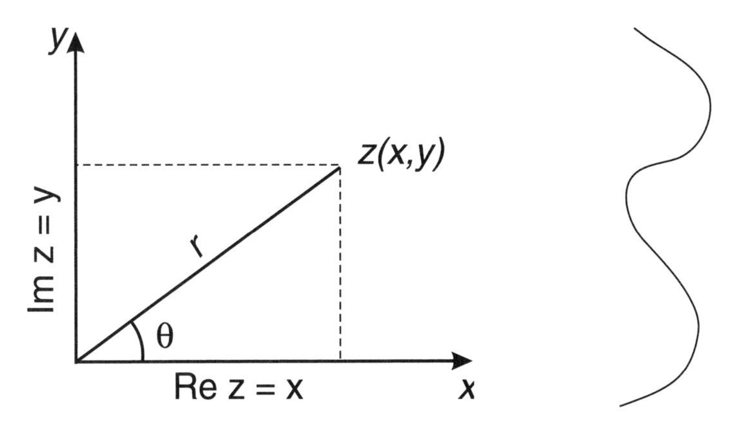

Figure 4.2 *Left:* Representation of a complex number as a vector in space. *Right:* An abstract drawing, or what?

Subtraction: $$z_1 - z_2 = (x_1 - x_2) + i(y_1 - y_2), \tag{4.7}$$

Multiplication: $$z_1 \times z_2 = (x_1 + iy_1) \times (x_2 + iy_2) \tag{4.8}$$

$$= (x_1 x_2 - y_1 y_2) + i(x_1 y_2 + x_2 y_1)$$

Division: $$\frac{z_1}{z_2} = \frac{x_1 + iy_1}{x_2 + iy_2} \times \frac{x_2 - iy_2}{x_2 - iy_2} \tag{4.9}$$

$$= \frac{(x_1 x_2 + y_1 y_2) + i(y_1 x_2 - x_1 y_2)}{x_2^2 + y_2^2}.$$

An amazing theorem by Euler relates the base of the natural logarithm system, complex numbers, and trigonometry:

$$e^{i\theta} = \cos\theta + i\sin\theta \quad \text{(Euler's theorem)}. \tag{4.10}$$

This leads to the polar representation of complex numbers (Figure 4.2 left),

$$z \equiv x + iy = re^{i\theta} = r\cos\theta + ir\sin\theta. \tag{4.11}$$

Likewise, Euler's theorem can be applied with a complex argument to obtain

$$e^z = e^{x+iy} = e^x e^{iy} = e^x(\cos y + i\sin y). \tag{4.12}$$

4.3 Resistance Becomes Impedance (Theory)

We apply Kirchhoff's laws to the RLC circuit in Figure 4.1 left by summing voltage drops as we work our way around the circuit. This gives the differential equation for the current $I(t)$ in the circuit:

$$\frac{dV(t)}{dt} = R\,\frac{dI}{dt} + L\,\frac{d^2I}{dt^2} + \frac{I}{C}, \tag{4.13}$$

where we have taken an extra time derivative to eliminate an integral over the current. The analytic solution follows by assuming that the voltage has the form $V(t) = V_0 \cos \omega t$ and by guessing that the resulting current $I(t) = I_0 e^{-i\omega t}$ will also be complex, with its real part the physical current. Because (4.13) is linear in I, the law of linear superposition holds, and so we can solve for the complex I and then extract its real and imaginary parts:

$$I(t) = \frac{1}{Z} V_0 e^{-i\omega t}, \quad Z = R + i\left(\frac{1}{\omega C} - \omega L\right), \tag{4.14}$$

$$\Rightarrow \quad I(t) = \frac{V_0}{|Z|} e^{-i(\omega t + \theta)} = \frac{V_0}{|Z|}\left[\cos(\omega t + \theta) - i \sin(\omega t + \theta)\right], \tag{4.15}$$

$$|Z| = \sqrt{R^2 + \left(\frac{1}{\omega C} - \omega L\right)^2}, \quad \theta = \tan^{-1}\left(\frac{1/\omega C - \omega L}{R}\right).$$

We see that the amplitude of the current equals the amplitude of the voltage divided by the magnitude of the complex impedance, and that the phase of the current relative to that of the voltage is given by θ.

The solution for the two *RLC* circuits in parallel (Figure 4.1 right) is analogous to that with ordinary resistors. Two impedances in series have the same current passing through them, and so we add voltages. Two impedances in parallel have the same voltage across them, and so we add currents:

$$Z_{\text{ser}} = Z_1 + Z_2, \quad \frac{1}{Z_{\text{par}}} = \frac{1}{Z_1} + \frac{1}{Z_2}. \tag{4.16}$$

4.4 Abstract Data Structures, Objects (CS)

What do you see when you look at the *abstract* object on the right of Figure 4.2? Some readers may see a face in profile, others may see some parts of human anatomy, and others may see a total absence of artistic ability. This figure is abstract in the sense that it does not try to present a true or realistic picture of the object but rather uses a symbol to suggest more than meets the eye. Abstract or formal concepts pervade mathematics and science because they make it easier to describe nature. For example, we may define $v(t)$ as the velocity of an object as a function of time. This is an abstract concept in the sense that we cannot see $v(t)$ but rather just infer it from the changes in the observable position. In computer science we create an abstract object by using a symbol to describe a collection of items. In Java we have built- in or *primitive data types* such as integers, floating-point numbers, Booleans, and strings. In addition, we may define *abstract data structures* of our own creation by combining primitive data types into more complicated structures called *objects*. These objects are abstract in the sense that they are named with a single symbol yet they contain multiple parts.

To distinguish between the general structure of objects we create and the set of data that its parts contain, the general object is called a *class*, while the object with specific values for its parts is called an *instance* of the class, or just an *object*. In this unit our objects will be complex numbers, while at other times they may be plots, vectors, or matrices. The classes that we form will not only contain objects (data structures) but also the associated methods for modifying the objects, with the entire class thought of as an object.

In computer science, abstract data structures must possess three properties:

1. **Typename:** Procedure to construct new data types from elementary pieces.
2. **Set values:** Mechanism for assigning values to the defined data type.
3. **Set operations:** Rules that permit operations on the new data type (you would not have gone to all the trouble of declaring a new data type unless you were interested in doing something with it).

In terms of these properties, when we declare a "complex" variable to have real and imaginary parts, we satisfy property 1. When we assign *doubles* to the parts, we satisfy property 2. And when we define addition and subtraction, we satisfy property 3.

Before we examine how these properties are applied in our programs, let us review the structure we have been using in our Java programs. When we start our programs with a declaration statement such as `double x`, this tells the Java compiler the kind of variable `x` is, so that Java will store it properly in memory and use proper operations on it. The general rule is that *every variable we use in a program must have its data type declared.* For primitive (built-in) data types, we declare them to be `double`, `float`, `int`, `char`, `long`, `short`, or `boolean`. If our program employs some user-defined abstract data structures, then they too must be declared. This declaration must occur even if we do not define the meaning of the data structure until later in the program (the compiler checks on that). Consequently, when our program refers to a number z as complex, the compiler must be told at some point that there are *both* a real part x and an imaginary part y that make up a complex number.

The actual process of creating objects in your Java program requires *nonstatic* class variables and methods. This means we leave out the word `static` in declaring the class and class variables. If the class and class variables are no longer static, they may be thought of as *dynamic*. Likewise, methods that deal with objects may be either static or dynamic. The static ones take objects as arguments, much like conventional mathematical functions. In contrast, dynamic methods accomplish the same end by modifying or interacting with the objects. Although you can deal with objects using static methods, you must use dynamic (nonstatic) methods to enjoy the full power of object-oriented programming.

4.4.1 Object Declaration and Construction

Though we may assign a name like x to an object, because objects have multiple components, you cannot assign one explicit *value* to the object. It follows then, that when Java deals with objects, it does so by *reference*; that is, the name of the variable *refers to the location in memory* where the object is stored and not to the explicit values of the object's parts. To see what this means in practice, the class file Complex.java in Listing 4.1 adds and multiplies complex numbers, with the complex numbers represented as objects.

```
// Complex.java:   Creates "Complex" class with static members

public class Complex {
  public double re, im;                                    // Nonstatic class variables

  public Complex ()  { re = 0;  im = 0; }                       // Default constructor

  public Complex(double x, double y)  { re = x;  im = y; }        // Full constructor

  public static Complex add(Complex a, Complex b) {              // Static add method
    Complex temp = new Complex();                              // Create Complex temp
    temp.re = a.re + b.re;
    temp.im = a.im + b.im;
    return temp;
  }

  public static Complex mult(Complex a, Complex b) {             // Static mult method
    Complex temp = new Complex();                              // Create Complex temp
    temp.re = a.re * b.re - a.im * b.im;
    temp.im = a.re * b.im + a.im * b.re;
    return temp;
  }

  public static void main(String[] argv) {                          // Main method
    Complex a, b;                                          // Declare 2 Complex objects
    a = new Complex();                                          // Create the objects
    b = new Complex(4.7, 3.2);
    Complex c = new Complex(3.1, 2.4);                          // Declare, create in 1
    System.out.println("a,b=("+a.re+","+a.im+"), ("+b.re+","+b.im+"),");
    System.out.println("c = ("+c.re+ ", " +c.im+ ")");
    a = add(b, c);                                              // Perform arithmetic
    System.out.println( "b + c = (" +a.re+ ", " +a.im+ "), ");
    a = mult(b, c);
    System.out.println( "b*c = (" +a.re+ ", " +a.im+ ")");
  }
}
```

Listing 4.1 Complex.java defines the object class Complex. This permits the use of complex data types (objects). Note the two types of Complex constructors.

4.4.2 Implementation in Java

1. Enter the program Complex.java by hand, trying to understand it as best you are able. (Yes, we know that you can just copy it, but then you do not become familiar with the constructs.)
2. Observe how the word Complex is a number of things in this program. It is the name of the *class* (line 3), as well as the name of two methods that create the object (lines 6 and 10). Methods, such as these, that create objects are called *constructors*. Although neophytes view these multiple uses of the name Complex as confusing, more experienced users often view it as elegant and efficient. Look closely and take note that this program has nonstatic variables (the word static is not on line 4).
3. Compile and execute this program and check that the output agrees with the results you obtain by doing a hand calculation. ▌

The first thing to notice about Complex.java is that the class is declared on line 3 with the statement

```
public class Complex {                                                    3
```

The main method is declared on line 24 with the statement

```
public static void main(String[] argv)                                    24
```

These are the same techniques you have seen before. However, on line 4 we see that the variables re and im are declared for the entire class with the statement

```
public double re, im;                                                     4
```

Although this is similar to the declaration we have seen before for class variables, observe that the word static is absent. This indicates that these variables are interactive or dynamic, that is, parts of an object. They are dynamic in the sense that they will be different for each specific object (*instance* of the class) created. That is, if we define z1 and z2 to be Complex objects, then the variables re and im each will be different for z1 and z2.

We extract the component parts of our complex object by *dot operations*, much like extracting the components of a physical vector by taking dot products of it with unit vectors along each axis:

z1.re	real part of object z1	**z1.im**	imaginary part of object z1
z2.re	real part of object z2	**z2.im**	imaginary part of object z2

This same dot notation is used to access the methods of objects, as we will see shortly. On line 6 in Listing 4.1 we see a method Complex declared with the statement

```
public Complex()                                                          6
```

On line 8 we see a method Complex declared yet again, but with a different statement:

```
public Complex(double x, double y)                                    8
```

Some explanation is clearly in order! First, notice that both of these methods are nonstatic (the word static is absent). In fact, they are the methods that construct our complex number object, which we call Complex. Second, notice that the name of each of these methods is the same as the name of the class, Complex.[1] They are spelled with their first letters capitalized, rather than with the lowercase letters usually used for methods, and because these objects have the same name as the class that contains them.

The two Complex methods are used to *construct* the object, and for this reason are called *constructors*. The first Complex constructor on line 6 is seen to be a method that takes no argument and returns no value (yes, this appears rather weird, but be patient). When Complex is called with no argument, as we see on line 18, the real and imaginary parts of the complex number (object) are automatically set to zero. This method is called the *default constructor* since it does what Java would otherwise do automatically (by default) when first creating an object, namely, set all its component parts initially to zero. We have explicitly included it for pedagogical purposes.

Exercise: Remove the default constructor (on line 6 with no argument) from Complex.java and check that you get the same result for the call to Complex().

The Complex method on line 8 implements the standard way to construct complex numbers. It is seen to take the two doubles x and y as input arguments, to set the real part of the complex number (object) to x, to set the imaginary part to y, and then to return to the place in the program that called it. This method is an additional constructor for complex numbers but differs from the default constructor by taking arguments. Inasmuch as the nondefault constructor takes arguments while the default constructor does not, Java does not confuse it with the default constructor even though both methods have the same name.

Okay, let us now take stock of what we have up to this point. On line 4 Complex.java has declared the variables re and im that will be the two separate parts of the created object. As each instance of each object created will have different values for the object's parts, these variables are referred to as *instance variables*. Because the name of the class file and the names of the objects it creates are all the same, it is sometimes useful to use yet another word to distinguish one from the other. Hence the phrase *instance of a class* is used to refer to the created objects (in our example, a, b, and c). This distinguishes them from the definition of the abstract data type.

[1] These two nonstatic methods are special as they permit the parameters characterizing the object that they construct to be passed via a new statement.

Look now at the main method to see how to go about creating objects using the constructor Complex. In the usual place for declaring variables (line 25 in the listing) we have the statement

```
Complex a, b;                                                        25
```

Because the compiler knows that Complex is not one of its primitive (built-in) data types, it assumes that it must be one that we have defined. In the present case, the class file contains the nonstatic class named Complex, as well as the constructors for Complex objects (data types). This means that the compiler does not have to look very far to know what you mean by a complex data type. For this reason, when the statement Complex a, b; on line 25 declares the variables a and b to be Complex objects, the compiler knows that they are manifestly objects since the constructors are *not* static.

Recall that declaring a variable type, such as double or int, does not assign a value to the variable but instead tells the compiler to add the name of the variable to the list of variables it will encounter. Likewise, the declaration statement Complex a, b; lets the compiler know what type of variables these are without assigning values to their parts. The actual *creation of objects* requires us to place numerical values in the memory locations that have been reserved for them. Seeing that an object has multiple parts, we cannot give all its parts initial values with a simple assignment statement like a = 0, so something fancier is called for. This is exactly why constructor methods are used. Specifically, on line 26 we have the object a created with the statement

```
a = new Complex();                                                   26
```

and on line 27 we have the object b created with the statement

```
b = new Complex(4.7, 3.2);                                           27
```

Look at how the creation of a new object requires the command new to precede the name of the constructor (Complex in this case). Also note that line 26 uses the default constructor method to set both the re and im parts of a to zero, while line 27 uses the second constructor method to set the re part of b to 4.7 and the im part of b to 3.2.

Just as we have done with the primitive data types of Java, it is possible to both declare and initialize an object in one statement. Indeed, line 28 does just that for object c with the statement

```
Complex c = new Complex(3.1, 2.4);                                   28
```

Notice how the data type Complex precedes the variable name c in line 28 because the variable c has not previously been declared; future uses of c should *not* declare or create it again.

4.4.3 Static and Nonstatic Methods

Once our complex-number objects have been declared (added to the variable list) and created (assigned values), it is easy to do arithmetic with them. First, we will get some experience with object arithmetic using the traditional static methods, and then in §4.4.4 we will show how to perform the same object arithmetic using *nonstatic* methods.

We know the rules of complex arithmetic and complex trigonometry and now will write Java methods to implement them. It makes sense to place these methods in the same class file that defines the data type since these associated methods are needed to manipulate objects of that data type. On line 31 in our *main* program we see the statement

```
a = add(b, c);                                                          31
```

This says to add the complex number b to the complex number c and then to store the result "as" (in the memory location reserved for) the complex number a. You may recall that we initially set the re and im parts of a to zero in line 26 using the default Complex constructor. This statement will replace the initial zero values with those computed in line 31. The method add that adds two complex numbers is defined on lines 10–15. It starts with the statement

```
public static Complex add(Complex a, Complex b)                         10
```

and is declared to be *static* with the two Complex objects (numbers) a and b as arguments. The fact that the word Complex precedes the method's name add signifies that the method will return a Complex number object as its result. We have the option of defining other names like complex_add or plus for this addition method.

The calculational part of the add method starts on line 11 by declaring and creating a temporary complex number temp that will contain the result of the addition of the complex numbers a and b. As indicated before, the *dot operator* convention means that temp.re will contain the re part of temp and that temp.im will contain the imaginary part. Thus the statements on lines 12 and 13,

```
temp.re = a.re + b.re;                                                  12
temp.im = a.im + b.im;                                                  13
```

add the complex numbers a and b by extracting the real parts of each, adding them together, and then storing the result as the re part of temp. Line 13 determines the imaginary part of the sum in an analogous manner. Finally, the statement

```
return temp;                                                            21
```

returns the object (complex number) temp as the value of add(Complex a, Complex b). Because a complex number has two parts, both parts must be returned to the calling program, and this is what return temp does.

4.4.4 Nonstatic Methods

The program ComplexDyn.java in Listing 4.2 also adds and multiplies complex numbers as objects, but it uses what are called *dynamic, nonstatic,* or *interactive* methods. This is more elegant and powerful, but less like, the traditional procedural programming. To avoid confusion and to permit you to run both the static and nonstatic versions without them interfering with each other, the nonstatic version is called ComplexDyn, in contrast to the Complex used for the static method. Notice how the names of the methods in ComplexDyn and Complex are the same, although they go about their business differently.

```
// ComplexDyn.java: Complex object class with nonstatic members

public class ComplexDyn {
  public double re; public double im;                    // Nonstatic class variables

  public ComplexDyn() { re = 0; im = 0; }                     // Default constructor

  public ComplexDyn(double x, double y) { re = x; im = y; }           // Constructor

  public void add(ComplexDyn other)                         // Dynamic other + this
    { this.re = this.re + other.re;  this.im = this.im + other.im; }

  public void mult(ComplexDyn other) {                       // Dynamic other*this
    ComplexDyn ans = new ComplexDyn();                             // Intermediate
    ans.re  = this.re * other.re - this.im * other.im;
    ans.im  = this.re * other.im + this.im * other.re;
    this.re = ans.re;                                 // Copy value into returned object
    this.im = ans.im;
  }

  public static void main(String[] argv) {                 // Indep static Main object
    ComplexDyn a, b;                                  // Declare 2 Complex objects
    a = new ComplexDyn();                                      // Create objects
    b = new ComplexDyn(4.7, 3.2);
    ComplexDyn c = new ComplexDyn(3.1, 2.4);                   // Declare, create
    System.out.println("a,b=("+a.re+", "+a.im+"),("+b.re+", "+ b.im+"),");
    System.out.println("c = (" +c.re+ ", " +c.im+ ")");
    c.add(b);                                                  // Nonstatic add
    a = c;
    System.out.println("b + c = (" + a.re + ", " + a.im + "), ");
    c = new ComplexDyn(3.1, 2.4);
    c.mult(b);                                                 // Nonstatic mult
    System.out.println("b*c = (" + c.re + ", " + c.im + ")");
  }
}
```

Listing 4.2 ComplexDyn.java defines the object class ComplexDyn. This permits the use of dynamic complex data objects, for example, as c.add(b).

Exercise

1. Enter the ComplexDyn.java class file by hand, trying to understand it in the process. If you have entered Complex.java by hand, you may modify that program to save some time (but be careful!).

2. Compile and execute this program and check that the output agrees with the results you obtained in the exercises in §4.2.

Nonstatic methods go about their business by modifying the properties of the objects to which they are attached (complex numbers in the present case). In fact, we shall see that nonstatic methods are literally appended to the names of objects much as the endings of verbs are modified when their tenses change. In a sense, when the method is appended to an object, it creates a new object. To cite an instance, on line 28 we see the operation

```
c.add(b);                                         // Nonstatic addition 28
```

Study the way this statement says to take the complex number object c and modify it using the add method that adds b to the object. This statement results in new values for the parts of object c. Because object c is modified by this action, line 28 is equivalent to our static operation

```
c = add(c,b);                                     // Static method equivalent 28
```

Regardless of the approach, since c now contains the sum c + b, if we want to use c again, we must redefine it, as we do on line 31. On line 32 we take object c and multiply it by b via

```
c.mult(b);                                        // Nonstatic multiplication 32
```

This method changes c to c * b. Thus line 32 has the static method equivalence

```
c = mult(c, b);                                   // Static method equivalent 32
```

We see from these two examples that nonstatic methods are called using the same dot operator used to refer to instance variables of an object. In contrast, static methods take the object as arguments and do not use the dot operator. Thus we called the static methods via add(c,b) and mult(c,b) and called the nonstatic methods via c.add(b) and c.mult(b).

The static methods here do not need a dot operator since they are called from within the class Complex or ComplexDyn that defined them. However, they would need a dot operator if called from another class. For example, you may have seen that the square root method is called Math.sqrt(x). This is actually the static method sqrt from the class Math. You could call the static add(c,b) method of class Complex by using Complex.add(c,b). This works within the program (class) Complex as well but is not required. It is required, however, if add is called from other classes.

Observe now how the object-oriented add method has the distinctive form

```
public void add(ComplexDyn other)                                        10
{this.re = this.re + other.re; this.im = this.im + other.im;}            11
```

Line 10 tells us that the method add is nonstatic (the word static is absent), that no value or object is returned (the void), and that there is one argument other of the type

ComplexDyn. What is unusual about this nonstatic method is that it is supposed to add two complex numbers together, yet there is only one argument given to the method and no object is returned! Indeed, the assumption is that since the method is nonstatic, it will be used to modify only the object to which it will be attached. Hence it literally goes without saying that there is an object around for this method to modify, and the this reference is used to refer to "this" calling object. In fact, the reason the argument to the method is conventionally called other is to distinguish it from the this object that the method will modify. (We are being verbose for clarity's sake: The word "this" may be left out of these statements without changing their actions.) Consequently, when the object addition is done in line 11 with

```
this.re = this.re + other.re;                // Addition of re parts of this and other 11
```

it is understood that re refers to the current object being modified (this), while other refers to the "other" object being used to make the modifications.

4.5 Complex Currents (Solution)

1. Extend the class Complex.java or ComplexDyn.java by adding new methods to subtract, take the modulus, take the complex conjugate, and determine the phase of complex numbers.
2. Test your methods by checking that the following identities hold for a variety of complex numbers:

$$\begin{aligned} z+z &= 2z, & z+z* &= 2\ \mathrm{Re}\ z \\ z-z &= 0, & z-z* &= 2\ \mathrm{Im}\ z \\ zz* &= |z|^2, & zz* &= r^2 \quad \text{(which is real)} \end{aligned} \tag{4.17}$$

   Hint: Compare your output to some cases of pure real, pure imaginary, and simple complex numbers that you are able to evaluate by hand.
3. Equation (4.14) gives the magnitude and phase of the current in a single *RLC* circuit. Modify the given complex arithmetic program so that it performs the required complex arithmetic.
4. Compute and then make a plot of the magnitude and phase of the current in the circuit as a function of frequency $0 \le \omega \le 2$.
5. Construct a $z(x, y)$ surface plot of the magnitude and phase of the current as functions of *both* the frequency of the external voltage ω and of the resistance R. Observe how the magnitude has a maximum when the external frequency $\omega = 1/\sqrt{LC}$. This is the *resonance* frequency.
6. Another approach is to make a 3-D visualization of the complex Z as a function of a complex argument (Figure 4.3). Do this by treating the frequency $\omega = x + iy$ as a complex number. You should find a sharp peak at $x = \mathrm{Re}(\omega) = 1$. Adjust the plot so that you tell where $\mathrm{Im}(1/Z)$ changes sign. If you look closely at the graph, you should also see that there is a maximum for a negative imaginary value of ω. This is related to the length of the lifetime of the resonance.

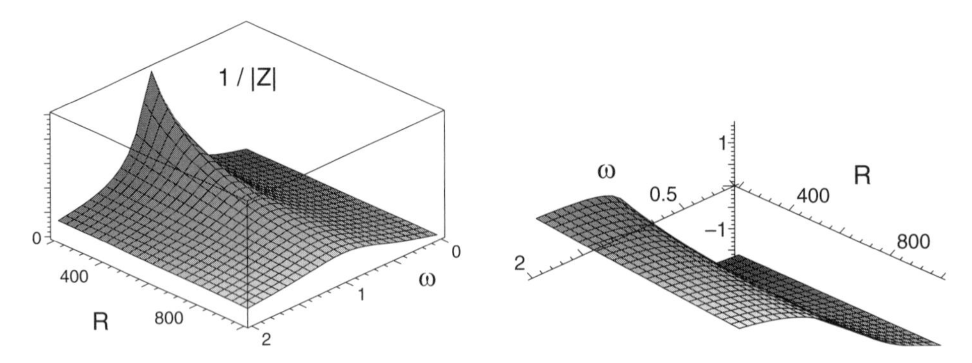

Figure 4.3 *Left:* A plot of $1/|Z|$ *versus* resistance R and frequency ω showing that the magnitude of the current has a maximum at $\omega = 1$. *Right:* A plot of the current's phase *versus* resistance R and frequency ω showing that below resonance, $\omega < 1$, the current lags the voltage, while above resonance the current leads the voltage.

7. **Assessment:** You should notice a resonance peak in the magnitude at the same frequency for which the phase vanishes. The smaller the resistance R, the more sharply the circuit should pass through resonance. These types of circuits were used in the early days of radio to tune to a specific frequency. The sharper the peak, the better the quality of reception.
8. The second part of the problem dealing with the two circuits in parallel is very similar to the first part. You need to change only the value of the impedance Z used. To do that, explicitly perform the complex arithmetic implied by (4.16), deduce a new value for the impedance, and then repeat the calculation of the current.

4.6 OOP Worked Examples

Creating object-oriented programs requires a transition from a procedural programming mindset, in which functions take arguments as input and produce answers as output, to one in which objects are created, probed, transferred, and modified. To assist you in the transition, we present here two sample procedural programs and their OOP counterparts. In both cases the OOP examples are longer but presumably easier to modify and extend.

4.6.1 OOP Beats

You obtain beats if you add together two sine functions y_1 and y_2 with nearly identical frequencies,

$$y_3(t) = y_1(t) + y_2(t) = A\,\sin(30\,t) + A\,\sin(33\,t). \tag{4.18}$$

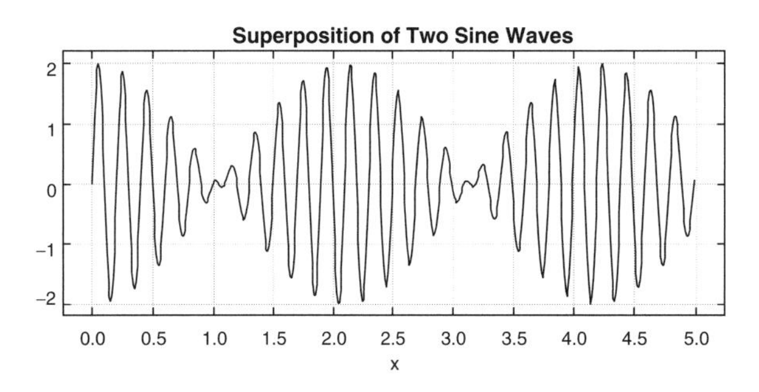

Figure 4.4 Superposition of two waves with similar wave numbers (a PtPlot).

Beats look like a single sine wave with a slowly varying amplitude (Figure 4.4). In Listing 4.3 we give Beats.java, a simple program that plots beats. You see here that all the computation is done in the main program, with no methods called other than those for plotting. On lines 14 and 15 the variables y1 and y2 are defined as the appropriate functions of time and then added together on line 16 to form the beats. Contrast this with the object-oriented program OOPBeats.java in Listing 4.4 that produces the same graph.

```
// Beats.java:     plots beats
import ptolemy.plot.*;                                                    2

public class Beats {                                                      4

  public static void main(String[] argv) {                                6
    double y1, y2, y3, x;
    int i;                                                                8
    x = 0.;                                        // Initial position
    Plot myPlot = new Plot();                                             10
    myPlot.setTitle("Superposition of Two Sine Waves");
    myPlot.setXLabel(" x");                                               12
    for ( i = 1;  i < 501;  i++ ) {
      y1 = Math.sin(30*x);                                 // Wave 1      14
      y2 = Math.sin(33*x);                                 // Wave 2
      y3 = y1 + y2;                                    // Sum of waves    16
      myPlot.addPoint(0, x, y3, true);
      x = x + 0.01;                          // Small increment in x      18
    }
   PlotApplication app = new  PlotApplication(myPlot);                    20
} }
```

Listing 4.3 Beats.java plots beats using procedural programming. Contrast this with the object-oriented program OOPBeats.java in Listing 4.4.

In this OOP version, the main program is at the very end, on lines 26–29. It is short because all it does is create an OOPbeats object named sumsines on line 27 with the appropriate parameters and then on line 28 sums the two waves by having the method sumwaves modify the object. The constructor for an OOPbeats object is given on line 7, followed by the sumwaves method. The sumwaves method takes no arguments and returns no value; the waves are summed on line 19, and the graph plotted all within the method.

```
//   OOPBeats.java:  OOP Superposition 2 Sine waves
import ptolemy.plot.*;

public class OOPBeats {
  public double A, k1, k2;
                                                          // Class Constructor
  public OOPBeats(double Ampl, double freq1, double freq2)
    { A = Ampl;  k1 = freq1;  k2 = freq2; }

  public void sumwaves() {                                   // Sums 2 waves
    int i;
    double y1, y2, y3, x = 0;
    Plot myPlot = new Plot();
    myPlot.setTitle("Superposition of two Sines");
    myPlot.setXLabel(" x");
    for ( i = 1; i < 501; i++ ) {
      y1 = A*Math.sin(k1*x);                                    // 1st Sine
      y2 = A*Math.sin(k2*x);                                    // 2nd Sine
      y3 = y1 + y2;                                          // Superpositon
      myPlot.addPoint(0, x, y3, true);
      x = x + 0.01;                                           // Increment x
    }
    PlotApplication app = new  PlotApplication(myPlot);
  }

  public static void main(String[] argv) {                  // Class instance
    OOPBeats sumsines = new OOPBeats(1., 30., 33.);             // Instance
    sumsines.sumwaves();                                // Call sumsins' method
  }
}
```

Listing 4.4 OOPBeats.java plots beats using OOP. Contrast this with the procedural program Beats.java in Listing 4.3.

4.6.2 OOP Planet

In our second example we add together periodic functions representing positions *versus* time. One set describes the position of the moon as it revolves around a planet, and the other set describes the position of the planet as it revolves about the sun. Specifically, the planet orbits the sun at a radius $R = 4$ units with an angular frequency $\omega_p = 1\,\text{rad/s}$, while the moon orbits Earth at a radius $r = 1$ unit from the planet and an angular velocity $\omega_s = 14\,\text{rad/s}$. The position of the planet at time t relative to the sun is described by

$$x_p = R\,\cos(\omega_p\,t), \quad y_p = R\,\sin(\omega_p\,t). \tag{4.19}$$

The position of the satellite, relative to the sun, is given by the sum of its position relative to the planet and the position of the planet relative to the sun:

$$\begin{aligned} x_s &= x_p + r\,\cos(\omega_s\,t) = R\,\cos(\omega_p\,t) + r\,\cos(\omega_s\,t), \\ y_s &= y_p + r\,\sin(\omega_s\,t) = R\,\sin(\omega_p\,t) + r\,\sin(\omega_s\,t). \end{aligned}$$

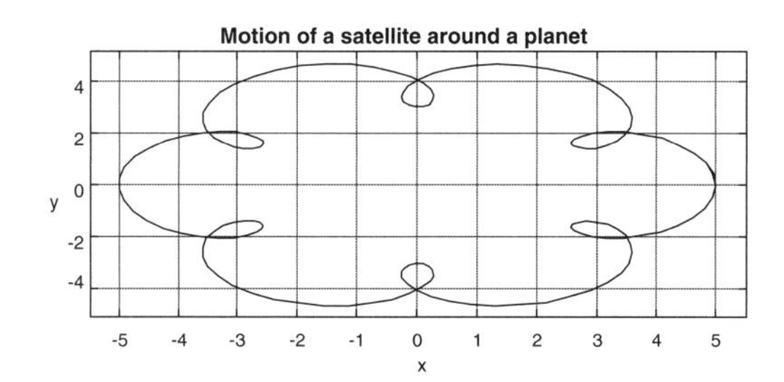

Figure 4.5 The trajectory of a satellite as seen from the sun.

So again this looks like beating (if $\omega_s \simeq \omega_p$ and if we plot x or y *versus* t), except that now we will make a parametric plot of $x(t)$ *versus* $y(t)$ to obtain a visualization of the orbit. A procedural program Moon.java to do this summation and to produce Figure 4.5 is in Listing 4.5.

```
//   Moon.java:  moon orbiting a planet
import ptolemy.plot.*;

public class Moon {

  public static void main(String[] argv) {
    double Radius, wplanet, radius, wmoon, time, x, y;
    Radius = 4. ;                                              // Planet
    wplanet = 2. ;                                    // Omega of planet
    radius = 1. ;                                      // Moon's orbit r
    wmoon = 14. ;                               // Oemga moon wrt planet
    Plot myPlot = new Plot();
    myPlot.setTitle("Motion of a moon around a planet ");
    myPlot.setXLabel(" x");
    myPlot.setYLabel(" y");
    for ( time = 0. ; time < 3.2; time = time + 0.02)   {
      x = Radius *Math.cos(wplanet*time) + radius*Math.cos(wmoon*time);
      y = Radius *Math.sin(wplanet*time) + radius*Math.sin(wmoon*time);
      myPlot.addPoint(0, x, y, true);
    }
    PlotApplication app = new  PlotApplication(myPlot);
  }
}
```

Listing 4.5 The procedural program **Moon.java** computes the trajectory of a satellite as seen from the sun. You need to write your own OOP version of this program.

Exercise: Rewrite the program using OOP.

1. Define a mother class OOPlanet containing:

Radius	Planet's orbit radius
wplanet	Planet's orbit ω
(xp, yp)	Planet's coordinates
getX(double t), getY(double t)	Planet coordinates methods
trajectory()	Method for planet's orbit

2. Define a daughter class OOPMoon containing:

radius	Radius of moon's orbit
wmoon	Frequency of moon in orbit
(xm, ym)	Moon's coordinates
trajectory()	Method for moon's orbit relative to sun

3. The main program must contain one instance of the class planet and another instance of the class Moon, that is, one planet object and one Moon object.
4. Have each instance call its own trajectory method to plot the appropriate orbit. For the planet this should be a circle, while for the moon it should be a circle with retrogrades (Figure 4.5).

One solution, which produces the same results as the previous program, is the program OOPlanet.java in Listing 4.6. As with OOPbeats.java, the main program for OOPlanet.java is at the end. It is short because all it does is create an OOPMoon object with the appropriate parameters and then have the moon's orbit plotted by applying the trajectory method to the object.

```
// OOPPlanet.java: Planet orbiting Sun
import ptolemy.plot.*;

public class OOPPlanet {
  double Radius, wplanet, xp, yp;                    // Orbit r, omega, Planet coordinates

  public OOPPlanet() { Radius = 0.; wplanet = 0.; }             // Default constructor

  public OOPPlanet(double Rad, double pomg) { Radius = Rad; wplanet = pomg; }

  public double getX( double time )                      // Get x of planet at time t
  { return Radius*Math.cos( wplanet*time ); }

  public double getY( double time )                  // Get y of the planet at time t
  { return Radius*Math.sin( wplanet*time ); }

  public void trajectory() {                              // Trajectory of the planet
    double time;
    Plot myPlot = new Plot();
    myPlot.setTitle( "Motion of a planet around the Sun" );
    myPlot.setXLabel(" x");
    myPlot.setYLabel(" y");
    for(time = 0.;time <3.2; time = time + 0.02)  {
      xp = getX( time );
      yp = getY( time );
      myPlot.addPoint(0, xp, yp, true);  }
    PlotApplication app = new  PlotApplication( myPlot );
    }
}                                                              // Planet class ends

class OOPMoon extends OOPPlanet {                  // OOPMoon = daughter class of planet
  double radius, wmoon, xm, ym;                     // R, omega satellite, moon wrt Sun

  public OOPMoon()  { radius = 0.;   wmoon = 0.; }

  public OOPMoon(double Rad, double pomg, double rad, double momg) {    // Full Constr
   Radius = Rad;   wplanet = pomg;  radius = rad;  wmoon = momg; }
                                              // Coordinates of moon relative to Sun
  public void trajectory() {
```

```
      double time;
      Plot myPlot = new Plot();
      myPlot.setTitle("Satellite orbit about planet");
      myPlot.setXLabel(" x");
      myPlot.setYLabel(" y");
      for(time = 0.; time<3.2; time = time+0.02){
        xm = getX(time) + radius*Math.cos(wmoon*time);
        ym = getY(time) + radius*Math.sin(wmoon*time);
        myPlot.addPoint(0, xm, ym, true); }
      PlotApplication app = new  PlotApplication(myPlot);
      }

    public static void main(String[] argv) {
      double Rad, pomg, rad,  momg;
      Rad = 4.;                                              // Planet
      pomg = 2.;                                  // Angular velocity of planet
      rad = 1.;                                        // Satellite orbit radius
      momg = 14.;                           // Ang. vel, satellite around planet
      // Uncomment next 2 lines for planet trajectory and
//      comment other two (Moon) lines
      //OOPlanet earth = new OOPlanet(Rad, pomg);
      //earth.trajectory();
      //next two lines if desirethe Moon trajectory
      //but previous two lines must be commented
      OOPMoon Selene = new OOPMoon(Rad, pomg, rad, momg);
      Selene.trajectory();
}}
```

Listing 4.6 OOPPlanet.java creates an OOPMoon object and then plots the moon's orbit by applying the trajectory method to the object.

What is new about this program is that it contains two classes, OOPlanet beginning on line 4 and OOPMoon beginning on line 31. This means that when you compile the program, you should obtain two class files, OOPlanet.class and OOPMoon.class. Yet because execution begins in the main method and the only main method is in OOPMoon, you need to execute OOPMoon.class to run the program:

```
% java OOPMoon                                    Execute main method
```

Scan the code to see how the class OOPMoon is within the class OOPlanet and is therefore a *subclass*. That being the case, OOPMoon is called a *daughter class* and OOPlanet is called a *mother class*. The daughter class inherits the properties of the mother class as well as having properties of its own. Thus, on lines 46 and 47, OOPMoon uses the getX(time) and getY(time) methods from the OOPlanet class without having to say OOPlanet.getX(time) to specify the class name.

4.7 Unit II. Advanced Objects: Baton Projectiles ⊙

In this unit we look at more advanced aspects of OOP. These aspects are designed to help make programming more efficient by making the reuse of already written components easier and more reliable. The ideal is to permit this even for entirely different future projects for which you will have no memory or knowledge of the internal workings of the already written components that you want to reuse. OOP concepts can be particularly helpful in complicated projects in which you need to add new features without "breaking" the old ones and in which you may be modifying code that you did not write.

4.8 Trajectory of a Thrown Baton (Problem)

We wish to describe the trajectory of a baton that spins as it travels through the air. On the left in Figure 4.6 the baton is shown as as two identical spheres joined by a massless bar. Each sphere has mass m and radius r, with the centers of the spheres separated by a distance L. The baton is thrown with the initial velocity (Figure 4.6 center) corresponding to a rotation about the center of the lower sphere.

Problem: Write an OOP program that computes the position and velocity of the baton as a function of time. The program should

1. plot the position of each end of the baton as a function of time;
2. plot the translational kinetic energy, the rotational kinetic energy, and the potential energy of the baton, all as functions of time;
3. use several classes as building blocks so that you may change one building block without affecting the rest of the program;
4. (optional) then be extended to solve for the motion of a baton with an additional lead weight at its center.

4.8.1 Combined Translation and Rotation (Theory)

Classical dynamics describes the motion of the baton as the motion of its center of mass (CM) (marked with an "X" in Figure 4.6), plus a rotation about the CM. Because the translational and rotational motions are independent, each may be determined separately, and because we ignore air resistance, the angular velocity ω about the CM is constant.

The baton is thrown with an initial velocity (Figure 4.6 center). The simplest way to view this is as a translation of the entire baton with a velocity $\mathbf{V}_0$ and a rotation of angular velocity ω about the CM (Figure 4.6 right). To determine ω, we note that the tangential velocity due to rotation is

$$v_t = \frac{1}{2}\omega L. \tag{4.20}$$

For the direction of rotation as indicated in Figure 4.6, this tangential velocity is added to the CM velocity at the top of the baton and is subtracted from the CM velocity at the bottom. Because the total velocity equals 0 at the bottom and $2\mathbf{V}_0$ at the top, we are able to solve for ω:

$$\frac{1}{2}\omega L - V_0 = 0 \;\Rightarrow\; V_0 = \frac{1}{2}\omega L, \quad \Rightarrow\; \omega = \frac{2V_0}{L}. \tag{4.21}$$

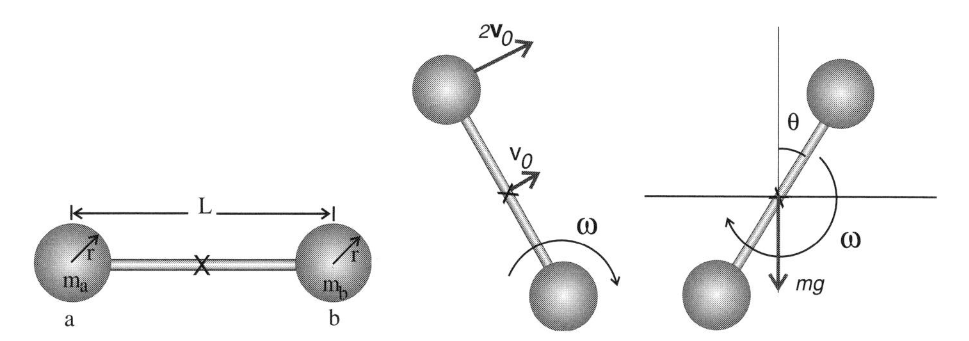

Figure 4.6 *Left:* The baton before it is thrown. "X" marks the CM. *Center:* The initial conditions for the baton as it is thrown. *Right:* The baton spinning in the air under the action of gravity.

If we ignore air resistance, the only force acing on the baton is gravity, and it acts at the CM of the baton (Figure 4.6 right). Figure 4.7 shows a plot of the trajectory $[x(t), y(t)]$ of the CM:

$$(x_{\rm cm},\, y_{\rm cm}) = \left(V_{0x}t,\, V_{0y}t - \frac{1}{2}gt^2\right), \quad (v_{x,\rm cm},\, v_{y,\rm cm}) = (\,V_{0x},\, V_{0y} - gt)\,,$$

where the horizontal and vertical components of the initial velocity are

$$V_{0x} = V_0 \cos\theta, \quad V_{0y} = V_0 \sin\theta.$$

Even though ω = constant, it is a constant about the CM, which itself travels along a parabolic trajectory. Consequently, the motion of the baton's ends may appear complicated to an observer on the ground (Figure 4.7 right). To describe the motion of the ends, we label one end of the baton a and the other end b (Figure 4.6 left). Then, for an angular orientation ϕ of the baton,

$$\phi(t) = \omega t + \phi_0 = \omega t, \tag{4.22}$$

where we have taken the initial $\phi = \phi_0 = 0$. Relative to the CM, the ends of the baton are described by the polar coordinates

$$(r_a, \phi_a) = \left(\frac{L}{2}, \phi(t)\right), \quad (r_b, \phi_b) = \left(\frac{L}{2}, \phi(t) + \pi\right). \tag{4.23}$$

The ends of the baton are also described by the Cartesian coordinates

$$(x'_a, y'_a) = \frac{L}{2}\left[\cos\omega t,\, \sin\omega t\right], \quad (x'_b, y'_b) = \frac{L}{2}\left[\cos(\omega t + \pi),\, \sin(\omega t + \pi)\right].$$

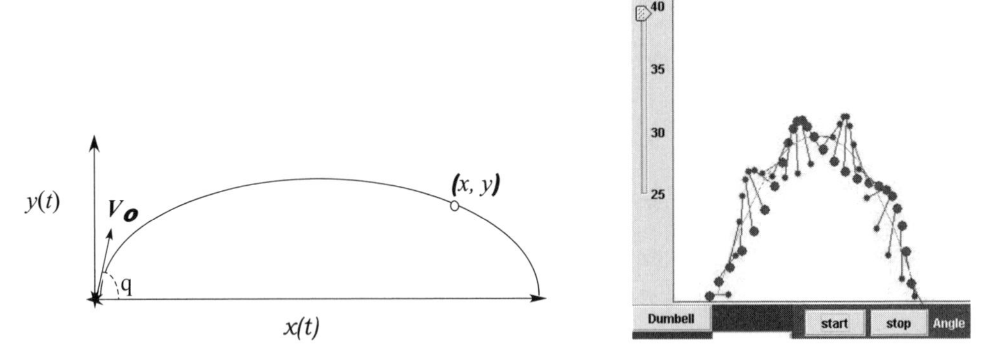

Figure 4.7 *Left:* The trajectory ($x(t)$, $y(t)$) followed by the baton's CM. *Right:* The applet `JParabola.java` showing the entire baton as its CM follows a parabola.

The baton's ends, as seen by a stationary observer, have the vector sum of the position of the CM plus the position relative to the CM:

$$(x_a,\ y_a) = \left[V_{0x}t + \frac{L}{2}\cos(\omega t),\ V_{0y}t - \frac{1}{2}gt^2 + \frac{L}{2}\sin(\omega t)\right], \tag{4.24}$$

$$(x_b,\ y_b) = \left[V_{0x}t + \frac{L}{2}\cos(\omega t + \pi),\ V_{0y}t - \frac{1}{2}gt^2 + \frac{L}{2}\sin(\omega t + \pi)\right].$$

If L_a and L_b are the distances of m_a and m_b from CM, then

$$L_a = \frac{m_b}{m_a + m_b}, \quad L_b = \frac{m_a}{m_a + m_b}, \quad \Rightarrow \quad m_a L_a = m_b L_b. \tag{4.25}$$

The moment of inertia of the masses (ignoring the bar connecting them) is

$$I_{\text{masses}} = m_a L_a^2 + m_b L_b^2. \tag{4.26}$$

If the bar connecting the masses is uniform with mass m and length L, then it has a moment of inertia about its CM of

$$I_{\text{bar}} = \frac{1}{12} m L^2. \tag{4.27}$$

Because the CM of the bar is at the same location as the CM of the masses, the total moment of inertia for the system is just the sum of the two:

$$I_{\text{tot}} = I_{\text{masses}} + I_{\text{bar}}. \tag{4.28}$$

The potential energy of the masses is

$$\mathrm{PE}_{\mathrm{masses}} = (m_a + m_b)gh = (m_a + m_b)g\left(V_0\, t\sin\theta - \frac{1}{2}gt^2\right), \tag{4.29}$$

while the potential energy of the bar just has $m_a + m_b$ replaced by m since both share the same CM location. The rotational kinetic energy of rotation is

$$\mathrm{KE}_{\mathrm{rot}} = \frac{1}{2}I\omega^2, \tag{4.30}$$

with ω the angular velocity and I the moment of inertia for either the masses or the bar (or the sum). The translational kinetic energy of the masses is

$$\mathrm{KE}_{\mathrm{trans}} = \frac{1}{2}m\left[(V_0\,\sin\theta - g\,t)^2 + (V_0\,\cos\theta)^2\right], \tag{4.31}$$

with $m_a + m_b$ replaced by m for the bar's translational KE.

To get a feel for the interestingly complicated motion of a baton, we recommend that the reader try out the applet JParabola on the CD (Fig. 4.7 right).

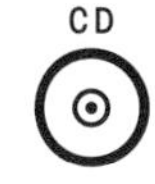


```
% appletviewer jcenterofmass.html
```

4.9 OOP Design Concepts (CS)

In accord with our belief that much of education is just an understanding of what the words mean, we start by defining OOP as programming containing component objects with four characteristics [Smi 91]:

Encapsulation: The data and the *methods* used to produce or access data are encapsulated into entities called *objects.* For our problem, the data are initial positions, velocities, and properties of a baton, and the objects are the baton and its path. As part of the OOP philosophy, data are manipulated only via distinct methods.

Abstraction: Operations applied to objects are assumed to yield standard results according to the nature of the objects. To illustrate, summing two matrices always gives another matrix. By incorporating abstraction into programming, we concentrate more on solving the problem and less on the details of the implementation.

Inheritance: Objects inherit characteristics (including code) from their ancestors yet may be different from their ancestors. A baton inherits the motion of a point particle, which in this case describes the motion of the CM, and extends that by permitting rotations about the CM. In addition, if we form a red baton, it inherits the characteristics of a colorless baton but with the property of color added to it.

Polymorphism: Methods with the same name may affect different objects differently. Child objects may have *member* functions with the same name but with properties differing from those of their ancestors (analogous to method overload, where the method used depends upon the method's arguments).

We now solve our baton problem using OOP techniques. Although it is also possible to use the more traditional techniques of procedural programming, this problem contains the successive layers of complexity that make it appropriate for OOP. We will use several source (.java) files for this problem, each yielding a different class file. Each class will correspond to a different physical aspect of the baton, with additional classes added as needed. There will be a class Path.java to represent the trajectory of the CM, a class Ball.java to represent the masses on the ends of the baton, and a class Baton.java to assemble the other classes into an object representing a projected and spinning baton. Ultimately we will combine the classes to solve our problem.

4.9.1 Including Multiple Classes

The codes Ball.java, Path.java, and Baton.java in Listings 4.7–4.9 produce three class files. You may think of each class as an object that may be created, manipulated, or destroyed as needed. (Recall that objects are abstract data types with multiple parts.) In addition, since these classes may be shared with other Java classes, more complicated objects may be constructed by using these classes as building blocks.

```
//      Ball.java: Isolated Ball with Mass and Radius
//      "Object" class, no main method, creates & probes objects                 2

public class Ball {                                                               4
  public double m, r;                         // Nonstatic variables unique to ball
                                                                                  6
  Ball(double mass, double radius) { m = mass;  r = radius; }     // Constructor
                                                                                  8
  public double getM() {return m;}                                    // Get mass
                                                                                  10
  public double getR() {return r;}                                  // Get radius
                                                                                  12
  public double getI() {return (2./5.)*m*r*r;}                           // Get I
}                                                                                 14
```

Listing 4.7 The class **Ball** representing the ball on the end of the baton.

To use class files that are not in your working directory, use the import command to tell Java where to find them.[2] For example, in Chapter 3, "Visualization Tools," we used import ptolemy.plot.* to tell Java to retrieve all class files found in the ptolemy/plot directory. A complication arises here in that multiple class files may contain more than one main method. That being the case, Java uses the first main it finds, starting in the directory from which you issued the java command.

[2] Actually, the Java compiler looks through all the directories in your classpath and imports the first instance of the needed class that it finds.

```java
// Path.java:          Parabolic Trajectory Class

public class Path {
  public static final double g = 9.8;                          // Static, same all Paths
  public double v0x, v0y;                                   // Non-static, unique each Path

  public Path(double v0, double theta)    {v0x = v0 * Math.cos(theta*Math.PI/180.);
                                           v0y = v0 * Math.sin(theta * Math.PI / 180.);  }

  public double getX(double t) { return v0x * t; }

  public double getY(double t) { return v0y * t - 0.5*g*t*t; }
}
```

Listing 4.8 The class **Path** creating an object that represents the trajectory of the center of mass.

```java
// Baton.java: Combines classes to form Baton
import ptolemy.plot.*;

public class Baton {
   public double L;                                   // Nonstatic variables unique ea baton
   public double w;                                                            // Omega
   public Path path;                                                     // Path object
   public Ball ball;                                                     // Ball object

   Baton(Path p, Ball b, double L1, double w1) {path = p; ball = b; L = L1; w = w1;}

   public double getM()      { return 2*ball.getM();}

   public double getI()    {return 2*ball.getI() + 1./2.*ball.getM()*L*L; }

   public double getXa(double t)    {return path.getX(t) + L/2*Math.cos(w*t);}

   public double getYa(double t)    {return path.getY(t) + L/2*Math.sin(w*t); }

   public double getXb(double t)    { return path.getX(t) - L/2* Math.cos(w*t); }

   public double getYb(double t)   { return path.getY(t) - L/2*Math.sin(w*t); }

   public static void main(String args[]) {                              // Main method
      double x, y;
      Plot myPlot = new Plot();                                          // Create Plot
      Ball myBall = new Ball(0.5, 0.4);                                  // Create Ball
      Path myPath = new Path(15., 34.);                                  // Create Path
      Baton myBaton = new Baton(myPath, myBall, 2.5, 15.);                     // Baton
      myPlot.setTitle("y vs x");
      myPlot.setXLabel("x");
      myPlot.setYLabel("y");
      for ( double t = 0.; myPath.getY(t) >= 0.;     t += 0.02)
      {   x = myBaton.getXa(t);
          y = myBaton.getYa(t);
          System.out.println("t = " + t + " x = " + x + " y = " + y);
          myPlot.addPoint(0, x, y, true);    }
      PlotApplication app = new PlotApplication(myPlot);
} }
```

Listing 4.9 Baton.java combines ball and path classes to form a baton.

In a project such as this where there are many different types of objects, it is a good idea to define each object inside its own .java file (and therefore its own .class file) and to place the main method in a file such as Main.java or ProjectName.java. This separates individual objects from the helper codes that glue the objects together. And since reading a well-written main method should give you a fair idea of what the entire program does, you want to be able to find the main methods easily.

4.9.2 Ball and Path Class Implementation

The Ball class in Listing 4.7 creates an object representing a sphere of mass m, radius r, and moment of inertia I. It is our basic building block. Scrutinize the length of the methods in Ball; most are short. Inasmuch as we will be using Ball as a building block, it is a good idea to keep the methods simple and just add more methods to create more complicated objects. Take stock of how similar Ball is to the Complex class in its employment of *dynamic (nonstatic)* variables. In the present case, m and r behave like the re and im dynamic variables in the Complex class in that they too act by being attached to the end of an object's name. As an example, myBall.m extracts the mass of a ball object.

In Ball.java we have defined three dynamic methods, getM, getR, and getI. When affixed to a particular ball object, these methods extract its mass, radius, and moment of inertia, respectively. Dynamic methods are like dynamic variables in that they behave differently depending on the object they modify. To cite an instance, ball1.getR() and ball2.getR() return different values if ball1 and ball2 have different radii. The getI method computes and returns the moment of inertia $I = \frac{2}{5}mr^2$ of a sphere for an axis passing through its center. The methods getM and getR are *template* methods; that is, they do not compute anything now but are included to facilitate future extensions. To name an instance, if the Ball class becomes more complex, you may need to sum the masses of its constituent parts in order to return the ball's total mass. With the template in place, you do that without having to reacquaint yourself with the rest of the code first.

Look back now and count all the methods in the Ball class. You should find four, none of which is a main method. This is fine because these methods are used by other classes, one of which will have a main method.

Exercise: Compile the Ball class. If the Java compiler does not complain, you know Ball.java contains valid code. Next try to run the byte code in Ball.class:

```
> java Ball                                        // Run Ball.class
```

You should get an error message of the type java.lang.NoSuchMethodError, with the word main at the end. This is Java's way of saying you need a main method to execute a class file. ▮

Exercise: Be adventurous and make a main method for the Ball class. Because we have not yet included Path and Baton objects, you will not be able to do much more than test that you have created the Ball object, but that is at least a step in the right direction:

```
public static void main(String[] args) {
    Ball myBall = new Ball(3.2, 0.8);
    System.out.println("M:  " + myBall.getM());
    System.out.println("R:  " + myBall.getR());
    System.out.println("I:  " + myBall.getI());  }
```

This testing code creates a Ball object and prints out its properties by affixing get methods to the object. Compile and run the modified Ball.java and thereby ensure that the Ball class still works properly. ▌

The class Path in Listing 4.8 creates an object that represents the trajectory $[(x(t), y(t)]$ of the center of mass. The class Path is another building block that we will use to construct the baton's trajectory. It computes the initial velocity components V_{0x} and V_{0y} and stores them as the dynamic class variables v0x and v0y. These variables need to be dynamic because each new path will have its own initial velocity. The acceleration resulting from gravity is the constant g, which being the same for all objects can be safely declared as a static variable independent of class instance. Survey how Path.java stores g not only as a static variable but also as a class variable so that its value is available to all methods in the class. The constructor method Path( ) of the class Path takes the polar coordinates (V_0, θ) as arguments and computes the components of the initial velocity, (v0x, v0y). This too is a building-block class, so it does not need a main method.

Exercise: Use the main method below to test the class Path. Make a Path object and find its properties at several different times. Remember, since this a test code, it does not need to do anything much. Just making an object and checking that it has the expected properties are enough. ▌

```
public static void main(String[] args) {
    Path myPath = new Path(3.0, 45.0);
    for (double t = 0.0; t <= 4.0; t += 1.0) {
        double x = myPath.getX(t);
        double y = myPath.getY(t);
        System.out.println("t = ", t, " x = ", x, " y = ", y);   } }
```

4.9.3 Composition, Objects Within Objects

A good way to build a complex system is to assemble it from simpler parts. By way of example, automobiles are built from wheels, engines, seats, and so forth, with each of these parts being built from simpler parts yet. OOP builds programs in much the same way. We start with the primitive data types of integers, floating-point numbers, and Boolean variables and combine them into more complicated data types called objects (what we did in combining two doubles into a Complex object). Then we build more complicated objects from simpler objects, and so forth.

The technique of constructing complex objects from simpler ones is called *composition.* As a consequence of the simple objects being contained within the more complex ones, the former are described by *nonstatic class variables*. This means that their properties change depending upon which object they are within. When you use composition to create more complex objects, you are working at a *higher level of abstraction.* Ideally, composition hides the distracting details of the simpler objects from your view so that you focus on the major task to be accomplished. This is analogous to first designing the general form of a bridge before worrying about the colors of the cables to be used.

4.9.4 Baton Class Implementation

Now that we have assembled the building-block classes, we combine them to create the baton's trajectory. We call the combined class Baton.java (Listing 4.9) and place the methods to compute the positions of the ends of the baton relative to the CM in it. Check first how the Baton class and its methods occupy lines 4–22, while the main method is on lines 24–39. Whether the main method is placed first or last is a matter of taste; Java does not care—but some programmers care very much. Look next at how the Baton class contains the four dynamic class variables, L, w, path, and ball. Being dynamic, their values differ for each baton, and since they are class variables (not within any methods), they may be used by all methods in the class without being passed as arguments.

The subobjects used to construct the baton object are created with the statements

```
public Path path;                                  // Path subobject 7
public Ball ball;                                  // Ball subobject 8
```

These statements tell Java that we are creating the variables path and ball to represent objects of the types Path and Ball. To do this, we must place the methods defining Ball and Path in the directory in which we are creating a Baton. The Java compiler is flexible enough for you to declare class variables in any order or even pass classes as arguments.

The constructor Baton(Path p, Ball b, . . .) on line 10 takes the Path and Ball objects as arguments and constructs the Baton object from them. On lines 7 and 8 it assigns these arguments to the appropriate class variables path and ball. We create a Baton from a Ball and a Path such that there is a Ball object at each end, with the CM following the Path object:

```
Baton myBaton = new Baton(myPath, myBall, 0.5, 15.);                29
```

Study how the Baton constructor stores the Ball and Path objects passed to it inside the Baton class even though Ball and Path belong to different classes.

On lines 12–22 we define the methods for manipulating baton objects. They all have a get as part of their name. This is the standard way of indicating that a method will retrieve or extract some property from the object to which the method is appended. For instance, Baton.getM returns 2m, that is, the sum of the masses of the two spheres. Likewise, the getI method uses the parallel axes theorem to determine the moment of inertia of the two spheres about the CM, $I = 2I_m + \frac{1}{2}mL^2$, where m and I_m are the mass and moment of inertia of the object about its center of mass. On lines 14–22 we define the methods getXa, getYa, getXb, and getYb. These take the time t as an argument and return the coordinates of the baton's ends. In each method we first determine the position of the CM by calling path.getX or path.getY and then add on the relative coordinates of the ends. On line 24 we get to the main method. It starts by creating a Plot object myPlot, a Path object myPath, and a Ball object myBall. In each case we set the initial conditions for the object by passing them as arguments to the constructor (what is called after the new command).

4.9.5 Composition Exercise

1. Compile and run the latest version of Baton.java. For this to be successful, you must tell Java to look in the current directory for the class files corresponding to Ball.java and Path.java. One way to do that is to issue the javac and java commands with the -classpath option, with the location of the classes following the option. Here the dot . is shorthand for "the current directory":

   ```
   > javac –classpath . Baton.java          // Include current directory classes
   > java –classpath . Baton                // Include current directory classes
   ```

   The program should run and plot the trajectory of one end of the baton as it travels through the air (Figure 4.7).
2. If you want to have the Java compiler automatically include the current directory in the classpath (and to avoid the –classpath . option), you need to change your CLASSPATH environment variable to include the present working directory.
3. On line 33 we see that the program executes a for loop over values of t for which the baton remains in the air:

   ```
   for (double t = 0.0; myPath.getY(t) >= 0.0; t += 0.02)                33
   ```

   This says to repeat the loop as long as $y(t)$ is positive, that is, as long as the baton is in the air. Of course we could have had the for loop remain active for times up to the hang time T, but then we would have had to calculate the hang time! The weakness in our approach is that the loop will be repeated indefinitely if $y(t)$ never becomes negative.
4. Plot the trajectory of end b of the baton on the same graph that shows the trajectory of end a. You may do this by copying and pasting the for loop for a and then modifying it for b (making sure to change the data set number in the call to PtPlot so that the two ends are plotted in different colors).
5. Use a PtPlot application to print out your graph.

6. Change the mass of the ball variable to some large number, for instance, 50 kg, in the Baton constructor method. Add print statements in the constructor and the main program to show how the ball class variable and the myBall object are affected by the new mass. You should find that ball and myBall both reference the same object since they both refer to the same memory location. In this way changes to one object are reflected in the other object. ▮

In Java, an object is passed between methods and manipulated by *reference*. This means that its address in memory is passed and not the actual values of all the component parts of it. On the other hand, primitive data types like int and double are manipulated by *value*:

```
Ball myBall = new Ball(1.0, 3.0); // Create object
Ball p = myBall;      // Now p refers to same object
Ball q = myBall;      // Create another reference
```

At times we may actually say that objects are references. This means that when one object is set equal to another, both objects *point* to the same location in memory (the start location of the first component of the object). Therefore all three variables myBall, p, and q in the above code fragment refer to the same object in memory. If we change the mass of the ball, all three variables will reflect the new mass value. This also works for object arguments: If you pass an object as an argument to a method and the method modifies the object, then the object in the calling program will also be modified.

4.9.6 Calculating the Baton's Energy (Extension)

Extend your classes so that they plot the energy of the baton as a function of time. Plot the kinetic energy of translation, the kinetic energy of rotation, and the potential energy as functions of time:

1. The translational kinetic energy of the baton is the energy associated with the motion of the center of mass. Write a getKEcm method in the Baton class that returns the kinetic energy of translation $\mathrm{KE_{cm}}(t) = mv_{\mathrm{cm}}(t)^2/2$. In terms of pseudocode the method is

```
Get present value of Vx.
Get present value of Vy.
Compute V^2 = Vx^2 + Vy^2.
Return mV^2/2.
```

   Before you program this method, write getVx and getVy methods that extract the CM velocity from a baton. Seeing as how the Path class already computes $x_{\mathrm{cm}}(t)$ and $y_{\mathrm{cm}}(t)$, it is the logical place for the velocity methods. As a guide, we suggest consulting the getX and getY methods.
2. Next we need the method getKEcm in the Baton class to compute $\mathrm{KE_{cm}}(t)$. Inasmuch as the method will be in the Baton class, we may call any of the

methods in Baton, as well as access the path and ball subobjects there (subobjects because they reside inside the Baton object or class). We obtain the velocity components by applying the getV methods to the path subobject within the Baton object:

```
public double getKEcm(double t) {
    double vx = path.getVx(t);
    double vy = path.getVy(t);
    double v2 = vx * vx + vy * vy;
    return getM() * v2 / 2; }
```

Even though the method is in a different class than the object, Java handles this. Study how getM(), being within getKEcm, acts on the same object as does getKEcm without explicitly specifying the object.

3. Compile the modified Baton and Path classes.
4. Modify Baton.java to plot the translational kinetic energy of the center of mass as a function of time. Comment out the old for loops used for plotting the positions of the baton's ends and add the code

```
for (double t = 0.0; myPath.getY(t)>= 0.0; t += 0.02) {
    double KEcm = myBaton.getKEcm(t);
    myPlot.addPoint(0, t, KEcm, true);        }
```

Compile the modified Baton.java and check that your plot is physically reasonable. The translational kinetic energy should decrease and then increase as the baton goes up and comes down.

5. Write a method in the Baton class that computes the kinetic energy of rotation about the CM, $\mathrm{KE}_{ro} = \frac{1}{2}I\omega^2$. Call getI to extract the moment of inertia of the baton and check that all classes still compile properly.
6. The potential energy of the baton $\mathrm{PE}(t) = mgy_{\mathrm{cm}}(t)$ is that of a point particle with the total mass of the baton located at the CM. Write a method in the Baton class that computes PE. Use getM to extract the mass of the baton and use path.g to extract the acceleration due to gravity g. To determine the height as a function of time, write a method path.getY(t) that accesses the path object. Make sure that the methods getKEcm, getKEr, and getPE are in the Baton class.
7. Plot on one graph the translational kinetic energy, the kinetic energy of rotation, the potential energy, and the total energy. The plots may be obtained with commands such as

```
for (double t =0.; myPath.getY(t) >= 0.; t += 0.02) {
    double KEcm = myBaton.getKEcm(t);      // KE of CM
    double KEr = myBaton.getKEr(t);        // KE of rotation
    double PE = myBaton.getPE(t); // Potential
    double total = KEcm + KEr + PE; // Total Energy
    myPlot.addPoint( 0, t, KEcm, true);      // To data set 0
    myPlot.addPoint( 1, t, KEr, true);       // To data set 1
    myPlot.addPoint( 2, t, PE, true);   // To data set 2
    myPlot.addPoint( 3, t, total, true);          } // To data set 3
```

Check that all the plotting commands are object-oriented, with myPlot being the plot object. Label your data sets with a myPlot.addLegend command outside the for loop and check that your graph is physically reasonable. The total energy and rotational energies should both remain constant in time. However, the gravitational potential energy should fluctuate. ▮

4.9.7 Examples of Inheritance and Object Hierarchies

Up until this point we have built up our Java classes via *composition*, that is, by placing objects inside other objects (using objects as arguments). As powerful as composition is, it is not appropriate in all circumstances. As a case in point, you may want to modify the Baton class to create similar, but not identical, objects such as 3-D batons. A direct approach to extending the program would be to copy and paste parts of the original code into a new class and then modify the new class. However, this is error-prone and leads to long, complicated, repetitive code. The OOP approach applies the concept of *inheritance* to allow us to create new objects that inherit the properties of old objects but have additional properties as well. This is how we create entire hierarchies of objects.

As an example, let us say we want to place red balls on the ends of the baton. We make a new class RedBall that inherits properties from Ball by using the extend command:

```
public class RedBall extends Ball {
    . . .
```

As written, this code creates a RedBall class that is identical to the original Ball class. The key word extends tells Java to copy all the methods and variables from Ball into RedBall. In OOP terminology, the Ball class is the *parent class* or *superclass*, and the RedBall class is the *child class* or *subclass*. It follows then that a class hierarchy is a sort of family tree for classes, with the parent classes at the top of the tree. As things go, children beget children of their own and trees often grow high.

To make RedBall different from Ball, we add the property of color:

```
public class RedBall extends Ball {

    String getColor()
        {return "Red"; }

    double getR()
        {return 1.0;}
}
```

Now we append the getColor method to a RedBall to find out its color. Consider what it means to have the getR method defined in both the Ball and RedBall classes. We do this because the Java compiler assumes that the getR method in RedBall is more specialized, so it ignores the original method in the Ball class. In computer science language, we would say that the new getR method *overrides* the original method.

4.9.8 Baton with a Lead Weight (Application)

As a second example, we employ inheritance to create a class of objects representing a baton with a weight at its center. We call this new class LeadBaton.java and make it a child of the parent class Baton.java. Consequently, the LeadBaton class inherits all the methods from the parent Baton class, in addition to having new ones of its own:

```
// LeadBaton: class inheritance of methods from Baton

public class LeadBaton {
    public double M;        // Non-static class variables

    LeadBaton(Path p, Ball b, double L1, double w1, double M1) {
        super(p, b, L1, w1);          // Baton constructor
        M = M1;
    }

public double getM()
    { return super.getM() + M;       }      // Call getM in Baton
```

Here the nondefault constructor LeadBaton(. . .) takes five arguments, while the Baton constructor takes only three. For the LeadBaton constructor to work, it must call the Baton constructor in order to inherit the properties of a Baton. This is accomplished by use of the key word super, which is shorthand for *look in the superclass* and which tells Java to look in the parent, or superclass, for the constructor. We may also call methods with the super key word; to illustrate, super.getM() will call the getM method from Baton in place of the getM method from LeadBaton. Finally, because the LeadBaton class assigns new values to the mass, LeadBaton overrides the getM method of Baton.

Exercise:

1. Run and create plots from the LeadBaton class. Start by removing the main method from Baton.java and placing it in the file Main.java. Instead of creating a Baton, now create a LeadBaton with

   LeadBaton myBaton = new LeadBaton(myPath, myBall, 2.5, 15., 10.);

   Here the argument "10." describes a 10-kg mass at the center of the baton.
2. Compile and run the main method, remembering to use the option "–classpath." if needed. You should get a plot of the energies of the lead baton *versus* time. Compare its energy to an ordinary baton's and comment on the differences.

3. You should see now how OOP permits us to create many types of batons with only slight modifications of the code. You can switch between a `Baton` and a `LeadBaton` object with only a single change to `main`, a modification that would be significantly more difficult with procedural programming. ∎

4.9.9 Encapsulation to Protect Classes

In the previous section we created the classes for `Ball`, `Path`, and `Baton` objects. In all cases the Java source code for each class had the same basic structure: class variables, constructors, and `get` methods. Yet classes do different things, and it is common to categorize the functions of classes as either of the following.

Interface: How the outside world manipulates an object; all methods that are applied to that object; or

Implementation: The actual internal workings of an object; how the methods make their changes.

As applied to our program, the interface for the `Ball` class includes a constructor `Ball` and the `getM`, `getR`, and `getI` methods. The interface for the `Path` class includes a constructor `Path` and the `getX` and `getY` methods.

Pure OOP strives to keep objects abstract and to manipulate them only through methods. This makes it easy to follow and to control where variables are changed and thereby makes modifying an existing program easier and less error-prone. With this purpose in mind, we separate methods into those that perform calculations and those that cause the object to do things. In addition, to protect our objects from being misused by outsiders, we invoke the **private** (in contrast to `public`) key word when declaring class variables. This ensures that these variables may be accessed and changed only from inside the class. Outside code may still manipulate our objects, but it will have to do so by calling the methods we have tested and know will not damage them.

Once we have constructed the methods and made the class variables private, we have objects that are protected by having their internal codes entirely hidden to outside users. As programmers, we may rewrite an object's code as we want and still have the same working object with a *fixed interface* for the rest of the world. Furthermore, since the object's interface is constant, even though we may change the object, there is no need to modify any code that uses the object. This is a great advance in the ability to reuse code and to use other people's codes properly.

This two-step process of creating and protecting abstract objects is known as *encapsulation*. An encapsulated object may be manipulated only in a general manner that keeps the irrelevant details of its internal workings safely hidden within. Just what constitutes an "irrelevant detail" is in the eye of the programmer. In general, you should place the `private` key word before every nonstatic class variable and then write the appropriate methods for accessing the relevant variables. This OOP process of hiding the object's variables is called *data hiding*.

4.9.10 Encapsulation Exercise

1. Place the `private` key word before all the class variables in `Ball.java`. This accomplishes the first step in encapsulating an object. Print out `myBall.m` and `myBall.r` from the main method. The Java compiler should complain because the variables are now private (visible to `Ball` class members only) and the main method is outside the `Ball` class.
2. Create methods that allow you to manipulate the object in an abstract way; for example, to modify the mass of a `Ball` object and assign it to the private class variable `m`, include the command `myBall.setM(5.0)`. This is the second step in encapsulation. We already have the methods `getM`, `getR`, and `getI`, and the object constructor `Ball`, but they do not assign a mass to the ball. Insofar as we have used a method to change the private variable `m`, we have kept our code as general as possible and still have our objects encapsulated.
3. When we write `getM()`, we are saying that `M` is the *property* to be retrieved from a `Ball` object. Inversely, the method `setM` sets the property `M` of an object equal to the argument that is given. This is part of encapsulation because with both `get` and `set` methods on hand, you do not need to access the class variables from outside the class. The use of `get` and `set` methods is standard practice in Java. You do not have to write `get` and `set` methods for every class you create, but you should create these methods for any class you want encapsulated. If you look back at Chapter 3, "Visualization Tools," you will see that the classes in the PtPlot library have many `get` and `set` methods, for example, `getTitle`, `setTitle`, `getXLabel`, and `setXLabel`.
4. Java's `interface` key word allows us to specify an interface. Here `BallInterface` defines an interface for Ball-like objects:

```
public interface BallInterface {
    public double getM();
    public double getR();
    public double getI();
}
```

   This interface does not do anything by itself, but if you modify `Ball.java` so that `public class Ball` is replaced by `public class Ball implements BallInterface`, then the Java compiler will check that the `Ball` class has all the methods specified in the interface. The Java commands `interface` and `implements` are useful in having the compiler check that your classes have all the required methods.
5. Add an arbitrary new method to the interface and compile `Ball`. If the method is found in `Ball.java`, then the `Ball` class will compile without error. ▌

4.9.11 Complex Object Interface (Extension)

In Listing 4.10 we display KomplexInterface.java, our design for an interface for complex numbers. To avoid confusion with the Complex objects, we call the new objects Komplex. We include methods for addition, subtraction, multiplication, division, negation, and conjugation, as well as get and set methods for the real, imaginary, modulus, and phase. We include all methods in the interface and check that javac compiles the interface without error. Remember, an interface must give the arguments and return type for each method.

```
//        KomplexInterface:   complex numbers via interface
                                                                    2
public interface KomplexInterface {
    public double getRe();                                          4
    public double getIm();
    public double setRe();                                          6
    public double setIm();
     // type = 0: polar representation;  other: rectangular         8
    public void add(Komplex  other, int type);
    public void sub(Komplex other, int type);                       10
    public void mult(Komplex other, int type);
    public void div(Komplex other, int type);                       12
    public void conj(int type);          }
```

Listing 4.10 KomplexInterface.java is an interface for complex numbers and is used in Komplex.java in Listing 4.11.

We still represent complex numbers in Cartesian or polar coordinates:

$$z = x + iy = re^{i\theta}. \tag{4.32}$$

Insofar as the complex number itself is independent of representation, we must be able to switch between the rectangular or polar representation. This is useful because certain manipulations are simpler in one representation than in the other; for example, division is easier in polar represenation:

$$\frac{z_1}{z_2} = \frac{a+ib}{c+id} = \frac{ac+bd+i(bc-ad)}{c^2+d^2} = \frac{r_1 e^{i\theta_1}}{r_2 e^{i\theta_2}} = \frac{r_1}{r_2} e^{i(\theta_1-\theta_2)}. \tag{4.33}$$

Listings 4.11 and 4.12 are our implementation of an interface that permits us to use either representation when manipulating complex numbers. There are three files, Komplex, KomplexInterface, and KomplexTest, all given in the listings. Because these classes call each other, each must be in a class by itself. However, for the compiler to find all the classes that it needs, all three classes must be compiled with the same javac command:

```
% javac Komplex.java KomplexInterface.java KomplexTest.java
% java KomplexTest                                  // Run test
```

```java
// Komplex: Cartesian/polar complex via interface
// 'type = 0' -> polar representation, else rectangular

public class Komplex implements KomplexInterface {
  public double mod, theta, re, im;

  public Komplex() {                                          // Default constructor
    mod = 0;   theta = 0;     re = 0;        im = 0;  }

  public Komplex(double x, double y, int type) {                      //Constructor
    if (type == 0) {mod = x;   theta = y; } else {re = x;    im = y; } }

  public double getRe() { return mod*Math.cos(theta); }

  public double getIm() { return mod*Math.sin(theta); }

  public double setRe() { re = mod*Math.cos(theta); return re; }

  public double setIm() { im = mod*Math.sin(theta); return im; }

  public void add(Komplex other, int type) {
    double tempMod = 0.;
    if (type == 0) {
      tempMod = Math.sqrt(this.mod*this.mod + other.mod*other.mod
            +   2*this.mod*other.mod*Math.cos(this.theta-other.theta));
      this.theta = Math.atan2(this.mod*Math.sin(this.theta)
        + other.mod*Math.sin(other.theta), this.mod*Math.cos(this.theta)
                                        + other.mod*Math.cos(other.theta));
      this.mod = tempMod;
    } else { this.re = this.re + other.re; this.im = this.im + other.im; }
  }

  public void sub(Komplex other, int type) {
    if (type == 0) {
      this.mod = Math.sqrt(this.mod*this.mod +  other.mod*other.mod -
       2*this.mod*other.mod*(Math.cos(this.theta)*Math.cos(other.theta)
                       + Math.sin(this.theta)*Math.sin(other.theta)));
      this.theta =  Math.atan((this.mod*Math.sin(this.theta)
        -other.mod*Math.sin(other.theta))/(this.mod*Math.cos(this.theta)
                                       -other.mod*Math.cos(other.theta)));
    } else { this.re = this.re-other.re; this.im = this.im-other.im; }
  }

   public void div(Komplex other, int type) {
    if (type == 0) { this.mod = this.mod/other.mod;
                     this.theta = this.theta-other.theta;
    } else{ this.re = (this.re*other.re + this.im*other.im)/
          (other.re*other.re + other.im*other.im);
           this.im = (this.im*other.re-this.re*other.im)/
          ( other.re*other.re +  other.im*other.im );
      }
  }

  public void mult(Komplex other, int type) {
    if (type == 0) {
      this.mod = this.mod*other.mod;
      this.theta = this.theta + other.theta;
    } else {
      Komplex ans = new Komplex();
      ans.re = this.re*other.re-this.im*other.im;
      ans.im = this.re*other.im + this.im*other.re;
      this.re = ans.re;
      this.im = ans.im;
      }
  }

  public void conj(int type) {
    if (type == 0)  { this.mod = this.mod; this.theta = -this.theta; }
    else  {this.re = this.re; this.im = -this.im; }   }
}
```

Listing 4.11 Komplex.java manipulates complex numbers using the interface `KomplexInterface` in Listing 4.10.

```
//  KomplexTest:   test KomplexInterface
public class KomplexTest  {

  public static void main(String[] argv)  {
    Komplex a, e;
    e = new Komplex();
    a = new Komplex(1., 1., 1);
    Komplex b = new Komplex(1., 2., 1);
    System.out.println ("Cartesian: Re a = " + a.re + ", Im a = " + a.im + "");
    System.out.println ("Cartesian: Re b = " + b.re + ", Im b = " + b.im + "");
    b.add(a, 1);
    e = b;
    System.out.println("Cartesian: e=b + a=" + e.re + " " + e.im + "");
     // Polar Version, uses get and set methods
    a = new Komplex(Math.sqrt(2.), Math.PI/4., 0);          // Polar via 0
    b = new Komplex(Math.sqrt(5.), Math.atan2(2., 1.), 0);
    System.out.println ("Polar: Re a = " + a.getRe() + ", Im a = " + a.getIm() + "");
    System.out.println ("Polar: Re b = " + b.getRe() + ", Im b = " + b.getIm() + "");
    b.add(a, 0);
    e = b;
    System.out.println ("Polar e=b + a = " + e.getRe() + " " + e.getIm() + "");
  }
}
```

Listing 4.12 KomplexTest.java tests Komplex and KomplexInterface. All three classes must be compiled with the same javac command.

You should observe how KomplexInterface requires us to have methods for getting and setting the real and imaginary parts of Komplex objects, as well as adding, subtracting, multiplying, dividing, and conjugating complex objects. (In the comments we see the suggestion that there should also be methods for getting and setting the modulus and phase.)

The class Komplex contains the constructors for Komplex objects. This differs from our previous implementation Complex by having the additional integer variable type. If type = 0, then the complex numbers are in polar representation, else they are in Cartesian representation. So, for example, the method for arithmetic, such as the add method on line 22, is actually two different methods depending upon the value of type. In contrast, the get and set methods for real and imaginary parts are needed only for the polar representation, and so the value of type is not needed.

4.9.12 Polymorphism, Variable Multityping

Polymorphism allows a variable name declared as one type to contain other types as needed while the program runs. The idea may be applied to both the class and the interface. Class polymorphism allows a variable that is declared as one type to contain types it inherits. To illustrate, if we declare myBaton of type Baton,

Baton myBaton;

then it will be valid to assign an object of type Baton to that variable, which is what we have been doing all along. However, it is also permissible to assign a

LeadBaton object to myBaton, and in fact it is permissible to assign any other class that it inherits from the Baton class to that variable:

```
myBaton = Baton(myPath, myBall, 0.5, 15.);                      // Usual
myBaton = LeadBaton(myPath, myBall, 0.9, 20., 15.);             // OK too
myBaton = LeadBaton(myPath, myBall, 0.1, 1.5, 80.);             // Also OK
```

Polymorphism applies to the arguments of methods as well. If we declare an argument as type Baton, we are saying that the class must be a Baton or else some class that is a child class of Baton. This is possible because the child classes will have the same methods as the original Baton class (a child class may override a method or leave it alone, but it may not eliminate it).

4.10 Supplementary Exercises

Use a Java *interface* to introduce another object corresponding to the polar representation of complex numbers:

$$r = \sqrt{x^2 + y^2}, \quad \theta = \tan^{-1}(y/x), \quad x = r\cos\theta, \quad y = r\sin\theta.$$

1. Define a constructor Complex (r, theta, 1) that constructs the polar representation of a complex number from r and θ. (The 1 is there just to add a third argument and thereby to make the constructor unique.)
2. Define a method (static or nonstatic) that permits conversion from the Cartesian to the polar representation of complex numbers.
3. Define a method (static or nonstatic) that permits conversion from the polar to the Cartesian representation of complex numbers.
4. Define methods (static or nonstatic) for addition, subtraction, multiplication, and division of complex numbers in polar representation. (*Hint:* Multiplication and division are a snap for complex numbers in polar representation, while addition and subtraction are easier for complex numbers in Cartesian representation.) ∎

4.11 OOP Example: Superposition of Motions

The isotropy of space implies that motion in one direction is independent of motion in other directions. So, when a soccer ball is kicked, we have acceleration in the vertical direction and simultaneous, yet independent, uniform motion in the horizontal direction. In addition, Galilean invariance (velocity independence of Newton's laws of motion) tells us that when an acceleration is added to uniform motion, the distance covered due to the acceleration adds to the distance covered due to uniform velocity.

Your **problem** is to describe motion in such a way that velocities and accelerations in each direction are treated as separate entities or objects independent of motion in other directions. In this way the problem is viewed consistently from both the programming philosophy and the basic physics.

4.12 Newton's Laws of Motion (Theory)

Newton's second law of motion relates the force vector **F** acting on a mass m to the acceleration vector **a** of the mass:

$$\mathbf{F} = m\mathbf{a}, \quad F_i = m\frac{d^2x_i}{dt^2}, \quad (i = 1, 2, 3). \tag{4.34}$$

If the force in the x direction vanishes, $F_x = 0$, the equation of motion (4.34) has a solution corresponding to uniform motion in the x direction with a constant velocity v_{0x}:

$$x = x_0 + v_{0x}t. \tag{4.35}$$

Equation (4.35) is the *base* or *parent* object in our example. If the force in the y direction also vanishes, then there will also be uniform y motion:

$$y = y_0 + v_{0y}t. \tag{4.36}$$

We consider uniform x motion as a parent and view uniform y motion as a child.

Equation (4.34) tells us that a constant force in the x direction causes a constant acceleration a_x in that direction. The solution of the x equation of motion with uniform acceleration is

$$x = x_0 + v_{0x}t + \tfrac{1}{2}a_x t^2. \tag{4.37}$$

For projectile motion without air resistance, we usually have $a_x = 0$ and $a_y = -g = -9.8\,\mathrm{m/s}^2$:

$$y = y_0 + v_{0y}t - \tfrac{1}{2}gt^2. \tag{4.38}$$

This y motion is a child of the parent x motion.

4.13 OOP Class Structure (Method)

The *class structure* we use to solve our problem contains the objects

Parent class Um1D: 1-D uniform motion for given initial conditions,
Child class Um2D: uniform 2-D motion; child class of Um1D,
Child class Am2d: 2-D accelerated motion; child class of Um2D.

The *member functions* include

x: position after time t,
archive: creator of a file of position *versus* time.

For our projectile motion, *encapsulation* is a combination of the initial conditions (x_0, v_{x0}) with the member functions used to compute $x(t)$. Our member functions are the creator of the class of uniform 1-D motion Um1D and the creator x(t) of a file of x as a function of time t. *Inheritance* is the child class Um2D for uniform motion in both the x and y directions, it being created from the parent class Um1D of 1-D uniform motion. *Abstraction* is present (although not used powerfully) by the simple addition of motion in the x and y directions. *Polymorphism* is present by having the member function that creates the output file different for 1-D and 2-D motions. In this implementation of OOP, the class Accm2D for accelerated motion in two dimensions inherits uniform motion in two dimensions (which in turn inherits uniform 1-D motion) and adds to it the attribute of acceleration.

4.14 Java Implementation

```
// Accm2D.java OOP accelerated motion in 2D
import java.io.*;

class Um1D {                                              // Um1D class created
  protected double delt;                          // So children may access data
  protected int steps;                             // Time steps for file output
  private double x00, vx, time;

  Um1D(double x0, double dt, double vx0, double tott) {
    x00   = 0;                                 // Constructor Um1D, initializes
    delt  = dt;
    vx    = vx0;
    time  = tott;
    steps = (int)(tott/delt);
  }

  protected double x(double tt)                           // Creates x = xo + v*dt
  { return (x00 + tt*vx); }                                            // Method x

  public void archive() throws IOException, FileNotFoundException { // Method archive
    PrintWriter w = new PrintWriter(new FileOutputStream("unimot1D.dat"), true);
    int i;
    double xx, tt;
    tt = 0. ;
    for ( i = 1 ;  i <= steps ;  i += 1) {
      xx = x(tt);                                        // Computes X = Xo + t*v
      w.println(" " + tt + " " + xx + " ");
      tt = tt + delt;
  } } }

class Um2D extends Um1D {                            // Class Um2D = 2D child of UmD1
  private double y00, vy;
                                                           // Um2D constructor
  Um2D(double x0, double dt, double vx0, double tott, double y0, double vy0)  {
    super(x0, dt, vx0, tott);
    y00  = y0;
    vy   = vy0;
  }
```

```
   protected double y(double tt) { return (y00 + tt*vy); }

  // Method archive: override Um1D.archive for 2D uniform motion
  public void archive()  throws IOException, FileNotFoundException {
    PrintWriter q = new PrintWriter(new FileOutputStream("unimot2D.dat"), true);
    int i;
    double xx, yy, tt;
    tt = 0.;
    for ( i = 1 ;  i <= steps ;  i += 1) {
      xx = x(tt);
      yy = y(tt);                                              // Data now x vs y
      q.println(" " + yy + " " + xx + " ");
      tt = tt + delt;
  } } }

public class Accm2D extends Um2D {                       // Class Accm2D: child of Um2D
  private double ax, ay;

  Accm2D(double x0, double dt, double vx0, double tott, double y0,
         double vy0, double accx, double accy ) {               // Constructor Accm2D
    super(x0, dt, vx0, tott, y0, vy0);
    ax = accx;
    ay = accy;
  }
  protected double xy(double tt, int i) {                                // Method xy
    double dt2, xxac, yyac;
    dt2  = 0.5*tt*tt;
    xxac = x(tt) + ax*dt2;
    yyac = y(tt) + ay*dt2;
    if (i==1) return xxac;  else return yyac;
  }
                                          // Method archive: override Um2D.archive
  public void archive() throws IOException, FileNotFoundException  {
    PrintWriter l = new PrintWriter(new FileOutputStream("accm2D.dat"), true);
    int i;
    double tt, xxac, yyac;
    tt = 0.;
    for ( i = 1 ;  i <= steps ;  i += 1) {
      xxac = xy(tt, 1);
      yyac = xy(tt, 2);
      l.println(" " + xxac + " " + yyac + " ");
      tt = tt + delt;
  } }

  public static void main(String[] argv) throws IOException, FileNotFoundException {
    double inix, iniy, inivx, inivy, aclx, acly, dtim, ttotal;
    inix  = 0.;       dtim   = 0.1;
    inivx = 14.;      ttotal = 4.;
    iniy  = 0.;       inivy  = 14.;
    aclx  = 0.;       acly   = -9.8;
    Accm2D acmo2d = new Accm2D(inix, dtim, inivx, ttotal, iniy, inivy, aclx, acly);
     acmo2d.archive();
} }
```

Listing 4.13 Accm2D.java is an OOP program for accelerated motion in two dimensions.

5

Monte Carlo Simulations (Nonthermal)

Unit I of this chapter addresses the problem of how computers generate numbers that appear random and how we can determine how random they are. Unit II shows how to use these random numbers to simulate physical processes. In Chapter 6, "Integration," we see show how to use these random numbers to evaluate integrals, and in Chapter 15, "Thermodynamic Simulations & Feynman Quantum Path Integration," *we investigate the use of random numbers to simulate thermal processes and the fluctuations in quantum systems.*

5.1 Unit I. Deterministic Randomness

Some people are attracted to computing because of its deterministic nature; it's nice to have a place in one's life where nothing is left to chance. Barring machine errors or undefined variables, you get the same output every time you feed your program the same input. Nevertheless, many computer cycles are used for *Monte Carlo* calculations that at their very core include elements of chance. These are calculations in which random numbers generated by the computer are used to *simulate* naturally random processes, such as thermal motion or radioactive decay, or to solve equations on the average. Indeed, much of computational physics' recognition has come about from the ability of computers to solve previously intractable problems using Monte Carlo techniques.

5.2 Random Sequences (Theory)

We define a sequence of numbers $r_1, r_2, \ldots$ as *random* if there are no correlations among the numbers. Yet being random does not mean that all the numbers in the sequence are equally likely to occur. If all the numbers in a sequence are equally likely to occur, then the sequence is said to be *uniform*, and the numbers can be random as well. To illustrate, 1, 2, 3, 4, ... is uniform but probably not random. Further, it is possible to have a sequence of numbers that, in some sense, are random but have very short-range correlations among themselves, for example,

$$r_1,\ (1-r_1),\ r_2,\ (1-r_2),\ r_3,\ (1-r_3), \ldots$$

have short-range but not long-range correlations.

Mathematically, the likelihood of a number occurring is described by a distribution function $P(r)$, where $P(r)\,dr$ is the probability of finding r in the interval $[r, r+dr]$. A *uniform* distribution means that $P(r) =$ a constant. The standard random-number generator on computers generates uniform distributions between 0 and 1. In other words, the standard random-number generator outputs numbers in this interval, each with an equal probability yet each independent of the previous number. As we shall see, numbers can also be generated nonuniformly and still be random.

By the nature of their construction, computers are deterministic and so cannot create a random sequence. By the nature of their creation, computed random number sequences must contain correlations and in this way are not truly random. Although it may be a bit of work, if we know r_m and its preceding elements, it is always possible to figure out r_{m+1}. For this reason, computers are said to generate *pseudorandom numbers* (yet with our incurable laziness we won't bother saying "pseudo" all the time). While more sophisticated generators do a better job at hiding the correlations, experience shows that if you look hard enough or use these numbers long enough, you will notice correlations. A primitive alternative to generating random numbers is to read in a table of true random numbers determined by naturally random processes such as radioactive decay or to connect the computer to an experimental device that measures random events. This alternative is not good for production work but may be a useful check in times of doubt.

5.2.1 Random-Number Generation (Algorithm)

The *linear congruent* or *power residue* method is the common way of generating a pseudorandom sequence of numbers $0 \le r_i \le M-1$ over the interval $[0, M-1]$. You multiply the previous random number r_{i-1} by the constant a, add another constant c, take the *modulus* by M, and then keep just the fractional part (remainder)[1] as the next random number r_{i+1}:

$$r_{i+1} \stackrel{\text{def}}{=} (a\,r_i + c) \bmod M = \text{remainder}\left(\frac{a\,r_i + c}{M}\right). \tag{5.1}$$

The value for r_1 (the *seed*) is frequently supplied by the user, and *mod* is a built-in function on your computer for *remaindering* (it may be called *amod* or *dmod*). This is essentially a bit-shift operation that ends up with the least significant part of the input number and thus counts on the randomness of round-off errors to generate a random sequence.

As an example, if $c = 1, a = 4, M = 9$, and you supply $r_1 = 3$, then you obtain the sequence

$$r_1 = 3, \tag{5.2}$$

[1] You may obtain the same result for the modulus operation by subtracting M until any further subtractions would leave a negative number; what remains is the *remainder*.

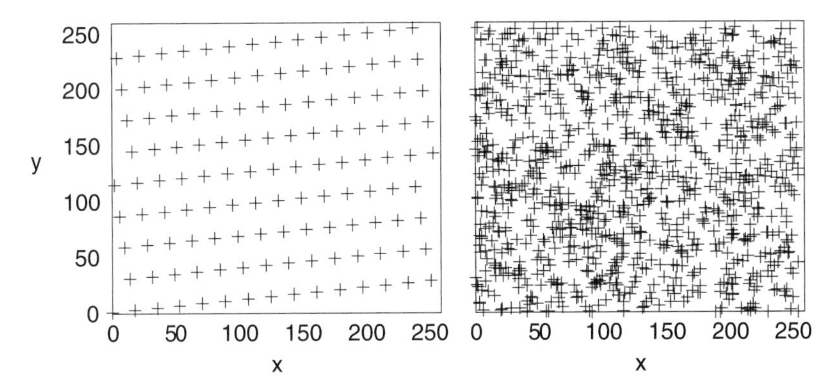

Figure 5.1 *Left:* A plot of successive random numbers $(x, y) = (r_i, r_{i+1})$ generated with a deliberately "bad" generator. *Right:* A plot generated with the library routine *drand48*.

$$r_2 = (4 \times 3 + 1)\text{mod } 9 = 13 \text{ mod } 9 = \text{rem } \frac{13}{9} = 4, \tag{5.3}$$

$$r_3 = (4 \times 4 + 1)\text{mod } 9 = 17 \text{ mod } 9 = \text{rem } \frac{17}{9} = 8, \tag{5.4}$$

$$r_4 = (4 \times 8 + 1)\text{mod } 9 = 33 \text{ mod } 9 = \text{rem } \frac{33}{9} = 6, \tag{5.5}$$

$$r_{5-10} = 7,\ 2,\ 0,\ 1,\ 5,\ 3. \tag{5.6}$$

We get a sequence of length $M = 9$, after which the entire sequence repeats. If we want numbers in the range $[0, 1]$, we divide the r's by $M = 9$:

$$0.333,\ 0.444,\ 0.889,\ 0.667,\ 0.778,\ 0.222,\ 0.000,\ 0.111,\ 0.555,\ 0.333.$$

This is still a sequence of length 9 but is no longer a sequence of integers. If random numbers in the range $[A, B]$ are needed, you only need to **scale**:

$$x_i = A + (B - A)r_i, \quad 0 \le r_i \le 1, \ \Rightarrow \ A \le x_i \le B. \tag{5.7}$$

As a rule of thumb: *Before using a random-number generator in your programs, you should check its range and that it produces numbers that "look" random.*

Although not a mathematical test, you should always make a graphical display of your random numbers. Your visual cortex is quite refined at recognizing patterns and will tell you immediately if there is one in your random numbers. For instance, Figure 5.1 shows generated sequences from "good" and "bad" generators, and it is clear which is not random (although if you look hard enough at the random points, your mind may well pick out patterns there too).

The linear congruent method (5.1) produces integers in the range $[0, M-1]$ and therefore becomes completely correlated if a particular integer comes up a second time (the whole cycle then repeats). In order to obtain a longer sequence, a and M should be large numbers but not so large that ar_{i-1} overflows. On a computer using 48-bit integer arithmetic, the built-in random-number generator may use M values as large as $2^{48} \simeq 3 \times 10^{14}$. A 32-bit generator may use $M = 2^{31} \simeq 2 \times 10^{9}$. If your program uses approximately this many random numbers, you may need to reseed the sequence during intermediate steps to avoid the cycle repeating.

Your computer probably has random-number generators that are better than the one you will compute with the power residue method. You may check this out in the manual or the help pages (try the *man* command in Unix) and then test the generated sequence. These routines may have names like *rand, rn, random, srand, erand, drand,* or *drand48*.

We recommend a version of *drand48* as a random-number generator. It generates random numbers in the range $[0, 1]$ with good spectral properties by using 48-bit integer arithmetic with the parameters[2]

$$M = 2^{48}, \quad c = B \text{ (base 16)} = 13 \text{ (base 8)}, \tag{5.8}$$

$$a = \text{5DEECE66D (base 16)} = 273673163155 \text{ (base 8)}. \tag{5.9}$$

To initialize the random sequence, you need to plant a seed in it. In Fortran you call the subroutine *srand48* to plant your seed, while in Java you issue the statement `Random randnum = new Random(seed);` (see `RandNum.java` in Listing 5.1 for details).

```java
// RandNum.java: random numbers via java.util.Random.class
import java.io.*;                                        // Location of PrintWriter
import java.util.*;                                         // Location of Random

public class RandNum {
  public static void main(String[] argv) throws IOException, FileNotFoundException {
  PrintWriter q = new PrintWriter(new FileOutputStream("RandNum.DAT"), true);
  long seed = 899432;                                  // Initialize 48 bit generator
  Random randnum = new Random(seed);
  int imax = 100; int i = 0;
 // generate random numbers and store in data file:
  for ( i=1; i <= imax; i++ ) q.println(randnum.nextDouble());
    System.out.println(" ");
    System.out.println("RandNum Program Complete.");
    System.out.println("Data stored in RandNum.DAT");
    System.out.println(" ");
  }
}                                                                   // End of class
```

Listing 5.1 RandNum.java calls the random-number generator from the Java utility class. Note that a different seed is needed for a different sequence.

[2] Unless you know how to do 48-bit arithmetic and how to input numbers in different bases, it may be better to enter large numbers like $M = 112233$ and $a = 9999$.

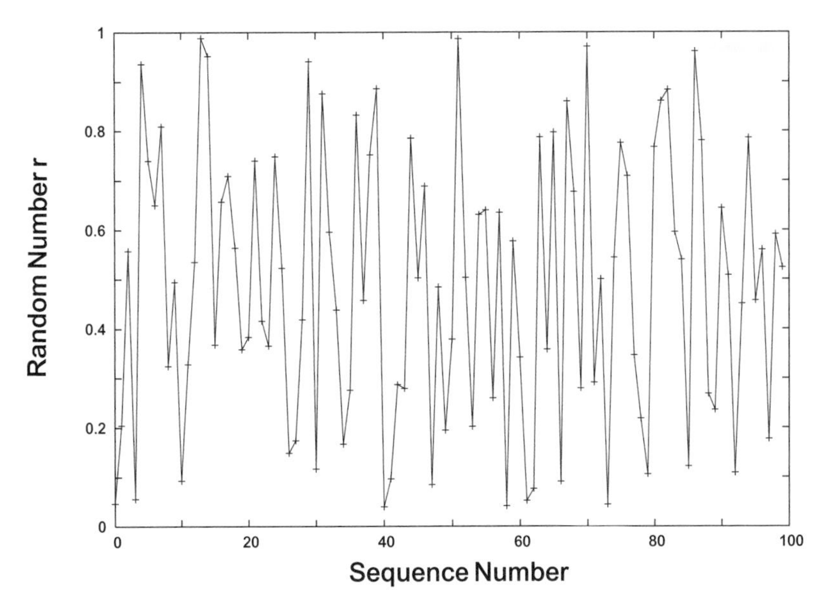

Figure 5.2 A plot of a uniform pseudorandom sequence r_i *versus* i. The points are connected to make it easier to follow the order.

5.2.2 Implementation: Random Sequence

1. Write a simple program to generate random numbers using the linear congruent method (5.1).
2. For pedagogical purposes, try the unwise choice: $(a, c, M, r_1) = (57, 1, 256, 10)$. Determine the *period*, that is, how many numbers are generated before the sequence repeats.
3. Take the pedagogical sequence of random numbers and look for correlations by observing clustering on a plot of successive pairs $(x_i, y_i) = (r_{2i-1}, r_{2i})$, $i = 1, 2, \ldots$. (Do *not* connect the points with lines.) You may "see" correlations (Figure 5.1), which means that you should not use this sequence for serious work.
4. Make your own version of Figure 5.2; that is, plot r_i *versus* i.
5. Test the built-in random-number generator on your computer for correlations by plotting the same pairs as above. (This should be good for serious work.)
6. Test the linear congruent method again with reasonable constants like those in (5.8) and (5.9). Compare the scatterplot you obtain with that of the built-in random-number generator. (This, too, should be good for serious work.) ▌

TABLE 5.1
A Table of a Uniform Pseudorandom Sequence r_i Generated by `RandNum.java`

0.04689502438508175	0.20458779675039795	0.5571907470797255	0.05634336673593088
0.9360668645897467	0.7399399139194867	0.6504153029899553	0.8096333704183057
0.3251217462543319	0.49447037101884717	0.14307712613141128	0.32858127644188206
0.5351001685588616	0.9880354395691023	0.9518097953073953	0.36810077925659423
0.6572443815038911	0.7090768515455671	0.5636787474592884	0.3586277378006649
0.38336910654033807	0.7400223756022649	0.4162083381184535	0.3658031553038087
0.7484798900468111	0.522694331447043	0.14865628292663913	0.1741881539527136
0.41872631012020123	0.9410026890120488	0.1167044926271289	0.8759009012786472
0.5962535409033703	0.4382385414974941	0.166837081276193	0.27572940246034305
0.832243048236776	0.45757242791790875	0.7520281492540815	0.8861881031774513
0.04040867417284555	0.14690149294881334	0.2869627609844023	0.27915054491588953
0.7854419848382436	0.502978394047627	0.688866810791863	0.08510414855949322
0.48437643825285326	0.19479360033700366	0.3791230234714642	0.9867371389465821

For scientific work we recommend using an industrial-strength random-number generator. To see why, here we assess how *bad* a careless application of the power residue method can be.

5.2.3 Assessing Randomness and Uniformity

Because the computer's random numbers are generated according to a definite rule, the numbers in the sequence must be correlated with each other. This can affect a simulation that assumes random events. Therefore it is wise for you to test a random-number generator to obtain a numerical measure of its uniformity and randomness before you stake your scientific reputation on it. In fact, some tests are simple enough for you to make it a habit to run them simultaneously with your simulation. In the examples to follow, we test for either randomness or uniformity.

1. Probably the most obvious, but often neglected, test for randomness and uniformity is to look at the numbers generated. For example, Table 5.1 presents some output from `RandNum.java`. If you just look at these numbers, you will know immediately that they all lie between 0 and 1, that they appear to differ from each other, and that there is no obvious pattern (like 0.3333).
2. As we have seen, a quick visual test (Figure 5.2) involves taking this same list and plotting it with r_i as ordinate and i as abscissa. Observe how there appears to be a uniform distribution between 0 and 1 and no particular correlation

between points (although your eye and brain will try to recognize some kind of pattern).

3. A simple test of uniformity evaluates the kth moment of a distribution:

$$\langle x^k \rangle = \frac{1}{N} \sum_{i=1}^{N} x_i^k. \tag{5.10}$$

If the numbers are distributed *uniformly*, then (5.10) is approximately the moment of the distribution function $P(x)$:

$$\frac{1}{N} \sum_{i=1}^{N} x_i^k \simeq \int_0^1 dx\, x^k P(x) \simeq \frac{1}{k+1} + O\left(\frac{1}{\sqrt{N}}\right). \tag{5.11}$$

If (5.11) holds for your generator, then you know that the distribution is uniform. If the deviation from (5.11) varies as $1/\sqrt{N}$, then you *also* know that the distribution is random.

4. Another simple test determines the near-neighbor correlation in your random sequence by taking sums of products for small k:

$$C(k) = \frac{1}{N} \sum_{i=1}^{N} x_i\, x_{i+k}, \quad (k = 1, 2, \ldots). \tag{5.12}$$

If your random numbers x_i and x_{i+k} are distributed with the joint probability distribution $P(x_i, x_{i+k})$ and are independent and uniform, then (5.12) can be approximated as an integral:

$$\frac{1}{N} \sum_{i=1}^{N} x_i\, x_{i+k} \simeq \int_0^1 dx \int_0^1 dy\, xy\, P(x, y) = \frac{1}{4}. \tag{5.13}$$

If (5.13) holds for your random numbers, then you know that they are uniform and independent. If the deviation from (5.13) varies as $1/\sqrt{N}$, then you *also* know that the distribution is random.

5. As we have seen, an effective test for randomness is performed by making a scatterplot of $(x_i = r_{2i}, y_i = r_{2i+1})$ for many i values. If your points have noticeable regularity, the sequence is not random. If the points are random, they should uniformly fill a square with no discernible pattern (a cloud) (Figure 5.1).

6. Test your random-number generator with (5.11) for $k = 1, 3, 7$ and $N = 100, 10,000, 100,000$. In each case print out

$$\sqrt{N} \left| \frac{1}{N} \sum_{i=1}^{N} x_i^k - \frac{1}{k+1} \right| \tag{5.14}$$

to check that it is of order 1. ■

5.3 Unit II. Monte Carlo Applications

Now that we have an idea of how to use the computer to generate pseudorandom numbers, we build some confidence that we can use these numbers to incorporate the element of chance into a simulation. We do this first by simulating a random walk and then by simulating an atom decaying spontaneously. After that, we show how knowing the statistics of random numbers leads to the best way to evaluate multidimensional integrals.

5.4 A Random Walk (Problem)

Consider a perfume molecule released in the front of a classroom. It collides randomly with other molecules in the air and eventually reaches your nose even though you are hidden in the last row. The **problem** is to determine how many collisions, on the average, a perfume molecule makes in traveling a distance R. You are given the fact that a molecule travels an average (*root-mean-square*) distance r_{rms} between collisions.

5.4.1 Random-Walk Simulation

CD

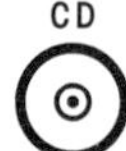

*There are a number of ways to simulate a random walk with (surprise, surprise) different assumptions yielding different physics. We will present the simplest approach for a 2-D walk, with a minimum of theory, and end up with a model for normal diffusion. The research literature is full of discussions of various versions of this problem. For example, Brownian motion corresponds to the limit in which the individual step lengths approach zero with no time delay between steps. Additional refinements include collisions within a moving medium (*abnormal diffusion*), including the velocities of the particles, or even pausing between steps. Models such as these are discussed in Chapter 13, "*Fractals & Statistical Growth,*" and demonstrated by some of the corresponding applets on the CD.*

In our random-walk simulation (Figure 5.3) an artificial *walker* takes sequential steps with the *direction* of each step *independent* of the direction of the previous step. For our model we start at the origin and take N steps in the xy plane of *lengths* (not coordinates)

$$(\Delta x_1, \Delta y_1),\ (\Delta x_2, \Delta y_2),\ (\Delta x_3, \Delta y_3), \ldots,\ (\Delta x_N, \Delta y_N). \tag{5.15}$$

Even though each step may be in a different direction, the distances along each Cartesian axis just add algebraically (aren't vectors great?). Accordingly, the radial distance R from the starting point after N steps is

$$\begin{aligned} R^2 &= (\Delta x_1 + \Delta x_2 + \cdots + \Delta x_N)^2 + (\Delta y_1 + \Delta y_2 + \cdots + \Delta y_N)^2 \\ &= \Delta x_1^2 + \Delta x_2^2 + \cdots + \Delta x_N^2 + 2\Delta x_1 \Delta x_2 + 2\Delta x_1 \Delta x_3 + 2\Delta x_2 \Delta x_1 + \cdots \\ &\quad + (x \rightarrow y). \end{aligned} \tag{5.16}$$

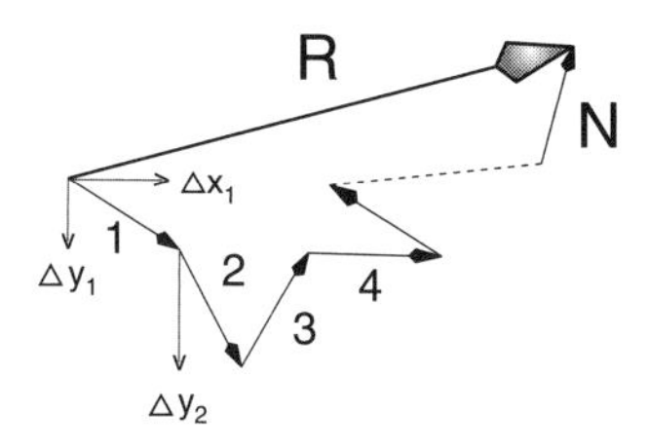

Figure 5.3 Some of the N steps in a random walk that end up a distance R from the origin. Notice how the Δx's for each step add algebraically.

If the walk is random, the particle is equally likely to travel in any direction at each step. If we take the average of a large number of such random steps, all the cross terms in (5.16) will vanish and we will be left with

$$\begin{aligned} R_{\rm rms}^2 &\simeq \langle \Delta x_1^2 + \Delta x_2^2 + \cdots + \Delta x_N^2 + \Delta y_1^2 + \Delta y_2^2 + \cdots + \Delta y_N^2 \rangle \\ &= \langle \Delta x_1^2 + \Delta y_1^2 \rangle + \langle \Delta x_2^2 + \Delta y_2^2 \rangle + \cdots \\ &= N\langle r^2 \rangle = N r_{\rm rms}^2, \\ \Rightarrow &\quad \boxed{R_{\rm rms} \simeq \sqrt{N} r_{\rm rms}}, \end{aligned} \tag{5.17}$$

where $r_{\rm rms} = \sqrt{\langle r^2 \rangle}$ is the *root-mean-square* step size.

To summarize, if the walk is random, then we expect that after a large number of steps the average *vector* distance from the origin will vanish:

$$\langle \vec{R} \rangle = \langle x \rangle \vec{i} + \langle y \rangle \vec{j} \simeq 0. \tag{5.18}$$

However, (5.17) indicates that the average *scalar* distance from the origin is $\sqrt{N} r_{\rm rms}$, where each step is of average length $r_{\rm rms}$. In other words, the vector endpoint will be distributed uniformly in all quadrants, and so the displacement vector averages to zero, but the length of that vector does not. For large N values, $\sqrt{N} r_{\rm rms} \ll N r_{\rm rms}$ but does not vanish. In our experience, practical simulations agree with this theory, but rarely perfectly, with the level of agreement depending upon the details of how the averages are taken and how the randomness is built into each step.

5.4.2 Implementation: Random Walk

The program `Walk.java` on the instructor's CD is our random-walk simulation. It's key element is random values for the x and y components of each step,

```
x += ( randnum.nextDouble() - 0.5 );
y += ( randnum.nextDouble() - 0.5 );
rsq[i] += x*x + y*y;                                    // Radius
```

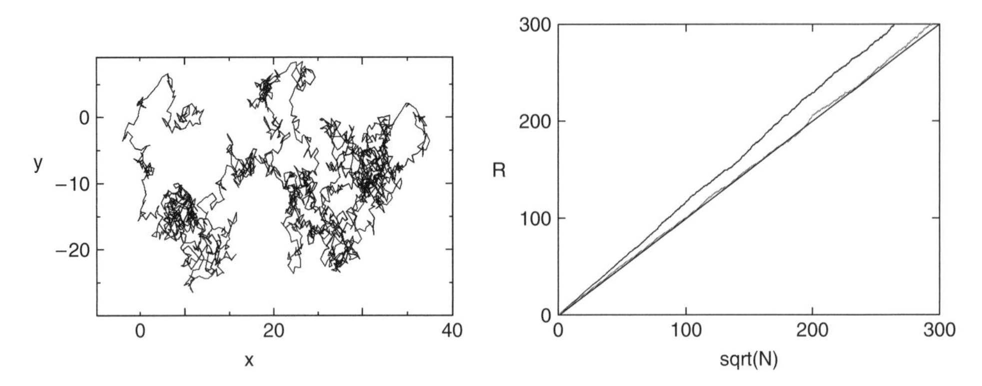

Figure 5.4 *Left:* A computer simulation of a random walk. *Right:* The distance covered in two simulated walks of *N* steps using different schemes for including randomness. The theoretical prediction (5.17) is the straight line.

Here we omit the scaling factor that normalizes each step to length 1. When using your computer to simulate a random walk, you should expect to obtain (5.17) only as the average displacement after many trials, not necessarily as the answer for each trial. You may get different answers depending on how you take your random steps (Figure 5.4 right).

Start at the origin and take a 2-D random walk with your computer.

1. To increase the amount of randomness, independently choose random values for $\Delta x'$ and $\Delta y'$ in the range $[-1, 1]$. Then normalize them so that each step is of unit length

$$\Delta x = \frac{1}{L}\Delta x', \quad \Delta y = \frac{1}{L}\Delta y', \quad L = \sqrt{\Delta x'^2 + \Delta y'^2}.$$

2. Use a plotting program to draw maps of several independent random walks, each of 1000 steps. Comment on whether these look like what you would expect of a random walk.
3. If you have your walker taking N steps in a single trial, then conduct a total number $K \simeq \sqrt{N}$ of trials. Each trial should have N steps and start with a different seed.
4. Calculate the mean square distance R^2 for each trial and then take the average of R^2 for all your K trials:

$$\langle R^2(N) \rangle = \frac{1}{K}\sum_{k=1}^{K} R^2_{(k)}(N).$$

5. Check the validity of the assumptions made in deriving the theoretical result (5.17) by checking how well

$$\frac{\langle \Delta x_i \Delta x_{j\neq i}\rangle}{R^2} \simeq \frac{\langle \Delta x_i \Delta y_j\rangle}{R^2} \simeq 0.$$

   Do your checking for both a single (long) run and for the average over trials.
6. Plot the root-mean-square distance $R_{\text{rms}} = \sqrt{\langle R^2(N)\rangle}$ as a function of $\sqrt{N}$. Values of N should start with a small number, where $R \simeq \sqrt{N}$ is not expected to be accurate, and end at a quite large value, where two or three places of accuracy should be expected on the average. ▌

5.5 Radioactive Decay (Problem)

Your **problem** is to simulate how a small number N of radioactive particles decay.[3] In particular, you are to determine when radioactive decay looks like exponential decay and when it looks *stochastic* (containing elements of chance). Because the exponential decay law is a large-number approximation to a natural process that always ends with small numbers, our simulation should be closer to nature than is the exponential decay law (Figure 5.5). In fact, if you go to the CD and "listen" to the output of the decay simulation code, what you will hear sounds very much like a Geiger counter, a convincing demonstration of the realism of the simulation.

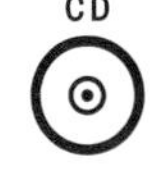

Spontaneous decay is a natural process in which a particle, with no external stimulation, decays into other particles. Even though the probability of decay of any one particle in any time interval is constant, just when it decays is a random event. Because the exact moment when any one particle decays is random, it does not matter how long the particle has been around or whether some other particles have decayed. In other words, the probability $\mathcal{P}$ of any one particle decaying per unit time interval is a constant, and when that particle decays, it is gone forever. Of course, as the total number of particles decreases with time, so will the number of decays, but the probability of any one particle decaying in some time interval is always the same constant as long as that particle exists.

5.5.1 Discrete Decay (Model)

Imagine having a sample of $N(t)$ radioactive nuclei at time t (Figure 5.5 inset). Let ΔN be the number of particles that decay in some small time interval Δt. We convert the statement "the probability $\mathcal{P}$ of any one particle decaying per unit time is a constant" into the equation

[3] Spontaneous decay is also discussed in Chapter 8, "Solving Systems of Equations with Matrices; Data Fitting," where we fit an exponential function to a decay spectrum.

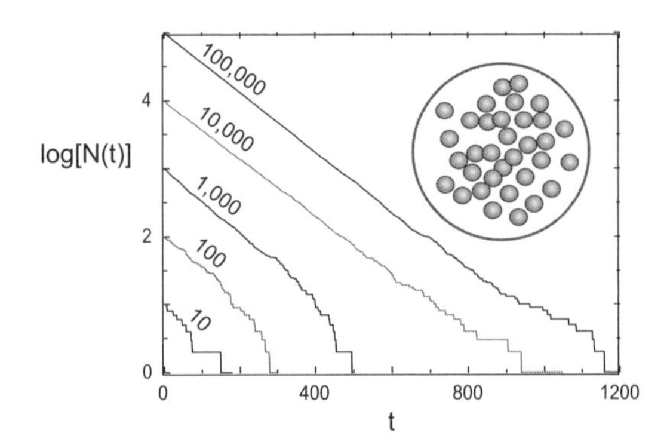

Figure 5.5 A sample containing *N* nuclei, each one of which has the same probability of decaying per unit time (circle). Semilog plots of the results from several decay simulations. Notice how the decay appears exponential (like a straight line) when the number of nuclei is large, but stochastic for *log N* $\leq$ *2.0*.

$$\mathcal{P} = \frac{\Delta N(t)/N(t)}{\Delta t} = -\lambda, \tag{5.19}$$

$$\Rightarrow \frac{\Delta N(t)}{\Delta t} = -\lambda N(t), \tag{5.20}$$

where the constant λ is called the *decay rate*. Of course the real decay rate or *activity* is $\Delta N(t)/\Delta t$, which varies with time. In fact, because the total activity is proportional to the total number of particles still present, it too is stochastic with an exponential-like decay in time. [Actually, because the number of decays $\Delta N(t)$ is proportional to the difference in random numbers, its stochastic nature becomes evident before that of $N(t)$.]

Equation (5.20) is a *finite-difference equation* in terms of the experimental measurables $N(t)$, $\Delta N(t)$, and Δt. Although it cannot be integrated the way a differential equation can, it can be solved numerically when we include the fact that the decay process is random. Because the process is random, we cannot predict a single value for $\Delta N(t)$, although we can predict the average number of decays when observations are made of many identical systems of N decaying particles.

5.5.2 Continuous Decay (Model)

When the number of particles $N \to \infty$ and the observation time interval $\Delta t \to 0$, an approximate form of the radioactive decay law (5.20) results:

$$\frac{\Delta N(t)}{\Delta t} \longrightarrow \frac{dN(t)}{dt} = -\lambda N(t). \tag{5.21}$$

This can be integrated to obtain the time dependence of the total number of particles and of the total activity:

$$N(t) = N(0)e^{-\lambda t} = N(0)e^{-t/\tau}, \tag{5.22}$$

$$\frac{dN(t)}{dt} = -\lambda N(0)e^{-\lambda t} = \frac{dN}{dt}e^{-\lambda t}(0). \tag{5.23}$$

We see that in this limit we obtain exponential decay, which leads to the identification of the decay rate λ with the inverse lifetime:

$$\lambda = \frac{1}{\tau}. \tag{5.24}$$

So we see from its derivation that exponential decay is a good description of nature for a large number of particles where $\Delta N/N \simeq 0$. The basic law of nature (5.19) is always valid, but as we will see in the simulation, exponential decay (5.23) becomes less and less accurate as the number of particles becomes smaller and smaller.

5.5.3 Decay Simulation

A program for simulating radioactive decay is surprisingly simple but not without its subtleties. We increase time in discrete steps of Δt, and for each time interval we count the number of nuclei that have decayed during that Δt. The simulation quits when there are no nuclei left to decay. Such being the case, we have an outer loop over the time steps Δt and an inner loop over the remaining nuclei for each time step. The pseudocode is simple:

```
input N, lambda
t=0
while N > 0
    DeltaN = 0
    for i = 1..N
        if (r_i < lambda) DeltaN = DeltaN + 1
    end for
    t = t +1
    N = N - DeltaN
    Output t, DeltaN, N
end while
```

When we pick a value for the decay rate $\lambda = 1/\tau$ to use in our simulation, we are setting the scale for times. If the actual decay rate is $\lambda = 0.3 \times 10^6\ \mathrm{s}^{-1}$ and if we decide to measure times in units of 10^{-6} s, then we will choose random numbers $0 \leq r_i \leq 1$, which leads to λ values lying someplace near the middle of the range (e.g., $\lambda \simeq 0.3$). Alternatively, we can use a value of $\lambda = 0.3 \times 10^6\ \mathrm{s}^{-1}$ in our simulation and then scale the random numbers to the range $0 \leq r_i \leq 10^6$. However, unless you plan to compare your simulation to experimental data, you do not have to worry

```
// Decay.java: Spontaneous decay simulation
import java.io.*;
import java.util.*;

public class Decay {
  static double lambda = 0.01;                                          // Decay constant
  static int max = 1000, time_max = 500, seed = 68111;                          // Params

  public static void main(String[] argv) throws IOException, FileNotFoundException {
    int atom, time, number, nloop;
    double decay;
    PrintWriter w = new PrintWriter(new FileOutputStream("decay.dat"), true);
    number = nloop = max;                                               // Initial value
    Random r = new Random(seed);                                  // Seed number generator
    for ( time = 0;  time <= time_max;  time++ ) {                            // Time loop
      for ( atom = 1;  atom <= number;  atom++ ) {                           // Decay loop
        decay = r.nextDouble();
        if (decay  <  lambda) nloop--;                                          // A decay
    }
      number = nloop;
      w.println( "  " + time + "  " + (double)number/max);
    }
   System.out.println("data stored in decay.dat");
} }
```

Listing 5.2 Decay.java simulates spontaneous decay in which a decay occurs if a random number is smaller than the decay parameter.

about the scale for time but instead should focus on the physics behind the slopes and relative magnitudes of the graphs.

5.6 Decay Implementation and Visualization

Write a program to simulate radioactive decay using the simple program in Listing 5.2 as a guide. You should obtain results like those in Figure 5.5.

1. Plot the logarithm of the number left $\ln N(t)$ and the logarithm of the decay rate $\ln \Delta N(t)$ *versus* time. Note that the simulation measures time in steps of Δt (generation number).
2. Check that you obtain what looks like exponential decay when you start with large values for $N(0)$, but that the decay displays its stochastic nature for small $N(0)$ [large $N(0)$ values are also stochastic; they just don't look like it].
3. Create two plots, one showing that the slopes of $N(t)$ *versus* t are *independent* of $N(0)$ and another showing that the slopes are proportional to λ.
4. Create a plot showing that within the expected statistical variations, $\ln N(t)$ and $\ln \Delta N(t)$ are proportional.
5. Explain in your own words how a process that is spontaneous and random at its very heart can lead to exponential decay.
6. How does your simulation show that the decay is exponential-like and not a power law such as $N = \beta t^{-\alpha}$?

6

Integration

In this chapter we discuss numerical integration, a basic tool of scientific computation. We derive the Simpson and trapezoid rules but just sketch the basis of Gaussian quadrature, which, though our standard workhorse, is long in derivation. We do discuss Gaussian quadrature in its various forms and indicate how to transform the Gauss points to a wide range of intervals. We end the chapter with a discussion of Monte Carlo integration, which is fundamentally different from other integration techniques.

6.1 Integrating a Spectrum (Problem)

Problem: An experiment has measured $dN(t)/dt$, the number of particles per unit time entering a counter. Your **problem** is to integrate this spectrum to obtain the number of particles $N(1)$ that entered the counter in the first second for an arbitrary decay rate

$$N(1) = \int_0^1 \frac{dN(t)}{dt}\, dt. \tag{6.1}$$

6.2 Quadrature as Box Counting (Math)

The integration of a function may require some cleverness to do analytically but is relatively straightforward on a computer. A traditional way to perform numerical integration by hand is to take a piece of graph paper and count the number of boxes or *quadrilaterals* lying below a curve of the integrand. For this reason numerical integration is also called *numerical quadrature* even when it becomes more sophisticated than simple box counting.

The Riemann definition of an integral is the limit of the sum over boxes as the width h of the box approaches zero (Figure 6.1):

$$\int_a^b f(x)\, dx = \lim_{h\to 0}\left[h \sum_{i=1}^{(b-a)/h} f(x_i)\right]. \tag{6.2}$$

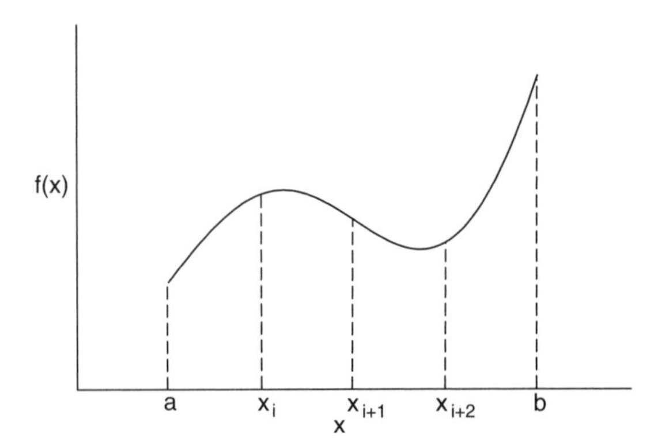

Figure 6.1 The integral $\int_a^b f(x)\,dx$ is the area under the graph of $f(x)$ from a to b. Here we break up the area into four regions of equal widths h.

The numerical integral of a function $f(x)$ is approximated as the equivalent of a finite sum over boxes of height $f(x)$ and width w_i:

$$\int_a^b f(x)\,dx \simeq \sum_{i=1}^{N} f(x_i)w_i, \tag{6.3}$$

which is similar to the Riemann definition (6.2) except that there is no limit to an infinitesimal box size. Equation (6.3) is the standard form for all integration algorithms; the function $f(x)$ is evaluated at N points in the interval $[a, b]$, and the function values $f_i \equiv f(x_i)$ are summed with each term in the sum weighted by w_i. While in general the sum in (6.3) gives the exact integral only when $N \to \infty$, it may be exact for finite N if the integrand is a polynomial. The different integration algorithms amount to different ways of choosing the points and weights. Generally, the precision increases as N gets larger, with round-off error eventually limiting the increase. Because the "best" approximation depends on the specific behavior of $f(x)$, there is no universally best approximation. In fact, some of the automated integration schemes found in subroutine libraries switch from one method to another and change the methods for different intervals until they find ones that work well for each interval.

In general you should not attempt a numerical integration of an integrand that contains a singularity without first removing the singularity by hand.[1] You may be able to do this very simply by breaking the interval down into several subintervals

[1] In Chapter 20, "Integral Equations in Quantum Mechanics," we show how to remove such a singularity even when the integrand is unknown.

so the singularity is at an endpoint where an integration point is not placed or by a change of variable:

$$\int_{-1}^{1} |x| f(x)\, dx = \int_{-1}^{0} f(-x)\, dx + \int_{0}^{1} f(x)\, dx, \tag{6.4}$$

$$\int_{0}^{1} x^{1/3}\, dx = \int_{0}^{1} 3y^3\, dy, \quad (y = x^{1/3}), \tag{6.5}$$

$$\int_{0}^{1} \frac{f(x)\, dx}{\sqrt{1-x^2}} = 2\int_{0}^{1} \frac{f(1-y^2)\, dy}{\sqrt{2-y^2}}, \quad (y^2 = 1-x). \tag{6.6}$$

Likewise, if your integrand has a very slow variation in some region, you can speed up the integration by changing to a variable that compresses that region and places few points there. Conversely, if your integrand has a very rapid variation in some region, you may want to change to variables that expand that region to ensure that no oscillations are missed.

6.2.1 Algorithm: Trapezoid Rule

The trapezoid and Simpson integration rules use values of $f(x)$ at evenly spaced values of x. They use N points $x_i (i = 1, N)$ evenly spaced at a distance h apart throughout the integration region $[a, b]$ and *include the endpoints*. This means that there are $(N-1)$ intervals of length h:

$$h = \frac{b-a}{N-1}, \quad x_i = a + (i-1)h, \quad i = 1, N, \tag{6.7}$$

where we start our counting at $i = 1$. The trapezoid rule takes each integration interval i and constructs a trapezoid of width h in it (Figure 6.2). This approximates $f(x)$ by a straight line in each interval i and uses the average height $(f_i + f_{i+1})/2$ as the value for f. The area of each such trapezoid is

$$\int_{x_i}^{x_i+h} f(x)\, dx \simeq \frac{h(f_i + f_{i+1})}{2} = \frac{1}{2} h f_i + \frac{1}{2} h f_{i+1}. \tag{6.8}$$

In terms of our standard integration formula (6.3), the "rule" in (6.8) is for $N = 2$ points with weights $w_i \equiv \frac{1}{2}$ (Table 6.1).

In order to apply the trapezoid rule to the entire region $[a, b]$, we add the contributions from each subinterval:

$$\int_{a}^{b} f(x)\, dx \simeq \frac{h}{2} f_1 + h f_2 + h f_3 + \cdots + h f_{N-1} + \frac{h}{2} f_N. \tag{6.9}$$

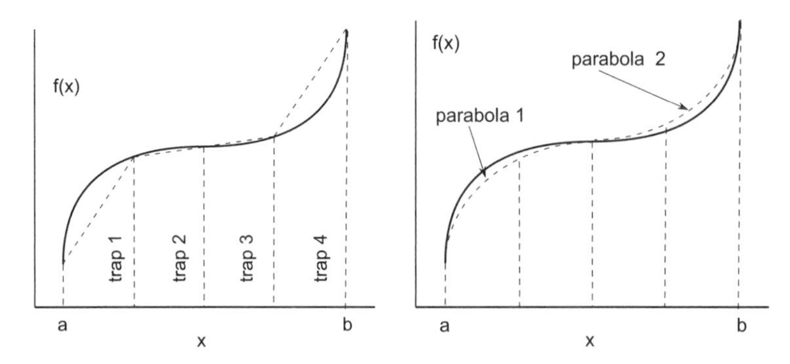

Figure 6.2 *Right:* Straight-line sections used for the trapezoid rule. *Left:* Two parabolas used in Simpson's rule.

You will notice that because the internal points are counted twice (as the end of one interval and as the beginning of the next), they have weights of $h/2 + h/2 = h$, whereas the endpoints are counted just once and on that account have weights of only $h/2$. In terms of our standard integration rule (6.32), we have

$$w_i = \left\{ \frac{h}{2}, h, \ldots, h, \frac{h}{2} \right\} \quad \text{(trapezoid rule).} \tag{6.10}$$

In Listing 6.1 we provide a simple implementation of the trapezoid rule.

6.2.2 Algorithm: Simpson's Rule

For each interval, Simpson's rule approximates the integrand $f(x)$ by a parabola (Figure 6.2 right):

$$f(x) \simeq \alpha x^2 + \beta x + \gamma, \tag{6.11}$$

TABLE 6.1
Elementary Weights for Uniform-Step Integration Rules

Name	Degree	Elementary Weights
Trapezoid	1	$(1, 1)\frac{h}{2}$
Simpson's	2	$(1, 4, 1)\frac{h}{3}$
$\frac{3}{8}$	3	$(1, 3, 3, 1)\frac{3}{8}h$
Milne	4	$(14, 64, 24, 64, 14)\frac{h}{45}$

```
// Trap.java trapezoid-rule integration of parabola

public class Trap {
  public static final double A = 0., B =3.;                        // Constant endpoints
  public static final int N = 100;                              // N points (not intervals)

  public static void main(String[] args) {                        // Main does summation
    double sum, h, t, w;
    int i;
    h = (B - A)/(N - 1);                                           // Initialization
    sum = 0.;
    for ( i=1 ;  i <= N;  i=i + 1)  {                                  // Trap rule
      t = A + (i-1) * h;
      if ( i==1 || i==N ) w = h/2.; else w = h;                       // End wt=h/2
      sum = sum + w  * t * t;
    }
    System.out.println(sum);
  }
}                                                // OUTPUT  9.000459136822773
```

Listing 6.1 Trap.java integrates the function t^2 via the trapezoid rule. Note how the step size h depends on the interval and how the weights at the ends and middle differ.

with all intervals equally spaced. The area under the parabola for each interval is

$$\int_{x_i}^{x_i+h} (\alpha x^2 + \beta x + \gamma)\, dx = \frac{\alpha x^3}{3} + \frac{\beta x^2}{2} + \gamma x \Bigg|_{x_i}^{x_i+h}. \tag{6.12}$$

In order to relate the parameters α, β, and γ to the function, we consider an interval from -1 to $+1$, in which case

$$\int_{-1}^{1} (\alpha x^2 + \beta x + \gamma)\, dx = \frac{2\alpha}{3} + 2\gamma. \tag{6.13}$$

But we notice that

$$\begin{aligned} & f(-1) = \alpha - \beta + \gamma, \qquad f(0) = \gamma, \qquad f(1) = \alpha + \beta + \gamma, \\ \Rightarrow \quad & \alpha = \frac{f(1) + f(-1)}{2} - f(0), \quad \beta = \frac{f(1) - f(-1)}{2}, \quad \gamma = f(0). \end{aligned} \tag{6.14}$$

In this way we can express the integral as the weighted sum over the values of the function at three points:

$$\int_{-1}^{1} (\alpha x^2 + \beta x + \gamma)\, dx = \frac{f(-1)}{3} + \frac{4f(0)}{3} + \frac{f(1)}{3}. \tag{6.15}$$

Because three values of the function are needed, we generalize this result to our problem by evaluating the integral over two adjacent intervals, in which case we evaluate the function at the two endpoints and in the middle (Table 6.1):

$$\int_{x_i-h}^{x_i+h} f(x)\,dx = \int_{x_i}^{x_i+h} f(x)\,dx + \int_{x_i-h}^{x_i} f(x)\,dx$$

$$\simeq \frac{h}{3}f_{i-1} + \frac{4h}{3}f_i + \frac{h}{3}f_{i+1}. \tag{6.16}$$

Simpson's rule requires the elementary integration to be over *pairs* of intervals, which in turn requires that the total number of intervals be even or that the number of points N be odd. In order to apply Simpson's rule to the entire interval, we add up the contributions from each pair of subintervals, counting all but the first and last endpoints twice:

$$\int_a^b f(x)dx \simeq \frac{h}{3}f_1 + \frac{4h}{3}f_2 + \frac{2h}{3}f_3 + \frac{4h}{3}f_4 + \cdots + \frac{4h}{3}f_{N-1} + \frac{h}{3}f_N. \tag{6.17}$$

In terms of our standard integration rule (6.3), we have

$$\boxed{w_i = \left\{\frac{h}{3}, \frac{4h}{3}, \frac{2h}{3}, \frac{4h}{3}, \ldots, \frac{4h}{3}, \frac{h}{3}\right\}} \quad \text{(Simpson's rule)}. \tag{6.18}$$

The sum of these weights provides a useful check on your integration:

$$\sum_{i=1}^{N} w_i = (N-1)h. \tag{6.19}$$

Remember, the number of points N must be odd for Simpson's rule.

6.2.3 Integration Error (Analytic Assessment)

In general, you should choose an integration rule that gives an accurate answer using the least number of integration points. We obtain a crude estimate of the *approximation* or *algorithmic error* $\mathcal{E}$ and the relative error ϵ by expanding $f(x)$ in a Taylor series around the midpoint of the integration interval. We then multiply that error by the number of intervals N to estimate the error for the entire region $[a, b]$. For the trapezoid and Simpson rules this yields

$$\mathcal{E}_t = O\left(\frac{[b-a]^3}{N^2}\right) f^{(2)}, \quad \mathcal{E}_s = O\left(\frac{[b-a]^5}{N^4}\right) f^{(4)}, \quad \epsilon_{t,s} = \frac{\mathcal{E}_{t,s}}{f}. \tag{6.20}$$

We see that the third-derivative term in Simpson's rule cancels (much like the central-difference method does in differentiation). Equations (6.20) are illuminating in showing how increasing the sophistication of an integration rule leads to an error that decreases with a higher inverse power of N, yet is also proportional to higher derivatives of f. Consequently, for small intervals and functions $f(x)$ with well-behaved derivatives, Simpson's rule should converge more rapidly than the trapezoid rule.

To model the error in integration, we assume that after N steps the *relative* round-off error is random and of the form

$$\epsilon_{\text{ro}} \simeq \sqrt{N}\epsilon_m, \tag{6.21}$$

where ϵ_m is the machine precision, $\epsilon \sim 10^{-7}$ for single precision and $\epsilon \sim 10^{-15}$ for double precision (the standard for science). Because most scientific computations are done with doubles, we will assume double precision. We want to determine an N that minimizes the total error, that is, the sum of the approximation and round-off errors:

$$\epsilon_{\text{tot}} \simeq \epsilon_{\text{ro}} + \epsilon_{\text{approx}}. \tag{6.22}$$

This occurs, approximately, when the two errors are of equal magnitude, which we approximate even further by assuming that the two errors are equal:

$$\epsilon_{\text{ro}} = \epsilon_{\text{approx}} = \frac{\mathcal{E}_{\text{trap,simp}}}{f}. \tag{6.23}$$

To continue the search for optimum N for a general function f, we set the scale of function size and the lengths by assuming

$$\frac{f^{(n)}}{f} \simeq 1, \quad b-a=1 \quad \Rightarrow \quad h=\frac{1}{N}. \tag{6.24}$$

The estimate (6.23), when applied to the **trapezoid rule**, yields

$$\sqrt{N}\epsilon_m \simeq \frac{f^{(2)}(b-a)^3}{fN^2} = \frac{1}{N^2}, \tag{6.25}$$

$$\Rightarrow \quad N \simeq \frac{1}{(\epsilon_m)^{2/5}} = \left(\frac{1}{10^{-15}}\right)^{2/5} = 10^6, \tag{6.26}$$

$$\Rightarrow \epsilon_{\text{ro}} \simeq \sqrt{N}\epsilon_m = 10^{-12}. \tag{6.27}$$

The estimate (6.23), when applied to **Simpson's rule**, yields

$$\sqrt{N}\epsilon_m = \frac{f^{(4)}(b-a)^5}{fN^4} = \frac{1}{N^4}, \tag{6.28}$$

$$\Rightarrow N = \frac{1}{(\epsilon_m)^{2/9}} = \left(\frac{1}{10^{-15}}\right)^{2/9} = 2154, \tag{6.29}$$

$$\Rightarrow \epsilon_{\text{ro}} \simeq \sqrt{N}\epsilon_m = 5 \times 10^{-14}. \tag{6.30}$$

These results are illuminating in that they show how

- Simpson's rule is an improvement over the trapezoid rule.
- It is possible to obtain an error close to machine precision with Simpson's rule (and with other higher-order integration algorithms).
- Obtaining the best numerical approximation to an integral is not achieved by letting $N \to \infty$ but with a relatively small $N \leq 1000$.

6.2.4 Algorithm: Gaussian Quadrature

It is often useful to rewrite the basic integration formula (6.3) such that we separate a weighting function $W(x)$ from the integrand:

$$\int_a^b f(x)\, dx \equiv \int_a^b W(x) g(x)\, dx \simeq \sum_{i=1}^{N} w_i g(x_i). \tag{6.31}$$

In the Gaussian quadrature approach to integration, the N points and weights are chosen to make the approximation error vanish if $g(x)$ were a $(2N-1)$-degree polynomial. To obtain this incredible optimization, the points x_i end up having a specific distribution over $[a, b]$. In general, if $g(x)$ is smooth or can be made smooth by factoring out some $W(x)$ (Table 6.2), Gaussian algorithms will produce higher accuracy than the trapezoid and Simpson rules for the same number of points. Sometimes the integrand may not be smooth because it has different behaviors in different regions. In these cases it makes sense to integrate each region separately and then add the answers together. In fact, some "smart" integration subroutines decide for themselves how many intervals to use and what rule to use in each.

All the rules indicated in Table 6.2 are Gaussian with the general form (6.31). We can see that in one case the weighting function is an exponential, in another a

TABLE 6.2
Types of Gaussian Integration Rules

Integral	*Name*	*Integral*	*Name*
$\int_{-1}^{1} f(y)\, dy$	Gauss	$\int_{-1}^{1} \frac{F(y)}{\sqrt{1-y^2}}\, dy$	Gauss–Chebyshev
$\int_{-\infty}^{\infty} e^{-y^2} F(y)\, dy$	Gauss–Hermite	$\int_0^{\infty} e^{-y} F(y)\, dy$	Gauss–Laguerre
$\int_0^{\infty} \frac{e^{-y}}{\sqrt{y}} F(y)\, dy$	Associated Gauss–Laguerre		

TABLE 6.3
Points and Weights for four-point Gaussian Quadrature

$\pm y_i$	w_i
0.33998 10435 84856	0.65214 51548 62546
0.86113 63115 94053	0.34785 48451 37454

Gaussian, and in several an integrable singularity. In contrast to the equally spaced rules, there is never an integration point at the extremes of the intervals, yet the values of the points and weights change as the number of points N changes. Although we will leave the derivation of the Gaussian points and weights to the references on numerical methods, we note here that for ordinary Gaussian (Gauss–Legendre) integration, the points y_i turn out to be the N zeros of the Legendre polynomials, with the weights related to the derivatives, $P_N(y_i) = 0$, and $w_i = 2/([(1-y_i^2)[P_N'(y_i)]^2]$. Subroutines to generate these points and weights are standard in mathematical function libraries, are found in tables such as those in [A&S 72], or can be computed. The *gauss* subroutines we provide on the CD also scale the points to span specified regions. As a check that your points are correct, you may want to compare them to the four-point set in Table 6.3.

6.2.4.1 MAPPING INTEGRATION POINTS

Our standard convention (6.3) for the general interval $[a, b]$ is

$$\int_a^b f(x)\,dx \simeq \sum_{i=1}^{N} f(x_i) w_i. \tag{6.32}$$

With Gaussian points and weights, the y interval $-1 < y_i \le 1$ must be *mapped* onto the x interval $a \le x \le b$. Here are some mappings we have found useful in our work. In all cases (y_i, w_i') are the elementary Gaussian points and weights for the interval $[-1, 1]$, and we want to scale to x with various ranges.

1. $[-1, 1] \to [a, b]$ uniformly, $(a+b)/2 =$ **midpoint:**

$$x_i = \frac{b+a}{2} + \frac{b-a}{2} y_i, \qquad w_i = \frac{b-a}{2} w_i', \tag{6.33}$$

$$\Rightarrow \quad \int_a^b f(x)\,dx = \frac{b-a}{2} \int_{-1}^{1} f[x(y)]\,dy. \tag{6.34}$$

2. $[0 \to \infty]$, $a =$ **midpoint:**

$$x_i = a\frac{1+y_i}{1-y_i}, \qquad w_i = \frac{2a}{(1-y_i)^2} w_i'. \tag{6.35}$$

3. $[-\infty \to \infty]$, **scale set by** a:

$$x_i = a\frac{y_i}{1-y_i^2}, \quad w_i = \frac{a(1+y_i^2)}{(1-y_i^2)^2} w_i'. \tag{6.36}$$

4. $[b \to \infty]$, $a+2b =$ **midpoint:**

$$x_i = \frac{a+2b+ay_i}{1-y_i}, \quad w_i = \frac{2(b+a)}{(1-y_i)^2} w_i'. \tag{6.37}$$

5. $[0 \to b]$, $ab/(b+a) =$ **midpoint:**

$$x_i = \frac{ab(1+y_i)}{b+a-(b-a)y_i}, \quad w_i = \frac{2ab^2}{(b+a-(b-a)y_i)^2} w_i'. \tag{6.38}$$

As you can see, even if your integration range extends out to infinity, there will be points at large but not infinite x. As you keep increasing the number of grid points N, the last x_i gets larger but always remains finite.

6.2.5 Integration Implementation and Error Assessment

1. Write a double-precision program to integrate an arbitrary function numerically using the trapezoid rule, the Simpson rule, and Gaussian quadrature. For our assumed **problem** there is an analytic answer:

$$\frac{dN(t)}{dt} = e^{-t} \quad \Rightarrow \quad N(1) = \int_0^1 e^{-t}\, dt = 1 - e^{-1}.$$

2. Compute the relative error $\epsilon = |(\text{numerical-exact})/\text{exact}|$ in each case. Present your data in the tabular form

N	ϵ_T	ϵ_S	ϵ_G
2	...	...	...
10	...	...	...

with spaces or tabs separating the fields. Try N values of 2, 10, 20, 40, 80, 160, ... (*Hint*: Even numbers may not be the assumption of every rule.)

3. Make a log-log plot of relative error *versus* N (Figure 6.3). You should observe that

$$\epsilon \simeq CN^{\alpha} \quad \Rightarrow \quad \log \epsilon = \alpha \log N + \text{constant}.$$

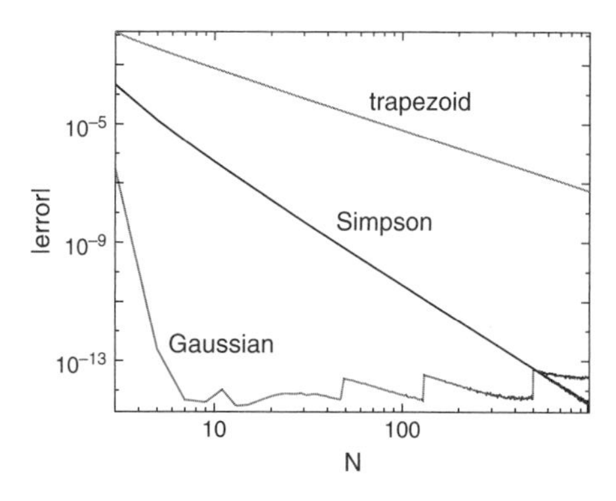

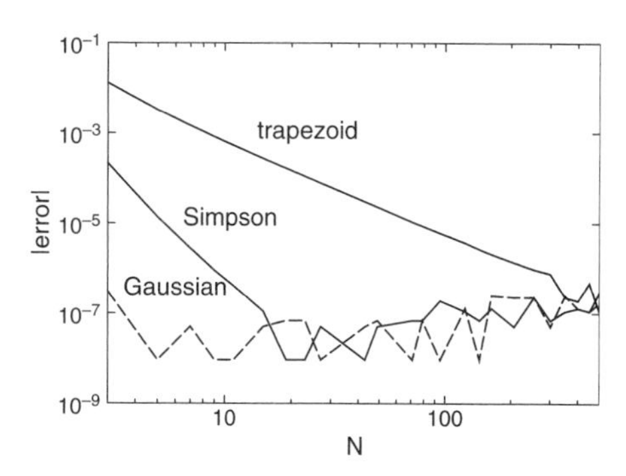

Figure 6.3 Log-log plot of the error in the integration of exponential decay using the trapezoid rule, Simpson's rule, and Gaussian quadrature *versus* the number of integration points *N*. Approximately 15 decimal places of precision are attainable with double precision (*left*), and 7 places with single precision (*right*).

This means that a power-law dependence appears as a straight line on a log-log plot, and that if you use $\log_{10}$, then the ordinate on your log-log plot will be the negative of the number of decimal places of precision in your calculation.

4. Use your plot or table to estimate the power-law dependence of the error ϵ on the number of points N and to determine the number of decimal places of precision in your calculation. Do this for both the trapezoid and Simpson rules and for both the algorithmic and round-off error regimes. (Note that it may be hard to reach the round-off error regime for the trapezoid rule because the approximation error is so large.)

In Listing 6.2 we give a sample program that performs an integration with Gaussian points. The method gauss generates the points and weights and may be useful in other applications as well.

```
// IntegGauss.java: integration via Gauss Quadrature
import java.io.*;                                          // Location of PrintWriter

public class IntegGauss {
  static final double max_in = 1001;                              // Numb intervals
  static final double vmin = 0., vmax = 1.;                          // Int ranges
  static final double ME = 2.7182818284590452354E0 ;                // Euler's const

 public static void main(String[] argv)  throws IOException, FileNotFoundException {
   int i;
   double result;
   PrintWriter t = new PrintWriter(new FileOutputStream("IntegGauss.dat"), true);
   for ( i=3;  i <= max_in;  i  += 2) {
     result = gaussint(i, vmin, vmax);
     t.println("" + i + "  " + Math.abs(result-1 + 1/ME));
   }
   System.out.println("Output in IntegGauss.dat");
```

```
  }

  public static double f (double x) {return (Math.exp(-x));}                    // f(x)

  public static double gaussint (int no, double min, double max)  {
    int n;
    double quadra = 0., w[] = new double[2001], x[] = new double[2001];
    gauss (no, 0, min, max, x, w);                                    // Returns pts & wts
    for ( n=0;  n <  no;  n++ )     quadra += f(x[n])*w[n];        // Calculate integral
    return (quadra);
  }

public static void gauss(int npts, int job, double a, double b,
                                                double x[], double w[]) {
      int m = 0, i = 0, j = 0;
      double t = 0., t1 = 0., pp = 0., p1 = 0., p2 = 0., p3 = 0. ;
      double xi, eps = 3.E-14;                                      // Accuracy: ADJUST!
     m = (npts + 1)/2;
      for ( i=1;  i <= m;  i++ )  {
        t = Math.cos(Math.PI*((double)i-0.25)/((double)npts + 0.5));
        t1 = 1;
        while((Math.abs(t-t1)) >= eps)  {
          p1 = 1. ;  p2 = 0. ;
          for ( j=1;  j <= npts;  j++ )  {
            p3 = p2;   p2 = p1;
            p1=((2.*(double)j-1)*t*p2-((double)j-1.)*p3)/((double)j);
          }
          pp = npts*(t*p1-p2)/(t*t-1.);
          t1 = t;          t = t1 - p1/pp;
        }
        x[i-1] = -t;   x[npts-i] = t;
        w[i-1]   = 2./((1.-t*t)*pp*pp);
        w[npts-i] = w[i-1];
        System.out.println(" x[i-1]"+ x[i-1] +" w " + w[npts-i]);
      }
      if (job==0)  {
        for ( i=0;  i < npts ;  i++ )  {
          x[i] = x[i]*(b-a)/2. + (b + a)/2.;
          w[i] = w[i]*(b-a)/2.;
        }
      }
      if (job==1)  {
        for ( i=0;  i < npts;  i++ )  {
          xi=x[i];
          x[i] = a*b*(1. + xi) / (b + a-(b-a)*xi);
          w[i] = w[i]*2.*a*b*b/((b + a-(b-a)*xi)*(b+a-(b-a)*xi));
        }
      }
      if (job==2) {
        for ( i=0;  i < npts;  i++ )  {
           xi=x[i];
           x[i] = (b*xi+ b + a + a) / (1.-xi);
           w[i] = w[i]*2.*(a + b)/((1.-xi)*(1.-xi));
        }
      }
      return;
} }
```

Listing 6.2 IntegGauss.java integrates the function $f(x)$ via Gaussian quadrature. The points and weights are generated in the method gauss, which remains fixed for all applications. Note that the parameter eps, which controls the level of precision desired, should be set by the user, as should the value for job, which controls the mapping of the Gaussian points onto arbitrary intervals (they are generated for $-1 \leq x \leq 1$).

6.3 Experimentation

Try two integrals for which the answers are less obvious:

$$F_1 = \int_0^{2\pi} \sin(1000x)\, dx, \quad F_2 = \int_0^{2\pi} \sin^x(100x)\, dx. \tag{6.39}$$

Explain why the computer may have trouble with these integrals. ∎

6.4 Higher-Order Rules (Algorithm)

As in numerical differentiation, we can use the known functional dependence of the error on interval size h to reduce the integration error. For simple rules like the trapezoid and Simpson rules, we have the analytic estimates (6.23), while for others you may have to experiment to determine the h dependence. To illustrate, if $A(h)$ and $A(h/2)$ are the values of the integral determined for intervals h and $h/2$, respectively, and we know that the integrals have expansions with a leading error term proportional to h^2,

$$A(h) \simeq \int_a^b f(x)\, dx + \alpha h^2 + \beta h^4 + \cdots, \tag{6.40}$$

$$A\left(\frac{h}{2}\right) \simeq \int_a^b f(x)\, dx + \frac{\alpha h^2}{4} + \frac{\beta h^4}{16} + \cdots. \tag{6.41}$$

Consequently, we make the h^2 term vanish by computing the combination

$$\frac{4}{3} A\left(\frac{h}{2}\right) - \frac{1}{3} A(h) \simeq \int_a^b f(x)\, dx - \frac{\beta h^4}{4} + \cdots. \tag{6.42}$$

Clearly this particular trick (Romberg's extrapolation) works only if the h^2 term dominates the error and then only if the derivatives of the function are well behaved. An analogous extrapolation can also be made for other algorithms.

In Table 6.1 we gave the weights for several equal-interval rules. Whereas the Simpson rule used two intervals, the three-eighths rule uses three, and the Milne[2] rule four. (These are single-interval rules and must be strung together to obtain a rule *extended* over the entire integration range. This means that the points that end one interval and begin the next are weighted twice.) You can easily determine the number of elementary intervals integrated over, and check whether you and we have written the weights right, by summing the weights for any rule. The sum is

[2] There is, not coincidentally, a Milne Computer Center at Oregon State University, although there no longer is a central computer there.

the integral of $f(x) = 1$ and must equal h times the number of intervals (which in turn equals $b - a$):

$$\sum_{i=1}^{N} w_i = h \times N_{\text{intervals}} = b - a. \tag{6.43}$$

6.5 Problem: Monte Carlo Integration by Stone Throwing

Imagine yourself as a farmer walking to your furthermost field to add algae-eating fish to a pond having an algae explosion. You get there only to read the instructions and discover that you need to know the area of the pond in order to determine the correct number of the fish to add. Your **problem** is to measure the area of this irregularly shaped pond with just the materials at hand [G,T&C 06].

It is hard to believe that Monte Carlo techniques can be used to evaluate integrals. After all, we do not want to gamble on the values! While it is true that other methods are preferable for single and double integrals, Monte Carlo techniques are best when the dimensionality of integrations gets large! For our pond problem, we will use a *sampling* technique (Figure 6.4):

1. Walk off a box that completely encloses the pond and remove any pebbles lying on the ground within the box.
2. Measure the lengths of the sides in natural units like *feet*. This tells you the area of the enclosing box A_{box}.
3. Grab a bunch of pebbles, count their number, and then throw them up in the air in random directions.
4. Count the number of splashes in the pond N_{pond} and the number of pebbles lying on the ground within your box N_{box}.
5. Assuming that you threw the pebbles uniformly and randomly, the number of pebbles falling into the pond should be proportional to the area of the pond A_{pond}. You determine that area from the simple ratio

$$\frac{N_{\text{pond}}}{N_{\text{pond}} + N_{\text{box}}} = \frac{A_{\text{pond}}}{A_{\text{box}}} \quad \Rightarrow \quad A_{\text{pond}} = \frac{N_{\text{pond}}}{N_{\text{pond}} + N_{\text{box}}} A_{\text{box}}. \tag{6.44}$$

6.5.1 Stone Throwing Implementation

Use sampling (Figure 6.4) to perform a 2-D integration and thereby determine π:

1. Imagine a circular pond enclosed in a square of side 2 ($r = 1$).
2. We know the analytic area of a circle $\oint dA = \pi$.
3. Generate a sequence of random numbers $-1 \le r_i \le +1$.

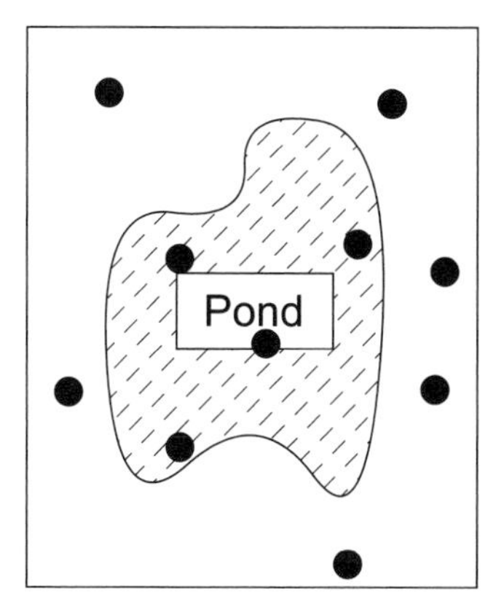

Figure 6.4 Throwing stones in a pond as a technique for measuring its area. There is a tutorial on this on the CD where you can see the actual "splashes" (the dark spots) used in an integration.

4. For $i = 1$ to N, pick $(x_i, y_i) = (r_{2i-1}, r_{2i})$.
5. If $x_i^2 + y_i^2 < 1$, let $N_{\text{pond}} = N_{\text{pond}} + 1$; otherwise let $N_{\text{box}} = N_{\text{box}} + 1$.
6. Use (6.44) to calculate the area, and in this way π.
7. Increase N until you get π to three significant figures (we don't ask much — that's only slide-rule accuracy).

6.5.2 Integration by Mean Value (Math)

The standard Monte Carlo technique for integration is based on the *mean value theorem* (presumably familiar from elementary calculus):

$$I = \int_a^b dx\, f(x) = (b-a)\langle f \rangle. \tag{6.45}$$

The theorem states the obvious if you think of integrals as areas: The value of the integral of some function $f(x)$ between a and b equals the length of the interval $(b-a)$ times the mean value of the function over that interval $\langle f \rangle$ (Figure 6.5). The integration algorithm uses Monte Carlo techniques to evaluate the mean in (6.45). With a sequence $a \le x_i \le b$ of N uniform random numbers, we want to determine the *sample mean* by *sampling* the function $f(x)$ at these points:

$$\langle f \rangle \simeq \frac{1}{N} \sum_{i=1}^{N} f(x_i). \tag{6.46}$$

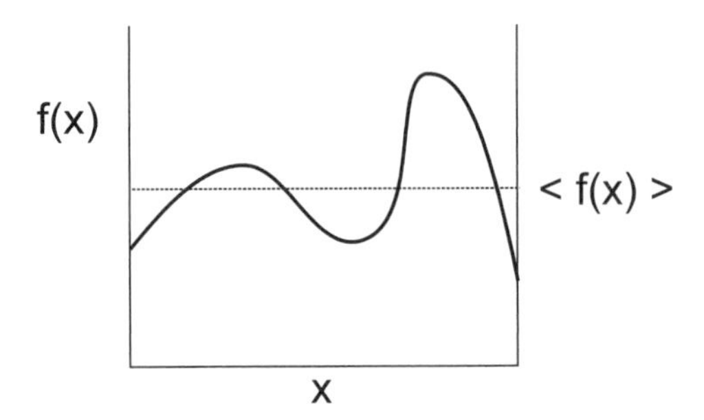

Figure 6.5 The area under the curve $f(x)$ is the same as that under the dashed line $y = \langle f \rangle$.

This gives us the very simple integration rule:

$$\int_a^b dx\, f(x) \simeq (b-a)\frac{1}{N}\sum_{i=1}^{N} f(x_i) = (b-a)\langle f \rangle. \tag{6.47}$$

Equation (6.47) looks much like our standard algorithm for integration (6.3) with the "points" x_i chosen randomly and with uniform weights $w_i = (b-a)/N$. Because no attempt has been made to obtain the best answer for a given value of N, this is by no means an optimized way to evaluate integrals; but you will admit it is simple. If we let the number of samples of $f(x)$ approach infinity $N \to \infty$ or if we keep the number of samples finite and take the average of infinitely many runs, the laws of statistics assure us that (6.47) will approach the correct answer, at least if there were no round-off errors.

For readers who are familiar with statistics, we remind you that the uncertainty in the value obtained for the integral I after N samples of $f(x)$ is measured by the standard deviation σ_I. If σ_f is the standard deviation of the integrand f in the sampling, then for normal distributions we have

$$\sigma_I \simeq \frac{1}{\sqrt{N}}\sigma_f. \tag{6.48}$$

So for large N, the error in the value obtained for the integral decreases as $1/\sqrt{N}$.

6.6 High-Dimensional Integration (Problem)

Let's say that we want to calculate some properties of a small atom such as magnesium with 12 electrons. To do that we need to integrate atomic wave functions over the three coordinates of each of 12 electrons. This amounts to a $3 \times 12 = 36$-D integral. If we use 64 points for each integration, this requires about $64^{36} \simeq 10^{65}$ evaluations of the integrand. If the computer were fast and could evaluate the integrand a billion times per second, this would take about 10^{56} s, which is significantly longer than the age of the universe ($\sim 10^{17}$ s).

Your **problem** is to find a way to perform multidimensional integrations so that you are still alive to savor the answers. Specifically, evaluate the 10-D integral

$$I = \int_0^1 dx_1 \int_0^1 dx_2 \cdots \int_0^1 dx_{10} \, (x_1 + x_2 + \cdots + x_{10})^2 . \tag{6.49}$$

Check your numerical answer against the analytic one, $\frac{155}{6}$. ▌

6.6.1 Multidimensional Monte Carlo

It is easy to generalize mean value integration to many dimensions by picking random points in a multidimensional space. For example,

$$\int_a^b dx \int_c^d dy \, f(x,y) \simeq (b-a)(d-c)\frac{1}{N}\sum_i^N f(\mathbf{x}_i) = (b-a)(d-c)\langle f \rangle . \tag{6.50}$$

6.6.2 Error in Multidimensional Integration (Assessment)

When we perform a multidimensional integration, the error in the Monte Carlo technique, being statistical, decreases as $1/\sqrt{N}$. This is valid even if the N points are distributed over D dimensions. In contrast, when we use these same N points to perform a D-dimensional integration as D 1-D integrals using a rule such as Simpson's, we use N/D points for each integration. For fixed N, this means that the number of points used for each integration decreases as the number of dimensions D increases, and so the error in each integration increases with D. Furthermore, the total error will be approximately N times the error in each integral. If we put these trends together and look at a particular integration rule, we will find that at a value of $D \simeq 3$–4 the error in Monte Carlo integration is similar to that of conventional schemes. For larger values of D, the Monte Carlo method is always more accurate!

6.6.3 Implementation: 10-D Monte Carlo Integration

Use a built-in random-number generator to perform the 10-D Monte Carlo integration in (6.49). Our program `Int10d.java` is available on the instructor's CD.

1. Conduct 16 trials and take the average as your answer.
2. Try sample sizes of $N = 2, 4, 8, \ldots, 8192$.
3. Plot the error *versus* $1/\sqrt{N}$ and see if linear behavior occurs.

6.7 Integrating Rapidly Varying Functions (Problem)

It is common in many physical applications to integrate a function with an approximately Gaussian dependence on x. The rapid falloff of the integrand means that our Monte Carlo integration technique would require an incredibly large number of points to obtain even modest accuracy. Your **problem** is to make Monte Carlo integration more efficient for rapidly varying integrands.

6.7.1 Variance Reduction (Method)

If the function being integrated never differs much from its average value, then the standard Monte Carlo mean value method (6.47) should work well with a large, but manageable, number of points. Yet for a function with a large *variance* (i.e., one that is not "flat"), many of the random evaluations of the function may occur where the function makes a slight contribution to the integral; this is, basically, a waste of time. The method can be improved by mapping the function f into a function g that has a smaller variance over the interval. We indicate two methods here and refer you to [Pres 00] and [Koon 86] for more details.

The first method is a *variance reduction* or *subtraction technique* in which we devise a flatter function on which to apply the Monte Carlo technique. Suppose we construct a function $g(x)$ with the following properties on $[a, b]$:

$$|f(x) - g(x)| \leq \epsilon, \qquad \int_a^b dx\, g(x) = J. \tag{6.51}$$

We now evaluate the integral of $f(x) - g(x)$ and add the result to J to obtain the required integral

$$\int_a^b dx\, f(x) = \int_a^b dx\, [f(x) - g(x)] + J. \tag{6.52}$$

If we are clever enough to find a simple $g(x)$ that makes the variance of $f(x) - g(x)$ less than that of $f(x)$ and that we can integrate analytically, we obtain more accurate answers in less time.

6.7.2 Importance Sampling (Method)

A second method for improving Monte Carlo integration is called *importance sampling* because it lets us sample the integrand in the most important regions. It derives from expressing the integral in the form

$$I = \int_a^b dx\, f(x) = \int_a^b dx\, w(x) \frac{f(x)}{w(x)}. \tag{6.53}$$

If we now use $w(x)$ as the *weighting function* or *probability distribution* for our random numbers, the integral can be approximated as

$$I = \left\langle \frac{f}{w} \right\rangle \simeq \frac{1}{N} \sum_{i=1}^{N} \frac{f(x_i)}{w(x_i)}. \tag{6.54}$$

The improvement from (6.54) is that a judicious choice of weighting function $w(x) \propto f(x)$ makes $f(x)/w(x)$ more constant and thus easier to integrate.

6.7.3 Von Neumann Rejection (Method)

A simple, ingenious method for generating random points with a probability distribution $w(x)$ was deduced by von Neumann. This method is essentially the same as the rejection or sampling method used to guess the area of a pond, only now the pond has been replaced by the weighting function $w(x)$, and the arbitrary box around the lake by the arbitrary constant W_0. Imagine a graph of $w(x)$ *versus* x (Figure 6.6). Walk off your box by placing the line $W = W_0$ on the graph, with the only condition being $W_0 \geq w(x)$. We next "throw stones" at this graph and count only those that fall into the $w(x)$ pond. That is, we generate uniform distributions in x and $y \equiv W$ with the maximum y value equal to the width of the box W_0:

$$(x_i, W_i) = (r_{2i-1}, W_0 r_{2i}). \tag{6.55}$$

We then reject all x_i that do not fall into the pond:

$$\text{If } W_i < w(x_i), \text{ accept}, \quad \text{If } W_i > w(x_i), \text{ reject}. \tag{6.56}$$

The x_i values so accepted will have the weighting $w(x)$ (Figure 6.6). The largest acceptance occurs where $w(x)$ is large, in this case for midrange x. In Chapter 15, "Thermodynamic Simulations & Feynman Quantum Path Integration," we apply a variation of the rejection technique known as the *Metropolis algorithm*. This algorithm has now become the cornerstone of computation thermodynamics.

6.7.4 Simple Gaussian Distribution

The central limit theorem can be used to deduce a Gaussian distribution via a simple summation. The theorem states, under rather general conditions, that if $\{r_i\}$ is a

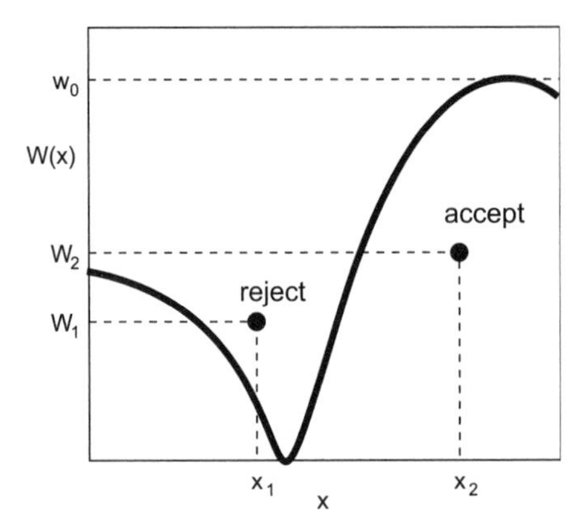

Figure 6.6 The Von Neumann rejection technique for generating random points with weight $w(x)$.

sequence of mutually independent random numbers, then the sum

$$x_N = \sum_{i=1}^{N} r_i \tag{6.57}$$

is distributed normally. This means that the generated x values have the distribution

$$P_N(x) = \frac{\exp\left[-\frac{(x-\mu)^2}{2\sigma^2}\right]}{\sqrt{2\pi\sigma^2}}, \quad \mu = N\langle r\rangle, \ \ \sigma^2 = N(\langle r^2\rangle - \langle r\rangle^2). \tag{6.58}$$

6.8 Nonuniform Assessment ⊙

Use the von Neumann rejection technique to generate a normal distribution of standard deviation 1 and compare to the simple Gaussian method.

6.8.1 Implementation: Nonuniform Randomness ⊙

In order for $w(x)$ to be the weighting function for random numbers over $[a, b]$, we want it to have the properties

$$\int_a^b dx\, w(x) = 1, \quad [w(x) > 0], \quad d\mathcal{P}(x \to x + dx) = w(x)\, dx, \tag{6.59}$$

where $d\mathcal{P}$ is the probability of obtaining an x in the range $x \to x + dx$. For the uniform distribution over $[a, b]$, $w(x) = 1/(b - a)$.

Inverse transform/change of variable method $\odot$: Let us consider a change of variables that takes our original integral I (6.53) to the form

$$I = \int_a^b dx\, f(x) = \int_0^1 dW \frac{f[x(W)]}{w[x(W)]}. \tag{6.60}$$

Our aim is to make this transformation such that there are equal contributions from all parts of the range in W; that is, we want to use a uniform sequence of random numbers for W. To determine the new variable, we start with $u(r)$, the uniform distribution over $[0, 1]$,

$$u(r) = \begin{cases} 1, & \text{for } 0 \leq r \leq 1, \\ 0, & \text{otherwise.} \end{cases} \tag{6.61}$$

We want to find a mapping $r \leftrightarrow x$ or probability function $w(x)$ for which probability is conserved:

$$w(x)\, dx = u(r)\, dr, \quad \Rightarrow \quad w(x) = \left| \frac{dr}{dx} \right| u(r). \tag{6.62}$$

This means that even though x and r are related by some (possibly) complicated mapping, x is also random with the probability of x lying in $x \to x + dx$ equal to that of r lying in $r \to r + dr$.

To find the mapping between x and r (the tricky part), we change variables to $W(x)$ defined by the integral

$$W(x) = \int_{-\infty}^{x} dx'\, w(x'). \tag{6.63}$$

We recognize $W(x)$ as the (incomplete) integral of the probability density $u(r)$ up to some point x. It is another type of distribution function, the integrated probability of finding a random number less than the value x. The function $W(x)$ is on that account called a *cumulative distribution function* and can also be thought of as the area to the left of $r = x$ on the plot of $u(r)$ *versus* r. It follows immediately from the definition (6.63) that $W(x)$ has the properties

$$W(-\infty) = 0; \quad W(\infty) = 1, \tag{6.64}$$

$$\frac{dW(x)}{dx} = w(x), \quad dW(x) = w(x)\, dx = u(r)\, dr. \tag{6.65}$$

Consequently, $W_i = \{r_i\}$ is a uniform sequence of random numbers, and we just need to invert (6.63) to obtain x values distributed with probability $w(x)$.

The crux of this technique is being able to invert (6.63) to obtain $x = W^{-1}(r)$. Let us look at some analytic examples to get a feel for these steps (numerical inversion is possible and frequent in realistic cases).

Uniform weight function w**:** We start with the familiar uniform distribution

$$w(x) = \begin{cases} \frac{1}{b-a}, & \text{if } a \leq x \leq b, \\ 0, & \text{otherwise.} \end{cases} \tag{6.66}$$

After following the rules, this leads to

$$W(x) = \int_a^x dx' \frac{1}{b-a} = \frac{x-a}{b-a} \tag{6.67}$$

$$\Rightarrow \quad x = a + (b-a)W \quad \Rightarrow \quad W^{-1}(r) = a + (b-a)r, \tag{6.68}$$

where $W(x)$ is always taken as uniform. In this way we generate uniform random $0 \leq r \leq 1$ and uniform random $a \leq x \leq b$.

Exponential weight: We want random points with an exponential distribution:

$$w(x) = \begin{cases} \frac{1}{\lambda} e^{-x/\lambda}, & \text{for } x > 0, \\ 0, & \text{for } x < 0, \end{cases} \quad W(x) = \int_0^x dx' \frac{1}{\lambda} e^{-x'/\lambda} = 1 - e^{-x/\lambda},$$

$$\Rightarrow \quad x = -\lambda \ln(1-W) \equiv -\lambda \ln(1-r). \tag{6.69}$$

In this way we generate uniform random r: $[0,1]$ and obtain $x = -\lambda \ln(1-r)$ distributed with an exponential probability distribution for $x > 0$. Notice that our prescription (6.53) and (6.54) tells us to use $w(x) = e^{-x/\lambda}/\lambda$ to remove the exponential-like behavior from an integrand and place it in the weights and scaled points ($0 \leq x_i \leq \infty$). Because the resulting integrand will vary less, it may be approximated better as a polynomial:

$$\int_0^\infty dx\, e^{-x/\lambda} f(x) \simeq \frac{\lambda}{N} \sum_{i=1}^N f(x_i), \quad x_i = -\lambda \ln(1-r_i). \tag{6.70}$$

Gaussian (normal) distribution: We want to generate points with a normal distribution:

$$w(x') = \frac{1}{\sqrt{2\pi}\sigma} e^{-(x'-\overline{x})^2/2\sigma^2}. \tag{6.71}$$

This by itself is rather hard but is made easier by generating uniform distributions in angles and then using trigonometric relations to convert them to a Gaussian distribution. But before doing that, we keep things simple by realizing that we can obtain (6.71) with mean $\overline{x}$ and standard deviation σ by scaling and a translation of a simpler $w(x)$:

$$w(x) = \frac{1}{\sqrt{2\pi}} e^{-x^2/2}, \quad x' = \sigma x + \overline{x}. \tag{6.72}$$

We start by generalizing the statement of probability conservation for two different distributions (6.62) to two dimensions [Pres 94]:

$$p(x, y)\, dx\, dy = u(r_1, r_2)\, dr_1\, dr_2 \quad \Rightarrow \quad p(x, y) = u(r_1, r_2) \left| \frac{\partial(r_1, r_2)}{\partial(x, y)} \right|.$$

We recognize the term in vertical bars as the Jacobian determinant:

$$J = \left| \frac{\partial(r_1, r_2)}{\partial(x, y)} \right| \stackrel{\text{def}}{=} \frac{\partial r_1}{\partial x}\frac{\partial r_2}{\partial y} - \frac{\partial r_2}{\partial x}\frac{\partial r_1}{\partial y}. \tag{6.73}$$

To specialize to a Gaussian distribution, we consider $2\pi r$ as angles obtained from a uniform random distribution r, and x and y as Cartesian coordinates that will have a Gaussian distribution. The two are related by

$$x = \sqrt{-2 \ln r_1} \cos 2\pi r_2, \quad y = \sqrt{-2 \ln r_1} \sin 2\pi r_2. \tag{6.74}$$

The inversion of this mapping produces the Gaussian distribution

$$r_1 = e^{-(x^2+y^2)/2}, \quad r_2 = \frac{1}{2\pi} \tan^{-1} \frac{y}{x}, \quad J = -\frac{e^{-(x^2+y^2)/2}}{2\pi}. \tag{6.75}$$

The solution to our problem is at hand. We use (6.74) with r_1 and r_2 uniform random distributions, and x and y are then Gaussian random distributions centered around $x = 0$.

7

Differentiation & Searching

In this chapter we add two more tools to our computational toolbox: numerical differentiation and trial-and-error searching. In Unit I we derive the forward-difference, central-difference, and extrapolated-difference methods for differentiation. They will be used throughout the book, especially for partial differential equations. In Unit II we devise ways to search for solutions to nonlinear equations by trial and error and apply our new-found numerical differentiation tools there. Although trial-and-error searching may not sound very precise, it is in fact widely used to solve problems where analytic solutions do not exist or are not practical. In Chapter 8, "Solving Systems of Equations with Matrices; Data Fitting," *we extend these search and differentiation techniques to the solution of simultaneous equations using matrix techniques. In Chapter 9,* "Differential Equation Applications," *we combine trial-and-error searching with the solution of ordinary differential equations to solve the quantum eigenvalue problem.*

7.1 Unit I. Numerical Differentiation

Problem: Figure 7.1 shows the trajectory of a projectile with air resistance. The dots indicate the times t at which measurements were made and tabulated. Your **problem** is to determine the velocity $dy/dt \equiv y'$ as a function of time. Note that since there is realistic air resistance present, there is no analytic function to differentiate, only this table of numbers.

You probably did rather well in your first calculus course and feel competent at taking derivatives. However, you may not ever have taken derivatives of a table of numbers using the elementary definition

$$\frac{dy(t)}{dt} \stackrel{\text{def}}{=} \lim_{h\to 0} \frac{y(t+h)-y(t)}{h}. \tag{7.1}$$

In fact, even a computer runs into errors with this kind of limit because it is wrought with subtractive cancellation; the computer's finite word length causes the numerator to fluctuate between 0 and the machine precision ϵ_m as the denominator approaches zero.

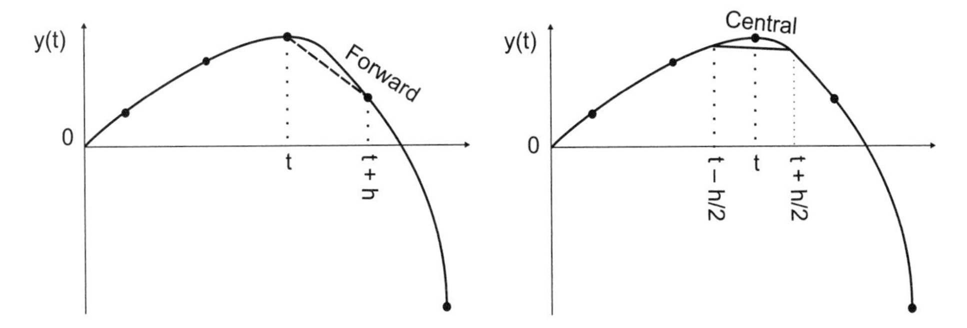

Figure 7.1 Forward-difference approximation (slanted dashed line) and central-difference approximation (horizontal line) for the numerical first derivative at point *t*. The central difference is seen to be more accurate. (The trajectory is that of a projectile with air resistance.)

7.2 Forward Difference (Algorithm)

The most direct method for numerical differentiation starts by expanding a function in a Taylor series to obtain its value a small step h away:

$$y(t+h) = y(t) + h\,\frac{dy(t)}{dt} + \frac{h^2}{2!}\frac{d^2y(t)}{dt^2} + \frac{h^3}{3!}\frac{dy^3(t)}{dt^3} + \cdots . \tag{7.2}$$

We obtain the *forward-difference* derivative algorithm by solving (7.2) for $y'(t)$:

$$\boxed{\left.\frac{dy(t)}{dt}\right|_{\text{fd}} \stackrel{\text{def}}{=} \frac{y(t+h)-y(t)}{h}.} \tag{7.3}$$

An approximation for the error follows from substituting the Taylor series:

$$\left.\frac{dy(t)}{dt}\right|_{\text{fd}} \simeq \frac{dy(t)}{dt} + \frac{h}{2}\frac{dy^2(t)}{dt^2} + \cdots . \tag{7.4}$$

You can think of this approximation as using two points to represent the function by a straight line in the interval from x to $x+h$ (Figure 7.1 left).

The approximation (7.3) has an error proportional to h (unless the heavens look down upon you kindly and make y'' vanish). We can make the approximation error smaller by making h smaller, yet precision will be lost through the subtractive cancellation on the left-hand side (LHS) of (7.3) for too small an h. To see how the forward-difference algorithm works, let $y(t) = a + bt^2$. The exact derivative is $y' = 2bt$, while the computed derivative is

$$\left.\frac{dy(t)}{dt}\right|_{\text{fd}} \simeq \frac{y(t+h) - y(t)}{h} = 2bt + bh. \tag{7.5}$$

This clearly becomes a good approximation only for small $h(h \ll 2t)$.

7.3 Central Difference (Algorithm)

An improved approximation to the derivative starts with the basic definition (7.1) or geometrically as shown in Figure 7.1 on the right. Now, rather than making a single step of h forward, we form a *central difference* by stepping forward half a step and backward half a step:

$$\boxed{\left.\frac{dy(t)}{dt}\right|_{\text{cd}} \stackrel{\text{def}}{=} \frac{y(t+h/2) - y(t-h/2)}{h}.} \tag{7.6}$$

We estimate the error in the central-difference algorithm by substituting the Taylor series for $y(t \pm h/2)$ into (7.6):

$$\begin{aligned} y\left(t+\frac{h}{2}\right) - y\left(t-\frac{h}{2}\right) &\simeq \left[y(t) + \frac{h}{2}y'(t) + \frac{h^2}{8}y''(t) + \frac{h^3}{48}y'''(t) + \mathcal{O}(h^4)\right] \\ &\quad - \left[y(t) - \frac{h}{2}y'(t) + \frac{h^2}{8}y''(t) - \frac{h^3}{48}y'''(t) + \mathcal{O}(h^4)\right] \\ &= hy'(t) + \frac{h^3}{24}y'''(t) + \mathcal{O}(h^5), \end{aligned}$$

$$\Rightarrow \quad \left.\frac{dy(t)}{dt}\right|_{\text{cd}} \simeq y'(t) + \frac{1}{24}h^2 y'''(t) + \mathcal{O}(h^4). \tag{7.7}$$

The important difference between this algorithm and the forward difference from (7.3) is that when $y(t-h/2)$ is subtracted from $y(t+h/2)$, all terms containing an even power of h in the two Taylor series cancel. This make the central-difference algorithm accurate to order h^2 (h^3 before division by h), while the forward difference is accurate only to order h. If the $y(t)$ is smooth, that is, if $y'''h^2/24 \ll y''h/2$, then you can expect the central-difference error to be smaller. If we now return to our parabola example (7.5), we will see that the central difference gives the exact derivative independent of h:

$$\left.\frac{dy(t)}{dt}\right|_{\text{cd}} \simeq \frac{y(t+h/2) - y(t-h/2)}{h} = 2bt. \tag{7.8}$$

7.4 Extrapolated Difference (Method)

Because a differentiation rule based on keeping a certain number of terms in a Taylor series also provides an expression for the error (the terms not included), we can reduce the theoretical error further by forming a combination of algorithms whose summed errors extrapolate to zero. One algorithm is the central-difference algorithm (7.6) using a half-step back and a half-step forward. The second algorithm is another central-difference approximation using quarter-steps:

$$\left.\frac{dy(t,h/2)}{dt}\right|_{\text{cd}} \stackrel{\text{def}}{=} \frac{y(t+h/4)-y(t-h/4)}{h/2} \simeq y'(t)+\frac{h^2}{96}\frac{d^3y(t)}{dt^3}+\cdots. \tag{7.9}$$

A combination of the two eliminates both the quadratic and linear error terms:

$$\left.\frac{dy(t)}{dt}\right|_{\text{ed}} \stackrel{\text{def}}{=} \frac{4D_{\text{cd}}y(t,h/2)-D_{\text{cd}}y(t,h)}{3} \tag{7.10}$$

$$\simeq \frac{dy(t)}{dt} - \frac{h^4 y^{(5)}(t)}{4\times 16\times 120}+\cdots. \tag{7.11}$$

Here (7.10) is the extended-difference algorithm, (7.11) gives its error, and D_{cd} represents the central-difference algorithm. If $h=0.4$ and $y^{(5)}\simeq 1$, then there will be only one place of round-off error and the truncation error will be approximately machine precision ϵ_m; this is the best you can hope for.

When working with these and similar higher-order methods, it is important to remember that while they may work as designed for well-behaved functions, they may fail badly for functions containing noise, as may data from computations or measurements. If noise is evident, it may be better to first smooth the data or fit them with some analytic function using the techniques of Chapter 8, "Solving Systems of Equations with Matrices; Data Fitting," and then differentiate.

7.5 Error Analysis (Assessment)

The approximation errors in numerical differentiation decrease with decreasing step size h, while round-off errors increase with decreasing step size (you have to take more steps and do more calculations). Remember from our discussion in Chapter 2, "Errors & Uncertainties in Computations," that the best approximation occurs for an h that makes the total error $\epsilon_{\text{approx}}+\epsilon_{\text{ro}}$ a minimum, and that as a rough guide this occurs when $\epsilon_{\text{ro}}\simeq\epsilon_{\text{approx}}$.

We have already estimated the approximation error in numerical differentiation rules by making a Taylor series expansion of $y(x+h)$. The approximation error with the forward-difference algorithm (7.3) is $\mathcal{O}(h)$, while that with the central-difference algorithm (7.7) is $\mathcal{O}(h^2)$:

$$\epsilon_{\text{approx}}^{\text{fd}} \simeq \frac{y''h}{2}, \quad \epsilon_{\text{approx}}^{\text{cd}} \simeq \frac{y'''h^2}{24}. \tag{7.12}$$

To obtain a rough estimate of the round-off error, we observe that differentiation essentially subtracts the value of a function at argument x from that of the same function at argument $x+h$ and then divides by h: $y' \simeq [y(t+h) - y(t)]/h$. As h is made continually smaller, we eventually reach the round-off error limit where $y(t+h)$ and $y(t)$ differ by just machine precision ϵ_m:

$$\epsilon_{\text{ro}} \simeq \frac{\epsilon_m}{h}. \tag{7.13}$$

Consequently, round-off and approximation errors become equal when

$$\epsilon_{\text{ro}} \simeq \epsilon_{\text{approx}},$$

$$\begin{aligned} \frac{\epsilon_m}{h} &\simeq \epsilon_{\text{approx}}^{\text{fd}} = \frac{y^{(2)}h}{2}, & \frac{\epsilon_m}{h} &\simeq \epsilon_{\text{approx}}^{\text{cd}} = \frac{y^{(3)}h^2}{24}, \\ \Rightarrow \quad h_{\text{fd}}^2 &= \frac{2\epsilon_m}{y^{(2)}}, & \Rightarrow \quad h_{\text{cd}}^3 &= \frac{24\epsilon_m}{y^{(3)}}. \end{aligned} \tag{7.14}$$

We take $y' \simeq y^{(2)} \simeq y^{(3)}$ (which may be crude in general, though not bad for e^t or $\cos t$) and assume double precision, $\epsilon_m \simeq 10^{-15}$:

$$\begin{aligned} h_{\text{fd}} &\simeq 4 \times 10^{-8}, & h_{\text{cd}} &\simeq 3 \times 10^{-5}, \\ \Rightarrow \quad \epsilon_{\text{fd}} &\simeq \frac{\epsilon_m}{h_{\text{fd}}} \simeq 3 \times 10^{-8}, & \Rightarrow \quad \epsilon_{\text{cd}} &\simeq \frac{\epsilon_m}{h_{\text{cd}}} \simeq 3 \times 10^{-11}. \end{aligned} \tag{7.15}$$

This may seem backward because the better algorithm leads to a larger h value. It is not. The ability to use a larger h means that the error in the central-difference method is about 1000 times smaller than the error in the forward-difference method.

We give a full program Diff.java on the instructor's disk, yet the programming for numerical differentiation is so simple that we need give only the lines

```
FD = ( y(t+h) - y(t) ) /h;                                    // forward diff
CD = ( y(t+h/2) - y(t-h/2) ) /h;                              // central diff
ED = (8*(y(t+h/4) - y(t-h/4)) - (y(t+h/2)-y(t-h/2))) /(3*h);  // extrap diff
```

1. Use forward-, central-, and extrapolated-difference algorithms to differentiate the functions $\cos t$ and e^t at $t = 0.1, 1.,$ and 100.
   a. Print out the derivative and its relative error $\mathcal{E}$ as functions of h. Reduce the step size h until it equals machine precision $h \simeq \epsilon_m$.
   b. Plot $\log_{10} |\mathcal{E}|$ *versus* $\log_{10} h$ and check whether the number of decimal places obtained agrees with the estimates in the text.
   c. See if you can identify regions where truncation error dominates at large h and round-off error at small h in your plot. Do the slopes agree with our model's predictions?

7.6 Second Derivatives (Problem)

Let's say that you have measured the position $y(t)$ *versus* time for a particle (Figure 7.1). Your **problem** now is to determine the force on the particle. Newton's second law tells us that force and acceleration are linearly related:

$$F = ma = m\frac{d^2y}{dt^2}, \tag{7.16}$$

where F is the force, m is the particle's mass, and a is the acceleration. So if we can determine the acceleration $a(t) = d^2y/dt^2$ from the $y(t)$ values, we can determine the force.

The concerns we expressed about errors in first derivatives are even more valid for second derivatives where additional subtractions may lead to additional cancellations. Let's look again at the central-difference method:

$$\frac{dy(t)}{dt} \simeq \frac{y(t+h/2) - y(t-h/2)}{h}. \tag{7.17}$$

This algorithm gives the derivative at t by moving forward and backward from t by $h/2$. We take the second derivative d^2y/dt^2 to be the central difference of the first derivative:

$$\frac{d^2y(t)}{dt^2} \simeq \frac{y'(t+h/2) - y'(t-h/2)}{h}$$

$$\simeq \frac{[y(t+h) - y(t)] - [y(t) - y(t-h)]}{h^2} \tag{7.18}$$

$$= \frac{y(t+h) + y(t-h) - 2y(t)}{h^2}. \tag{7.19}$$

As we did for first derivatives, we determine the second derivative at t by evaluating the function in the region surrounding t. Although the form (7.19) is more compact and requires fewer steps than (7.18), it may increase subtractive cancellation by first storing the "large" number $y(t+h) + y(t-h)$ and then subtracting another large number $2y(t)$ from it. We ask you to explore this difference as an exercise.

7.6.1 Second-Derivative Assessment

Write a program to calculate the second derivative of $\cos t$ using the central-difference algorithms (7.18) and (7.19). Test it over four cycles. Start with $h \simeq \pi/10$ and keep reducing h until you reach machine precision. ▌

7.7 Unit II. Trial-and-Error Searching

Many computer techniques are well-defined sets of procedures leading to definite outcomes. In contrast, some computational techniques are trial-and-error algorithms in which decisions on what steps to follow are made based on the current values of variables, and the program quits only when it thinks it has solved the problem. (We already did some of this when we summed a power series until the terms became small.) Writing this type of program is usually interesting because we must foresee how to have the computer act intelligently in all possible situations, and running them is very much like an experiment in which it is hard to predict what the computer will come up with.

7.8 Quantum States in a Square Well (Problem)

Probably the most standard problem in quantum mechanics[1] is to solve for the energies of a particle of mass m bound within a 1-D square well of radius a:

$$V(x) = \begin{cases} -V_0, & \text{for } |x| \le a, \\ 0, & \text{for } |x| \ge a. \end{cases} \tag{7.20}$$

As shown in quantum mechanics texts [Gott 66], the energies of the bound states $E = -E_B < 0$ within this well are solutions of the transcendental equations

$$\sqrt{10 - E_B}\, \tan\left(\sqrt{10 - E_B}\right) = \sqrt{E_B} \quad \text{(even)}, \tag{7.21}$$

$$\sqrt{10 - E_B}\, \text{cotan}\left(\sqrt{10 - E_B}\right) = \sqrt{E_B} \quad \text{(odd)}, \tag{7.22}$$

where even and odd refer to the symmetry of the wave function. Here we have chosen units such that $\hbar = 1$, $2m = 1$, $a = 1$, and $V_0 = 10$. Your **problem** is to

1. Find several bound-state energies E_B for even wave functions, that is, the solution of (7.21).
2. See if making the potential deeper, say, by changing the 10 to a 20 or a 30, produces a larger number of, or deeper bound states.

7.9 Trial-and-Error Roots via the Bisection Algorithm

Trial-and-error root finding looks for a value of x at which

$$f(x) = 0,$$

where the 0 on the right-hand side (RHS) is conventional (an equation such as $10 \sin x = 3x^3$ can easily be written as $10 \sin x - 3x^3 = 0$). The search procedure starts

[1] We solve this same problem in §9.9 using an approach that is applicable to almost any potential and which also provides the wave functions. The approach of this section works only for the eigenenergies of a square well.

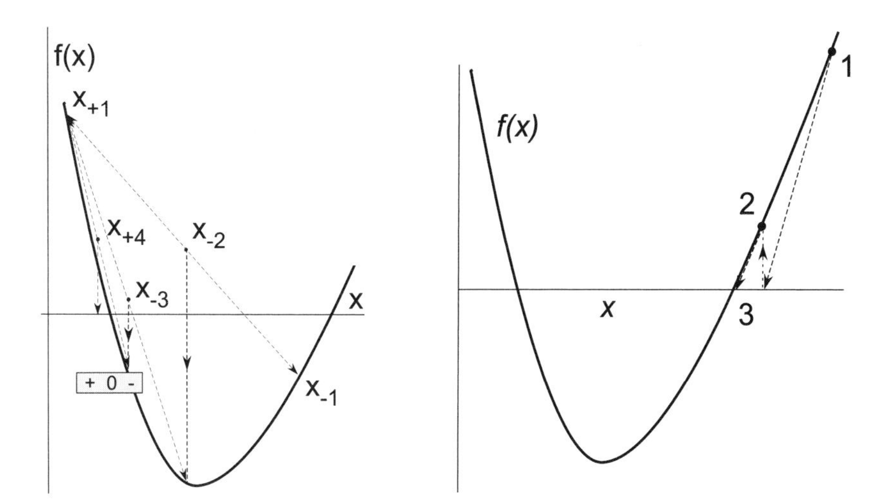

Figure 7.2 A graphical representation of the steps involved in solving for a zero of $f(x)$ using the bisection algorithm (*left*) and the Newton–Raphson method (*right*). The bisection algorithm takes the midpoint of the interval as the new guess for x, and so each step reduces the interval size by one-half. The Newton–Raphson method takes the new guess as the zero of the line tangent to $f(x)$ at the old guess. Four steps are shown for the bisection algorithm, but only two for the more rapidly convergent Newton–Raphson method.

with a guessed value for x, substitutes that guess into $f(x)$ (the "trial"), and then sees how far the LHS is from zero (the "error"). The program then changes x based on the error and tries out the new guess in $f(x)$. The procedure continues until $f(x) \simeq 0$ to some desired level of precision, until the changes in x are insignificant, or until it appears that progress is not being made.

The most elementary trial-and-error technique is the *bisection algorithm*. It is reliable but slow. If you know some interval in which $f(x)$ changes sign, then the bisection algorithm will always converge to the root by finding progressively smaller and smaller intervals in which the zero occurs. Other techniques, such as the Newton–Raphson method we describe next, may converge more quickly, but if the initial guess is not close, it may become unstable and fail completely.

The basis of the bisection algorithm is shown on the left in Figure 7.2. We start with two values of x between which we know a zero occurs. (You can determine these by making a graph or by stepping through different x values and looking for a sign change.) To be specific, let us say that $f(x)$ is negative at x_- and positive at x_+:

$$f(x_-) < 0, \quad f(x_+) > 0. \tag{7.23}$$

(Note that it may well be that $x_- > x_+$ if the function changes from positive to negative as x increases.) Thus we start with the interval $x_+ \leq x \leq x_-$ within which

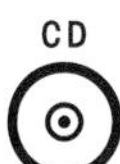

we know a zero occurs. The algorithm (given on the CD as Bisection.java) then picks the new x as the bisection of the interval and selects as its new interval the half in which the sign change occurs:

```
x = ( xPlus + xMinus ) / 2
if ( f(x) f(xPlus) > 0 ) xPlus = x
    else xMinus = x
```

This process continues until the value of $f(x)$ is less than a predefined level of precision or until a predefined (large) number of subdivisions occurs.

The example in Figure 7.2 on the left shows the first interval extending from $x_- = x_{+1}$ to $x_+ = x_{-1}$. We then bisect that interval at x, and since $f(x) < 0$ at the midpoint, we set $x_- \equiv x_{-2} = x$ and label it x_{-2} to indicate the second step. We then use $x_{+2} \equiv x_{+1}$ and x_{-2} as the next interval and continue the process. We see that only x_- changes for the first three steps in this example, but that for the fourth step x_+ finally changes. The changes then become too small for us to show.

7.9.1 Bisection Algorithm Implementation

1. The first step in implementing any search algorithm is to get an idea of what your function looks like. For the present problem you do this by making a plot of $f(E) = \sqrt{10 - E_B}\ \tan(\sqrt{10 - E_B}) - \sqrt{E_B}$ *versus* E_B. Note from your plot some approximate values at which $f(E_B) = 0$. Your program should be able to find more exact values for these zeros.
2. Write a program that implements the bisection algorithm and use it to find some solutions of (7.21).
3. *Warning:* Because the tan function has singularities, you have to be careful. In fact, your graphics program (or Maple) may not function accurately near these singularities. One cure is to use a different but equivalent form of the equation. Show that an equivalent form of (7.21) is

$$\sqrt{E}\ \cot(\sqrt{10 - E}) - \sqrt{10 - E} = 0. \tag{7.24}$$

4. Make a second plot of (7.24), which also has singularities but at different places. Choose some approximate locations for zeros from this plot.
5. Compare the roots you find with those given by Maple or Mathematica. ▌

7.10 Newton–Raphson Searching (A Faster Algorithm)

The Newton–Raphson algorithm finds approximate roots of the equation

$$f(x) = 0$$

more quickly than the bisection method. As we see graphically in Figure 7.2 on the right, this algorithm is the equivalent of drawing a straight line $f(x) \simeq mx + b$ tangent to the curve at an x value for which $f(x) \simeq 0$ and then using the intercept of the line with the x axis at $x = -b/m$ as an improved guess for the root. If the "curve" were a straight line, the answer would be exact; otherwise, it is a good approximation if the guess is close enough to the root for $f(x)$ to be nearly linear. The process continues until some set level of precision is reached. If a guess is in a region where $f(x)$ is nearly linear (Figure 7.2), then the convergence is much more rapid than for the bisection algorithm.

The analytic formulation of the Newton–Raphson algorithm starts with an old guess x_0 and expresses a new guess x as a correction Δx to the old guess:

$$x_0 = \text{old guess}, \quad \Delta x = \text{unknown correction} \tag{7.25}$$

$$\Rightarrow \quad x = x_0 + \Delta x = \text{(unknown) new guess}. \tag{7.26}$$

We next expand the known function $f(x)$ in a Taylor series around x_0 and keep only the linear terms:

$$f(x = x_0 + \Delta x) \simeq f(x_0) + \left.\frac{df}{dx}\right|_{x_0} \Delta x. \tag{7.27}$$

We then determine the correction Δx by calculating the point at which this linear approximation to $f(x)$ crosses the x axis:

$$f(x_0) + \left.\frac{df}{dx}\right|_{x_0} \Delta x = 0, \tag{7.28}$$

$$\Rightarrow \quad \boxed{\Delta x = -\frac{f(x_0)}{df/dx|_{x_0}}.} \tag{7.29}$$

The procedure is repeated starting at the improved x until some set level of precision is obtained.

The Newton–Raphson algorithm (7.29) requires evaluation of the derivative df/dx at each value of x_0. In many cases you may have an analytic expression for the derivative and can build it into the algorithm. However, especially for more complicated problems, it is simpler and less error-prone to use a numerical forward-difference approximation to the derivative:[2]

$$\frac{df}{dx} \simeq \frac{f(x+\delta x) - f(x)}{\delta x}, \tag{7.30}$$

where δx is some small change in x that you just chose [different from the Δ used for searching in (7.29)]. While a central-difference approximation for the derivative would be more accurate, it would require additional evaluations of the f's, and once you find a zero, it does not matter how you got there. On the CD we give the

[2] We discuss numerical differentiation in Chapter 7, "Differentiation & Searching."

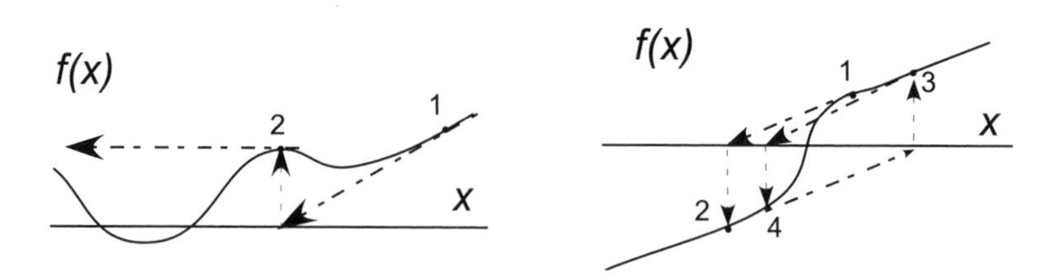

Figure 7.3 Two examples of how the Newton-Raphson algorithm may fail if the initial guess is not in the region where $f(x)$ can be approximated by a straight line. *Left:* A guess lands at a local minimum/maximum, that is, a place where the derivative vanishes, and so the next guess ends up at $x=\infty$. *Right:* The search has fallen into an infinite loop. Backtracking would help here.

programs Newton_cd.java (also Listing 7.1) and Newton_fd.java, which implement the derivative both ways.

```
// Newton_cd.java: Newton-Raphson root finder, central diff derivative

public class Newton_cd {

  public static double f(double x) { return 2*Math.cos(x) - x; }          // function

  public static void main(String[] argv) {
    double x = 2., dx = 1e-2, F= f(x), eps = 1e-6, df;
    int it, imax = 100;                           // Max no of iterations permitted
    for ( it = 0;  it <= imax;  it++ )  {
      System.out.println("Iteration # = "+it+" x = "+x+" f(x) = "+F);
      df = ( f(x + dx/2) - f(x-dx/2) )/dx;                   // Central diff deriv
      dx = -F/df;
      x += dx;                                                      // New guess
      F = f(x);                                                   // Save for use
      if ( Math.abs(F) <= eps )  {                       // Check for convergence
        System.out.println("Root found, tolerance eps = " + eps);
        break;
} } } }
```

Listing 7.1 Newton_cd.java uses the Newton-Raphson method to search for a zero of the function $f(x)$. A central-difference approximation is used to determine df/dx.

7.10.1 Newton-Raphson Algorithm with Backtracking

Two examples of possible problems with the Newton–Raphson algorithm are shown in Figure 7.3. On the left we see a case where the search takes us to an x value where the function has a local minimum or maximum, that is, where $df/dx=0$. Because $\Delta x=-f/f'$, this leads to a horizontal tangent (division by zero), and so the next guess is $x=\infty$, from where it is hard to return. When this happens, you need to start your search with a different guess and pray that you do not fall into this trap again. In cases where the correction is very large but maybe not infinite, you may want to try backtracking (described below) and hope that by taking a smaller step you will not get into as much trouble.

In Figure 7.3 on the right we see a case where a search falls into an infinite loop surrounding the zero without ever getting there. A solution to this problem is called *backtracking*. As the name implies, in cases where the new guess $x_0 + \Delta x$ leads to an increase in the magnitude of the function, $|f(x_0 + \Delta x)|^2 > |f(x_0)|^2$, you should backtrack somewhat and try a smaller guess, say, $x_0 + \Delta x/2$. If the magnitude of f still increases, then you just need to backtrack some more, say, by trying $x_0 + \Delta x/4$ as your next guess, and so forth. Because you know that the tangent line leads to a local decrease in $|f|$, eventually an acceptable small enough step should be found.

The problem in both these cases is that the initial guesses were not close enough to the regions where $f(x)$ is approximately linear. So again, a good plot may help produce a good first guess. Alternatively, you may want to start your search with the bisection algorithm and then switch to the faster Newton–Raphson algorithm when you get closer to the zero.

7.10.2 Newton–Raphson Algorithm Implementation

1. Use the Newton–Raphson algorithm to find some energies E_B that are solutions of (7.21). Compare this solution with the one found with the bisection algorithm.
2. Again, notice that the 10 in this equation is proportional to the strength of the potential that causes the binding. See if making the potential deeper, say, by changing the 10 to a 20 or a 30, produces more or deeper bound states. (Note that in contrast to the bisection algorithm, your initial guess must be closer to the answer for the Newton–Raphson algorithm to work.)
3. Modify your algorithm to include backtracking and then try it out on some problem cases. ∎

8

Solving Systems of Equations with Matrices; Data Fitting

Unit I of this chapter applies the trial-and-error techniques developed in Chapter 7, "Differentiation & Searching," *to solve a set of simultaneous nonlinear equations. This leads us into general matrix computing using scientific libraries. In Unit II we look at several ways in which theoretical formulas are fit to data and see that these often require the matrix techniques of Unit I.*

8.1 Unit I. Systems of Equations and Matrix Computing

Physical systems are often modeled by systems of simultaneous equations written in matrix form. As the models are made more realistic, the matrices often become large, and computers become an excellent tool for solving such problems. What makes computers so good is that matrix manipulations intrinsically involve the continued repetition of a small number of simple instructions, and algorithms exist to do this quite efficiently. Further speedup may be achieved by *tuning* the codes to the computer's architecture, as discussed in Chapter 14, "High-Performance Computing Hardware, Tuning, & Parallel Computing."

Industrial-strength subroutines for matrix computing are found in well-established scientific libraries. These subroutines are usually an order of magnitude or more faster than the elementary methods found in linear algebra texts,[1] are usually designed to minimize round-off error, and are often "robust," that is, have a high chance of being successful for a broad class of problems. For these reasons we recommend that you *do not write your own matrix subroutines* but instead get them from a library. An additional value of library routines is that you can often run the same program either on a desktop machine or on a parallel supercomputer, with matrix routines automatically adapting to the local architecture.

The thoughtful reader may be wondering when a matrix is "large" enough to require the use of a library routine. While in the past large may have meant a

[1] Although we prize the book [Pres 94] and what it has accomplished, we cannot recommend taking subroutines from it. They are neither optimized nor documented for easy, standalone use, whereas the subroutine libraries recommended in this chapter are.

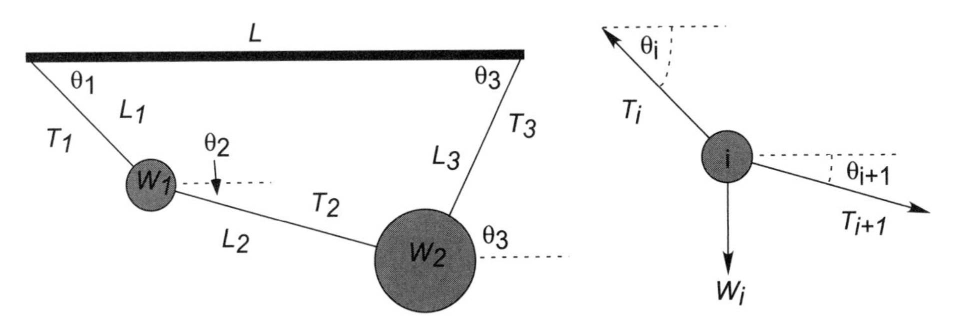

Figure 8.1 *Left:* Two weights connected by three pieces of string and suspended from a horizontal bar of length *L*. The angles and the tensions in the strings are unknown. *Right:* A free body diagram for one weight in equilibrium.

good fraction of your computer's random-access memory (RAM), we now advise that a library routine be used whenever the matrix computations are so numerically intensive that you must wait for results. In fact, even if the sizes of your matrices are small, as may occur in graphical processing, there may be library routines designed just for that which speed up your computation.

Now that you have heard the sales pitch, you may be asking, "What's the cost?" In the later part of this chapter we pay the costs of having to find what libraries are available, of having to find the name of the routine in that library, of having to find the names of the subroutines your routine calls, and then of having to figure out how to call all these routines properly. And because some of the libraries are in Fortran, if you are a C programmer you may also be taxed by having to call a Fortran routine from your C program. However, there are now libraries available in most languages.

8.2 Two Masses on a String

Two weights $(W_1, W_2) = (10, 20)$ are hung from three pieces of string with lengths $(L_1, L_2, L_3) = (3, 4, 4)$ and a horizontal bar of length $L = 8$ (Figure 8.1). The **problem** is to find the angles assumed by the strings and the tensions exerted by the strings.

In spite of the fact that this is a simple problem requiring no more than first-year physics to formulate, the coupled transcendental equations that result are inhumanely painful to solve analytically. However, we will show you how the computer can solve this problem, but even then only by a trial-and-error technique with no guarantee of success. Your **problem** is to test this solution for a variety of weights and lengths and then to extend it to the three-weight problem (not as easy as it may seem). In either case check the physical reasonableness of your solution; the deduced tensions should be positive and of similar magnitude to the weights of the spheres, and the deduced angles should correspond to a physically realizable geometry, as confirmed with a sketch. Some of the exploration you should do is to see at what point your initial guess gets so bad that the computer is unable to find a physical solution.

8.2.1 Statics (Theory)

We start with the geometric constraints that the horizontal length of the structure is L and that the strings begin and end at the same height (Figure 8.1 left):

$$L_1 \cos\theta_1 + L_2 \cos\theta_2 + L_3 \cos\theta_3 = L, \tag{8.1}$$

$$L_1 \sin\theta_1 + L_2 \sin\theta_2 - L_3 \sin\theta_3 = 0, \tag{8.2}$$

$$\sin^2\theta_1 + \cos^2\theta_1 = 1, \tag{8.3}$$

$$\sin^2\theta_2 + \cos^2\theta_2 = 1, \tag{8.4}$$

$$\sin^2\theta_3 + \cos^2\theta_3 = 1. \tag{8.5}$$

Observe that the last three equations include trigonometric identities as independent equations because we are treating $\sin\theta$ and $\cos\theta$ as independent variables; this makes the search procedure easier to implement. The basics physics says that since there are no accelerations, the sum of the forces in the horizontal and vertical directions must equal zero (Figure 8.1 right):

$$T_1 \sin\theta_1 - T_2 \sin\theta_2 - W_1 = 0, \tag{8.6}$$

$$T_1 \cos\theta_1 - T_2 \cos\theta_2 = 0, \tag{8.7}$$

$$T_2 \sin\theta_2 + T_3 \sin\theta_3 - W_2 = 0, \tag{8.8}$$

$$T_2 \cos\theta_2 - T_3 \cos\theta_3 = 0. \tag{8.9}$$

Here W_i is the weight of mass i and T_i is the tension in string i. Note that since we do not have a rigid structure, we cannot assume the equilibrium of torques.

8.2.2 Multidimensional Newton–Raphson Searching

Equations (8.1)–(8.9) are nine simultaneous nonlinear equations. While linear equations can be solved directly, nonlinear equations cannot [Pres 00]. You can use the computer to search for a solution by guessing, but there is no guarantee of finding one. We apply to our set the same Newton–Raphson algorithm as used to solve a single equation by renaming the nine unknown angles and tensions as the subscripted variable y_i and placing the variables together as a vector:

$$\mathbf{y} = \begin{pmatrix} x_1 \\ x_2 \\ x_3 \\ x_4 \\ x_5 \\ x_6 \\ x_7 \\ x_8 \\ x_9 \end{pmatrix} = \begin{pmatrix} \sin\theta_1 \\ \sin\theta_2 \\ \sin\theta_3 \\ \cos\theta_1 \\ \cos\theta_2 \\ \cos\theta_3 \\ T_1 \\ T_2 \\ T_3 \end{pmatrix}. \tag{8.10}$$

The nine equations to be solved are written in a general form with zeros on the right-hand sides and placed in a vector:

$$f_i(x_1, x_2, \ldots, x_N) = 0, \quad i = 1, N, \tag{8.11}$$

$$\mathbf{f}(\mathbf{y}) = \begin{pmatrix} f_1(\mathbf{y}) \\ f_2(\mathbf{y}) \\ f_3(\mathbf{y}) \\ f_4(\mathbf{y}) \\ f_5(\mathbf{y}) \\ f_6(\mathbf{y}) \\ f_7(\mathbf{y}) \\ f_8(\mathbf{y}) \\ f_9(\mathbf{y}) \end{pmatrix} = \begin{pmatrix} 3x_4 + 4x_5 + 4x_6 - 8 \\ 3x_1 + 4x_2 - 4x_3 \\ x_7x_1 - x_8x_2 - 10 \\ x_7x_4 - x_8x_5 \\ x_8x_2 + x_9x_3 - 20 \\ x_8x_5 - x_9x_6 \\ x_1^2 + x_4^2 - 1 \\ x_2^2 + x_5^2 - 1 \\ x_3^2 + x_6^2 - 1 \end{pmatrix} = \mathbf{0}. \tag{8.12}$$

The solution to these equations requires a set of nine x_i values that make all nine f_i's vanish simultaneously. Although these equations are not very complicated (the physics after all is elementary), the terms quadratic in x make them nonlinear, and this makes it hard or impossible to find an analytic solution. The search algorithm is to guess a solution, expand the nonlinear equations into linear form, solve the resulting linear equations, and continue to improve the guesses based on how close the previous one was to making $\mathbf{f} = 0$.

Explicitly, let the approximate solution at any one stage be the set $\{x_i\}$ and let us assume that there is an (unknown) set of corrections $\{\Delta x_i\}$ for which

$$f_i(x_1 + \Delta x_1, x_2 + \Delta x_2, \ldots, x_9 + \Delta x_9) = 0, \quad i = 1, 9. \tag{8.13}$$

We solve for the approximate Δx_i's by assuming that our previous solution is close enough to the actual one for two terms in the Taylor series to be accurate:

$$f_i(x_1 + \Delta x_1, \ldots, x_9 + \Delta x_9) \simeq f_i(x_1, \ldots, x_9) + \sum_{j=1}^{9} \frac{\partial f_i}{\partial x_j} \Delta x_j = 0. \quad i = 1, 9. \tag{8.14}$$

We now have a solvable set of nine linear equations in the nine unknowns Δx_i, which we express as a single matrix equation

$$f_1 + \partial f_1/\partial x_1\, \Delta x_1 + \partial f_1/\partial x_2\, \Delta x_2 + \cdots + \partial f_1/\partial x_9\, \Delta x_9 = 0,$$

$$f_2 + \partial f_2/\partial x_1\, \Delta x_1 + \partial f_2/\partial x_2\, \Delta x_2 + \cdots + \partial f_2/\partial x_9\, \Delta x_9 = 0,$$

$$\ddots$$

$$f_9 + \partial f_9/\partial x_1\, \Delta x_1 + \partial f_9/\partial x_2\, \Delta x_2 + \cdots + \partial f_9/\partial x_9\, \Delta x_9 = 0,$$

$$\begin{pmatrix} f_1 \\ f_2 \\ \ddots \\ f_9 \end{pmatrix} + \begin{pmatrix} \partial f_1/\partial x_1 & \partial f_1/\partial x_2 & \cdots & \partial f_1/\partial x_9 \\ \partial f_2/\partial x_1 & \partial f_2/\partial x_2 & \cdots & \partial f_2/\partial x_9 \\ & \ddots & & \\ \partial f_9/\partial x_1 & \partial f_9/\partial x_2 & \cdots & \partial f_9/\partial x_9 \end{pmatrix} \begin{pmatrix} \Delta x_1 \\ \Delta x_2 \\ \ddots \\ \Delta x_9 \end{pmatrix} = 0. \tag{8.15}$$

Note now that the derivatives and the f's are all evaluated at known values of the x_i's, so only the vector of the Δx_i values is unknown. We write this equation in matrix notation as

$$\mathbf{f} + \mathbf{F}'\mathbf{\Delta x} = 0, \quad \Rightarrow \quad \mathbf{F}'\mathbf{\Delta x} = -\mathbf{f}, \tag{8.16}$$

$$\mathbf{\Delta x} = \begin{pmatrix} \Delta x_1 \\ \Delta x_2 \\ \ddots \\ \Delta x_9 \end{pmatrix}, \quad \mathbf{f} = \begin{pmatrix} f_1 \\ f_2 \\ \ddots \\ f_9 \end{pmatrix}, \quad \mathbf{F}' = \begin{pmatrix} \partial f_1/\partial x_1 & \cdots & \partial f_1/\partial x_9 \\ \partial f_2/\partial x_1 & \cdots & \partial f_2/\partial x_9 \\ & \ddots & \\ \partial f_9/\partial x_1 & \cdots & \partial f_9/\partial x_9 \end{pmatrix}.$$

Here we use **bold** to emphasize the vector nature of the columns of f_i and Δx_i values and call the matrix of the derivatives $\mathbf{F}'$ (it is also sometimes called $\mathbf{J}$ because it is the *Jacobian* matrix).

The equation $\mathbf{F}'\mathbf{\Delta x} = -\mathbf{f}$ is in the standard form for the solution of a linear equation (often written $\mathbf{A}\mathbf{x} = \mathbf{b}$), where $\mathbf{\Delta x}$ is the vector of unknowns and $\mathbf{b} = -\mathbf{f}$. Matrix equations are solved using the techniques of linear algebra, and in the sections to follow we shall show how to do that on a computer. In a formal (and sometimes practical) sense, the solution of (8.16) is obtained by multiplying both sides of the equation by the inverse of the $\mathbf{F}'$ matrix:

$$\mathbf{\Delta x} = -\mathbf{F}'^{-1}\mathbf{f}, \tag{8.17}$$

where the inverse must exist if there is to be a unique solution. Although we are dealing with matrices now, this solution is identical in form to that of the 1-D problem, $\Delta x = -(1/f')f$. In fact, one of the reasons we use formal or abstract notation for matrices is to reveal the simplicity that lies within.

As we indicated for the single-equation Newton–Raphson method, while for our two-mass problem we can derive analytic expressions for the derivatives $\partial f_i/\partial x_j$, there are $9 \times 9 = 81$ such derivatives for this (small) problem, and entering them all would be both time-consuming and error-prone. In contrast, especially for more complicated problems, it is straightforward to program a forward-difference approximation for the derivatives,

$$\frac{\partial f_i}{\partial x_j} \simeq \frac{f_i(x_j + \Delta x_j) - f_i(x_j)}{\delta x_j}, \tag{8.18}$$

where each individual x_j is varied independently since these are partial derivatives and δx_j are some arbitrary changes you input. While a central-difference approximation for the derivative would be more accurate, it would also require more evaluations of the f's, and once we find a solution it does not matter how accurate our algorithm for the derivative was.

As also discussed for the 1-D Newton–Raphson method (§7.10.1), the method can fail if the initial guess is not close enough to the zero of f (here all N of them) for the f's to be approximated as linear. The *backtracking* technique may be applied here as well, in the present case, progressively decreasing the corrections Δx_i until $|f|^2 = |f_1|^2 + |f_2|^2 + \cdots + |f_N|^2$ decreases.

8.3 Classes of Matrix Problems (Math)

It helps to remember that the rules of mathematics apply even to the world's most powerful computers. For example, you *should* have problems solving equations if you have more unknowns than equations or if your equations are not linearly independent. But do not fret. While you cannot obtain a unique solution when there are not enough equations, you may still be able to map out a space of allowable solutions. At the other extreme, if you have more equations than unknowns, you have an *overdetermined* problem, which may not have a unique solution. An overdetermined problem is sometimes treated using data fitting in which a solution to a sufficient set of equations is found, tested on the unused equations, and then improved if needed. Not surprisingly, this latter technique is known as the *linear least-squares method* because it finds the best solution "on the average."

The most basic matrix problem is the system of linear equations you have to solve for the two-mass **problem**:

$$\mathbf{A}\mathbf{x} = \mathbf{b}, \quad \mathbf{A}_{N\times N}\,\mathbf{x}_{N\times 1} = \mathbf{b}_{N\times 1}, \tag{8.19}$$

where $\mathbf{A}$ is a known $N \times N$ matrix, $\mathbf{x}$ is an unknown vector of length N, and $\mathbf{b}$ is a known vector of length N. The best way to solve this equation is by Gaussian elimination or lower-upper (LU) decomposition. This yields the vector $\mathbf{x}$ without explicitly calculating $\mathbf{A}^{-1}$. Another, albeit slower and less robust, method is to determine the inverse of $\mathbf{A}$ and then form the solution by multiplying both sides of (8.19) by $\mathbf{A}^{-1}$:

$$\mathbf{x} = \mathbf{A}^{-1}\mathbf{b}. \tag{8.20}$$

Both the direct solution of (8.19) and the determination of a matrix's inverse are standards in a matrix subroutine library.

If you have to solve the matrix equation

$$\mathbf{A}\mathbf{x} = \lambda\mathbf{x}, \tag{8.21}$$

with $\mathbf{x}$ an unknown vector and λ an unknown parameter, then the direct solution (8.20) will not be of much help because the matrix $\mathbf{b} = \lambda\mathbf{x}$ contains the unknowns λ and $\mathbf{x}$. Equation (8.21) is the *eigenvalue problem*. It is harder to solve than (8.19) because solutions exist for only certain λ values (or possibly none depending on $\mathbf{A}$). We use the identity matrix to rewrite (8.21) as

$$[\mathbf{A} - \lambda\mathbf{I}]\mathbf{x} = 0, \tag{8.22}$$

and we see that multiplication by $[\mathbf{A} - \lambda\mathbf{I}]^{-1}$ yields the *trivial solution*

$$\mathbf{x} = 0 \quad \text{(trivial solution)}. \tag{8.23}$$

While the trivial solution is a bona fide solution, it is trivial. A more interesting solution requires the existence of a condition that forbids us from multiplying both sides of (8.22) by $[\mathbf{A} - \lambda\mathbf{I}]^{-1}$. That condition is the nonexistence of the inverse, and if you recall that Cramer's rule for the inverse requires division by $\det[\mathbf{A} - \lambda\mathbf{I}]$, it is clear that the inverse fails to exist (and in this way eigenvalues *do* exist) when

$$\det[\mathbf{A} - \lambda\mathbf{I}] = 0. \tag{8.24}$$

The λ values that satisfy this *secular equation* are the eigenvalues of (8.21).

If you are interested in only the eigenvalues, you should look for a matrix routine that solves (8.24). To do that, first you need a subroutine to calculate the determinant of a matrix, and then a search routine to zero in on the solution of (8.24). Such routines are available in libraries. The traditional way to solve the eigenvalue problem (8.21) for both eigenvalues and eigenvectors is by *diagonalization*. This is equivalent to successive changes of basis vectors, each change leaving the eigenvalues unchanged while continually decreasing the values of the off-diagonal elements of $\mathbf{A}$. The sequence of transformations is equivalent to continually operating on the original equation with a matrix $\mathbf{U}$:

$$\mathbf{U}\mathbf{A}(\mathbf{U}^{-1}\mathbf{U})\mathbf{x} = \lambda\mathbf{U}\mathbf{x}, \tag{8.25}$$

$$(\mathbf{U}\mathbf{A}\mathbf{U}^{-1})(\mathbf{U}\mathbf{x}) = \lambda\mathbf{U}\mathbf{x}, \tag{8.26}$$

until one is found for which $\mathbf{UAU}^{-1}$ is diagonal:

$$\mathbf{UAU}^{-1} = \begin{pmatrix} \lambda_1^{'} & & \cdots & 0 \\ 0 & \lambda_2^{'} & \cdots & 0 \\ 0 & 0 & \lambda_3^{'} & \cdots \\ 0 & \cdots & & \lambda_N^{'} \end{pmatrix}. \tag{8.27}$$

The diagonal values of $\mathbf{UAU}^{-1}$ are the eigenvalues with eigenvectors

$$\mathbf{x}_i = \mathbf{U}^{-1}\hat{\mathbf{e}}_i; \tag{8.28}$$

that is, the eigenvectors are the columns of the matrix $\mathbf{U}^{-1}$. A number of routines of this type are found in subroutine libraries.

8.3.1 Practical Aspects of Matrix Computing

Many scientific programming bugs arise from the improper use of arrays.[2] This may be due to the extensive use of matrices in scientific computing or to the complexity of keeping track of indices and dimensions. In any case, here are some rules of thumb to observe.

Computers are finite: Unless you are careful, your matrices will use so much memory that your computation will slow down significantly, especially if it starts to use virtual memory. As a case in point, let's say that you store data in a 4-D array with each index having a *physical dimension* of 100: `A[100] [100] [100] [100]`. This array of $(100)^4$ 64-byte words occupies $\simeq$1 GB of memory.

Processing time: Matrix operations such as inversion require on the order of N^3 steps for a square matrix of dimension N. Therefore, doubling the dimensions of a 2-D square matrix (as happens when the number of integration steps is doubled) leads to an *eightfold* increase in processing time.

Paging: Many operating systems have *virtual memory* in which disk space is used when a program runs out of RAM (see Chapter 14, "High-Performance Computing Hardware, Tuning, and Parallel Computing," for a discussion of how computers arrange memory). This is a slow process that requires writing a full *page* of words to the disk. If your program is near the memory limit at which paging occurs, even a slight increase in a matrix's dimension may lead to an order-of-magnitude increase in execution time.

Matrix storage: While we may think of matrices as multidimensional blocks of stored numbers, the computer stores them as linear strings. For instance, a matrix `a[3][3]` in Java and C is stored in *row-major order* (Figure 8.2 left):

$$a_{1,1}\ a_{1,2}\ a_{1,3}\ a_{2,1}\ a_{2,2}\ a_{2,3}\ a_{3,1}\ a_{3,2}\ a_{3,3}\ \ldots,$$

[2] Even a vector $V(N)$ is called an array, albeit a 1-D one.

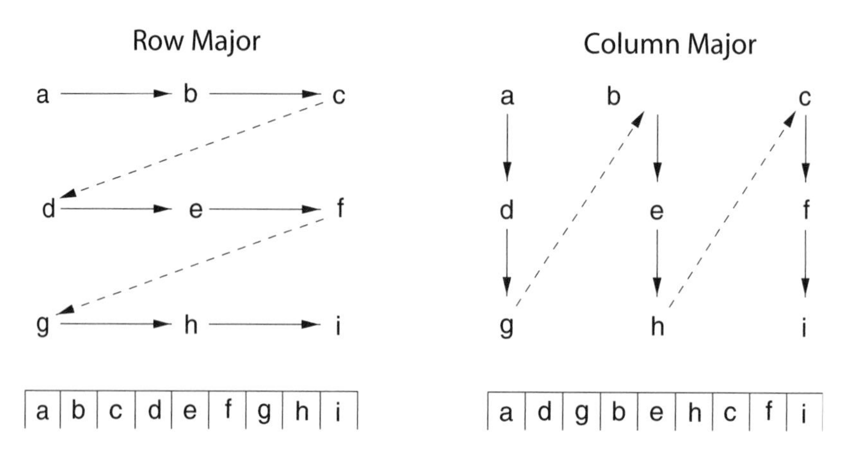

Figure 8.2 *Left:* Row-major order used for matrix storage in C and Java.
Right: Column-major order used for matrix storage in Fortran. How successive matrix elements are stored in a linear fashion in memory is shown at the bottom.

while in Fortran it is stored in *column-major order* (Figure 8.2 right):

$$a_{1,1}\ a_{2,1}\ a_{3,1}\ a_{1,2}\ a_{2,2}\ a_{3,2}\ a_{1,3}\ a_{2,3}\ a_{3,3}\ \ \ldots.$$

It is important to keep this linear storage scheme in mind in order to write proper code and to permit the mixing of Fortran and C programs.

When dealing with matrices, you have to balance the clarity of the operations being performed against the efficiency with which the computer performs them. For example, having one matrix with many indices such as `V[L,Nre,Nspin,k,kp,Z,A]` may be neat packaging, but it may require the computer to jump through large blocks of memory to get to the particular values needed (large *strides*) as you vary `k`, `kp`, and `Nre`. The solution would be to have several matrices such as `V1[Nre,Nspin,k,kp,Z,A]`, `V2[Nre,Nspin,k,kp,Z,A]`, and `V3[Nre,Nspin,k,kp,Z,A]`.

Subscript 0: It is standard in C and Java to have array indices begin with the value 0. While this is now permitted in Fortran, the standard has been to start indices at 1. On that account, in addition to the different locations in memory due to row-major and column-major ordering, the same matrix element is referenced differently in the different languages:

Location	Java/C Element	Fortran Element
Lowest	a[0][0]	a(1,1)
	a[0][1]	a(2,1)
	a[1][0]	a(3,1)
	a[1][1]	a(1,2)
	a[2][0]	a(2,2)
Highest	a[2][1]	a(3,2)

Physical and logical dimensions: When you run a program, you issue commands such as `double a[3][3]` or `Dimension a(3,3)` that tell the compiler how much memory it needs to set aside for the array a. This is called *physical memory*. Sometimes you may use arrays without the full complement of values declared in the declaration statements, for example, as a test case. The amount of memory you actually use to store numbers is the matrix's *logical size*.

Modern programming techniques, such as those used in Java, C, and Fortran90, permit *dynamic memory allocation*; that is, you may use variables as the dimension of your arrays and read in the values of the variables at run time. With these languages you should read in the sizes of your arrays at run time and thus give them the same physical and logical sizes. However, Fortran77, which is the language used for many library routines, requires the dimensions to be specified at compile time, and so the physical and logical sizes may well differ. To see why care is needed if the physical and logical sizes of the arrays differ, imagine that you declared a[3][3] but defined elements only up to a[2][2]. Then the a in storage would look like

a[1][1]' a[1][2]' a[1][3] a[2][1]' a[2][2]' a[2][3] a[3][1] a[3][2] a[3][3],

where only the elements with primes have values assigned to them. Clearly, the defined a values do not occupy sequential locations in memory, and so an algorithm processing this matrix cannot assume that the next element in memory is the next element in your array. This is the reason why subroutines from a library often need to know *both* the physical and logical sizes of your arrays.

Passing sizes to subprograms ⊙: This is needed when the logical and physical dimensions of arrays differ, as is true with some library routines but probably not with the programs you write. In cases such as those using external libraries, you must also watch that the sizes of your matrices do not exceed the bounds that have been declared in the subprograms. This may occur *without* an error message and probably will give you the wrong answers. In addition, if you are running a C program that calls a Fortran subroutine, you will need to pass *pointers* to variables and not the actual values of the variables to the Fortran subprograms (Fortran makes *reference calls*, which means it deals with pointers only as subprogram arguments). Here we have a program possibly running some data stored nearby:

```
main                                              // In main program
  dimension a(100), b(400)

  function Sample(a)                              // In subroutine
    dimension a(10)                          // Smaller dimension
    a(300) = 12                    // Out of bounds, but no message
```

One way to ensure size compatibility among main programs and subroutines is to declare array sizes only in your main program and then pass those sizes along to your subprograms as arguments.

Equivalence, pointers, references manipulations ⊙: Once upon a time computers had such limited memories that programmers conserved memory by having different variables occupy the *same* memory location, the theory being that this would cause no harm as long as these variables were not being used at the same time. This was done by the use of `Common` and `Equivalence` statements in Fortran and by manipulations using pointers and references in other languages. These types of manipulations are now obsolete (the bane of object-oriented programming) and can cause endless grief; do not use them unless it is a matter of "life or death"!

Say what's happening: You decrease programming errors by using self-explanatory labels for your indices (subscripts), stating what your variables mean, and describing your storage schemes.

Tests: Always test a library routine on a small problem whose answer you know (such as the exercises in §8.3.4). Then you'll know if you are supplying it with the right arguments and if you have all the links working.

8.3.2 Implementation: Scientific Libraries, World Wide Web

Some major scientific and mathematical libraries available include the following.

NETLIB	A WWW metalib of free math libraries	ScaLAPACK	Distributed memory LAPACK
LAPACK	Linear Algebra Pack	JLAPACK	LAPACK library in Java
SLATEC	Comprehensive math and statistical pack	ESSL	Engineering and Science Subroutine Library (IBM)
IMSL	International Math and Statistical Libraries	CERNLIB	European Centre for Nuclear Research Library
BLAS	Basic Linear Algebra Subprograms	JAMA	Java Matrix Library
NAG	Numerical Algorithms Group (UK Labs)	LAPACK ++	Linear algebra in C++
TNT	C++ Template Numerical Toolkit	GNU Scientific GSL	Full scientific libraries in C and C++

Except for ESSL, IMSL, and NAG, all these libraries are in the public domain. However, even the proprietary ones are frequently available on a central computer or via an institutionwide site license. General subroutine libraries are treasures to possess because they typically contain optimized routines for almost everything you might want to do, such as

Linear algebra manipulations	Matrix operations	Interpolation, fitting
Eigensystem analysis	Signal processing	Sorting and searching
Solutions of linear equations	Differential equations	Roots, zeros, and extrema
Random-number operations	Statistical functions	Numerical quadrature

You can search the Web to find out about these libraries or to download one if it is not already on your computer. Alternatively, an excellent place to start looking for a library is Netlib, a repository of free software, documents, and databases of interest to computational scientists.

Linear Algebra Package (LAPACK) is a free, portable, modern (1990) library of Fortran77 routines for solving the most common problems in numerical linear algebra. It is designed to be efficient on a wide range of high-performance computers under the proviso that the hardware vendor has implemented an efficient set of Basic Linear Algebra Subroutines (BLAS). In contrast to LAPACK, the Sandia, Los Alamos, Air Force Weapons Laboratory Technical Exchange Committee (SLATEC) library contains general-purpose mathematical and statistical Fortran routines and is consequently more general. Nonetheless, it is not as tuned to the architecture of a particular machine as is LAPACK.

Sometimes a subroutine library supplies only Fortran routines, and this requires a C programmer to call a Fortran routine (we describe how to do that in Appendix E). In some cases, C-language routines may also be available, but they may not be optimized for a particular machine.

As an example of what may be involved in using a scientific library, consider the SLATEC library, which we recommend. The full library contains a guide, a table of contents, and documentation via comments in the source code. The subroutines are classified by the Guide to Available Mathematical Software (GAMS) system. For our masses-on-strings **problem** we have found the needed routines:

snsq-s, dnsq-d	Find zero of n-variable, nonlinear function
snsqe-s, dnsqe-d	Easy-to-use snsq

If you extract these routines, you will find that they need the following:

enorm.f	j4save.f	r1mach.f	xerprn.f	fdjac1.f	r1mpyq.f
xercnt.f	xersve.f	fdump.f	qform.f	r1updt.f	xerhlt.f
xgetua.f	dogleg.f	i1mach.f	qrfac.f	snsq.f	xermsg.f

Of particular interest in these "helper" routines, are *i1mach.f*, *r1mach.f*, and *d1mach.f*. They tell LAPACK the characteristic of your particular machine when the library is first installed. Without that knowledge, LAPACK does not know when convergence is obtained or what step sizes to use.

8.3.3 JAMA: Java Matrix Library

JAMA is a basic linear algebra package for Java developed at the U.S. National Institute of Science (NIST) (see reference [Jama] for documentation). We recommend it because it works well, is natural and understandable to nonexperts, is free, and helps make scientific codes more universal and portable, and because not much else is available. JAMA provides object-oriented classes that construct true

Matrix objects, add and multiply matrices, solve matrix equations, and print out entire matrices in an aligned row-by-row format. JAMA is intended to serve as the standard matrix class for Java.[3] Because this book uses Java for its examples, we now give some JAMA examples.

The first example is the matrix equation $\mathbf{A}\mathbf{x} = \mathbf{b}$ for the solution of a set of linear equations with $\mathbf{x}$ unknown. We take $\mathbf{A}$ to be 3×3, $\mathbf{x}$ to be 3×1, and $\mathbf{b}$ to be 3×1:

```
double[][] array = { {1.,2.,3}, {4.,5.,6.}, {7.,8.,10.} };
Matrix A = new Matrix(array);
Matrix b = Matrix.random(3,1);
Matrix x = A.solve(b);
Matrix Residual = A.times(x).minus(b);
```

Here the vectors and matrices are declared and created as Matrix variables, with $\mathbf{b}$ given random values. We then solve the 3×3 linear system of equations $\mathbf{A}\mathbf{x} = \mathbf{b}$ with the single command Matrix x = A.solve(b) and compute the residual $\mathbf{A}\mathbf{x} - \mathbf{b}$ with the command Residual = A.times(x).minus(b).

Our second JAMA example arises in the solution for the principal-axes system for a cube and requires us to find a coordinate system in which the inertia tensor is diagonal. This entails solving the eigenvalue problem

$$\mathbf{I}\,\overrightarrow{\omega} = \lambda\vec{\omega}, \tag{8.29}$$

where $\mathbf{I}$ is the original inertia matrix, $\vec{\omega}$ is an eigenvector, λ is an eigenvalue, and we use arrows to indicate vectors. The program JamaEigen.java in Listing 8.1 solves for the eigenvalues and vectors and produces output of the form

```
Input Matrix
 0.66667                  --0.25000            --0.25000
                --0.25000 0.66667              --0.25000
                --0.25000            --0.25000 0.66667

Eigenvalue: lambda.Re[] = 0.1666666666666665, 0.9166666666666666,
0.9166666666666666

 Matrix with column eigenvectors                       First Eigenvector, Vec
                --0.57735                 --0.70711              --0.40825       -0.57735
                --0.57735 0.70711                   --0.40825      --0.57735
                --0.57735 0.00000 0.81650             --0.57735

Does LHS = RHS?
 --0.146225044865 --0.146225044865
 --0.146225044865 --0.146225044865
 --0.146225044865 --0.146225044865
```

Look at JamaEigen and notice how on line 9 we first set up the array I with all the elements of the inertia tensor and then on line 10 create a matrix MatI with the same elements as the array. On line 13 the eigenvalue problem is solved with the creation of an eigenvalue object E via the JAMA command:

[3] A sibling matrix package, *Jampack* [Jama] has also been developed at NIST and at the University of Maryland, and it works for complex matrices as well.

```
/* JamaEigen.java: eigenvalue problem with JAMA. JAMA must be in same directory or
   included in CLASSPATH. Uses Matrix.class; see Matrix.java or documentation    */
import Jama.*;
import java.io.*;

public class JamaEigen {

  public static void main(String[] argv) {
  double[][] I = { {2./3,-1./4,-1./4}, {-1./4,2./3,-1./4},  {-1./4,-1./4,2./3}};
  Matrix MatI = new Matrix(I);                          // Form Matrix from 2D arrays
  System.out.print( "Input Matrix" );
  MatI.print (10, 5);                                          // Jama Matrix print
  EigenvalueDecomposition E = new EigenvalueDecomposition(MatI); // Eigenvalue finder
  double[] lambdaRe = E.getRealEigenvalues();                  // Real, Imag eigens
  double[] lambdaIm =  E.getImagEigenvalues();                       // Imag eigens
  System.out.println("Eigenvalues: \t lambda.Re[]="
                    + lambdaRe[0]+","+lambdaRe[1]+", "+lambdaRe[2]);
  Matrix V = E.getV();                                  // Get matrix of eigenvectors
  System.out.print("\n Matrix with column eigenvectors ");
  V.print (10, 5);
  Matrix Vec = new Matrix(3,1);                         // Extract single eigenvector
  Vec.set( 0, 0, V.get(0, 0) );
  Vec.set( 1, 0, V.get(1, 0) );
  Vec.set( 2, 0, V.get(2, 0) );
  System.out.print( "First Eigenvector, Vec" );
  Vec.print (10,5);
  Matrix LHS = MatI.times(Vec);                           // Should get Vec as answer
  Matrix RHS = Vec.times(lambdaRe[0]);
  System.out.print( "Does LHS = RHS?" );
  LHS.print (18, 12);
  RHS.print (18, 12);
}}
```

Listing 8.1 JamaEigen.java uses the JAMA matrix library to solve eigenvalue problems. Note that JAMA defines and manipulates the new data type (object) `Matrix`, which differs from an array but can be created from one.

```
EigenDecomposition E = new EigenDecomposition(MatI);    13
```

Then on line 14 we extract (get) a vector `lambdaRe` of length 3 containing the three (real) eigenvalues `lambdaRe[0]`, `lambdaRe[1]`, `lambdaRe[2]`:

```
double[] lambdaRe = E.getRealEigenvalues();    14
```

On line 18 we create a 3×3 matrix `V` containing the eigenvectors in the three columns of the matrix with the JAMA command:

```
Matrix V = E.getV();    18
```

which takes the eigenvector object `E` and gets the vectors from it. Then, on lines 22–24 we form a vector `Vec` (a 3×1 `Matrix`) containing a single eigenvector by extracting the elements from `V` with a `get` method and assigning them with a `set` method:

```
Vec.set(0,0,V.get(0,0));    22
```

```
/* JamaFit: JAMA matrix libe least-squares parabola fit of y(x) = b0 + b1 x + b2 x^2
            JAMA must be in same directory as program, or included in CLASSPATH     */

import Jama.*;
import java.io.*;

public class JamaFit {

  public static void main(String[] argv)throws IOException, FileNotFoundException {
    PrintWriter w =  new PrintWriter(new FileOutputStream("jamafit.dat"), true);
    double []   x = {1., 1.05, 1.15, 1.32,  1.51, 1.68, 1.92};                   // Data
    double []   y = {0.52, 0.73, 1.08, 1.44, 1.39, 1.46, 1.58};
    double [] sig = {0.1, 0.1, 0.2, 0.3, 0.2, 0.1, 0.1};
    double sig2, s, sx, sxx, sy, sxxx, sxxxx, sxy, sxxy, rhl, xx, yy;
    double [][] Sx = new double[3][3];                                 // Create 3x3 array
    double [][] Sy = new double[3][1];                                 // Create 3x1 array
    int Nd = 7, i;                                                   // Number of data points
    s = sx = sxx = sy = sxxx = sxxxx = sxy = sxy = sxxy = 0;

    for ( i=0; i <= Nd-1; i++ )  {                                // Generate matrix elements
      sig2   = sig[i]*sig[i];    s    += 1./sig2;          sx    += x[i]/sig2;
      sy     += y[i]/sig2;       rhl   = x[i]*x[i];        sxx   += rhl/sig2;
      sxxy   += rhl*y[i]/sig2;   sxy += x[i]*y[i]/sig2;    sxxx  += rhl*x[i]/sig2;
      sxxxx += rhl*rhl/sig2;
    }
    Sx[0][0] = s;                                                          // Assign arrays
    Sx[0][1] = Sx[1][0] = sx;
    Sx[0][2] = Sx[2][0] = Sx[1][1] = sxx;
    Sx[1][2] = Sx[2][1] = sxxx;
    Sx[2][2] = sxxxx;   Sy[0][0] = sy;    Sy[1][0] = sxy;   Sy[2][0] = sxxy;
    Matrix MatSx = new Matrix(Sx);                                   // Form Jama Matrices
    Matrix MatSy = new Matrix(3, 1);
    MatSy.set(0, 0, sy);
    MatSy.set(1, 0, sxy);
    MatSy.set(2, 0, sxxy);
    Matrix B = MatSx.inverse().times(MatSy);                            // Determine inverse
    Matrix Itest = MatSx.inverse().times(MatSx);                            // Test inverse
    System.out.print( "B Matrix via inverse" );                               // Jama print
    B.print (16, 14);
    System.out.print( "MatSx.inverse().times(MatSx) " );
    Itest.print (16, 14);
    B = MatSx.solve(MatSy);                                             // Direct solution too
    System.out.print( "B Matrix via direct" );
    B.print (16,14);
                                  // Extract via Jama get & Print parabola coefficients
    System.out.println("FitParabola2 Final Results");
    System.out.println("\n");
    System.out.println("y(x) = b0 + b1 x + b2 x^2");
    System.out.println("\n");
    System.out.println("b0 = "+B.get(0,0));
    System.out.println("b1 = "+B.get(1,0));
    System.out.println("b2 = "+B.get(2,0));
    System.out.println("\n");
    for ( i=0; i <= Nd-1; i++ ) {                                               // Test fit
      s = B.get(0,0) + B.get(1,0) * x[i] + B.get(2,0) * x[i] * x[i];
      System.out.println("i, x, y, yfit = "+i+", "+x[i]+", "+y[i]+", "+s);
    }
    for ( i=0; i < Nd; i++)  {
      yy=B.get(0,0)+B.get(1,0)*x[i]+B.get(2,0)*x[i]*x[i];
      w.println(" "+x[i] + "  " +yy + "  "+y[i]);
    }
    System.out.println("Output in jamafit.dat");
} }
```

Listing 8.2 JamaFit.java performs a least-squares fit of a parabola to data using the JAMA matrix library to solve the set of linear equations $ax = b$.

Our final JAMA example, JamaFit.java in Listing 8.2, demonstrates many of the features of JAMA. It arises in the context of least-squares fitting, as discussed in §8.7 where we give the equations being used to fit the parabola $y(x) = b_0 + b_1 x + b_2 x^2$ to a set of N_D measured data points $(y_i, y_i \pm \sigma_i)$. For illustration, the equation is solved both directly and by matrix inversion, and several techniques for assigning values to JAMA's Matrix are used.

8.3.4 Exercises for Testing Matrix Calls

Before you direct the computer to go off crunching numbers on a million elements of some matrix, it's a good idea for you to try out your procedures on a small matrix, especially one for which you know the right answer. In this way it will take you only a short time to realize how hard it is to get the calling procedure perfectly right! Here are some exercises.

1. Find the inverse of $\mathbf{A} = \begin{pmatrix} +4 & -2 & +1 \\ +3 & +6 & -4 \\ +2 & +1 & +8 \end{pmatrix}$.
   a. As a general procedure, applicable even if you do not know the analytic answer, check your inverse in both directions; that is, check that $\mathbf{A}\mathbf{A}^{-1} = \mathbf{A}^{-1}\mathbf{A} = \mathbf{I}$.
   b. Verify that $\mathbf{A}^{-1} = \frac{1}{263}\begin{pmatrix} +52 & +17 & +2 \\ -32 & +30 & +19 \\ -9 & -8 & +30 \end{pmatrix}$.
2. Consider the same matrix $\mathbf{A}$ as before, now used to describe three simultaneous linear equations, $\mathbf{Ax} = \mathbf{b}$, or explicitly,

$$\begin{pmatrix} a_{11} & a_{12} & a_{13} \\ a_{21} & a_{22} & a_{23} \\ a_{31} & a_{32} & a_{33} \end{pmatrix} \begin{pmatrix} x_1 \\ x_2 \\ x_3 \end{pmatrix} = \begin{pmatrix} b_1 \\ b_2 \\ b_3 \end{pmatrix}.$$

   Here the vector $\mathbf{b}$ on the RHS is assumed known, and the problem is to solve for the vector $\mathbf{x}$. Use an appropriate subroutine to solve these equations for the three different $\mathbf{x}$ vectors appropriate to these three different $\mathbf{b}$ values on the RHS:

$$b_1 = \begin{pmatrix} +12 \\ -25 \\ +32 \end{pmatrix}, \quad b_2 = \begin{pmatrix} +4 \\ -10 \\ +22 \end{pmatrix}, \quad b_3 = \begin{pmatrix} +20 \\ -30 \\ +40 \end{pmatrix}.$$

   The solutions should be

$$\mathbf{x}_1 = \begin{pmatrix} +1 \\ -2 \\ +4 \end{pmatrix}, \quad \mathbf{x}_2 = \begin{pmatrix} +0.312 \\ -0.038 \\ +2.677 \end{pmatrix}, \quad \mathbf{x}_3 = \begin{pmatrix} +2.319 \\ -2.965 \\ +4.790 \end{pmatrix}.$$

3. Consider the matrix $\mathbf{A} = \begin{pmatrix} \alpha & \beta \\ -\beta & \alpha \end{pmatrix}$, where you are free to use any values you want for α and β. Use a numerical eigenproblem solver to show that the eigenvalues and eigenvectors are the complex conjugates

$$\mathbf{x}_{1,2} = \begin{pmatrix} +1 \\ \mp i \end{pmatrix}, \qquad \lambda_{1,2} = \alpha \mp i\beta.$$

4. Use your eigenproblem solver to find the eigenvalues of the matrix

$$\mathbf{A} = \begin{pmatrix} -2 & +2 & -3 \\ +2 & +1 & -6 \\ -1 & -2 & +0 \end{pmatrix}.$$

   a. Verify that you obtain the eigenvalues $\lambda_1 = 5$, $\lambda_2 = \lambda_3 = -3$. Notice that double roots can cause problems. In particular, there is a uniqueness problem with their eigenvectors because any combination of these eigenvectors is also an eigenvector.
   b. Verify that the eigenvector for $\lambda_1 = 5$ is proportional to

$$\mathbf{x}_1 = \frac{1}{\sqrt{6}} \begin{pmatrix} -1 \\ -2 \\ +1 \end{pmatrix}.$$

   c. The eigenvalue -3 corresponds to a double root. This means that the corresponding eigenvectors are degenerate, which in turn means that they are not unique. Two linearly independent ones are

$$\mathbf{x}_2 = \frac{1}{\sqrt{5}} \begin{pmatrix} -2 \\ +1 \\ +0 \end{pmatrix}, \qquad \mathbf{x}_3 = \frac{1}{\sqrt{10}} \begin{pmatrix} 3 \\ 0 \\ 1 \end{pmatrix}.$$

   In this case it's not clear what your eigenproblem solver will give for the eigenvectors. Try to find a relationship between your computed eigenvectors with the eigenvalue -3 and these two linearly independent ones.

5. Your model of some physical system results in $N = 100$ coupled linear equations in N unknowns:

$$a_{11}y_1 + a_{12}y_2 + \cdots + a_{1N}y_N = b_1,$$
$$a_{21}y_1 + a_{22}y_2 + \cdots + a_{2N}y_N = b_2,$$
$$\cdots$$
$$a_{N1}y_1 + a_{N2}y_2 + \cdots + a_{NN}y_N = b_N.$$

In many cases, the a and b values are known, so your exercise is to solve for all the x values, taking **a** as the *Hilbert* matrix and **b** as its first row:

$$[a_{ij}] = \mathbf{a} = \left[\frac{1}{i+j-1}\right] = \begin{pmatrix} 1 & \frac{1}{2} & \frac{1}{3} & \frac{1}{4} & \cdots & \frac{1}{100} \\ \frac{1}{2} & \frac{1}{3} & \frac{1}{4} & \frac{1}{5} & \cdots & \frac{1}{101} \\ \ddots & & & & & \\ \frac{1}{100} & \frac{1}{101} & \cdots & & \cdots & \frac{1}{199} \end{pmatrix},$$

$$[b_i] = \mathbf{b} = \left[\frac{1}{i}\right] = \begin{pmatrix} 1 \\ \frac{1}{2} \\ \frac{1}{3} \\ \ddots \\ \frac{1}{100} \end{pmatrix}.$$

Compare to the analytic solution

$$\begin{pmatrix} y_1 \\ y_2 \\ \ddots \\ y_N \end{pmatrix} = \begin{pmatrix} 1 \\ 0 \\ \ddots \\ 0 \end{pmatrix}.$$

8.3.5 Matrix Solution of the String Problem

We have now set up the solution to our problem of two masses on a string and have the matrix tools needed to solve it. Your **problem** is to check out the physical reasonableness of the solution for a variety of weights and lengths. You should check that the deduced tensions are positive and that the deduced angles correspond to a physical geometry (e.g., with a sketch). Since this is a physics-based problem, we know that the sine and cosine functions must be less than 1 in magnitude and that the tensions should be similar in magnitude to the weights of the spheres.

8.3.6 Explorations

1. See at what point your initial guess gets so bad that the computer is unable to find a physical solution.
2. A possible problem with the formalism we have just laid out is that by incorporating the identity $\sin^2\theta_i + \cos^2\theta_i = 1$ into the equations we may be

discarding some information about the sign of $\sin\theta$ or $\cos\theta$. If you look at Figure 8.1, you can observe that for some values of the weights and lengths, θ_2 may turn out to be negative, yet $\cos\theta$ should remain positive. We can build this condition into our equations by replacing $f_7 - f_9$ with f's based on the form

$$f_7 = x_4 - \sqrt{1-x_1^2}, \quad f_8 = x_5 - \sqrt{1-x_2^2}, \quad f_9 = x_6 - \sqrt{1-x_3^2}. \tag{8.30}$$

See if this makes any difference in the solutions obtained.

2. ⊙ Solve the similar three-mass problem. The approach is the same, but the number of equations gets larger.

8.4 Unit II. Data Fitting

Data fitting is an art worthy of serious study by all scientists. In this unit we just scratch the surface by examining how to interpolate within a table of numbers and how to do a least-squares fit to data. We also show how to go about making a least-squares fit to nonlinear functions using some of the search techniques and subroutine libraries we have already discussed.

8.5 Fitting an Experimental Spectrum (Problem)

Problem: The cross sections measured for the resonant scattering of a neutron from a nucleus are given in Table 8.1 along with the measurement number (index), the energy, and the experimental error. Your **problem** is to determine values for the cross sections at energy values lying between those measured by experiment.

You can solve this **problem** in a number of ways. The simplest is to numerically *interpolate* between the values of the experimental $f(E_i)$ given in Table 8.1. This is direct and easy but does not account for there being experimental noise in the data. A more appropriate way to solve this problem (discussed in §8.7) is to find the *best fit* of a theoretical function to the data. We start with what we believe to be the "correct" theoretical description of the data,

$$f(E) = \frac{f_r}{(E-E_r)^2 + \Gamma^2/4}, \tag{8.31}$$

where f_r, E_r, and Γ are unknown parameters. We then adjust the parameters to obtain the best fit. This is a best fit in a statistical sense but in fact may not pass through all (or any) of the data points. For an easy, yet effective, introduction to statistical data analysis, we recommend [B&R 02].

These two techniques of interpolation and least-squares fitting are powerful tools that let you treat tables of numbers as if they were analytic functions and sometimes let you deduce statistically meaningful constants or conclusions from measurements. In general, you can view data fitting as *global* or *local*. In global fits,

TABLE 8.1
Experimental Values for a Scattering Cross Section $g(E)$ as a Function of Energy

i	1	2	3	4	5	6	7	8	9
E_i (MeV) $[\equiv x_i]$	0	25	50	75	100	125	150	175	200
$g(E_i)$ (mb)	10.6	16.0	45.0	83.5	52.8	19.9	10.8	8.25	4.7
Error = $\pm\sigma_i$ (mb)	9.34	17.9	41.5	85.5	51.5	21.5	10.8	6.29	4.14

a single function in x is used to represent the entire set of numbers in a table like Table 8.1. While it may be spiritually satisfying to find a single function that passes through all the data points, if that function is not the correct function for describing the data, the fit may show nonphysical behavior (such as large oscillations) between the data points. The rule of thumb is that if you must interpolate, keep it local and view global interpolations with a critical eye.

8.5.1 Lagrange Interpolation (Method)

Consider Table 8.1 as ordered data that we wish to interpolate. We call the independent variable x and its tabulated values $x_i (i = 1, 2, \ldots)$, and we assume that the dependent variable is the function $g(x)$, with tabulated values $g_i = g(x_i)$. We assume that $g(x)$ can be approximated as a $(n-1)$-degree polynomial in each interval i:

$$g_i(x) \simeq a_0 + a_1 x + a_2 x^2 + \cdots + a_{n-1} x^{n-1}, \quad (x \simeq x_i). \tag{8.32}$$

Because our fit is local, we do not assume that one $g(x)$ can fit all the data in the table but instead use a different polynomial, that is, a different set of a_i values, for each region of the table. While each polynomial is of low degree, multiple polynomials are used to span the entire table. If some care is taken, the set of polynomials so obtained will behave well enough to be used in further calculations without introducing much unwanted noise or discontinuities.

The classic interpolation formula was created by Lagrange. He figured out a closed-form one that directly fits the $(n-1)$-order polynomial (8.32) to n values of the function $g(x)$ evaluated at the points x_i. The formula is written as the sum of polynomials:

$$g(x) \simeq g_1 \lambda_1(x) + g_2 \lambda_2(x) + \cdots + g_n \lambda_n(x), \tag{8.33}$$

$$\lambda_i(x) = \prod_{j(\neq i)=1}^{n} \frac{x - x_j}{x_i - x_j} = \frac{x - x_1}{x_i - x_1} \frac{x - x_2}{x_i - x_2} \cdots \frac{x - x_n}{x_i - x_n}. \tag{8.34}$$

For three points, (8.33) provides a second-degree polynomial, while for eight points it gives a seventh-degree polynomial. For example, here we use a four-point

Lagrange interpolation to determine a third-order polynomial that reproduces the values $x_{1-4} = (0, 1, 2, 4)$, $f_{1-4} = (-12, -12, -24, -60)$:

$$g(x) = \frac{(x-1)(x-2)(x-4)}{(0-1)(0-2)(0-4)}(-12) + \frac{x(x-2)(x-4)}{(1-0)(1-2)(1-4)}(-12)$$

$$+ \frac{x(x-1)(x-4)}{(2-0)(2-1)(2-4)}(-24) + \frac{x(x-1)(x-2)}{(4-0)(4-1)(4-2)}(-60),$$

$$\Rightarrow \quad g(x) = x^3 - 9x^2 + 8x - 12. \tag{8.35}$$

As a check we see that

$$g(4) = 4^3 - 9(4^2) + 32 - 12 = -60, \qquad g(0.5) = -10.125. \tag{8.36}$$

If the data contain little noise, this polynomial can be used with some confidence within the range of the data, but with risk beyond the range of the data.

Notice that Lagrange interpolation makes no restriction that the points in the table be evenly spaced. As a check, it is also worth noting that the sum of the Lagrange multipliers equals one, $\sum_{i=1}^{n} \lambda_i = 1$. Usually the Lagrange fit is made to only a small region of the table with a small value of n, even though the formula works perfectly well for fitting a high-degree polynomial to the entire table. The difference between the value of the polynomial evaluated at some x and that of the actual function is equal to the *remainder*

$$R_n \simeq \frac{(x-x_1)(x-x_2)\cdots(x-x_n)}{n!} g^{(n)}(\zeta), \tag{8.37}$$

where ζ lies somewhere in the interpolation interval but is otherwise undetermined. This shows that if significant high derivatives exist in $g(x)$, then it cannot be approximated well by a polynomial. In particular, if $g(x)$ is a table of experimental data, it is likely to contain noise, and then it is a bad idea to fit a curve through all the data points.

8.5.2 Lagrange Implementation and Assessment

Consider the experimental neutron scattering data in Table 8.1. The expected theoretical functional form that describes these data is (8.31), and our empirical fits to these data are shown in Figure 8.3.

1. Write a subroutine to perform an n-point Lagrange interpolation using (8.33). Treat n as an arbitrary input parameter. (You can also do this exercise with the spline fits discussed in § 8.5.4.)
2. Use the Lagrange interpolation formula to fit the entire experimental spectrum with one polynomial. (This means that you must fit all nine data points

with an eight-degree polynomial.) Then use this fit to plot the cross section in steps of 5 MeV.

3. Use your graph to deduce the resonance energy E_r (your peak position) and Γ (the full width at half-maximum). Compare your results with those predicted by our theorist friend, $(E_r, \Gamma) = (78, 55)$ MeV.
4. A more realistic use of Lagrange interpolation is for local interpolation with a small number of points, such as three. Interpolate the preceding cross-sectional data in 5-MeV steps using three-point Lagrange interpolation. (Note that the end intervals may be special cases.) ▮

This example shows how easy it is to go wrong with a high-degree-polynomial fit. Although the polynomial is guaranteed to pass through all the data points, the representation of the function away from these points can be quite unrealistic. Using a low-order interpolation formula, say, $n = 2$ or 3, in each interval usually eliminates the wild oscillations. If these local fits are then matched together, as we discuss in the next section, a rather continuous curve results. Nonetheless, you must recall that if the data contain errors, a curve that actually passes through them may lead you astray. We discuss how to do this properly in §8.7.

8.5.3 Explore Extrapolation

We deliberately have not discussed *extrapolation* of data because it can lead to serious *systematic* errors; the answer you get may well depend more on the function you assume than on the data you input. Add some adventure to your life and use the programs you have written to extrapolate to values outside Table 8.1. Compare your results to the theoretical Breit–Wigner shape (8.31). ▮

8.5.4 Cubic Splines (Method)

If you tried to interpolate the resonant cross section with Lagrange interpolation, then you saw that fitting parabolas (three-point interpolation) within a table may avoid the erroneous and possibly catastrophic deviations of a high-order formula. (A two-point interpolation, which connects the points with straight lines, may not lead you far astray, but it is rarely pleasing to the eye or precise.) A sophisticated variation of an $n = 4$ interpolation, known as *cubic splines*, often leads to surprisingly eye-pleasing fits. In this approach (Figure 8.3), cubic polynomials are fit to the function in each interval, with the additional constraint that the first and second derivatives of the polynomials be continuous from one interval to the next. This continuity of slope and curvature is what makes the spline fit particularly eye-pleasing. It is analogous to what happens when you use the flexible spline drafting tool (a lead wire within a rubber sheath) from which the method draws its name.

The series of cubic polynomials obtained by spline-fitting a table of data can be integrated and differentiated and is guaranteed to have well-behaved derivatives.

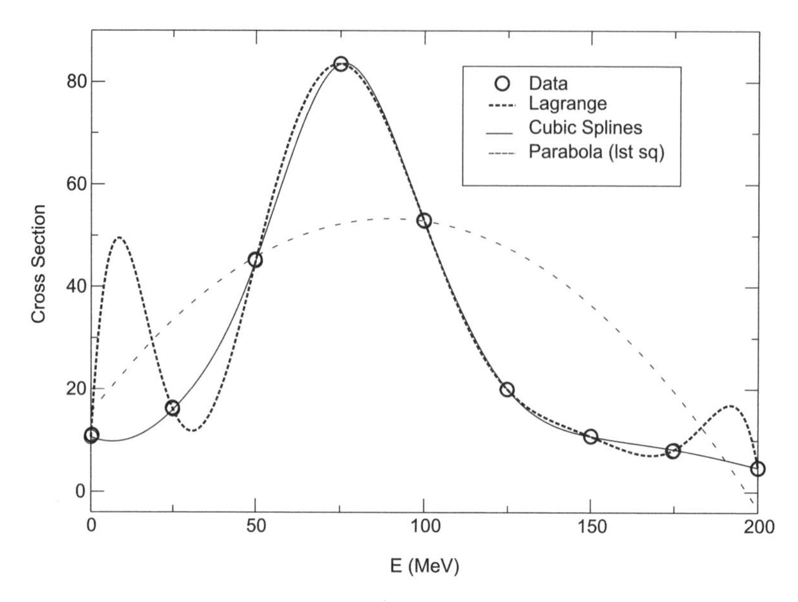

Figure 8.3 Three fits to cross-section data. Short dashed line: Lagrange interpolation using an eight-degree polynomial that passes through all the data points but has nonphysical oscillations between points; solid line: cubic splines (smooth but not accurate); dashed line: Least-squares parabola fit (a best fit with a bad theory). The best approach is to do a least-squares fit of the correct theoretical function, the Breit–Wigner method (8.31).

The existence of meaningful derivatives is an important consideration. As a case in point, if the interpolated function is a potential, you can take the derivative to obtain the force. The complexity of simultaneously matching polynomials and their derivatives over all the interpolation points leads to many simultaneous linear equations to be solved. This makes splines unattractive for hand calculation, yet easy for computers and, not surprisingly, popular in both calculations and graphics. To illustrate, the smooth curves connecting points in most "draw" programs are usually splines, as is the solid curve in Figure 8.3.

The basic approximation of splines is the representation of the function $g(x)$ in the subinterval $[x_i, x_{i+1}]$ with a cubic polynomial:

$$g(x) \simeq g_i(x), \quad \text{for } x_i \le x \le x_{i+1}, \tag{8.38}$$

$$g_i(x) = g_i + g_i'(x - x_i) + \frac{1}{2}g_i''(x - x_i)^2 + \frac{1}{6}g_i'''(x - x_i)^3. \tag{8.39}$$

This representation makes it clear that the coefficients in the polynomial equal the values of $g(x)$ and its first, second, and third derivatives at the tabulated points x_i. Derivatives beyond the third vanish for a cubic. The computational chore is to determine these derivatives in terms of the N tabulated g_i values. The matching of g_i at the *nodes* that connect one interval to the next provides the equations

$$g_i(x_{i+1}) = g_{i+1}(x_{i+1}), \quad i = 1, N-1. \tag{8.40}$$

The matching of the first *and* second derivatives at each interval's boundaries provides the equations

$$g'_{i-1}(x_i) = g'_i(x_i), \quad g''_{i-1}(x_i) = g''_i(x_i). \tag{8.41}$$

The additional equations needed to determine all constants is obtained by matching the third derivatives at adjacent nodes. Values for the third derivatives are found by approximating them in terms of the second derivatives:

$$g'''_i \simeq \frac{g''_{i+1} - g''_i}{x_{i+1} - x_i}. \tag{8.42}$$

As discussed in Chapter 7, "Differentiation & Searching," a *central-difference approximation* would be better than a forward-difference approximation, yet (8.42) keeps the equations simpler.

It is straightforward though complicated to solve for all the parameters in (8.39). We leave that to other reference sources [Thom 92, Pres 94]. We can see, however, that matching at the boundaries of the intervals results in only $(N-2)$ linear equations for N unknowns. Further input is required. It usually is taken to be the boundary conditions at the endpoints $a = x_1$ and $b = x_N$, specifically, the second derivatives $g''(a)$ and $g''(b)$. There are several ways to determine these second derivatives:

Natural spline: Set $g''(a) = g''(b) = 0$; that is, permit the function to have a slope at the endpoints but no curvature. This is "natural" because the derivative vanishes for the flexible spline drafting tool (its ends being free).

Input values for g' at the boundaries: The computer uses $g'(a)$ to approximate $g''(a)$. If you do not know the first derivatives, you can calculate them numerically from the table of g_i values.

Input values for g'' at the boundaries: Knowing values is of course better than approximating values, but it requires the user to input information. If the values of g'' are not known, they can be approximated by applying a forward-difference approximation to the tabulated values:

$$g''(x) \simeq \frac{[g(x_3) - g(x_2)]/[x_3 - x_2] - [g(x_2) - g(x_1)]/[x_2 - x_1]}{[x_3 - x_1]/2}. \tag{8.43}$$

8.5.4.1 CUBIC SPLINE QUADRATURE (EXPLORATION)

A powerful integration scheme is to fit an integrand with splines and then integrate the cubic polynomials analytically. If the integrand $g(x)$ is known only at its tabulated values, then this is about as good an integration scheme as is possible;

if you have the ability to calculate the function directly for arbitrary x, Gaussian quadrature may be preferable. We know that the spline fit to g in each interval is the cubic (8.39)

$$g(x) \simeq g_i + g_i'(x - x_i) + \frac{1}{2}g_i''(x - x_i)^2 + \frac{1}{6}g_i'''(x - x_i)^3. \quad (8.44)$$

It is easy to integrate this to obtain the integral of g for this interval and then to sum over all intervals:

$$\int_{x_i}^{x_{i+1}} g(x)\, dx \simeq \left(g_i x + \frac{1}{2}g_i' x_i^2 + \frac{1}{6}g_i'' x^3 + \frac{1}{24}g_i''' x^4 \right)\Bigg|_{x_i}^{x_{i+1}}, \quad (8.45)$$

$$\int_{x_j}^{x_k} g(x)\, dx = \sum_{i=j}^{k} \left(g_i x + \frac{1}{2}g_i' x_i^2 + \frac{1}{6}g_i'' x^3 + \frac{1}{24}g_i''' x^4 \right)\Bigg|_{x_i}^{x_{i+1}}. \quad (8.46)$$

Making the intervals smaller does not necessarily increase precision, as subtractive cancellations in (8.45) may get large.

8.5.5 Spline Fit of Cross Section (Implementation)

CD

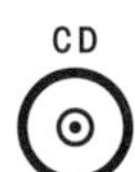

Fitting a series of cubics to data is a little complicated to program yourself, so we recommend using a library routine. While we have found quite a few Java-based spline applications available on the internet, none seemed appropriate for interpreting a simple set of numbers. That being the case, we have adapted the splint.c and the spline.c functions from [Pres 94] to produce the SplineAppl.java program shown in Listing 8.3 (there is also an applet version on the CD). Your **problem** now is to carry out the assessment in § 8.5.2 using cubic spline interpolation rather than Lagrange interpolation.

8.6 Fitting Exponential Decay (Problem)

Figure 8.4 presents actual experimental data on the number of decays ΔN of the π meson as a function of time [Stez 73]. Notice that the time has been "binned" into $\Delta t = 10$-ns intervals and that the smooth curve is the theoretical exponential decay expected for very large numbers. Your **problem** is to deduce the lifetime τ of the π meson from these data (the tabulated lifetime of the pion is 2.6×10^{-8} s).

8.6.1 Theory to Fit

Assume that we start with N_0 particles at time $t = 0$ that can decay to other particles.[4] If we wait a short time Δt, then a small number ΔN of the particles will decay *spontaneously*, that is, with no external influences. This decay is a stochastic

[4] Spontaneous decay is discussed further and simulated in § 5.5.

process, which means that there is an element of chance involved in just when a decay will occur, and so no two experiments are expected to give exactly the same results. The basic law of nature for spontaneous decay is that the number of decays ΔN in a time interval Δt is proportional to the number of particles $N(t)$ present at that time and to the time interval

$$\Delta N(t) = -\frac{1}{\tau} N(t) \Delta t \quad \Rightarrow \quad \frac{\Delta N(t)}{\Delta t} = -\lambda N(t). \tag{8.47}$$

```
/* SplineAppl.java: Application version of cubic spline fitting. Interpolates
   array x[n], y[n], x0 < x1 ...  < x(n-1). yp1, ypn: y' at ends evaluated internally
   y2[]: y" array; yp1, ypn > e30 for natural spline    */
import java.io.*;

public class SplineAppl  {

  public static void main(String[] argv) throws IOException, FileNotFoundException {
    PrintWriter w = new PrintWriter(new FileOutputStream("Spline.dat"), true);
    PrintWriter q = new PrintWriter(new FileOutputStream("Input.dat"), true);
    double x[] = {0.,1.2,2.5,3.7,5.,6.2,7.5,8.7,9.9};                      // input
    double y[] = {0.,0.93,.6,-0.53,-0.96,-0.08,0.94,0.66,-0.46};
    int  i,n = x.length,  np = 15, klo, khi, k;
    double y2[] = new double[9], u[] = new double[n];
    double h, b, a, Nfit, p, qn, sig, un, yp1, ypn, xout, yout;
    for ( i=0; i < n; i++ ) q.println (" " + x[i] + " " + y[i] + " ");
    Nfit = 30;                                                            // N output pts
    yp1 = (y[1]-y[0])/(x[1]-x[0]) - (y[2]-y[1])/(x[2]-x[1]) +(y[2]-y[0])/(x[2]-x[0]);
    ypn = (y[n-1]-y[n-2])/(x[n-1]-x[n-2]) - (y[n-2]-y[n-3])
                  /(x[n-2]-x[n-3]) + (y[n-1]-y[n-3])/(x[n-1]-x[n-3]);
    if (yp1 > 0.99e30) y2[0] = u[0] = 0. ; // Natural
      else {y2[0] = (-0.5); u[0] = (3/(x[1]-x[0]))*((y[1]-y[0])/(x[1]-x[0])-yp1);}
    for ( i=1;  i <= n-2;  i++ )  {                              // Decomposition loop
      sig = (x[i]-x[i-1])/(x[i + 1]-x[i-1]);
      p = sig*y2[i-1] + 2. ;
      y2[i] = (sig-1.)/p;
      u[i] = (y[i+1]-y[i])/(x[i + 1]-x[i])-(y[i]-y[i-1])/(x[i]-x[i-1]);
      u[i] = (6.*u[i]/(x[i+1]-x[i-1])-sig*u[i-1])/p;
    }
    if (ypn > 0.99e30) qn = un = 0. ;                            // Test for natural
     else {qn = 0.5; un = (3/(x[n-1]-x[n-2]))*(ypn-(y[n-1]-y[n-2])/(x[n-1]-x[n-2]));}
    y2[n-1] = (un-qn*u[n-2])/(qn*y2[n-2] + 1.);
    for ( k = n-2;  k>= 0; k--)  y2[k] = y2[k]*y2[k + 1] + u[k];
    for ( i=1;  i <= Nfit;  i++ ) {                      // initialization ends, begin fit
      xout = x[0] + (x[n-1]-x[0])*(i-1)/(Nfit);
      klo = 0;                                                      // Bisection algor
      khi = n-1;                                          // klo, khi bracket xout value
      while (khi-klo >1) {k = (khi+klo) >> 1; if (x[k] > xout) khi =k; else klo = k;}
      h = x[khi]-x[klo];
      if (x[k] > xout) khi = k;    else klo = k;
      h = x[khi]-x[klo];  a = (x[khi]-xout)/h;
      b = (xout-x[klo])/h;
      yout = (a*y[klo] + b*y[khi] +((a*a*a-a)*y2[klo] + (b*b*b-b)*y2[khi])*(h*h)/6.);
      w.println (" " + xout + " " + yout + " ");
    }
    System.out.println("data stored in Spline.dat");
} }
```

Listing 8.3 SplineAppl.java is an application version of an applet given on the CD that performs a cubic spline fit to data. The arrays `x[]` and `y[]` are the data to fit, and the values of the fit at `Nfit` points are output into the file `Spline.dat`.

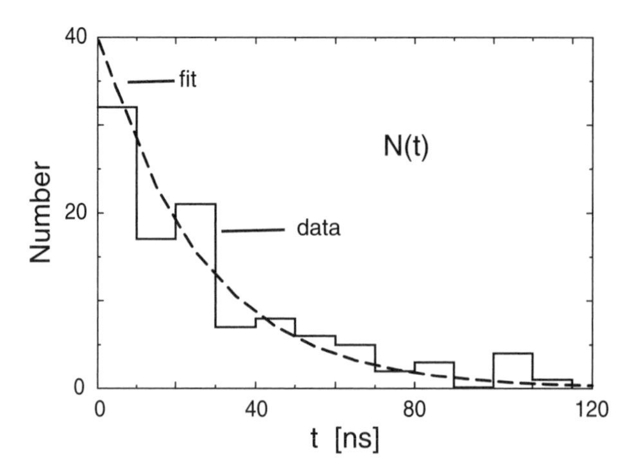

Figure 8.4 A reproduction of the experimental measurement in (Stez 73) of the number of decays of a π meson as a function of time. Measurements are made during time intervals of 10-ns length. Each "event" corresponds to a single decay.

Here $\tau = 1/\lambda$ is the *lifetime* of the particle, with λ the rate parameter. The actual decay *rate* is given by the second equation in (8.47). If the number of decays ΔN is very small compared to the number of particles N, and if we look at vanishingly small time intervals, then the difference equation (8.47) becomes the differential equation

$$\frac{dN(t)}{dt} \simeq -\lambda N(t) = \frac{1}{\tau} N(t). \tag{8.48}$$

This differential equation has an exponential solution for the number as well as for the decay rate:

$$N(t) = N_0 e^{-t/\tau}, \qquad \frac{dN(t)}{dt} = -\frac{N_0}{\tau} e^{-t/\tau} = \frac{dN(0)}{dt} e^{-t/\tau}. \tag{8.49}$$

Equation (8.49) is the theoretical formula we wish to "fit" to the data in Figure 8.4. The output of such a fit is a "best value" for the lifetime τ.

8.7 Least-Squares Fitting (Method)

Books have been written and careers have been spent discussing what is meant by a "good fit" to experimental data. We cannot do justice to the subject here and refer the reader to [B&R 02, Pres 94, M&W 65, Thom 92]. However, we will emphasize three points:

1. If the data being fit contain errors, then the "best fit" in a statistical sense should not pass through all the data points.
2. If the theory is not an appropriate one for the data (e.g., the parabola in Figure 8.3), then its best fit to the data may not be a good fit at all. This is good, for it indicates that this is not the right theory.
3. Only for the simplest case of a linear least-squares fit can we write down a closed-form solution to evaluate and obtain the fit. More realistic problems are usually solved by *trial-and-error* search procedures, sometimes using sophisticated subroutine libraries. However, in §8.7.6 we show how to conduct such a nonlinear search using familiar tools.

Imagine that you have measured N_D data values of the independent variable y as a function of the dependent variable x:

$$(x_i, y_i \pm \sigma_i), \quad i = 1, N_D, \tag{8.50}$$

where $\pm\sigma_i$ is the uncertainty in the ith value of y. (For simplicity we assume that all the errors σ_i occur in the dependent variable, although this is hardly ever true [Thom 92]). For our problem, y is the number of decays as a function of time, and x_i are the times. Our goal is to determine how well a mathematical function $y = g(x)$ (also called a *theory* or a *model*) can describe these data. Alternatively, if the theory contains some parameters or constants, our goal can be viewed as determining the best values for these parameters. We assume that the model function $g(x)$ contains, in addition to the functional dependence on x, an additional dependence upon M_P parameters $\{a_1, a_2, \ldots, a_{M_P}\}$. Notice that the parameters $\{a_m\}$ are not variables, in the sense of numbers read from a meter, but rather are parts of the theoretical model, such as the size of a box, the mass of a particle, or the depth of a potential well. For the exponential decay function (8.49), the parameters are the lifetime τ and the initial decay rate $dN(0)/dt$. We indicate this as

$$g(x) = g(x; \{a_1, a_2, \ldots, a_{M_P}\}) = g(x; \{a_m\}). \tag{8.51}$$

We use the chi-square (χ^2) measure as a gauge of how well a theoretical function g reproduces data:

$$\boxed{\chi^2 \stackrel{\text{def}}{=} \sum_{i=1}^{N_D} \left(\frac{y_i - g(x_i; \{a_m\})}{\sigma_i} \right)^2,} \tag{8.52}$$

where the sum is over the N_D experimental points $(x_i, y_i \pm \sigma_i)$. The definition (8.52) is such that smaller values of χ^2 are better fits, with $\chi^2 = 0$ occurring if the theoretical curve went through the center of every data point. Notice also that the $1/\sigma_i^2$ weighting means that measurements with larger errors[5] contribute less to χ^2.

Least-squares fitting refers to adjusting the parameters in the theory until a minimum in χ^2 is found, that is, finding a curve that produces the least value for the summed squares of the deviations of the data from the function $g(x)$. In general, this is the best fit possible or the best way to determine the parameters in a theory. The M_P parameters $\{a_m, m=1, M_P\}$ that make χ^2 an extremum are found by solving the M_P equations:

$$\frac{\partial \chi^2}{\partial a_m} = 0, \quad \Rightarrow \sum_{i=1}^{N_D} \frac{[y_i - g(x_i)]}{\sigma_i^2} \frac{\partial g(x_i)}{\partial a_m} = 0, \quad (m=1, M_P). \tag{8.53}$$

More usually, the function $g(x; \{a_m\})$ has a sufficiently complicated dependence on the a_m values for (8.53) to produce M_P simultaneous nonlinear equations in the a_m values. In these cases, solutions are found by a trial-and-error search through the M_P-dimensional parameter space, as we do in §8.7.6. To be safe, when such a search is completed, you need to check that the minimum χ^2 you found is *global* and not *local*. One way to do that is to repeat the search for a whole grid of starting values, and if different minima are found, to pick the one with the lowest χ^2.

8.7.1 Least-Squares Fitting: Theory and Implementation

When the deviations from theory are due to random errors and when these errors are described by a Gaussian distribution, there are some useful rules of thumb to remember [B&R 02]. You know that your fit is good if the value of χ^2 calculated via the definition (8.52) is approximately equal to the number of degrees of freedom $\chi^2 \simeq N_D - M_P$, where N_D is the number of data points and M_P is the number of parameters in the theoretical function. If your χ^2 is much less than $N_D - M_P$, it doesn't mean that you have a "great" theory or a really precise measurement; instead, you probably have too many parameters or have assigned errors (σ_i values) that are too large. In fact, too small a χ^2 may indicate that you are fitting the random scatter in the data rather than missing approximately one-third of the error bars, as expected for a normal distribution. If your χ^2 is significantly greater than $N_D - M_P$, the theory may not be good, you may have significantly underestimated your errors, or you may have errors that are not random.

The M_P simultaneous equations (8.53) can be simplified considerably if the functions $g(x; \{a_m\})$ depend *linearly* on the parameter values a_i, e.g.,

$$g(x; \{a_1, a_2\}) = a_1 + a_2 x. \tag{8.54}$$

[5] If you are not given the errors, you can guess them on the basis of the apparent deviation of the data from a smooth curve, or you can weigh all points equally by setting $\sigma_i \equiv 1$ and continue with the fitting.

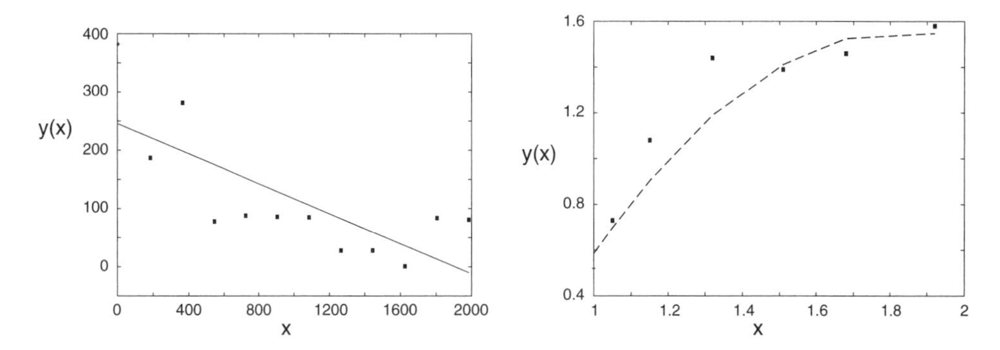

Figure 8.5 *Left:* A linear least-squares best fit of data to a straight line. Here the deviation of theory from experiment is greater than would be expected from statistics, or in other words, a straight line is not a good theory for these data. *Right:* A linear least-squares best fit of different data to a parabola. Here we see that the fit misses approximately one-third of the points, as expected from the statistics for a good fit.

In this case (also known as *linear regression* and shown on the left in Figure 8.5) there are $M_P = 2$ parameters, the slope a_2, and the y intercept a_1. Notice that while there are only two parameters to determine, there still may be an arbitrary number N_D of data points to fit. Remember, a unique solution is not possible unless the number of data points is equal to or greater than the number of parameters. For this linear case, there are just two derivatives,

$$\frac{\partial g(x_i)}{\partial a_1} = 1, \quad \frac{\partial g(x_i)}{\partial a_2} = x_i, \tag{8.55}$$

and after substitution, the χ^2 minimization equations (8.53) can be solved [Pres 94]:

$$a_1 = \frac{S_{xx}S_y - S_x S_{xy}}{\Delta}, \qquad a_2 = \frac{S S_{xy} - S_x S_y}{\Delta}, \tag{8.56}$$

$$S = \sum_{i=1}^{N_D} \frac{1}{\sigma_i^2}, \qquad S_x = \sum_{i=1}^{N_D} \frac{x_i}{\sigma_i^2}, \quad S_y = \sum_{i=1}^{N_D} \frac{y_i}{\sigma_i^2}, \tag{8.57}$$

$$S_{xx} = \sum_{i=1}^{N_D} \frac{x_i^2}{\sigma_i^2}, \qquad S_{xy} = \sum_{i=1}^{N_D} \frac{x_i y_i}{\sigma_i^2}, \quad \Delta = S S_{xx} - S_x^2. \tag{8.58}$$

Statistics also gives you an expression for the *variance* or uncertainty in the deduced parameters:

$$\sigma_{a_1}^2 = \frac{S_{xx}}{\Delta}, \quad \sigma_{a_2}^2 = \frac{S}{\Delta}. \tag{8.59}$$

This is a measure of the uncertainties in the values of the fitted parameters arising from the uncertainties σ_i in the measured y_i values. A measure of the dependence of the parameters on each other is given by the *correlation coefficient*:

$$\rho(a_1, a_2) = \frac{\mathrm{cov}(a_1, a_2)}{\sigma_{a_1}\sigma_{a_2}}, \quad \mathrm{cov}(a_1, a_2) = \frac{-S_x}{\Delta}. \tag{8.60}$$

Here $\mathrm{cov}(a_1, a_2)$ is the *covariance* of a_1 and a_2 and vanishes if a_1 and a_2 are independent. The correlation coefficient $\rho(a_1, a_2)$ lies in the range $-1 \le \rho \le 1$, with a positive ρ indicating that the errors in a_1 and a_2 are likely to have the same sign, and a negative ρ indicating opposite signs.

The preceding analytic solutions for the parameters are of the form found in statistics books but are not optimal for numerical calculations because subtractive cancellation can make the answers unstable. As discussed in Chapter 2, "Errors & Uncertainties in Computations," a rearrangement of the equations can decrease this type of error. For example, [Thom 92] gives improved expressions that measure the data relative to their averages:

$$a_1 = \overline{y} - a_2\overline{x}, \quad a_2 = \frac{S_{xy}}{S_{xx}}, \quad \overline{x} = \frac{1}{N}\sum_{i=1}^{N_d} x_i, \quad \overline{y} = \frac{1}{N}\sum_{i=1}^{N_d} y_i$$

$$S_{xy} = \sum_{i=1}^{N_d} \frac{(x_i - \overline{x})(y_i - \overline{y})}{\sigma_i^2}, \quad S_{xx} = \sum_{i=1}^{N_d} \frac{(x_i - \overline{x})^2}{\sigma_i^2}. \tag{8.61}$$

In `JamaFit.java` in Listing 8.2 and on the CD, we give a program that fits a parabola to some data. You can use it as a model for fitting a line to data, although you can use our closed-form expressions for a straight-line fit. In `Fit.java` on the instructor's CD we give a program for fitting to the decay data.

8.7.2 Exponential Decay Fit Assessment

Fit the exponential decay law (8.49) to the data in Figure 8.4. This means finding values for τ and $\Delta N(0)/\Delta t$ that provide a best fit to the data and then judging how good the fit is.

1. Construct a table $(\Delta N/\Delta t_i, t_i)$, for $i = 1, N_D$ from Figure 8.4. Because time was measured in bins, t_i should correspond to the middle of a bin.
2. Add an estimate of the error σ_i to obtain a table of the form $(\Delta N/\Delta t_i \pm \sigma_i, t_i)$. You can estimate the errors by eye, say, by estimating how much the histogram values appear to fluctuate about a smooth curve, or you can take $\sigma_i \simeq \sqrt{\text{events}}$. (This last approximation is reasonable for large numbers, which this is not.)

3. In the limit of very large numbers, we would expect a plot of $\ln |dN/dt|$ *versus* t to be a straight line:

$$\ln \left| \frac{\Delta N(t)}{\Delta t} \right| \simeq \ln \left| \frac{\Delta N_0}{\Delta t} \right| - \frac{1}{\tau} \Delta t.$$

This means that if we treat $\ln |\Delta N(t)/\Delta t|$ as the dependent variable and time Δt as the independent variable, we can use our linear fit results. Plot $\ln |\Delta N/\Delta t|$ *versus* Δt.
4. Make a least-squares fit of a straight line to your data and use it to determine the lifetime τ of the π meson. Compare your deduction to the tabulated lifetime of 2.6×10^{-8} s and comment on the difference.
5. Plot your best fit on the same graph as the data and comment on the agreement.
6. Deduce the goodness of fit of your straight line and the approximate error in your deduced lifetime. Do these agree with what your "eye" tells you? ▌

8.7.3 Exercise: Fitting Heat Flow

The table below gives the temperature T along a metal rod whose ends are kept at a fixed constant temperature. The temperature is a function of the distance x along the rod.

x_i (cm)	1.0	2.0	3.0	4.0	5.0	6.0	7.0	8.0	9.0
T_i (C)	14.6	18.5	36.6	30.8	59.2	60.1	62.2	79.4	99.9

1. Plot the data to verify the appropriateness of a linear relation

$$T(x) \simeq a + bx. \tag{8.62}$$

2. Because you are not given the errors for each measurement, assume that the least significant figure has been rounded off and so $\sigma \geq 0.05$. Use that to compute a least-squares straight-line fit to these data.
3. Plot your best $a + bx$ on the curve with the data.
4. After fitting the data, compute the variance and compare it to the deviation of your fit from the data. Verify that about one-third of the points miss the σ error band (that's what is expected for a normal distribution of errors).
5. Use your computed variance to determine the χ^2 of the fit. Comment on the value obtained.
6. Determine the variances σ_a and σ_b and check whether it makes sense to use them as the errors in the deduced values for a and b. ▌

8.7.4 Linear Quadratic Fit (Extension)

As indicated earlier, as long as the function being fitted depends *linearly* on the unknown parameters a_i, the condition of minimum χ^2 leads to a set of simultaneous linear equations for the a's that can be solved on the computer using matrix techniques. To illustrate, suppose we want to fit the quadratic polynomial

$$g(x) = a_1 + a_2 x + a_3 x^2 \tag{8.63}$$

to the experimental measurements $(x_i, y_i, i = 1, N_D)$ (Figure 8.5 right). Because this $g(x)$ is linear in all the parameters a_i, we can still make a linear fit even though x is raised to the second power. [However, if we tried to a fit a function of the form $g(x) = (a_1 + a_2 x)\, \exp(-a_3\, x)$ to the data, then we would not be able to make a linear fit because one of the a's appears in the exponent.]

The best fit of this quadratic to the data is obtained by applying the minimum χ^2 condition (8.53) for $M_p = 3$ parameters and N_D (still arbitrary) data points. A solution represents the maximum likelihood that the deduced parameters provide a correct description of the data for the theoretical function $g(x)$. Equation (8.53) leads to the three simultaneous equations for a_1, a_2, and a_3:

$$\sum_{i=1}^{N_D} \frac{[y_i - g(x_i)]}{\sigma_i^2} \frac{\partial g(x_i)}{\partial a_1} = 0, \qquad \frac{\partial g}{\partial a_1} = 1, \tag{8.64}$$

$$\sum_{i=1}^{N_D} \frac{[y_i - g(x_i)]}{\sigma_i^2} \frac{\partial g(x_i)}{\partial a_2} = 0, \qquad \frac{\partial g}{\partial a_2} = x, \tag{8.65}$$

$$\sum_{i=1}^{N_D} \frac{[y_i - g(x_i)]}{\sigma_i^2} \frac{\partial g(x_i)}{\partial a_3} = 0, \qquad \frac{\partial g}{\partial a_3} = x^2. \tag{8.66}$$

Note: Because the derivatives are independent of the parameters (the a's), the a dependence arises only from the term in square brackets in the sums, and because that term has only a linear dependence on the a's, these equations are linear equations in the a's.

Exercise: Show that after some rearrangement, (8.64)–(8.66) can be written as

$$S a_1 + S_x a_2 + S_{xx} a_3 = S_y, \tag{8.67}$$

$$S_x a_1 + S_{xx} a_2 + S_{xxx} a_3 = S_{xy},$$

$$S_{xx} a_1 + S_{xxx} a_2 + S_{xxxx} a_3 = S_{xxy}.$$

Here the definitions of the S's are simple extensions of those used in (8.56)–(8.58) and are programmed in JamaFit.java shown in Listing 8.2. After placing the three parameters into a vector **a** and the three RHS terms in (8.67) into a vector **S**, these equations assume the matrix form:

$$[\alpha]\mathbf{a} = \mathbf{S}, \tag{8.68}$$

$$[\alpha] = \begin{bmatrix} S & S_x & S_{xx} \\ S_x & S_{xx} & S_{xxx} \\ S_{xx} & S_{xxx} & S_{xxxx} \end{bmatrix}, \quad \mathbf{a} = \begin{bmatrix} a_1 \\ a_2 \\ a_3 \end{bmatrix}, \quad \mathbf{S} = \begin{bmatrix} S_y \\ S_{xy} \\ S_{xxy} \end{bmatrix}.$$

This is the exactly the matrix problem we solved in §8.3.3 with the code JamaFit.java given in Listing 8.2. The solution for the parameter vector **a** is obtained by solving the matrix equations. Although for 3×3 matrices we can write out the solution in closed form, for larger problems the numerical solution requires matrix methods.

8.7.5 Linear Quadratic Fit Assessment

1. Fit the quadratic (8.63) to the following data sets [given as (x_1, y_1), $(x_2, y_2), \ldots$]. In each case indicate the values found for the a's, the number of degrees of freedom, *and* the value of χ^2.
   a. $(0, 1)$
   b. $(0, 1), (1, 3)$
   c. $(0, 1), (1, 3), (2, 7)$
   d. $(0, 1), (1, 3), (2, 7), (3, 15)$
2. Find a fit to the last set of data to the function

$$y = Ae^{-bx^2}. \tag{8.69}$$

   Hint: A judicious change of variables will permit you to convert this to a linear fit. Does a minimum χ^2 still have meaning here?

8.7.6 Nonlinear Fit of the Breit-Wigner Formula to a Cross Section

Problem: Remember how we started Unit II of this chapter by interpolating the values in Table 8.1, which gave the experimental cross section Σ as a function of energy. Although we did not use it, we also gave the theory describing these data, namely, the Breit–Wigner resonance formula (8.31):

$$f(E) = \frac{f_r}{(E - E_r)^2 + \Gamma^2/4}. \tag{8.70}$$

Your **problem** here is to determine what values for the parameters E_r, f_r, and Γ in (8.70) provide the best fit to the data in Table 8.1.

Because (8.70) is not a linear function of the parameters (E_r, Σ_0, Γ), the three equations that result from minimizing χ^2 are not linear equations and so cannot be solved by the techniques of *linear* algebra (matrix methods). However, in our study of the masses on a string problem in Unit I, we showed how to use the Newton–Raphson algorithm to search for solutions of simultaneous nonlinear equations. That technique involved expansion of the equations about the previous guess to obtain a set of linear equations and then solving the linear equations with the matrix libraries. We now use this same combination of fitting, trial-and-error searching, and matrix algebra to conduct a nonlinear least-squares fit of (8.70) to the data in Table 8.1.

Recollect that the condition for a best fit is to find values of the M_P parameters a_m in the theory $g(x, a_m)$ that minimize $\chi^2 = \sum_i [(y_i - g_i)/\sigma_i]^2$. This leads to the M_P equations (8.53) to solve

$$\sum_{i=1}^{N_D} \frac{[y_i - g(x_i)]}{\sigma_i^2} \frac{\partial g(x_i)}{\partial a_m} = 0, \quad (m = 1, M_P). \tag{8.71}$$

To find the form of these equations appropriate to our problem, we rewrite our theory function (8.70) in the notation of (8.71):

$$a_1 = f_r, \quad a_2 = E_R, \quad a_3 = \Gamma^2/4, \quad x = E, \tag{8.72}$$

$$\Rightarrow \quad g(x) = \frac{a_1}{(x - a_2)^2 + a_3}. \tag{8.73}$$

The three derivatives required in (8.71) are then

$$\frac{\partial g}{\partial a_1} = \frac{1}{(x - a_2)^2 + a_3}, \quad \frac{\partial g}{\partial a_2} = \frac{-2a_1(x - a_2)}{[(x - a_2)^2 + a_3]^2}, \quad \frac{\partial g}{\partial a_3} = \frac{-a_1}{[(x - a_2)^2 + a_3]^2}.$$

Substitution of these derivatives into the best-fit condition (8.71) yields three simultaneous equations in a_1, a_2, and a_3 that we need to solve in order to fit the $N_D = 9$ data points (x_i, y_i) in Table 8.1:

$$\sum_{i=1}^{9} \frac{y_i - g(x_i, a)}{(x_i - a_2)^2 + a_3} = 0, \quad \sum_{i=1}^{9} \frac{y_i - g(x_i, a)}{[(x_i - a_2)^2 + a_3]^2} = 0,$$

$$\sum_{i=1}^{9} \frac{\{y_i - g(x_i, a)\}\,(x_i - a_2)}{[(x_i - a_2)^2 + a_3]^2} = 0. \tag{8.74}$$

Even without the substitution of (8.70) for $g(x, a)$, it is clear that these three equations depend on the a's in a nonlinear fashion. That's okay because in §8.2.2 we derived the N-dimensional Newton–Raphson search for the roots of

$$f_i(a_1, a_2, \ldots, a_N) = 0, \quad i = 1, N, \tag{8.75}$$

where we have made the change of variable $y_i \to a_i$ for the present problem. We use that same formalism here for the $N = 3$ equations (8.74) by writing them as

$$f_1(a_1, a_2, a_3) = \sum_{i=1}^{9} \frac{y_i - g(x_i, a)}{(x_i - a_2)^2 + a_3} = 0, \tag{8.76}$$

$$f_2(a_1, a_2, a_3) = \sum_{i=1}^{9} \frac{\{y_i - g(x_i, a)\}\,(x_i - a_2)}{[(x_i - a_2)^2 + a_3]^2} = 0, \tag{8.77}$$

$$f_3(a_1, a_2, a_3) = \sum_{i=1}^{9} \frac{y_i - g(x_i, a)}{[(x_i - a_2)^2 + a_3]^2} = 0. \tag{8.78}$$

Because $f_r \equiv a_1$ is the peak value of the cross section, $E_R \equiv a_2$ is the energy at which the peak occurs, and $\Gamma = 2\sqrt{a_3}$ is the full width of the peak at half-maximum, good guesses for the a's can be extracted from a graph of the data. To obtain the nine derivatives of the three f's with respect to the three unknown a's, we use two nested loops over i and j, along with the forward-difference approximation for the derivative

$$\frac{\partial f_i}{\partial a_j} \simeq \frac{f_i(a_j + \Delta a_j) - f_i(a_j)}{\Delta a_j}, \tag{8.79}$$

where Δa_j corresponds to a small, say $\leq 1\%$, change in the parameter value.

8.7.6.1 NONLINEAR FIT IMPLEMENTATION

Use the Newton–Raphson algorithm as outlined in §8.7.6 to conduct a nonlinear search for the best-fit parameters of the Breit–Wigner theory (8.70) to the data in Table 8.1. Compare the deduced values of (f_r, E_R, Γ) to that obtained by inspection of the graph. The program `Newton_Jama2.java` on the instructor's CD solves this problem.

9

Differential Equation Applications

Part of the attraction of computational problem solving is that it is easy to solve almost every differential equation. Consequently, while most traditional (read "analytic") treatments of oscillations are limited to the small displacements about equilibrium where the restoring forces are linear, we eliminate those restrictions here and reveal some interesting nonlinear physics. In Unit I we look at oscillators that may be harmonic for certain parameter values but then become anharmonic. We start with simple systems that have analytic solutions, use them to test various differential-equation solvers, and then include time-dependent forces and investigate nonlinear resonances and beating.[1] *In Unit II we examine how a differential-equation solver may be combined with a search algorithm to solve the eigenvalue problem. In Unit III we investigate how to solve the simultaneous ordinary differential equations (ODEs) that arise in scattering, projectile motion, and planetary orbits.*

9.1 Unit I. Free Nonlinear Oscillations

Problem: In Figure 9.1 we show a mass m attached to a spring that exerts a restoring force toward the origin, as well as a hand that exerts a time-dependent external force on the mass. We are told that the restoring force exerted by the spring is nonharmonic, that is, not simply proportional to displacement from equilibrium, but we are not given details as to how this is nonharmonic. Your **problem** is to solve for the motion of the mass as a function of time. You may assume the motion is constrained to one dimension.

9.2 Nonlinear Oscillators (Models)

This is a problem in classical mechanics for which Newton's second law provides us with the equation of motion

$$F_k(x) + F_{\text{ext}}(x,t) = m\frac{d^2x}{dt^2}, \qquad (9.1)$$

[1] In Chapter 12, "Discrete & Continuous Nonlinear Dynamics," we make a related study of the realistic pendulum and its chaotic behavior. Some special properties of nonlinear equations are discussed in Chapter 19, "Solitons & Computational Fluid Dynamics."

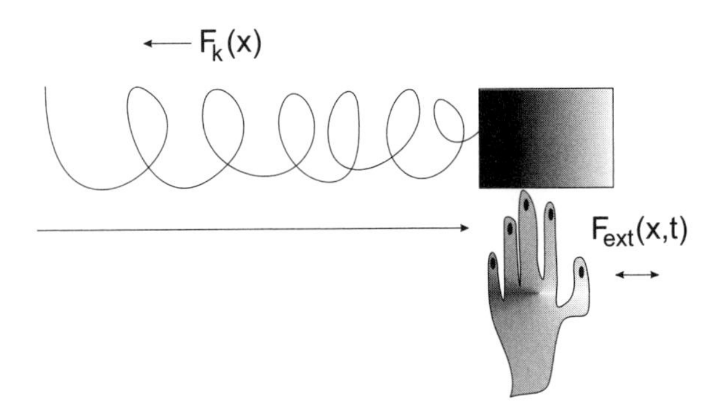

Figure 9.1 A mass m attached to a spring with restoring force $F_k(x)$ and with an external agency (a hand) subjecting the mass to a time-dependent driving force as well.

where $F_k(x)$ is the restoring force exerted by the spring and $F_{\rm ext}(x,t)$ is the external force. Equation (9.1) is the differential equation we must solve for arbitrary forces. Because we are not told just how the spring departs from being linear, we are free to try out some different models. As our first model, we try a potential that is a harmonic oscillator for small displacements x and also contains a perturbation that introduces a nonlinear term to the force for large x values:

$$V(x) \simeq \frac{1}{2}kx^2\left(1-\frac{2}{3}\alpha x\right), \tag{9.2}$$

$$\Rightarrow \quad F_k(x) = -\frac{dV(x)}{dx} = -kx(1-\alpha x) = m\frac{d^2x}{dt^2}, \tag{9.3}$$

where we have omitted the time-dependent external force. Equation (9.3) is the second-order ODE we need to solve. If $\alpha x \ll 1$, we should have essentially harmonic motion.

We can understand the basic physics of this model by looking at the curves on the left in Figure 9.2. As long as $x < 1/\alpha$, there will be a *restoring force* and the motion will be periodic (repeated exactly and indefinitely in time), even though it is harmonic (linear) only for small-amplitude oscillations. Yet, as the amplitude of oscillation gets larger, there will be an asymmetry in the motion to the right and left of the equilibrium position. And if $x > 1/\alpha$, the force will become repulsive and the mass will "roll" down the potential hill.

As a second model of a nonlinear oscillator, we assume that the spring's potential function is proportional to some arbitrary *even* power p of the displacement x from equilibrium:

$$V(x) = \frac{1}{p}kx^p, \quad (p\,\text{even}). \tag{9.4}$$

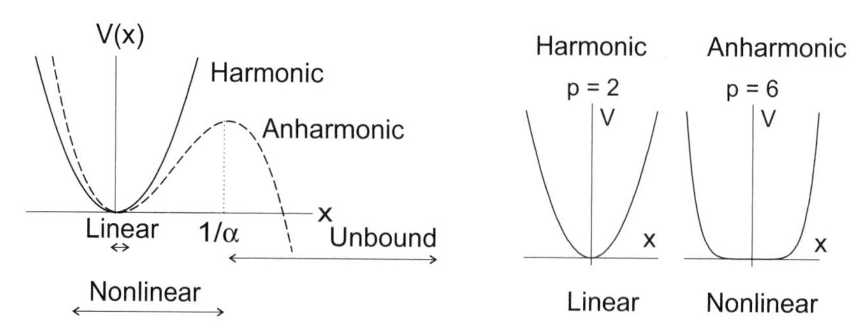

Figure 9.2 *Left:* The potential of an harmonic oscillator (solid curve) and of an oscillator with an anharmonic correction (dashed curve). *Right:* The shape of the potential energy function $V(x) \propto |x|^p$ for different p values. The linear and nonlinear labels refer to restoring force derived from these potentials.

We require an even p to ensure that the force,

$$F_k(x) = -\frac{dV(x)}{dx} = -kx^{p-1}, \tag{9.5}$$

contains an odd power of p, which guarantees that it is a *restoring* force for positive or negative x values. We display some characteristics of this potential on the right in Figure 9.2. We see that $p = 2$ is the harmonic oscillator and that $p = 6$ is nearly a square well with the mass moving almost freely until it hits the wall at $x \simeq \pm 1$. Regardless of the p value, the motion will be periodic, but it will be harmonic only for $p = 2$. Newton's law (9.1) gives the second-order ODE we need to solve:

$$F_{\text{ext}}(x,t) - kx^{p-1} = m\frac{d^2x}{dt^2}. \tag{9.6}$$

9.3 Types of Differential Equations (Math)

The background material in this section is presented to avoid confusion about semantics. The well-versed reader may want to skim or skip it.

Order: A general form for a *first-order* differential equation is

$$\frac{dy}{dt} = f(t,y), \tag{9.7}$$

where the "order" refers to the degree of the derivative on the LHS. The derivative or force function $f(t,y)$ on the RHS, is arbitrary. For instance, even if $f(t,y)$ is a nasty function of y and t such as

$$\frac{dy}{dt} = -3t^2 y + t^9 + y^7, \tag{9.8}$$

this is still first order in the derivative. A general form for a *second-order* differential equation is

$$\frac{d^2y}{dt^2} + \lambda\frac{dy}{dt} = f\left(t, \frac{dy}{dt}, y\right). \tag{9.9}$$

The derivative function f on the RHS is arbitrary and may involve any power of the first derivative as well. To illustrate,

$$\frac{d^2y}{dt^2} + \lambda\frac{dy}{dt} = -3t^2\left(\frac{dy}{dt}\right)^4 + t^9 y(t) \tag{9.10}$$

is a second-order differential equation, as is Newton's law (9.1).

In the differential equations (9.7) and (9.9), the time t is the *independent* variable and the position y is the *dependent* variable. This means that we are free to vary the time at which we want a solution, but not the value of the solution y at that time. Note that we often use the symbol y or Y for the dependent variable but that this is just a symbol. In some applications we use y to describe a position that is an independent variable instead of t.

Ordinary and partial: Differential equations such as (9.1) and (9.7) are *ordinary* differential equations because they contain only *one* independent variable, in these cases t. In contrast, an equation such as the Schrödinger equation

$$i\frac{\partial\psi(\mathbf{x},t)}{\partial t} = -\frac{1}{2m}\left[\frac{\partial^2\psi}{\partial x^2} + \frac{\partial^2\psi}{\partial y^2} + \frac{\partial^2\psi}{\partial z^2}\right] + V(\mathbf{x})\psi(\mathbf{x},t) \tag{9.11}$$

(where we have set $\hbar = 1$) contains several independent variables, and this makes it a *partial differential equation* (PDE). The partial derivative symbol ∂ is used to indicate that the dependent variable ψ depends simultaneously on several independent variables. In the early parts of this book we limit ourselves to ordinary differential equations. In Chapters 17–19 we examine a variety of partial differential equations.

Linear and nonlinear: Part of the liberation of computational science is that we are no longer limited to solving *linear equations*. A *linear* equation is one in which only the first power of y or $d^n y/d^n t$ appears; a *nonlinear* equation may contain higher powers. For example,

$$\frac{dy}{dt} = g^3(t)y(t) \quad \text{(linear)}, \qquad \frac{dy}{dt} = \lambda y(t) - \lambda^2 y^2(t) \quad \text{(nonlinear)}. \tag{9.12}$$

An important property of linear equations is the *law of linear superposition* that lets us add solutions together to form new ones. As a case in point, if $A(t)$ and

$B(t)$ are solutions of the linear equation in (9.12), then

$$y(t) = \alpha A(t) + \beta B(t) \tag{9.13}$$

is also a solution for arbitrary values of the constants α and β. In contrast, even if we were clever enough to guess that the solution of the nonlinear equation in (9.12) is

$$y(t) = \frac{a}{1 + be^{-\lambda t}} \tag{9.14}$$

(which you can verify by substitution), things would be amiss if we tried to obtain a more general solution by adding together two such solutions:

$$y_1(t) = \frac{a}{1 + be^{-\lambda t}} + \frac{a'}{1 + b'e^{-\lambda t}} \tag{9.15}$$

(which you can verify by substitution).

Initial and boundary conditions: The general solution of a first-order differential equation always contains one arbitrary constant. The general solution of a second-order differential equation contains two such constants, and so forth. For any specific problem, these constants are fixed by the *initial conditions*. For a first-order equation, the sole initial condition may be the position $y(t)$ at some time. For a second-order equation, the two initial conditions may be the position and velocity at some time. Regardless of how powerful the hardware and software that you employ, the mathematics remains valid, and so you must know the initial conditions in order to solve the problem uniquely.

In addition to the initial conditions, it is possible to further restrict the solutions of differential equations. One such way is by *boundary conditions* that constrain the solution to have fixed values at the boundaries of the solution space. Problems of this sort are called *eigenvalue problems*, and they are so demanding that solutions do not always exist, and even when they do exist, a trial-and-error search may be required to find them. In Unit II we discuss how to extend the techniques of the present unit to boundary-value problems.

9.4 Dynamic Form for ODEs (Theory)

A standard form for ODEs, which has found use both in numerical analysis [Pres 94, Pres 00] and in classical dynamics [Schk 94, Tab 89, J&S 98], is to express ODEs of *any order* as N simultaneous first-order ODEs:

$$\frac{dy^{(0)}}{dt} = f^{(0)}(t, y^{(i)}), \tag{9.16}$$

$$\frac{dy^{(1)}}{dt} = f^{(1)}(t, y^{(i)}), \tag{9.17}$$

$$\ddots \tag{9.18}$$

$$\frac{dy^{(N-1)}}{dt} = f^{(N-1)}(t, y^{(i)}), \tag{9.19}$$

where $y^{(i)}$ dependence in f is allowed but not any dependence on derivatives $dy^{(i)}/dt$. These equations can be expressed more compactly by use of the N-dimensional vectors (indicated here in **boldface**) $\mathbf{y}$ and $\mathbf{f}$:

$$d\mathbf{y}(t)/dt = \mathbf{f}(t, \mathbf{y}),$$

$$\mathbf{y} = \begin{pmatrix} y^{(0)}(t) \\ y^{(1)}(t) \\ \ddots \\ y^{(N-1)}(t) \end{pmatrix}, \quad \mathbf{f} = \begin{pmatrix} f^{(0)}(t, \mathbf{y}) \\ f^{(1)}(t, \mathbf{y}) \\ \ddots \\ f^{(N-1)}(t, \mathbf{y}) \end{pmatrix}. \tag{9.19}$$

The utility of such compact notation is that we can study the properties of the ODEs, as well as develop algorithms to solve them, by dealing with the single equation (9.20) without having to worry about the individual components. To see how this works, let us convert Newton's law

$$\frac{d^2x}{dt^2} = \frac{1}{m} F\left(t, \frac{dx}{dt}, x\right) \tag{9.20}$$

to standard dynamic form. The rule is that the RHS may not contain any explicit derivatives, although individual components of $y^{(i)}$ may represent derivatives. To pull this off, we define the position x as the dependent variable $y^{(0)}$ and the velocity dx/dt as the dependent variable $y^{(1)}$:

$$y^{(0)}(t) \stackrel{\text{def}}{=} x(t), \quad y^{(1)}(t) \stackrel{\text{def}}{=} \frac{dx}{dt} = \frac{d(0)}{dt}. \tag{9.21}$$

The second-order ODE (9.20) now becomes two simultaneous first-order ODEs:

$$\frac{dy^{(0)}}{dt} = y^{(1)}(t), \quad \frac{dy^{(1)}}{dt} = \frac{1}{m} F(t, y^{(0)}, y^{(1)}). \tag{9.22}$$

This expresses the acceleration [the second derivative in (9.20)] as the first derivative of the velocity $[y^{(2)}]$. These equations are now in the standard form (9.20), with the derivative or force function **f** having the two components

$$f^{(0)} = y^{(1)}(t), \quad f^{(1)} = \frac{1}{m} F(t, y^{(0)}, y^{(1)}), \tag{9.23}$$

where F may be an explicit function of time as well as of position and velocity.

To be even more specific, applying these definitions to our spring problem (9.6), we obtain the coupled first-order equations

$$\frac{dy^{(0)}}{dt} = y^{(1)}(t), \quad \frac{dy^{(1)}}{dt} = \frac{1}{m}\left[F_{\text{ext}}(x,t) - k y^{(0)}(t)^{p-1}\right], \tag{9.24}$$

where $y^{(0)}(t)$ is the position of the mass at time t and $y^{(1)}(t)$ is its velocity. In the standard form, the components of the force function and the initial conditions are

$$\begin{aligned} f^{(0)}(t,\mathbf{y}) &= y^{(1)}(t), \quad f^{(1)}(t,\mathbf{y}) = \frac{1}{m}\left[F_{\text{ext}}(x,t) - k(y^{(0)})^{p-1}\right], \\ y^{(0)}(0) &= x_0, \qquad\qquad y^{(1)}(0) = v_0. \end{aligned} \tag{9.25}$$

Breaking a second-order differential equation into two first-order ones is not just an arcane mathematical maneuver. In classical dynamics it occurs when transforming the single Newtonian equation of motion involving position and acceleration (9.1), into two *Hamiltonian* equations involving position and momentum:

$$\frac{dp_i}{dt} = F_i, \quad m\frac{dy_i}{dt} = p_i. \tag{9.26}$$

9.5 ODE Algorithms

The classic way to solve a differential equation is to start with the known initial value of the dependent variable, $y_0 \equiv y(t=0)$, and then use the derivative function $f(t,y)$ to find an approximate value for y at a small step $\Delta t = h$ forward in time; that is, $y(t=h) \equiv y_1$. Once you can do that, you can solve the ODE for all t values by just continuing stepping to larger times one small h at a time (Figure 9.3).[2] Error is always a concern when integrating differential equations because derivatives require small differences, and small differences are prone to subtractive cancellations and round-off error accumulation. In addition, because our stepping procedure for solving the differential equation is a continuous extrapolation of the initial conditions, with each step building on a previous extrapolation,

[2] To avoid confusion, notice that $y^{(n)}$ is the nth component of the y vector, while y_n is the value of y after n time steps. (Yes, there is a price to pay for elegance in notation.)

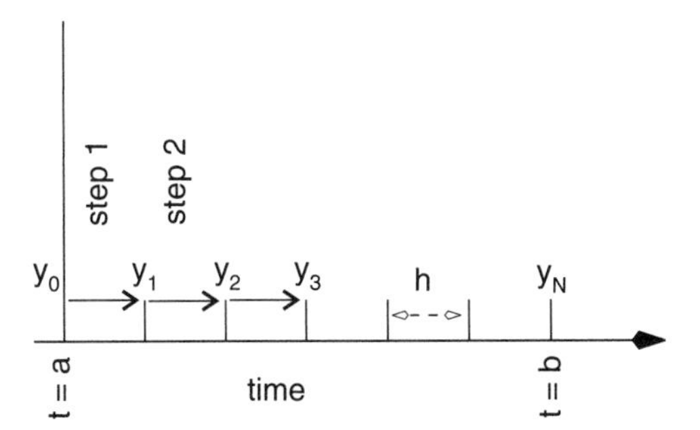

Figure 9.3 The steps of length h taken in solving a differential equation. The solution starts at time $t = a$ and is integrated to $t = b$.

this is somewhat like a castle built on sand; in contrast to interpolation, there are no tabulated values on which to anchor your solution.

It is simplest if the time steps used throughout the integration remain constant in size, and that is mostly what we shall do. Industrial-strength algorithms, such as the one we discuss in §9.5.2, adapt the step size by making h larger in regions where y varies slowly (this speeds up the integration and cuts down on round-off error) and making h smaller in regions where y varies rapidly.

9.5.1 Euler's Rule

Euler's rule (Figure 9.4 left) is a simple algorithm for integrating the differential equation (9.7) by one step and is just the forward-difference algorithm for the

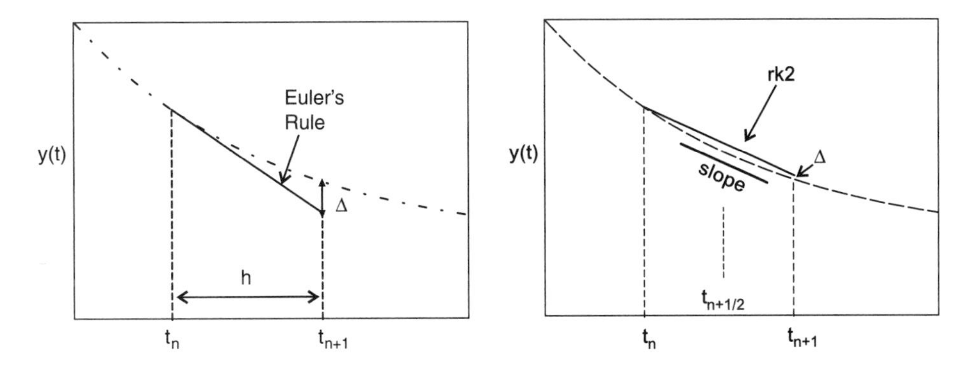

Figure 9.4 *Left:* Euler's algorithm for the forward integration of a differential equation for one time step. The linear extrapolation with the initial slope is seen to cause the error Δ. *Right:* The `rk2` algorithm is also a linear extrapolation of the solution y_n to y_{n+1}, but with the slope (bold line segment) at the interval's midpoint. The error is seen to be much smaller.

derivative:

$$\frac{d\mathbf{y}(t)}{dt} \simeq \frac{\mathbf{y}(t_{n+1}) - \mathbf{y}(t_n)}{h} = \mathbf{f}(t, \mathbf{y}), \tag{9.27}$$

$$\Rightarrow \quad \boxed{\mathbf{y}_{n+1} \simeq \mathbf{y}_n + h\mathbf{f}(t_n, \mathbf{y}_n),} \tag{9.28}$$

where $y_n \stackrel{\text{def}}{=} y(t_n)$ is the value of y at time t_n. We know from our discussion of differentiation that the error in the forward-difference algorithm is $\mathcal{O}(h^2)$, and so this too is the error in Euler's rule.

To indicate the simplicity of this algorithm, we apply it to our oscillator problem for the first time step:

$$y_1^{(0)} = x_0 + v_0 h, \quad y_1^{(1)} = v_0 + h\frac{1}{m}\left[F_{\text{ext}}(t=0) + F_k(t=0)\right]. \tag{9.29}$$

Compare these to the projectile equations familiar from first-year physics,

$$x = x_0 + v_0 h + \frac{1}{2}ah^2, \quad v = v_0 + ah, \tag{9.30}$$

and we see that the acceleration does not contribute to the distance covered (no h^2 term), yet it does contribute to the velocity (and so will contribute belatedly to the distance in the next time step). This is clearly a simple algorithm that requires very small h values to obtain precision. Yet using small values for h increases the number of steps and the accumulation of round-off error, which may lead to instability.[3] Whereas we do not recommend Euler's algorithm for general use, it is commonly used to start some of the more precise algorithms.

9.5.2 Runge–Kutta Algorithm

Even though no one algorithm will be good for solving all ODEs, the fourth-order Runge–Kutta algorithm `rk4`, or its extension with adaptive step size, `rk45`, has proven to be robust and capable of industrial-strength work. In spite of `rk4` being our recommended standard method, we derive the simpler `rk2` here and just give the results for `rk4`.

The Runge–Kutta algorithms for integrating a differential equation are based upon the formal (exact) integral of our differential equation:

$$\frac{dy}{dt} = f(t, y) \quad \Rightarrow \quad y(t) = \int f(t, y)\, dt \tag{9.31}$$

[3] Instability is often a problem when you integrate a $y(t)$ that decreases as the integration proceeds, analogous to upward recursion of spherical Bessel functions. In that case, and if you have a linear problem, you are best off integrating *inward* from large times to small times and then scaling the answer to agree with the initial conditions.

$$\Rightarrow \quad y_{n+1} = y_n + \int_{t_n}^{t_{n+1}} f(t, y)\, dt. \tag{9.32}$$

To derive the second-order Runge–Kutta algorithm rk2 (Figure 9.4 right, rk2.java on the CD), we expand $f(t, y)$ in a Taylor series about the *midpoint* of the integration interval and retain two terms:

$$f(t, y) \simeq f(t_{n+1/2}, y_{n+1/2}) + (t - t_{n+1/2}) \frac{df}{dt}(t_{n+1/2}) + \mathcal{O}(h^2). \tag{9.33}$$

Because $(t - t_{n+1/2})$ to any odd power is equally positive and negative over the interval $t_n \leq t \leq t_{n+1}$, the integral of $(t - t_{n+1/2})$ in (9.32) vanishes and we obtain our algorithm:

$$\int_{t_n}^{t_{n+1}} f(t, y)\, dt \simeq f(t_{n+1/2},\ y_{n+1/2})h + \mathcal{O}(h^3), \tag{9.34}$$

$$\Rightarrow \quad y_{n+1} \simeq y_n + hf(t_{n+1/2},\ y_{n+1/2}) + \mathcal{O}(h^3) \quad (\texttt{rk2}). \tag{9.35}$$

We see that while rk2 contains the same number of terms as Euler's rule, it obtains a higher level of precision by taking advantage of the cancellation of the $\mathcal{O}(h)$ terms [likewise, rk4 has the integral of the $t - t_{n+1/2}$ and $(t - t_{n+1/2})^3$ terms vanish]. Yet the price for improved precision is having to evaluate the derivative function and y at the middle of the time interval, $t = t_n + h/2$. And there's the rub, for we do not know the value of $y_{n+1/2}$ and cannot use this algorithm to determine it. The way out of this quandary is to use Euler's algorithm for $y_{n+1/2}$:

$$y_{n+1/2} \simeq y_n + \frac{1}{2}h\frac{dy}{dt} = y_n + \frac{1}{2}hf(t_n,\ y_n). \tag{9.36}$$

Putting the pieces all together gives the complete rk2 algorithm:

$$\boxed{\mathbf{y}_{n+1} \simeq \mathbf{y}_n + \mathbf{k}_2, \quad (\texttt{rk2})} \tag{9.37}$$

$$\boxed{\mathbf{k}_2 = h\mathbf{f}\left(t_n + \frac{h}{2},\ \mathbf{y}_n + \frac{\mathbf{k}_1}{2}\right), \quad \mathbf{k}_1 = h\,\mathbf{f}(t_n,\ \mathbf{y}_n),} \tag{9.38}$$

where we use boldface to indicate the vector nature of y and f. We see that the known derivative function **f** is evaluated at the ends and the midpoint of the interval, but that only the (known) initial value of the dependent variable **y** is required. This makes the algorithm self-starting.

As an example of the use of rk2, we apply it to our spring problem:

$$y_1^{(0)} = y_0^{(0)} + hf^{(0)}\left(\frac{h}{2}, y_0^{(0)} + k_1\right) \simeq x_0 + h\left[v_0 + \frac{h}{2}F_k(0)\right],$$

$$y_1^{(1)} = y_0^{(1)} + hf^{(1)}\left[\frac{h}{2}, y_0 + \frac{h}{2}f(0, y_0)\right] \simeq v_0 + \frac{h}{m}\left[F_{\text{ext}}\left(\frac{h}{2}\right) + F_k\left(y_0^{(1)} + \frac{k_1}{2}\right)\right].$$

These equations say that the position $y^{(0)}$ changes because of the initial velocity and force, while the velocity changes because of the external force at $t = h/2$ and the internal force at two intermediate positions. We see that the position now has an h^2 time dependence, which at last brings us up to the level of first-year physics.

rk4: The fourth-order Runge–Kutta method rk4 (Listing 9.1) obtains $\mathcal{O}(h^4)$ precision by approximating y as a Taylor series up to h^2 (a parabola) at the midpoint of the interval. This approximation provides an excellent balance of power, precision, and programming simplicity. There are now four gradient (k) terms to evaluate with four subroutine calls needed to provide a better approximation to $f(t, y)$ near the midpoint. This is computationally more expensive than the Euler method, but its precision is much better, and the step size h can be made larger. Explicitly, rk4 requires the evaluation of four intermediate slopes, and these are approximated with the Euler algorithm:

$$\boxed{\begin{aligned} &\mathbf{y}_{n+1} = \mathbf{y}_n + \frac{1}{6}(\mathbf{k}_1 + 2\mathbf{k}_2 + 2\mathbf{k}_3 + \mathbf{k}_4), \\ &\mathbf{k}_1 = h\mathbf{f}(t_n, \mathbf{y}_n), \qquad \mathbf{k}_2 = h\mathbf{f}\left(t_n + \frac{h}{2}, \mathbf{y}_n + \frac{\mathbf{k}_1}{2}\right), \\ &\mathbf{k}_3 = h\mathbf{f}\left(t_n + \frac{h}{2}, \mathbf{y}_n + \frac{\mathbf{k}_2}{2}\right), \qquad \mathbf{k}_4 = h\mathbf{f}(t_n + h, \mathbf{y}_n + \mathbf{k}_3). \end{aligned}} \tag{9.39}$$

rk45: (rk45.java on the CD) A variation of rk4, known as the Runge–Kutta–Fehlberg method or rk45 [Math 02], automatically doubles the step size and tests to see how an estimate of the error changes. If the error is still within acceptable bounds, the algorithm will continue to use the larger step size and thus speed up the computation; if the error is too large, the algorithm will decrease the step size until an acceptable error is found. As a consequence of the extra information obtained in the testing, the algorithm obtains $\mathcal{O}(h^5)$ precision but often at the expense of extra computing time. Whether that extra time is recovered by being able to use a larger step size depends upon the application.

9.5.3 Adams–Bashforth–Moulton Predictor-Corrector

Another approach for obtaining high precision in an ODE algorithm uses the solution from previous steps, say, y_{n-2} and y_{n-1}, in addition to y_n, to predict y_{n+1}.

```
// rk4.java: 4th order Runge-Kutta ODE Solver for arbitrary y(t)
import java.io.*;

public class rk4 {

  public static void main(String[] argv) throws IOException, FileNotFoundException {
    PrintWriter w = new PrintWriter(new FileOutputStream("rk4.dat"), true);
    double h, t, a = 0., b = 10.;                          // Step size, time, endpoints
    double ydumb[] = new double[2], y[] = new double[2], fReturn[] = new double[2];
    double k1[]    = new double[2], k2[] = new double[2], k3[]      = new double[2];
    double k4[]    = new double[2];
    int i, n = 100;
    y[0] = 3. ;    y[1] = -5. ;                                          // Initialize
    h = (b-a)/n;
    t = a;
    w.println(t + " " + y[0] + " " + y[1]);                             // File output
    while (t < b) {                                                       // Time loop
      if ( (t + h) > b ) h = b - t;                                       // Last step
      f(t, y, fReturn);                            // Evaluate RHS's, return in fReturn
      k1[0] = h*fReturn[0];  k1[1] = h*fReturn[1];          // Compute function values
      for ( i=0; i <= 1; i++ )  ydumb[i] = y[i] + k1[i]/2;
      f(t + h/2, ydumb, fReturn);
      k2[0] = h*fReturn[0];  k2[1] = h*fReturn[1];
      for ( i=0; i <= 1; i++ )  ydumb[i] = y[i] + k2[i]/2;
      f(t + h/2, ydumb, fReturn);
      k3[0] = h*fReturn[0];  k3[1] = h*fReturn[1];
      for ( i=0; i <= 1; i++ )  ydumb[i] = y[i] + k3[i];
      f(t + h, ydumb, fReturn);
      k4[0] = h*fReturn[0];    k4[1] = h*fReturn[1];
      for (i=0;i <= 1; i++)y[i]=y[i]+(k1[i]+2*(k2[i]+k3[i])+k4[i])/6;
      t = t + h;
      w.println(t + " " + y[0] + " " + y[1]);                             // File output
    }                                                                  // End while loop
    System.out.println("Output in rk4.dat");
  }
                                                                 // YOUR FUNCTION here
  public static void f( double t, double y[], double fReturn[] )
    { fReturn[0] = y[1];                                                   // RHS 1st eq
      fReturn[1] = -100*y[0]-2*y[1] + 10*Math.sin(3*t); }                  // RHS 2nd
}
```

Listing 9.1 rk4.java solves an ODE with the RHS given by the method f() using a fourth-order Runge–Kutta algorithm. Note that the method f(), which you will need to change for each problem, is kept separate from the algorithm, which it is best not to change.

(The Euler and rk methods use just the previous step.) Many of these methods tend to be like a Newton's search method; we start with a guess or *prediction* for the next step and then use an algorithm such as rk4 to check on the prediction. This yields a *correction*. As with rk45, one can use the difference between prediction and correction as a measure of the error and then adjust the step size to obtain improved precision [Math 92, Pres 00]. For those readers who may want to explore such methods, ABM.java on the CD gives our implementation of the *Adams–Bashforth–Moulton* predictor-corrector scheme.

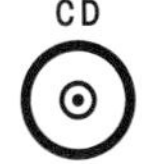

9.5.4 Assessment: rk2 *versus* rk4 *versus* rk45

While you are free to do as you please, we do *not* recommend that you write your own rk4 method unless you are very careful. We will be using rk4 for some

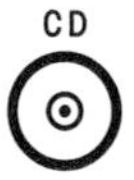

high-precision work, and unless you get every fraction and method call just right, your `rk4` may appear to work well but still not give all the precision that you should have. Regardless, we recommend that you write your own `rk2`, as doing so will make it clearer as to how the Runge–Kutta methods work, but without all the pain. We give `rk2`, `rk4`, and `rk45` codes on the CD and list `rk4.java` in Listing 9.1.

1. Write your own `rk2` method. Design your method for a general ODE; this means making the derivative function $f(t, x)$ a separate method.
2. Use your `rk2` solver in a program that solves the equation of motion (9.6) or (9.24). Use double precision to help control subtractive cancellation and plot both the position $x(t)$ and velocity dx/dt as functions of time.
3. Once your ODE solver compiles and executes, do a number of things to check that it is working well and that you know what h values to use.
   a. Adjust the parameters in your potential so that it corresponds to a pure harmonic oscillator (set $p = 2$ or $\alpha = 0$). For this case we have an analytic result with which to compare:

$$x(t) = A\ \sin(\omega_0 t + \phi), \quad v(t) = \omega_0 A\ \cos(\omega_0 t + \phi), \quad \omega_0 = \sqrt{k/m}.$$

   b. Pick values of k and m such that the period $T = 2\pi/\omega$ is a nice number with which to work (something like $T = 1$).
   c. Start with a step size $h \simeq T/5$ and make h smaller until the solution looks smooth, has a period that remains constant for a large number of cycles, and agrees with the analytic result. As a general rule of thumb, we suggest that you start with $h \simeq T/100$, where T is a characteristic time for the problem at hand. Here we want you to start with a large h so that you can see a bad solution turn good.
   d. Make sure that you have exactly the same initial conditions for the analytic and numerical solutions (zero displacement, nonzero velocity) and then plot the two together. It is good if you cannot tell them apart, yet that only ensures that there are approximately two places of agreement.
   e. Try different initial velocities and verify that a *harmonic* oscillator is *isochronous*, that is, that its period does *not* change as the amplitude varies.
4. Now that you know you can get a good solution of an ODE with `rk2`, **compare** the solutions obtained with the `rk2`, `rk4`, and `rk45` solvers.

TABLE 9.1
Comparison of ODE Solvers for Different Equations

Equation No.	*Method*	*Initial h*	*No. of Flops*	*Time* (ms)	*Relative Error*
(9.40)	`rk4`	0.01	1000	5.2	2.2×10^{-8}
	`rk45`	1.00	72	1.5	1.8×10^{-8}
(9.41)	`rk4`	0.01	227	8.9	1.8×10^{-8}
	`rk45`	0.1	3143	36.7	5.7×10^{-11}

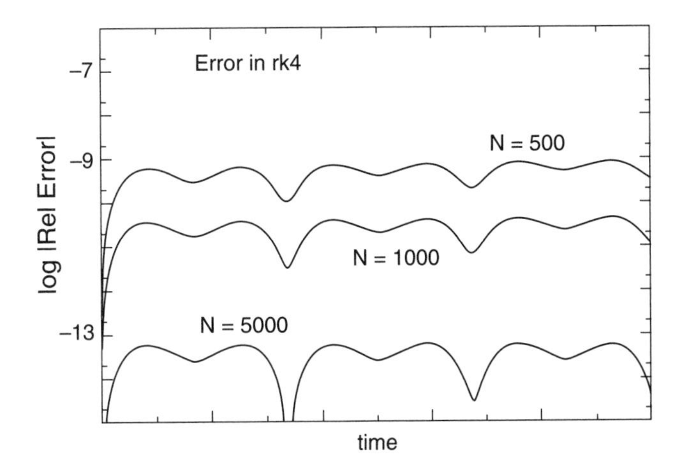

Figure 9.5 The log of the relative error (number of places of precision) obtained with rk4 using a differing number N of steps over the same time interval.

5. Make a table of comparisons similar to Table 9.1. There we compare rk4 and rk45 for the two equations

$$2yy'' + y^2 - y'^2 = 0, \tag{9.40}$$

$$y'' + 6y^5 = 0, \tag{9.41}$$

with initial conditions $([y(0), y'(0)] = (1, 1)$. Equation (9.40) yields oscillations with variable frequency and has an analytic solution with which to compare. Equation (9.41) corresponds to our standard potential (9.4), with $p = 6$. Although we have not tuned rk45, the table shows that by setting its tolerance parameter to a small enough number, rk45 will obtain better precision than rk4 (Figure 9.5) but that it requires ~10 times more floating-point operations and takes ~5 times longer. ▮

9.6 Solution for Nonlinear Oscillations (Assessment)

Use your rk4 program to study anharmonic oscillations by trying powers in the range $p = 2$–12 or anharmonic strengths in the range $0 \le \alpha x \le 2$. Do *not* include any explicit time-dependent forces yet. Note that for large values of p you may need to decrease the step size h from the value used for the harmonic oscillator because the forces and accelerations get large near the turning points.

1. Check that the solution remains periodic with constant amplitude and period for a given initial condition and value of p or α regardless of how nonlinear

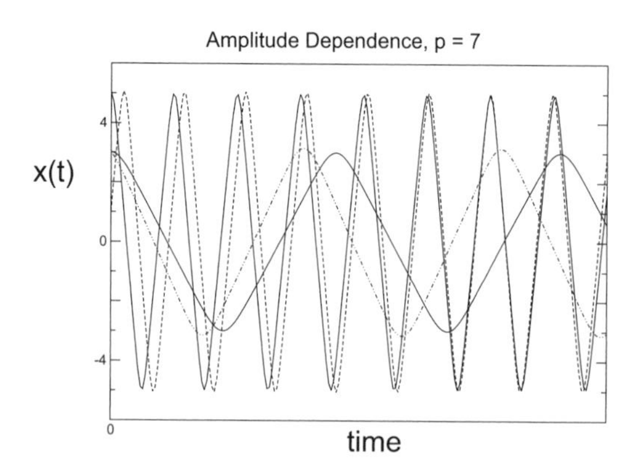

Figure 9.6 Different periods for oscillations within the restoring potential $V \propto x^7$ depending upon the amplitude.

you make the force. In particular, check that the maximum speed occurs at $x = 0$ and that the minimum speed occurs at maximum x. The latter is a consequence of energy conservation.

2. Verify that nonharmonic oscillators are *nonisochronous*, that is, that different initial conditions (amplitudes) lead to different periods (Figure 9.6).
3. Explain why the shapes of the oscillations change for different p's or α's.
4. Devise an algorithm to determine the period T of the oscillation by recording times at which the mass passes through the origin. Note that because the motion may be asymmetric, you must record at least three times.
5. Construct a graph of the deduced period as a function of initial amplitude.
6. Verify that the motion is oscillatory but not harmonic as the initial energy approaches $k/6\alpha^2$ or for $p > 6$.
7. Verify that for the anharmonic oscillator with $E = k/6\alpha^2$, the motion changes from oscillatory to translational. See how close you can get to the *separatrix* where a single oscillation takes an infinite time. (There is no separatrix for the power-law potential.)

9.6.1 Precision Assessment: Energy Conservation

We have not explicitly built energy conservation into our ODE solvers. Nonetheless, unless you have explicitly included a frictional force, it follows from the form of the equations of motion that energy must be a constant for all values of p or α. That being the case, the constancy of energy is a demanding test of the numerics.

1. Plot the potential energy $\text{PE}(t) = V[x(t)]$, the kinetic energy $\text{KE}(t) = mv^2(t)/2$, and the total energy $E(t) = \text{KE}(t) + \text{PE}(t)$, for 50 periods. Comment on the

correlation between PE(t) and KE(t) and how it depends on the potential's parameters.

2. Check the long-term *stability* of your solution by plotting

$$-\log_{10}\left|\frac{E(t)-E(t=0)}{E(t=0)}\right| \simeq \text{number of places of precision}$$

for a large number of periods (Figure 9.5). Because $E(t)$ should be independent of time, the numerator is the absolute error in your solution and when divided by $E(0)$, becomes the relative error (say 10^{-11}). If you cannot achieve 11 or more places, then you need to decrease the value of h or debug.

3. Because a particle bound by a large-p oscillator is essentially "free" most of the time, you should observe that the average of its kinetic energy over time exceeds its average potential energy. This is actually the physics behind the Virial theorem for a power-law potential:

$$\langle \text{KE} \rangle = \frac{p}{2}\langle \text{PE} \rangle. \tag{9.42}$$

Verify that your solution satisfies the Virial theorem. (Those readers who have worked the perturbed oscillator problem can use this relation to deduce an effective p value, which should be between 2 and 3.) ▌

9.7 Extensions: Nonlinear Resonances, Beats, and Friction

Problem: So far our oscillations have been rather simple. We have ignored friction and assumed that there are no external forces (hands) influencing the system's natural oscillations. Determine

1. How the oscillations change when friction is included.
2. How the resonances and beats of nonlinear oscillators differ from those of linear oscillators.
3. How introducing friction affects resonances.

9.7.1 Friction: Model and Implementation

The world is full of friction, and it is not all bad. For while friction may make it harder to pedal a bike through the wind, it also tends to stabilize motion. The simplest models for frictional force are called *static*, *kinetic*, and *viscous* friction:

$$F_f^{(\text{static})} \leq -\mu_s N, \quad F_f^{(\text{kinetic})} = -\mu_k N\frac{v}{|v|}, \quad F_f^{(\text{viscous})} = -bv. \tag{9.43}$$

Here N is the *normal force* on the object under consideration, μ and b are parameters, and v is the velocity. This model for static friction is appropriate for objects at rest, while the model for kinetic friction is appropriate for an object sliding on a dry surface. If the surface is lubricated, or if the object is moving through a viscous medium, then a frictional force proportional to velocity is a better model.[4]

1. Extend your harmonic oscillator code to include the three types of friction in (9.43) and observe how the motion differs for each.
2. *Hint:* For the simulation with static plus kinetic friction, each time the oscillator has $v = 0$ you need to check that the restoring force exceeds the static force of friction. If not, the oscillation must end at that instant. Check that your simulation terminates at nonzero x values.
3. For your simulations with viscous friction, investigate the qualitative changes that occur for increasing b values:

Underdamped: $b < 2m\omega_0$	Oscillation within a decaying envelope
Critically damped: $b = 2m\omega_0$	Nonoscillatory, finite time decay
Over damped: $b > 2m\omega_0$	Nonoscillatory, infinite time decay

9.7.2 Resonances and Beats: Model and Implementation

All stable physical systems will oscillate if displaced slightly from their rest positions. The frequency ω_0 with which such a system executes small oscillations about its rest positions is called its *natural frequency*. If an external sinusoidal force is applied to this system, and if the frequency of the external force equals the natural frequency ω_0, then a *resonance* may occur in which the oscillator absorbs energy from the external force and the amplitude of oscillation increases with time. If the oscillator and the driving force remain in phase over time, the amplitude will increase continuously unless there is some mechanism, such as friction or nonlinearities, to limit the growth.

If the frequency of the driving force is close to the natural frequency of the oscillator, then a related phenomena, known as *beating*, may occur. In this situation there is interference between the natural amplitude that is independent of the driving force plus an amplitude due to the external force. If the frequency of the driver is very close to the natural frequency, then the resulting motion,

$$x \simeq x_0 \sin \omega t + x_0 \sin \omega_0 t = \left(2x_0 \cos \frac{\omega - \omega_0}{2} t\right) \sin \frac{\omega + \omega_0}{2} t, \tag{9.44}$$

looks like the natural vibration of the oscillator at the average frequency $\frac{\omega + \omega_0}{2}$, yet with an amplitude $2x_0 \cos \frac{\omega - \omega_0}{2} t$ that varies with the slow *beat frequency* $\frac{\omega - \omega_0}{2}$.

[4] The effect of air resistance on projectile motion is studied in Unit III of this chapter.

9.8 Implementation: Inclusion of Time-Dependent Force

To extend our simulation to include an external force,

$$F_{\text{ext}}(t) = F_0 \sin \omega t, \tag{9.45}$$

we need to include some time dependence in the force function $\mathbf{f}(t, \mathbf{y})$ that occurs in our ODE solver.

1. Add the sinusoidal time-dependent external force (9.45) to the space-dependent restoring force in your program (do not include friction yet).
2. Start by using a very large value for the magnitude of the driving force F_0. This should lead to *mode locking* (the 500-pound-gorilla effect), where the system is overwhelmed by the driving force and, after the transients die out, the system oscillates in phase with the driver regardless of its frequency.
3. Now lower F_0 until it is close to the magnitude of the natural restoring force of the system. You need to have this near equality for beating to occur.
4. Verify that for the harmonic oscillator, the beat frequency, that is, the number of variations in intensity per unit time, equals the frequency difference $(\omega - \omega_0)/2\pi$ in cycles per second, where $\omega \simeq \omega_0$.
5. Once you have a value for F_0 matched well with your system, make a series of runs in which you progressively increase the frequency of the driving force for the range $\omega_0/10 \leq \omega \leq 10\omega_0$.
6. Make of plot of the maximum amplitude of oscillation that occurs as a function of the frequency of the driving force.
7. Explore what happens when you make nonlinear systems resonate. If the nonlinear system is close to being harmonic, you should get beating in place of the blowup that occurs for the linear system. Beating occurs because the natural frequency changes as the amplitude increases, and thus the natural and forced oscillations fall out of phase. Yet once out of phase, the external force stops feeding energy into the system, and the amplitude decreases; with the decrease in amplitude, the frequency of the oscillator returns to its natural frequency, the driver and oscillator get back in phase, and the entire cycle repeats.
8. Investigate now how the inclusion of viscous friction modifies the curve of amplitude *versus* driver frequency. You should find that friction broadens the curve.
9. Explain how the character of the resonance changes as the exponent p in the potential $V(x) = k|x|^p/p$ is made larger and larger. At large p, the mass effectively "hits" the wall and falls out of phase with the driver, and so the driver is less effective. ▌

9.9 Unit II. Binding A Quantum Particle

Problem: In this unit is we want to determine whether the rules of quantum mechanics are applicable inside a nucleus. More specifically, you are told that nuclei contain neutrons and protons (nucleons) with mass $mc^2 \simeq 940$ MeV and that a nucleus has a size of about 2 fm.[5] Your explicit **problem** is to see if these experimental facts are compatible, first, with quantum mechanics and, second, with the observation that there is a typical spacing of several million electron volts (MeV) between the ground and excited states in nuclei.

This problem requires us to solve the bound-state eigenvalue problem for the 1-D, time-dependent Schrödinger equation. Even though this equation is an ODE, which we know how to solve quite well by now, the extra requirement that we need to solve for bound states makes this an eigenvalue problem. Specifically, the bound-state requirement imposes boundary conditions on the form of the solution, which in turn means that a solution exists only for certain energies, the eigenenergies or eigenvalues.

If this all sounds a bit much for you now, rest assured that you do not need to understand all the physics behind these statements. What we want is for you to gain experience with the technique of conducting a numerical search for the eigenvalue in conjunction with solving an ODE numerically. This is how one solves the numerical ODE eigenvalue problem. In §20.2.1, we discuss how to solve the equivalent, but more advanced, momentum-space eigenvalue problem as a matrix problem. In Chapter 18, PDE Waves: String, Quantum Packet, *and we study the related problem of the motion of a quantum wave packet confined to a potential well. Further discussions of the numerical bound-state problem are found in [Schd 00, Koon 86].*

9.10 Theory: The Quantum Eigenvalue Problem

Quantum mechanics describes phenomena that occur on atomic or subatomic scales (an elementary particle is subatomic). It is a statistical theory in which the probability that a particle is located in a region dx around the point x is $\mathcal{P} = |\psi(x)|^2\,dx$, where $\psi(x)$ is called the *wave function*. If a particle of energy E moving in one dimension experiences a potential $V(x)$, its wave function is determined by an ODE (a PDE if greater than 1-D) known as the time-independent Schrödinger equation[6]:

$$\frac{-\hbar^2}{2m}\frac{d^2\psi(x)}{dx^2} + V(x)\psi(x) = E\psi(x). \tag{9.46}$$

[5] A fermi (fm) equals 10^{-13} cm $= 10^{-15}$ m, and $\hbar c \simeq 197.32$ MeV fm.

[6] The time-dependent equation requires the solution of a partial differential equation, as discussed in Chapter 18, "PDE Waves: String, Quantum Packet, and E&M."

Although we say we are solving for the energy E, in practice we solve for the wave vector κ. The energy is negative for bound states, and so we relate the two by

$$\kappa^2 = -\frac{2m}{\hbar^2}E = \frac{2m}{\hbar^2}|E|. \tag{9.47}$$

The Schrödinger equation then takes the form

$$\frac{d^2\psi(x)}{dx^2} - \frac{2m}{\hbar^2}V(x)\psi(x) = \kappa^2\psi(x). \tag{9.48}$$

When our problem tells us that the particle is bound, we are being told that it is confined to some finite region of space. The only way to have a $\psi(x)$ with a finite integral is to have it decay exponentially as $x \to \pm\infty$ (where the potential vanishes):

$$\psi(x) \to \begin{cases} e^{-\kappa x}, & \text{for } x \to +\infty, \\ e^{+\kappa x}, & \text{for } x \to -\infty. \end{cases} \tag{9.49}$$

In summary, although it is straightforward to solve the ODE (9.46) with the techniques we have learned so far, we must also require that the solution $\psi(x)$ simultaneously satisfies the boundary conditions (9.49). This extra condition turns the ODE problem into an *eigenvalue problem* that has solutions (*eigenvalues*) for only certain values of the energy E. The ground-state energy corresponds to the smallest (most negative) eigenvalue. The ground-state wave function (eigenfunction), which we must determine in order to find its energy, must be nodeless and even (symmetric) about $x = 0$. The excited states have higher (less negative) energies and wave functions that may be odd (antisymmetric).

9.10.1 Model: Nucleon in a Box

The numerical methods we describe are capable of handling the most realistic potential shapes. Yet to make a connection with the standard textbook case and to permit some analytic checking, we will use a simple model in which the potential $V(x)$ in (9.46) is a finite square well (Figure 9.7):

$$V(x) = \begin{cases} -V_0 = -83\,\text{MeV}, & \text{for } |x| \le a = 2\,\text{fm}, \\ 0, & \text{for } |x| > a = 2\,\text{fm}, \end{cases} \tag{9.50}$$

where values of 83 MeV for the depth and 2 fm for the radius are typical for nuclei (these are the units in which we solve the problem). With this potential the Schrödinger equation (9.48) becomes

$$\frac{d^2\psi(x)}{dx^2} + \left(\frac{2m}{\hbar^2}V_0 - \kappa^2\right)\psi(x) = 0, \quad \text{for } |x| \le a, \tag{9.51}$$

$$\frac{d^2\psi(x)}{dx^2} - \kappa^2\psi(x) = 0, \quad \text{for } |x| > a. \tag{9.52}$$

To evaluate the ratio of constants here, we insert c^2, the speed of light squared, into both the numerator and the denominator [L 96, Appendix A.1]:

$$\frac{2m}{\hbar^2} = \frac{2mc^2}{(\hbar c)^2} \simeq \frac{2 \times 940\,\text{MeV}}{(197.32\,\text{MeV fm})^2} = 0.0483\,\text{MeV}^{-1}\,\text{fm}^{-2}. \tag{9.53}$$

9.11 Combined Algorithms: Eigenvalues via ODE Solver Plus Search

The solution of the eigenvalue problem combines the numerical solution of the ordinary differential equation (9.48) with a trial-and-error search for a wave function that satisfies the boundary conditions (9.49). This is done in several steps:

1. Start on the very far *left* at $x = -X_{\text{max}} \simeq -\infty$, where $X_{\text{max}} \gg a$. Assume that the wave function there satisfies the left-hand boundary condition:

$$\psi_L(x = -X_{\text{max}}) = e^{+\kappa x} = e^{-\kappa X_{\text{max}}}.$$

2. Use your favorite ODE solver to step $\psi_L(x)$ in toward the origin (to the right) from $x = -X_{\text{max}}$ until you reach the *matching radius* x_{match}. The exact value of this matching radius is not important, and our final solution should be independent of it. On the left in Figure 9.7, we show a sample solution with $x_{\text{match}} = -a$; that is, we match at the left edge of the potential well. In the middle and on the right in Figure 9.7 we see some guesses that do not match.
3. Start on the very far *right*, that is, at $x = +X_{\text{max}} \simeq +\infty$, with a wave function that satisfies the right-hand boundary condition:

$$\psi_R(x = +\kappa X_{\text{max}}) = e^{-\kappa x} = e^{-\kappa X_{\text{max}}}.$$

4. Use your favorite ODE solver (e.g., rk4) to step $\psi_R(x)$ in toward the origin (to the left) from $x = +X_{\text{max}}$ until you reach the *matching radius* x_{match}. This means that we have stepped through the potential well (Figure 9.7).
5. In order for probability and current to be continuous at $x = x_{\text{match}}$, $\psi(x)$ and $\psi'(x)$ must be continuous there. Requiring the ratio $\psi'(x)/\psi(x)$, called the

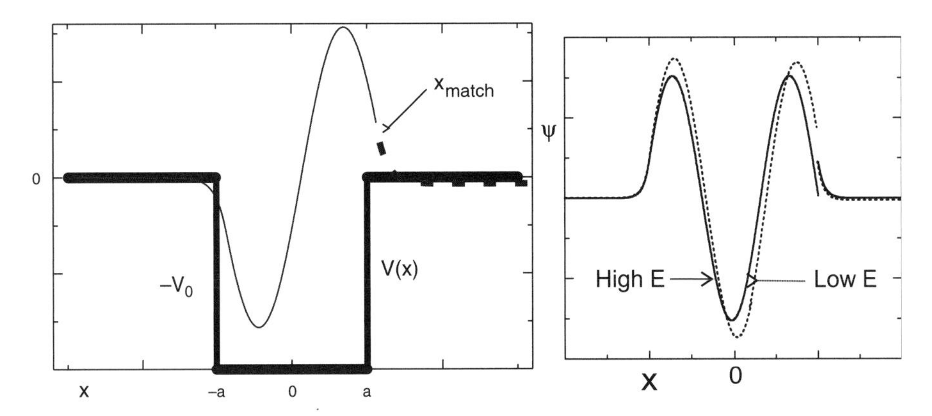

Figure 9.7 *Left:* Computed wave function and the square-well potential (bold lines). The wave function computed by integration in from the left is matched to the one computed by integration in from the right (dashed curve) at a point near the left edge of the well. Note how the wave function decays rapidly outside the well. *Right:* A first guess at a wave function with energy E that is 0.5% too low (dotted curve). We see that the left wave function does not vary rapidly enough to match the right one at $x = 500$. The solid curve shows a second guess at a wave function with energy E that is 0.5% too high. We see that now the left wave function varies too rapidly.

logarithmic derivative, to be continuous encapsulates both continuity conditions into a single condition and is independent of ψ's normalization.

6. Even though we do not know ahead of time which energies E or κ values are eigenvalues, we still need a starting value for the energy in order to use our ODE solver. Such being the case, we start the solution with a guess for the energy. A good guess for ground-state energy would be a value somewhat up from that at the bottom of the well, $E > -V_0$.
7. Because it is unlikely that any guess will be correct, the left- and right-wave functions will not quite match at $x = x_{\text{match}}$ (Figure 9.7). This is okay because we can use the amount of mismatch to improve the next guess. We measure how well the right and left wave functions match by calculating the difference

$$\Delta(E, x) = \left. \frac{\psi'_L(x)/\psi_L(x) - \psi'_R(x)/\psi_R(x)}{\psi'_L(x)/\psi_L(x) + \psi'_R(x)/\psi_R(x)} \right|_{x=x_{\text{match}}}, \tag{9.54}$$

where the denominator is used to avoid overly large or small numbers. Next we try a different energy, note how much $\Delta(E)$ has changed, and use this to deduce an intelligent guess at the next energy. The search continues until the left- and right-wave ψ'/ψ match within some tolerance.

9.11.1 Numerov Algorithm for the Schrödinger ODE ⊙

We generally recommend the fourth-order Runge–Kutta method for solving ODEs, and its combination with a search routine for solving the eigenvalue problem. In this section we present the Numerov method, an algorithm that is specialized for ODEs not containing any first derivatives (such as our Schrödinger equation). While this algorithm is not as general as `rk4`, *it is of $\mathcal{O}(h^6)$ and thus speeds up the calculation by providing additional precision.*

We start by rewriting the Schrödinger equation (9.48) in the compact form,

$$\frac{d^2\psi}{dx^2} + k^2(x)\psi = 0, \quad k^2(x) \stackrel{\text{def}}{=} \frac{2m}{\hbar^2} \begin{cases} E + V_0, & \text{for } |x| < a, \\ E, & \text{for } |x| > a, \end{cases} \tag{9.55}$$

where $k^2 = -\kappa^2$ in potential-free regions. Observe that although (9.55) is specialized to a square well, other potentials would have $V(x)$ in place of $-V_0$. The trick in the Numerov method is to get extra precision in the second derivative by taking advantage of there being no first derivative $d\psi/dx$ in (9.55). We start with the Taylor expansions of the wave functions,

$$\begin{aligned} \psi(x+h) &\simeq \psi(x) + h\psi^{(1)}(x) + \frac{h^2}{2}\psi^{(2)}(x) + \frac{h^3}{3!}\psi^{(3)}(x) + \frac{h^4}{4!}\psi^{(4)}(x) + \cdots \\ \psi(x-h) &\simeq \psi(x) - h\psi^{(1)}(x) + \frac{h^2}{2}\psi^{(2)}(x) - \frac{h^3}{3!}\psi^{(3)}(x) + \frac{h^4}{4!}\psi^{(4)}(x) + \cdots, \end{aligned}$$

where $\psi^{(n)}$ signifies the nth derivative $d^n\psi/dx^n$. Because the expansion of $\psi(x-h)$ has odd powers of h appearing with negative signs, all odd powers cancel when we add $\psi(x+h)$ and $\psi(x-h)$ together:

$$\begin{aligned} &\psi(x+h) + \psi(x-h) \simeq 2\psi(x) + h^2\psi^{(2)}(x) + \frac{h^4}{12}\psi^{(4)}(x) + \mathcal{O}(h^6), \\ \Rightarrow \quad &\psi^{(2)}(x) \simeq \frac{\psi(x+h) + \psi(x-h) - 2\psi(x)}{h^2} - \frac{h^2}{12}\psi^{(4)}(x) + \mathcal{O}(h^4). \end{aligned}$$

To obtain an algorithm for the second derivative we eliminate the fourth-derivative term by applying the operator $1 + \frac{h^2}{12}\frac{d^2}{dx^2}$ to the Schrödinger equation (9.55):

$$\psi^{(2)}(x) + \frac{h^2}{12}\psi^{(4)}(x) + k^2(x)\psi + \frac{h^2}{12}\frac{d^2}{dx^2}[k^2(x)\psi^{(4)}(x)] = 0.$$

```
// QuantumNumerov.java: solves Schroed eq via Numerov + Bisection Algor

import java.io.*;
import ptolemy.plot.*;

public class QuantumNumerov  {

  public static void main(String[] argv) {
    Plot myPlot = new Plot();
    myPlot.setTitle("Schrodinger Eqn Numerov-Eigenfunction");
    myPlot.setXLabel("x");
    myPlot.setYLabel("Y(x)");
    myPlot.setXRange(-1000,1000);
    myPlot.setColor(false);
    PlotApplication app = new PlotApplication(myPlot);
    Plot myPlot1 = new Plot();
    myPlot1.setTitle("Schrodinger Eqn Numerov-Probability Density");
    myPlot1.setXLabel("x");
    myPlot1.setYLabel("Y(x)^2");
    myPlot1.setXRange(-1000,1000);
    myPlot1.setYRange(-0.004,0.004);
    myPlot1.setColor(false);
    PlotApplication app1 = new PlotApplication(myPlot1);

    int i, istep, im, n, m, imax, nl, nr;
    double dl=1e-8, h, min, max, e, e1, de, xl0, xr0, xl, xr, f0, f1, sum, v, fact;
    double ul[] = new double[1501], ur[]  = new double[1501], k2l[]=new double[1501];
    double s[]  = new double[1501], k2r[] = new double[1501];
    n = 1501;    m = 5;       im = 0;     imax = 100;    xl0 = -1000;   xr0 =  1000;
    h   =  (xr0-xl0)/(n-1);
    min  = -0.001;      max = -0.00085;      e    = min;      de  = 0.01;
    ul[0] = 0.0;         ul[1] =  0.00001;  ur[0] =  0.0;  ur[1] =  0.00001;
    for (i = 0; i < n; ++i)  {                                // Set up the potential k2
      xl = xl0+i*h;
      xr = xr0-i*h;
      k2l[i] = (e-v(xl));
      k2r[i] = (e-v(xr));
    }
    im = 500;                                                      // The matching point
    nl = im+2;                       nr = n-im+1;
    numerov (nl,h,k2l,ul);       numerov (nr,h,k2r,ur);
    fact= ur[nr-2]/ul[im];                                          // Rescale the solution
    for (i = 0; i < nl; ++i)   ul[i] = fact*ul[i];
    f0 = (ur[nr-1]+ul[nl-1]-ur[nr-3]-ul[nl-3])/(2*h*ur[nr-2]);              // Log deriv
    istep = 0;                                             // Bisection algor for the root
    while ( Math.abs(de) > dl && istep < imax )  {
      e1 = e;
      e  = (min+max)/2;
      for (i = 0; i < n; ++i)  {
        k2l[i] = k2l[i]+(e-e1);
        k2r[i] = k2r[i]+(e-e1);
      }
      im=500;
      nl = im+2;                    nr = n-im+1;
      numerov (nl,h,k2l,ul);   numerov (nr,h,k2r,ur);
      fact=ur[nr-2]/ul[im];                                          // Rescale solution
      for (i = 0; i < nl; ++i)   ul[i] = fact*ul[i];
      f1 = (ur[nr-1]+ul[nl-1]-ur[nr-3]-ul[nl-3])/(2*h*ur[nr-2]);               // Log der
      if ((f0*f1) < 0)  { max = e; de = max-min; }
      else  { min = e;  de = max-min;  f0 = f1;}
      istep = istep+1;
    }
    sum = 0;
    for (i = 0; i < n; ++i) { if (i > im) ul[i] = ur[n-i-1]; sum = sum+ul[i]*ul[i]; }
    sum = Math.sqrt(h*sum);
    System.out.println("istep=" +istep);
    System.out.println("e= "+ e+" de= "+de);
    for (i = 0; i < n; i = i+m)  {
```

```
      xl = xl0+i*h;
      ul[i] = ul[i]/sum;
      myPlot.addPoint(1, xl, ul[i], true);
      myPlot1.addPoint(1, xl, ul[i]*ul[i], true);
    }
  }

  public static void numerov (int n,double h,double k2[],double u[]){
    int i;
    for (i = 1; i <n-1; ++i)  { u[i+1]=(2*u[i]*(1-5.*h*h/12.*k2[i])
                               -(1.+h*h/12.*k2[i-1])*u[i-1])/(1.+h*h/12.*k2[i+1]); }
  }

  public static double v(double x)  {
    double v;
    if (Math.abs(x)<=500)   v=-0.001;
    else   v=0;
    return v;
  }
}
```

Listing 9.2 QuantumNumerov.java solves the 1-D time-independent Scrödinger equation for bound-state energies using a Numerov method (rk4 also works, as we show in Listing 9.3).

We eliminate the $\psi^{(4)}$ terms by substituting the derived expression for the $\psi^{(2)}$:

$$\frac{\psi(x+h)+\psi(x-h)-2\psi(x)}{h^2}+k^2(x)\psi(x)+\frac{h^2}{12}\frac{d^2}{dx^2}[k^2(x)\psi(x)]\simeq 0.$$

Now we use a central-difference approximation for the second derivative of $k^2(x)\psi(x)$

$$h^2\frac{d^2[k^2(x)\psi(x)]}{dx^2}\simeq[(k^2\psi)_{x+h}-(k^2\psi)_x]+[(k^2\psi)_{x-h}-(k^2\psi)_x].$$

After this substitution we obtain the Numerov algorithm:

$$\boxed{\psi(x+h)\simeq\frac{2[1-\frac{5}{12}h^2k^2(x)]\psi(x)-[1+\frac{h^2}{12}k^2(x-h)]\psi(x-h)}{1+h^2k^2(x+h)/12}.}\tag{9.56}$$

We see that the Numerov algorithm uses the values of ψ at the two previous steps x and $x-h$ to move ψ forward to $x+h$. To step backward in x, we need only to reverse the sign of h. Our implementation of this algorithm, `Numerov.java`, is given in Listing 9.2.

9.11.2 Implementation: Eigenvalues via an ODE Solver Plus Bisection Algorithm

1. Combine your bisection algorithm search program with your `rk4` or Numerov ODE solver program to create an eigenvalue solver. Start with a step size $h=0.04$.

2. Write a subroutine that calculates the matching function $\Delta(E,x)$ as a function of energy and matching radius. This subroutine will be called by the bisection algorithm program to search for the energy at which $\Delta(E,x=2)$ vanishes.
3. As a first guess, take $E \simeq 65\,\text{MeV}$.
4. Search until $\Delta(E,x)$ changes in only the fourth decimal place. We do this in the code QuantumEigen.java shown in Listing 9.3.
5. Print out the value of the energy for each iteration. This will give you a feel as to how well the procedure converges, as well as a measure of the precision obtained. Try different values for the tolerance until you are confident that you are obtaining three good decimal places in the energy.
6. Build in a limit to the number of energy iterations you permit and print out when the iteration scheme fails.
7. As we have done, plot the wave function and potential on the same graph (you will have to scale the potential to get them both to fit).
8. Deduce, by counting the number of nodes in the wave function, whether the solution found is a ground state (no nodes) or an excited state (with nodes) and whether the solution is even or odd (the ground state must be even).
9. Include in your version of Figure 9.7 a horizontal line within the potential indicating the energy of the ground state relative to the potential's depth.
10. Increase the value of the initial energy guess and search for excited states. Make sure to examine the wave function for each state found to ensure that it is continuous and to count the number of nodes.
11. Add each new state found as another horizontal bar within the potential.
12. Verify that you have solved the **problem**, that is, that the spacing between levels is on the order of MeV for a nucleon bound in a several-fermi well. ▮

```
// QuantumEigen.java: solves Schroedinger eq via rk4 + Bisection Algor
import java.io.*;

public class QuantumEigen {
  static double eps = 1E-6 ;                          // Class variables; precision
  static int n_steps = 501;                                  // Number int steps

  public static void main(String[] argv) throws IOException, FileNotFoundException {
    double E = -17., h = 0.04;                    // Initial E in MeV, step size in fm
    double Emax, Emin, Diff;
    int count, count_max = 100;
    Emax = 1.1*E;        Emin = E/1.1;
    for ( count=0; count <= count_max; count++ ) {                 // Iteration loop
      E = (Emax + Emin)/2. ;                                       // Divide E range
      Diff = diff(E, h);
      System.out.println("E = " + E + ", L-R Log deriv(E) = " + Diff);
      if (diff(Emax, h)*Diff > 0) Emax = E;                    // Bisection algorithm
      else Emin = E;
      if ( Math.abs(Diff) < eps ) break;
    }
    plot(E, h);
    System.out.println("Final eigenvalue E = " + E);
    System.out.println("iterations, max = " + count + ", " + count_max);
    System.out.println("WF in QunatumL/R.dat, V in QuantumV.dat ");
  }                                                                  // End main
                                                      // Returns L-R log deriv
```

```
public static double diff(double E, double h)
                                    throws IOException, FileNotFoundException {
  double left, right, x;
  int ix, nL, nR, i_match;
  double y[] = new double[2];
  i_match = n_steps/3;                                          // Matching radius
  nL = i_match + 1;
  y[0] = 1.E-15;                                                // Initial wf on left
  y[1] = y[0]*Math.sqrt(-E*0.4829);
  for (ix = 0; ix < nL + 1; ix++) { x = h * (ix  -n_steps/2); rk4(x, y, h, 2, E); }
  left = y[1]/y[0];                                             // Log  derivative
  y[0] = 1.E-15;                               // - slope for even;  reverse for odd
  y[1] = -y[0]*Math.sqrt(-E*0.4829);                            // Initialize R wf
  for (ix = n_steps; ix > nL+1; ix--){x = h*(ix+1-n_steps/2); rk4(x, y, -h, 2, E);}
  right = y[1]/y[0];                                            // Log derivative
  return( (left - right)/(left + right) );
}

public static void plot(double E, double h)          // Repeat integrations for plot
                                    throws IOException, FileNotFoundException {
  PrintWriter L =  new PrintWriter(new FileOutputStream("QuantumL.dat"), true);
  PrintWriter R =  new PrintWriter(new FileOutputStream("QuantumR.dat"), true);
  PrintWriter Vx = new PrintWriter(new FileOutputStream("QuantumV.dat"), true);
  double left, right, normL, x = 0.;
  int ix, nL, nR, i_match, n_steps = 1501;          // Total no integration steps
  double y[] = new double[2], yL[][] = new double [2][505];
  i_match = 500;                                                // Matching point
  nL = i_match + 1;
  y[0] = 1.E-40;                                            // Initial wf on the left
  y[1] = -Math.sqrt(-E*0.4829) *y[0];
  for ( ix = 0;  ix <= nL;  ix++ ) {
    yL[0][ix] = y[0];  yL[1][ix] = y[1];
    x = h * (ix  -n_steps/2);
    rk4(x, y, h, 2, E);
  }                                                        // Integrate to the left
  y[0] = -1.E-15;                                // - slope: even;  reverse for odd
  y[1] = -Math.sqrt(-E*0.4829)*y[0];
  for ( ix = n_steps -1;   ix >= nL + 1;  ix--)  {                  // Integrate in
    x = h * (ix + 1 -n_steps/2);
    R.println(x + "  " + y[0] + "  " + y[1]);                       // File print
    Vx.println(x + "  " + 1.7E9*V(x));                              // Scaled V
    rk4(x, y, -h, 2, E);
  }
  x = x  - h;
  R.println(x + "  " + y[0] + "  " + y[1]);                         // File print
  normL = y[0]/yL[0][nL];                         // Renormalize L wf & derivative
  for ( ix = 0;  ix <= nL;  ix++ )  {
    x = h * (ix-n_steps/2 + 1);
    y[0] = yL[0][ix]*normL;
    y[1] = yL[1][ix]*normL;
    L.println(x + "  " + y[0] + "  " + y[1]);                       // File print
    Vx.println(x + "  " + 1.7E9*V(x));                              // Print V
  }
    return;
}

 public static void f(double x, double y[], double F[], double E)
   { F[0] = y[1];    F[1] = -(0.4829)*(E-V(x))*y[0];  }

public static double V(double x)
  { if (Math.abs(x)  <  10.)  return ( -16.);    else return(0.) ; }

public static void rk4(double t, double y[],double h,int Neqs,double E){
  int i;
  double F[]  = new double[Neqs], ydumb[]     = new double[Neqs];
  double k1[] = new double[Neqs]; double k2[] = new double[Neqs];
  double k3[] = new double[Neqs]; double k4[] = new double[Neqs];
  f(t, y, F,E);
```

```
    for (i=0; i<Neqs; i++) { k1[i] = h*F[i]; ydumb[i] = y[i] + k1[i]/2;}
    f(t + h/2, ydumb, F,E);
    for (i=0; i<Neqs; i++)  { k2[i] = h*F[i]; ydumb[i] = y[i] + k2[i]/2;}
    f(t + h/2, ydumb, F,E);
    for (i=0; i<Neqs; i++)  { k3[i]=  h*F[i]; ydumb[i] = y[i] + k3[i];}
    f(t + h, ydumb, F,E);
    for (i=0; i<Neqs; i++) {k4[i]=h*F[i]; y[i]=y[i]+(k1[i]+2*(k2[i]+k3[i])+k4[i])/6;}
  }
}
```

Listing 9.3 QuantumEigen.java solves the 1-D time-independent Schrödinger equation for bound-state energies using the rk4 algorithm.

9.12 Explorations

1. Check to see how well your search procedure works by using arbitrary values for the starting energy. For example, because no bound-state energies can lie below the bottom of the well, try $E \geq -V_0$, as well as some arbitrary fractions of V_0. In every case examine the resulting wave function and check that it is both symmetric and continuous.
2. Increase the depth of your potential progressively until you find several bound states. Look at the wave function in each case and correlate the number of nodes in the wave function and the position of the bound state in the well.
3. Explore how a bound-state energy changes as you change the depth V_0 of the well. In particular, as you keep decreasing the depth, watch the eigenenergy move closer to $E = 0$ and see if you can find the potential depth at which the bound state has $E \simeq 0$.
4. For a fixed well depth V_0, explore how the energy of a bound state changes as the well radius a is varied.
5. ⊙ Conduct some explorations in which you discover different combinations of (V_0, a) that give the same ground-state energies (discrete ambiguities). The existence of several different combinations means that a knowledge of ground-state energy is not enough to determine a unique depth of the well.
6. Modify the procedures to solve for the eigenvalue and eigenfunction for odd wave functions.
7. Solve for the wave function of a linear potential:
$$V(x) = -V_0 \begin{cases} |x|, & \text{for } |x| < a, \\ 0, & \text{for } |x| > a. \end{cases}$$
There is less potential here than for a square well, so you may expect smaller binding energies and a less confined wave function. (For this potential, there are no analytic results with which to compare.)
8. Compare the results obtained, and the time the computer took to get them, using both the Numerov and `rk4` methods.
9. **Newton–Raphson extension:** Extend the eigenvalue search by using the Newton–Raphson method in place of the bisection algorithm. Determine how much faster and more precise it is.

9.13 Unit III. Scattering, Projectiles, and Planetary Orbits

9.14 Problem 1: Classical Chaotic Scattering

Problem: One expects the classical scattering of a projectile from a barrier to be a continuous process. Yet it has been observed in experiments conducted on pinball machines (Figure 9.8 left) that for certain conditions, the projectile undergoes multiple internal scatterings and ends up with a final trajectory that is apparently unrelated to the initial one. Your **problem** is to determine if this process can be modeled as scattering from a static potential or if there must be active mechanisms built into the pinball machines that cause chaotic scattering.

Although this problem is easy to solve on the computer, the results have some chaotic features that are surprising (chaos is discussed further in Chapter 12, "Discrete & Continuous Nonlinear Dynamics"). In fact, the applet Disper2e.html *on the CD (created by Jaime Zuluaga) that simulates this problem continues to be a source of wonderment for readers as well as authors.*

9.14.1 Model and Theory

Our model for balls bouncing off the electrically activated bumpers in pinball machines is a point particle scattering from the stationary 2-D potential [Bleh 90]

$$V(x,y) = \pm x^2 y^2 e^{-(x^2+y^2)}. \tag{9.57}$$

This potential has four circularly symmetric peaks in the xy plane (Figure 9.8 right). The two signs correspond to repulsive and attractive potentials, respectively (the pinball machine contains only repulsive interactions). Because there are four peaks in this potential, we suspect that it may be possible to have multiple scatterings in which the projectile bounces back and forth among the peaks, somewhat as in a pinball machine.

The **theory** for this problem is classical dynamics. Visualize a scattering experiment in which a projectile starting out at an infinite distance from a target with a definite velocity **v** and an impact parameter b (Figure 9.8 right) is incident on a target. After interacting with the target and moving a nearly infinite distance from it, the scattered particle is observed at the scattering angle θ. Because the potential cannot recoil, the speed of the projectile does not change, but its direction does. An experiment typically measures the number of particles scattered and then converts this to a function, the differential cross section $\sigma(\theta)$, which is independent of the details of the experimental apparatus:

$$\sigma(\theta) = \lim \frac{N_{\rm scatt}(\theta)/\Delta\Omega}{N_{\rm in}/\Delta A_{\rm in}}. \tag{9.58}$$

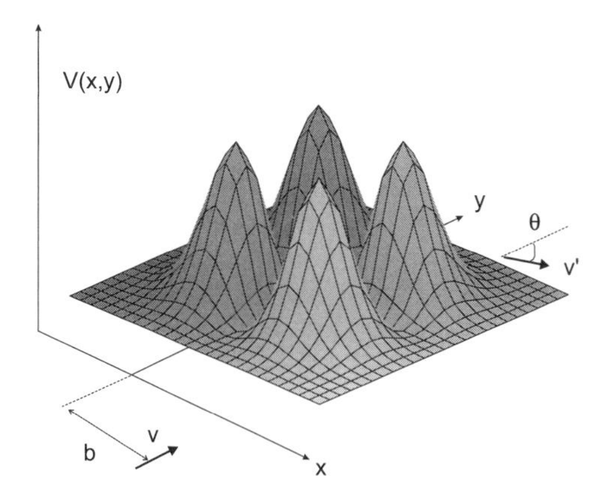

Figure 9.8 *Left:* A classic pinball machine in which multiple scatterings occur from the round structures on the top. *Right:* Scattering from the potential $V(x,y) = x^2 y^2 e^{-(x^2+y^2)}$. The incident velocity is v in the y direction, and the impact parameter (x value) is b. The velocity of the scattered particle is v', and its scattering angle is θ.

Here $N_{\rm scatt}(\theta)$ is the number of particles per unit time scattered into the detector at angle θ subtending a solid angle $\Delta\Omega$, $N_{\rm in}$ is the number of particle per unit time incident on the target of cross-sectional area $\Delta A_{\rm in}$, and the limit in (9.58) is for infinitesimal detector and area sizes.

The definition (9.58) for the cross section is the one that experimentalists use to convert their measurements to a function that can be calculated by theory. We as theorists solve for the trajectories of particles scattered from the potential (9.57) and from them deduce the scattering angle θ. Once we have the scattering angle, we predict the differential cross section from the dependence of the scattering angle upon the classical impact parameter b [M&T 03]:

$$\sigma(\theta) = \frac{b}{\left|\frac{d\theta}{db}\right| \sin\theta(b)}. \tag{9.59}$$

The surprise you should find in the simulation is that for certain parameters $d\theta/db$ has zeros and discontinuities, and this leads to a highly variable, large cross section.

The dynamical equations to solve are just Newton's law for the x and y motions with the potential (9.57):

$$\mathbf{F} = m\mathbf{a}$$

$$-\frac{\partial V}{\partial x}\hat{\imath} - \frac{\partial V}{\partial y}\hat{\jmath} = m\frac{d^2\mathbf{x}}{dt^2}, \tag{9.60}$$

$$\mp 2xye^{-(x^2+y^2)}\left[y(1-x^2)\hat{\imath} + x(1-y^2)\hat{\jmath}\right] = m\frac{d^2x}{dt^2}\hat{\imath} + m\frac{d^2y}{dt^2}\hat{\jmath}. \tag{9.61}$$

The equations for the x and y motions are simultaneous second-order ODEs:

$$m\frac{d^2x}{dt^2} = \mp 2y^2x(1-x^2)e^{-(x^2+y^2)}, \tag{9.62}$$

$$m\frac{d^2y}{dt^2} = \mp 2x^2y(1-y^2)e^{-(x^2+y^2)}. \tag{9.63}$$

Because the force vanishes at the peaks in Figure 9.8, these equations tell us that the peaks are at $x = \pm 1$ and $y = \pm 1$. Substituting these values into the potential (9.57) yields $V_{\text{max}} = \pm e^{-2}$, which sets the energy scale for the problem.

9.14.2 Implementation

In §9.16.1 we will also describe how to express simultaneous ODEs such as (9.62) and (9.63) in the standard `rk4` form. Even though both equations are independent, we solve them simultaneously to determine the scattering trajectory $[x(t), y(t)]$. We use the same `rk4` algorithm we used for a single second-order ODE, only now the arrays will be 4-D rather than 2-D:

$$\frac{d\mathbf{y}(t)}{dt} = \mathbf{f}(t, \mathbf{y}), \tag{9.64}$$

$$y^{(0)} \stackrel{\text{def}}{=} x(t), \quad y^{(1)} \stackrel{\text{def}}{=} y(t), \tag{9.65}$$

$$y^{(2)} \stackrel{\text{def}}{=} \frac{dx}{dt}, \quad y^{(3)} \stackrel{\text{def}}{=} \frac{dy}{dt}, \tag{9.66}$$

where the order in which the $y^{(i)}$s are assigned is arbitrary. With these definitions and equations (9.62) and (9.63), we can assign values for the force function:

$$f^{(0)} = y^{(2)}, \qquad f^{(1)} = y^{(3)}, \tag{9.67}$$

$$f^{(2)} = \frac{\mp 1}{m} 2y^2x(1-x^2)e^{-(x^2+y^2)} = \frac{\mp 1}{m} 2y^{(1)^2}y^{(0)}(1-y^{(0)^2})e^{-(y^{(0)^2}+y^{(1)^2})},$$

$$f^{(3)} = \frac{\mp 1}{m} 2x^2y(1-y^2)e^{-(x^2+y^2)} = \frac{\mp 1}{m} 2y^{(0)^2}y^{(1)}(1-y^{(1)^2})e^{-(y^{(0)^2}+y^{(1)^2})}.$$

To deduce the scattering angle from our simulation, we need to examine the trajectory of the scattered particle at an "infinite" separation from the target. To approximate that, we wait until the scattered particle no longer feels the potential (say $|\text{PE}|/\text{KE} \le 10^{-10}$) and call this infinity. The scattering angle is then deduced from the components of velocity,

$$\theta = \tan^{-1}\left(\frac{v_y}{v_x}\right) = \texttt{atan2(Vx, Vy)}. \tag{9.68}$$

Here atan2 is a function in most computer languages that computes the arctangent in the correct quadrant without requiring any explicit divisions (that can blow up).

9.14.3 Assessment

1. Apply the rk4 method to solve the simultaneous second-order ODEs (9.62) and (9.63) with a 4-D force function.
2. The **initial conditions** are (a) an incident particle with only a y component of velocity and (b) an impact parameter b (the initial x value). You do not need to vary the initial y, but it should be large enough such that $\mathrm{PE}/\mathrm{KE} \leq 10^{-10}$, which means that the $\mathrm{KE} \simeq E$.
3. Good parameters are $m = 0.5$, $v_y(0) = 0.5$, $v_x(0) = 0.0$, $\Delta b = 0.05$, $-1 \leq b \leq 1$. You may want to lower the energy and use a finer step size once you have found regions of rapid variation.
4. Plot a number of trajectories $[x(t), y(t)]$ that show usual and unusual behaviors. In particular, plot those for which backward scattering occurs, and consequently for which there is much multiple scattering.
5. Plot a number of phase space trajectories $[x(t), \dot{x}(t)]$ and $[y(t), \dot{y}(t)]$. How do these differ from those of bound states?
6. Determine the scattering angle $\theta =$ atan2(Vx,Vy) by determining the velocity of the scattered particle after it has left the interaction region, that is, $\mathrm{PE}/\mathrm{KE} \leq 10^{-10}$.
7. Identify which characteristics of a trajectory lead to discontinuities in $d\theta/db$ and thus $\sigma(\theta)$.
8. Run the simulations for both attractive and repulsive potentials and for a range of energies less than and greater than $V_{\mathrm{max}} = \exp(-2)$.
9. **Time delay:** Another way to find unusual behavior in scattering is to compute the *time delay* $T(b)$ as a function of the impact parameter b. The time delay is the increase in the time it takes a particle to travel through the interaction region after the interaction is turned on. Look for highly oscillatory regions in the semilog plot of $T(b)$, and once you find some, repeat the simulation at a finer scale by setting $b \simeq b/10$ (the structures are fractals, see Chapter 13, "Fractals & Statistical Growth").

9.15 Problem 2: Balls Falling Out of the Sky

Golf and baseball players claim that hit balls appear to fall straight down out of the sky at the end of their trajectories (the solid curve in Figure 9.9). Your **problem** is to determine whether there is a simple physics explanation for this effect or whether it is "all in the mind's eye." And while you are wondering why things fall out of the sky, see if you can use your new-found numerical tools to explain why planets do not fall out of the sky.

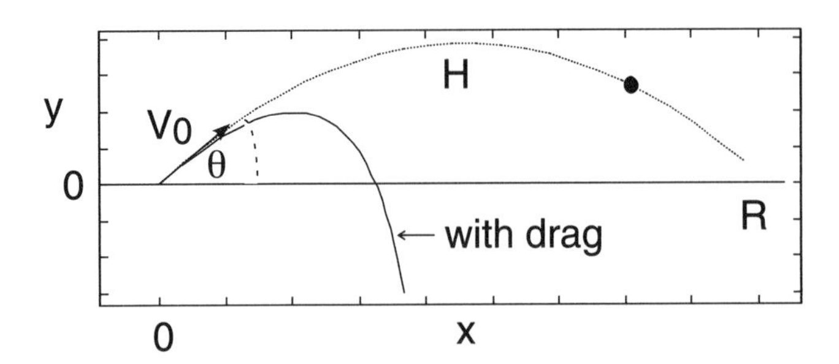

Figure 9.9 The trajectory of a projectile fired with initial velocity V_0 in the θ direction. The solid curve includes air resistance.

9.16 Theory: Projectile Motion with Drag

Figure 9.9 shows the initial velocity V_0 and inclination θ for a projectile launched from the origin. If we ignore air resistance, the projectile has only the force of gravity acting on it and therefore has a constant acceleration $g = 9.8\,\mathrm{m/s^2}$ in the negative y direction. The analytic solutions to the equations of motion are

$$x(t) = V_{0x}t, \qquad y(t) = V_{0y}t - \tfrac{1}{2}gt^2, \tag{9.69}$$

$$v_x(t) = V_{0x}, \qquad v_y(t) = V_{0y} - gt, \tag{9.70}$$

where $(V_{0x}, V_{0y}) = V_0(\cos\theta, \sin\theta)$. Solving for t as a function of x and substituting it into the $y(t)$ equation show that the trajectory is a parabola:

$$y = \frac{V_{0y}}{V_{0x}}x - \frac{g}{2V_{0x}^2}x^2. \tag{9.71}$$

Likewise, it is easy to show (dashed curve in Figure 9.9) that without friction the range $R = 2V_0^2 \sin\theta\cos\theta/g$ and the maximum height $H = \frac{1}{2}V_0^2\sin^2\theta/g$.

The parabola of frictionless motion is symmetric about its midpoint and so does not describe a ball dropping out of the sky. We want to determine if the inclusion of air resistance leads to trajectories that are much steeper at their ends than at their beginnings (solid curve in Figure 9.9). The basic physics is Newton's second law in two dimensions for a frictional force $\mathbf{F}^{(f)}$ opposing motion, and a vertical gravitational force $-mg\hat{\mathbf{e}}_y$:

$$\mathbf{F}^{(f)} - mg\hat{\mathbf{e}}_y = m\frac{d^2\mathbf{x}(t)}{dt^2}, \tag{9.72}$$

$$\Rightarrow \quad F_x^{(f)} = m\frac{d^2x}{dt^2}, F_y^{(f)} - mg = m\frac{d^2y}{dt^2}, \tag{9.73}$$

where the bold symbols indicate vector quantities.

The frictional force $\mathbf{F}^{(f)}$ is not a basic force of nature but rather a simple model of a complicated phenomenon. We do know that friction always opposes motion, which means it is in the direction opposite to velocity. One model assumes that the frictional force is proportional to a power n of the projectile's speed [M&T 03]:

$$\mathbf{F}^{(f)} = -k\, m\, |v|^n \, \frac{\mathbf{v}}{|v|}, \tag{9.74}$$

where the $-\mathbf{v}/|v|$ factor ensures that the frictional force is always in a direction opposite that of the velocity. Physical measurements indicate that the power n is noninteger and varies with velocity, and so a more accurate model would be a numerical one that uses the empirical velocity dependence $n(v)$. With a constant power law for friction, the equations of motion are

$$\frac{d^2x}{dt^2} = -k\, v_x^n \, \frac{v_x}{|v|}, \quad \frac{d^2y}{dt^2} = -g - k\, v_y^n \, \frac{v_y}{|v|}, \quad |v| = \sqrt{v_x^2 + v_y^2}. \tag{9.75}$$

We shall consider three values for n, each of which represents a different model for the air resistance: (1) $n = 1$ for low velocities; (2) $n = \frac{3}{2}$, for medium velocities; and (3) $n = 2$ for high velocities.

9.16.1 Simultaneous Second-Order ODEs

Even though (9.75) are simultaneous second-order ODEs, we can still use our regular ODE solver on them after expressing them in standard form

$$\frac{d\mathbf{y}}{dt} = \mathbf{y}(t, \mathbf{y}) \quad \text{(standard form)}. \tag{9.76}$$

We pick $\mathbf{y}$ to be the 4-D vector of dependent variables:

$$y^{(0)} = x(t), \quad y^{(1)} = \frac{dx}{dt}, \quad y^{(2)} = y(t), \quad y^{(3)} = \frac{dy}{dt}. \tag{9.77}$$

We express the equations of motion in terms of $\mathbf{y}$ to obtain the standard form:

$$\frac{dy^{(0)}}{dt} \left(\equiv \frac{dx}{dt}\right) = y^{(1)}, \quad \frac{dy^{(1)}}{dt} \left(\equiv \frac{d^2x}{dt^2}\right) = \frac{1}{m} F_x^{(f)}(\mathbf{y})$$

$$\frac{dy^{(2)}}{dt} \left(\equiv \frac{dy}{dt}\right) = y^{(3)}, \quad \frac{dy^{(3)}}{dt} \left(\equiv \frac{d^2y}{dt^2}\right) = \frac{1}{m} F_y^{(f)}(\mathbf{y}) - g.$$

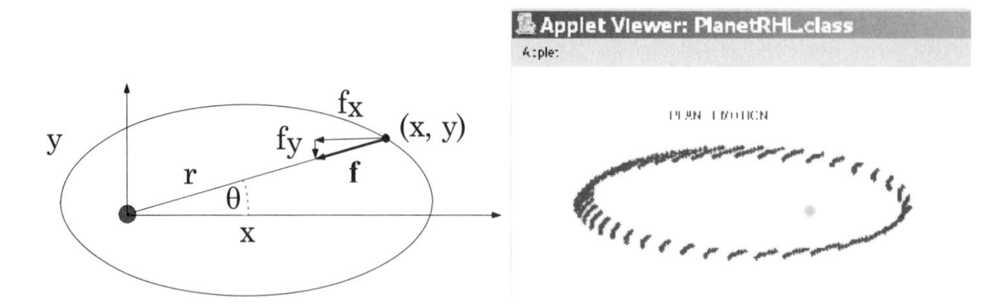

Figure 9.10 *Left:* The gravitational force on a planet a distance r from the sun. The x and y components of the force are indicated. *Right:* Output from the applet **Planet** (on the CD) showing the precession of a planet's orbit when the gravitational force $\propto 1/r^4$.

And now we just read off the components of the force function $\mathbf{f}(t, \mathbf{y})$:

$$f^{(0)} = y^{(1)}, \quad f^{(1)} = \frac{1}{m} F_x^{(f)}, \quad f^{(2)} = y^{(3)}, \quad f^{(3)} = \frac{1}{m} F_y^{(f)} - g.$$

9.16.2 Assessment

1. Modify your `rk4` program so that it solves the simultaneous ODEs for projectile motion (9.75) with friction ($n = 1$).
2. Check that you obtain graphs similar to those in Figure 9.9.
3. The model (9.74) with $n = 1$ is okay for low velocities. Now modify your program to handle $n = \frac{3}{2}$ (medium-velocity friction) and $n = 2$ (high-velocity friction). Adjust the value of k for the latter two cases such that the initial force of friction $k\, v_0^n$ is the same for all three cases.
4. What is your conclusion about balls falling out of the sky?

9.17 Problem 3: Planetary Motion

Newton's explanation of the motion of the planets in terms of a universal law of gravitation is one of the great achievements of science. He was able to prove that the planets traveled along elliptical paths with the sun at one vertex and to predict periods of the motion accurately. All Newton needed to postulate was that the force between a planet of mass m and the sun of mass M is

$$F^{(g)} = -\frac{GmM}{r^2}. \tag{9.78}$$

Here r is the planet-CM distance, G is the universal gravitational constant, and the attractive force lies along the line connecting the planet and the sun (Figure 9.10 left). The hard part for Newton was solving the resulting differential equations because he had to invent calculus to do it and then had go through numerous analytic manipulations. The numerical solution is straightforward since even for planets the equation of motion is still

$$\mathbf{f} = m\mathbf{a} = m\frac{d^2\mathbf{x}}{dt^2}, \tag{9.79}$$

with the force (9.78) having components (Figure 9.10)

$$f_x = F^{(g)} \cos\theta = F^{(g)}\frac{x}{r}, \tag{9.80}$$

$$f_y = F^{(g)} \sin\theta = F^{(g)}\frac{y}{r}, \tag{9.81}$$

$$r = \sqrt{x^2 + y^2}. \tag{9.82}$$

The equation of motion yields two simultaneous second-order ODEs:

$$\frac{d^2x}{dt^2} = -GM\frac{x}{r^3}, \quad \frac{d^2y}{dt^2} = -GM\frac{y}{r^3}. \tag{9.83}$$

9.17.1 Implementation: Planetary Motion

1. Assume units such that $GM = 1$ and use the initial conditions

$$x(0) = 0.5, \quad y(0) = 0, \quad v_x(0) = 0.0, \quad v_y(0) = 1.63.$$

2. Modify your ODE solver program to solve (9.83).
3. You may need to make the time step small enough so that the elliptical orbit closes upon itself, as it should, and the number of steps large enough such that the orbits just repeat.
4. Experiment with the initial conditions until you find the ones that produce a circular orbit (a special case of an ellipse).
5. Once you have obtained good precision, note the effect of progressively increasing the initial velocity until the orbits open up and become hyperbolic.
6. Using the same initial conditions that produced the ellipse, investigate the effect of the power in (9.78) being $1/r^4$ rather than $1/r^2$. You should find that the orbital ellipse now rotates or precesses (Figure 9.10). In fact, as you should verify, even a slight variation from an inverse square power law (as arises from general relativity) causes the orbit to precess. ▌

9.17.1.1 EXPLORATION: RESTRICTED THREE-BODY PROBLEM

Extend the previous solution for planetary motion to one in which a satellite of tiny mass moves under the influence of two planets of equal mass $M = 1$. Consider the planets as rotating about their center of mass in circular orbits and of such large mass that they are uninfluenced by the satellite. Assume that all motions remain in the xy plane and that the units are such that $G = 1$. ▮

10

Fourier Analysis: Signals and Filters

In Unit I of this chapter we examine Fourier series and Fourier transforms, two standard tools for decomposing periodic and nonperiodic motions. We find that as implemented for numerical computation, both the series and the integral become a discrete Fourier transform (DFT), which is simple to program. In Unit II we discuss the subject of signal filtering and see that various Fourier tools can be used to reduce noise in measured or simulated signals. In Unit III we present a discussion of the fast Fourier transform (FFT), a technique that is so efficient that it permits evaluations of DFTs in real time.

10.1 Unit I. Fourier Analysis of Nonlinear Oscillations

Consider a particle oscillating in the nonharmonic potential of equation (9.4):

$$V(x) = \frac{1}{p} k|x|^p, \tag{10.1}$$

for $p \neq 2$, or for the perturbed harmonic oscillator (9.2),

$$V(x) = \frac{1}{2} k x^2 \left(1 - \frac{2}{3}\alpha x\right). \tag{10.2}$$

While free oscillations in these potentials are always periodic, they are not truly sinusoidal. Your **problem** is to take the solution of one of these nonlinear oscillators and relate it to the solution

$$x(t) = A_0 \sin(\omega t + \phi_0) \tag{10.3}$$

of the linear harmonic oscillator. If your oscillator is sufficiently nonlinear to behave like the sawtooth function (Figure 10.1 left), then the Fourier spectrum you obtain should be similar to that shown on the right in Figure 10.1.

In general, when we undertake such a spectral analysis, we want to analyze the steady-state behavior of a system. This means that the initial transient behavior has had a chance to die out. It is easy to identify just what the initial transient is for linear systems but may be less so for nonlinear systems in which the "steady state" jumps among a number of configurations.

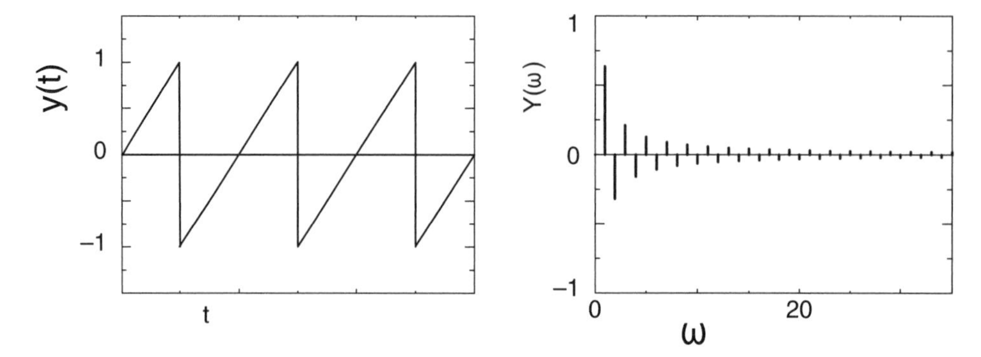

Figure 10.1 *Left:* A sawtooth function in time. *Right:* The Fourier spectrum of frequencies in natural units contained in this sawtooth function.

10.2 Fourier Series (Math)

Part of our interest in nonlinear oscillations arises from their lack of study in traditional physics courses even though linear oscillations are just the first approximation to a naturally oscillatory system. If the force on a particle is always toward its equilibrium position (a restoring force), then the resulting motion will be *periodic* but not necessarily *harmonic*. A good example is the motion in a highly anharmonic well $p \simeq 10$ that produces an $x(t)$ looking like a series of pyramids; this motion is periodic but not harmonic.

In numerical analysis there really is no distinction between a Fourier integral and a Fourier series because the integral is always approximated as a finite series. We will illustrate both methods. In a sense, our approach is the inverse of the traditional one in which the *fundamental* oscillation is determined analytically and the higher-frequency *overtones* are determined by perturbation theory [L&L,M 76]. We start with the full (numerical) periodic solution and then decompose it into what may be called *harmonics*. When we speak of fundamentals, overtones, and harmonics, we speak of solutions to the *linear boundary-value problem*, for example, of waves on a plucked violin string. In this latter case, and when given the correct conditions (enough musical skill), it is possible to excite individual harmonics or sums of them in the series

$$y(t) = b_0 \sin \omega_0 t + b_1 \sin 2\omega_0 t + \cdots . \tag{10.4}$$

Anharmonic oscillators vibrate at a single frequency (which may vary with amplitude) but not with a sinusoidal waveform. Expanding the oscillations in a Fourier series does not imply that the individual harmonics can be excited (played).

You may recall from classical mechanics that the general solution for a vibrating system can be expressed as the sum of the *normal modes* of that system. These expansions are possible because we have *linear operators* and, subsequently, the *principle of superposition*: If $y_1(t)$ and $y_2(t)$ are solutions of some linear equation, then $\alpha_1 y_1(t) + \alpha_2 y_2(t)$ is also a solution. The principle of linear superposition does not hold when we solve nonlinear problems. Nevertheless, it is always possible to expand a *periodic* solution of a *nonlinear* problem in terms of trigonometric functions with frequencies that are integer multiples of the true frequency of the nonlinear oscillator.[1] This is a consequence of *Fourier's theorem* being applicable to any single-valued periodic function with only a finite number of discontinuities. We assume we know the period T, that is, that

$$y(t+T) = y(t). \tag{10.5}$$

This tells us the *true frequency* ω:

$$\omega \equiv \omega_1 = \frac{2\pi}{T}. \tag{10.6}$$

A periodic function (usually called the *signal*) can be expanded as a series of harmonic functions with frequencies that are multiples of the true frequency:

$$y(t) = \frac{a_0}{2} + \sum_{n=1}^{\infty} (a_n \cos n\omega t + b_n \sin n\omega t). \tag{10.7}$$

This equation represents the signal $y(t)$ as the simultaneous sum of pure tones of frequency $n\omega$. The coefficients a_n and b_n are measures of the amount of $\cos n\omega t$ and $\sin n\omega t$ present in $y(t)$, specifically, the intensity or power at each frequency is proportional to $a_n^2 + b_n^2$.

The Fourier series (10.7) is a best fit in the least-squares sense of Chapter 8, "Solving Systems of Equations with Matrices; Data Fitting," because it minimizes $\sum_i [y(t_i) - y_i]^2$, where i denotes different measurements of the signal. This means that the series converges to the average behavior of the function but misses the function at discontinuities (at which points it converges to the mean) or at sharp corners (where it overshoots). A general function $y(t)$ may contain an infinite number of Fourier components, although low-accuracy reproduction is usually possible with a small number of harmonics.

The coefficients a_n and b_n are determined by the standard techniques for function expansion. To find them, multiply both sides of (10.7) by $\cos n\omega t$ or $\sin n\omega t$, integrate over one period, and project a single a_n or b_n:

$$\begin{pmatrix} a_n \\ b_n \end{pmatrix} = \frac{2}{T} \int_0^T dt \begin{pmatrix} \cos n\omega t \\ \sin n\omega t \end{pmatrix} y(t), \qquad \omega \stackrel{\text{def}}{=} \frac{2\pi}{T}. \tag{10.8}$$

[1] We remind the reader that every periodic system by definition has a period T and consequently a true frequency ω. Nonetheless, this does not imply that the system behaves like $\sin \omega t$. Only harmonic oscillators do that.

As seen in the b_n coefficients (Figure 10.1 right), these coefficients usually decrease in magnitude as the frequency increases and can occur with a negative sign, the negative sign indicating relative phase.

Awareness of the *symmetry* of the function $y(t)$ may eliminate the need to evaluate all the expansion coefficients. For example,

- a_0 is twice the average value of y:

$$a_0 = 2\langle y(t)\rangle . \tag{10.9}$$

- For an *odd function*, that is, one for which $y(-t) = -y(t)$, all of the coefficients $a_n \equiv 0$ and only half of the integration range is needed to determine b_n:

$$b_n = \frac{4}{T}\int_0^{T/2} dt\, y(t)\sin n\omega t. \tag{10.10}$$

  However, if there is no input signal for $t < 0$, we do not have a truly odd function, and so small values of a_n may occur.
- For an *even function*, that is, one for which $y(-t) = y(t)$, the coefficient $b_n \equiv 0$ and only half the integration range is needed to determine a_n:

$$a_n = \frac{4}{T}\int_0^{T/2} dt\, y(t)\cos n\omega t. \tag{10.11}$$

10.2.1 Example 1: Sawtooth Function

The sawtooth function (Figure 10.1) is described mathematically as

$$y(t) = \begin{cases} \frac{t}{T/2}, & \text{for } 0 \le t \le \frac{T}{2}, \\ \frac{t-T}{T/2}, & \text{for } \frac{T}{2} \le t \le T. \end{cases} \tag{10.12}$$

It is clearly periodic, nonharmonic, and discontinuous. Yet it is also odd and so can be represented more simply by shifting the signal to the left:

$$y(t) = \frac{t}{T/2}, \quad -\frac{T}{2} \le t \le \frac{T}{2}. \tag{10.13}$$

Even though the general shape of this function can be reproduced with only a few terms of the Fourier components, many components are needed to reproduce the sharp corners. Because the function is odd, the Fourier series is a sine series and

(10.8) determines the values

$$b_n = \frac{2}{T}\int_{-T/2}^{+T/2} dt\,\sin n\omega t \frac{t}{T/2} = \frac{\omega}{\pi}\int_{-\pi/\omega}^{+\pi/\omega} dt\,\sin n\omega t \frac{\omega t}{\pi} = \frac{2}{n\pi}(-1)^{n+1},$$

$$\Rightarrow \quad y(t) = \frac{2}{\pi}\left[\sin\omega t - \frac{1}{2}\sin 2\omega t + \frac{1}{3}\sin 3\omega t - \cdots\right]. \tag{10.14}$$

10.2.2 Example 2: Half-wave Function

The half-wave function

$$y(t) = \begin{cases} \sin\omega t, & \text{for } 0 < t < T/2, \\ 0, & \text{for } T/2 < t < T, \end{cases} \tag{10.15}$$

is periodic, nonharmonic (the upper half of a sine wave), and continuous, but with discontinuous derivatives. Because it lacks the sharp corners of the sawtooth function, it is easier to reproduce with a finite Fourier series. Equation (10.8) determines

$$a_n = \begin{cases} \frac{-2}{\pi(n^2-1)}, & n \text{ even or } 0, \\ 0, & n \text{ odd}, \end{cases} \qquad b_n = \begin{cases} \frac{1}{2}, & n = 1, \\ 0, & n \neq 1, \end{cases}$$

$$\Rightarrow \quad y(t) = \frac{1}{2}\sin\omega t + \frac{1}{\pi} - \frac{2}{3\pi}\cos 2\omega t - \frac{2}{15\pi}\cos 4\omega t + \cdots. \tag{10.16}$$

10.3 Exercise: Summation of Fourier Series

1. **Sawtooth function:** Sum the Fourier series for the *sawtooth function* up to order $n = 2, 4, 10, 20$ and plot the results over two periods.
   a. Check that in each case the series gives the mean value of the function *at* the points of discontinuity.
   b. Check that in each case the series *overshoots* by about 9% the value of the function on either side of the discontinuity (the *Gibbs phenomenon*).
2. **Half-wave function:** Sum the Fourier series for the *half-wave function* up to order $n = 2, 4, 10$ and plot the results over two periods. (The series converges quite well, doesn't it?)

10.4 Fourier Transforms (Theory)

Although a Fourier *series* is the right tool for approximating or analyzing periodic functions, the Fourier *transform* or *integral* is the right tool for nonperiodic functions. We convert from series to transform by imagining a system described by a continuum of "fundamental" frequencies. We thereby deal with *wave packets* containing continuous rather than discrete frequencies.[2] While the difference between series and transform methods may appear clear mathematically, when we approximate the Fourier integral as a finite sum, the two become equivalent.

By analogy with (10.7), we now imagine our function or signal $y(t)$ expressed in terms of a continuous series of harmonics:

$$y(t) = \int_{-\infty}^{+\infty} d\omega\, Y(\omega)\, \frac{e^{i\omega t}}{\sqrt{2\pi}}, \tag{10.17}$$

where for compactness we use a complex exponential function.[3] The expansion amplitude $Y(\omega)$ is analogous to the Fourier coefficients (a_n, b_n) and is called the *Fourier transform* of $y(t)$. The integral (10.17) is the inverse transform since it converts the transform to the signal. The *Fourier transform* converts $y(t)$ to its transform $Y(\omega)$:

$$Y(\omega) = \int_{-\infty}^{+\infty} dt\, \frac{e^{-i\omega t}}{\sqrt{2\pi}} y(t). \tag{10.18}$$

The $1/\sqrt{2\pi}$ factor in both these integrals is a common normalization in quantum mechanics but maybe not in engineering where only a single $1/2\pi$ factor is used. Likewise, the signs in the exponents are also conventions that do not matter as long as you maintain consistency.

If $y(t)$ is the measured response of a system (signal) as a function of time, then $Y(\omega)$ is the *spectral function* that measures the amount of frequency ω present in the signal. [However, some experiments may measure $Y(\omega)$ directly, in which case an inverse transform is needed to obtain $y(t)$.] In many cases $Y(\omega)$ is a complex function with positive and negative values and with significant variation in magnitude. Accordingly, it is customary to eliminate some of the complexity of $Y(\omega)$ by making a semilog plot of the squared modulus $|Y(\omega)|^2$ *versus* ω. This is called a

[2] We follow convention and consider time t the function's variable and frequency ω the transform's variable. Nonetheless, these can be reversed or other variables such as position x and wave vector k may also be used.

[3] Recollect the principle of linear superposition and that $\exp(i\omega t) = \cos\omega t + i\sin\omega t$. This means that the real part of y gives the cosine series, and the imaginary part the sine series.

power spectrum and provides you with an immediate view of the amount of power or strength in each component.

If the Fourier transform and its inverse are consistent with each other, we should be able to substitute (10.17) into (10.18) and obtain an identity:

$$Y(\omega)=\int_{-\infty}^{+\infty} dt\,\frac{e^{-i\omega t}}{\sqrt{2\pi}}\int_{-\infty}^{+\infty} d\omega'\,\frac{e^{i\omega' t}}{\sqrt{2\pi}}Y(\omega')=\int_{-\infty}^{+\infty} d\omega'\left\{\int_{-\infty}^{+\infty} dt\,\frac{e^{i(\omega'-\omega)t}}{2\pi}\right\}Y(\omega').$$

For this to be an identity, the term in braces must be the *Dirac delta function*:

$$\int_{-\infty}^{+\infty} dt\, e^{i(\omega'-\omega)t} = 2\pi\delta(\omega'-\omega). \tag{10.19}$$

While the delta function is one of the most common and useful functions in theoretical physics, it is not well behaved in a mathematical sense and misbehaves terribly in a computational sense. While it is possible to create numerical approximations to $\delta(\omega'-\omega)$, they may well be borderline pathological. It is certainly better for you to do the delta function part of an integration analytically and leave the nonsingular leftovers to the computer.

10.4.1 Discrete Fourier Transform Algorithm

If $y(t)$ or $Y(\omega)$ is known analytically or numerically, the integral (10.17) or (10.18) can be evaluated using the integration techniques studied earlier. In practice, the signal $y(t)$ is measured at just a finite number N of times t, and these are what we must use to approximate the transform. The resultant *discrete Fourier transform* is an approximation both because the signal is not known for all times and because we integrate numerically.[4] Once we have a discrete set of transforms, they can be used to reconstruct the signal for any value of the time. In this way the DFT can be thought of as a technique for interpolating, compressing and extrapolating data.

We assume that the signal $y(t)$ is sampled at $(N+1)$ discrete times (N time intervals), with a constant spacing h between times:

$$y_k \stackrel{\text{def}}{=} y(t_k), \qquad k=0,1,2,\ldots,N, \tag{10.20}$$

$$t_k \stackrel{\text{def}}{=} kh, \qquad h=\Delta t. \tag{10.21}$$

In other words, we measure the signal $y(t)$ once every hth of a second for a total time T. This corresponds to period T and *sampling rate* s:

$$T \stackrel{\text{def}}{=} Nh, \quad s=\frac{N}{T}=\frac{1}{h}. \tag{10.22}$$

[4] More discussion can be found in [B&H 95], which is devoted to just this topic.

Regardless of the actual periodicity of the signal, when we choose a period T over which to sample, the mathematics produces a $y(t)$ that is periodic with period T,

$$y(t+T) = y(t). \tag{10.23}$$

We recognize this periodicity, and ensure that there are only N independent measurements used in the transform, by requiring the first and last y's to be the same:

$$y_0 = y_N. \tag{10.24}$$

If we are analyzing a truly periodic function, then the first N points should all be within one period to guarantee their independence. Unless we make further assumptions, these N independent input data $y(t_k)$ can determine no more than N independent output Fourier components $Y(\omega_k)$.

The time interval T (which should be the period for periodic functions) is the largest time over which we consider the variation of $y(t)$. Consequently, it determines the lowest frequency,

$$\omega_1 = \frac{2\pi}{T}, \tag{10.25}$$

contained in our Fourier representation of $y(t)$. The frequencies ω_n are determined by the number of samples taken and by the total sampling time $T = Nh$ as

$$\omega_n = n\omega_1 = n\frac{2\pi}{Nh}, \qquad n = 0, 1, \ldots, N. \tag{10.26}$$

Here $\omega_0 = 0$ corresponds to the zero-frequency or DC component.

The DFT algorithm results from (1) evaluating the integral in (10.18) not from $-\infty$ to $+\infty$ but rather from time 0 to time T over which the signal is measured, and from (2) using the trapezoid rule for the integration[5]

$$Y(\omega_n) \stackrel{\text{def}}{=} \int_{-\infty}^{+\infty} dt\, \frac{e^{-i\omega_n t}}{\sqrt{2\pi}} y(t) \simeq \int_0^T dt\, \frac{e^{-i\omega_n t}}{\sqrt{2\pi}} y(t), \tag{10.27}$$

$$\simeq \sum_{k=1}^{N} h\, y(t_k) \frac{e^{-i\omega_n t_k}}{\sqrt{2\pi}} = h \sum_{k=1}^{N} y_k \frac{e^{-2\pi i k n/N}}{\sqrt{2\pi}}. \tag{10.28}$$

[5] The alert reader may be wondering what has happened to the $h/2$ with which the trapezoid rule weights the initial and final points. Actually, they are there, but because we have set $y_0 \equiv y_N$, two $h/2$ terms have been added to produce one h term.

To keep the final notation more symmetric, the step size h is factored from the transform Y and a discrete function Y_n is defined:

$$Y_n \stackrel{\text{def}}{=} \frac{1}{h} Y(\omega_n) = \sum_{k=1}^{N} y_k \frac{e^{-2\pi ikn/N}}{\sqrt{2\pi}}. \tag{10.29}$$

With this same care in accounting, and with $d\omega \to 2\pi/Nh$, we invert the Y_n's:

$$y(t) \stackrel{\text{def}}{=} \int_{-\infty}^{+\infty} d\omega \frac{e^{i\omega t}}{\sqrt{2\pi}} Y(\omega) \simeq \sum_{n=1}^{N} \frac{2\pi}{Nh} \frac{e^{i\omega_n t}}{\sqrt{2\pi}} Y(\omega_n). \tag{10.30}$$

Once we know the N values of the transform, we can use (10.30) to evaluate $y(t)$ for any time t. There is nothing illegal about evaluating Y_n and y_k for arbitrarily large values of n and k, yet there is also nothing to be gained. Because the trigonometric functions are periodic, we just get the old answer:

$$y(t_{k+N}) = y(t_k), \quad Y(\omega_{n+N}) = Y(\omega_n). \tag{10.31}$$

Another way of stating this is to observe that none of the equations change if we replace $\omega_n t$ by $\omega_n t + 2\pi n$. There are still just N independent output numbers for N independent inputs.

We see from (10.26) that the larger we make the time $T = Nh$ over which we sample the function, the smaller will be the frequency steps or resolution.[6] Accordingly, if you want a smooth frequency spectrum, you need to have a small frequency step $2\pi/T$. This means you need a large value for the total observation time T. While the best approach would be to measure the input signal for longer times, in practice a measured signal $y(t)$ is often extended in time ("padded") by adding zeros for times beyond the last measured signal, which thereby increases the value of T. Although this does not add new information to the analysis, it does build in the experimentalist's belief that the signal has no existence at times after the measurements are stopped.

While periodicity is expected for Fourier *series*, it is somewhat surprising for Fourier *integrals*, which have been touted as the right tool for nonperiodic functions. Clearly, if we input values of the signal for longer lengths of time, then the inherent period becomes longer, and if the repeat period is very long, it may be of little consequence for times short compared to the period. If $y(t)$ is actually periodic with period Nh, then the DFT is an excellent way of obtaining Fourier series. If the input function is not periodic, then the DFT can be a bad approximation near the endpoints of the time interval (after which the function will repeat) or, correspondingly, for the lowest frequencies.

The discrete Fourier transform and its inverse can be written in a concise and insightful way, and be evaluated efficiently, by introducing a complex variable Z

[6] See also §10.4.2 where we discuss the related phenomenon of aliasing.

for the exponential and then raising Z to various powers:

$$\boxed{y_k = \frac{\sqrt{2\pi}}{N}\sum_{n=1}^{N} Z^{-nk} Y_n,} \qquad Z = e^{-2\pi i/N}, \tag{10.32}$$

$$\boxed{Y_n = \frac{1}{\sqrt{2\pi}}\sum_{k=1}^{N} Z^{nk} y_k,} \qquad n = 0, 1, \ldots, N, \tag{10.33}$$

where $Z^{nk} \equiv [(Z)^n]^k$. With this formulation, the computer needs to compute only powers of Z. We give our DFT code in Listing 10.1. If your preference is to avoid complex numbers, we can rewrite (10.32) in terms of separate real and imaginary parts by applying Euler's theorem:

$$Z = e^{-i\theta}, \quad \Rightarrow \quad Z^{\pm nk} = e^{\mp ink\theta} = \cos nk\theta \mp i \sin nk\theta, \tag{10.34}$$

where $\theta \stackrel{\text{def}}{=} 2\pi/N$. In terms of the explicit real and imaginary parts,

$$\begin{aligned} Y_n = \frac{1}{\sqrt{2\pi}} \sum_{k=1}^{N} & [(\cos(nk\theta)\mathrm{Re}\, y_k + \sin(nk\theta)\, \mathrm{Im}\, y_k \\ & + \mathbf{i}(\cos(nk\theta)\, \mathrm{Im}\, y_k - \sin(nk\theta)\mathrm{Re}\, y_k)], \end{aligned} \tag{10.35}$$

$$\begin{aligned} y_k = \frac{\sqrt{2\pi}}{N} \sum_{n=1}^{N} & [(\cos(nk\theta)\, \mathrm{Re}\, Y_n - \sin(nk\theta)\mathrm{Im}\, Y_n \\ & + \mathbf{i}(\cos(nk\theta)\mathrm{Im}\, Y_n + \sin(nk\theta)\, \mathrm{Re}\, Y_n\,)]. \end{aligned} \tag{10.36}$$

Readers new to DFTs are often surprised when they apply these equations to practical situations and end up with transforms Y having imaginary parts, even though the signal y is real. Equation (10.35) shows that a real signal ($\mathrm{Im}\, y_k \equiv 0$) will yield an imaginary transform unless $\sum_{k=1}^{N} \sin(nk\theta)\, \mathrm{Re}\, y_k = 0$. This occurs only if $y(t)$ is an *even* function over $-\infty \le t \le +\infty$ *and* we integrate exactly. Because neither condition holds, the DFTs of real, even functions may have small imaginary parts. This is not due to an error in programming and in fact is a good measure of the approximation error in the entire procedure.

The computation time for a discrete Fourier transform can be reduced even further by use of the *fast Fourier transform* algorithm. An examination of (10.32) shows that the DFT is evaluated as a matrix multiplication of a vector of length N

containing the Z values by a vector of length N of y value. The time for this DFT scales like N^2, while the time for the FFT algorithm scales as $N \log_2 N$. Although this may not seem like much of a difference, for $N = 10^{2-3}$, the difference of 10^{3-5} is the difference between a minute and a week. For this reason, FFT is often used for on-line analysis of data. We discuss FFT techniques in §10.8.

```
//DFT.java: Discrete Fourier Transform
import java.io.*;

public class DFT {
  static final int N = 1000, Np = N;                          // Global constants
  static double [] signal = new double[N + 1];
  static double twopi = 2.*Math.PI, sq2pi = 1./Math.sqrt(twopi);
  static double h = twopi/N;

  public static void main(String[] argv) {
    double dftreal[] = new double[Np], dftimag[] = new double[Np];
    f(signal);
    fourier(dftreal, dftimag);
  }

  public static void fourier(double dftreal[], double dftimag[]) {
    double real, imag;                                         // Calc & plot Y(w)
    int n, k;
    for ( n = 0; n < Np; n++ ) {                               // Loop on frequency
      real = imag = 0. ;                                        // Clear variables
      for ( k = 0; k < N; k++ ){                                 // Loop for sums
        real += signal[k]*Math.cos((twopi*k*n)/N);
        imag += signal[k]*Math.sin((twopi*k*n)/N);
      }
      dftreal[n] = real*sq2pi;
      dftimag[n] = -imag*sq2pi;
    }
  }

  public static void f(double [] signal) {                     // Initial function
    int i;
    double step = twopi/N, x = 0.;
    for ( i=0; i <= N; i++ ) { signal[i] = 5.+10*Math.sin(x+2.); x += step; }
  }
}
```

Listing 10.1 DFT.java computes the discrete Fourier transform for the signal given in the method `f(signal[ ])`. You will have to add output and plotting to see the results. (The instructor's version also does an inverse transform and plots the results with PtPlot.)

10.4.2 Aliasing and Anti-aliasing (Assessment) ⊙

The sampling of a signal by DFT for only a finite number of times limits the accuracy of the deduced high-frequency components present in the signal. Clearly, good information about very high frequencies requires sampling the signal with small time steps so that all the wiggles can be included. While a poor deduction of the high-frequency components may be tolerable if all we care about are the low-frequency ones, the high-frequency components remain present in the signal

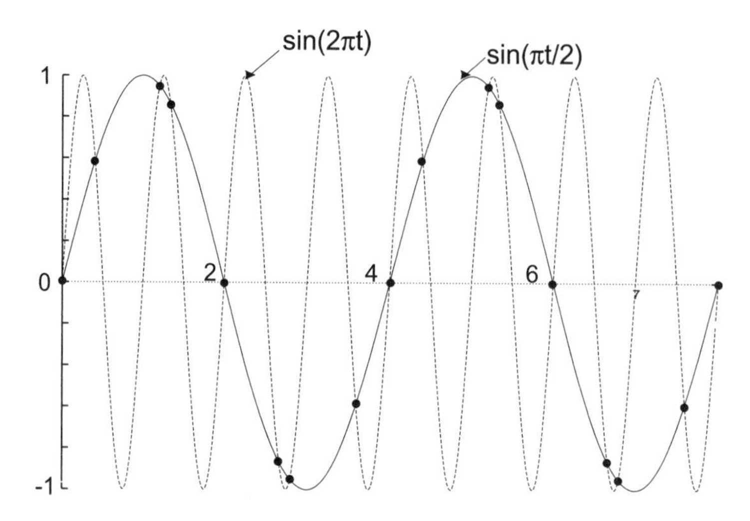

Figure 10.2 A plot of the functions $\sin(\pi t/2)$ and $\sin(2\pi t)$. If the sampling rate is not high enough, these signals will appear indistinguishable. If both are present in a signal (e.g., as a signal that is the sum of the two) and if the signal is not sampled at a high enough rate, the deduced low-frequency component will be contaminated by the higher-frequency component.

and may contaminate the low-frequency components that we deduce. This effect is called *aliasing* and is the cause of the moiré pattern distortion in digital images.

As an example, consider Figure 10.2 showing the two functions $\sin(\pi t/2)$ and $\sin(2\pi t)$ for $0 \leq t \leq 8$, with their points of overlap in bold. If we were unfortunate enough to sample a signal containing these functions at the times $t = 0, 2, 4, 6, 8$, then we would measure $y \equiv 0$ and assume that there was no signal at all. However, if we were unfortunate enough to measure the signal at the filled dots in Figure 10.2 where $\sin(\pi t/2) = \sin(2\pi t)$, specifically, $t = 0, \frac{12}{10}, \frac{4}{3}, \ldots$, then our Fourier analysis would completely miss the high-frequency component. In DFT jargon, we would say that the high-frequency component has been *aliased* by the low-frequency component. In other cases, some high-frequency values may be included in our sampling of the signal, but our sampling rate may not be high enough to include enough of them to separate the high-frequency component properly. In this case some high-frequency signals would be included spuriously as part of the low-frequency spectrum, and this would lead to spurious low-frequency oscillations when the signal is synthesized from its Fourier components.

More precisely, aliasing occurs when a signal containing frequency f is sampled at a rate of $s = N/T$ measurements per unit time, with $s \leq f/2$. In this case, the frequencies f and $f - 2s$ yield the same DFT, and we would not be able to determine that there are two frequencies present. That being the case, to avoid aliasing we want no frequencies $f > s/2$ to be present in our input signal. This is known as the *Nyquist criterion*. In practice, some applications avoid the effects of aliasing by filtering out the high frequencies from the signal and then analyzing the remaining low-frequency part. (The low-frequency *sinc filter* discussed in §10.7.1 is often used for this.) Even though this approach eliminates some high-frequency information, it lessens the distortion of the low-frequency components and so may lead to improved reproduction of the signal.

If accurate values for the high frequencies are required, then we will need to increase the sampling rate s by increasing the number N of samples taken within the fixed sampling time $T = Nh$. By keeping the sampling time constant and increasing the number of samples taken, we make the time step h smaller, and this picks up the higher frequencies. By increasing the number N of frequencies that you compute, you move the higher-frequency components you are interested in closer to the middle of the spectrum and thus away from the error-prone ends.

If we vary the the total time sampling time $T = Nh$ but not the sampling rate $s = N/T = 1/h$, we make ω_1 smaller because the discrete frequencies

$$\omega_n = n\omega_1 = n\frac{2\pi}{T} \tag{10.37}$$

are measured in steps of ω_1. This leads to a smoother frequency spectrum. However, to keep the time step the same and thus not lose high-frequency information, we would also need to increase the number of N samples. And as we said, this is often done, after the fact, by padding the end of the data set with zeros.

10.4.3 DFT for Fourier Series (Algorithm)

For simplicity let us consider the Fourier cosine series:

$$y(t) = \sum_{n=0}^{\infty} a_n \cos(n\omega t), \quad a_k = \frac{2}{T}\int_0^T dt \cos(k\omega t) y(t). \tag{10.38}$$

Here $T \stackrel{\text{def}}{=} 2\pi/\omega$ is the actual period of the system (not necessarily the period of the simple harmonic motion occurring for a small amplitude). We assume that the function $y(t)$ is sampled for a discrete set of times

$$y(t = t_k) \equiv y_k, \quad k = 0, 1, \ldots, N. \tag{10.39}$$

Because we are analyzing a periodic function, we retain the conventions used in the DFT and require the function to repeat itself with period $T = Nh$; that is, we assume that the amplitude is the same at the first and last points:

$$y_0 = y_N. \tag{10.40}$$

This means that there are only N independent values of y being used as input. For these N independent y_k values, we can determine uniquely only N expansion coefficients a_k. If we use the trapezoid rule to approximate the integration in (10.38), we determine the N independent Fourier components as

$$a_n \simeq \frac{2h}{T}\sum_{k=1}^{N}\cos\left(n\omega t_k\right)y(t_k) = \frac{2}{N}\sum_{k=1}^{N}\cos\left(\frac{2\pi nk}{N}\right)y_k, \quad n=0,\ldots,N. \tag{10.41}$$

Because we can determine only N Fourier components from N independent $y(t)$ values, our Fourier series for the $y(t)$ must be in terms of only these components:

$$y(t) \simeq \sum_{n=0}^{N} a_n \cos(n\omega t) = \sum_{n=0}^{N} a_n \cos\left(\frac{2\pi nt}{Nh}\right). \tag{10.42}$$

In summary, we sample the function $y(t)$ at N times, $t_1, \ldots, t_N$. We see that all the values of y sampled contribute to each a_k. Consequently, if we increase N in order to determine more coefficients, we must recompute all the a_n values. In the wavelet analysis in Chapter 11, "Wavelet Analysis & Data Compression," the theory is reformulated so that additional samplings determine higher Fourier components without affecting lower ones.

10.4.4 Assessments

Simple analytic input: It is always good to do these simple checks before examining more complex problems. If your system has some Fourier analysis packages (such as the graphing package *Ace/gr*), you may want to compare your results with those from the packages. Once you understand how the packages work, it makes sense to use them.

1. Sample the even signal

$$y(t) = 3\cos(\omega t) + 2\cos(3\omega t) + \cos(5\omega t).$$

   Decompose this into its components; then check that they are essentially real and in the ratio 3:2:1 (or 9:4:1 for the power spectrum) and that they resum to give the input signal.
2. Experiment on the separate effects of picking different values of the step size h and of enlarging the measurement period $T = Nh$.
3. Sample the odd signal

$$y(t) = \sin(\omega t) + 2\sin(3\omega t) + 3\sin(5\omega t).$$

   Decompose this into its components; then check that they are essentially imaginary and in the ratio 1:2:3 (or 1:4:9 if a power spectrum is plotted) and that they resum to give the input signal.
4. Sample the mixed-symmetry signal

$$y(t) = 5\sin(\omega t) + 2\cos(3\omega t) + \sin(5\omega t).$$

Decompose this into its components; then check that there are three of them in the ratio 5:2:1 (or 25:4:1 if a power spectrum is plotted) and that they resum to give the input signal.

5. Sample the signal

$$y(t) = 5 + 10\sin(t + 2).$$

Compare and explain the results obtained by sampling (a) without the 5, (b) as given but without the 2, and (c) without the 5 and without the 2.

6. In our discussion of aliasing, we examined Figure 10.2 showing the functions $\sin(\pi t/2)$ and $\sin(2\pi t)$. Sample the function

$$y(t) = \sin\left(\frac{\pi}{2}t\right) + \sin(2\pi t)$$

and explore how aliasing occurs. Explicitly, we know that the true transform contains peaks at $\omega = \pi/2$ and $\omega = 2\pi$. Sample the signal at a rate that leads to aliasing, as well as at a higher sampling rate at which there is no aliasing. Compare the resulting DFTs in each case and check if your conclusions agree with the Nyquist criterion. ∎

Highly nonlinear oscillator: Recall the numerical solution for oscillations of a spring with power $p = 12$ [see (10.1)]. Decompose the solution into a Fourier series and determine the number of higher harmonics that contribute at least 10%; for example, determine the n for which $|b_n/b_1| < 0.1$. Check that resuming the components reproduces the signal.

Nonlinearly perturbed oscillator: Remember the harmonic oscillator with a nonlinear perturbation (9.2):

$$V(x) = \frac{1}{2}kx^2\left(1 - \frac{2}{3}\alpha x\right), \quad F(x) = -kx(1 - \alpha x). \tag{10.43}$$

For very small amplitudes of oscillation ($x \ll 1/\alpha$), the solution $x(t)$ will essentially be only the first term of a Fourier series.

1. We want the signal to contain "approximately 10% nonlinearity." This being the case, fix your value of α so that $\alpha x_{\text{max}} \simeq 10\%$, where x_{max} is the maximum amplitude of oscillation. For the rest of the problem, keep the value of α fixed.
2. Decompose your numerical solution into a discrete Fourier spectrum.
3. Plot a graph of the percentage of importance of the first *two*, non-DC Fourier components as a function of the initial displacement for $0 < x_0 < 1/2\alpha$. You should find that higher harmonics are more important as the amplitude increases. Because both even and odd components are present, Y_n should be complex. Because a 10% effect in amplitude becomes a 1% effect in power, make sure that you make a semilog plot of the power spectrum.

4. As always, check that resumations of your transforms reproduce the signal.

(*Warning:* The ω you use in your series must correspond to the *true* frequency of the system, not just the ω of small oscillations.)

10.4.5 DFT of Nonperiodic Functions (Exploration)

Consider a simple model (a wave packet) of a "localized" electron moving through space and time. A good model for an electron initially localized around $x = 5$ is a Gaussian multiplying a plane wave:

$$\psi(x, t=0) = \exp\left[-\frac{1}{2}\left(\frac{x-5.0}{\sigma_0}\right)^2\right] e^{ik_0 x}. \tag{10.44}$$

This wave packet is not an eigenstate of the momentum operator[7] $p = id/dx$ and in fact contains a spread of momenta. Your **problem** is to evaluate the Fourier transform,

$$\psi(p) = \int_{-\infty}^{+\infty} dx \frac{e^{ipx}}{\sqrt{2\pi}} \psi(x), \tag{10.45}$$

as a way of determining the momenta components in (10.44).

10.5 Unit II. Filtering Noisy Signals

You measure a signal $y(t)$ that obviously contains noise. Your **problem** is to determine the frequencies that would be present in the signal if it did not contain noise. Of course, once you have a Fourier transform from which the noise has been removed, you can transform it to obtain a signal $s(t)$ with no noise.

In the process of solving this problem we examine two simple approaches: the use of autocorrelation functions and the use of filters. Both approaches find wide applications in science, with our discussion not doing the subjects justice. However, we will see filters again in the discussion of wavelets in Chapter 11, "Wavelet Analysis & Data Compression."

10.6 Noise Reduction via Autocorrelation (Theory)

We assume that the measured signal is the sum of the true signal $s(t)$, which we wish to determine, plus the *noise* $n(t)$:

$$y(t) = s(t) + n(t). \tag{10.46}$$

[7] We use natural units in which $\hbar = 1$.

Our first approach to separating the signal from the noise relies on that fact that noise is a random process and thus should not be correlated with the signal. Yet what do we mean when we say that two functions are *correlated?* Well, if the two tend to oscillate with their nodes and peaks in much the same places, then the two functions are clearly correlated. An analytic measure of the correlation of two arbitrary functions $y(t)$ and $x(t)$ is the *correlation function*

$$c(\tau) = \int_{-\infty}^{+\infty} dt\, y^*(t)\, x(t+\tau) \equiv \int_{-\infty}^{+\infty} dt\, y^*(t-\tau)\, x(t), \tag{10.47}$$

where τ, the *lag time*, is a variable. Even if the two signals have different magnitudes, if they have similar time dependences except for one lagging or leading the other, then for certain values of τ the integrand in (10.47) will be positive for all values of t. In this case the two signals interfere constructively and produce a large value for the correlation function. In contrast, if both functions oscillate independently, then it is just as likely for the integrand to be positive as to be negative, in which case the two signals interfere destructively and produce a small value for the integral.

Before we apply the correlation function to our problem, let us study some of its properties. We use (10.17) to express c, y^*, and x in terms of their Fourier transforms:

$$c(\tau) = \int_{-\infty}^{+\infty} d\omega''\, C(\omega'') \frac{e^{i\omega'' t}}{\sqrt{2\pi}}, \quad y^*(t) = \int_{-\infty}^{+\infty} d\omega\, Y^*(\omega) \frac{e^{-i\omega t}}{\sqrt{2\pi}},$$

$$x(t+\tau) = \int_{-\infty}^{+\infty} d\omega'\, X(\omega') \frac{e^{+i\omega t}}{\sqrt{2\pi}}. \tag{10.48}$$

Because ω, ω', and ω'' are dummy variables, other names may be used for these variables without changing any results. When we substitute these representations into the definition (10.47) and assume that the resulting integrals converge well enough to be rearranged, we obtain

$$\int_{-\infty}^{+\infty} d\omega''\, C(\omega'') e^{i\omega'' t} = \frac{1}{2\pi} \int_{-\infty}^{+\infty} d\omega \int_{-\infty}^{+\infty} d\omega'\, Y^*(\omega) X(\omega') e^{i\omega\tau} 2\pi \delta(\omega' - \omega)$$

$$= \int_{-\infty}^{+\infty} d\omega Y^*(\omega) X(\omega) e^{i\omega\tau},$$

$$\Rightarrow \quad C(\omega) = \sqrt{2\pi}\, Y^*(\omega) X(\omega), \tag{10.49}$$

where the last line follows because ω'' and ω are equivalent dummy variables. Equation (10.49) says that the Fourier transform of the correlation function between two signals is proportional to the product of the transform of one signal and the complex conjugate of the transform of the other. (We shall see a related convolution theorem for filters.)

A special case of the correlation function $c(\tau)$ is the *autocorrelation function* $A(\tau)$. It measures the correlation of a time signal with itself:

$$A(\tau) \stackrel{\text{def}}{=} \int_{-\infty}^{+\infty} dt\, y^*(t)\, y(t+\tau) \equiv \int_{-\infty}^{+\infty} dt\, y(t)\, y^*(t-\tau). \tag{10.50}$$

This function is computed by taking a signal $y(t)$ that has been measured over some time period and then averaging it over time using $y(t+\tau)$ as a weighting function. This process is also called *folding* a function onto itself (as might be done with dough) or a *convolution*. To see how this folding removes noise from a signal, we go back to the measured signal (10.46), which was the sum of pure signal plus noise $s(t)+n(t)$. As an example, on the upper left in Figure 10.3 we show a signal that was constructed by adding random noise to a smooth signal. When we compute the autocorrelation function for this signal, we obtain a function (upper right in Figure 10.3) that looks like a broadened, smoothed version of the signal $y(t)$. We can understand how the noise is removed by taking the Fourier transform of $s(t)+n(t)$ to obtain a simple sum of transforms:

$$Y(\omega) = S(\omega) + N(\omega), \tag{10.51}$$

$$\left\{ \begin{array}{c} S(\omega) \\ N(\omega) \end{array} \right\} = \int_{-\infty}^{+\infty} dt \left\{ \begin{array}{c} s(t) \\ n(t) \end{array} \right\} \frac{e^{-i\omega t}}{\sqrt{2\pi}}. \tag{10.52}$$

Because the autocorrelation function (10.50) for $y(t) = s(t)+n(t)$ involves the second power of y, is not a linear function, that is, $A_y \neq A_s + A_n$, but instead,

$$A_y(\tau) = \int_{-\infty}^{+\infty} dt\, [s(t)s(t+\tau) + s(t)n(t+\tau) + n(t)n(t+\tau)]\,. \tag{10.53}$$

If we assume that the noise $n(t)$ in the measured signal is truly random, then it should average to zero over long times and be uncorrelated at times t and $t+\tau$. This being the case, both integrals involving the noise vanish, and so

$$A_y(\tau) \simeq \int_{-\infty}^{+\infty} dt\, s(t)\, s(t+\tau) = A_s(\tau). \tag{10.54}$$

Thus, the part of the noise that is random tends to be averaged out of the autocorrelation function, and we are left with the autocorrelation function of approximately the pure signal.

This is all very interesting but is not the transform $S(\omega)$ of the pure signal that we need to solve our problem. However, application of (10.49) with $Y(\omega) = X(\omega) = S(\omega)$ tells us that the Fourier transform $A(\omega)$ of the autocorrelation function is proportional to $|S(\omega)|^2$:

$$A(\omega) = \sqrt{2\pi}\, |S(\omega)|^2. \tag{10.55}$$

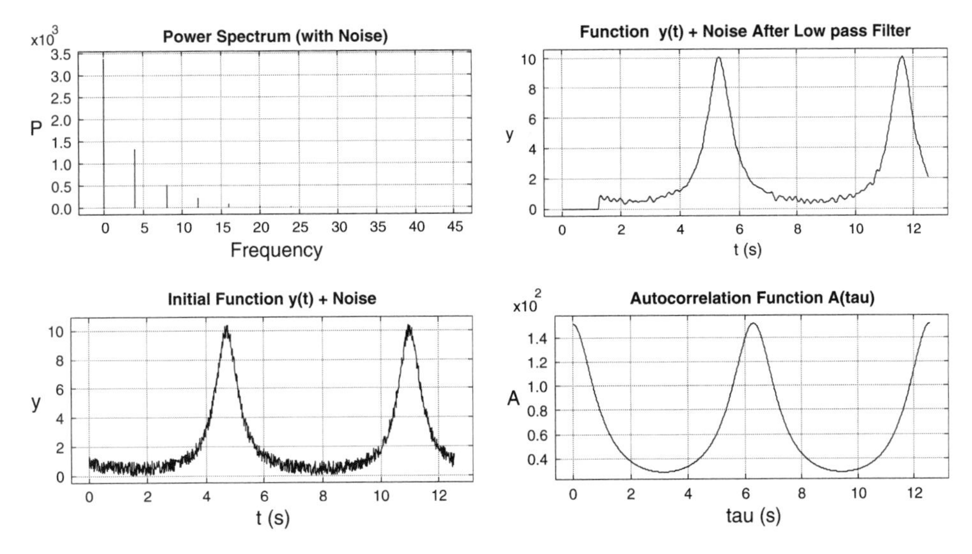

Figure 10.3 *From bottom left to right:* The function plus noise $s(t) + n(t)$, the autocorrelation function *versus* time, the power spectrum obtained from autocorrelation function, and the noisy signal after passage through a lowpass filter.

The function $|S(\omega)|^2$ is the *power spectrum* we discussed in §10.4. For practical purposes, knowing the power spectrum is often all that is needed and is easier to understand than a complex $S(\omega)$; in any case it is all that we can calculate.

As a procedure for analyzing data, we (1) start with the noisy measured signal and (2) compute its autocorrelation function $A(t)$ via the integral (10.50). Because this is just folding the signal onto itself, no additional functions or input is needed. We then (3) perform a DFT on the autocorrelation function $A(t)$ to obtain the power spectrum. For example, in Figure 10.3 we see a noisy signal (lower left), the autocorrelation function (lower right), which clearly is smoother than the signal, and finally, the deduced power spectrum (upper left). Notice that the broadband high-frequency components characteristic of noise are absent from the power spectrum. You can easily modify the sample program `DFT.java` in Listing 10.1 to compute the autocorrelation function and then the power spectrum from $A(\tau)$. We present a program `NoiseSincFilter` or `Filter.java` that does this on the CD.

10.6.1 Autocorrelation Function Exercises

1. Imagine that you have sampled the pure signal

$$s(t) = \frac{1}{1 - 0.9 \sin t}. \tag{10.56}$$

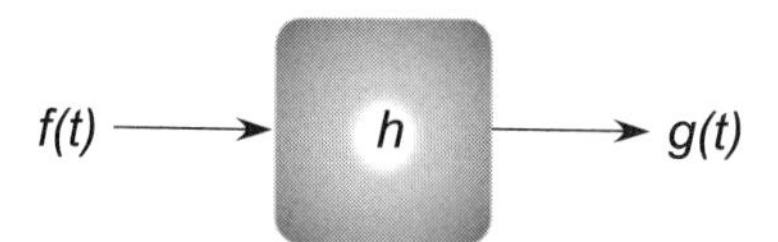

Figure 10.4 Input signal f is filtered by h, and the output is g.

Although there is just a single sine function in the denominator, there is an infinite number of overtones as follows from the expansion

$$s(t) \simeq 1 + 0.9 \sin t + (0.9 \sin t)^2 + (0.9 \sin t)^3 + \cdots . \tag{10.57}$$

a. Compute the DFT $S(\omega)$. Make sure to sample just one period but to cover the entire period. Make sure to sample at enough times (fine scale) to obtain good sensitivity to the high-frequency components.
b. Make a semilog plot of the power spectrum $|S(\omega)|^2$.
c. Take your input signal $s(t)$ and compute its autocorrelation function $A(\tau)$ for a full range of τ values (an analytic solution is okay too).
d. Compute the power spectrum indirectly by performing a DFT on the autocorrelation function. Compare your results to the spectrum obtained by computing $|S(\omega)|^2$ directly.

2. Add some random noise to the signal using a random number generator:

$$y(t_i) = s(t_i) + \alpha(2r_i - 1), \quad 0 \leq r_i \leq 1, \tag{10.58}$$

where α is an adjustable parameter. Try several values of α, from small values that just add some fuzz to the signal to large values that nearly hide the signal.
a. Plot your noisy data, their Fourier transform, and their power spectrum obtained directly from the transform with noise.
b. Compute the autocorrelation function $A(t)$ and its Fourier transform.
c. Compare the DFT of $A(\tau)$ to the power spectrum and comment on the effectiveness of reducing noise by use of the autocorrelation function.
d. For what value of α do you essentially lose all the information in the input? ∎

The code `Noise.java` that performs similar steps is available on the CD.

10.7 Filtering with Transforms (Theory)

A filter (Figure 10.4) is a device that converts an input signal $f(t)$ to an output signal $g(t)$ with some specific property for the latter. More specifically, an *analog filter* is defined [Hart 98] as integration over the input function:

$$g(t) = \int_{-\infty}^{+\infty} d\tau\, f(\tau)\, h(t-\tau) \stackrel{\text{def}}{=} f(t) * h(t). \tag{10.59}$$

The operation indicated in (10.59) occurs often enough that it is given the name *convolution* and is denoted by an asterisk $*$. The function $h(t)$ is called the *response* or *transfer function* of the filter because it is the response of the filter to a unit impulse:

$$h(t) = \int_{-\infty}^{+\infty} d\tau\, \delta(\tau)\, h(t-\tau). \tag{10.60}$$

Such being the case, $h(t)$ is also called the *unit impulse response function* or *Green's function*. Equation (10.59) states that the output $g(t)$ of a filter equals the input $f(t)$ convoluted with the transfer function $h(t-\tau)$. Because the argument of the response function is delayed by a time τ relative to that of the signal in the integral (10.59), τ is called the *lag time*. While the integration is over all times, the response of a good detector usually peaks around zero time. In any case, the response must equal zero for $\tau > t$ because events in the future cannot affect the present (causality).

The *convolution theorem* states that the Fourier transform of the convolution $g(t)$ is proportional to the product of the transforms of $f(t)$ and $h(t)$:

$$G(\omega) = \sqrt{2\pi}\, F(\omega)\, H(\omega). \tag{10.61}$$

The theorem results from expressing the functions in (10.59) by their transforms and using the resulting Dirac delta function to evaluate an integral (essentially what we did in our discussion of correlation functions). This is an example of how some relations are simpler in transform space than in time space.

Regardless of the domain used, filtering as we have defined it is a linear process involving just the first powers of f. This means that the output at one frequency is proportional to the input at that frequency. The constant of proportionality between the two may change with frequency and thus suppress specific frequencies relative to others, but that constant remains fixed in time. Because the law of linear superposition is valid for filters, if the input to a filter is represented as the sum of various functions, then the transform of the output will be the sum of the functions' Fourier transforms. Because the transfer function may be complex, $H(\omega) = |H(\omega)| \exp[i\phi(\omega)]$, the filter may also shift the phase of the input at frequency ω by an amount ϕ.

Filters that remove or decrease high-frequency components more than they do low-frequency components, are called *lowpass* filters. Those that filter out the low frequencies are called *highpass filters*. A simple lowpass filter is the RC circuit on the left in Figure 10.5, and it produces the transfer function

$$H(\omega) = \frac{1}{1+i\omega\tau} = \frac{1-i\omega\tau}{1+\omega^2\tau^2}, \tag{10.62}$$

where $\tau = RC$ is the time constant. The ω^2 in the denominator leads to a decrease in the response at high frequencies and therefore makes this a lowpass filter (the

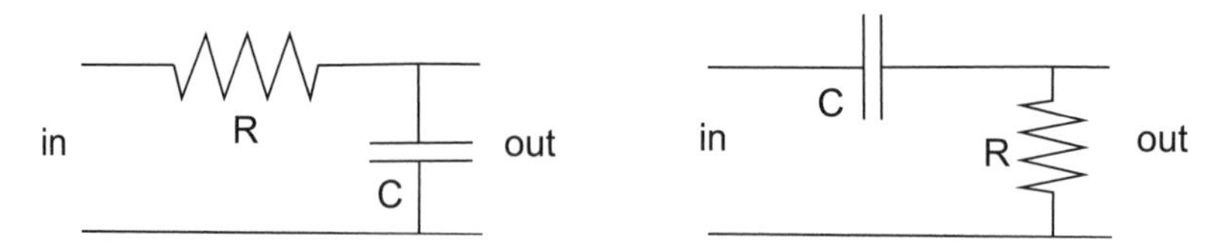

Figure 10.5 *Left:* An *RC* circuit arranged as a lowpass filter. *Right:* An *RC* circuit arranged as a highpass filter.

$i\omega$ affects only the phase). A simple highpass filter is the RC circuit on the right in Figure 10.5, and it produces the transfer function

$$H(\omega) = \frac{i\omega\tau}{1 + i\omega\tau} = \frac{i\omega\tau + \omega^2\tau^2}{1 + \omega^2\tau^2}. \tag{10.63}$$

$H = 1$ at large ω, yet H vanishes as $\omega \to 0$, which makes this a highpass filter.

Filters composed of resistors and capacitors are fine for analog signal processing. For digital processing we want a *digital filter* that has a specific response function for each frequency range. A physical model for a digital filter may be constructed from a delay line with taps at various spacing along the line (Figure 10.6) [Hart 98]. The signal read from tap n is just the input signal delayed by time $n\tau$, where the delay time τ is a characteristic of the particular filter. The output from each tap is described by the transfer function $\delta(t - n\tau)$, possibly with scaling factor c_n. As represented by the triangle on the right in Figure 10.6, the signals from all taps are ultimately summed together to form the total response function:

$$h(t) = \sum_{n=0}^{N} c_n\, \delta(t - n\tau). \tag{10.64}$$

In the frequency domain, the Fourier transform of a delta function is an exponential, and so (10.64) results in the transfer function

$$H(\omega) = \sum_{n=0}^{N} c_n\, e^{-i\, n\omega\tau}, \tag{10.65}$$

where the exponential indicates the phase shift from each tap.

If a digital filter is given a continuous time signal $f(t)$ as input, its output will be the discrete sum

$$g(t) = \int_{-\infty}^{+\infty} dt'\, f(t') \sum_{n=0}^{N} c_n\, \delta(t - t' - n\tau) = \sum_{n=0}^{N} c_n\, f(t - n\tau). \tag{10.66}$$

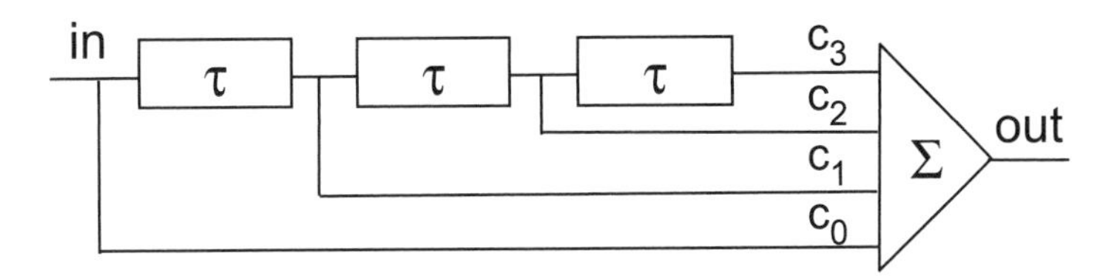

Figure 10.6 A delay-line filter in which the signal at different time translations is scaled by different amounts c_i.

And of course, if the signal's input is a discrete sum, its output will remain a discrete sum. (We restrict ourselves to nonrecursive filters [Pres 94].) In either case, we see that knowledge of the filter coefficients c_i provides us with all we need to know about a digital filter. If we look back at our work on the discrete Fourier transform in §10.4.1, we can view a digital filter (10.66) as a Fourier transform in which we use an N-point approximation to the Fourier integral. The c_n's then contain both the integration weights and the values of the response function at the integration points. The transform itself can be viewed as a filter of the signal into specific frequencies.

10.7.1 Digital Filters: Windowed Sinc Filters (Exploration) ⊙

Problem: Construct digital versions of highpass and lowpass filters and determine which filter works better at removing noise from a signal.

A popular way to separate the bands of frequencies in a signal is with a *windowed sinc filter* [Smi 99]. This filter is based on the observation that an ideal *lowpass* filter passes all frequencies below a cutoff frequency ω_c and blocks all frequencies above this frequency. And because there tends to be more noise at high frequencies than at low frequencies, removing the high frequencies tends to remove more noise than signal, although some signal is inevitably lost. One use for windowed sinc filters is in reducing aliasing by removing the high-frequency component of a signal before determining its Fourier components. The graph on the lower right in Figure 10.7 was obtained by passing our noisy signal through a sinc filter (using the program `Filter.java` given on the CD).

If both positive and negative frequencies are included, an ideal low-frequency filter will look like the rectangular pulse in frequency space:

$$H(\omega,\omega_c) = \text{rect}\left(\frac{\omega}{2\omega_c}\right) \qquad \text{rect}(\omega) = \begin{cases} 1, & \text{if } |\omega| \le \frac{1}{2}, \\ 0, & \text{otherwise.} \end{cases} \tag{10.67}$$

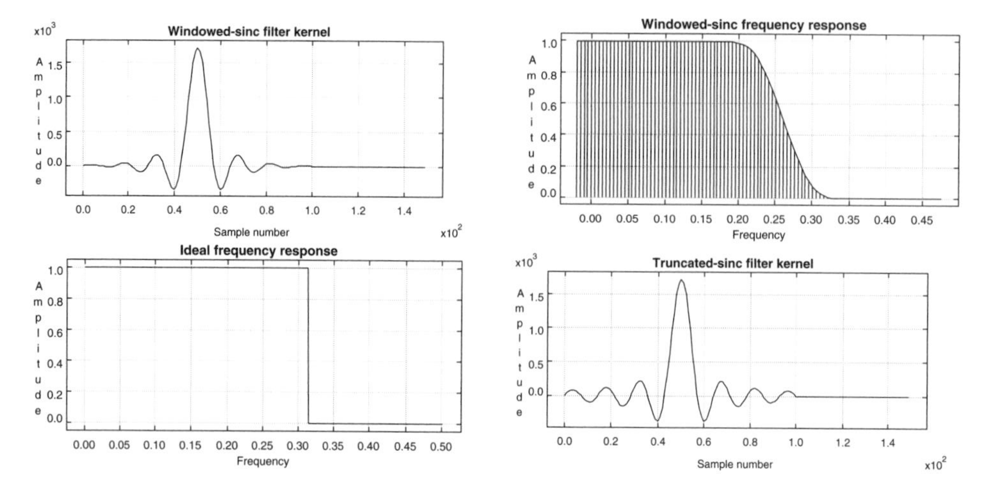

Figure 10.7 *Lower left:* Frequency response for an ideal lowpass filter. *Lower right:* Truncated-sinc filter kernel (time domain). *Upper left:* Windowed-sinc filter kernel. *Upper right:* windowed-sinc filter frequency response.

Here rect(ω) is the rectangular function (Figure 10.8). Although maybe not obvious, a rectangular pulse in the frequency domain has a Fourier transform that is proportional to the *sinc function* in the time domain [Smi 91, Wiki]

$$\int_{-\infty}^{+\infty} d\omega \, e^{-i\omega t} \mathrm{rect}(\omega) = \mathrm{sinc}\left(\frac{t}{2}\right) \stackrel{\mathrm{def}}{=} \frac{\sin(\pi t/2)}{\pi t/2}, \tag{10.68}$$

where the π's are sometimes omitted. Consequently, we can filter out the high-frequency components of a signal by convoluting it with $\sin(\omega_c t)/(\omega_c t)$, a technique also known as the *Nyquist–Shannon* interpolation formula. In terms of discrete transforms, the time-domain representation of the sinc filter is

$$h[i] = \frac{\sin(\omega_c i)}{i\pi}. \tag{10.69}$$

All frequencies below the cutoff frequency ω_c are passed with unit amplitude, while all higher frequencies are blocked.

In practice, there are a number of problems in using this function as the filter. First, as formulated, the filter is *noncausal*; that is, there are coefficients at negative times, which is nonphysical because we do not start measuring the signal until $t = 0$. Second, in order to produce a perfect rectangular response, we would have to sample the signal at an infinite number of times. In practice, we sample at $(M + 1)$

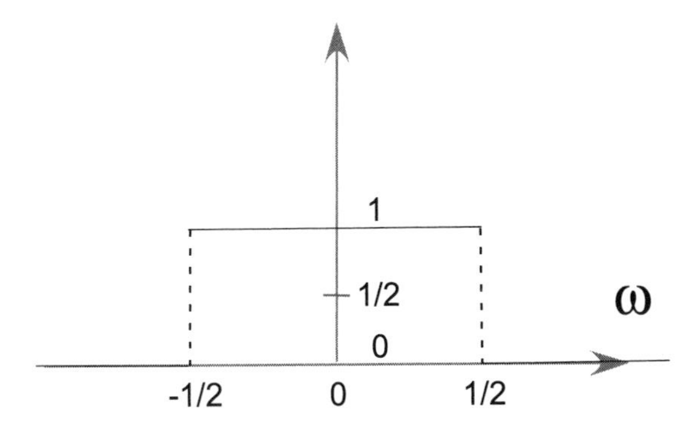

Figure 10.8 The rectangle function rect(ω) whose Fourier transform is sinc(t).

points (M even) placed symmetrically around the main lobe of $\sin(\pi t)/\pi t$ and then shift times to purely positive values,

$$h[i] = \frac{\sin[2\pi\omega_c(i - M/2)]}{i - M/2}, \quad 0 \leq t \leq M. \tag{10.70}$$

As might be expected, a penalty is incurred for making the filter discrete; instead of the ideal rectangular response, we obtain a *Gibbs overshoot*, with rounded corners and oscillations beyond the corner.

There are two ways to reduce the departures from the ideal filter. The first is to increase the length of times for which the filter is sampled, which inevitably leads to longer compute times. The other way is to smooth out the truncation of the sinc function by multiplying it with a smoothly tapered curve like the *Hamming window function*:

$$w[i] = 0.54 - 0.46\ \cos(2\pi i/M). \tag{10.71}$$

In this way the filter's kernel becomes

$$h[i] = \frac{\sin[2\pi\omega_c(i - M/2)]}{i - M/2}\left[0.54 - 0.46\ \cos(\frac{2\pi i}{M})\right]. \tag{10.72}$$

The cutoff frequency ω_c should be a fraction of the sampling rate. The time length M determines the *bandwidth* over which the filter changes from 1 to 0.

Exercise: Repeat the exercise that added random noise to a known signal, this time using the sinc filter to reduce the noise. See how small you can make the signal and still be able to separate it from the noise. (The code `Noise.java` that performs these steps is on the instructor's CD.)

10.8 Unit III. Fast Fourier Transform Algorithm ⊙

We have seen in (10.32) that a discrete Fourier transform can be written in the compact form

$$Y_n = \frac{1}{\sqrt{2\pi}} \sum_{k=1}^{N} Z^{nk}\, y_k, \quad Z = e^{-2\pi i/N}, \quad n = 0, 1, \ldots, N-1. \tag{10.73}$$

Even if the signal elements y_k to be transformed are real, Z is always complex, and therefore we must process both real and imaginary parts when computing transforms. Because both n and k range over N integer values, the $(Z^n)^k\, y_k$ multiplications in (10.73) require some N^2 multiplications and additions of complex numbers. As N gets large, as happens in realistic applications, this geometric increase in the number of steps slows down the algorithm.

In 1965, Cooley and Tukey discovered an algorithm[8] that reduces the number of operations necessary to perform a DFT from N^2 to roughly $N \log_2 N$ [Co 65, Donn 05]. Even though this may not seem like such a big difference, it represents a 100-fold speedup for 1000 data points, which changes a full day of processing into 15 min of work. Because of its widespread use (including cell phones), the fast Fourier transform algorithm is considered one of the 10 most important algorithms of all time.

The idea behind the FFT is to utilize the periodicity inherent in the definition of the DFT (10.73) to reduce the total number of computational steps. Essentially, the algorithm divides the input data into two equal groups and transforms only one group, which requires $\sim (N/2)^2$ multiplications. It then divides the remaining (nontransformed) group of data in half and transforms them, continuing the process until all the data have been transformed. The total number of multiplications required with this approach is approximately $N \log_2 N$.

Specifically, the FFT's time economy arises from the computationally expensive complex factor $Z^{nk} [= ((Z)^n)^k]$ being equal to the same cyclically repeated value as the integers n and k vary sequentially. For instance, for $N = 8$,

$$\begin{aligned}
Y_0 &= Z^0 y_0 + Z^0 y_1 + Z^0\, y_2 + Z^0\, y_3 + Z^0\, y_4 + Z^0\, y_5 + Z^0\, y_6 + Z^0\, y_7,\\
Y_1 &= Z^0 y_0 + Z^1 y_1 + Z^2 y_2 + Z^3\, y_3 + Z^4\, y_4 + Z^5\, y_5 + Z^6\, y_6 + Z^7\, y_7,\\
Y_2 &= Z^0 y_0 + Z^2 y_1 + Z^4\, y_2 + Z^6\, y_3 + Z^8\, y_4 + Z^{10} y_5 + Z^{12} y_6 + Z^{14} y_7,\\
Y_3 &= Z^0 y_0 + Z^3 y_1 + Z^6\, y_2 + Z^9\, y_3 + Z^{12} y_4 + Z^{15} y_5 + Z^{18} y_6 + Z^{21} y_7,
\end{aligned}$$

[8] Actually, this algorithm has been discovered a number of times, for instance, in 1942 by Danielson and Lanczos [Da 42], as well as much earlier by Gauss.

$$Y_4 = Z^0 y_0 + Z^4 y_1 + Z^8\, y_2 + Z^{12} y_3 + Z^{16} y_4 + Z^{20} y_5 + Z^{24} y_6 + Z^{28} y_7,$$
$$Y_5 = Z^0 y_0 + Z^5 y_1 + Z^{10} y_2 + Z^{15} y_3 + Z^{20} y_4 + Z^{25} y_5 + Z^{30} y_6 + Z^{35} y_7,$$
$$Y_6 = Z^0 y_0 + Z^6 y_1 + Z^{12} y_2 + Z^{18} y_3 + Z^{24} y_4 + Z^{30} y_5 + Z^{36} y_6 + Z^{42} y_7,$$
$$Y_7 = Z^0 y_0 + Z^7 y_1 + Z^{14} y_2 + Z^{21} y_3 + Z^{28} y_4 + Z^{35} y_5 + Z^{42} y_6 + Z^{49} y_7,$$

where we include $Z^0(\equiv 1)$ for clarity. When we actually evaluate these powers of Z, we find only four independent values:

$$\begin{aligned}
Z^0 &= \exp(0) = +1, & Z^1 &= \exp(-\tfrac{2\pi}{8} i) = +\tfrac{\sqrt{2}}{2} - i\,\tfrac{\sqrt{2}}{2},\\
Z^2 &= \exp(-\tfrac{2\pi}{8} 2i) = -i, & Z^3 &= \exp(-\tfrac{2\pi}{8} 3i) = -\tfrac{\sqrt{2}}{2} - i\,\tfrac{\sqrt{2}}{2},\\
Z^4 &= \exp(-\tfrac{2\pi}{8} 4i) = -Z^0, & Z^5 &= \exp(-\tfrac{2\pi}{8} 5i) = -Z^1,\\
Z^6 &= \exp(-\tfrac{2\pi}{8} 6i) = -Z^2, & Z^7 &= \exp(-\tfrac{2\pi}{8} 7i) = -Z^3,\\
Z^8 &= \exp(-\tfrac{2\pi}{8} 8i) = +Z^0, & Z^9 &= \exp(-\tfrac{2\pi}{8} 9i) = +Z^1,\\
Z^{10} &= \exp(-\tfrac{2\pi}{8} 10i) = +Z^2, & Z^{11} &= \exp(-\tfrac{2\pi}{8} 11i) = +Z^3,\\
Z^{12} &= \exp(-\tfrac{2\pi}{8} 11i) = -Z^0, & &\cdots.
\end{aligned} \tag{10.74}$$

When substituted into the definitions of the transforms, we obtain

$$\begin{aligned}
Y_0 &= Z^0 y_0 + Z^0 y_1 + Z^0 y_2 + Z^0 y_3 + Z^0 y_4 + Z^0 y_5 + Z^0 y_6 + Z^0 y_7,\\
Y_1 &= Z^0 y_0 + Z^1 y_1 + Z^2 y_2 + Z^3 y_3 - Z^0 y_4 - Z^1 y_5 - Z^2 y_6 - Z^3 y_7,\\
Y_2 &= Z^0 y_0 + Z^2 y_1 - Z^0 y_2 - Z^2 y_3 + Z^0 y_4 + Z^2 y_5 - Z^0 y_6 - Z^2 y_7,\\
Y_3 &= Z^0 y_0 + Z^3 y_1 - Z^2 y_2 + Z^1 y_3 - Z^0 y_4 - Z^3 y_5 + Z^2 y_6 - Z^1 y_7,\\
Y_4 &= Z^0 y_0 - Z^0 y_1 + Z^0 y_2 - Z^0 y_3 + Z^0 y_4 - Z^0 y_5 + Z^0 y_6 - Z^0 y_7,\\
Y_5 &= Z^0 y_0 - Z^1 y_1 + Z^2 y_2 - Z^3 y_3 - Z^0 y_4 + Z^1 y_5 - Z^2 y_6 + Z^3 y_7,\\
Y_6 &= Z^0 y_0 - Z^2 y_1 - Z^0 y_2 + Z^2 y_3 + Z^0 y_4 - Z^2 y_5 - Z^0 y_6 + Z^2 y_7,\\
Y_7 &= Z^0 y_0 - Z^3 y_1 - Z^2 y_2 - Z^1 y_3 - Z^0 y_4 + Z^3 y_5 + Z^2 y_6 + Z^1 y_7,\\
Y_8 &= Y_0.
\end{aligned}$$

We see that these transforms now require $8 \times 8 = 64$ multiplications of complex numbers, in addition to some less time-consuming additions. We place these equations in an appropriate form for computing by regrouping the terms into sums and differences of the y's:

$$Y_0 = Z^0(y_0 + y_4) + Z^0(y_1 + y_5) + Z^0(y_2 + y_6) + Z^0(y_3 + y_7),$$

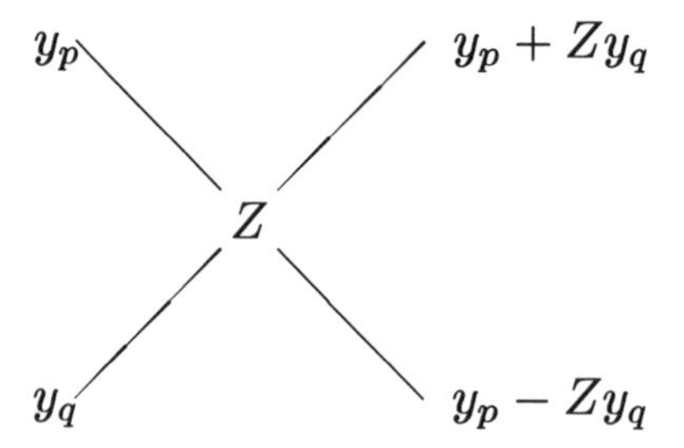

Figure 10.9 The basic butterfly operation in which elements y_p and y_q are transformed into $y_p + Z\,y_q$ and $y_p - Z\,y_q$.

$$
\begin{aligned}
Y_1 &= Z^0(y_0 - y_4) + Z^1(y_1 - y_5) + Z^2(y_2 - y_6) + Z^3(y_3 - y_7),\\
Y_2 &= Z^0(y_0 + y_4) + Z^2(y_1 + y_5) - Z^0(y_2 + y_6) - Z^2(y_3 + y_7),\\
Y_3 &= Z^0(y_0 - y_4) + Z^3(y_1 - y_5) - Z^2(y_2 - y_6) + Z^1(y_3 - y_7),\\
Y_4 &= Z^0(y_0 + y_4) - Z^0(y_1 + y_5) + Z^0(y_2 + y_6) - Z^0(y_3 + y_7),\\
Y_5 &= Z^0(y_0 - y_4) - Z^1(y_1 - y_5) + Z^2(y_2 - y_6) - Z^3(y_3 - y_7),\\
Y_6 &= Z^0(y_0 + y_4) - Z^2(y_1 + y_5) - Z^0(y_2 + y_6) + Z^2(y_3 + y_7),\\
Y_7 &= Z^0(y_0 - y_4) - Z^3(y_1 - y_5) - Z^2(y_2 - y_6) - Z^1(y_3 - y_7),\\
Y_8 &= Y_0.
\end{aligned}
$$

Note the repeating factors inside the parentheses, with combinations of the form $y_p \pm y_q$. These symmetries are systematized by introducing the *butterfly operation* (Figure 10.9). This operation takes the y_p and y_q data elements from the left wing and converts them to the $y_p + Zy_q$ elements in the upper- and lower-right wings. In Figure 10.10 we show what happens when we apply the butterfly operations to an entire FFT process, specifically to the pairs (y_0, y_4), (y_1, y_5), (y_2, y_6), and (y_3, y_7). Notice how the number of multiplications of complex numbers has been reduced: For the first butterfly operation there are 8 multiplications by Z^0; for the second butterfly operation there are 8 multiplications, and so forth, until a total of 24 multiplications is made in four butterflies. In contrast, 64 multiplications are required in the original DFT (10.8).

10.8.1 Bit Reversal

The reader may have observed that in Figure 10.10 we started with 8 data elements in the order 0–7 and that after three butterfly operators we obtained transforms in the order 0, 4, 2, 6, 1, 5, 3, 7. The astute reader may may also have observed that these numbers correspond to the bit-reversed order of 0–7.

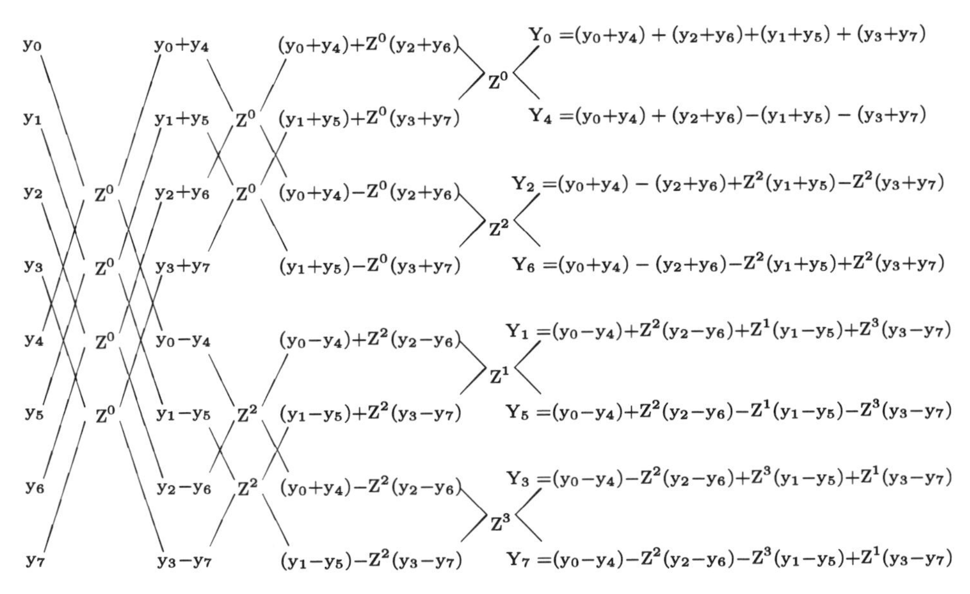

Figure 10.10 The butterfly operations performing a FFT on four pairs of data.

Let us look into this further. We need 3 bits to give the order of each of the 8 input data elements (the numbers 0–7). Explicitly, on the left in Table 10.1 we give the binary representation for decimal numbers 0–7, their bit reversals, and the corresponding decimal numbers. On the right we give the ordering for 16 input data elements, where we need 4 bits to enumerate their order. Notice that the order of the first 8 elements differs in the two cases because the number of bits being reversed differs. Notice too that after the reordering, the first half of the numbers are all even and the second half are all odd.

The fact that the Fourier transforms are produced in an order corresponding to the bit-reversed order of the numbers 0–7 suggests that if we process the data in the bit-reversed order 0, 4, 6, 2, 1, 5, 3, 7, then the output Fourier transforms will be ordered. We demonstrate this conjecture in Figure 10.11, where we see that to obtain the Fourier transform for the 8 input data, the butterfly operation had to be applied 3 times. The number 3 occurs here because it is the power of 2 that gives the number of data; that is, $2^3 = 8$. In general, in order for a FFT algorithm to produce transforms in the proper order, it must reshuffle the input data into bit-reversed order. As a case in point, our sample program starts by reordering the 16 (2^4) data elements given in Table 10.2. Now the 4 butterfly operations produce sequentially ordered output.

10.9 FFT Implementation

The first FFT program we are aware of was written in 1967 in Fortran IV by Norman Brenner at MIT's Lincoln Laboratory [Hi,76] and was hard for us to follow. Our

TABLE 10.1

Binary-Reversed 0–7			
Decimal	*Binary*	*Reversal*	*Decimal Reversal*
0	000	000	0
1	001	100	4
2	010	010	2
3	011	110	6
4	100	001	1
5	101	101	5
6	110	011	3
7	111	111	7

Binary-Reversed 0–16			
Decimal	*Binary*	*Reversal*	*Decimal Reversal*
0	0000	0000	0
1	0001	1000	8
2	0010	0100	4
3	0011	1100	12
4	0100	0010	2
5	0101	1010	10
6	0110	0110	6
7	0111	1110	14
8	1000	0001	1
9	1001	1001	9
10	1010	0101	5
11	1011	1101	13
12	1100	0011	3
13	1101	1011	11
14	1110	0111	7
15	1111	1111	15

(easier-to-follow) Java version of it is in Listing 10.2. Its input is $N = 2^n$ data to be transformed (FFTs always require 2^N input data). If the number of your input data is not a power of 2, then you can make it so by concatenating some of the initial data to the end of your input until a power of 2 is obtained; since a DFT is always periodic, this just starts the period a little earlier. This program assigns complex numbers at the 16 data points

$$y_m = m + mi, \quad m = 0, \ldots, 15, \tag{10.75}$$

reorders the data via bit reversal, and then makes four butterfly operations. The data are stored in the array `dtr[max][2]`, with the second subscript denoting real and imaginary parts. We increase speed further by using the 1-D array `data` to make memory access more direct:

$$\texttt{data}[1] = \texttt{dtr}[0][1], \quad \texttt{data}[2] = \texttt{dtr}[1][1], \quad \texttt{data}[3] = \texttt{dtr}[1][0], \ldots,$$

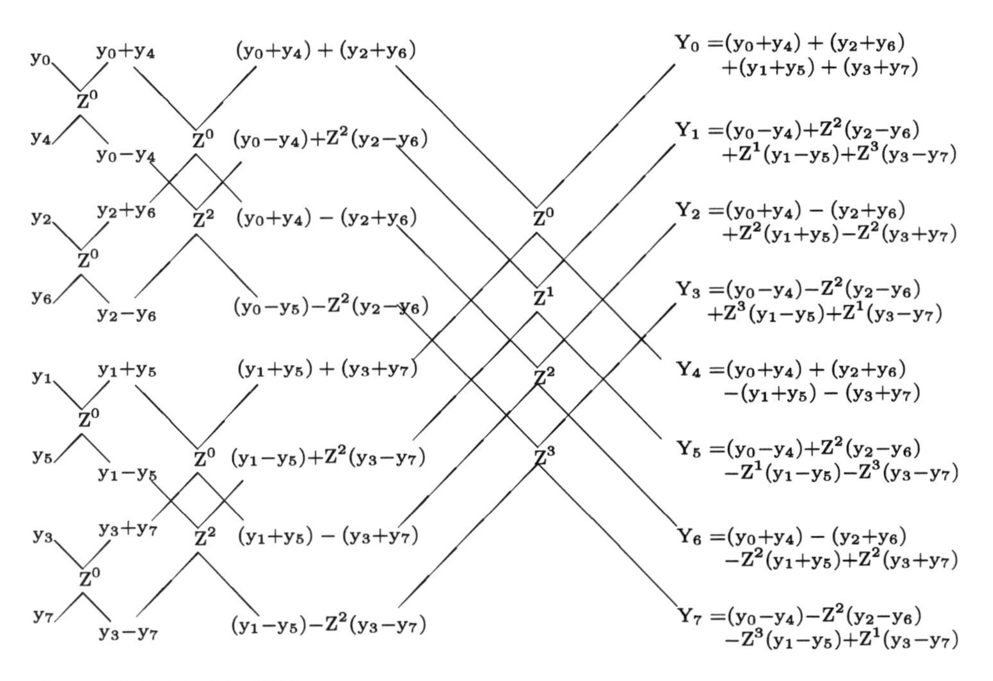

Figure 10.11 Modified FFT.

which also provides storage for the output. The FFT transforms data using the butterfly operation and stores the results back in dtr[][], where the input data were originally.

TABLE 10.2
Reordering for 16 Data Complex Points

Order	Input Data	New Order	Order	Input Data	New Order
0	0.0 + 0.0i	0.0 + 0.0i	8	8.0 + 8.0i	1.0 + 1.0i
1	1.0 + 1.0i	8.0 + 8.0i	9	9.0 + 9.0i	9.0 + 9.0i
2	2.0 + 2.0i	4.0 + 4.0i	10	10.0 + 10.i	5.0 + 5.0i
3	3.0 + 3.0i	12.0 + 12.0i	11	11.0 + 11.0i	13.0 + 13.0i
4	4.0 + 4.0i	2.0 + 2.0i	12	12.0 + 12.0i	3.0 + 3.0i
5	5.0 + 5.0i	10.0 + 10.i	13	13.0 + 13.0i	11.0 + 11.0i
6	6.0 + 6.0i	6.0 + 6.0i	14	14.0 + 14.i	7.0 + 7.0i
7	7.0 + 7.0i	14.0 + 14.0i	15	15.0 + 15.0i	15.0 + 15.0i

```
/* FFT.java:  FFT for complex numbers dtr[][]
              data1[i][0], data1[i][1] = Re, Im parts of point [i].
              When done, Re, Im Fourier Transforms placed in same array
              Required max = 2^m < 1024
              dtr[][] placed in array data[]:  */
import java.util.*;
import java.io.*;

public class FFT {
  public static int max = 2100;                                    // Global variables
  public static int points = 1026;                                 // Can be increased
  public static double data[] = new double[max];
  public static double dtr[][] = new double[points][2];

  public static void main(String[] argv) {
    int isign, i, nn = 16;                                          // Power of 2
    isign = -1;                                    // -1 transform, +1 inverse transform
    for ( i=0; i<nn; i++ ) {                                        // Form array
      dtr[i][0] = (double)(i);                                      // Re part
      dtr[i][1] = (double)(i);                                      // Im part
      System.out.println("dtr " + dtr[i][0] + " im " + dtr[i][1]);
    }
    fft(nn, isign);                                     // Call FFT, use global dtr[][]
    for(i=0; i <nn; i++) System.out.println("i "+i+" FT " +dtr[i][0]+"  "+dtr[i][1]);
  }

  public static void fft(int nn, int isign) {                     // FFT of dtr[n][2]
    int i, j, m, n, mmax, istep;
    Double tempr, tempi, wr, wi, wstpr, wstpi, theta, sinth;
    n = 2*nn;
    for( i = 0; i <= nn; i++ ) {                       // Original data in dtr to data
      j = 2*i + 1;
      data[j] = dtr[i][0];                                   // Real dtr, odd data[j]
      data[j+1] = dtr[i][1];                              // Imag dtr, even data[j+1]
      System.out.println(" Input data ");
      System.out.println(" dt "+j+"d  "+data[j]+"  "+ data[j+1]);
    }
    j = 1;                                            // Place data in bit reverse order
    for ( i = 1; i <= n; i=i+2 ) {
      if( (i-j)<0 ) {                               // Reorder equivalent to bit reverse
        tempr = data[j];
        tempi = data[j+1];
        data[j] = data[i];
        data[j+1] = data[i+1];
        data[i] = tempr;
        data[i+1] = tempi;
      }
      m = n/2;
      do { if ( (j-m)<=0 )  break; j = j-m; m = m/2; }
      while ( m-2>0 ); j = j+m;
    }                                           //  Print data and data to see reorder
    System.out.println(" Bit-reversed data ");
    for (i=1; i<=n; i=i+2) System.out.println (" i " + i + " data[i] "+data[i]);
    mmax = 2;
    while( (mmax-n)<0 ) {                                         // Begin transform
      istep = 2*mmax;
      theta = 6.2831853/(float)(isign*mmax);
      sinth = Math.sin(theta/2.0);
      wstpr = -2.0*sinth*sinth;
      wstpi = Math.sin(theta);
      wr = 1.0;      wi = 0.0;
      for (m=1; m <= mmax; m=m+2) {
        for(i=m; i<=n; i=i+istep) {
          j = i+mmax;
          tempr = wr*data[j]-wi*data[j+1];
          tempi = wr*data[j+1]+wi*data[j];
          data[j] = data[i]-tempr;
```

```
          data[j+1] = data[i+1]-tempi;
          data[i] = data[i]+tempr;
          data[i+1] = data[i+1]+tempi;
        }                                                                          // For i
        tempr = wr;
        wr = wr*wstpr-wi*wstpi+wr;
        wi = wi*wstpr+tempr*wstpi+wi;
      }                                                                            // For m
      mmax = istep;
    }                                                                              // While
    for(i=0; i<nn; i++)  {j = 2*i+1; dtr[i][0] = data[j]; dtr[i][1] = data[j+1]; }
} }
```

Listing 10.2 FFT.java computes the FFT or inverse transform depending upon the sign of isign.

10.10 FFT Assessment

1. Compile and execute FFT.java. Make sure you understand the output.
2. Take the output from FFT.java, inverse-transform it back to signal space, and compare it to your input. [Checking that the double transform is proportional to itself is adequate, although the normalization factors in (10.32) should make the two equal.]
3. Compare the transforms obtained with a FFT to those obtained with a DFT (you may choose any of the functions studied before). Make sure to compare both precision and execution times.

11

Wavelet Analysis & Data Compression

Problem: You have sampled the signal in Figure 11.1 that seems to contain an increasing number of frequencies as time increases. Your problem is to undertake a spectral analysis of this signal that tells you, in the most compact way possible, how much of each frequency is present at each instant in time. *Hint:* Although we want the method to be general enough to work with numerical data, for pedagogical purposes it is useful to know that the signal is

$$y(t) = \begin{cases} \sin 2\pi t, & \text{for } 0 \le t \le 2, \\ 5\sin 2\pi t + 10\sin 4\pi t, & \text{for } 2 \le t \le 8, \\ 2.5\sin 2\pi t + 6\sin 4\pi t + 10\sin 6\pi t, & \text{for } 8 \le t \le 12. \end{cases} \tag{11.1}$$

11.1 Unit I. Wavelet Basics

The Fourier analysis we used in §10.4.1 reveals the amount of the harmonic functions $\sin(\omega t)$ and $\cos(\omega t)$ and their overtones that are present in a signal. An expansion in periodic functions is fine for *stationary* signals (those whose forms do not change in time) but has shortcomings for the variable form of our **problem** signal (11.1). One such problem is that the Fourier reconstruction has all constituent frequencies occurring simultaneously and so does not contain *time resolution* information indicating when each frequency occurs. Another shortcoming is that all the Fourier components are correlated, which results in more information being stored than may be needed and no convenient way to remove the excess storage.

There are a number of techniques that extend simple Fourier analysis to nonstationary signals. In this chapter we include an introduction to *wavelet analysis*, a field that has seen extensive development and application in the last decade in areas as diverse as brain waves and gravitational waves. The idea behind wavelet analysis is to expand a signal in a complete set of functions (wavelets), each of which oscillates for a finite period of time and each of which is centered at a different time. To give you a preview before we get into the details, we show four sample wavelets in Figure 11.2. Because each wavelet is local in time, it is a wave packet[1]

[1] We discuss wave packets further in §11.2.

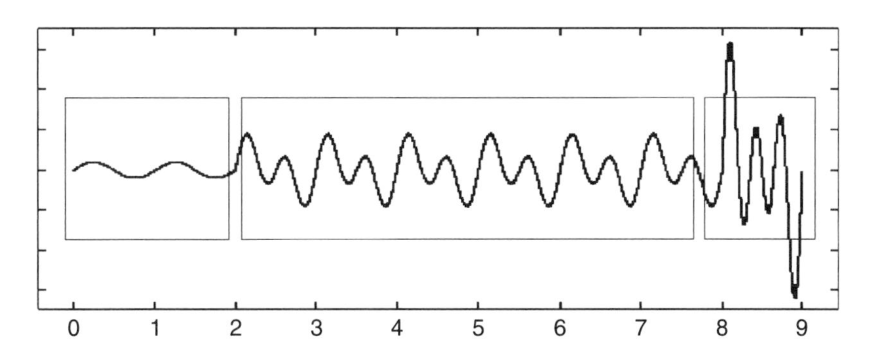

Figure 11.1 The time signal (11.1) containing an increasing number of frequencies as time increases. The boxes are possible placements of windows for short-time Fourier transforms.

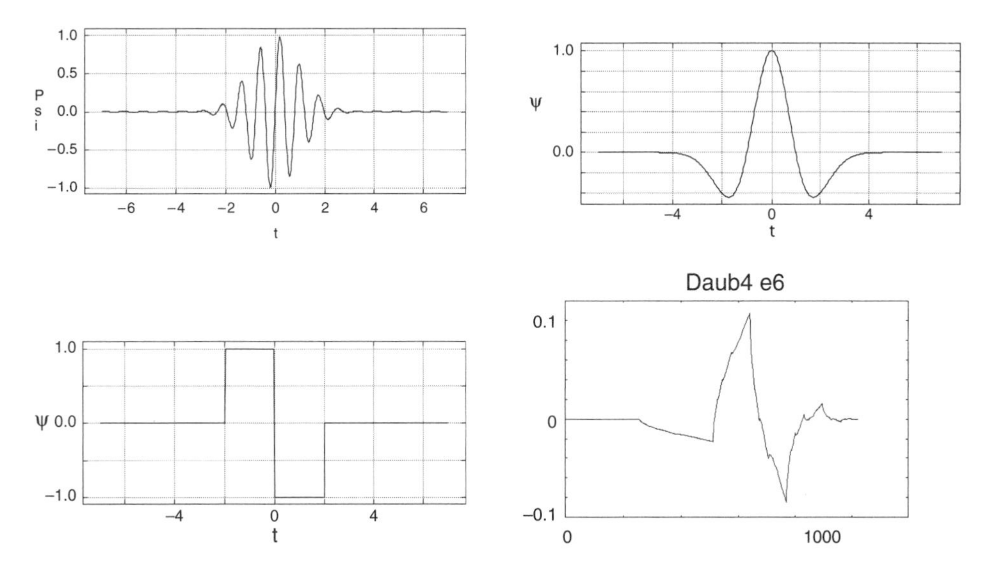

Figure 11.2 Four sample mother wavelets. *Clockwise from top:* Morlet (real part), Mexican hat, Daub4 e6 (explained later), and Haar. Wavelet basis functions are generated by scaling and translating these mother wavelets.

containing a range of frequencies. These wave packets are called wavelets because they are small and do not extend for long times.

Although wavelets are required to oscillate in time, they are not restricted to a particular functional form [Add 02]. As a case in point, they may be oscillating Gaussians (Morlet: top left in Figure 11.2),

$$\Psi(t) = e^{2\pi i t} e^{-t^2/2\sigma^2} = (\cos 2\pi t + i \sin 2\pi t) e^{-t^2/2\sigma^2} \quad \text{(Morlet)}, \qquad (11.2)$$

the second derivative of a Gaussian (Mexican hat, top right),

$$\Psi(t) = -\sigma^2 \frac{d^2}{dt^2} e^{-t^2/2\sigma^2} = \left(1 - \frac{t^2}{\sigma^2}\right) e^{-t^2/2\sigma^2}, \tag{11.3}$$

an up-and-down step function (lower left), or a fractal shape (bottom right). Such wavelets are *localized* in both time and frequency; that is, they are large for just a finite time and contain a finite range of frequencies. As we shall see, translating and scaling one of these *mother wavelets* generates an entire set of *child wavelet* basis functions, with each individual function covering a different frequency range at a different time.

11.2 Wave Packets and the Uncertainty Principle (Theory)

A *wave packet* or *wave train* is a collection of waves added together in such a way as to produce a pulse of width Δt. As we shall see, the Fourier transform of a wave packet is a pulse in the frequency domain of width $\Delta\omega$. We will first study such wave packets analytically and then use others numerically. An example of a simple wave packet is just a sine wave that oscillates at frequency ω_0 for N periods (Figure 11.3 left) [A&W 01]:

$$y(t) = \begin{cases} \sin\omega_0 t, & \text{for} \quad |t| < N\frac{\pi}{\omega_0} \equiv N\frac{T}{2}, \\ 0, & \text{for} \quad |t| > N\frac{\pi}{\omega_0} \equiv N\frac{T}{2}, \end{cases} \tag{11.4}$$

where we relate the frequency to the period via the usual $\omega_0 = 2\pi/T$. In terms of these parameters, the width of the wave packet is

$$\Delta t = NT = N\frac{2\pi}{\omega_0}. \tag{11.5}$$

The Fourier transform of the wave packet (11.4) is a straight-forward application of the transform formula (10.18):

$$Y(\omega) = \int_{-\infty}^{+\infty} dt \, \frac{e^{-i\omega t}}{\sqrt{2\pi}} y(t) = \frac{-i}{\sqrt{2\pi}} \int_0^{N\pi/\omega_0} dt \, \sin\omega_0 t \, \sin\omega t$$

$$= \frac{(\omega_0 + \omega)\sin\left[(\omega_0 - \omega)\frac{N\pi}{\omega_0}\right] - (\omega_0 - \omega)\sin\left[(\omega_0 + \omega)\frac{N\pi}{\omega_0}\right]}{\sqrt{2\pi}(\omega_0^2 - \omega^2)}, \tag{11.6}$$

where we have dropped a factor of $-i$ that affects only the phase. While at first glance (11.6) appears to be singular at $\omega = \omega_0$, it just peaks there (Figure 11.3 right), reflecting the predominance of frequency ω_0. However, there are sharp corners in the signal $y(t)$ (Figure 11.3 left), and these give $Y(\omega)$ a width $\Delta\omega$.

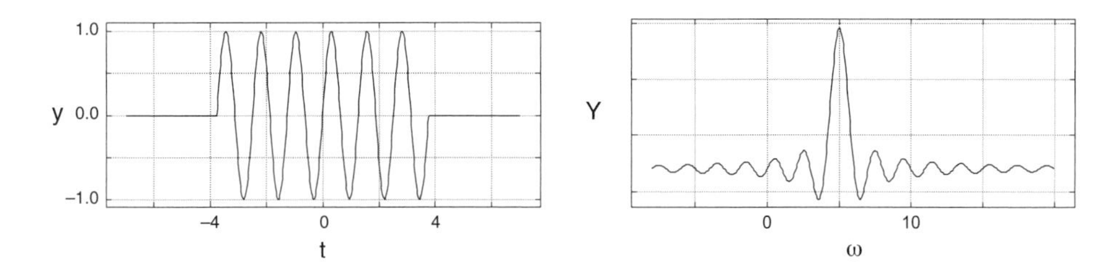

Figure 11.3 *Left:* A wave packet in time corresponding to the functional form (11.4) with $\omega_0 = 5$ and $N = 6$. *Right:* The Fourier transform in frequency of this same wave packet.

There is a fundamental relation between the widths Δt and $\Delta\omega$ of a wave packet. Although we use a specific example to determine that relation, it is true in general. While there may not be a precise definition of "width" for all functions, one can usually deduce a good measure of the width (say, within 25%). To illustrate, if we look at the right of Figure 11.3, it makes sense to use the distance between the first zeros of the transform $Y(\omega)$ (11.6) as the width $\Delta\omega$. The zeros occur at

$$\frac{\omega - \omega_0}{\omega_0} = \pm\frac{1}{N} \Rightarrow \quad \Delta\omega \simeq \omega - \omega_0 = \frac{\omega_0}{N}, \tag{11.7}$$

where N is the number of cycles in our original wave packet. Because the wave packet in time makes N oscillations each of period T, a reasonable measure of the width Δt of the signal $y(t)$ is

$$\Delta t = NT = N\frac{2\pi}{\omega_0}. \tag{11.8}$$

When the products of the frequency width (11.7) and the time width (11.8) are combined, we obtain

$$\boxed{\Delta t\,\Delta\omega \geq 2\pi.} \tag{11.9}$$

The greater-than sign is used to indicate that this is a minimum, that is, that $y(t)$ and $Y(\omega)$ extend beyond Δt and $\Delta\omega$, respectively. Nonetheless, most of the signal and transform should lie within the bound (11.9).

A relation of the form (11.9) also occurs in quantum mechanics, where it is known as the *Heisenberg uncertainty principle*, with Δt and $\Delta\omega$ being called the uncertainties in t and ω. It is true for transforms in general and states that as a signal is made more localized in time (smaller Δt) the transform becomes less localized (larger $\Delta\omega$). Conversely, the signal $y(t) = \sin\omega_0 t$ is completely localized in frequency and so has an infinite extent in time, $\Delta t \simeq \infty$.

11.2.1 Wave Packet Assessment

Consider the following wave packets:

$$y_1(t) = e^{-t^2/2}, \quad y_2(t) = \sin(8t)e^{-t^2/2}, \quad y_3(t) = (1-t^2)\, e^{-t^2/2}.$$

For each wave packet:

1. Estimate the width Δt. A good measure might be the *full width at half-maxima* (FWHM) of $|y(t)|$.
2. Evaluate and plot the Fourier transform $Y(\omega)$.
3. Estimate the width $\Delta\omega$ of the transform. A good measure might be the *full width at half-maxima* of $|Y(\omega)|$.
4. Determine the constant C for the uncertainty principle

$$\Delta t\, \Delta\omega \geq 2\pi C.$$

11.3 Short-Time Fourier Transforms (Math)

The constant amplitude of the functions $\sin\omega t$ and $\cos\omega t$ for all times can limit the usefulness of Fourier transforms. Because these functions and their overtones extend over all times with that constant amplitude, there is considerable overlap among them, and thus the information present in various Fourier components is correlated. This is undesirable for compressed data storage, where you want to store a minimum number of data and want to be able to cut out some of these data with just a minimal effect on the signal reconstruction.[2] In *lossless compression*, which reproduces the original signal exactly, you save space by storing how many times each data element is repeated and where each element is located. In *lossy compression*, in addition to removing repeated elements, you also eliminate some transform components, consistent with the uncertainty relation (11.9) and with the level of resolution required in the reproduction. This leads to a greater compression.

In §10.4.1 we defined the Fourier transform $Y(\omega)$ of a signal $y(t)$ as

$$Y(\omega) = \int_{-\infty}^{+\infty} dt\, \frac{e^{-i\omega t}}{\sqrt{2\pi}}\, y(t) = \langle \omega\, | y \rangle . \tag{11.10}$$

As is true for simple vectors, you can think of (11.10) as giving the overlap or scalar product of the basis function $\exp(i\omega t)/\sqrt{2\pi}$ and the signal $y(t)$ [notice that the

[2] Wavelets have also proven to be a highly effective approach to data compression, with the Joint Photographic Experts Group (JPEG) 2000 standard being based on wavelets. In Appendix G we give a full example of image compression with wavelets.

complex conjugate of the exponential basis function appears in (11.10)]. Another view of (11.10) is as the mapping or projection of the signal into ω space. In this case the overlap projects the amount of the periodic function $\exp(i\omega t)/\sqrt{2\pi}$ in the signal $y(t)$. In other words, the Fourier component $Y(\omega)$ is also the correlation between the signal $y(t)$ and the basis function $\exp(i\omega t)/\sqrt{2\pi}$, which is what results from filtering the signal $y(t)$ through a frequency-ω filter. If there is no $\exp(i\omega t)$ in the signal, then the integral vanishes and there is no output. If $y(t) = \exp(i\omega t)$, the signal is at only one frequency, and the integral is accordingly singular.

The problem signal in Figure 11.1 clearly has different frequencies present at different times and for different lengths of time. In the past this signal might have been analyzed with a precursor of wavelet analysis known as the *short-time Fourier transform*. With that technique, the signal $y(t)$ is "chopped up" into different segments along the time axis, with successive segments centered about successive times $\tau_1, \tau_2, \ldots, \tau_N$. For instance, we show three such segments in Figure 11.1. Once we have the dissected signal, a Fourier analysis is made of each segment. We are then left with a sequence of transforms $\left(Y_{\tau_1}^{(\mathrm{ST})}, Y_{\tau_2}^{(\mathrm{ST})}, \ldots, Y_{\tau_N}^{(\mathrm{ST})}\right)$, one for each short-time interval, where the superscript $^{(\mathrm{ST})}$ indicates short time.

Rather than chopping up a signal, we express short-time Fourier transforming mathematically by imagining translating a *window function* $w(t-\tau)$ by a time τ over the signal in Figure 11.1:

$$Y^{(\mathrm{ST})}(\omega, \tau) = \int_{-\infty}^{+\infty} dt \, \frac{e^{i\omega t}}{\sqrt{2\pi}} \, w(t-\tau)\, y(t). \tag{11.11}$$

Here the values of the translation time τ correspond to different locations of window w over the signal, and the window function is essentially a transparent box of small size on an opaque background. Any signal within the width of the window is transformed, while the signal lying outside the window is not seen. Note that in (11.11) the extra variable τ in the Fourier transform indicating the location of the time around which the window was placed. Clearly, since the short-time transform is a function of two variables, a surface or 3-D plot is needed to view the amplitude as a function of both ω and τ.

11.4 The Wavelet Transform

The wavelet transform of a time signal $y(t)$,

$$\boxed{Y(s, \tau) = \int_{-\infty}^{+\infty} dt \, \psi_{s,\tau}^{*}(t) y(t)} \quad \text{(wavelet transform)}, \tag{11.12}$$

is similar in concept and notation to a short-time Fourier transform. Rather than using $\exp(i\omega t)$ as the basis functions, we use wave packets or wavelets $\psi_{s,\tau}(t)$

localized in time, such as the those shown in Figure 11.2. Because each wavelet is localized in time, each acts as its own window function. Because each wavelet is oscillatory, each contains its own small range of frequencies.

Equation (11.12) says that the wavelet transform $Y(s,\tau)$ is a measure of the amount of basis function $\psi_{s,\tau}(t)$ present in the signal $y(t)$. The τ variable indicates the time portion of the signal being decomposed, while the s variable is equivalent to the frequency present during that time:

$$\omega = \frac{2\pi}{s}, \qquad s = \frac{2\pi}{\omega} \quad \text{(scale–frequency relation)}. \tag{11.13}$$

Because it is key to much that follows, it is a good idea to think about (11.13) for a while. If we are interested in the time *details* of a signal, then this is another way of saying that we are interested in what is happening at small values of the *scale* s. Equation (11.13) indicates that small values of s correspond to high-frequency components of the signal. That being the case, the time details of the signal are in the high-frequency, or low-scale, components.

11.4.1 Generating Wavelet Basis Functions

The conceptual discussion of wavelets is over, and it is time to get to work. We first need a technique for generating wavelet bases, and then we need to discretize this technique. As is often the case, the final formulation will turn out to be simple and short, but it will be a while before we get there.

Just as the expansion of a function in a complete orthogonal set is not restricted to any particular set, so the theory of wavelets is not restricted to any particular wavelet basis, although there is some art involved in choosing the most appropriate wavelets for a given signal. The standard way to generate a family of wavelet basis functions starts with $\Psi(t)$, a *mother* or *analyzing* function of the real variable t, and then use this to generate daughter wavelets. As a case in point, let us start with the mother wavelet

$$\Psi(t) = \sin(8t)e^{-t^2/2}. \tag{11.14}$$

We then generate the four wavelet basis functions displayed in Figure 11.4 by scaling, translating, and normalizing this mother wavelet:

$$\psi_{s,\tau}(t) \stackrel{\text{def}}{=} \frac{1}{\sqrt{s}}\Psi\left(\frac{t-\tau}{s}\right) = \frac{1}{\sqrt{s}}\sin\left[\frac{8(t-\tau)}{s}\right] e^{-(t-\tau)^2/2s^2}. \tag{11.15}$$

We see that larger or smaller values of s, respectively, expand or contract the mother wavelet, while different values of τ shift the center of the wavelet. Because the wavelets are inherently oscillatory, the scaling leads to the same number of oscillations occurring in different time spans, which is equivalent to having basis states

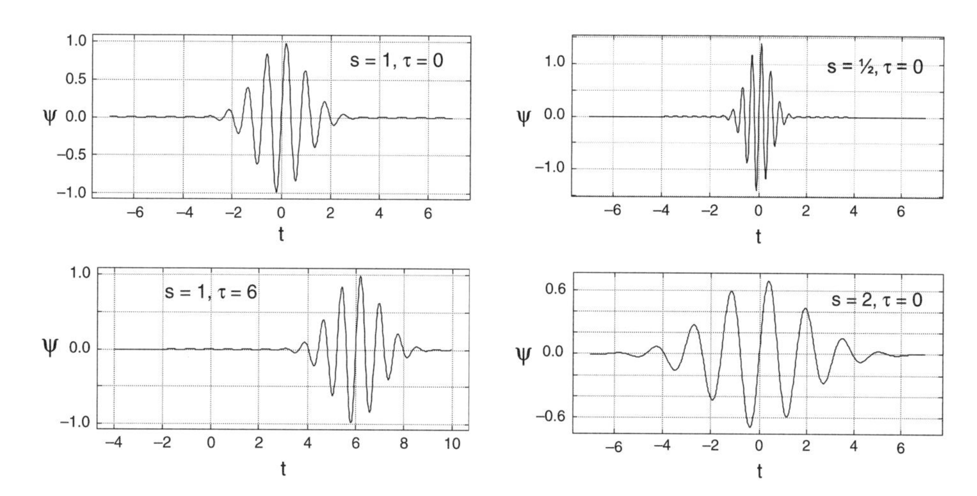

Figure 11.4 Four wavelet basis functions generated by scaling (s) and translating (τ) the oscillating Gaussian mother wavelet. *Clockwise from top:* ($s = 1, \tau = 0$), ($s = 1/2, \tau = 0$), ($s = 1, \tau = 6$), and ($s = 2, \tau = 60$). Note how $s < 1$ is a wavelet with higher frequency, while $s > 1$ has a lower frequency than the $s = 1$ mother. Likewise, the $\tau = 6$ wavelet is just a translated version of the $\tau = 0$ one directly above it.

with differing frequencies. We see that $s < 1$ produces a higher-frequency wavelet, while $s > 1$ produces a lower-frequency one, both of the same shape. As we shall see, we do not need to store much information to outline the large-time-scale s behavior of a signal (its *smooth envelope*), but we do need more information to specify its short-time-scale s behavior (*details*). And if we want to resolve finer features in the signal, then we will need to have more information on yet finer details. Here the division by $\sqrt{s}$ is made to ensure that there is equal "power" (or energy or intensity) in each region of s, although other normalizations can also be found in the literature. After substituting in the daughters, the wavelet transform (11.12) and its inverse [VdB 99] are

$$\boxed{Y(s,\tau) = \frac{1}{\sqrt{s}} \int_{-\infty}^{+\infty} dt\, \Psi^*\left(\frac{t-\tau}{s}\right) y(t)} \quad \textbf{(wavelet transform)}, \qquad (11.16)$$

$$\boxed{y(t) = \frac{1}{C} \int_{-\infty}^{+\infty} d\tau \int_{0}^{+\infty} ds \frac{\psi^*_{s,\tau}(t)}{s^{3/2}} Y(s,\tau)} \quad \textbf{(inverse transform)}, \qquad (11.17)$$

where the normalization constant C depends on the wavelet used.

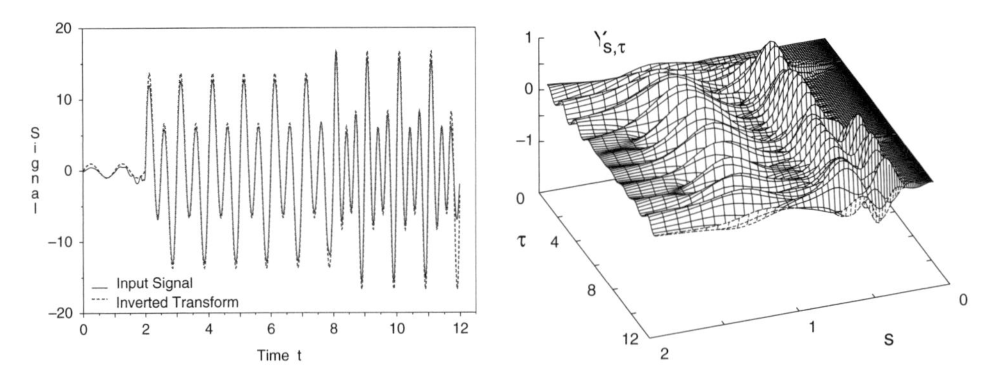

Figure 11.5 *Left:* Comparison of the input and reconstituted signal using Morlet wavelets. As expected for Fourier transforms, the reconstruction is least accurate near the endpoints. *Right:* The continuous wavelet spectrum obtained by analyzing the input signal (11.18) with Morelet wavelets. Observe how at small values of time τ there is predominantly one frequency present, how a second, higher-frequency (smaller-scale) component enters at intermediate times, and how at larger times a still higher-frequency component enters. Further observation indicates that the large-s component has an amplitude consistent with the input. (Figure courtesy of Z. Dimcovic.)

The general requirements for a mother wavelet Ψ are [Add 02, VdB 99]

1. $\Psi(t)$ is real.
2. $\Psi(t)$ oscillates around zero such that its average is zero:

$$\int_{-\infty}^{+\infty} \Psi(t)\, dt = 0.$$

3. $\Psi(t)$ is local, that is, a wave packet, and is square-integrable:

$$\Psi(|t| \to \infty) \to 0 \quad \text{(rapidly)}, \qquad \int_{-\infty}^{+\infty} |\Psi(t)|^2\, dt < \infty.$$

4. The transforms of low powers of t vanish, that is, the first p moments:

$$\int_{-\infty}^{+\infty} t^0\, \Psi(t)\, dt = \int_{-\infty}^{+\infty} t^1\, \Psi(t)\, dt = \cdots = \int_{-\infty}^{+\infty} t^{p-1}\, \Psi(t)\, dt = 0.$$

This makes the transform more sensitive to details than to general shape.

You can think of scale as being like the scale on a map (also discussed in §13.5.2 with reference to fractal analysis) or in terms of *resolution*, as might occur in photographic images. Regardless of the words, we will see in Chapter 12, "Discrete & Continuous Nonlinear Dynamics," that if we have a fractal, then we have a self-similar object that looks the same at all scales or resolutions. Similarly, each

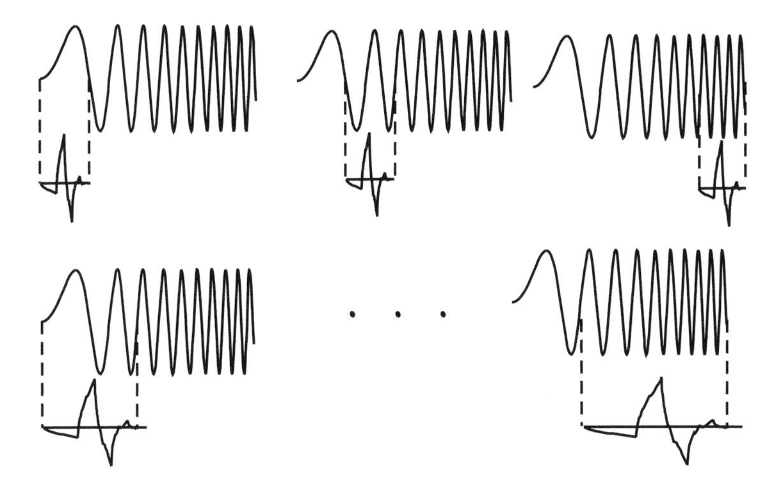

Figure 11.6 A signal is analyzed by starting with a narrow wavelet at the signal's beginning and producing a coefficient that measures the similarity of the signal to the wavelet. The wavelet is successively shifted over the length of the signal. Then the wavelet is expanded and the analysis repeated.

wavelet basis function in a set is self-similar to the others, but at a different scale or location.

In summary, wavelet bases are functions of the time variable t, as well as of the two parameters s and τ. The t variable is integrated over to yield a transform that is a function of the time scale s (frequency $2\pi/s$) and window location τ.

As an example of how we use the two degrees of freedom, consider the analysis of a chirp signal $\sin(60t^2)$ (Figure 11.6). We see that a slice at the beginning of the signal is compared to our first basis function. (The comparison is done via the *convolution* of the wavelet with the signal.) This first comparison is with a narrow version of the wavelet, that is, at low scale, and yields a single coefficient. The comparison at this scale continues with the next signal slice and ends when the entire signal has been covered (the top row in Figure 11.6). Then the wavelet is expanded, and comparisons are repeated. Eventually, the data are processed at all scales and at all time intervals. The narrow signals correspond to a high-resolution analysis, while the broad signals correspond to low resolution. As the scales get larger (lower frequencies, lower resolution), fewer details of the time signal remain visible, but the overall shape or gross features of the signal become clearer.

11.4.2 Continuous Wavelet Transform Implementation

We want to develop some intuition as to what wavelet transforms look like before going on to apply them in unknown situations. Accordingly, modify the program you have been using for the discrete Fourier transform so that it now computes the continuous wavelet transform.

1. You will want to see the effect of using different mother wavelets. Accordingly, write a method that calculates the mother wavelet for
   a. a Morlet wavelet (11.2),
   b. a Mexican hat wavelet (11.3),
   c. a Haar wavelet (the square wave in Figure 11.2).
2. Try out your transform for the following input signals and see if the results make sense:
   a. A pure sine wave $y(t) = \sin 2\pi t$,
   b. A sum of sine waves $y(t) = 2.5 \sin 2\pi t + 6 \sin 4\pi t + 10 \sin 6\pi t$,
   c. The nonstationary signal for our problem (11.1)

$$y(t) = \begin{cases} \sin 2\pi t, & \text{for } 0 \le t \le 2, \\ 5 \sin 2\pi t + 10 \sin 4\pi t, & \text{for } 2 \le t \le 8, \\ 2.5 \sin 2\pi t + 6 \sin 4\pi t + 10 \sin 6\pi t, & \text{for } 8 \le t \le 12. \end{cases} \tag{11.18}$$

   d. The half-wave function

$$y(t) = \begin{cases} \sin \omega t, & \text{for } 0 < t < T/2, \\ 0, & \text{for } T/2 < t < T. \end{cases}$$

3. ⊙ Use (11.17) to invert your wavelet transform and compare the reconstructed signal to the input signal (you can normalize the two to each other). On the right in Figure 11.5 we show our comparison. ▌

In Listing 11.1 we give our *continuous wavelet transformation* `CWTzd.java` [Lang]. Because wavelets, with their transforms in two variables, are somewhat hard to grasp at first, we suggest that you write your own code and include a portion that does the inverse transform as a check (this is in fact what we have done in the instructor's code `Wavelets.java`). In the next section we will describe the *discrete wavelet transformation* that makes optimal discrete choices for the scale and time translation parameters s and τ. Figure 11.5 shows the spectrum produced for the input signal (11.1) in Figure 11.1. As was our goal, we see predominantly one frequency at short times, two frequencies at intermediate times, and three frequencies at longer times.

11.5 Unit II. Discrete Wavelet Transforms and Multiresolution Analysis⊙

As was true for DFTs, if a time signal is measured at only N discrete times,

$$y(t_m) \equiv y_m, \quad m = 1, \ldots, N, \tag{11.19}$$

```java
// CWT_zd.java  Continuous Wavelet Transform. Written by Ziatko Dimcovic
// Outputs transform.dat = TF, recSigNorm.dat =TF^{-1}
import java.io.*;

public class CWTzd {                                          // N.B. many class variables
  public static final double PI = Math.PI;
  public static double iT = 0.0, fT = 12.0, W = fT - iT;                        // i,f times
  public static int N = 1000; public static double h = W/N;                       //Steps
  public static int noPtsSig = N, noS = 100, noTau = 100;                      // # of pts
  public static double iTau = 0., iS = 0.1, tau = iTau, s = iS;
  // Need *very* small s steps for high-frequency, but only if s is small
  // Thus increment s by multiplying by number close enough to 1
  public static double dTau = W/noTau, dS = Math.pow(W/iS, 1./noS);

  public static void main(String[] args) throws IOException, FileNotFoundException {
    System.out.printf("\nUsing:\n\ttau + dTau, dTau = % 3.3f (noTau = % d)"
      + "\n\t s*dS, dS = % 3.3f (noS = % d)% n % n", dTau, noTau, dS, noS);
    String transformData = "transform.dat";                                   // Data file
    double[] sig = new double[noPtsSig];                                         // Signal
    signal(noPtsSig, sig, false);
    double[][] Y = new double[noS][noTau];                                    // Transform
    for (int i = 0; i < noS; i++, s *= dS) {                                    // Scaling
      tau = iT;
      for (int j = 0; j < noTau; j++, tau+=dTau)                            // Translation
        Y[i][j] = transform(s, tau, sig);
    }                                                    // Print normalized TF to file
    PrintWriter wd = new PrintWriter( new FileWriter(transformData), true);
    double maxY = 0.001;
    for (int i = 0; i < noS; i++)
      for (int j = 0; j < noTau; j++)
        if( Y[i][j]>maxY || Y[i][j]<-1*maxY )
          maxY = Math.abs( Y[i][j] );                                        // Find max Y
    tau = iT; s = iS;
    for (int i = 0; i < noS; i++, s*=dS) {                                   // Write data
      for (int j = 0; j < noTau; j++, tau+=dTau) {                            // Transform
        wd.println(s + " " + tau + " " + Y[i][j]/maxY); }                   // Norm to max
      tau = iT;
      wd.println();                                                      // For gnuplot 3D
    }
    wd.close();
                                                                       // Find inverse TF
    String recSigData = "recSig.dat";
    PrintWriter wdRecSig=new PrintWriter(new FileWriter(recSigData),true);
    double[] recSig = new double[sig.length];                          // Same resolution
    double t = 0.0;
    for (int rs = 0; rs < recSig.length; rs++, t += h) {
      recSig[rs] = invTransform(t, Y);
      wdRecSig.println(t + " " + recSig[rs]);                                // Write data
    }
    wdRecSig.close();
    System.out.println("\nDone.\n");  }                                        // End main

  public static double transform(double s, double tau, double[] sig) {
    double integral = 0., t = iT;                     // "initial time" = class variable
    for (int i = 0; i < sig.length; i++, t+=h) integral += sig[i]*morlet(t,s,tau)*h;
    return integral / Math.sqrt(s);  }

  public static double invTransform(double t, double[][] Y)  {
    double s = iS, tau = iTau, recSig_t = 0;                    // Don't change static's
    for (int i = 0; i < noS; i++, s *= dS) {
      tau = iTau;
      for (int j = 0; j < noTau; j++, tau += dTau)
        recSig_t += dTau*dS * Math.pow(s,-1.5)* Y[i][j] * morlet(t,s,tau);
    }
    return recSig_t; }

  public static double morlet(double t, double s, double tau) {                  // Mother
    double T = ((t-tau)/s);
```

```
    return Math.sin(8*T) * Math.exp( -T*T/2 );
  }

  public static void signal(int noPtsSig, double[] y, boolean plotIt) {
    double t = 0.0; double hs = W / noPtsSig;
    for (int i = 0; i < noPtsSig; i++, t+=hs) {
      double t1 = W/6., t2 = 4.*W/6.;
      if ( t>= iT && t<=t1 ) y[i] = Math.sin(2*PI*t);
      else if ( t>=t1 && t<=t2 ) y[i] = 5.*Math.sin(2*PI*t) + 10.*Math.sin(4*PI*t);
      else if ( t>=t2 && t<=fT )
        y[i]=2.5*Math.sin(2*PI*t)+6.*Math.sin(4*PI*t)+10*Math.sin(6*PI*t);
      else {
        System.err.println("\n\tIn signal(...) : t out of range.\n");
        System.exit(1);
      } } }

  public static void setParameters(int ptsTransf) {N=ptsTransf; h=W/N; noPtsSig = N;}

  public static void setParameters(int ptsTau, int ptsS) {noTau = ptsTau; noS = ptsS;
    dTau = W/noTau; dS = Math.pow(W/iS, 1./noS); }

  public static void setParameters(int ptsTransf, int ptsTau, int ptsS) {
    N = ptsTransf; h = W/N; noPtsSig = N;
    noTau = ptsTau; noS = ptsS;
    dTau = W/noTau; dS = Math.pow(W/iS, 1./noS); }
  }                                                              // End class
```

Listing 11.1 CWTzd.java computes a normalized continuous wavelet transform of the signal data in input[ ] (here assigned as a sum of sine functions) using Morlet wavelets (courtesy of Z. Dimcovic). The discrete wavelet transform (DWT) in Listing 11.2 is faster and yields a compressed transform but is less transparent.

then we can determine only N independent components of the transform Y. The trick is to compute only the N independent components required to reproduce the input signal, consistent with the uncertainty principle. The *discrete wavelet transform* (DWT) technique evaluates the transforms with discrete values for the scaling parameter s and the time translation parameter τ:

$$\psi_{j,k}(t) = \frac{\Psi\left[(t-k2^j)/2^j\right]}{\sqrt{2^j}} \equiv \frac{\Psi\left(t/2^j - k\right)}{\sqrt{2^j}} \quad \text{(DWT)}, \tag{11.20}$$

$$s = 2^j, \quad \tau = \frac{k}{2^j}, \quad k,\ j = 0, 1, \ldots. \tag{11.21}$$

Here j and k are integers whose maximum values are yet to be determined, and we have assumed that the total time interval $T = 1$, so that time is always measured in integer values. This choice of s and τ, based on powers of 2, is called a *dyadic grid* arrangement and is seen to automatically perform the scalings and translations at the different time scales that are at the heart of wavelet analysis.[3] The discrete

[3] Note that some references scale down with increasing j, in contrast to our scaling up.

wavelet transform now becomes

$$Y_{j,k} = \int_{-\infty}^{+\infty} dt\, \psi_{j,k}(t)\, y(t) \simeq \sum_m \psi_{j,k}(t_m) y(t_m) h \quad \text{(DWT)}, \tag{11.22}$$

where the discreteness here refers to the wavelet basis set and *not* the time variable. For an orthonormal wavelet basis, the inverse discrete transform is then

$$y(t) = \sum_{j,\,k=-\infty}^{+\infty} Y_{j,k}\, \psi_{j,k}(t) \quad \text{(inverse DWT)}. \tag{11.23}$$

This inversion will exactly reproduce the input signal at the N input points if we sum over an infinite number of terms [Add 02]. Practical calculations will be less exact.

Notice in (11.20) and (11.22) that we have kept the time variable t in the wavelet basis functions continuous, even though s and τ are made discrete. This is useful in establishing the orthonormality of the basis functions,

$$\int_{-\infty}^{+\infty} dt\, \psi^*_{j,k}(t)\, \psi_{j',k'}(t) = \delta_{jj'}\, \delta_{kk'}, \tag{11.24}$$

where $\delta_{m,n}$ is the Kronecker delta function. Being normalized to 1 means that each wavelet basis has "unit energy"; being orthogonal means that each basis function is independent of the others. And because wavelets are localized in time, the different transform components have low levels of correlation with each other. Altogether, this leads to efficient and flexible data storage.

The use of a discrete wavelet basis makes it clear that we sample the input signal at the discrete values of time determined by the integers j and k. In general, you want time steps that sample the signal at enough times in each interval to obtain the desired level of precision. A rule of thumb is to start with 100 steps to cover each major feature. Ideally, the needed times correspond to the times at which the signal was sampled, although this may require some forethought.

Consider Figure 11.7. We measure a signal at a number of discrete times within the intervals (k or τ values) corresponding to the vertical columns of fixed width along the time axis. For each time interval, we want to sample the signal at a number of scales (frequencies or j values). However, as discussed in §11.2, the basic mathematics of Fourier transforms indicates that the width Δt of a wave packet $\psi(t)$ and the width $\Delta\omega$ of its Fourier transform $Y(\omega)$ are related by an uncertainty principle

$$\Delta\omega\, \Delta t \geq 2\pi.$$

This relation constrains the number of times we can meaningfully sample a signal in order to determine a number of Fourier components. So while we may want a high-resolution reproduction of our signal, we do not want to store more data than are

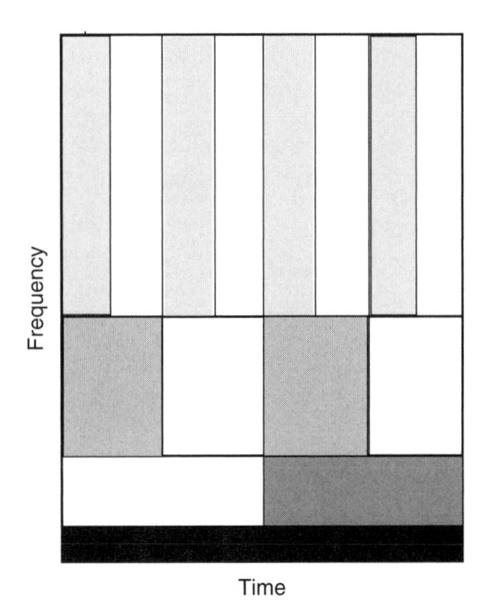

Figure 11.7 Time and frequency resolutions. Each box represents an equal portion of the time–frequency plane but with different proportions of time and frequency.

needed to obtain that reproduction. If we sample the signal for times centered about some τ in an interval of width $\Delta\tau$ (Figure 11.7) and then compute the transform at a number of scales s or frequencies $\omega = 2\pi/s$ covering a range of height $\Delta\omega$, then the relation between the height and width is restricted by the uncertainty relation, which means that each of the rectangles in Figure 11.7 has the same area $\Delta\omega\,\Delta t = 2\pi$. The increasing heights of the rectangles at higher frequencies means that a larger range of frequencies should be sampled as the frequency increases. The premise here is that the low-frequency components provide the gross or *smooth* outline of the signal which, being smooth, does not require much detail, while the high-frequency components give the details of the signal over a short time interval and so require many components in order to record these details with high resolution.

Industrial-strength wavelet analyses do not compute explicit integrals but instead apply a technique known as *multiresolution analysis* (MRA). We give an example of this technique in Figure 11.8 and in the code `DWT.java` in Listing 11.2. It is based on a *pyramid algorithm* that samples the signal at a finite number of times and then passes it successively through a number of *filters*, with each filter representing a digital version of a wavelet.

Filters were discussed in §10.7, where in (10.59) we defined the action of a linear filter as a convolution of the filter response function with the signal. A comparison of the definition of a filter to the definition of a wavelet transform (11.12) shows that the two are essentially the same. Such being the case, the result of the transform operation is a weighted sum over the input signal values, with each weight the product of the integration weight times the value of the wavelet function at the

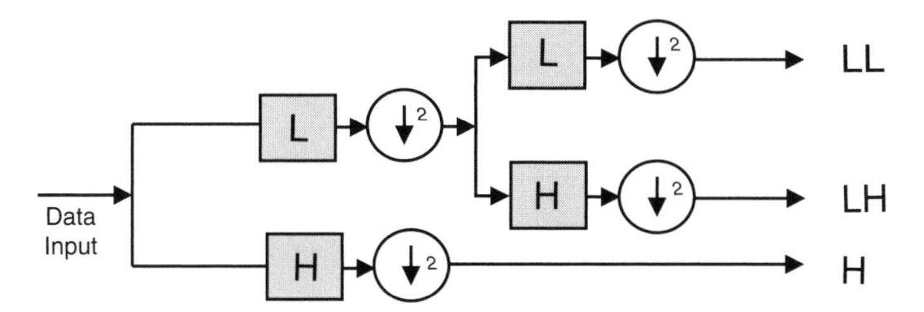

Figure 11.8 A multifrequency dyadic (power-of-2) filter tree used for discrete wavelet transformation. The L boxes represent lowpass filters, while the H boxes represent highpass filters, each of which performs a convolution (transform). The circles containing ↓2 filter out half of the signal that enters them, which is called *subsampling* or *factor-of-2 decimation*.

integration point. Therefore, *rather than tabulate explicit wavelet functions, a set of filter coefficients is all that is needed for discrete wavelet transforms*.

Because each filter in Figure 11.8 changes the relative strengths of the different frequency components, passing the signal through a series of filters is equivalent, in the wavelet sense, to analyzing the signal at different scales. This is the origin of the name "multiresolution analysis." Figure 11.8 shows how the pyramid algorithm passes the signal through a series of highpass filters (H) and then through a series of lowpass filters (L). Each filter changes the scale to that of the level below. Notice too, the circles containing ↓2 in Figure 11.8. This operation filters out half of the signal and so is called *subsampling* or *factor-of-2 decimation*. It is the way we keep the areas of each box in Figure 11.7 constant as we vary the scale and translation times. We consider subsampling further when we discuss the pyramid algorithm.

In summary, the DWT process decomposes the signal into *smooth* information stored in the low-frequency components and *detailed* information stored in the high-frequency components. Because *high-resolution* reproductions of signals require more information about details than about gross shape, the pyramid algorithm is an effective way to compress data while still maintaining high resolution (we implement compression in Appendix G). In addition, because components of different resolution are independent of each other, it is possible to lower the number of data stored by systematically eliminating higher-resolution components. The use of wavelet filters builds in progressive scaling, which is particularly appropriate for fractal-like reproductions.

11.5.1 Pyramid Scheme Implementation ⊙

We now wish to implement the pyramid scheme outlined in Figure 11.8. The filters L and H will be represented by matrices, which is an approximate way to perform

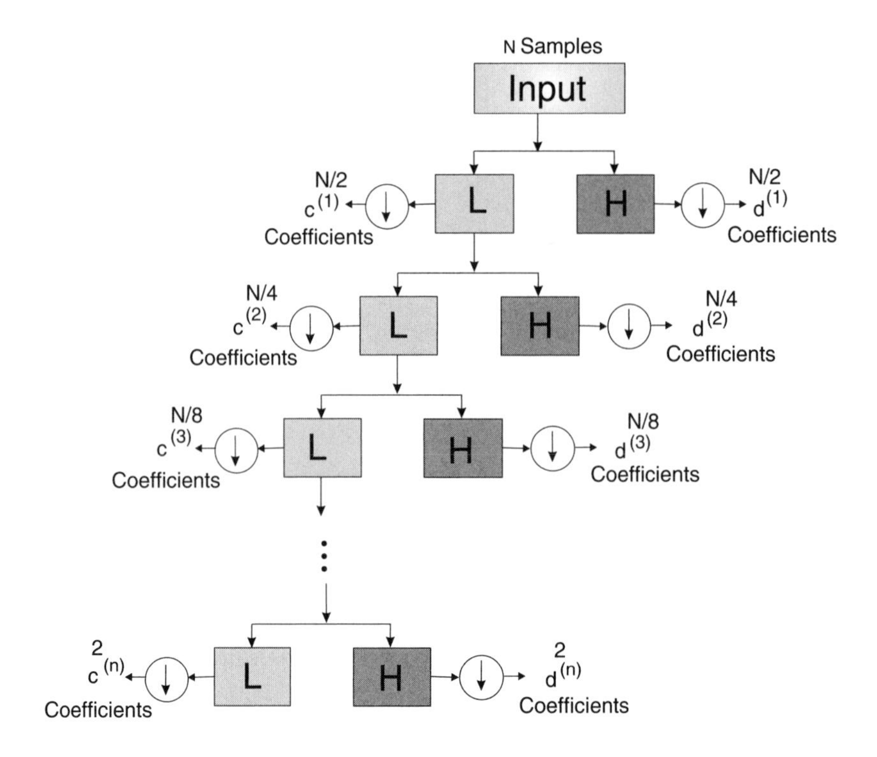

Figure 11.9 The original signal is processed by high- and low-band filters, and the outputs are downsampled with every other point kept. The process continues until there are only two output points of high-band filtering and two points of low-band filtering. The total number of output data equals the total number of signal points. It may be easier to understand the output of such an analysis as the signal passed through various filters rather than as a set of Fourier-like coefficients.

the integrations or convolutions. Then there is a decimation of the output by one-half, and finally an interleaving of the output for further filtering. This process simultaneously cuts down on the number of points in the data set and changes the scale and the resolution. The decimation reduces the number of values of the remaining signal by one half, with the low-frequency part discarded because the details are in the high-frequency parts.

As indicated in Figure 11.9, the pyramid algorithm's DWT successively (1) applies the (soon-to-be-derived) c matrix (11.35) to the whole N-length vector,

$$\begin{pmatrix} Y_0 \\ Y_1 \\ Y_2 \\ Y_3 \end{pmatrix} = \begin{pmatrix} c_0 & c_1 & c_2 & c_3 \\ c_3 & -c_2 & c_1 & -c_0 \\ c_2 & c_3 & c_0 & c_1 \\ c_1 & -c_0 & c_3 & -c_2 \end{pmatrix} \begin{pmatrix} y_0 \\ y_1 \\ y_2 \\ y_3 \end{pmatrix}, \tag{11.25}$$

(2) applies it to the $(N/2)$-length smooth vector, (3) and then repeats until two smooth components remain. (4) After each filtering, the elements are ordered, with the newest two smooth elements on top, the newest detailed elements below, and the older detailed elements below that. (5) The process continues until there are just two smooth elements left.

To illustrate, here we filter and reorder an initial vector of length $N = 8$:

$$\begin{pmatrix} y_1 \\ y_2 \\ y_3 \\ y_4 \\ y_5 \\ y_6 \\ y_7 \\ y_8 \end{pmatrix} \overset{\text{filter}}{\longrightarrow} \begin{pmatrix} s_1^{(1)} \\ d_1^{(1)} \\ s_2^{(1)} \\ d_2^{(1)} \\ s_3^{(1)} \\ d_3^{(1)} \\ s_4^{(1)} \\ d_4^{(1)} \end{pmatrix} \overset{\text{order}}{\longrightarrow} \begin{pmatrix} s_1^{(1)} \\ s_2^{(1)} \\ s_3^{(1)} \\ s_4^{(1)} \\ \hline d_1^{(1)} \\ d_2^{(1)} \\ d_3^{(1)} \\ d_4^{(1)} \end{pmatrix} \overset{\text{filter}}{\longrightarrow} \begin{pmatrix} s_1^{(2)} \\ d_1^{(2)} \\ s_2^{(2)} \\ d_2^{(2)} \\ \hline d_1^{(1)} \\ d_2^{(1)} \\ d_3^{(1)} \\ d_4^{(1)} \end{pmatrix} \overset{\text{order}}{\longrightarrow} \begin{pmatrix} s_1^{(2)} \\ s_2^{(2)} \\ d_1^{(2)} \\ d_2^{(2)} \\ \hline d_1^{(1)} \\ d_2^{(1)} \\ d_3^{(1)} \\ d_4^{(1)} \end{pmatrix}. \tag{11.26}$$

The discrete inversion of a transform vector back to a signal vector is made using the transpose (inverse) of the transfer matrix at each stage. For instance,

$$\begin{pmatrix} y_0 \\ y_1 \\ y_2 \\ y_3 \end{pmatrix} = \begin{pmatrix} c_0 & c_3 & c_2 & c_1 \\ c_1 & -c_2 & c_3 & -c_0 \\ c_2 & c_1 & c_0 & c_3 \\ c_3 & -c_0 & c_1 & -c_2 \end{pmatrix} \begin{pmatrix} Y_0 \\ Y_1 \\ Y_2 \\ Y_3 \end{pmatrix}. \tag{11.27}$$

As a more realistic example, imagine that we have sampled the chirp signal $y(t) = \sin(60t^2)$ for 1024 times. The filtering process through which we place this signal is illustrated as a passage from the top to the bottom in Figure 11.9. First the original 1024 samples are passed through a single low band and a single high band (which is mathematically equivalent to performing a series of convolutions). As indicated by the down arrows, the output of the first stage is then downsampled (the number reduced by a factor of 2). This results in 512 points from the high-band filter as well as 512 points from the low-band filter. This produces the first-level output. The output coefficients from the high-band filters are called $\{d_i^{(1)}\}$ to indicate that they show details, and $\{s_i^{(1)}\}$ to indicate that they show smooth features. The superscript indicates that this is the first level of processing. The detail coefficients $\{d^{(1)}\}$ are stored to become part of the final output.

In the next level down, the 512 smooth data $\{s_i^{(1)}\}$ are passed through new low- and high-band filters using a broader wavelet. The 512 outputs from each are downsampled to form a smooth sequence $\{s_i^{(2)}\}$ of size 256 and a detailed sequence $\{d_i^{(2)}\}$ of size 256. Again the detail coefficients $\{d^{(2)}\}$ are stored to become part of the final output. (Note that this is only half the size of the previously stored details.) The process continues until there are only two numbers left for the detail coefficients and two numbers left for the smooth coefficients. Since this last filtering is done

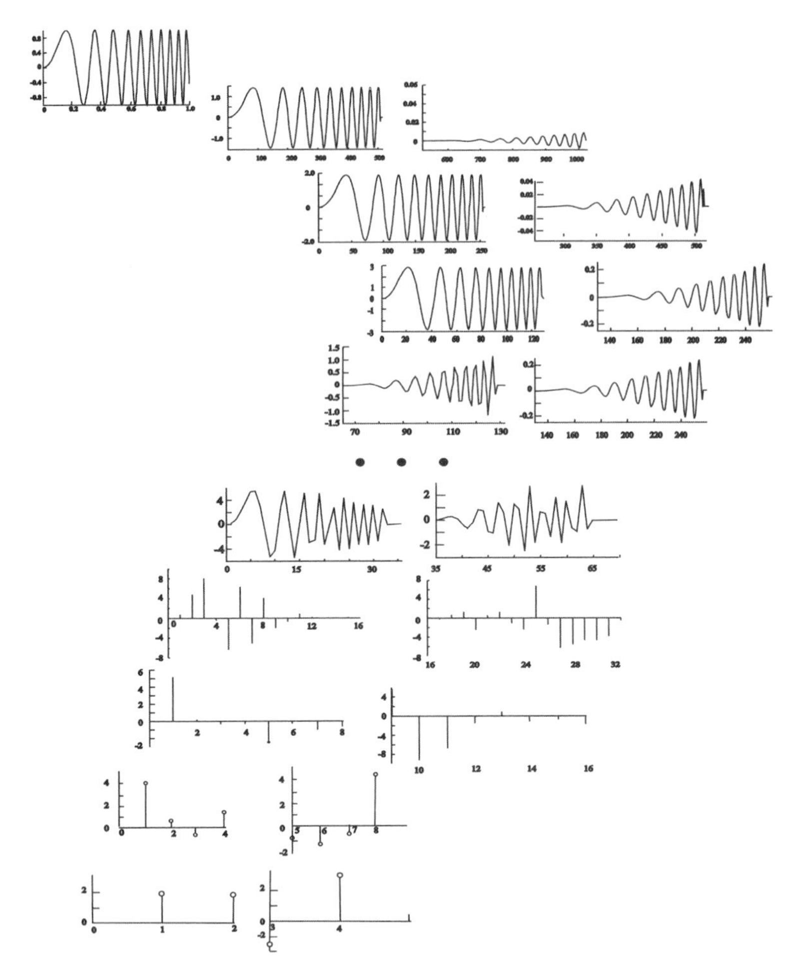

Figure 11.10 The filtering of the original signal at the top goes through the pyramid algorithm and produces the outputs shown, in successive passes. The sampling is reduced by a factor of 2 in each step. Note that in the upper graphs we have connected the points to make the output look continuous, while in the lower graphs, with fewer points, we have plotted the output as histograms to make the points more evident.

with the broadest wavelet, it is of the lowest resolution and therefore requires the least information.

In Figure 11.10 we show the actual effects on the chirp signal of pyramid filtering for various levels in the processing. (The processing is done with four-coefficient *Daub4* wavelets, which we will discuss soon.) At the uppermost level, the Daub4 wavelet is narrow, and so convoluting this wavelet with successive sections of the signal results in smooth components that still contain many large high-frequency parts. The detail components, in contrast, are much smaller in magnitude. In the next stage, the wavelet is dilated to a lower frequency and the analysis is repeated *on just the smooth (low-band) part.* The resulting output is similar, but with coarser features for the smooth coefficients and larger values for the details. Note that in the upper graphs we have connected the points to make the output look continuous, while in the lower graphs, with fewer points, we have plotted the output as histograms to make the points more evident. Eventually the downsampling leads to just two coefficients output from each filter, at which point the filtering ends.

To reconstruct the original signal (called *synthesis* or *transformation*) a reversed process is followed: Begin with the last sequence of four coefficients, upsample them, pass them through low- and high-band filters to obtain new levels of coefficients, and repeat until all the N values of the original signal are recovered. The inverse scheme is the same as the processing scheme (Figure 11.9), only now the direction of all the arrows is reversed.

11.5.2 Daubechies Wavelets via Filtering

We should now be able to understand that digital wavelet analysis has been standardized to the point where classes of wavelet basis functions are specified not by their analytic forms but rather by their *wavelet filter coefficients*. In 1988, the Belgian mathematician Ingrid Daubechies discovered an important class of such filter coefficients [Daub 95]. We will study just the Daub4 class containing the four coefficients c_0, c_1, c_2, and c_3.

Imagine that our input contains the four elements $\{y_1, y_2, y_3, y_4\}$ corresponding to measurements of a signal at four times. We represent a lowpass filter L and a highpass filter H in terms of the four filter coefficients as

$$L = \begin{pmatrix} +c_0 & +c_1 & +c_2 & +c_3 \end{pmatrix} \tag{11.28}$$

$$H = \begin{pmatrix} +c_3 & -c_2 & +c_1 & -c_0 \end{pmatrix}. \tag{11.29}$$

To see how this works, we form an input vector by placing the four signal elements in a column and then multiply the input by L and H:

$$L \begin{pmatrix} y_0 \\ y_1 \\ y_2 \\ y_3 \end{pmatrix} = \begin{pmatrix} +c_0 & +c_1 & +c_2 & +c_3 \end{pmatrix} \begin{pmatrix} y_0 \\ y_1 \\ y_2 \\ y_3 \end{pmatrix} = c_0 y_0 + c_1 y_1 + c_2 y_2 + c_3 y_3,$$

$$H \begin{pmatrix} y_0 \\ y_1 \\ y_2 \\ y_3 \end{pmatrix} = \begin{pmatrix} +c_3 & -c_2 & +c_1 & -c_0 \end{pmatrix} \begin{pmatrix} y_0 \\ y_1 \\ y_2 \\ y_3 \end{pmatrix} = c_3 y_0 - c_2 y_1 + c_1 y_2 - c_0 y_3.$$

We see that if we choose the values of the c_i's carefully, the result of L acting on the signal vector is a single number that may be viewed as a weighted average of the four input signal elements. Since an averaging process tends to smooth out data, the lowpass filter may be thought of as a *smoothing filter* that outputs the general shape of the signal.

In turn, we see that if we choose the c_i values carefully, the result of H acting on the signal vector is a single number that may be viewed as the weighted differences of the input signal. Since a differencing process tends to emphasize the variation in the data, the highpass filter may be thought of as a *detail* filter that produces a large output when the signal varies considerably, and a small output when the signal is smooth.

We have just seen how the individual L and H filters, each represented by a single row, output one number when acting upon an input signal containing four elements in a column. If we want the output of the filtering process Y to contain the same number of elements as the input (four y's in this case), we just stack the L and H filters together:

$$\begin{pmatrix} Y_0 \\ Y_1 \\ Y_2 \\ Y_3 \end{pmatrix} = \begin{pmatrix} L \\ H \\ L \\ H \end{pmatrix} \begin{pmatrix} y_0 \\ y_1 \\ y_2 \\ y_3 \end{pmatrix} = \begin{pmatrix} c_0 & c_1 & c_2 & c_3 \\ c_3 & -c_2 & c_1 & -c_0 \\ c_2 & c_3 & c_0 & c_1 \\ c_1 & -c_0 & c_3 & -c_2 \end{pmatrix} \begin{pmatrix} y_0 \\ y_1 \\ y_2 \\ y_3 \end{pmatrix}. \tag{11.30}$$

Of course the first and third rows of the Y vector will be identical, as will the second and fourth, but we will get to that soon.

Now we go about determining the values of the filter coefficients c_i by placing specific demands upon the output of the filter. We start by recalling that in our discussion of discrete Fourier transforms we observed that a transform is equivalent to a rotation from the time domain to the frequency domain. Yet we know from our study of linear algebra that rotations are described by orthogonal matrices, that is, matrices whose inverses are equal to their transposes. In order for the inverse transform to return us to the input signal, the transfer matrix must be orthogonal. For our wavelet transformation to be orthogonal, we must have the 4×4 filter matrix times its transpose equal to the identity matrix:

$$\begin{pmatrix} c_0 & c_1 & c_2 & c_3 \\ c_3 & -c_2 & c_1 & -c_0 \\ c_2 & c_3 & c_0 & c_1 \\ c_1 & -c_0 & c_3 & -c_2 \end{pmatrix} \begin{pmatrix} c_0 & c_3 & c_2 & c_1 \\ c_1 & -c_2 & c_3 & -c_0 \\ c_2 & c_1 & c_0 & c_3 \\ c_3 & -c_0 & c_1 & -c_2 \end{pmatrix} = \begin{pmatrix} 1 & 0 & 0 & 0 \\ 0 & 1 & 0 & 0 \\ 0 & 0 & 1 & 0 \\ 0 & 0 & 0 & 1 \end{pmatrix},$$

$$\Rightarrow \quad c_0^2 + c_1^2 + c_2^2 + c_3^2 = 1, \quad c_2 c_0 + c_3 c_1 = 0. \tag{11.31}$$

Two equations in four unknowns are not enough for a unique solution, so we now include the further requirement that the detail filter $H = (c_3,\ -c_0,\ c_1,\ -c_2)$ must output a zero if the input is smooth. We define "smooth" to mean that the input is constant or linearly increasing:

$$\begin{pmatrix} y_0 & y_1 & y_2 & y_3 \end{pmatrix} = \begin{pmatrix} 1 & 1 & 1 & 1 \end{pmatrix} \quad \text{or} \quad \begin{pmatrix} 0 & 1 & 2 & 3 \end{pmatrix}. \tag{11.32}$$

This is equivalent to demanding that the moments up to order p are zero, that is, that we have an "approximation of order p." Explicitly,

$$H\left(y_0 \quad y_1 \quad y_2 \quad y_3\right) = H\left(1 \quad 1 \quad 1 \quad 1\right) = H\left(0 \quad 1 \quad 2 \quad 3\right) = 0,$$

$$\Rightarrow \quad c_3 - c_2 + c_1 - c_0 = 0, \quad 0 \times c_3 - 1 \times c_2 + 2 \times c_1 - 3 \times c_0 = 0,$$

$$\Rightarrow \quad c_0 = \frac{1+\sqrt{3}}{4\sqrt{2}} \simeq 0.483, \quad c_1 = \frac{3+\sqrt{3}}{4\sqrt{2}} \simeq 0.836, \tag{11.33}$$

$$c_2 = \frac{3-\sqrt{3}}{4\sqrt{2}} \simeq 0.224, \quad c_3 = \frac{1-\sqrt{3}}{4\sqrt{2}} \simeq -0.129. \tag{11.34}$$

We now have our basic Daub4 filter coefficients. They can be used to process signals with more than four elements by creating a square filter matrix of the needed dimension that repeats these elements by placing the row versions of L and H along the diagonal, with successive pairs displaced two columns to the right. For example, for eight elements,

$$\begin{pmatrix} Y_0 \\ Y_1 \\ Y_2 \\ Y_3 \\ Y_4 \\ Y_5 \\ Y_6 \\ Y_7 \end{pmatrix} = \begin{pmatrix} c_0 & c_1 & c_2 & c_3 & 0 & 0 & 0 & 0 \\ c_3 & -c_2 & c_1 & -c_0 & 0 & 0 & 0 & 0 \\ 0 & 0 & c_0 & c_1 & c_2 & c_3 & 0 & 0 \\ 0 & 0 & c_3 & -c_2 & c_1 & -c_0 & 0 & 0 \\ 0 & 0 & 0 & 0 & c_0 & c_1 & c_2 & c_3 \\ 0 & 0 & 0 & 0 & c_3 & -c_2 & c_1 & -c_0 \\ c_2 & c_3 & 0 & 0 & 0 & 0 & c_0 & c_1 \\ c_1 & -c_0 & 0 & 0 & 0 & 0 & c_3 & -c_2 \end{pmatrix} \begin{pmatrix} y_0 \\ y_1 \\ y_2 \\ y_3 \\ y_4 \\ y_5 \\ y_6 \\ y_7 \end{pmatrix}. \tag{11.35}$$

Note that in order not to lose any information, the last pair on the bottom two rows is wrapped over to the left. If you perform the actual multiplications indicated in (11.35), you will note that the output has successive *smooth* and *detailed* information. The output is processed with the pyramid scheme.

The time dependences of two Daub4 wavelets is displayed in Figure 11.11. To obtain these from our filter coefficients, first imagine that an elementary wavelet $y_{1,1}(t) \equiv \psi_{1,1}(t)$ is input into the filter. This should result in a transform $Y_{1,1} = 1$. Inversely, we obtain $y_{1,1}(t)$ by applying the inverse transform to a Y vector with a 1 in the first position and zeros in all the other positions. Likewise, the ith member of the Daubechies class is obtained by applying the inverse transform to a Y vector with a 1 in the ith position and zeros in all the other positions. Our code for doing this is `Scale.java` and is on the CD.

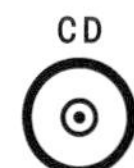

On the left in Figure 11.11 is the wavelet for coefficient 6 (thus the e6 notation). On the right in Figure 11.11 is the sum of two wavelets corresponding to the coefficients 10 and 58. We see that the two wavelets have different levels of scale as well as

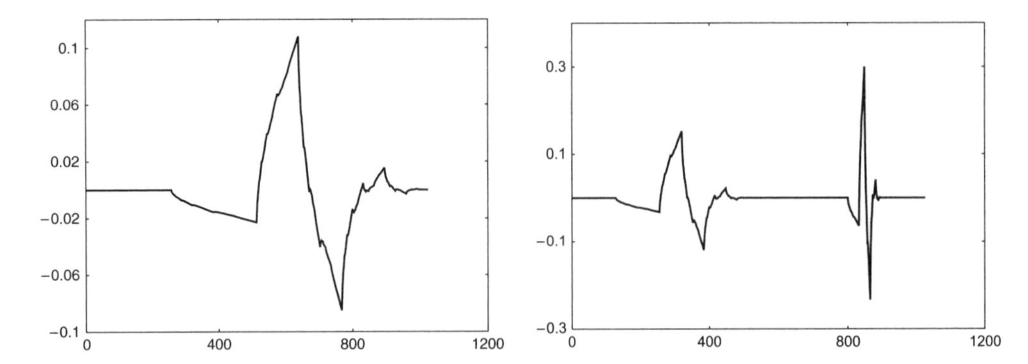

Figure 11.11 *Left:* The Daub4 e6 wavelet constructed by inverse transformation of the wavelet coefficients. *Right:* The sum of Daub4 e10 and Daub4 1e58 wavelets of different scale and time displacements.

different time positions. So even though the time dependence of the wavelets is not evident when wavelet (filter) coefficients are used, it is there.

11.5.3 DWT Implementation and Exercise

Listing 11.2 gives our program for performing a DWT on the chirp signal $y(t) = \sin(60t^2)$. The method pyram calls the daube4 method to perform the DWT or inverse DWT, depending upon the value of sign.

```
// DWT.java: DUAB4 Wavelet TF; input=indata.dat, output=outdata.dat
// sign = + 1: DWT, -1: InvDWT,  2**n input
import java.io.*;

public class DWT {
  static double c0, c1, c2, c3;                                  // Global variables
  static int N = 1024, n;                                        // 2^n  data points

  public static void main(String[] argv) throws IOException, FileNotFoundException {
    PrintWriter w = new PrintWriter (new FileOutputStream("indata.dat"), true);
    PrintWriter q = new PrintWriter (new FileOutputStream("outdata.dat"), true);
    System.out.println("DWT: input = indata.dat, output = outdata.dat");
    int i, dn;
    double xi, inxi;  double f[] = new double[N + 1];                    // Data vector
    inxi = 1./(double)(N);                                    // For chirp 0 <= t <= 1
    xi = 0. ;
    for ( i=1; i <= N; i++ ) {
      f[i] = chirp(xi);                                     // Change for other signal
      xi = xi + inxi;
      w.println(" " + xi + " " + f[i] + " ");                          // Indata.dat
    }
    n = N;                                          // number datapoints must = power of 2
    pyram(f, n, 1);                                          // DWTF (1 -> -1 for inverse)
    // pyram(f, n, -1);                                           // (1 -> -1 for inverse)
```

```
    for (i=1; i <= N; i++ ) q.println(" " +i+ " " +f[i]+ " ");
  }                                                              // Main

  public static void pyram(double f[], int n, int sign) {        // Pyramid
    int nd, nend;
    double sq3, fsq2;                                    // sqrt(3), 4sqrt(2)
    if (n < 4) return;                                       // Too few data
    sq3 = Math.sqrt(3);
    fsq2 = 4.*Math.sqrt(2);                              // DAUB4 coefficients
    c0 = (1. +sq3)/fsq2; c1 = (3. +sq3)/fsq2; c2 = (3.-sq3)/fsq2; c3 = (1.-sq3)/fsq2;
    nend = 4;                        // Controls output (1:1024, 2:512, etc)
    if (sign >= 0) for (nd=n; nd >= nend; nd /= 2) daube4(f, nd, sign);
      else for (nd=4; nd <= n; nd *= 2) daube4(f, nd, sign);
  }

  public static void daube4(double f[], int n, int sign) {
    /** Daubechies 4 DWT or 1/DWT
     * @param f[]: data containing DWT or 1/DWT
     * @param n: decimation after pyramidal algorithm
     * @param sign: >=0 for DWT, @param sign:  < 0 for inverse TF     */
    double tr[] = new double [n + 1];                  // Temporary variable
    int i, j, mp, mp1;
    if (n < 4) return;
    mp = n/2; mp1 = mp + 1;                   // Midpoint Midpoint +1 of array
    if (sign >= 0) {                                                // DWT
      j = 1;
      for ( i = 1; j <= n-3; i++ ) {            // Smooth then detailed filters
        tr[i]      = c0*f[j] + c1*f[j + 1] + c2*f[j + 2] + c3*f[j + 3];
        tr[i + mp] = c3*f[j] - c2*f[j + 1] + c1*f[j + 2] - c0*f[j + 3];
        j += 2;                                              // Downsampling
      }
      tr[i] = c0*f[n-1] + c1*f[n] + c2*f[1] + c3*f[2];                 // Low
      tr[i + mp] = c3*f[n-1]-c2*f[n] + c1*f[1]-c0*f[2];                // High
    } else {                                                   // inverse DWT
      tr[1] = c2*f[mp] + c1*f[n] + c0*f[1] + c3*f[mp1];                // Low
      tr[2] = c3*f[mp]-c0*f[n] + c1*f[1]-c2*f[mp1];                    // High
      for ( i=1, j=3; i < mp; i++ ) {
        tr[j] = c2*f[i] + c1*f[i + mp] + c0*f[i + 1] + c3*f[i + mp1];
        j += 1;                                  // upsamplig c coefficients
        tr[j] = c3*f[i]-c0*f[i + mp] + c1*f[i + 1]-c2*f[i + mp1];
        j += 1;                                 // upsampling d coefficients
      }                                                                // For
      }                                                               // Else
    for ( i=1; i <= n; i++ )    f[i] = tr[i];               // Copy TF in array
  }                                                                 // Daube4

  public static double chirp(double xi) {double y; y = Math.sin(60*xi*xi); return y;}
}                                                                    // Class
```

Listing 11.2 DWT.java computes the discrete wavelet transform using the pyramid algorithm for the 2^n signal values stored in f[ ] (here assigned as the chirp signal $\sin 60t^2$). The Daub4 digital wavelets are the basis functions, and sign $= \pm 1$ for transform/inverse.

1. Modify the program so that you output to a file the values for the input signal that your code has read in. It is always important to check your input.
2. Try to reproduce the left of Figure 11.10 by using various values for the variable nend that controls when the filtering ends. A value nend=1024 should produce just the first step in the downsampling (top row in Figure 11.10). Selecting nend=512 should produce the next row, while nend=4 should output just two smooth and detailed coefficients.

3. Reproduce the scale–time diagram shown on the right in Figure 11.10. This diagram shows the output at different scales and serves to interpret the main components of the signal and the time in which they appear. The time line at the bottom of the figure corresponds to a signal of length 1 over which 256 samples were recorded. The low-band (smooth) components are shown on the left, and the high-band components on the right.
   a. The bottommost figure results when `nend` $= 256$.
   b. The figure in the second row up results from `nend` $= 128$, and we have the output from two filterings. The output contains 256 coefficients but divides time into four intervals and shows the frequency components of the original signal in more detail.
   c. Continue with the subdivisions for `nend` $= 64, 32, 16, 8$, and 4.
4. For each of these choices except the topmost, divide the time by 2 and separate the intervals by vertical lines.
5. The topmost spectrum is your final output. Can you see any relation between it and the chirp signal?
6. Change the sign of `sign` and check that the inverse DWT reproduces the original signal.
7. Use the code to visualize the time dependence of the Daubechies mother function at different scales.
   a. Start by performing an inverse transformation on the eight-component signal `[0,0,0,0,1,0,0,0]`. This should yield a function with a width of about 5 units.
   b. Next perform an inverse transformation on a unit vector with $N = 32$ but with all components except the fifth equal to zero. The width should now be about 25 units, a larger scale but still covering the same time interval.
   c. Continue this procedure until you obtain wavelets of 800 units.
   d. Finally, with $N = 1024$, select a portion of the mother wavelet with data in the horizontal interval [590,800]. This should show self-similarity similar to that at the bottom of Figure 11.11.

12

Discrete & Continuous Nonlinear Dynamics

Nonlinear dynamics is one of the success stories of computational science. It has been explored by mathematicians, scientists, and engineers, with computers as an essential tool. The computations have led to the discovery of new phenomena such as solitons, chaos, and fractals, as you will discover on your own. In addition, because biological systems often have complex interactions and may not be in thermodynamic equilibrium states, models of them are often nonlinear, with properties similar to those of other complex systems.

In Unit I we develop the logistic map as a model for how bug populations achieve dynamic equilibrium. It is an example of a very simple but nonlinear equation producing surprising complex behavior. In Unit II we explore chaos for a continuous system, the driven realistic pendulum. Our emphasis there is on using phase space as an example of the usefulness of an abstract space to display the simplicity underlying complex behavior. In Unit III we extend the discrete logistic map to nonlinear differential models of coupled predator–prey populations and their corresponding phase space plots.

12.1 Unit I. Bug Population Dynamics (Discrete)

Problem: The populations of insects and the patterns of weather do not appear to follow any simple laws.[1] At times they appear stable, at other times they vary periodically, and at other times they appear chaotic, only to settle down to something simple again. Your **problem** is to deduce if a simple, discrete law can produce such complicated behavior.

12.2 The Logistic Map (Model)

Imagine a bunch of insects reproducing generation after generation. We start with N_0 bugs, then in the next generation we have to live with N_1 of them, and after i generations there are N_i bugs to bug us. We want to define a model of how N_n varies with the discrete generation number n. For guidance, we look to the radioactive decay simulation in Chapter 5, "Monte Carlo Simulations", where the

[1] Except maybe in Oregon, where storm clouds come to spend their weekends.

discrete decay law, $\Delta N/\Delta t = -\lambda N$, led to exponential-like decay. Likewise, if we reverse the sign of λ, we should get exponential-like growth, which is a good place to start our modelling. We assume that the bug-breeding rate is proportional to the number of bugs:

$$\frac{\Delta N_i}{\Delta t} = \lambda\, N_i. \tag{12.1}$$

Because we know as an empirical fact that exponential growth usually tapers off, we improve the model by incorporating the observation that bugs do not live on love alone; they must also eat. But bugs, not being farmers, must compete for the available food supply, and this might limit their number to a maximum N_* (called the *carrying capacity*). Consequently, we modify the exponential growth model (12.1) by introducing a growth rate λ' that decreases as the population N_i approaches N_*:

$$\lambda = \lambda'(N_* - N_i) \;\Rightarrow\; \frac{\Delta N_i}{\Delta t} = \lambda'(N_* - N_i)N_i. \tag{12.2}$$

We expect that when N_i is small compared to N_*, the population will grow exponentially, but that as N_i approaches N_*, the growth rate will decrease, eventually becoming negative if N_i exceeds N_*, the carrying capacity.

Equation (12.2) is one form of the *logistic map*. It is usually written as a relation between the number of bugs in future and present generations:

$$N_{i+1} = N_i + \lambda'\,\Delta t(N_* - N_i)N_i, \tag{12.3}$$

$$= N_i\,(1 + \lambda'\,\Delta t N_*)\left[1 - \frac{\lambda'\,\Delta t}{1 + \lambda'\,\Delta t N_*} N_i\right]. \tag{12.4}$$

The map looks simple when expressed in terms of natural variables:

$$\boxed{x_{i+1} = \mu x_i(1 - x_i),} \tag{12.5}$$

$$\mu \stackrel{\text{def}}{=} 1 + \lambda'\,\Delta t N_*, \quad x_i \stackrel{\text{def}}{=} \frac{\lambda'\,\Delta t}{\mu} N_i \simeq \frac{N_i}{N_*}, \tag{12.6}$$

where μ is a dimensionless growth parameter and x_i is a dimensionless population variable. Observe that the *growth rate* μ equals 1 when the breeding rate λ' equals 0, and is otherwise expected to be larger than 1. If the number of bugs born per generation $\lambda'\,\Delta t$ is large, then $\mu \simeq \lambda'\,\Delta t N_*$ and $x_i \simeq N_i/N_*$. That is, x_i is essentially a fraction of the carrying capacity N_*. Consequently, we consider x values in the range $0 \le x_i \le 1$, where $x = 0$ corresponds to no bugs and $x = 1$ to the maximum population. Note that there is clearly a linear, quadratic dependence of the RHS of (12.5) on x_i. In general, a map uses a function $f(x)$ to map one number in a sequence to another,

$$x_{i+1} = f(x_i). \tag{12.7}$$

For the logistic map, $f(x) = \mu x(1-x)$, with the quadratic dependence of f on x making this a nonlinear map, while the dependence on only the one variable x_i makes it a *one-dimensional* map.

12.3 Properties of Nonlinear Maps (Theory)

Rather than do some fancy mathematical analysis to determine the properties of the logistic map [Rash 90], we prefer to have you study it directly on the computer by plotting x_i *versus* generation number i. Some typical behaviors are shown in Figure 12.1. In Figure 12.1A we see equilibration into a single population; in Figure 12.1B we see oscillation between two population levels; in Figure 12.1C we see oscillation among four levels; and in Figure 12.1D we see a chaotic system. The initial population x_0 is known as the *seed*, and as long as it is not equal to zero, its exact value usually has little effect on the population dynamics (similar to what we found when generating pseudorandom numbers). In contrast, the dynamics are unusually sensitive to the value of the growth parameter μ. For those values of μ at which the dynamics are complex, there may be extreme sensitivity to the initial condition x_0 as well as to the exact value of μ.

12.3.1 Fixed Points

An important property of the map (12.5) is the possibility of the sequence x_i reaching a *fixed* point at which x_i remains or fluctuates about. We denote such fixed points as x_*. At a *one-cycle* fixed point, there is no change in the population from generation i to generation $i+1$; that is,

$$x_{i+1} = x_i = x_*. \tag{12.8}$$

Using the logistic map (12.5) to relate x_{i+1} to x_i yields the algebraic equation

$$\mu x_*(1-x_*) = x_* \quad \Rightarrow \quad x_* = 0 \quad \text{or } x_* = \frac{\mu - 1}{\mu}. \tag{12.9}$$

The nonzero fixed point $x_* = (\mu-1)/\mu$ corresponds to a stable population with a balance between birth and death that is reached regardless of the initial population (Figure 12.1A). In contrast, the $x_* = 0$ point is unstable and the population remains static only as long as no bugs exist; if even a few bugs are introduced, exponential growth occurs. Further analysis (§12.8) tells us that the stability of a population is determined by the magnitude of the derivative of the mapping function $f(x_i)$ at

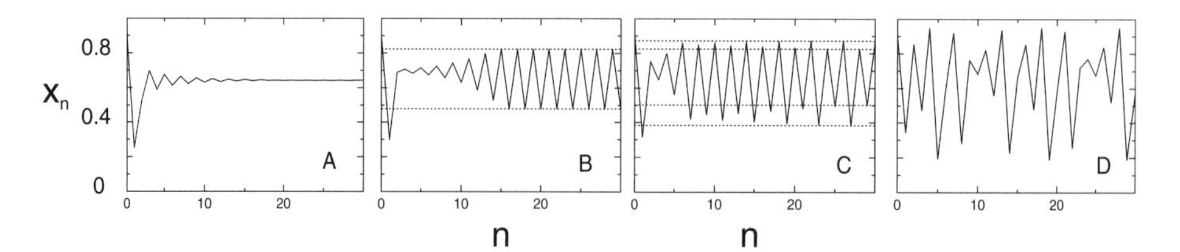

Figure 12.1 The insect population x_n *versus* the generation number n for various growth rates. (A) $\mu = 2.8$, a single attractor. If the fixed point is $x_n = 0$, the system becomes extinct. (B) $\mu = 3.3$, a double attractor. (C) $\mu = 3.5$, a quadruple attractor. (D) $\mu = 3.8$, a chaotic regime. If $\mu < 1$, the population goes extinct.

the fixed point [Rash 90]:

$$\left| \frac{df}{dx} \right|_{x_*} < 1 \quad \text{(stable)}. \tag{12.10}$$

For the one cycle of the logistic map (12.5), we have

$$\left. \frac{df}{dx} \right|_{x_*} = \mu - 2\mu x_* = \begin{cases} \mu, & \text{stable at } x_* = 0 \text{ if } \mu < 1, \\ 2 - \mu, & \text{stable at } x_* = \frac{\mu - 1}{\mu} \text{ if } \mu < 3. \end{cases} \tag{12.11}$$

12.3.2 Period Doubling, Attractors

Equation (12.11) tells us that while the equation for fixed points (12.9) may be satisfied for all values of μ, the populations will not be stable if $\mu > 3$. For $\mu > 3$, the system's long-term population *bifurcates* into two populations (a *two-cycle*), an effect known as *period doubling* (Figure 12.1B). Because the system now acts as if it were attracted to two populations, these populations are called *attractors* or *cycle points*. We can easily predict the x values for these two-cycle attractors by demanding that generation $i + 2$ have the same population as generation i:

$$x_i = x_{i+2} = \mu x_{i+1}(1 - x_{i+1}) \;\Rightarrow\; x_* = \frac{1 + \mu \pm \sqrt{\mu^2 - 2\mu - 3}}{2\mu}. \tag{12.12}$$

We see that as long as $\mu > 3$, the square root produces a real number and thus that physical solutions exist (complex or negative x_* values are nonphysical). We leave it to your computer explorations to discover how the system continues to double periods as μ continues to increase. In all cases the pattern is the same: One of the populations bifurcates into two.

12.4 Mapping Implementation

Program the logistic map to produce a sequence of population values x_i as a function of the generation number i. These are called *map orbits*. The assessment consists of confirmation of Feigenbaum's observations [Feig 79] of the different behavior patterns shown in Figure 12.1. These occur for growth parameter $\mu = (0.4, 2.4, 3.2, 3.6, 3.8304)$ and seed population $x_0 = 0.75$. Identify the following on your graphs:

1. **Transients:** Irregular behaviors before reaching a steady state that differ for different seeds.
2. **Asymptotes:** In some cases the steady state is reached after only 20 generations, while for larger μ values, hundreds of generations may be needed. These steady-state populations are independent of the seed.
3. **Extinction:** If the growth rate is too low, $\mu \leq 1$, the population dies off.
4. **Stable states:** The stable single-population states attained for $\mu < 3$ should agree with the prediction (12.9).
5. **Multiple cycles:** Examine the map orbits for a growth parameter μ increasing continuously through 3. Observe how the system continues to double periods as μ increases. To illustrate, in Figure 12.1C with $\mu = 3.5$, we notice a steady state in which the population alternates among four attractors (a *four-cycle*).
6. **Intermittency:** Observe simulations for $3.8264 < \mu < 3.8304$. Here the system appears stable for a finite number of generations and then jumps all around, only to become stable again.
7. **Chaos:** We define *chaos* as the deterministic behavior of a system displaying no discernible regularity. This may seem contradictory; if a system is deterministic, it must have step-to-step correlations (which, when added up, mean long-range correlations); but if it is chaotic, the complexity of the behavior may hide the simplicity within. In an operational sense, a chaotic system is one with an extremely high sensitivity to parameters or initial conditions. This sensitivity to even minuscule changes is so high that it is impossible to predict the long-range behavior unless the parameters are known to infinite precision (a physical impossibility).

   The system's behavior in the chaotic region is critically dependent on the exact values of μ and x_0. Systems may start out with nearly identical values for μ and x_0 but end up with quite different ones. In some cases the complicated behaviors of nonlinear systems will be *chaotic*, but unless you have a bug in your program, they will not be random.[2]

   a. Compare the long-term behaviors of starting with the two essentially identical seeds $x_0 = 0.75$ and $x'_0 = 0.75(1+\epsilon)$, where $\epsilon \simeq 2 \times 10^{-14}$.

[2] You may recall from Chapter 5, "Monte Carlo Simulations," that a random sequence of events does not even have step-by-step correlations.

b. Repeat the simulation with $x_0 = 0.75$ and two essentially identical survival parameters, $\mu = 4.0$ and $\mu' = 4.0(1-\epsilon)$. Both simulations should start off the same but eventually diverge. ∎

12.5 Bifurcation Diagram (Assessment)

Computing and watching the population change with generation number gives a good idea of the basic dynamics, at least until it gets too complicated to discern patterns. In particular, as the number of bifurcations keeps increasing and the system becomes chaotic, it is hard for us to see a simple underlying structure within the complicated behavior. One way to visualize what is going on is to concentrate on the attractors, that is, those populations that appear to attract the solutions and to which the solutions continuously return. A plot of these attractors (long-term iterates) of the logistic map as a function of the growth parameter μ is an elegant way to summarize the results of extensive computer simulations.

A *bifurcation diagram* for the logistic map is given in Figure 12.2, while one for a Gaussian map is given in Figure 12.3. For each value of μ, hundreds of iterations are made to make sure that all transients essentially die out, and then the values (μ, x_*) are written to a file for hundreds of iterations after that. If the system falls into an n cycle for this μ value, then there should predominantly be n different values written to the file. Next, the value of the initial populations x_0 is changed slightly, and the entire procedure is repeated to ensure that no fixed points are missed. When finished, your program will have stepped through all the values of growth parameter μ, and for each value of μ it will have stepped through all the values of the initial population x_0. Our sample program Bugs.java is shown in Listing 12.1.

```
// Bugs.java: Bifurcation diagram for logistic map
import java.io.*;

public class Bugs {
  static double m_min =0.0, m_max =4., step =0.01 ;                    // Class variables

  public static void main(String[] argv) throws IOException, FileNotFoundException {
    double m, y;
    int i;                                                        // Output data to Bugs.dat
    PrintWriter w = new PrintWriter(new FileOutputStream("Bugs.dat"), true);
    for ( m = m_min; m <= m_max; m += step) {                              // mu loop
      y = 0.5;                                                      // Arbitrary seed
      for (i=1;   i <=200; i++ ) y = m*y*(1-y);                          // Transients
      for (i=201; i <=401; i++ ) {
      y = m*y*(1-y);
      w.println( ""+ m+" "+ y);}
    }
  System.out.println("sorted data stored in Bugs.dat.");
} }
```

Listing 12.1 Bugs.java, the basis of a program for producing the bifurcation diagram of the logistic map. A finished program requires finer grids, a scan over initial values, and removal of duplicates.

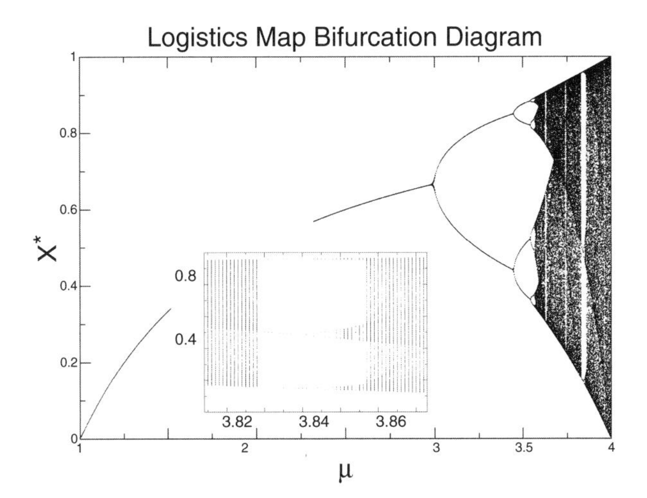

Figure 12.2 The bifurcation plot, attractor populations *versus* growth rate, for the logistic map. The inset shows some details of a three-cycle window.

12.5.1 Bifurcation Diagram Implementation

The last part of this problem is to reproduce Figure 12.2 at various levels of detail. (You can listen to a sonification of this diagram on the CD or use one of the applet there to create your own sonification.) While the best way to make a visualization of this sort would be with visualization software that permits you to vary the intensity of each individual point on the screen, we simply plot individual points and have the density in each region determined by the number of points plotted there. When thinking about plotting many individual points to draw a figure, it is important to keep in mind that your screen resolution is $\sim$100 dots per inch and your laser printer resolution may be 300 dots per inch. This means that if you plot a point at each pixel, you will be plotting $\sim 3000 \times 3000 \simeq 10$ million elements. *Beware:* This can require some time and may choke a printer. In any case, printing at a finer resolution is a waste of time.

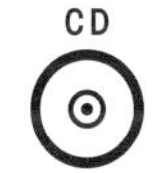

12.5.2 Visualization Algorithm: Binning

1. Break up the range $1 \leq \mu \leq 4$ into 1000 steps and loop through them. These are the "bins" into which we will place the x_* values.
2. In order not to miss any structures in your bifurcation diagram, loop through a range of initial x_0 values as well.
3. Wait at least 200 generations for the transients to die out and then print the next several hundred values of (μ, x_*) to a file.

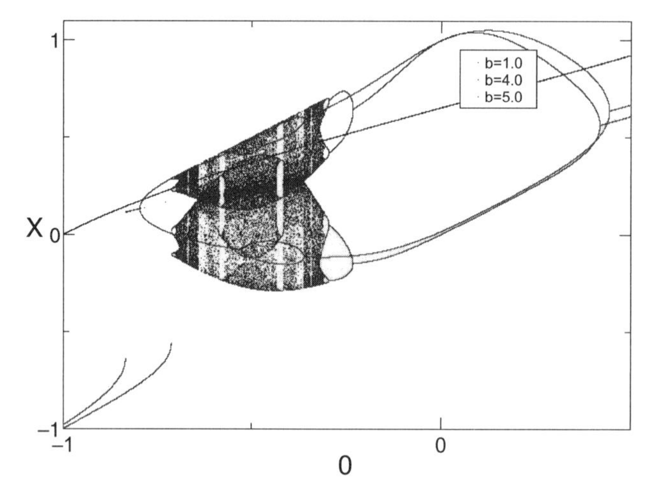

Figure 12.3 A bifurcation plot for the Gaussian map. (Courtesy of W. Hager.)

4. Print your x_* values to no more than three or four decimal places. You will not be able to resolve more places than this on your plot, and this restriction will keep your output files smaller by permitting you to remove duplicates. It is hard to control the number of decimal places in the output with Java's standard print commands (although `printf` and `DecimalFormat` do permit control). A simple approach is to multiply the x_i values by 1000 and then throw away the part to the right of the decimal point. Because $0 \le x_n \le 1$, this means that $0 \le 100 * x_n \le 1000$, and you can throw away the decimal part by casting the resulting numbers as integers:

```
Ix[i]= (int)(1000*x[i])                    // Convert to 0 ≤ ints ≤ 1000
```

   You may then divide by 1000 if you want floating-point numbers.
5. You also need to remove duplicate values of (x, μ) from your file (they just take up space and plot on top of each other). You can do that in Unix/Linux with the *sort -u* command.
6. Plot your file of x_* *versus* μ. Use small symbols for the points and do not connect them.
7. Enlarge sections of your plot and notice that a similar bifurcation diagram tends to be contained within each magnified portion (this is called *self-similarity*).
8. Look over the series of bifurcations occurring at

$$\mu_k \simeq 3,\ 3.449,\ 3.544,\ 3.5644,\ 3.5688,\ 3.569692,\ 3.56989,\ \ldots . \qquad (12.13)$$

   The end of this series is a region of chaotic behavior.

9. Inspect the way this and other sequences begin and then end in chaos. The changes sometimes occur quickly, and so you may have to make plots over a very small range of μ values to see the structures.
10. A close examination of Figure 12.2 shows regions where, with a slight increase in μ, a very large number of populations suddenly change to very few populations. Whereas these may appear to be artifacts of the video display, this is a real effect and these regions are called *windows*. Check that at $\mu = 3.828427$, chaos turns into a three-cycle population.

12.5.3 Feigenbaum Constants (Exploration)

Feigenbaum discovered that the sequence of μ_k values (12.13) at which bifurcations occur follows a regular pattern [Feig 79]. Specifically, it converges geometrically when expressed in terms of the distance between bifurcations δ:

$$\mu_k \quad \rightarrow \quad \mu_\infty - \frac{c}{\delta^k}, \qquad \delta = \lim_{k\rightarrow\infty} \frac{\mu_k - \mu_{k-1}}{\mu_{k+1} - \mu_k}. \tag{12.14}$$

Use your sequence of μ_k values to determine the constants in (12.14) and compare them to those found by Feigenbaum:

$$\mu_\infty \simeq 3.56995, \quad c \simeq 2.637, \quad \delta \simeq 4.6692. \tag{12.15}$$

Amazingly, the value of the *Feigenbaum constant* δ is universal for all second-order maps.

12.6 Random Numbers via Logistic Map (Exploration) ⊙

There are claims that the logistic map in the chaotic region ($\mu \geq 4$),

$$x_{i+1} \simeq 4x_i(1 - x_i), \tag{12.16}$$

can be used to generate random numbers [P&R 95]. Although successive x_i's are correlated, if the population for approximately every sixth generation is examined, the correlation is effectively gone and random numbers result. To make the sequence more uniform, a trigonometric transformation is used:

$$y_i = \frac{1}{\pi}\cos^{-1}(1 - 2x_i). \tag{12.17}$$

Use the random-number tests discussed in Chapter 5, "Monte Carlo Simulation," to test this claim.

TABLE 12.1
Several Nonlinear Maps to Explore

Name	$f(x)$	Name	$f(x)$
Logistic	$\mu x(1-x)$	Tent	$\mu(1-2\|x-1/2\|)$
Ecology	$xe^{\mu(1-x)}$	Quartic	$\mu[1-(2x-1)^4]$
Gaussian	$e^{bx^2}+\mu$		

12.7 Other Maps (Exploration)

Bifurcations and chaos are characteristic properties of nonlinear systems. Yet systems can be nonlinear in a number of ways. Table 12.1 lists four maps that generate x_i sequences containing bifurcations. The tent map derives its nonlinear dependence from the absolute value operator, while the logistic map is a subclass of the ecology map. Explore the properties of these other maps and note the similarities and differences.

12.8 Signals of Chaos: Lyapunov Coefficients ⊙

The Lyapunov coefficient λ_i provides an analytic measure of whether a system is chaotic [Wolf 85, Ram 00, Will 97]. Physically, the coefficient is a measure of the growth rate of the solution near an attractor. For 1-D maps there is only one such coefficient, whereas in general there is a coefficient for each direction in space. The essential assumption is that neighboring paths x_n near an attractor have an n (or time) dependence $L \propto \exp(\lambda t)$. Consequently, orbits that have $\lambda > 0$ diverge and are chaotic; orbits that have $\lambda = 0$ remain marginally stable, while orbits with $\lambda < 0$ are periodic and stable. Mathematically, the Lyapunov coefficient or exponent is defined as

$$\lambda = \lim_{t\to\infty} \frac{1}{t} \log \frac{L(t)}{L(t_0)}, \tag{12.18}$$

where $L(t)$ is the distance between neighboring phase space trajectories at time t.

We calculate the Lyapunov exponent for a general 1-D map,

$$x_{n+1} = f(x_n), \tag{12.19}$$

and then apply the result to the logistic map. To determine stability, we examine perturbations about a reference trajectory x_0 by adding a small perturbation and iterating once [Mann 90, Ram 00]:

$$\hat{x}_0 = x_0 + \delta x_0, \quad \hat{x}_1 = x_1 + \delta x_1. \tag{12.20}$$

We substitute this into (12.19) and expand f in a Taylor series around x_0:

$$x_1 + \delta x_1 = f(x_0 + \delta x_0) \simeq f(x_0) + \left.\frac{\delta f}{\delta x}\right|_{x_0} \delta x_0 = x_1 + \left.\frac{\delta f}{\delta x}\right|_{x_0} \delta x_0,$$

$$\Rightarrow \quad \delta x_1 \simeq \left(\frac{\delta f}{\delta x}\right)_{x_0} \delta x_0. \tag{12.21}$$

(This is the proof of our earlier statement about the stability of maps.) To deduce the general result we examine one more iteration:

$$\delta x_2 \simeq \left(\frac{\delta f}{\delta x}\right)_{x_1} \delta x_1 = \left(\frac{\delta f}{\delta x}\right)_{x_0} \left(\frac{\delta f}{\delta x}\right)_{x_1} \delta x_0, \tag{12.22}$$

$$\Rightarrow \quad \delta x_n = \prod_{i=0}^{n-1} \left(\frac{\delta f}{\delta x}\right)_{x_i} \delta x_0. \tag{12.23}$$

This last relation tells us how trajectories differ on the average after n steps:

$$|\delta x_n| = L^n |\delta x_0|, \quad L^n = \prod_{i=0}^{n-1} \left| \left(\frac{\delta f}{\delta x}\right)_{x_i} \right|. \tag{12.24}$$

We now solve for the Lyapunov number L and take its logarithm to obtain the Lyapunov exponent:

$$\lambda = \ln(L) = \lim_{n\to\infty} \frac{1}{n} \sum_{i=0}^{n-1} \ln \left| \left(\frac{\delta f}{\delta x}\right)_{x_i} \right|. \tag{12.25}$$

For the logistic map we obtain

$$\lambda = \frac{1}{n} \sum_{i=0}^{n-1} \ln |\mu - 2\mu x_i|, \tag{12.26}$$

where the sum is over iterations.

The code `LyapLog.java` in Listing 12.2 computes the Lyapunov exponents for the bifurcation plot of the logistic map. In Figure 12.4 left we show its output and note the sign changes in λ where the system becomes chaotic, and abrupt changes in slope at bifurcations. (A similar curve is obtained for the fractal dimension of the logistic map, as indeed the two are proportional.)

12.8.1 Shannon Entropy ⊙

Shannon entropy, like the Lyapunov exponent, is another analytic measure of chaos. It is a measure of uncertainty that has proven useful in communication theory

```
// LyapLog.java: Lyapunov coef for logistic map
import java.io.*;

public class LyapLog {
  static double m_min = 2.8, m_max = 4., step = 0.002 ;

  public static void main(String[] argv) throws IOException, FileNotFoundException {
    double m, y, suma, lyap[] = new double[100];
    int i;
    PrintWriter w = new PrintWriter(new FileOutputStream("logistic.dat"), true);
    PrintWriter q = new PrintWriter(new FileOutputStream("lyap.dat"), true);
    System.out.println("data stored in logistic.dat and lyap.dat");
    for ( m = m_min; m <= m_max; m += step) {                       // m loop
      y = 0.5;
      for ( i=1; i <= 200; i++ ) y = m*y*(1-y);
      suma = 0. ;                                                   // Skip transients
      for ( i=201; i <= 401; i++ ) {
        y = m*y*(1-y);
        suma = suma + Math.log(Math.abs(m*(1.-2.*y)));              // Lyapunov
        w.println( m + "  " + y);
      }
      lyap[(int)m] = suma/401;                        // Normalize Lyapunov exponent
      q.println( m + "   " + lyap[(int)m]) ;
} } }
```

Listing 12.2 LyapLog.java computes Lyapunov exponents for the bifurcation plot of the logistic map as a function of growth rate. Note the fineness of the μ grid.

[Shannon 48, Ott 02, G,T&C 06] and led to the concept of information entropy. Imagine that an experiment has N possible outcomes. If the probability of each is $p_1, p_2, \ldots, p_N$, with normalization such that $\sum_{i=1}^{N} p_i = 1$, the Shannon entropy is defined as

$$S_{\text{Shannon}} = -\sum_{i=1}^{N} p_i \, ln \, p_i. \tag{12.27}$$

If $p_i \equiv 0$, there is no uncertainty and $S_{\text{Shannon}} = 0$, as you might expect. If all N outcomes have equal probability, $p_i \equiv 1/N$, we obtain $S_{\text{Shannon}} = \ln N$, which diverges slowly as $N \to \infty$.

The code Entropy.java in Listing 12.3 computes the Shannon entropy for the the logistic map as a function of the growth parameter μ. The results (Figure 12.4 left) are seen to be quite similar to the Lyapunov exponent, again with discontinuities occurring at the bifurcations.

12.9 Unit I Quiz

1. Consider the logistic map.
   a. Make sketches of what a graph of population x_i *versus* generation number i would look like for extinction and for a period-two cycle.
   b. Describe in words and possibly a diagram the relation between the preceding two sketches and the bifurcation plot of x_i *versus* i.

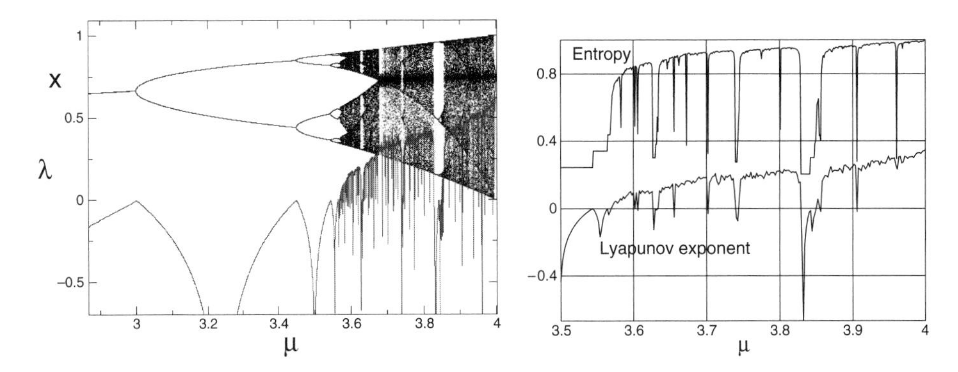

Figure 12.4 *Left:* Lyapunov exponent and bifurcation values for the logistic map as functions of the growth rate μ. *Right:* Shannon entropy (reduced by a factor of 5) and the Lyapunov coefficient for the logistic map.

2. Consider the *tent map*. Rather than compute this map, study it with just a piece of paper.
   a. Make sketches of what a graph of population x_i *versus* generation number i would look like for extinction, for a period-one cycle, and for a period-two cycle.
   b. Show that there is a single fixed point for $\mu > 1/2$ and a period-two cycle for $\mu > 1$.

```
// Entropy.java, Shannon entropy of logistic map
import java.io.*;
import java.util.*;

public class Entropy {

  public static void main(String[] argv) throws IOException, FileNotFoundException {
    PrintWriter w = new PrintWriter (new FileOutputStream("Entropy.dat"), true);
    double prob[] = new double[1001];                                    //
        Probabilities
    int nbin = 1000, nmax = 100000, j, n, ibin;
    double entropy, x, mu;
    System.out.println("Entropy output in Entropy.dat");
    for ( mu = 3.5; mu <= 4. ; mu = mu + 0.001) {                        // Values of mu
      for ( j=1; j < nbin; j++ ) prob[j] = 0;
      x = 0.5;
      for ( n=1; n <= nmax; n++ ) {
        x = mu*x*(1.-x);                                    // Logistic map, Skip transients
        if (n > 30000) { ibin = (int)(x*nbin) + 1;   prob[ibin] = prob[ibin] + 1;}
      }
      entropy = 0. ;
      for ( ibin = 1; ibin <= nbin; ibin++ ) if (prob[ibin]>0)
          entropy = entropy-(prob[ibin]/nmax)*Math.log(prob[ibin]/nmax);
      w.println(" " + mu + " " + entropy);
    }
  }
}
```

Listing 12.3 Entropy.java computes the Shannon entropy for the logistic map as a function of growth parameter μ.

12.10 Unit II. Pendulums Become Chaotic (Continuous)

In Unit I on bugs we discovered that a simple nonlinear difference equation yields solutions that may be simple, complicated, or chaotic. Unit III will extend that model to the differential form, which also exhibits complex behaviors. Now we search for similar nonlinear, complex behaviors in the differential equation describing a realistic pendulum. Because chaotic behavior may resemble noise, it is important to be confident that the unusual behaviors arise from physics and not numerics. Before we explore the solutions, we provide some theoretical background on the use of phase space plots for revealing the beauty and simplicity underlying complicated behaviors. We also provide two chaotic pendulum applets (Figure 12.5) for assistance in understanding the new concepts. Our study is based on the description in [Rash 90], on the analytic discussion of the parametric oscillator in [L&L,M 76], and on a similar study of the vibrating pivot pendulum in [G,T&C 06].

CD

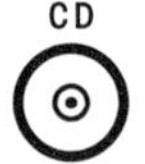

Consider the pendulum on the left in Figure 12.5. We see a pendulum of length l *driven* by an external sinusoidal torque **f** through air with a coefficient of drag α. Because there is no restriction that the angular displacement θ be small, we call this a *realistic* pendulum. Your **problem** is to describe the motion of this pendulum, first when the driving torque is turned off but the initial velocity is large enough to send the pendulum over the top, and then when the driving torque is turned on.

12.11 Chaotic Pendulum ODE

What we call a *chaotic pendulum* is just a pendulum with friction and a driving torque (Figure 12.5 left) but with no small-deflection-angle approximation. Newton's laws of rotational motion tell us that the sum of the gravitational torque $-mgl\sin\theta$, the frictional torque $-\beta\dot{\theta}$, and the external torque $\tau_0\cos\omega t$ equals the moment of inertia of the pendulum times its angular acceleration [Rash 90]:

$$I\,\frac{d^2\theta}{dt^2} = -mgl\,\sin\theta - \beta\,\frac{d\theta}{dt} + \tau_0\cos\omega t, \tag{12.28}$$

$$\Rightarrow\quad \frac{d^2\theta}{dt^2} = -\omega_0^2\,\sin\theta - \alpha\,\frac{d\theta}{dt} + f\,\cos\omega t, \tag{12.29}$$

$$\text{where}\quad \omega_0 = \frac{mgl}{I},\quad \alpha = \frac{\beta}{I},\quad f = \frac{\tau_0}{I}. \tag{12.30}$$

Equation (12.29) is a second-order time-dependent nonlinear differential equation. Its nonlinearity arises from the $\sin\theta$, as opposed to the θ, dependence of

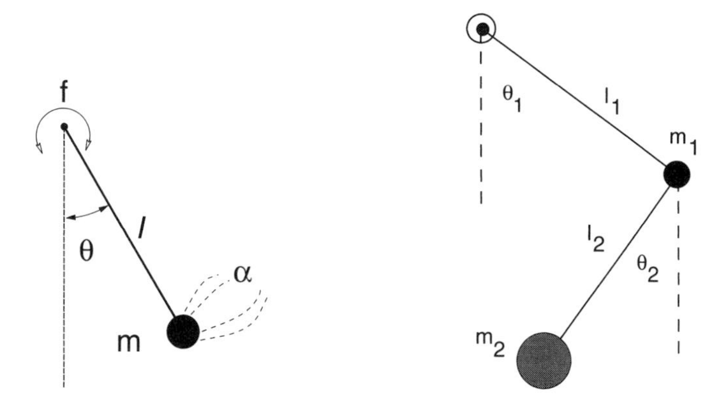

Figure 12.5 *Left:* A pendulum of length l driven through air by an external sinusoidal torque. The strength of the torque is given by f and that of air resistance by α. *Right:* A double pendulum.

the gravitational torque. The parameter ω_0 is the natural frequency of the system arising from the restoring torque, α is a measure of the strength of friction, and f is a measure of the strength of the driving torque. In our standard ODE form, $d\mathbf{y}/dt = \mathbf{y}$ (Chapter 9, "Differential Equation Applications"), we have two simultaneous first-order equations:

$$\frac{dy^{(0)}}{dt} = y^{(1)}, \tag{12.31}$$

$$\frac{dy^{(1)}}{dt} = -\omega_0^2 \sin y^{(0)} - \alpha y^{(1)} + f \cos \omega t,$$

$$\text{where} \quad y^{(0)} = \theta(t), \quad y^{(1)} = \frac{d\theta(t)}{dt}. \tag{12.32}$$

12.11.1 Free Pendulum Oscillations

If we ignore friction and external torques, (12.29) takes the simple form

$$\frac{d^2\theta}{dt^2} = -\omega_0^2 \sin \theta \tag{12.33}$$

If the displacements are small, we can approximate $\sin\theta$ by θ and obtain the linear equation of simple harmonic motion with frequency ω_0:

$$\frac{d^2\theta}{dt^2} \simeq -\omega_0^2 \theta \;\Rightarrow\; \theta(t) = \theta_0 \sin(\omega_0 t + \phi). \tag{12.34}$$

In Chapter 9, "Differential Equation Applications," we studied how nonlinearities produce anharmonic oscillations, and indeed (12.33) is another good candidate for such studies. As before, we expect solutions of (12.33) for the free realistic pendulum to be periodic, but with a frequency $\omega \simeq \omega_0$ only for small oscillations. Furthermore, because the restoring torque, $mgl \sin\theta \simeq mgl(\theta - \theta^3/3)$, is less than the $mgl\theta$ assumed in a harmonic oscillator, realistic pendulums swing slower (have longer periods) as their angular displacements are made larger.

12.11.2 Solution as Elliptic Integrals

The analytic solution to the realistic pendulum is a textbook problem [L&L,M 76, M&T 03, Schk 94], except that it is hardly a solution and hardly analytic. The "solution" is based on energy being a constant (integral) of the motion. For simplicity, we start the pendulum off at rest from its maximum displacement θ_m. Because the initial energy is all potential, we know that the total energy of the system equals its initial potential energy (Figure 12.5),

$$E = \mathrm{PE}(0) = mgl - mgl\cos\theta_m = 2mgl\sin^2\left(\frac{\theta_m}{2}\right). \tag{12.35}$$

Yet since $E = \mathrm{KE} + \mathrm{PE}$ is a constant, we can write for any value of θ

$$2mgl\sin^2\frac{\theta_m}{2} = \frac{1}{2}I\left(\frac{d\theta}{dt}\right)^2 + 2mgl\sin^2\frac{\theta}{2},$$

$$\Rightarrow \quad \frac{d\theta}{dt} = 2\omega_0\left[\sin^2\frac{\theta_m}{2} - \sin^2\frac{\theta}{2}\right]^{1/2} \quad \Rightarrow \quad \frac{dt}{d\theta} = \frac{T_0/\pi}{\left[\sin^2(\theta_m/2) - \sin^2(\theta/2)\right]^{1/2}},$$

$$\Rightarrow \quad \frac{T}{4} = \frac{T_0}{4\pi}\int_0^{\theta_m}\frac{d\theta}{\left[\sin^2(\theta_m/2) - \sin^2(\theta/2)\right]^{1/2}} = \frac{T_0}{4\pi\sin\theta_m}F\left(\frac{\theta_m}{2},\frac{\theta}{2}\right), \tag{12.36}$$

$$\Rightarrow \quad T \simeq T_0\left[1 + \frac{1}{4}\sin^2\frac{\theta_m}{2} + \frac{9}{64}\sin^4\frac{\theta_m}{2} + \cdots\right], \tag{12.37}$$

where we have assumed that it takes $T/4$ for the pendulum to travel from $\theta = 0$ to θ_m. The integral in (12.36) is an *elliptic integral of the first kind*. If you think of an elliptic integral as a generalization of a trigonometric function, then this is a closed-form solution; otherwise, it's an integral needing computation. The series expansion of the period (12.37) is obtained by expanding the denominator and integrating it term by term. It tells us, for example, that an amplitude of 80° leads to a 10% slowdown of the pendulum relative to the small θ result. In contrast, we will determine the period empirically without the need for any expansions.

12.11.3 Implementation and Test: Free Pendulum

As a preliminary to the solution of the full equation (12.29), modify your `rk4` program to solve (12.33) for the free oscillations of a realistic pendulum.

1. Start your pendulum at $\theta = 0$ with $\dot{\theta}(0) \neq 0$. Gradually increase $\dot{\theta}(0)$ to increase the importance of nonlinear effects.
2. Test your program for the linear case ($\sin\theta \rightarrow \theta$) and verify that
   a. your solution is harmonic with frequency $\omega_0 = 2\pi/T_0$, and
   b. the frequency of oscillation is independent of the amplitude.
3. Devise an algorithm to determine the period T of the oscillation by counting the time it takes for three successive passes of the amplitude through $\theta = 0$. (You need *three* passes because a general oscillation may not be symmetric about the origin.) Test your algorithm for simple harmonic motion where you know T_0.
4. For the realistic pendulum, observe the change in period as a function of increasing initial energy or displacement. Plot your observations along with (12.37).
5. Verify that as the initial KE approaches $2mgl$, the motion remains oscillatory but not harmonic (Figure 12.8).
6. At $E = 2\,\text{mgl}$ (the *separatrix*), the motion changes from oscillatory to rotational ("over the top" or "running"). See how close you can get to the separatrix and to its infinite period.
7. ⊙ Use the applet `HearData` (Figure 12.6) to convert your different oscillations to sound and hear the difference between harmonic motion (boring) and anharmonic motion containing overtones (interesting).

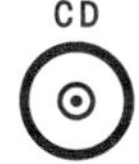

12.12 Visualization: Phase Space Orbits

The conventional solution to an equation of motion is the position $x(t)$ and the velocity $v(t)$ as functions of time. Often behaviors that appear complicated as functions of time appear simpler when viewed in an abstract space called *phase space*, where the ordinate is the velocity $v(t)$ and the abscissa is the position $x(t)$ (Figure 12.7). As we see from the phase space figures, the solutions form geometric objects that are easy to recognize. (We provide two applets on the CD, `Pend1` and `Pend2`, to help the reader make the connections between phase space shapes and the corresponding physical motion.)

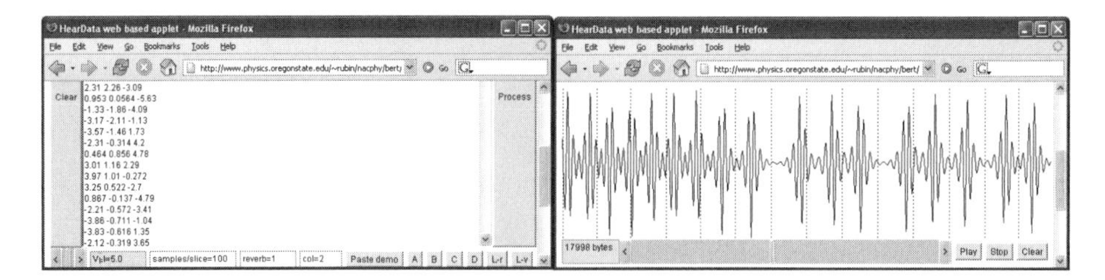

Figure 12.6 The data screen (*left*) and output screen (*right*) of the applet HearData on the CD. Columns of $(t_i, x(t_i))$ data are pasted into the data window, processed into the graph in the output window, and then converted to sound that is played by Java.

The position and velocity of a free harmonic oscillator are given by the trigonometric functions

$$x(t) = A\sin(\omega t), \quad v(t) = \frac{dx}{dt} = \omega A\cos(\omega t). \tag{12.38}$$

When substituted into the total energy, we obtain two important results:

$$E = \text{KE} + \text{PE} = \left(\frac{1}{2}m\right)v^2 + \left(\frac{1}{2}\omega^2 m^2\right)x^2 \tag{12.39}$$

$$= \frac{\omega^2 m^2 A^2}{2m}\cos^2(\omega t) + \frac{1}{2}\omega^2 m^2 A^2 \sin^2(\omega t) = \frac{1}{2}m\omega^2 A^2. \tag{12.40}$$

The first equation, being that of an ellipse, proves that the harmonic oscillator follows closed elliptical orbits in phase space, with the size of the ellipse increasing with the system's energy. The second equation proves that the total energy is a constant of the motion. Different initial conditions having the same energy start at different places on the same ellipse and transverse the same orbits.

In Figures 12.7–12.10 we show various phase space structures. *Study these figures and their captions* and note the following:

- The orbits of anharmonic oscillations will still be ellipselike but with angular corners that become more distinct with increasing nonlinearity.
- Closed trajectories describe periodic oscillations [the same (x, v) occur again and again], with clockwise motion.
- Open orbits correspond to nonperiodic or "running" motion (a pendulum rotating like a propeller).
- Regions of space where the potential is repulsive lead to open trajectories in phase space (Figure12.7 left).

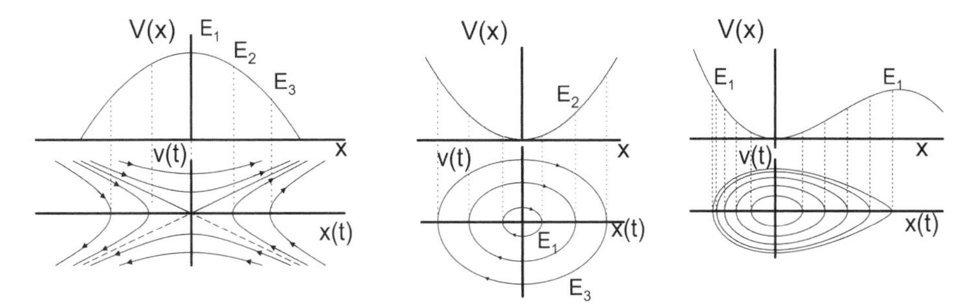

Figure 12.7 Three potentials and their characteristic behaviors in phase space. The different orbits correspond to different energies, as indicated by the limits within the potentials (dashed lines). Notice the absence of trajectories in regions forbidden by energy conservation. *Left:* A repulsive potential leads to open orbits in phase space characteristic of nonperiodic motion. The phase space trajectories cross at the hyperbolic point in the middle, an unstable equilibrium point. *Middle:* The harmonic oscillator leads to symmetric ellipses in phase space; the closed orbits indicate periodic behavior, and the symmetric trajectories indicate a symmetric potential. *Right:* A nonharmonic oscillator. Notice that the ellipselike trajectories neither are ellipses nor are symmetric with respect to the $v(t)$ axis.

- As seen in Figure 12.8 left, the separatrix corresponds to the trajectory in phase space that separates open and closed orbits. Motion on the separatrix is indeterminant, as the pendulum may balance at the maxima of $V(\theta)$.
- Friction may cause the energy to decrease with time and the phase space orbit to spiral into a *fixed point*.
- For certain parameters, a closed *limit cycle* occurs in which the energy pumped in by the external torque exactly balances that lost by friction (Figure 12.8 right).
- Because solutions for different initial conditions are unique, different orbits do not cross. Nonetheless, open orbits join at points of unstable equilibrium (*hyperbolic points* in Figure 12.7 left) where an indeterminacy exists. ▌

12.12.1 Chaos in Phase Space

It is easy to solve the nonlinear ODE (12.31) on the computer using our usual techniques. However, it is not so easy to understand the solutions because they are so rich in complexity. The solutions are easier to understand in phase space, particularly if you learn to recognize some characteristic structures there. Actually, there are a number of "tools" that can be used to decide if a system is chaotic in contrast to just complex. Geometric structures in phase space is one of them,

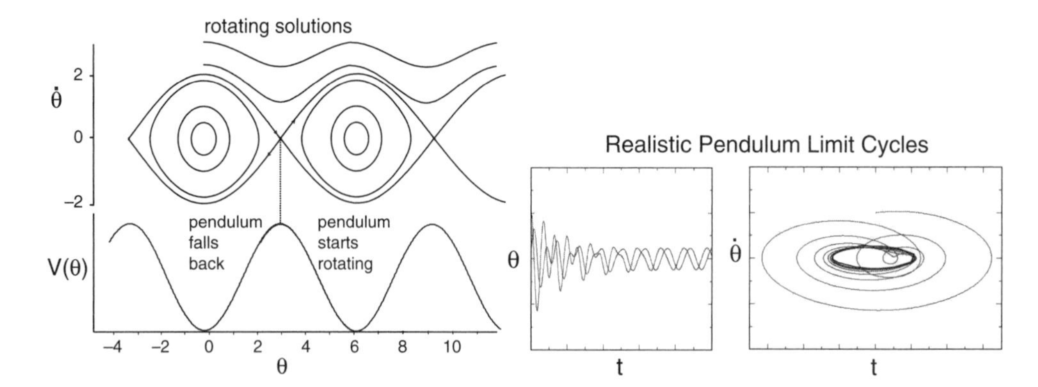

Figure 12.8 *Left:* Phase space trajectories for a plane (*i.e.* 2-D) pendulum including "over the top" or rotating solutions. The trajectories are symmetric with respect to vertical and horizontal reflections through the origin. At the bottom of the figure is shown the corresponding θ dependence of the potential. *Right:* Position *versus* time for two initial conditions of a chaotic pendulum that end up with the same limit cycle, and the corresponding phase space orbits. (Courtesy of W. Hager.)

and determination of the Lyupanov coefficient (discussed in §12.8) is another. Both signal the simplicity lying within the complexity.

What may be surprising is that even though the ellipselike figures in phase space were observed originally for free systems with no friction and no driving torque, similar structures continue to exist for driven systems with friction. The actual trajectories may not remain on a single structure for all times, but they are *attracted* to them and return to them often. In contrast to periodic motion, which corresponds to closed figures in phase space, random motion appears as a diffuse cloud filling an entire energetically accessible region. Complex or chaotic motion falls someplace in between (Figure 12.9 middle). If viewed for long times and for many initial conditions, chaotic trajectories (flows) through phase space, while resembling the familiar geometric figures, they may contain dark or diffuse *bands* in places rather than single lines. The continuity of trajectories within bands means that a continuum of solutions is possible and that the system flows continuously among the different trajectories forming the band. The transitions among solutions cause the coordinate space solutions to appear chaotic and are what makes them hypersensitive to the initial conditions (the slightest change in which causes the system to flow to nearby trajectories).

Pick out the following phase space structures in your simulations.

Limit cycles: When a chaotic pendulum is driven by a not-too-large driving torque, it is possible to pick the magnitude for this torque such that after the initial transients die off, the average energy put into the system during one period exactly balances the average energy dissipated by friction during that

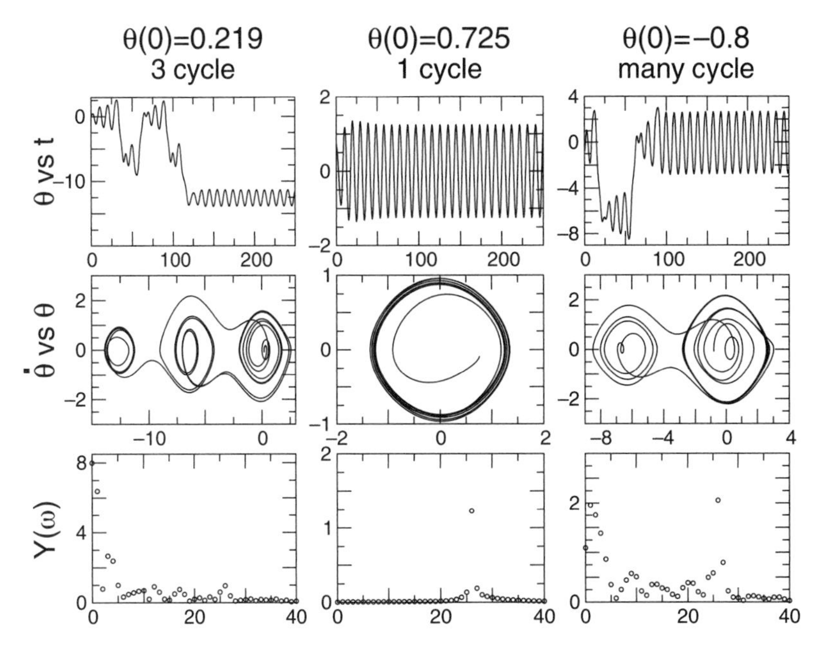

Figure 12.9 Position and phase space plots for a chaotic pendulum with $\omega_0 = 1$, $\alpha = 0.2$, $f = 0.52$, and $\omega = 0.666$. The rightmost initial condition displays more of the broadband Fourier spectrum characteristic of chaos. (Examples of chaotic behavior can be seen in Figure 12.10.)

period (Figure 12.8 right):

$$\langle f \cos \omega t \rangle = \left\langle \alpha \frac{d\theta}{dt} \right\rangle = \left\langle \alpha \frac{d\theta(0)}{dt} \cos \omega t \right\rangle \quad \Rightarrow \quad f = \alpha \frac{d\theta(0)}{dt}. \tag{12.41}$$

This leads to *limit cycles* that appear as closed ellipselike figures, yet the solution may be unstable and make sporadic jumps between limit cycles.

Predictable attractors: Well-defined, fairly simple periodic behaviors that are not particularly sensitive to the initial conditions. These are orbits, such as fixed points and limit cycles, into which the system settles or returns to often. If your location in phase space is near a predictable attractor, ensuing times will bring you to it.

Strange attractors: Well-defined, yet complicated, semiperiodic behaviors that appear to be uncorrelated with the motion at an earlier time. They are distinguished from predicable attractors by being fractal (Chapter 13, "Fractals & Statistical Growth") chaotic, and highly sensitive to the initial conditions [J&S 98]. Even after millions of oscillations, the motion remains *attracted* to them.

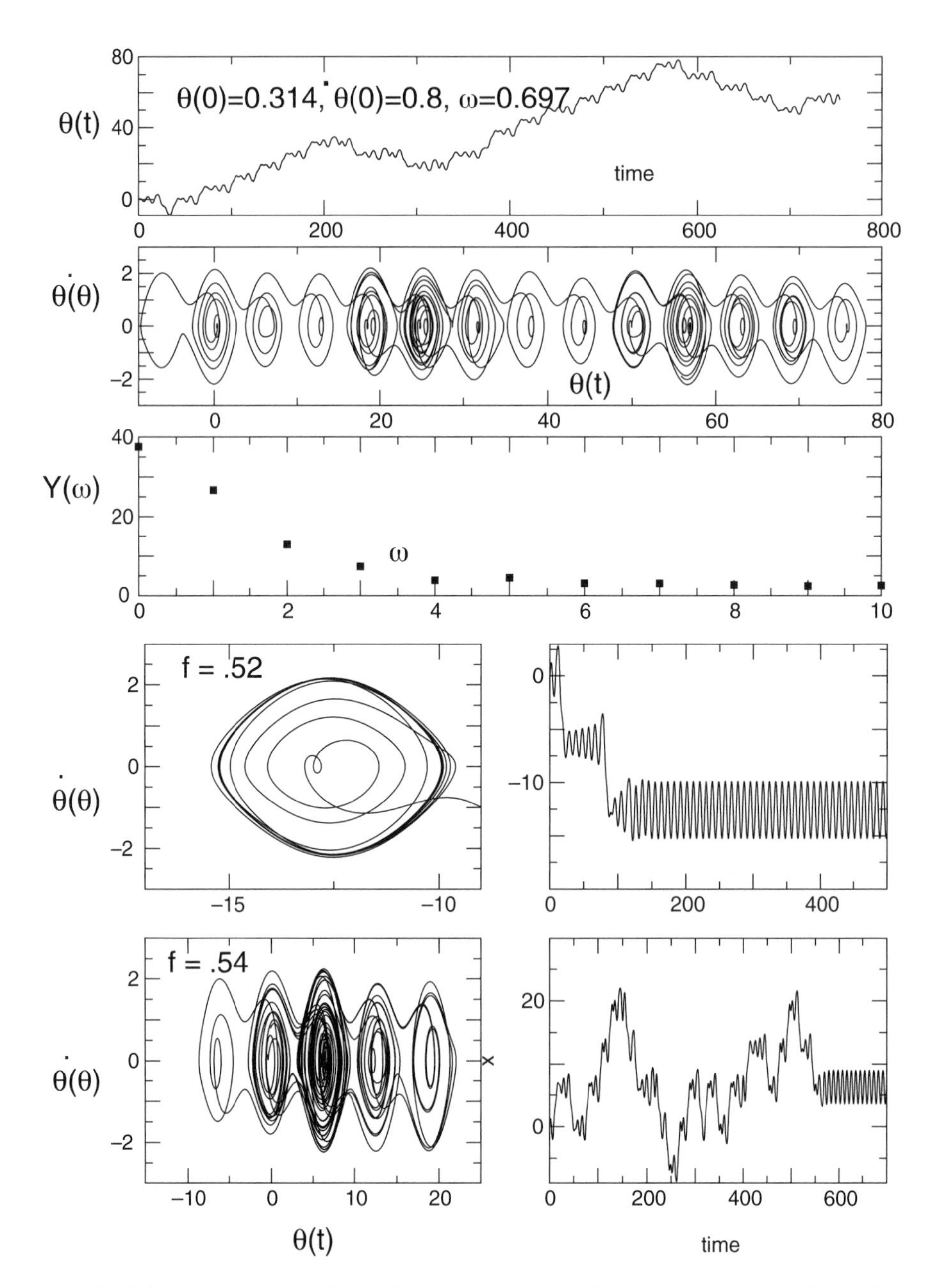

Figure 12.10 Some examples of complicated behaviors of a realistic pendulum. In the top three rows sequentially we see $\theta(t)$ behaviors, a phase space diagram containing regular patterns with dark bands, and a broad Fourier spectrum. These features are characteristic of chaos. In the bottom two rows we see how the behavior changes abruptly after a slight change in the magnitude of the force and that for $f = 0.54$ there occur the characteristic broad bands of chaos.

Chaotic paths: Regions of phase space that appear as filled-in bands rather than lines. Continuity within the bands implies complicated behaviors, yet still with simple underlying structure.

Mode locking: When the magnitude f of the driving torque is larger than that for a limit cycle (12.41), the driving torque can overpower the natural oscillations, resulting in a steady-state motion at the frequency of the driver. This is called *mode locking*. While mode locking can occur for linear or nonlinear systems, for nonlinear systems the driving torque may lock onto the system by exciting its overtones, leading to a rational relation between the driving frequency and the natural frequency:

$$\frac{\omega}{\omega_0} = \frac{n}{m}, \quad n, m = \text{integers}. \tag{12.42}$$

Butterfly effects: One of the classic quips about the hypersensitivity of chaotic systems to the initial conditions is that the weather pattern in North America is hard to predict well because it is sensitive to the flapping of butterfly wings in South America. Although this appears to be counterintuitive because we know that systems with essentially identical initial conditions should behave the same, eventually the systems diverge. The applet `pend2` (Figure 12.11 bottom) lets you compare two simulations of nearly identical initial conditions.As seen on the right in Figure 12.11, the initial conditions for both pendulums differ by only 1 part in 917, and so the initial paths in phase space are the same. Nonetheless, at just the time shown here, the pendulums balance in the vertical position, and then one falls before the other, leading to differing oscillations and differing phase space plots from this time onward.

12.12.2 Assessment in Phase Space

The challenge in understanding simulations of the chaotic pendulum (12.31) is that the 4-D parameter space $(\omega_0, \alpha, f, \omega)$ is so immense that only sections of it can be studied systematically. We expect that sweeping through driving frequency ω should show resonances and beating; sweeping through the frictional force α should show underdamping, critical damping, and overdamping; and sweeping through the driving torque f should show mode locking (for the right values of ω). All these behaviors can be found in the solution of your differential equation, yet they are mixed together in complex ways.

In this assessment you should try to reproduce the behaviors shown in the phase space diagrams in Figure 12.9 and in the applets in Figure 12.11. *Beware*: Because the system is chaotic, you should expect that your results will be sensitive to the exact values of the initial conditions and to the details of your integration routine. We suggest that you experiment; start with the parameter values we used to produce our plots and then observe the effects of making *very small* changes in parameters

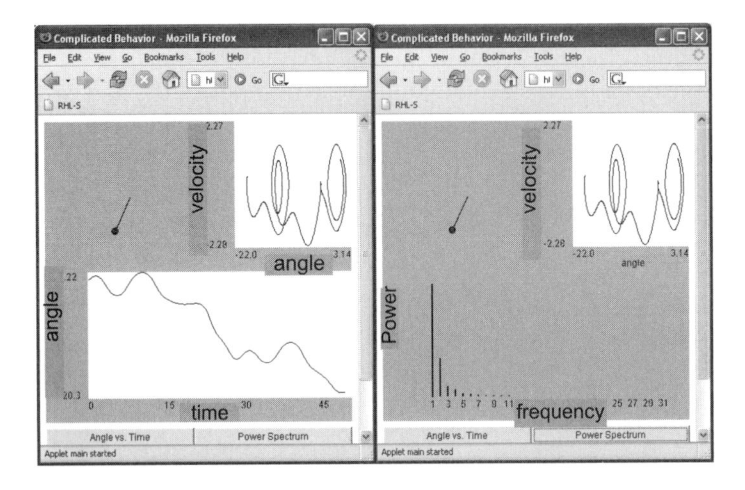

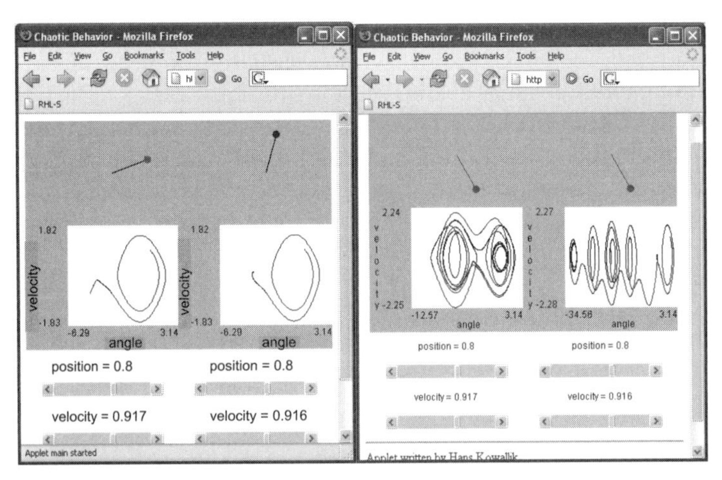

Figure 12.11 *Top row:* Output from the applet `Pend1` producing an animation of a chaotic pendulum, along with the corresponding position *versus* time and phase space plots. *Right:* The resulting Fourier spectrum produced by `Pend1`, *Bottom row:* Output from the applet `Pend2` producing an animation of two chaotic pendula, along with the corresponding phase space plots, and the final output with limit cycles (dark bands).

until you obtain different modes of behavior. Consequently, an inability to reproduce our results for the parameter values does not necessarily imply that something is "wrong."

1. Take your solution to the realistic pendulum and include friction, making α an input parameter. Run it for a variety of initial conditions, including over-the-top ones. Since no energy is fed to the system, you should see spirals.

2. Next, verify that with no friction, but with a very small driving torque, you obtain a perturbed ellipse in phase space.
3. Set the driving torque's frequency to be close to the natural frequency ω_0 of the pendulum and search for beats (Figure 12.6 right). Note that you may need to adjust the magnitude and phase of the driving torque to avoid an "impedance mismatch" between the pendulum and driver.
4. Finally, scan the frequency ω of the driving torque and search for nonlinear resonance (it looks like beating).
5. **Explore chaos**: Start off with the initial conditions we used in Figure 12.9,

$$(x_0, v_0) = (-0.0885, 0.8), \quad (-0.0883, 0.8), \quad (-0.0888, 0.8).$$

   To save time and storage, you may want to use a larger time step for plotting than the one used to solve the differential equations.
6. Indicate which parts of the phase space plots correspond to transients. (The applets on the CD may help you with this, especially if you watch the phase space features being built up in time.)
7. Ensure that you have found:
   a. a period-3 limit cycle where the pendulum jumps between three major orbits in phase space,
   b. a running solution where the pendulum keeps going over the top,
   c. chaotic motion in which some paths in the phase space appear as bands.
8. Look for the "butterfly effect" (Figure 12.11 bottom). Start two pendulums off with identical positions but with velocities that differ by 1 part in 1000. Notice that the initial motions are essentially identical but that at some later time the motions diverge. ∎

12.13 Exploration: Bifurcations of Chaotic Pendulums

We have seen that a chaotic system contains a number of dominant frequencies and that the system tends to "jump" from one to another. This means that the dominant frequencies occur sequentially, in contrast to linear systems where they occur simultaneously. We now want to explore this jumping as a computer experiment. If we sample the instantaneous angular velocity $\dot{\theta} = d\theta/dt$ of our chaotic simulation at various instances in time, we should obtain a series of values for the frequency, with the major Fourier components occurring more often than others.[3] These are the frequencies to which the system is *attracted*. That being the case, if we make a scatterplot of the sampled $\dot{\theta}$s for many times at one particular value of the driving force and then change the magnitude of the driving force slightly and sample the frequencies again, the resulting plot may show distinctive patterns of frequencies. That a bifurcation diagram similar to the one for bug populations results is one of the mysteries of life.

[3] We refer to this angular velocity as $\dot{\theta}$ since we have already used ω for the frequency of the driver and ω_0 for the natural frequency.

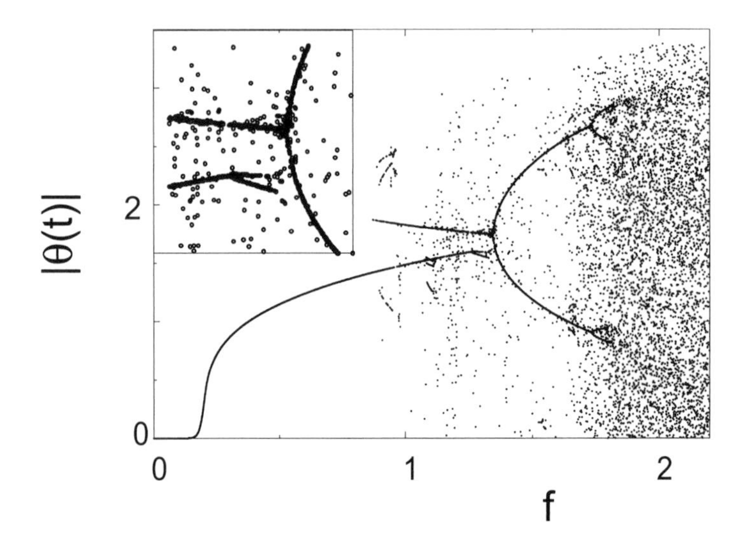

Figure 12.12 A bifurcation diagram for the damped pendulum with a vibrating pivot (see also the similar diagram for a double pendulum, Figure 12.14). The ordinate is $|d\theta/dt|$, the absolute value of the instantaneous angular velocity at the beginning of the period of the driver, and the abscissa is the magnitude of the driving force f. Note that the heavy line results from the overlapping of points, not from connecting the points (see enlargement in the inset).

In the scatterplot in Figure 12.12, we sample $\dot{\theta}$ for the motion of a chaotic pendulum with a vibrating pivot point (in contrast to our usual vibrating external torque). This pendulum is similar to our chaotic one (12.29), but with the driving force depending on $\sin\theta$:

$$\frac{d^2\theta}{dt^2} = -\alpha\,\frac{d\theta}{dt} - \left(\omega_0^2 + f\cos\omega t\right)\sin\theta. \tag{12.43}$$

Essentially, the acceleration of the pivot is equivalent to a sinusoidal variation of g or ω_0^2. Analytic [L&L,M 76, § 25–30] and numeric [DeJ 92, G,T&C 06] studies of this system exist. To obtain the bifurcation diagram in Figure 12.12:

1. Use the initial conditions $\theta(0) = 1$ and $\dot{\theta}(0) = 1$.
2. Set $\alpha = 0.1$, $\omega_0 = 1$, and $\omega = 2$, and vary $0 \le f \le 2.25$.
3. For each value of f, wait 150 periods of the driver before sampling to permit transients to die off. Sample $\dot{\theta}$ for 150 times at the instant the driving force passes through zero.
4. Plot the 150 values of $|\dot{\theta}|$ *versus* f. ▮

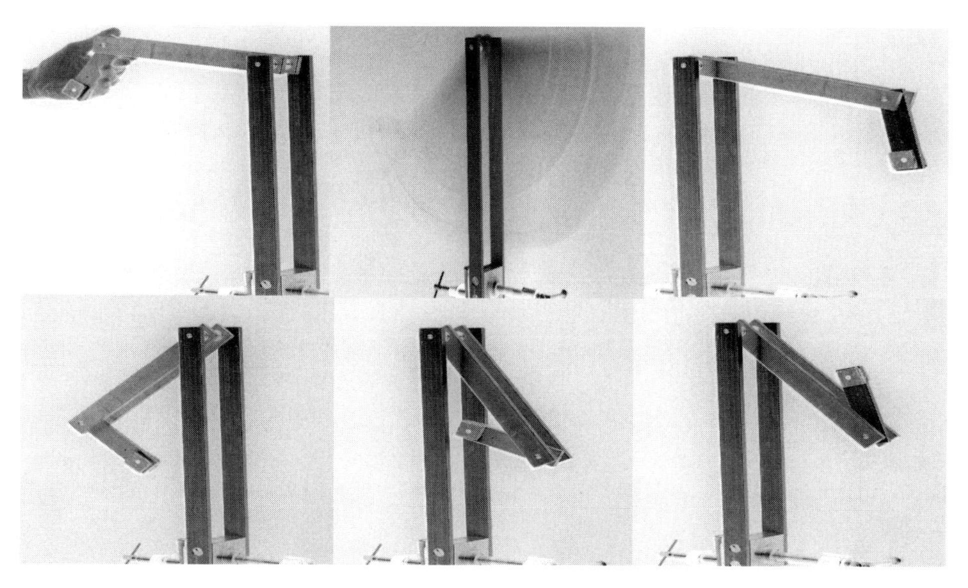

Figure 12.13 Photographs of a double pendulum built by a student in the OSU Physics Department. The longer pendulum consists of two separated shafts so that the shorter one can rotate through it. Both pendula can go over the top. We see the pendulum released from rest and then moving quickly. The flash photography stops the motion in various stages. (Photograph, R. Landau.)

12.14 Alternative Problem: The Double Pendulum

For those of you who have already studied a chaotic pendulum, an alternative is to study a double pendulum without any small-angle approximation (Figure 12.5 right and Fig. 12.13, and animation DoublePend.mpg on the CD). A double pendulum has a second pendulum connected to the first, and because each pendulum acts as a driving force for the other, we need not include an external driving torque to produce a chaotic system (there are enough degrees of freedom without it).

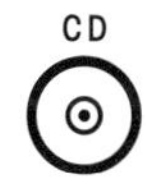

The equations of motions for the double pendulum are derived most directly from the Lagrangian formulation of mechanics. The Lagrangian is fairly simple but has the θ_1 and θ_2 motions innately coupled:

$$L = \mathrm{KE} - \mathrm{PE} = \frac{1}{2}(m_1+m_2)l_1^2\dot{\theta_1}^2 + \frac{1}{2}m_2 l_2^2 \dot{\theta_2}^2 \tag{12.44}$$

$$+ m_2 l_1 l_2 \dot{\theta_1}\dot{\theta_2}\cos(\theta_1-\theta_2) + (m_1+m_2) g l_1 \cos\theta_1 + m_2 g l_2 \cos\theta_2.$$

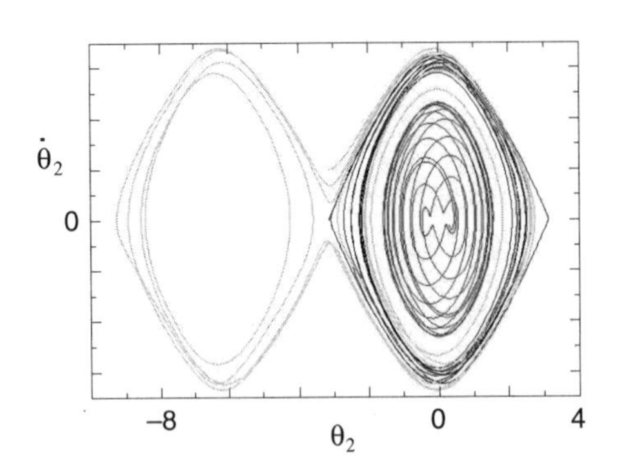

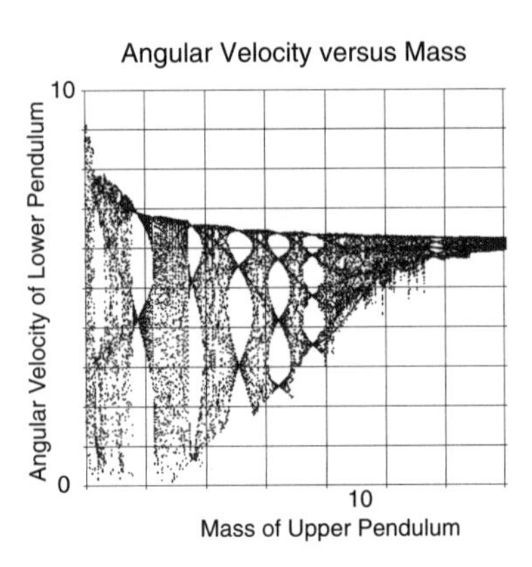

Figure 12.14 *Left:* Phase space trajectories for a double pendulum with $m_1 = 10m_2$, showing two dominant attractors. *Right:* A bifurcation diagram for the double pendulum displaying the instantaneous velocity of the lower pendulum as a function of the mass of the upper pendulum. (Both plots are courtesy of J. Danielson.)

The resulting dynamic equations couple the θ_1 and θ_2 motions:

$$(m_1 + m_2)l_1\ddot{\theta}_1 + m_2 l_2 \ddot{\theta}_2 \cos(\theta_1 - \theta_2) + m_2 l_2 \dot{\theta}_2^{\,2} \sin(\theta_1 - \theta_2) \tag{12.45}$$
$$+g(m_1 + m_2)\sin\theta_1 = 0,$$

$$m_2 l_2 \ddot{\theta}_2 + m_2 l_1 \ddot{\theta}_1 \cos(\theta_1 - \theta_2) - m_2 l_1 \dot{\theta}_1^{\,2} \sin(\theta_1 - \theta_2) + mg \sin\theta_2 = 0. \tag{12.46}$$

Usually textbooks approximate these equations for small oscillations, which diminish the effects of the coupling. "Slow" and "fast" modes then result for in-phase and antiphase oscillations, respectively, that look much like regular harmonic motions. What's more interesting is the motion that results without any small-angle restrictions, particularly when the pendulums have enough initial energy to go over the top (Figure 12.13). On the left in Figure 12.14 we see several phase space plots for the lower pendulum with $m_1 = 10m_2$. When given enough initial kinetic energy to go over the top, the trajectories are seen to flow between two major attractors as energy is transferred back and forth to the upper pendulum.

On the right in Figure 12.14 is a bifurcation diagram for the double pendulum. This was created by sampling and plotting the instantaneous angular velocity $\dot{\theta}_2$ of the lower pendulum at 70 times as the pendulum passed through its equilibrium position. The mass of the upper pendulum (a convenient parameter) was then changed, and the process repeated. The resulting structure is fractal and indicates bifurcations in the number of dominant frequencies in the motion. A plot of the Fourier or wavelet spectrum as a function of mass is expected to show similar characteristic frequencies.

12.15 Assessment: Fourier/Wavelet Analysis of Chaos

We have seen that a realistic pendulum experiences a restoring torque, $\tau_g \propto \sin\theta \simeq \theta - \theta^3/3! + \theta^5/5! + \cdots$, that contains nonlinear terms that lead to nonharmonic behavior. In addition, when a realistic pendulum is driven by an external sinusoidal torque, the pendulum may mode-lock with the driver and so oscillate at a frequency that is rationally related to the driver's. Consequently, the behavior of the realistic pendulum is expected to be a combination of various periodic behaviors, with discrete jumps between modes.

In this assessment you should determine the Fourier components present in the pendulum's complicated and chaotic behaviors. You should show that a three-cycle structure, for example, contains three major Fourier components, while a five-cycle structure has five. You should also notice that when the pendulum goes over the top, its spectrum contains a steady-state (DC) component.

1. Dust off your program for analyzing a $y(t)$ into Fourier components. Alternatively, you may use a Fourier analysis tool contained in your graphics program or system library (e.g., Grace and OpenDX).
2. Apply your analyzer to the solution of the chaotic pendulum for the cases where there are one-, three-, and five-cycle structures in phase space. Deduce the major frequencies contained in these structures.
3. Compare your results with the output of the `Pend1` applet (Figure 12.11 top).
4. Try to deduce a relation among the Fourier components, the natural frequency ω_0, and the driving frequency ω.
5. A classic signal of chaos is a broadband, although not necessarily flat, Fourier spectrum. Examine your system for parameters that give chaotic behavior and verify this statement. ▮

Wavelet Exploration: We saw in Chapter 11, "Wavelet Analysis & Data Compression", that a wavelet expansion is more appropriate than a Fourier expansion for a signal containing components that occur for finite periods of time. Because chaotic oscillations are just such signals, repeat the Fourier analysis of this section using wavelets instead of sines and cosines. Can you discern the temporal sequence of the various components?

12.16 Exploration: Another Type of Phase Space Plot

Imagine that you have measured the displacement of some system as a function of time. Your measurements appear to indicate characteristic nonlinear behaviors, and you would like to check this by making a phase space plot but without going to the trouble of measuring the conjugate momenta to plot *versus* displacement. Amazingly enough, one may also plot $x(t+\tau)$ *versus* $x(t)$ as a function of time to obtain a phase space plot [Abar 93]. Here τ is a *lag time* and should be chosen as some fraction of a characteristic time for the system under study. While this may not seem like a valid way to make a phase space plot, recall the forward difference approximation for the derivative,

$$v(t) = \frac{dx(t)}{dt} \simeq \frac{x(t+\tau) - x(t)}{\tau}. \tag{12.47}$$

We see that plotting $x(t+\tau)$ *versus* $x(t)$ is equivalent to plotting $v(t)$ *versus* $x(t)$.

Exercise: Create a phase space plot from the output of your chaotic pendulum by plotting $\theta(t+\tau)$ *versus* $\theta(t)$ for a large range of t values. Explore how the graphs change for different values of the lag time τ. Compare your results to the conventional phase space plots you obtained previously. ▌

12.17 Further Explorations

1. The nonlinear behavior in once-common objects such as vacuum tubes and metronomes is described by the **van der Pool equation**,

$$\frac{d^2x}{dt^2} + \mu(x^2 - x_0^2)\frac{dx}{dt} + \omega_0^2 x = 0. \tag{12.48}$$

The behavior predicted for these systems is *self-limiting* because the equation contains a limit cycle that is also a predictable attractor. You can think of (12.48) as describing an oscillator with x-dependent damping (the μ term). If $x > x_0$, friction slows the system down; if $x < x_0$, friction speeds the system up. Orbits internal to the limit cycle spiral out until they reach the limit cycle; orbits external to it spiral in.

2. **Duffing oscillator:** Another damped, driven nonlinear oscillator is

$$\frac{d^2\theta}{dt^2} - \frac{1}{2}\theta(1-\theta^2) = -\alpha\frac{d\theta}{dt} + f\cos\omega t. \tag{12.49}$$

While similar to the chaotic pendulum, it is easier to find multiple attractors for this oscillator [M&L 85].

3. **Lorenz attractor:** In 1962 Lorenz [Tab 89] was looking for a simple model for weather prediction and simplified the heat transport equations to

$$\frac{dx}{dt} = 10(y - x), \quad \frac{dy}{dt} = -xz + 28x - y, \quad \frac{dz}{dt} = xy - \frac{8}{3}z. \tag{12.50}$$

The solution of these simultaneous first-order nonlinear equations gave the complicated behavior that has led to the modern interest in chaos (after considerable doubt regarding the reliability of the numerical solutions).

4. **A 3-D computer fly:** Make $x+y$, $x+z$, and $y+z$ plots of the equations

$$x = \sin ay - z\cos bx, \quad y = z\sin cx - \cos dy, \quad z = e\sin x. \tag{12.51}$$

Here the parameter e controls the degree of apparent randomness.

5. **Hénon–Heiles potential:** The potential and Hamiltonian

$$V(x,y) = \frac{1}{2}x^2 + \frac{1}{2}y^2 + x^2y - \frac{1}{3}y^3, \quad H = \frac{1}{2}p_x^2 + \frac{1}{2}p_y^2 + V(x,y), \tag{12.52}$$

are used to describe three interacting astronomical objects. The potential binds the objects near the origin but releases them if they move far out. The equations of motion follow from the Hamiltonian equations:

$$\frac{dp_x}{dt} = -x - 2xy, \quad \frac{dp_y}{dt} = -y - x^2 + y^2, \quad \frac{dx}{dt} = p_x, \quad \frac{dy}{dt} = p_y.$$

a. Numerically solve for the position $[x(t), y(t)]$ for a particle in the Hénon–Heiles potential.
b. Plot $[x(t), y(t)]$ for a number of initial conditions. Check that the initial condition $E < \frac{1}{6}$ leads to a bounded orbit.
c. Produce a Poincaré section in the (y, p_y) plane by plotting (y, p_y) each time an orbit passes through $x = 0$.

12.18 Unit III. Coupled Predator–Prey Models ⊙

In Unit I we saw complicated behavior arising from a model of bug population dynamics in which we imposed a maximum population. We described that system with a discrete logistic map. In Unit II we saw complex behaviors arising from differential equations and learned how to use phase space plots to understand them. In this unit we study the differential equation model describing predator–prey population dynamics proposed by the American physical chemist Lotka [Lot 25] and the Italian mathematician Volterra [Volt 26]. Differential equations are easy to solve numerically and should be a good approximation if populations are large. However, there are equivalent discrete map versions of the model as well. Though simple, versions of these equations are used to model biological systems and neural networks.

Problem: Is it possible to use a small number of predators to control a population of pests so that the number of pests remains approximately constant? Include in your considerations the interaction between the populations as well as the competition for food and predation time.

12.19 Lotka-Volterra Model

We extend the logistic map to the Lotka–Volterra Model (LVM) to describe two populations coexisting in the same geographical region. Let

$$p(t) = \text{prey density}, \quad P(t) = \text{predator density}. \tag{12.53}$$

In the absence of interactions between the species, we assume that the prey population p breeds at a per-capita rate of a, which would lead to exponential growth:

$$\frac{dp}{dt} = ap, \quad \Rightarrow \quad p(t) = p(0)e^{at}. \tag{12.54}$$

Yet exponential growth does not occur because the predators P eat more prey as the prey numbers increase. The interaction rate between predator and prey requires both to be present, with the simplest assumption being that it is proportional to their joint probability:

$$\text{Interaction rate} = bpP.$$

This leads to a prey growth rate including both predation and breeding:

$$\boxed{\frac{dp}{dt} = a\,p - b\,p\,P,} \quad \text{(LVM-I for prey)}. \tag{12.55}$$

If left to themselves, predators P will also breed and increase their population. Yet predators need animals to eat, and if there are no other populations to prey upon, they will eat each other (or their young) at a per-capita mortality rate m:

$$\left.\frac{dP}{dt}\right|_{\text{competition}} = -mP, \Rightarrow P(t) = P(0)e^{-mt}. \tag{12.56}$$

However, once there are prey to interact with (read "eat") at the rate bpP, the predator population will grow at the rate

$$\boxed{\frac{dP}{dt} = \epsilon\,b\,p\,P - m\,P} \quad \text{(LVM-I for predators)}, \tag{12.57}$$

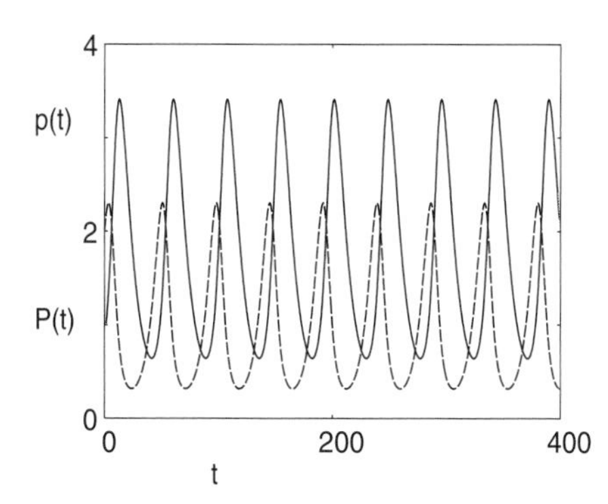

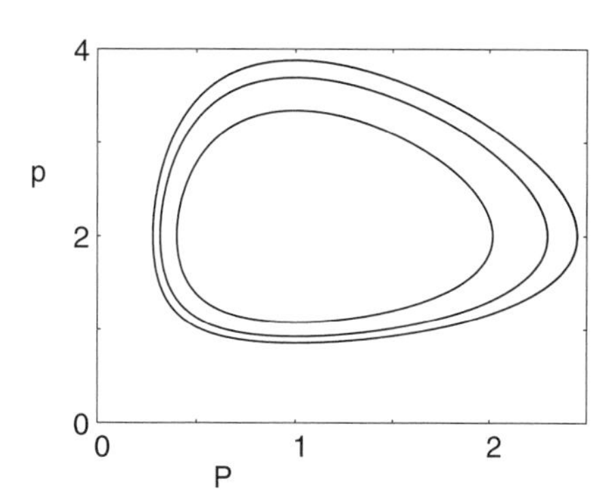

Figure 12.15 The populations of prey *p* and predator *P* from the Lotka–Volterra model. *Left:* The time dependences of the prey *p(t)* (solid) and the predators *P(t)* (dashed). *Right:* Prey population *p versus* predator population *P*. The different orbits in this "phase space plot" correspond to different initial conditions.

where ϵ is a constant that measures the efficiency with which predators convert prey interactions into food.

Equations (12.55) and (12.57) are two simultaneous ODEs and are our first model. We solve them with the `rk4` algorithm of Chapter 9, "Differential Equation Applications", after placing them in the standard dynamic form,

$$d\mathbf{y}/dt = \mathbf{f}(\mathbf{y}, t),$$

$$y_0 = p, \qquad f_0 = a\, y_0 - b\, y_0\, y_1, \tag{12.58}$$

$$y_1 = P, \qquad f_1 = \epsilon\, b\, y_0\, y_1 - m\, y_1.$$

A sample code to do this is `PredatorPrey.java` and is given on the CD. Results from our solution are shown in Figure 12.15. On the left we see that the two populations oscillate out of phase with each other in time; when there are many prey, the predator population eats them and grows; yet then the predators face a decreased food supply and so their population decreases; that in turn permits the prey population to grow, and so forth. On the right in Figure 12.15 we plot a phase space plot (phase space plots are discussed in Unit II) of $P(t)$ *versus* $p(t)$. A closed orbit here indicates a limit cycle that repeats indefinitely. Although increasing the initial number of predators does decrease the maximum number of pests, it is not a satisfactory solution to our **problem**, as the large variation in the number of pests cannot be called control.

12.19.1 LVM with Prey Limit

The initial assumption in the LVM that prey grow without limit in the absence of predators is clearly unrealistic. As with the logistic map, we include a limit on prey numbers that accounts for depletion of the food supply as the prey population grows. Accordingly, we modify the constant growth rate $a \rightarrow a(1-p/K)$ so that growth vanishes when the population reaches a limit K, the *carrying capacity*,

$$\boxed{\frac{dp}{dt} = a\,p\left(1-\frac{p}{K}\right) - b\,p\,P, \quad \frac{dP}{dt} = \epsilon\,b\,p\,P - m\,P} \quad \text{(LVM-II)}. \qquad (12.59)$$

The behavior of this model with prey limitations is shown in Figure 12.16. We see that both populations exhibit damped oscillations as they approach their equilibrium values. In addition, and as hoped for, the equilibrium populations are independent of the initial conditions. Note how the phase space plot spirals inward to a single close limit cycle, on which it remains, with little variation in prey number. This is control, and we may use it to predict the expected pest population.

12.19.2 LVM with Predation Efficiency

An additional unrealistic assumption in the original LVM is that the predators immediately eat all the prey with which they interact. As anyone who has watched a cat hunt a mouse knows, predators spend time finding prey and also chasing, killing, eating, and digesting it (all together called *handling*). This extra time decreases the rate bpP at which prey are eliminated. We define the *functional response* p_a as the probability of one predator finding one prey. If a single predator spends time t_{search} searching for prey, then

$$p_a = b\,t_{\text{search}}\,p \;\Rightarrow\; t_{\text{search}} = \frac{p_a}{bp}. \qquad (12.60)$$

If we call t_h the time a predator spends handling a single prey, then the effective time a predator spends handling a prey is $p_a\,t_h$. Such being the case, the total time T that a predator spends finding and handling a single prey is

$$T = t_{\text{search}} + t_{\text{handling}} = \frac{p_a}{bp} + p_a t_h \;\Rightarrow\; \frac{p_a}{T} = \frac{bp}{1+bpt_h},$$

where p_a/T is the effective *rate* of eating prey. We see that as the number of prey $p \rightarrow \infty$, the efficiency in eating them $\rightarrow 1$. We include this efficiency in (12.59) by modifying the rate b at which a predator eliminates prey to $b/(1+bpt_h)$:

$$\boxed{\frac{dp}{dt} = ap\left(1-\frac{p}{K}\right) - \frac{bpP}{1+bpt_h},} \quad \text{(LVM-III)}. \qquad (12.61)$$

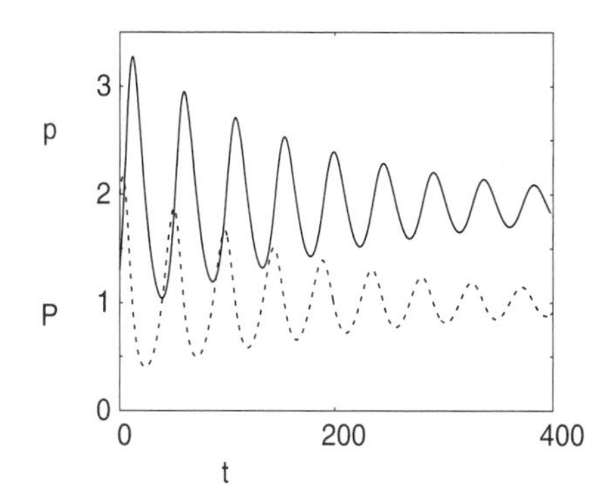

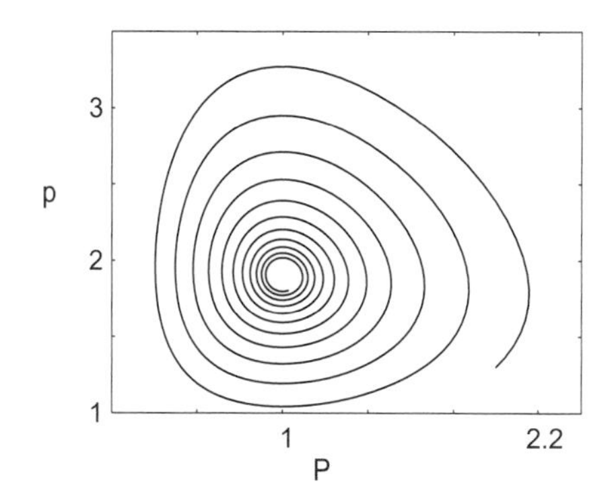

Figure 12.16 The Lotka–Volterra model of prey population *p* and predator population *P* with a prey population limit. *Left:* The time dependences of the prey *p(t)* (solid) and the predators *P(t)* (dashed). *Right:* Prey population *p versus* predator population *P*.

To be more realistic about the predator growth, we also place a limit on the predator carrying capacity but make it proportional to the number of prey:

$$\boxed{\frac{dP}{dt} = mP\left(1 - \frac{P}{kp}\right),} \quad \text{(LVM-III)}. \qquad (12.62)$$

Solutions for the extended model (12.61) and (12.62) are shown in Figure 12.17. Observe the existence of three dynamic regimes as a function of b:

- small b: no oscillations, no overdamping,
- medium b: damped oscillations that converge to a stable equilibrium,
- large b: limit cycle.

The transition from equilibrium to a limit cycle is called a *phase transition*.

We finally have a satisfactory solution to our **problem**. Although the prey population is not eliminated, it can be kept from getting too large and from fluctuating widely. Nonetheless, changes in the parameters can lead to large fluctuations or to nearly vanishing predators.

12.19.3 LVM Implementation and Assessment

1. Write a program to solve (12.61) and (12.62) using the `rk4` algorithm and the following parameter values.

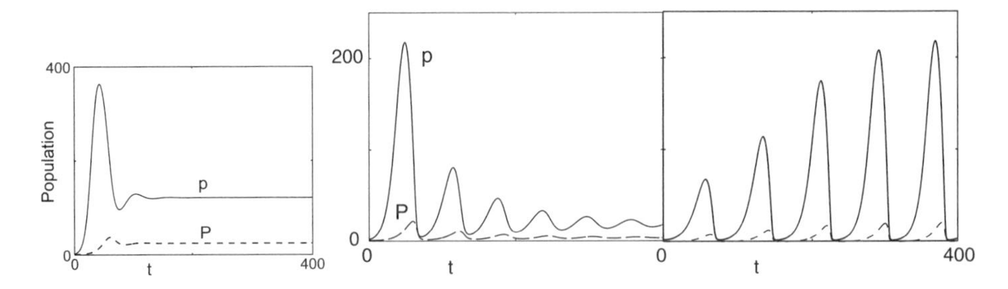

Figure 12.17 Lotka-Volterra model with predation efficiency and prey limitations. From left to right: overdamping, $b = 0.01$; damped oscillations, $b = 0.1$, and limit cycle, $b = 0.3$.

Model	a	b	ϵ	m	K	k
LVM-I	0.2	0.1	1	0.1	0	—
LVM-II	0.2	0.1	1	0.1	20.0	—
LVM-III	0.2	0.1	—	0.1	500.0	0.2

2. For each of the three models, construct
   a. a time series for prey and predator populations,
   b. phase space plots of predator *versus* prey populations.
3. **LVM-I**: Compute the equilibrium values for the prey and predator populations. Do you think that a model in which the cycle amplitude depends on the initial conditions can be realistic? Explain.
4. **LVM-II**: Calculate numerical values for the equilibrium values of the prey and predator populations. Make a series of runs for different values of prey carrying capacity K. Can you deduce how the equilibrium populations vary with prey carrying capacity?
5. Make a series of runs for different initial conditions for predator and prey populations. Do the cycle amplitudes depend on the initial conditions?
6. **LVM-III**: Make a series of runs for different values of b and reproduce the three regimes present in Figure 12.17.
7. Calculate the critical value for b corresponding to a phase transition between the stable equilibrium and the limit cycle. ▌

12.19.4 Two Predators, One Prey (Exploration)

1. Another version of the LVM includes the possibility that two populations of predators P_1 and P_2 may "share" the same prey population p. Investigate the behavior of a system in which the prey population grows logistically in the

absence of predators:

$$\frac{dp}{dt} = ap\left(1 - \frac{p}{K}\right) - (b_1 P_1 + b_2 P_2)\, p, \tag{12.63}$$

$$\frac{dP}{dt} = \epsilon_1 b_1 p P_1 - m_1 P_1, \qquad \frac{dP_2}{dt} = \epsilon_2 b_2 p P_2 - m_2 P_2. \tag{12.64}$$

a. Use the following values for the model parameters and initial conditions: $a = 0.2$, $K = 1.7$, $b_1 = 0.1$, $b_2 = 0.2$, $m_1 = m_2 = 0.1$, $\epsilon_1 = 1.0$, $\epsilon_2 = 2.0$, $p(0) = P_2(0) = 1.7$, and $P_1(0) = 1.0$.
b. Determine the time dependences for each population.
c. Vary the characteristics of the second predator and calculate the equilibrium population for the three components.
d. What is your answer to the question, "Can two predators that share the same prey coexist?"

2. The nonlinear nature of the Lotka–Volterra model can lead to chaos and fractal behavior. Search for chaos by varying the growth rates. ▌

13

Fractals & Statistical Growth

It is common to notice regular and eye-pleasing natural objects, such as plants and sea shells, that do not have well-defined geometric shapes. When analyzed mathematically, some of these objects have a dimension that is a fractional number, rather than an integer, and so are called fractals. *In this chapter we implement simple, statistical models that generate fractals. To the extent that these models generate structures that look like those in nature, it is reasonable to assume that the natural processes must be following similar rules arising from the basic physics or biology that creates the objects. Detailed applications of fractals can be found in [Mand 82, Arm 91, E&P 88, Sand 94, PhT 88].*

13.1 Fractional Dimension (Math)

Benoit Mandelbrot, who first studied fractional-dimension figures with supercomputers at IBM Research, gave them the name *fractals* [Mand 82]. Some geometric objects, such as Koch curves, are exact fractals with the same dimension for all their parts. Other objects, such as bifurcation curves, are statistical fractals in which elements of randomness occur and the dimension can be defined only locally or on the average.

Consider an abstract object such as the density of charge within an atom. There are an infinite number of ways to measure the "size" of this object. For example, each moment $\langle r^n \rangle$ is a measure of the size, and there is an infinite number of moments. Likewise, when we deal with complicated objects, there are different definitions of dimension and each may give a somewhat different value. In addition, the fractal dimension is often defined by using a measuring box whose size approaches zero, which is not practical for realistic applications.

Our first definition of the fractional dimension d_f (or *Hausdorff–Besicovitch dimension*) is based on our knowledge that a line has dimension 1, a triangle has dimension 2, and a cube has dimension 3. It seems perfectly reasonable to ask if there is some mathematical formula that agrees with our experience with regular objects yet can also be used for determining fractional dimensions. For simplicity, let us consider objects that have the same length L on each side, as do equilateral triangles and squares, and that have uniform density. We postulate that the dimension of an object is determined by the dependence of its total mass upon its length:

$$M(L) \propto L^{d_f}, \tag{13.1}$$

where the power d_f is the *fractal dimension*. As you may verify, this rule works with the 1-D, 2-D, and 3-D regular figures in our experience, so it is a reasonable definition. When we apply (13.1) to fractal objects, we end up with fractional values for d_f. Actually, we will find it easier to determine the fractal dimension not from an object's mass, which is *extensive* (depends on size), but rather from its density, which is *intensive*. The density is defined as mass/length for a linear object, as mass/area for a planar object, and as mass/volume for a solid object. That being the case, for a planar object we hypothesize that

$$\rho = \frac{M(L)}{\text{area}} \propto \frac{L^{d_f}}{L^2} \propto L^{d_f - 2}. \tag{13.2}$$

13.2 The Sierpiński Gasket (Problem 1)

To generate our first fractal (Figure 13.1), we play a game of chance in which we place dots at points picked randomly within a triangle. Here are the rules (which you should try out in the margins now).

1. Draw an equilateral triangle with vertices and coordinates:

$$\text{vertex } 1\text{: } (a_1, b_1); \quad \text{vertex } 2\text{: } (a_2, b_2); \quad \text{vertex } 3\text{: } (a_3, b_3).$$

2. Place a dot at an arbitrary point $P = (x_0, y_0)$ within this triangle.
3. Find the next point by selecting randomly the integer 1, 2, or 3:
   a. If 1, place a dot halfway between P and vertex 1.
   b. If 2, place a dot halfway between P and vertex 2.
   c. If 3, place a dot halfway between P and vertex 3.
4. Repeat the process using the last dot as the new P.

Mathematically, the coordinates of successive points are given by the formulas

$$(x_{k+1}, y_{k+1}) = \frac{(x_k, y_k) + (a_n, b_n)}{2}, \qquad n = \text{integer}\,(1 + 3r_i), \tag{13.3}$$

where r_i is a random number between 0 and 1 and where the *integer* function outputs the closest integer smaller than or equal to the argument. After 15,000 points, you should obtain a collection of dots like those on the left in Figure 13.1.

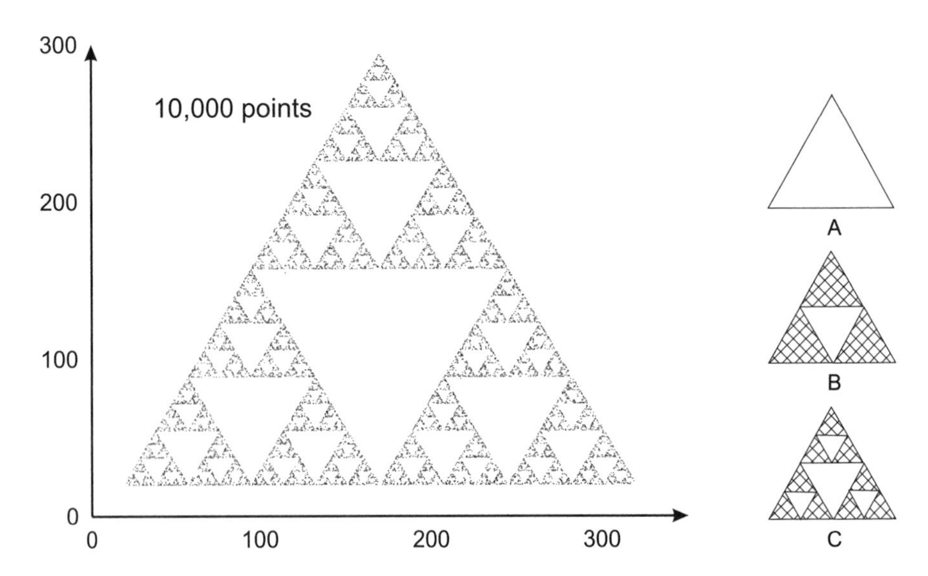

Figure 13.1 *Left:* A Sierpiński gasket containing 10,000 points constructed as a statistical fractal. Each filled part of this figure is self-similar. *Right:* A Sierpiński gasket constructed by successively connecting the midpoints of the sides of each equilateral triangle. (A–C) The first three steps in the process.

13.2.1 Sierpiński Implementation

Write a program to produce a Sierpiński gasket. Determine empirically the fractal dimension of your figure. Assume that each dot has mass 1 and that $\rho = CL^{\alpha}$. (You can have the computer do the counting by defining an array *box* of all 0 values and then change a 0 to a 1 when a dot is placed there.)

13.2.2 Assessing Fractal Dimension

The topology in Figure 13.1 was first analyzed by the Polish mathematician Sierpiński. Observe that there is the same structure in a small region as there is in the entire figure. In other words, if the figure had infinite resolution, any part of the figure could be scaled up in size and would be similar to the whole. This property is called *self-similarity*.

We construct a regular form of the Sierpiński gasket by removing an inverted equilateral triangle from the center of all filled equilateral triangles to create the next figure (Figure 13.1 right). We then repeat the process ad infinitum, scaling up the triangles so each one has side $r = 1$ after each step. To see what is unusual about this type of object, we look at how its density (mass/area) changes with size and then apply (13.2) to determine its fractal dimension. Assume that each triangle has mass m and assign unit density to the single triangle:

$$\rho(L=r) \propto \frac{M}{r^2} = \frac{m}{r^2} \stackrel{\text{def}}{=} \rho_0 \qquad \text{(Figure 13.1A)}$$

Next, for the equilateral triangle with side $L = 2$, the density

$$\rho(L=2r) \propto \frac{(M=3m)}{(2r)^2} = 34mr^2 = \frac{3}{4}\rho_0 \qquad \text{(Figure 13.1B)}$$

We see that the extra white space in Figure 13.1B leads to a density that is $\frac{3}{4}$ that of the previous stage. For the structure in Figure 13.1C, we obtain

$$\rho(L=4r) \propto \frac{(M=9m)}{(4r)^2} = (34)^2 \frac{m}{r^2} = \left(\frac{3}{4}\right)^2 \rho_0. \qquad \text{(Figure 13.1C)}$$

We see that as we continue the construction process, the density of each new structure is $\frac{3}{4}$ that of the previous one. Interesting. Yet in (13.2) we derived that

$$\rho \propto CL^{d_f-2}. \tag{13.4}$$

Equation (13.4) implies that a plot of the logarithm of the density ρ *versus* the logarithm of the length L for successive structures yields a straight line of slope

$$d_f - 2 = \frac{\Delta \log \rho}{\Delta \log L}. \tag{13.5}$$

As applied to our problem,

$$d_f = 2 + \frac{\Delta \log \rho(L)}{\Delta \log L} = 2 + \frac{\log 1 - \log \frac{3}{4}}{log 1 - \log 2} \simeq 1.58496. \tag{13.6}$$

As is evident in Figure 13.1, as the gasket grows larger (and consequently more massive), it contains more open space. So even though its mass approaches infinity as $L \to \infty$, its density approaches zero! And since a 2-D figure like a solid triangle has a constant density as its length increases, a 2-D figure has a slope equal to 0. Since the Sierpiński gasket has a slope $d_f - 2 \simeq -0.41504$, it fills space to a lesser extent than a 2-D object but more than a 1-D object does; it is a fractal with dimension ≤ 1.6.

13.3 Beautiful Plants (Problem 2)

It seems paradoxical that natural processes subject to chance can produce objects of high regularity and symmetry. For example, it is hard to believe that something as beautiful and graceful as a fern (Figure 13.2 left) has random elements in it. Nonetheless, there is a clue here in that much of the fern's beauty arises from the similarity of each part to the whole (self-similarity), with different ferns similar but not identical to each other. These are characteristics of fractals. Your **problem** is to discover if a simple algorithm including some randomness can draw regular ferns.

Figure 13.2 *Left:* A fern after 30,000 iterations of the algorithm (13.10). If you enlarge this, you will see that each frond has a similar structure. *Right:* A fractal tree created with the simple algorithm (13.13)

If the algorithm produces objects that resemble ferns, then presumably you have uncovered mathematics similar to that responsible for the shapes of ferns.

13.3.1 Self-affine Connection (Theory)

In (13.3), which defines mathematically how a Sierpiński gasket is constructed, a *scaling factor* of $\frac{1}{2}$ is part of the relation of one point to the next. A more general transformation of a point $P = (x, y)$ into another point $P' = (x', y')$ via *scaling* is

$$(x', y') = s(x, y) = (sx, sy) \quad \text{(scaling)}. \tag{13.7}$$

If the scale factor $s > 0$, an amplification occurs, whereas if $s < 0$, a reduction occurs. In our definition (13.3) of the Sierpiński gasket, we also added in a constant a_n. This is a *translation operation*, which has the general form

$$(x', y') = (x, y) + (a_x, a_y) \quad \text{(translation)}. \tag{13.8}$$

Another operation, not used in the Sierpiński gasket, is a *rotation* by angle θ:

$$x' = x\cos\theta - y\sin\theta, \quad y' = x\ \sin\theta + y\cos\theta \quad \text{(rotation)}. \tag{13.9}$$

The entire set of transformations, scalings, rotations, and translations defines an *affine transformation* (affine denotes a close relation between successive points). The transformation is still considered affine even if it is a more general linear transformation with the coefficients not all related by a single θ (in that case, we can have contractions and reflections). What is important is that the object created with these rules turns out to be self-similar; each step leads to new parts of the object that bear the same relation to the ancestor parts as the ancestors did to theirs. This is what makes the object look similar at all scales.

13.3.2 Barnsley's Fern Implementation

We obtain a Barnsley's fern [Barns 93] by extending the dots game to one in which new points are selected using an affine connection with some elements of chance mixed in:

$$(x,y)_{n+1} = \begin{cases} (0.5, 0.27y_n), & \text{with 2\% probability,} \\ (-0.139x_n + 0.263y_n + 0.57 \\ \quad 0.246x_n + 0.224y_n - 0.036), & \text{with 15\% probability,} \\ (0.17x_n - 0.215y_n + 0.408 \\ \quad 0.222x_n + 0.176y_n + 0.0893), & \text{with 13\% probability,} \\ (0.781x_n + 0.034y_n + 0.1075 \\ \quad -0.032x_n + 0.739y_n + 0.27), & \text{with 70\% probability.} \end{cases} \tag{13.10}$$

To select a transformation with probability $\mathcal{P}$, we select a uniform random number $0 \le r \le 1$ and perform the transformation if r is in a range proportional to $\mathcal{P}$:

$$\mathcal{P} = \begin{cases} 2\%, & r < 0.02, \\ 15\%, & 0.02 \le r \le 0.17, \\ 13\%, & 0.17 < r \le 0.3, \\ 70\%, & 0.3 < r < 1. \end{cases} \tag{13.11}$$

The rules (13.10) and (13.11) can be combined into one:

$$(x,y)_{n+1} = \begin{cases} (0.5, 0.27y_n), & r < 0.02, \\ (-0.139x_n + 0.263y_n + 0.57 \\ \quad 0.246x_n + 0.224y_n - 0.036), & 0.02 \le r \le 0.17, \\ (0.17x_n - 0.215y_n + 0.408 \\ \quad 0.222x_n + 0.176y_n + 0.0893), & 0.17 < r \le 0.3, \\ (0.781x_n + 0.034y_n + 0.1075, \\ \quad -0.032x_n + 0.739y_n + 0.27), & 0.3 < r < 1. \end{cases} \tag{13.12}$$

Although (13.10) makes the basic idea clearer, (13.12) is easier to program.

The starting point in Barnsley's fern (Figure 13.2) is $(x_1, y_1) = (0.5, 0.0)$, and the points are generated by repeated iterations. An important property of this fern is that it is not completely self-similar, as you can see by noting how different the stems and the fronds are. Nevertheless, the stem can be viewed as a compressed copy of a frond, and the fractal obtained with (13.10) is still *self-affine*, yet with a dimension that varies in different parts of the figure.

13.3.3 Self-affinity in Trees Implementation

Now that you know how to grow ferns, look around and notice the regularity in trees (such as in Figure 13.2 right). Can it be that this also arises from a self-affine structure? Write a program, similar to the one for the fern, starting at $(x_1, y_1) = (0.5, 0.0)$ and iterating the following self-affine transformation:

$$(x_{n+1}, y_{n+1}) = \begin{cases} (0.05x_n, 0.6y_n), & 10\% \text{ probability}, \\ (0.05x_n, -0.5y_n + 1.0), & 10\% \text{ probability}, \\ (0.46x_n - 0.15y_n, 0.39x_n + 0.38y_n + 0.6), & 20\% \text{ probability}, \\ (0.47x_n - 0.15y_n, 0.17x_n + 0.42y_n + 1.1), & 20\% \text{ probability}, \\ (0.43x_n + 0.28y_n, -0.25x_n + 0.45y_n + 1.0), & 20\% \text{ probability}, \\ (0.42x_n + 0.26y_n, -0.35x_n + 0.31y_n + 0.7), & 20\% \text{ probability}. \end{cases} \tag{13.13}$$

13.4 Ballistic Deposition (Problem 3)

There are a number of physical and manufacturing processes in which particles are deposited on a surface and form a film. Because the particles are evaporated from a hot filament, there is randomness in the emission process yet the produced films turn out to have well-defined, regular structures. Again we suspect fractals. Your **problem** is to develop a model that simulates this growth process and compare your produced structures to those observed.

13.4.1 Random Deposition Algorithm

The idea of simulating random depositions was first reported in [Vold 59], which includes tables of random numbers used to simulate the sedimentation of moist spheres in hydrocarbons. We shall examine a method of simulation [Fam 85] that results in the deposition shown in Figure 13.3. Consider particles falling onto and sticking to a horizontal line of length L composed of 200 deposition sites. All particles start from the same height, but to simulate their different velocities, we assume they start at random distances from the left side of the line. The simulation consists of generating uniform random sites between 0 and L and having a particle stick to the site on which it lands. Because a realistic situation may have columns of aggregates of different heights, the particle may be stopped before it makes it to the line, or it may bounce around until it falls into a hole. We therefore assume that if the column height at which the particle lands is greater than that of both its neighbors,

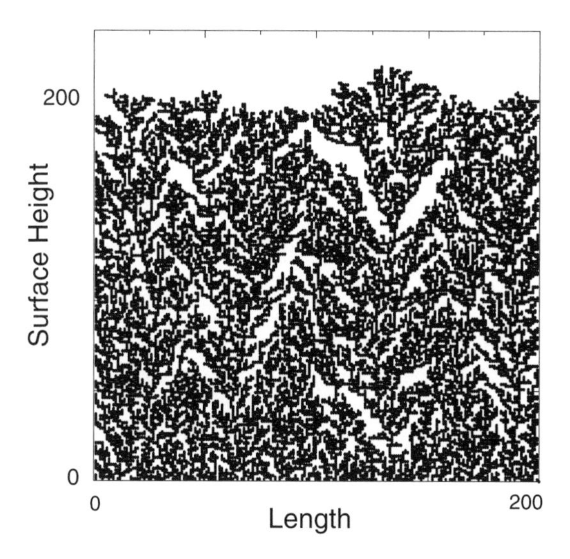

Figure 13.3 A simulation of the ballistic deposition of 20,000 particles on a substrate of length 200. The vertical height increases with the length of deposition time so that the top is the final surface.

it will add to that height. If the particle lands in a hole, or if there is an adjacent hole, it will fill up the hole. We speed up the simulation by setting the height of the hole equal to the maximum of its neighbors:

1. Choose a random site r.
2. Let the array h_r be the height of the column at site r.
3. Make the decision:

$$h_r = \begin{cases} h_r + 1, & \text{if } h_r \geq h_{r-1}, \;\; h_r > h_{r+1}, \\ \max[h_{r-1}, h_{r+1}], & \text{if } h_r < h_{r-1}, \;\; h_r < h_{r+1}. \end{cases} \tag{13.14}$$

Our sample simulation is Fractals/Film.java on the CD (FilmDim.java on the instructor's CD), with the essential loop:

```
int spot = random.nextInt(200); if (spot == 0)  {
    if ( coast[spot] < coast[spot+1] ) coast[spot] = coast[spot+1];
        else coast[spot]++;
} else if (spot == coast.length - 1) {
    if (coast[spot] < coast[spot-1]) coast[spot] = coast[spot-1];
    else coast[spot]++;
} else if ( coast[spot]<coast[spot-1] && coast[spot]<coast[spot+1] )
{
    if ( coast[spot-1] > coast[spot+1] ) coast[spot] = coast[spot-1];
        else coast[spot] = coast[spot+1];
} else coast[spot]++;
```

The results of this type of simulation show several empty regions scattered throughout the line (Figure 13.3), which is an indication of the statistical nature of the process while the film is growing. Simulations by Fereydoon reproduced the experimental observation that the average height increases linearly with time and produced fractal surfaces. (You will be asked to determine the fractal dimension of a similar surface as an exercise.)

13.5 Length of the British Coastline (Problem 4)

In 1967 Mandelbrot [Mand 67] asked a classic question, "How long is the coast of Britain?" If Britain had the shape of Colorado or Wyoming, both of which have straight-line boundaries, its perimeter would be a curve of dimension 1 with finite length. However, coastlines are geographic not geometric curves, with each portion of the coast often statistically self-similar to the entire coast yet on a reduced scale. In the latter cases the perimeter may be modeled as a fractal, in which case the length is either infinite or meaningless. Mandelbrot deduced the dimension of the west coast of Great Britain to be $d_f = 1.25$. In your **problem** we ask you to determine the dimension of the perimeter of one of our fractal simulations.

13.5.1 Coastlines as Fractals (Model)

The length of the coastline of an island is the perimeter of that island. While the concept of perimeter is clear for regular geometric figures, some thought is required to give it meaning for an object that may be infinitely self-similar. Let us assume that a map maker has a ruler of length r. If she walks along the coastline and counts the number of times N that she must place the ruler down in order to *cover* the coastline, she will obtain a value for the length L of the coast as Nr. Imagine now that the map maker keeps repeating her walk with smaller and smaller rulers. If the coast was a geometric figure or a *rectifiable curve*, at some point the length L would become essentially independent of r and would approach a constant. Nonetheless, as discovered empirically by Richardson [Rich 61] for natural coastlines, such as those of South Africa and Britain, the perimeter appears to be a function of r:

$$L(r) = Mr^{1-d_f}, \tag{13.15}$$

where M and d_f are empirical constants. For a geometric figure or for Colorado, $d_f = 1$ and the length approaches a constant as $r \to 0$. Yet for a fractal with $d_f > 1$, the perimeter $L \to \infty$ as $r \to 0$. This means that as a consequence of self-similarity, fractals may be of finite size but have infinite perimeters. Physically, at some point there may be no more details to discern as $r \to 0$ (say, at the quantum or Compton size limit), and so the limit may not be meaningful.

13.5.2 Box Counting Algorithm

Consider a line of length L broken up into segments of length r (Figure 13.4 left). The number of segments or "boxes" needed to cover the line is related to the size r of the box by

$$N(r) = \frac{L}{r} = \frac{C}{r}, \tag{13.16}$$

where C is a constant. A proposed definition of fractional dimension is the power of r in this expression as $r \to 0$. In our example, it tells us that the line has dimension $d_f = 1$. If we now ask how many little circles of radius r it would take to *cover* or fill a circle of area A (Figure 13.4 middle), we will find that

$$N(r) = \lim_{r\to 0} \frac{A}{\pi r^2} \quad \Rightarrow \quad d_f = 2, \tag{13.17}$$

as expected. Likewise, counting the number of little spheres or cubes that can be packed within a large sphere tells us that a sphere has dimension $d_f = 3$. In general, if it takes N little spheres or cubes of side $r \to 0$ to cover some object, then the fractal dimension d_f can be deduced as

$$N(r) = C \left(\frac{1}{r}\right)^{d_f} = C' \, s^{d_f} \quad (\text{as } r \to 0), \tag{13.18}$$

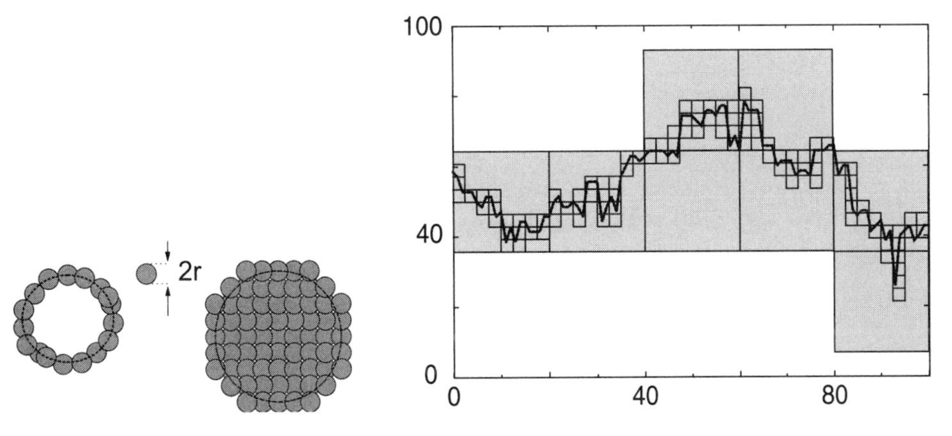

Figure 13.4 Examples of the use of box counting to determine fractal dimension. On the left the perimeter is being covered, in the middle an entire figure is being covered, and on the right a "coastline" is being covered by boxes of two different sizes (scales).

$$\log N(r) = \log C - d_f \log(r) \quad (\text{as } r \to 0), \tag{13.19}$$

$$\Rightarrow \quad d_f = -\lim_{r\to 0} \frac{\Delta N(r)}{\Delta r}. \tag{13.20}$$

Here $s \propto 1/r$ is called the *scale* in geography, so $r \to 0$ corresponds to an infinite scale. To illustrate, you may be familiar with the low scale on a map being 10,000 m to a centimeter, while the high scale is 100 m to a centimeter. If we want the map to show small details (sizes), we need a map of high scale.

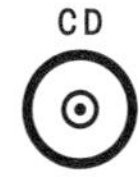

We will use box counting to determine the dimension of a perimeter, not of an entire figure. Once we have a value for d_f, we can determine a value for the length of the perimeter via (13.15). (If you cannot wait to see box counting in action, on the CD you will find an applet `Jfracdim` that goes through all the steps of box counting before your eyes and even plots the results.)

13.5.3 Coastline Implementation and Exercise

Rather than ruin our eyes using a geographic map, we use a mathematical one. Specifically, with a little imagination you will see that the top portion of Figure 13.3

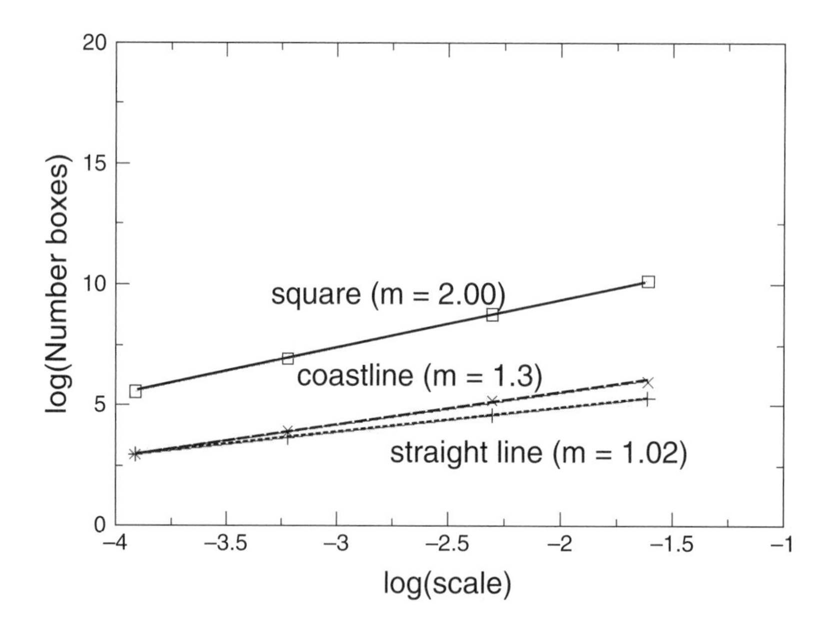

Figure 13.5 Dimensions of a line, box, and coastline determined by box counting.

looks like a natural coastline. Determine d_f by covering this figure, or one you have generated, with a semitransparent piece of graph paper[1], and counting the number of boxes containing any part of the coastline (Figures 13.4 and 13.5).

1. Print your coastline graph with the same physical scale (*aspect ratio*) for the vertical and horizontal axes. This is required because the graph paper you will use for box counting has square boxes and so you want your graph to also have the same vertical and horizontal scales. Place a piece of graph paper over your printout and look though the graph paper at your coastline. If you do not have a piece of graph paper available, or if you are unable to obtain a printout with the same aspect ratio for the horizontal and vertical axes, add a series of closely spaced horizontal and vertical lines to your coastline printout and use these lines as your graph paper. (Box counting should still be accurate if both your coastline and your graph paper are have the same aspect ratios.)
2. The vertical height in our printout was 17 cm, and the largest division on our graph paper was 1 cm. This sets the scale of the graph as 1:17, or $s = 17$ for the largest divisions (lowest scale). Measure the vertical height of your fractal, compare it to the size of the biggest boxes on your "piece" of graph paper, and thus determine your lowest scale.
3. With our largest boxes of $1\,\mathrm{cm} \times 1\,\mathrm{cm}$, we found that the coastline passed through $N = 24$ large boxes, that is, that 24 large boxes covered the coastline at $s = 17$. Determine how many of the largest boxes (lowest scale) are needed to cover your coastline.
4. With our next smaller boxes of $0.5\,\mathrm{cm} \times 0.5\,\mathrm{cm}$, we found that 51 boxes covered the coastline at a scale of $s = 34$. Determine how many of the midsize boxes (midrange scale) are needed to cover your coastline.
5. With our smallest boxes of $1\,\mathrm{mm} \times 1\,\mathrm{mm}$, we found that 406 boxes covered the coastline at a scale of $s = 170$. Determine how many of the smallest boxes (highest scale) are needed to cover your coastline.
6. Equation (13.20) tells us that as the box sizes get progressively smaller, we have

$$\log N \simeq \log A + d_f \log s,$$

$$\Rightarrow \quad d_f \simeq \frac{\Delta \log N}{\Delta \log s} = \frac{\log N_2 - \log N_1}{\log s_2 - \log s_1} = \frac{\log(N_2/N_1)}{\log(s_2/s_1)}.$$

Clearly, only the relative scales matter because the proportionality constants cancel out in the ratio. A plot of $\log N$ *versus* $\log s$ should yield a straight line. In our example we found a slope of $d_f = 1.23$. Determine the slope and thus the fractal dimension for your coastline. Although only two points are needed

[1] Yes, we are suggesting a painfully analog technique based on the theory that trauma leaves a lasting impression. If you prefer, you can store your output as a matrix of 1 and 0 values and let the computer do the counting, but this will take more of your time!

to determine the slope, use your lowest scale point as an important check. (Because the fractal dimension is defined as a limit for infinitesimal box sizes, the highest scale points are more significant.)

7. As given by (13.15), the perimeter of the coastline

$$L \propto s^{1.23-1} = s^{0.23}. \tag{13.21}$$

If we keep making the boxes smaller and smaller so that we are looking at the coastline at higher and higher scale *and* if the coastline is self-similar at all levels, then the scale s will keep getting larger and larger with no limits (or at least until we get down to some quantum limits). This means

$$L \propto \lim_{s \to \infty} s^{0.23} = \infty. \tag{13.22}$$

Does your fractal lead to an infinite coastline? Does it make sense that a small island like Britain, which you can walk around, has an infinite perimeter? ▮

13.6 Correlated Growth, Forests, and Films (Problem 5)

It is an empirical fact that in nature there is increased likelihood that a plant will grow if there is another one nearby (Figure 13.6 left). This *correlation* is also valid for the "growing" of surface films, as in the previous algorithm. Your **problem** is to include correlations in the surface simulation.

13.6.1 Correlated Ballistic Deposition Algorithm

A variation of the ballistic deposition algorithm, known as the *correlated ballistic deposition algorithm*, simulates mineral deposition onto substrates on which dendrites form [Tait 90]. We extend the previous algorithm to include the likelihood that a freshly deposited particle will attract another particle. We assume that the probability of sticking $\mathcal{P}$ depends on the distance d that the added particle is from the last one (Figure 13.6 right):

$$\mathcal{P} = c\, d^{\eta}. \tag{13.23}$$

Here η is a parameter and c is a constant that sets the probability scale.[2] For our implementation we choose $\eta = -2$, which means that there is an inverse square attraction between the particles (decreased probability as they get farther apart).

[2] The absolute probability, of course, must be less than one, but it is nice to choose c so that the relative probabilities produce a graph with easily seen variations.

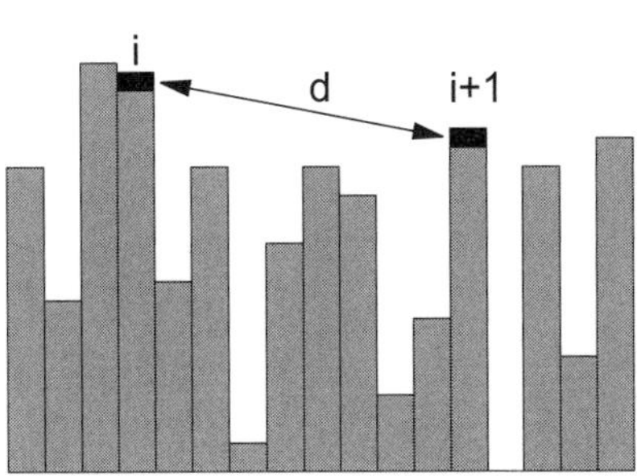

Figure 13.6 *Left:* A view that might be seen in the undergrowth of a forest or a correlated ballistic deposition. *Right:* The probability of particle $i+1$ sticking in some column depends on the distance d from the previously deposited particle i.

As in our study of uncorrelated deposition, a uniform random number in the interval $[0, L]$ determines the column in which the particle will be deposited. We use the same rules about the heights as before, but now a second random number is used in conjunction with (13.23) to decide if the particle will stick. For instance, if the computed probability is 0.6 and if $r < 0.6$, the particle will be accepted (sticks); if $r > 0.6$, the particle will be rejected.

13.7 Globular Cluster (Problem 6)

Consider a bunch of grapes on an overhead vine. Your **problem** is to determine how its tantalizing shape arises. In a flash of divine insight, you realize that these shapes, as well as others such as those of dendrites, colloids, and thin-film structure, appear to arise from an aggregation process that is limited by diffusion.

13.7.1 Diffusion-Limited Aggregation Algorithm

A model of diffusion-limited aggregation (DLA) has successfully explained the relation between a cluster's perimeter and mass [W&S 83]. We start with a 2-D lattice containing a seed particle in the middle, draw a circle around the particle, and place another particle on the circumference of the circle at some random angle. We then release the second particle and have it execute a random walk, much like the one we studied in Chapter 5, "Monte Carlo Simulations," but restricted to vertical or horizontal jumps between lattice sites. This is a type of *Brownian motion* that simulates diffusion. To make the model more realistic, we let the length of each

```
import java.io.*;           //Location of PrintWriter
import java.util.*;         //Location of Random
import java.lang.*;         //Location of Math

public class COLUMN {

  public static void main(String[] argv) throws IOException, FileNotFoundException {
    PrintWriter q = new PrintWriter( new FileOutputStream("COLUMN.DAT"),true);
    long seed = 971761;                        // Initialize 48 bit random number generator
    Random randnum = new Random(seed);                  // Next random: randnum.nextDouble()
    int max = 100000, npoints = 200;                          // Number iterations, spaces
    int i = 0, dist = 0, r = 0, x = 0, y = 0, oldx = 0, oldy = 0;
    double pp = 0.0, prob = 0.0;
    int hit[] = new int[200];
    for (i = 0; i< npoints; i++) hit[i] = 0;                                  // Clear array
    oldx = 100;
    oldy = 0;
    for( i = 1; i <= max; i++) {
    r = (int)(npoints*randnum.nextDouble());
    x = r-oldx;
    y = hit[r]-oldy;
    dist = x*x + y*y;
    if (dist == 0) prob = 1.0;             // Sticking prob depends on x to last particle
      else prob = 9.0/dist;                                       // nu = -2.0, c = 0.9
      pp = randnum.nextDouble();
        if (pp < prob) {
          if(r>0 && r<(npoints-1)) {
            if((hit[r] >= hit[r-1]) && (hit[r] >= hit[r+1])) hit[r]++;
                else if (hit[r-1] > hit[r+1]) hit[r] = hit[r-1];
                else hit[r] = hit[r+1];
            oldx = r;
            oldy = hit[r];
            q.println(r+"  "+hit[r]);
            }
          }
    }
    System.out.println(" ");
    System.out.println("COLUMN Program Complete.");
    System.out.println("Data stored in COLUMN.DAT");
    System.out.println(" ");
  }
}   // End of class
```

Listing 13.1 Column.java simulates correlated ballistic deposition of minerals onto substrates on which dendrites form.

step vary according to a random Gaussian distribution. If at some point during its random walk, the particle encounters another particle within one lattice spacing, they stick together and the walk terminates. If the particle passes outside the circle from which it was released, it is lost forever. The process is repeated as often as desired and results in clusters (Figure 13.7 and applet dla).

1. Write a subroutine that generates random numbers with a Gaussian distribution.[3]
2. Define a 2-D lattice of points represented by the array grid[400][400] with all elements initially zero.

[3] We indicated how to do this in § 6.8.1.

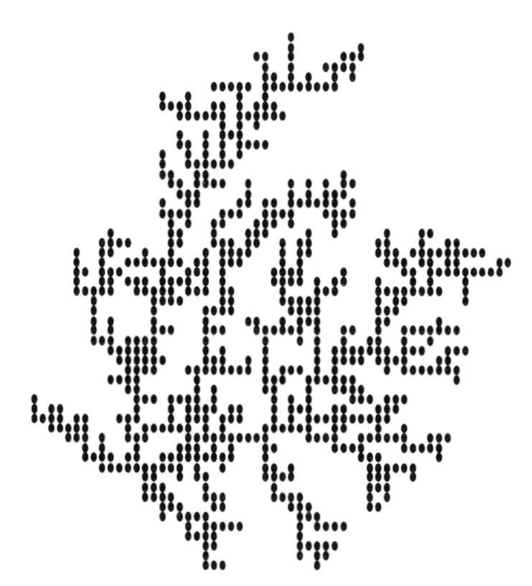

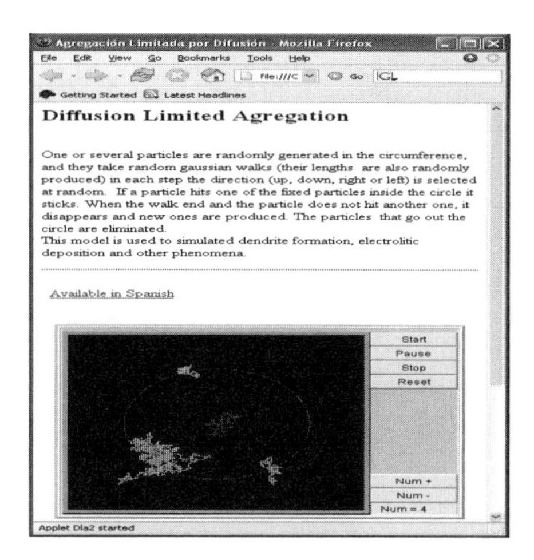

Figure 13.7 *Left:* A globular cluster of particles of the type that might occur in a colloid. *Right:* The applet **Dla2en.html** on the CD lets you watch these clusters grow. Here the cluster is at the center of the circle, and random walkers are started at random points around the circle.

3. Place the seed at the center of the lattice; that is, set grid[199][199]=1.
4. Imagine a circle of radius 180 lattice spacings centered at grid[199][199]. This is the circle from which we release particles.
5. Determine the angular position of the new particle on the circle's circumference by generating a uniform random angle between 0 and 2π.
6. Compute the x and y positions of the new particle on the circle.
7. Determine whether the particle moves horizontally or vertically by generating a uniform random number $0 < r_{xy} < 1$ and applying the rule

$$\text{if} \begin{cases} r_{xy} < 0.5, & \text{motion is vertical,} \\ r_{xy} > 0.5, & \text{motion is horizontal.} \end{cases} \tag{13.24}$$

8. Generate a Gaussian-weighted random number in the interval $[-\infty, \infty]$. This is the size of the step, with the sign indicating direction.
9. We now know the total distance and direction the particle will move. It jumps one lattice spacing at a time until this total distance is covered.
10. Before a jump, check whether a nearest-neighbor site is occupied:
    a. If occupied, the particle stays at its present position and the walk is over.
    b. If unoccupied, the particle jumps one lattice spacing.
    c. Continue the checking and jumping until the total distance is covered, until the particle sticks, or until it leaves the circle.

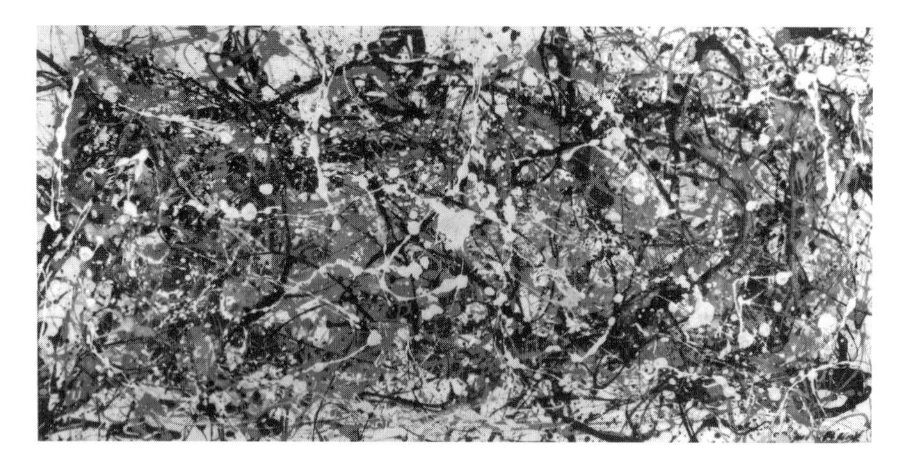

Figure 13.8 Number 8 by the American painter Jackson Pollock. (Used with permission, Neuberger Museum, State University of New York.) It has been found that Pollock's paintings exhibit a characteristic fractal structure. See if you can determine the fractal dimensions within this painting.

11. Once one random walk is over, another particle can be released and the process repeated. This is how the cluster grows. ▮

Because many particles are lost, you may need to generate hundreds of thousands of particles to form a cluster of several hundred particles.

13.7.2 Fractal Analysis of a DLA (or Pollock) Graph; Assessment

A cluster generated with the DLA technique is shown in Figure 13.7. We wish to analyze it to see if the structure is a fractal and, if so, to determine its dimension. (As an alternative, you may analyze the fractal nature of the Pollock painting in Figure 13.8, a technique used to determine the authenticity of this sort of art.) As a control, *simultaneously* analyze a geometric figure, such as a square or circle, whose dimension is known. The analysis is a variation of the one used to determine the length of the coastline of Britain.

1. If you have not already done so, use the box counting method to determine the fractal dimension of a simple square.
2. Draw a square of length L, small relative to the size of the cluster, around the seed particle. (Small might be seven lattice spacings to a side.)
3. Count the number of particles within the square.

4. Compute the density ρ by dividing the number of particles by the number of sites available in the box (49 in our example).
5. Repeat the procedure using larger and larger squares.
6. Stop when the cluster is covered.
7. The (box counting) fractal dimension d_f is estimated from a log-log plot of the density ρ *versus* L. If the cluster is a fractal, then (13.2) tells us that $\rho \propto L^{d_f-2}$, and the graph should be a straight line of slope $d_f - 2$.

The graph we generated had a slope of -0.36, which corresponds to a fractal dimension of 1.66. Because random numbers are involved, the graph you generate will be different, but the fractal dimension should be similar. (Actually, the structure is multifractal, and so the dimension varies with position.)

13.8 Fractal Structures in a Bifurcation Graph (Problem 7)

Recollect the project involving the logistics map where we plotted the values of the stable population numbers *versus* the growth parameter μ. Take one of the bifurcation graphs you produced and determine the fractal dimension of different parts of the graph by using the same technique that was applied to the coastline of Britain.

13.9 Fractals from Cellular Automata

We have already indicated in places how statistical models may lead to fractals. There is a class of statistical models known as cellular automata *that produce complex behaviors from very simple systems. Here we study some.*

Cellular automata were developed by von Neumann and Ulam in the early 1940s (von Neumann was also working on the theory behind modern computers then). Though very simple, cellular automata have found applications in many branches of science [Peit 94, Sipp 96]. Their classic definition is [Barns 93]:

> A cellular automaton is a discrete dynamical system in which space, time, and the states of the system are discrete. Each point in a regular spatial lattice, called a cell, can have any one of a finite number of states, and the states of the cells in the lattice are updated according to a local rule. That is, the state of a cell at a given time depends only on its won state one time step previously, and the states of its nearby neighbors at the previous time step. All cells on the lattice are updated synchronously, and so the state of the entice lattice advances in discrete time steps.

The program `CellAut.java` given on the CD creates a simple 1-D cellular automaton that grows on your screen. A cellular automaton in two dimensions consists of a number of square cells that grow upon each other. A famous one, invented by

Conway in the 1970s, is Conway's Game of Life. In this, cells with value 1 are alive, while cells with value 0 are dead. Cells grow according to the following rules:

1. If a cell is alive and if two or three of its eight neighbors are alive, then the cell remains alive.
2. If a cell is alive and if more than three of its eight neighbors are alive, then the cell dies because of overcrowding.
3. If a cell is alive and only one of its eight neighbors is alive, then the cell dies of loneliness.
4. If a cell is dead and more than three of its neighbors are alive, then the cell revives.

A variation on the Game of Life is to include a "rule one out of eight" that a cell will be alive if exactly one of its neighbors is alive, otherwise the cell will remain unchanged. The program OutofEight.java (Listing 13.2) starts with one live cell at the center of the 3-D array cell[34][34][2] and grows on your screen from there.

```
// Game of the life with 1 out of 8 rule
import java.io.*;
public class OutofEight {

  public static void main (String[] argv)throws IOException     {
    int cell[][][]=new int[34][34][2],i, r, j, alive;
      for ( j = 0; j < 33; j++ )  {                      // Initial state of cells
        for ( i = 0; i < 33; i++ ) cell[j][i][0] = 0;
        cell[16][16][0] = 1;
        for (i = 0; i < 33; i++) System.out.print(" "+cell[j][i][0]);
        System.out.println("");
      }                                                                      // j
      for ( r = 0; r < 10; r++ ) {
        for ( j = 1; j < 32; j++ ) {
          for ( i = 1; i < 32; i++ ) {
            alive=0;
            if (    cell[i-1][j][0] == 1 ) alive = alive + 1;
            if (    cell[i+1][j][0] == 1 ) alive = alive + 1;
            if (    cell[i][j-1][0] == 1 ) alive = alive + 1;
            if (    cell[i][j+1][0] == 1 ) alive = alive + 1;
            if ( cell[i-1][j-1][0] == 1 ) alive = alive + 1;
            if ( cell[i+1][j-1][0] == 1 ) alive = alive + 1;
            if ( cell[i-1][j+1][0] == 1 ) alive = alive + 1;
            if ( cell[i+1][j+1][0] == 1 ) alive = alive + 1;
            if ( cell[i][j][0] == 0 && alive == 1 ) cell[i][j][1]=1;
          }                                                                  // i
        }                                                                    // j
      for(j=0;j<33;j++) {for(i=0;i<33;i++)cell[j][i][0]=cell[j][i][1];}
      System.out.println("Press any key to continue ");
      for ( j = 0; j < 33; j++ )   {
        for ( i = 0; i<33; i++ )   {
        if ( cell[j][i][1] == 1)  System.out.print("X");
          else System.out.print(" ");
        }
        System.out.println("");
      }                                                                      // j
      System.in.read();
      }                                                                      // r
} }
```

Listing 13.2 OutofEight.java is an extension of Conway's Game of Life in which cells always revive if one out of eight neighbors is alive.

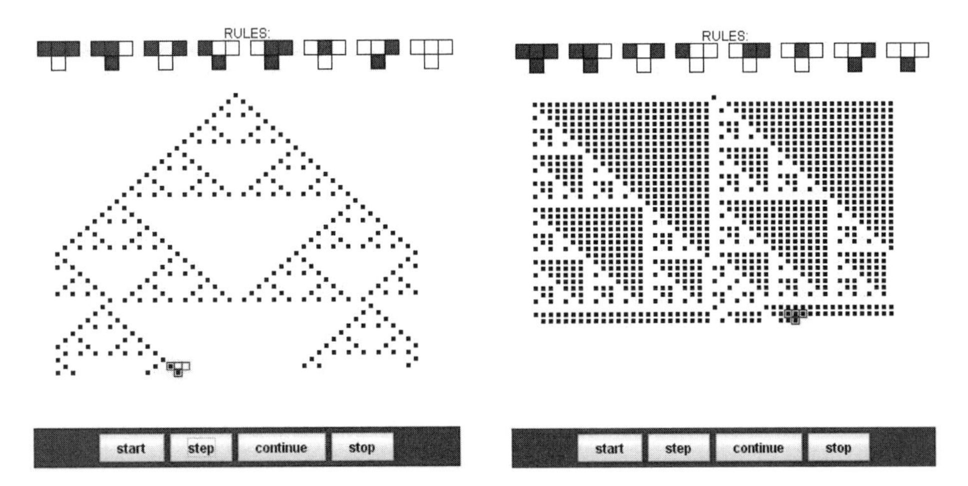

Figure 13.9 The rules for two versions of the Game of Life, given graphically on the top row, create the gaskets below. (Output of this sort is obtained from the applet JCellAut on the CD.)

In 1983 Wolfram developed the statistical mechanics of cellular automata and indicated how one can be used to generate a Sierpiński gasket [Wolf 83]. Since we have already seen that a Sierpiński gasket exhibits fractal geometry (§13.2), this represents a microscopic model of how fractals may occur in nature. This model uses eight rules, given graphically at the top of Figure 13.9, to generate new cells from old. We see all possible configurations for three cells in the top row, and the begetted next generation in the row below. At the bottom of Figure 13.9 is a Sierpiński gasket of the type created by the applet JCellAut on the CD (under Applets). This plays the game and lets you watch and control the growth of the gasket.

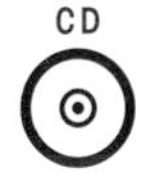

13.10 Perlin Noise Adds Realism ⊙

We have already seen in this chapter how statistical fractals are able to generate objects with a striking resemblance to those in nature. This appearance of realism may be further enhanced by including a type of coherent randomness known as Perlin noise. The resulting textures so resemble those of clouds, smoke, and fire that one cannot help but wonder if a similar mechanism might also be occurring in nature. The technique we are about to discuss was developed by Ken Perlin of New York University, who won an Academy Award (an Oscar) in 1997 for it and has continued to improve it [Perlin]. This type of coherent noise has found use in important physics simulations of stochastic media [Tick 04], as well as in video games and motion pictures (Tron).

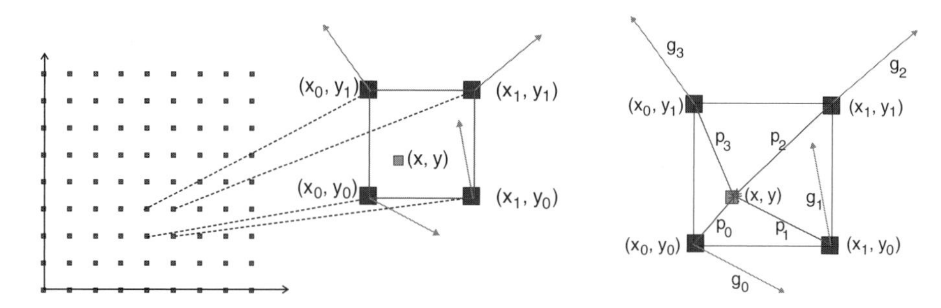

Figure 13.10 *Left:* The rectangular grid used to locate a square in space and a corresponding point within the square. As shown with the arrows, unit vectors $\mathbf{g}_i$ with random orientation are assigned at each grid point. *Right:* A point within each square is located by drawing the four $\mathbf{p}_i$. The $\mathbf{g}_i$ vectors are the same as on the left.

The inclusion of Perlin noise in a simulation adds both randomness and a type of coherence among points in space that tends to make dense regions denser and sparse regions sparser. This is similar to our correlated ballistic deposition simulations (§13.6.1) and related to chaos in its long-range randomness and short-range correlations. We start with some known function of x and y and add noise to it. For this purpose Perlin used the mapping or *ease* function (Figure 13.11 right)

$$f(p) = 3p^2 - 2p^3. \tag{13.25}$$

As a consequence of its S shape, this mapping makes regions close to 0 even closer to 0, while making regions close to 1 even closer (in other words, it increases the tendency to clump, which shows up as higher contrast). We then break space up into a uniform rectangular grid of points (Figure 13.10 left) and consider a point (x, y) within a square with vertices (x_0, y_0), (x_1, y_0), (x_0, y_1), and (x_1, y_1). We next assign unit gradients vectors $\mathbf{g}_0$–$\mathbf{g}_3$ with random orientation at each grid point. A point within each square is located by drawing the four $\mathbf{p}_i$ vectors (Figure13.10 right):

$$\mathbf{p}_0 = (x - x_0)\mathbf{i} + (y - y_0)\mathbf{j}, \quad \mathbf{p}_1 = (x - x_1)\mathbf{i} + (y - y_0)\mathbf{j}, \tag{13.26}$$

$$\mathbf{p}_2 = (x - x_1)\mathbf{i} + (y - y_1)\mathbf{j}, \quad \mathbf{p}_3 = (x - x_0)\mathbf{i} + (y - y_1)\mathbf{j}. \tag{13.27}$$

Next the scalar products of the $\mathbf{p}$'s and the $\mathbf{g}$'s are formed:

$$s = \mathbf{p}_0 \cdot \mathbf{g}_0, \quad t = \mathbf{p}_1 \cdot \mathbf{g}_1, \quad v = \mathbf{p}_2 \cdot \mathbf{g}_2, \quad u = \mathbf{p}_3 \cdot \mathbf{g}_3. \tag{13.28}$$

As shown on the left in Figure 13.11, the numbers s, t, u, and v are assigned to the four vertices of the square and represented there by lines perpendicular to the square with lengths proportional to the values of s, t, u, and v (which can be positive or negative).

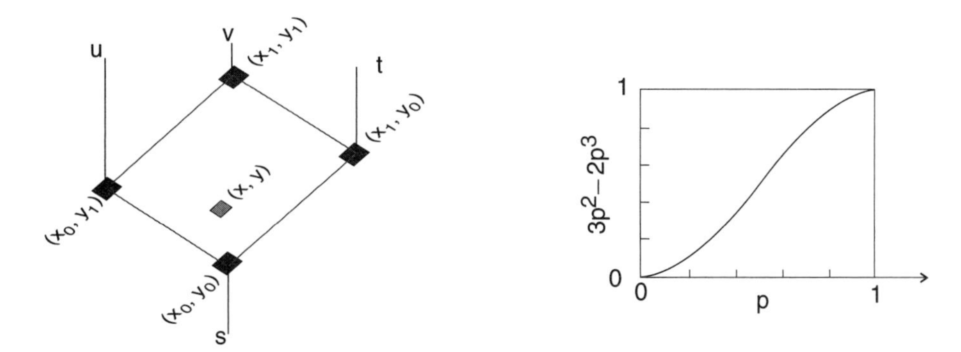

Figure 13.11 *Left:* The numbers *s, t, u,* and *v* are represented by perpendiculars to the four vertices, with lengths proportional to their values. *Right:* The function $3p^2 - 2p^3$ is used as a map of the noise at a point like (x, y) to others close by.

The actual mapping proceeds via a number of steps (Figure 13.12):

1. Transform the point (x, y) to (s_x, s_y),

$$s_x = 3x^2 - 2x^3, \quad s_y = 3y^2 - 2y^3. \tag{13.29}$$

2. Assign the lengths s, t, u, and v to the vertices in the mapped square.
3. Obtain the height a (Figure 13.12) via linear interpolation between s and t.

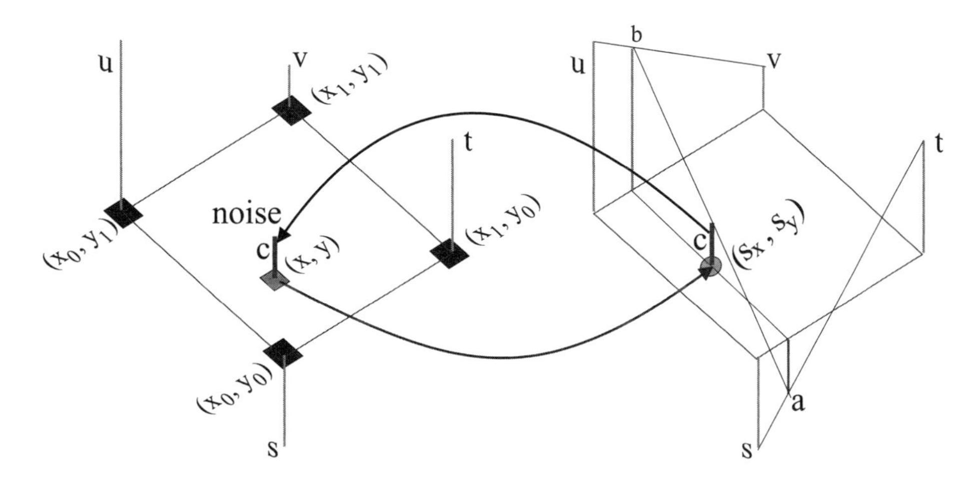

Figure 13.12 *Left*: The point (x, y) is mapped to point (s_x, x_y). *Right*: Using (13.29). Then three linear interpolations are performed to find *c*, the noise at (x, y).

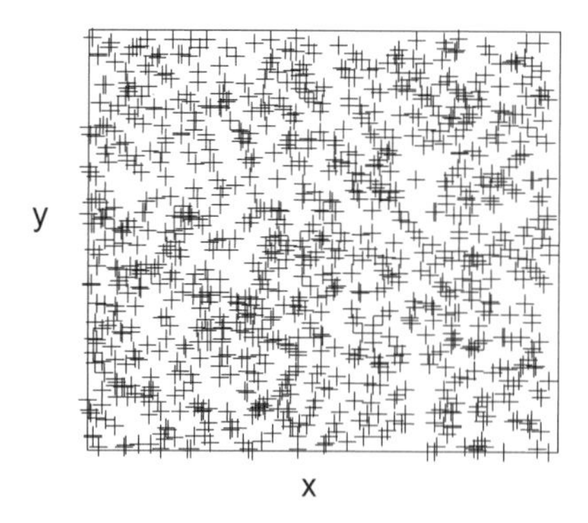

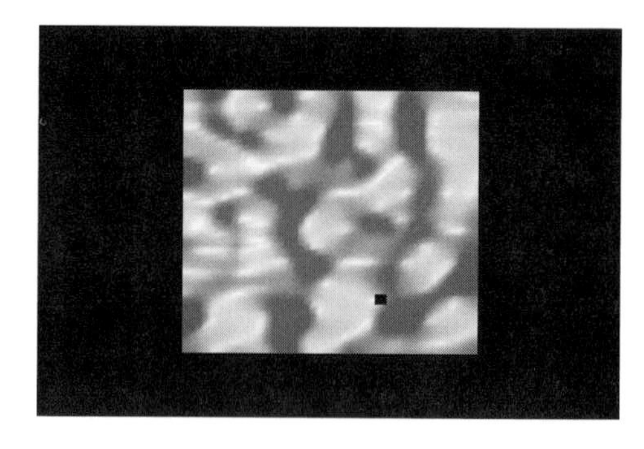

Figure 13.13 After the addition of Perlin noise, the random scatterplot on the left becomes the clusters on the right.

4. Obtain the height b via linear interpolation between u and v.
5. Obtain s_y as a linear interpolation between a and b.
6. The vector c so obtained is now the two components of the noise at (x, y).

Perlin's original C code to accomplish this mapping (along with other goodies) is found in [Perlin]. It takes as input the plot of random points (r_{2i}, r_{2i+1}) on the left in Figure 13.13 (which is the same as Figure 5.1) and by adding coherent noise produces the image on the right in Figure 13.13. The changes we made from the original program are (1) including an int before the variables p[], start, i, and j, and (2) adding # include <time.h> and the line srand(time(NULL)); at the beginning of method init( ) in order to obtain different random numbers each time the program runs. The main method of the C program we used is below. The program outputs a data file that we visualized with OpenDX to produce the image montania.tiff on the right in Figure 13.13.

13.10.1 Including Ray Tracing

Ray tracing is a technique that renders an image of a scene by simulating the way rays of light actually travel [Pov-Ray]. To avoid tracing rays that do not contribute to the final image, ray-tracing programs start at the viewer, trace rays backward onto the scene, and then back again onto the light sources. You can vary the location of the viewer and light sources and the properties of the objects being viewed, as well as atmospheric conditions such as fog, haze, and fire.

As an example of what can be done, on the left in Figure 13.14 we show the output from the ray-tracing program Pov-Ray [Pov-Ray], using as input the coherent random noise on the right in Figure 13.13. The program options we used are given in Listing 13.3 and are seen to include commands to color the islands, to include

Figure 13.14 *Left:* The output from the Pov-Ray ray-tracing program that took as input the 2-D coherent random noise plot in Figure 13.13 and added height and fog. *Right:* An image of a surface of revolution produced by Pov-Ray in which the marblelike texture is created by Perlin noise.

waves, and to give textures to the sky and the sea. Pov-Ray also allows the possibility of using Perlin noise to give textures to the objects to be created. For example, the stone cup on the right in Figure 13.14 has a marblelike texture produced by Perlin noise.

```
main() {
    int m, n;
    float hol, x, ym, ve[2];
    FILE *pf;
    y = 0.2;
    pf = fopen("mountain1.dat", "w");
    for ( n=0; n<50; n++ )  {
        x = 0.1;
        ve[1] = y*n;
        for ( m=0; m<50; m++ )  {
    ve[0] = x*m;                                    // Coordinates point between 0 1
    hol = noise2(ve);
    fprintf( pf, "%f\n", hol );
        }
    }
    fclose(pf);
}
```

```
// Islands.pov  Pov-Ray program to create Islands, by Manuel J Paez
plane {
  <0, 1, 0>, 0                                                         // Sky
  pigment { color rgb <0, 0, 1> }
  scale 1
  rotate <0, 0, 0>
  translate y*0.2
}
global_settings {
  adc_bailout 0.00392157
  assumed_gamma 1.5
  noise_generator 2
}
#declare Island_texture = texture {
  pigment {
```

```
      gradient <0, 1, 0>                                              // Vertical direction
      color_map {                                                     // Color the islands
        [ 0.15 color rgb <1, 0.968627, 0> ]
        [ 0.2  color rgb <0.886275, 0.733333, 0.180392>   ]
        [ 0.3  color rgb <0.372549, 0.643137, 0.0823529>  ]
        [ 0.4  color rgb <0.101961, 0.588235, 0.184314>   ]
        [ 0.5  color rgb <0.223529, 0.666667, 0.301961>   ]
        [ 0.6  color rgb <0.611765, 0.886275, 0.0196078>  ]
        [ 0.69 color rgb <0.678431, 0.921569, 0.0117647>  ]
        [ 0.74 color rgb <0.886275, 0.886275, 0.317647>   ]
        [ 0.86 color rgb <0.823529, 0.796078, 0.0196078>  ]
        [ 0.93 color rgb <0.905882, 0.545098, 0.00392157> ]
        }
    }
    finish {
      ambient rgbft <0.2, 0.2, 0.2, 0.2, 0.2>
      diffuse 0.8
    }
}
camera {                                              // Camera characteristics and location
  perspective
  location <-15, 6, -20>                                                        // Location
  sky <0, 1, 0>
  direction <0, 0, 1>
  right <1.3333, 0, 0>
  up <0, 1, 0>
  look_at <-0.5, 0, 4>                                                     // Look at point
  angle 36
}
light_source {<-10, 20, -25>, rgb <1, 0.733333, 0.00392157>}                      // Light

#declare Islands = height_field {                                // Takes  gif, finds heights
  gif "d:\pov\montania.gif"                                           // Windows directory
  scale <50, 2, 50>
  translate <-25, 0, -25>
}
object {                                                                         // Islands
  Islands
  texture {
    Island_texture
    scale 2
  }
}
box {                                                               // Upper box face = sea
  <-50, 0, -50>, <50, 0.3, 50>                                      // 2 opposite vertices
  translate <-25, 0, -25>
  texture {                                                               // Simulate waves
    normal {
      spotted
      0.4
      scale <0.1, 1, 0.1>
    }
    pigment { color rgb <0.164706, 0.556863, 0.901961> }
  }
}
fog {                                                                     // A constant fog
   fog_type 1
   distance 30
   rgb <0.984314, 1, 0.964706>
}
```

Listing 13.3 Islands.pov in the Codes/Animations/Fractals/ directory gives the Pov-Ray ray-tracing commands needed to convert the coherent noise random plot of Figure 13.13 into the mountainlike image on the left in Figure 13.14.

13.11 Quiz

1. Recall how box counting is used to determine the fractal dimension of an object. Imagine that the result of some experiment or simulation is an interesting geometric figure.
   a. What might be the physical/theoretical importance of determining that this object is a fractal?
   b. What might be the importance of determining its fractal dimension?
   c. Why is it important to use *more* than two sizes of boxes?
   d. Below is a figure composed of boxes of side 1.

1	1	1
1	1	1
1	1	1

   Use box counting to determine the fractal dimension of this figure. ∎

14

High-Performance Computing Hardware, Tuning, and Parallel Computing

In this chapter and in Appendix D we discuss a number of topics associated with high-performance computing (HPC). If history can be our guide, today's HPC hardware and software will be on desktop machines a decade from now. In Unit I we discuss the theory of a high-performance computer's memory and central processor design. In Unit II we examine parallel computers. In Appendix D we extend Unit II by giving a detailed tutorial on use of the message-passing interface (MPI) package, while a tutorial on an earlier package, Parallel virtual machine (PVM), is provided on the CD. In Unit III we discuss some techniques for writing programs that are optimized for HPC hardware, such as virtual memory and cache. By running the short implementations given in Unit III, you will experiment with your computer's memory and experience some of the concerns, techniques, rewards, and shortcomings of HPC.

CD

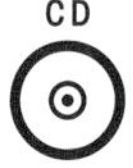

HPC is a broad subject, and our presentation is brief and given from a practitioner's point of view. The text [Quinn 04] surveys parallel computing and MPI from a computer science point of view. References on parallel computing include [Sterl 99, Quinn 04, Pan 96, VdeV 94, Fox 94]. References on MPI include Web resources [MPI, MPI2, MPImis] and the texts [Quinn 04, Pach 97, Lusk 99]. More recent developments, such as programming for multicore computers, cell computers, and field-programmable gate accelerators, will be discussed in future books.

14.1 Unit I. High-Performance Computers (CS)

By definition, supercomputers are the fastest and most powerful computers available, and at this instant, the term "supercomputers" almost always refers to parallel machines. They are the superstars of the high-performance class of computers. Unix workstations and modern personal computers (PCs), which are small enough in size and cost to be used by a small group or an individual, yet powerful enough for large-scale scientific and engineering applications, can also be high-performance computers. We define *high-performance computers* as machines with a good balance among the following major elements:

- Multistaged (pipelined) functional units.
- Multiple central processing units (CPUs) (parallel machines).

- Multiple cores.
- Fast central registers.
- Very large, fast memories.
- Very fast communication among functional units.
- Vector, video, or array processors.
- Software that integrates the above effectively.

As a simple example, it makes little sense to have a CPU of incredibly high speed coupled to a memory system and software that cannot keep up with it (the present state of affairs).

14.2 Memory Hierarchy

An idealized model of computer architecture is a CPU sequentially executing a stream of instructions and reading from a continuous block of memory. To illustrate, in Figure 14.1 we see a vector `A[ ]` and an array `M[ ][ ]` loaded in memory and about to be processed. The real world is more complicated than this. First, matrices are not stored in blocks but rather in linear order. For instance, in Fortran it is in *column-major* order:

```
M(1,1) M(2,1) M(3,1) M(1,2) M(2,2) M(3,2) M(1,3) M(2,3) M(3,3),
```

while in Java and C it is in *row-major* order:

```
M(0,0) M(0,1) M(0,2) M(1,0) M(1,1) M(1,2) M(2,0) M(2,1) M(2,2).
```

Second, the values for the matrix elements may not even be in the same physical place. Some may be in RAM, some on the disk, some in cache, and some in the CPU. To give some of these words more meaning, in Figures 14.2 and 14.3 we show simple models of the memory architecture of a high-performance computer. This hierarchical arrangement arises from an effort to balance speed and cost with fast, expensive memory supplemented by slow, less expensive memory. The memory architecture may include the following elements:

CPU: Central processing unit, the fastest part of the computer. The CPU consists of a number of very-high-speed memory units called *registers* containing the *instructions* sent to the hardware to do things like fetch, store, and operate on data. There are usually separate registers for instructions, addresses, and *operands* (current data). In many cases the CPU also contains some specialized parts for accelerating the processing of floating-point numbers.

Cache (high-speed buffer): A small, very fast bit of memory that holds instructions, addresses, and data in their passage between the very fast CPU registers and the slower RAM. This is seen in the next level down the pyramid in Figure 14.3. The main memory is also called *dynamic RAM* (DRAM), while the

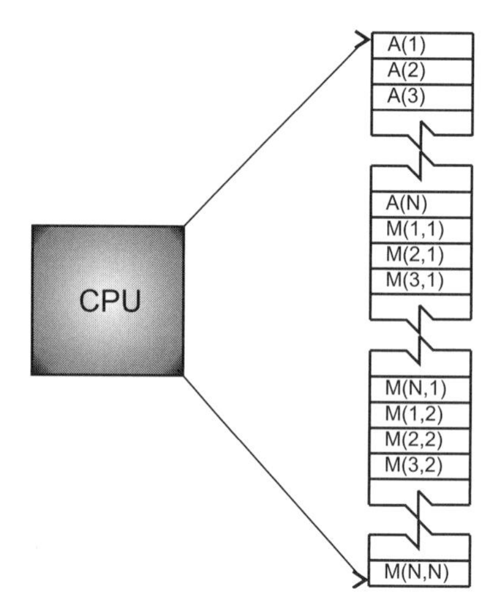

Figure 14.1 The logical arrangement of the CPU and memory showing a Fortran array *A*(*N*) and matrix *M*(*N*, *N*) loaded into memory.

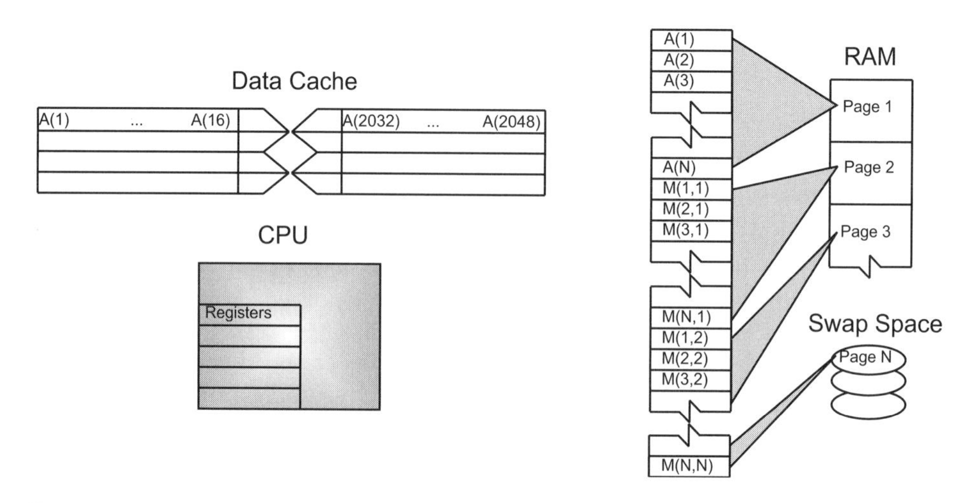

Figure 14.2 The elements of a computer's memory architecture in the process of handling matrix storage.

cache is called *static RAM* (SRAM). If the cache is used properly, it eliminates the need for the CPU to wait for data to be fetched from memory.

Cache and data lines: The data transferred to and from the cache or CPU are grouped into cache lines or data lines. The time it takes to bring data from memory into the cache is called *latency*.

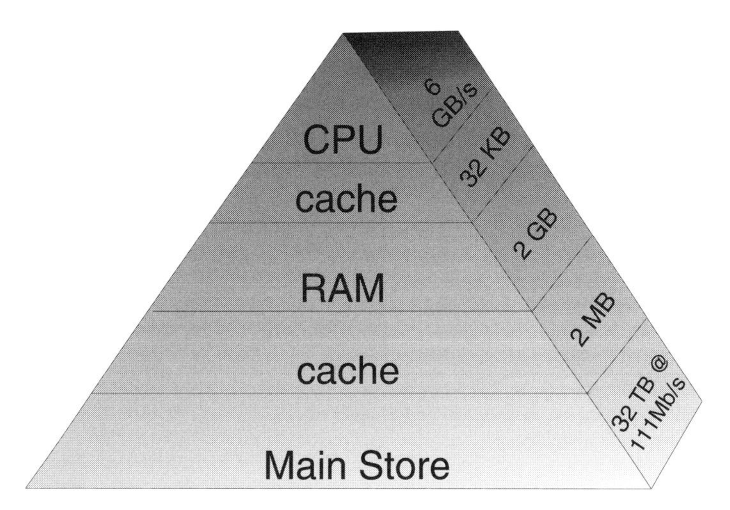

Figure 14.3 Typical memory hierarchy for a single-processor, high-performance computer (B = bytes, K, M, G, T = kilo, mega, giga, tera).

RAM: Random-access memory or central memory is in the middle memory in the hierarchy in Figure 14.3. RAM can be accessed directly, that is, in random order, and it can be accessed quickly, that is, without mechanical devices. It is where your program resides while it is being processed.

Pages: Central memory is organized into pages, which are blocks of memory of fixed length. The operating system labels and organizes its memory pages much like we do the pages of a book; they are numbered and kept track of with a *table of contents*. Typical page sizes are from 4 to 16 kB.

Hard disk: Finally, at the bottom of the memory pyramid is permanent storage on magnetic disks or optical devices. Although disks are very slow compared to RAM, they can store vast amounts of data and sometimes compensate for their slower speeds by using a cache of their own, the *paging storage controller*.

Virtual memory: True to its name, this is a part of memory you will not find in our figures because it is *virtual*. It acts like RAM but resides on the disk.

When we speak of fast and slow memory we are using a time scale set by the clock in the CPU. To be specific, if your computer has a clock speed or cycle time of 1 ns, this means that it could perform a billion operations per second if it could get its hands on the needed data quickly enough (typically, more than 10 cycles are needed to execute a single instruction). While it usually takes 1 cycle to transfer data from the cache to the CPU, the other memories are much slower, and so you can speed your program up by not having the CPU wait for transfers among different levels of memory. Compilers try to do this for you, but their success is affected by your programming style.

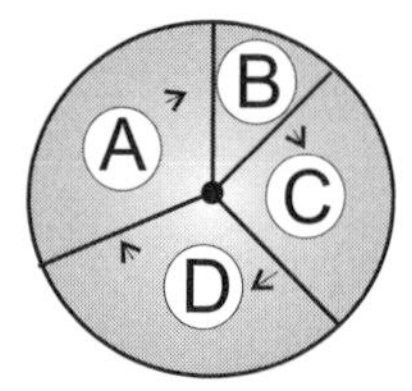

Figure 14.4 Multitasking of four programs in memory at one time in which the programs are executed in round-robin order.

As shown in Figure 14.2 for our example, virtual memory permits your program to use more pages of memory than can physically fit into RAM at one time. A combination of operating system and hardware *maps* this virtual memory into pages with typical lengths of 4–16 KB. Pages not currently in use are stored in the slower memory on the hard disk and brought into fast memory only when needed. The separate memory location for this switching is known as *swap space* (Figure 14.2). Observe that when the application accesses the memory location for M[i][j], the number of the page of memory holding this address is determined by the computer, and the location of M[i][j] within this page is also determined. A *page fault* occurs if the needed page resides on disk rather than in RAM. In this case the entire page must be read into memory while the least recently used page in RAM is swapped onto the disk. Thanks to virtual memory, it is possible to run programs on small computers that otherwise would require larger machines (or extensive reprogramming). The price you pay for virtual memory is an order-of-magnitude slowdown of your program's speed when virtual memory is actually invoked. But this may be cheap compared to the time you would have to spend to rewrite your program so it fits into RAM or the money you would have to spend to buy a computer with enough RAM for your problem.

Virtual memory also allows *multitasking*, the simultaneous loading into memory of more programs than can physically fit into RAM (Figure 14.4). Although the ensuing switching among applications uses computing cycles, by avoiding long waits while an application is loaded into memory, multitasking increases the total throughout and permits an improved computing environment for users. For example, it is multitasking that permits a windows system to provide us with multiple windows. Even though each window application uses a fair amount of memory, only the single application currently receiving input must actually reside in memory; the rest are *paged out* to disk. This explains why you may notice a slight delay when switching to an idle window; the pages for the now active program are being placed into RAM and the least used application still in memory is simultaneously being paged out.

TABLE 14.1
Computation of $c = (a + b)/(d * f)$

Arithmetic Unit	*Step 1*	*Step 2*	*Step 3*	*Step 4*
A1	Fetch a	Fetch b	Add	—
A2	Fetch d	Fetch f	Multiply	—
A3	—	—	—	Divide

14.3 The Central Processing Unit

How does the CPU get to be so fast? Often, it employs *prefetching* and *pipelining*; that is, it has the ability to prepare for the next instruction before the current one has finished. It is like an assembly line or a bucket brigade in which the person filling the buckets at one end of the line does not wait for each bucket to arrive at the other end before filling another bucket. In the same way a processor fetches, reads, and decodes an instruction while another instruction is executing. Consequently, even though it may take more than one cycle to perform some operations, it is possible for data to be entering and leaving the CPU on each cycle. To illustrate, Table 14.1 indicates how the operation $c = (a + b)/(d * f)$ is handled. Here the pipelined arithmetic units A1 and A2 are simultaneously doing their jobs of fetching and operating on operands, yet arithmetic unit A3 must wait for the first two units to complete their tasks before it has something to do (during which time the other two sit idle).

14.4 CPU Design: Reduced Instruction Set Computer

Reduced instruction set computer (RISC) architecture (also called *superscalar*) is a design philosophy for CPUs developed for high-performance computers and now used broadly. It increases the arithmetic speed of the CPU by decreasing the number of instructions the CPU must follow. To understand RISC we contrast it with *complex instruction set computer* (CISC), architecture. In the late 1970s, processor designers began to take advantage of *very-large-scale integration* (VLSI) which allowed the placement of hundreds of thousands of devices on a single CPU chip. Much of the space on these early chips was dedicated to *microcode* programs written by chip designers and containing machine language instructions that set the operating characteristics of the computer. There were more than 1000 instructions available, and many were similar to higher-level programming languages like *Pascal* and *Forth*. The price paid for the large number of complex instructions was slow speed, with a typical instruction taking more than 10 clock cycles. Furthermore, a 1975 study by Alexander and Wortman of the *XLP* compiler of the IBM System/360 showed that about 30 low-level instructions accounted for 99% of the use with only 10 of these instructions accounting for 80% of the use.

The RISC philosophy is to have just a small number of instructions available at the chip level but to have the regular programmer's high level-language, such as Fortran or C, translate them into efficient machine instructions for a particular computer's architecture. This simpler scheme is cheaper to design and produce, lets the processor run faster, and uses the space saved on the chip by cutting down on microcode to increase arithmetic power. Specifically, RISC increases the number of internal CPU registers, thus making it possible to obtain longer pipelines (cache) for the data flow, a significantly lower probability of memory conflict, and some instruction-level parallelism.

The theory behind this philosophy for RISC design is the simple equation describing the execution time of a program:

$$\text{CPU time} = \text{no.instructions} \times \text{cycles/instruction} \times \text{cycle time}. \tag{14.1}$$

Here "CPU time" is the time required by a program, "no. instructions" is the total number of machine-level instructions the program requires (sometimes called the *path length*), "cycles/instruction" is the number of CPU clock cycles each instruction requires, and "cycle time" is the actual time it takes for one CPU cycle. After viewing (14.1) we can understand the CISC philosophy, which tries to reduce CPU time by reducing no. instructions, as well as the RISC philosophy, which tries to reduce CPU time by reducing cycles/instruction (preferably to one). For RISC to achieve an increase in performance requires a greater decrease in cycle time and cycles/instruction than the increase in the number of instructions.

In summary, the elements of RISC are the following.

Single-cycle execution for most machine-level instructions.
Small instruction set of less than 100 instructions.
Register-based instructions operating on values in registers, with memory access confined to load and store to and from registers.
Many registers, usually more than 32.
Pipelining, that is, concurrent processing of several instructions.
High-level compilers to improve performance.

14.5 CPU Design: Multiple-Core Processors

The year preceding the publication of this book has seen a rapid increase in the inclusion of dual-core, or even quad-core, chips as the computational engine of computers. As seen in Figure 14.5, a dual-core chip has two CPUs in one integrated circuit with a shared interconnect and a shared level-2 cache. This type of configuration with two or more identical processors connected to a single shared main memory is called *symmetric multiprocessing*, or SMP. It is likely that by the time you read this book, 16-core or greater chips will be available.

Although multicore chips were designed for game playing and single precision, they should also be useful in scientific computing if new tools, algorithms, and programming methods are employed. These chips attain more speed with less heat and more energy efficiency than single-core chips, whose heat generation limits them to clock speeds of less than 4 GHz. In contrast to multiple single-core

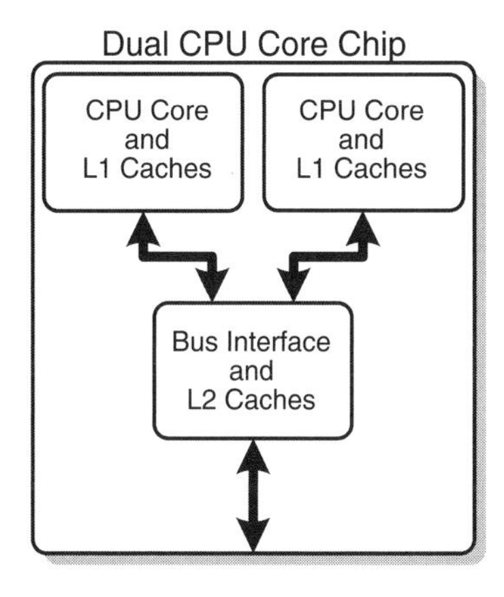

Figure 14.5 *Left:* A generic view of the Intel core-2 dual-core processor, with CPU-local level-1 caches and a shared, on-die level-2 cache (courtesy of D. Schmitz). *Right:* The AMD Athlon 64 X2 3600 dual-core CPU (Wikimedia Commons).

chips, multicore chips use fewer transistors per CPU and are thus simpler to make and cooler to run.

Parallelism is built into a multicore chip because each core can run a different task. However, since the cores usually share the same communication channel and level-2 cache, there is the possibility of a communication bottleneck if both CPUs use the bus at the same time. Usually the user need not worry about this, but the writers of compilers and software must so that your code will run in parallel. As indicated in our MPI tutorial in Appendix D, modern Intel compilers make use of each multiple core and even have MPI treat each core as a separate processor.

14.6 CPU Design: Vector Processor

Often the most demanding part of a scientific computation involves matrix operations. On a classic (von Neumann) scalar computer, the addition of two vectors of physical length 99 to form a third ultimately requires 99 sequential additions (Table 14.2). There is actually much behind-the-scenes work here. For each element i there is the *fetch* of $a(i)$ from its location in memory, the *fetch* of $b(i)$ from its location in memory, the *addition* of the numerical values of these two elements in a CPU register, and the *storage* in memory of the sum in $c(i)$. This fetching uses up time and is wasteful in the sense that the computer is being told again and again to do the same thing.

When we speak of a computer doing *vector processing*, we mean that there are hardware components that perform mathematical operations on entire rows or columns of matrices as opposed to individual elements. (This hardware can also handle single-subscripted matrices, that is, mathematical vectors.) In the vector

TABLE 14.2
Computation of Matrix $[C] = [A] + [B]$

Step 1	*Step 2*	...	*Step 99*
$c(1) = a(1) + b(1)$	$c(2) = a(2) + b(2)$	...	$c(99) = a(99) + b(99)$

TABLE 14.3
Vector Processing of Matrix $[A] + [B] = [C]$

Step 1	Step 2	Step 3	...	Step Z
$c(1) = a(1) + b(1)$				
	$c(2) = a(2) + b(2)$			
		$c(3) = a(3) + b(3)$		
			...	
				$c(Z) = a(Z) + b(Z)$

processing of $[A] + [B] = [C]$, the successive fetching of and addition of the elements A and B are grouped together and overlaid, and $Z \simeq 64$–256 elements (the *section size*) are processed with one command, as seen in Table 14.3. Depending on the array size, this method may speed up the processing of vectors by a factor of about 10. If all Z elements were truly processed in the same step, then the speedup would be ~ 64–256.

Vector processing probably had its heyday during the time when computer manufacturers produced large mainframe computers designed for the scientific and military communities. These computers had proprietary hardware and software and were often so expensive that only corporate or military laboratories could afford them. While the Unix and then PC revolutions have nearly eliminated these large vector machines, some do exist, as well as PCs that use vector processing in their video cards. Who is to say what the future holds in store?

14.7 Unit II. Parallel Computing

There is little question that advances in the hardware for parallel computing are impressive. Unfortunately, the software that accompanies the hardware often seems stuck in the 1960s. In our view, message passing has too many details for application scientists to worry about and requires coding at a much, or more, elementary level than we prefer. However, the increasing occurrence of clusters in which the nodes are symmetric multiprocessors has led to the development of sophisticated compilers that follow simpler programming models; for example, *partitioned global address space* compilers such as *Co-Array Fortran*, *Unified Parallel C*, and *Titanium*. In these approaches the programmer views a global array of data and then manipulates these data as if they were contiguous. Of course the data really are distributed, but the software takes care of that outside the programmer's view. Although the program may not be as efficient a use of the processors as hand coding, it is a lot easier, and as the number of processors becomes very large, one can live with a greater degree of inefficiency. In any case, if each node of the computer has a number of processors with a shared memory and there are a number of nodes, then some type of a hybrid programming model will be needed.

Problem: Start with the program you wrote to generate the bifurcation plot for bug dynamics in Chapter 12, "Discrete & Continuous Nonlinear Dynamics," and modify it so that different ranges for the growth parameter μ are computed simultaneously on multiple CPUs. Although this small a problem is not worth investing your time in to obtain a shorter turnaround time, it is worth investing your time into it gain some experience in parallel computing. In general, parallel computing holds the promise of permitting you to obtain faster results, to solve bigger problems, to run simulations at finer resolutions, or to model physical phenomena more realistically; but it takes some work to accomplish this.

14.8 Parallel Semantics (Theory)

We saw earlier that many of the tasks undertaken by a high-performance computer are run in parallel by making use of internal structures such as pipelined and segmented CPUs, hierarchical memory, and separate I/O processors. While these tasks are run "in parallel," the modern use of *parallel computing* or *parallelism* denotes applying multiple processors to a single problem [Quinn 04]. It is a computing environment in which some number of CPUs are running asynchronously and communicating with each other in order to exchange intermediate results and coordinate their activities.

For instance, consider matrix multiplication in terms of its elements:

$$[B] = [A][B] \quad \Rightarrow \quad B_{i,j} = \sum_{k=1}^{N} A_{i,k} B_{k,j}. \tag{14.2}$$

Because the computation of $B_{i,j}$ for particular values of i and j is independent of the computation of all the other values, each $B_{i,j}$ can be computed in parallel, or each row or column of $[B]$ can be computed in parallel. However, because $B_{k,j}$ on the RHS of (14.2) must be the "old" values that existed before the matrix multiplication, some communication among the parallel processors is required to ensure that they do not store the "new" values of $B_{k,j}$ before all the multiplications are complete. This $[B] = [A][B]$ multiplication is an example of *data dependency*, in which the data elements used in the computation depend on the order in which they are used. In contrast, the matrix multiplication $[C] = [A][B]$ is a *data parallel* operation in which the data can be used in any order. So already we see the importance of communication, synchronization, and understanding of the mathematics behind an algorithm for parallel computation.

The processors in a parallel computer are placed at the *nodes* of a communication network. Each node may contain one CPU or a small number of CPUs, and the communication network may be internal to or external to the computer. One way of categorizing parallel computers is by the approach they employ in handling instructions and data. From this viewpoint there are three types of machines:

- **Single-instruction, single-data (SISD):** These are the classic (von Neumann) serial computers executing a single instruction on a single data stream before the next instruction and next data stream are encountered.
- **Single-instruction, multiple-data (SIMD):** Here instructions are processed from a single stream, but the instructions act concurrently on multiple data elements. Generally the nodes are simple and relatively slow but are large in number.
- **Multiple instructions, multiple data (MIMD):** In this category each processor runs independently of the others with independent instructions and data. These are the types of machines that employ *message-passing* packages, such as MPI, to communicate among processors. They may be a collection of workstations linked via a network, or more integrated machines with thousands of processors on internal boards, such as the Blue Gene computer described in §14.13. These computers, which do not have a shared memory space, are also called *multicomputers*. Although these types of computers are some of the most difficult to program, their low cost and effectiveness for certain classes of problems have led to their being the dominant type of parallel computer at present.

The running of independent programs on a parallel computer is similar to the multitasking feature used by Unix and PCs. In multitasking (Figure 14.6 left) several independent programs reside in the computer's memory simultaneously and share the processing time in a round robin or priority order. On a SISD computer, only one program runs at a single time, but if other programs are in memory, then it does not take long to switch to them. In multiprocessing (Figure 14.6 right) these jobs may all run at the same time, either in different parts of memory or in the memory of different computers. Clearly, multiprocessing becomes complicated if separate processors are operating on different parts of the *same* program because

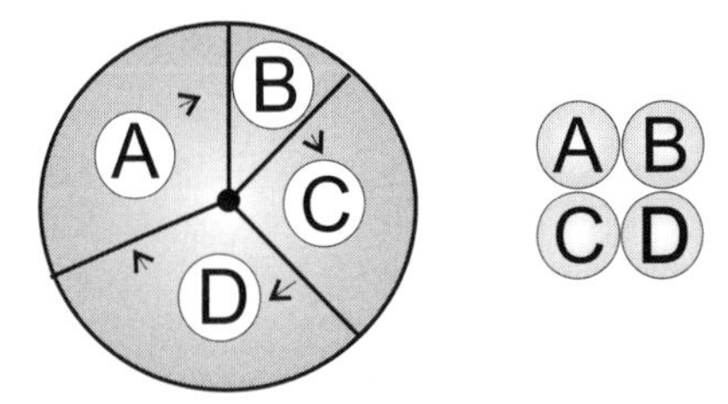

Figure 14.6 *Left:* Multitasking of four programs in memory at one time. On a SISD computer the programs are executed in round robin order. *Right:* Four programs in the four separate memories of a MIMD computer.

then synchronization and load balance (keeping all the processors equally busy) are concerns.

In addition to instructions and data streams, another way of categorizing parallel computation is by *granularity*. A *grain* is defined as a measure of the computational work to be done, more specifically, the ratio of computation work to communication work.

- **Coarse-grain parallel:** Separate programs running on separate computer systems with the systems coupled via a conventional communication network. An illustration is six Linux PCs sharing the same files across a network but with a different central memory system for each PC. Each computer can be operating on a different, independent part of one problem at the same time.
- **Medium-grain parallel:** Several processors executing (possibly different) programs simultaneously while accessing a common memory. The processors are usually placed on a common *bus* (communication channel) and communicate with each other through the memory system. Medium-grain programs have different, independent, *parallel subroutines* running on different processors. Because the compilers are seldom smart enough to figure out which parts of the program to run where, the user must include the multitasking routines in the program.[1]
- **Fine-grain parallel:** As the granularity decreases and the number of nodes increases, there is an increased requirement for fast communication among the nodes. For this reason fine-grain systems tend to be custom-designed machines. The communication may be via a central bus or via shared memory for a small number of nodes, or through some form of high-speed network for massively parallel machines. In the latter case, the compiler divides the work among the processing nodes. For example, different `for` loops of a program may be run on different nodes.

14.9 Distributed Memory Programming

An approach to concurrent processing that, because it is built from commodity PCs, has gained dominant acceptance for coarse- and medium-grain systems is *distributed memory*. In it, each processor has its own memory and the processors exchange data among themselves through a high-speed switch and network. The data exchanged or *passed* among processors have encoded *to* and *from* addresses and are called *messages*. The *clusters* of PCs or workstations that constitute a *Beowulf*[2] are

[1] Some experts define our medium grain as coarse grain yet this distinction changes with time.

[2] Presumably there is an analogy between the heroic exploits of the son of Ecgtheow and the nephew of Hygelac in the 1000 C.E. poem *Beowulf* and the adventures of us

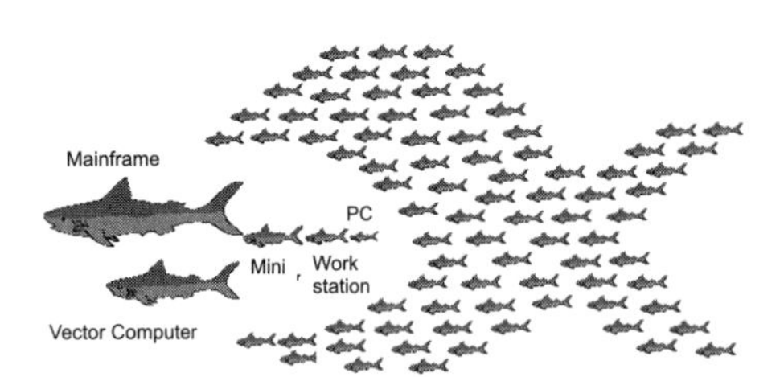

Figure 14.7 Two views of modern parallel computing (courtesy of Yuefan Deng).

examples of distributed memory computers (Figure 14.7). The unifying characteristic of a cluster is the integration of highly replicated compute and communication components into a single system, with each node still able to operate independently. In a Beowulf cluster, the components are commodity ones designed for a general market, as are the communication network and its high-speed switch (special interconnects are used by major commercial manufacturers, but they do not come cheaply). *Note*: A group of computers connected by a network may also be called a cluster but unless they are designed for parallel processing, with the same type of processor used repeatedly and with only a limited number of processors (the *front end*) onto which users may log in, they are not usually called a Beowulf.

The literature contains frequent arguments concerning the differences among clusters, commodity clusters, Beowulfs, constellations, massively parallel systems, and so forth [Dong 05]. Even though we recognize that there are major differences between the clusters on the top 500 list of computers and the ones that a university researcher may set up in his or her lab, we will not distinguish these fine points in the introductory materials we present here.

For a message-passing program to be successful, the data must be divided among nodes so that, at least for a while, each node has all the data it needs to run an independent subtask. When a program begins execution, data are sent to all the nodes. When all the nodes have completed their subtasks, they exchange data again in order for each node to have a complete new set of data to perform the next subtask. This repeated cycle of data exchange followed by processing continues until the full task is completed. Message-passing MIMD programs are also *single-program, multiple-data* programs, which means that the programmer writes a single program that is executed on all the nodes. Often a separate host program,

---

common folk assembling parallel computers from common elements that have surpassed the performance of major corporations and their proprietary, multi-million-dollar supercomputers.

which starts the programs on the nodes, reads the input files and organizes the output.

14.10 Parallel Performance

Imagine a cafeteria line in which all the servers appear to be working hard and fast yet the ketchup dispenser has some relish partially blocking its output and so everyone in line must wait for the ketchup lovers up front to ruin their food before moving on. This is an example of the slowest step in a complex process determining the overall rate. An analogous situation holds for parallel processing, where the "relish" may be the issuing and communicating of instructions. Because the computation cannot advance until all the instructions have been received, this one step may slow down or stop the entire process.

As we soon will demonstrate, the speedup of a program will not be significant unless you can get $\sim$90% of it to run in parallel, and even then most of the speedup will probably be obtained with only a small number of processors. This means that you need to have a computationally intense problem to make parallelization worthwhile, and that is one of the reasons why some proponents of parallel computers with thousands of processors suggest that you not apply the new machines to old problems but rather look for new problems that are both big enough and well-suited for massively parallel processing to make the effort worthwhile.

The equation describing the effect on speedup of the balance between serial and parallel parts of a program is known as Amdahl's law [Amd 67, Quinn 04]. Let

$$p = \text{no. of CPUs} \quad T_1 = 1\text{-CPU time}, \quad T_p = p\text{-CPU time}. \tag{14.3}$$

The maximum speedup S_p attainable with parallel processing is thus

$$S_p^{\max} = \frac{T_1}{T_p} \to p. \tag{14.4}$$

This limit is never met for a number of reasons: Some of the program is serial, data and memory conflicts occur, communication and synchronization of the processors take time, and it is rare to attain a perfect load balance among all the processors. For the moment we ignore these complications and concentrate on how the *serial* part of the code affects the speedup. Let f be the fraction of the program that potentially may run on multiple processors. The fraction $1 - f$ of the code that cannot be run in parallel must be run via serial processing and thus takes time:

$$T_s = (1 - f)T_1 \quad \text{(serial time)}. \tag{14.5}$$

The time T_p spent on the p parallel processors is related to T_s by

$$T_p = f\,\frac{T_1}{p}. \tag{14.6}$$

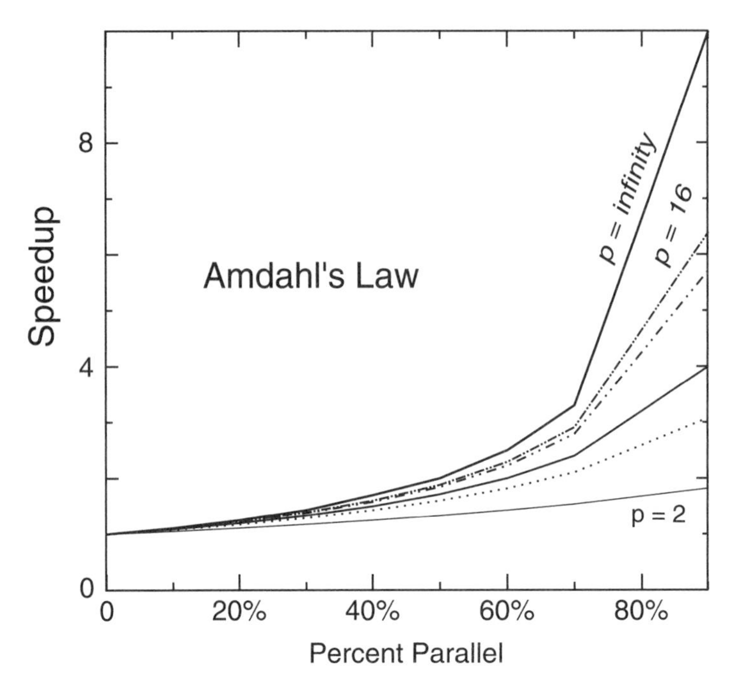

Figure 14.8 The theoretical speedup of a program as a function of the fraction of the program that potentially may be run in parallel. The different curves correspond to different numbers of processors.

That being so, the speedup S_p as a function of f and the number of processors is

$$\boxed{S_p = \frac{T_1}{T_s + T_p} = \frac{1}{1 - f + f/p}} \quad \text{(Amdahl's law).} \tag{14.7}$$

Some theoretical speedups are shown in Figure 14.8 for different numbers p of processors. Clearly the speedup will not be significant enough to be worth the trouble unless most of the code is run in parallel (this is where the 90% of your in-parallel figure comes from). Even an infinite number of processors cannot increase the speed of running the serial parts of the code, and so it runs at one processor speed. In practice this means many problems are limited to a small number of processors and that often for realistic applications only 10%–20% of the computer's peak performance may be obtained.

14.10.1 Communication Overhead

As discouraging as Amdahl's law may seem, it actually *overestimates* speedup because it ignores the *overhead* for parallel computation. Here we look at communication overhead. Assume a completely parallel code so that its speedup is

$$S_p = \frac{T_1}{T_p} = \frac{T_1}{T_1/p} = p. \tag{14.8}$$

The denominator assumes that it takes no time for the processors to communicate. However, it take a finite time, called *latency*, to get data out of memory and into the cache or onto the communication network. When we add in this latency, as well as other times that make up the *communication time* T_c, the speedup decreases to

$$S_p \simeq \frac{T_1}{T_1/p + T_c} < p \quad \text{(with communication time)}. \tag{14.9}$$

For the speedup to be unaffected by communication time, we need to have

$$\frac{T_1}{p} \gg T_c \quad \Rightarrow \quad p \ll \frac{T_1}{T_c}. \tag{14.10}$$

This means that as you keep increasing the number of processors p, at some point the time spent on computation T_1/p must equal the time T_c needed for communication, and adding more processors leads to greater execution time as the processors wait around more to communicate. This is another limit, then, on the maximum number of processors that may be used on any one problem, as well as on the effectiveness of increasing processor speed without a commensurate increase in communication speed.

The continual and dramatic increases in CPU speed, along with the widespread adoption of computer clusters, is leading to a changing view as to how to judge the speed of an algorithm. Specifically, the slowest step in a process is usually the rate-determining step, and the increasing speed of CPUs means that this slowest step is more and more often access to or communication among processors. Such being the case, while the number of computational steps is still important for determining an algorithm's speed, the number and amount of memory access and interprocessor communication must also be mixed into the formula. This is currently an active area of research in algorithm development.

14.11 Parallelization Strategy

A typical organization of a program for both serial and parallel tasks is

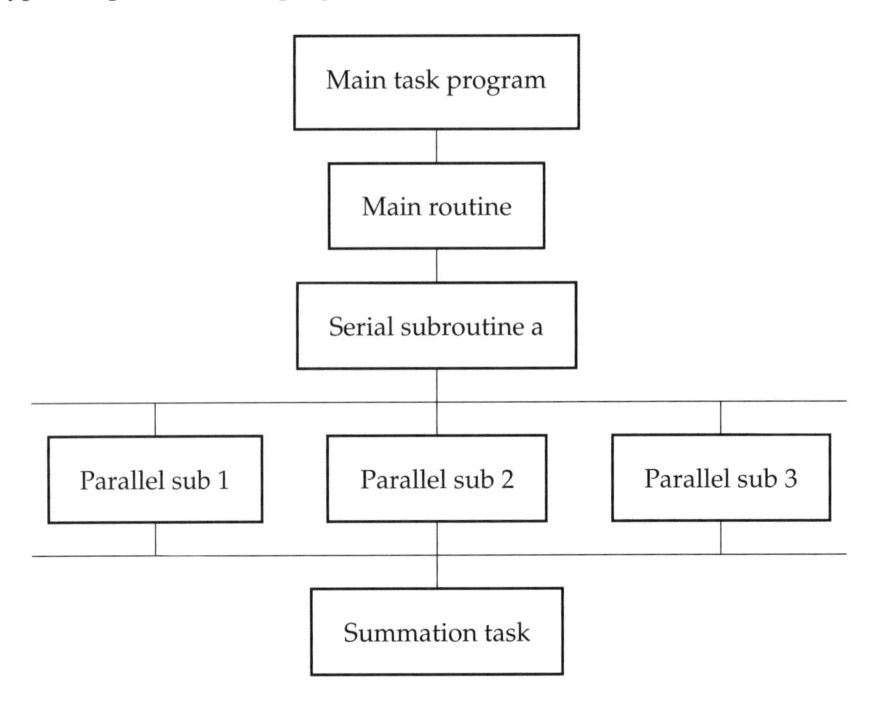

The user organizes the work into units called *tasks,* with each task assigning work (*threads*) to a processor. The main task controls the overall execution as well as the subtasks that run independent parts of the program (called *parallel subroutines, slaves, guests,* or *subtasks*). These parallel subroutines can be distinctive subprograms, multiple copies of the same subprogram, or even `for` loops.

It is the programmer's responsibility to ensure that the breakup of a code into parallel subroutines is mathematically and scientifically valid and is an equivalent formulation of the original program. As a case in point, if the most intensive part of a program is the evaluation of a large Hamiltonian matrix, you may want to evaluate each row on a different processor. Consequently, the key to parallel programming is to identify the parts of the program that may benefit from parallel execution. To do that the programmer should understand the program's data structures (discussed below), know in what order the steps in the computation must be performed, and know how to coordinate the results generated by different processors.

The programmer helps speed up the execution by keeping many processors simultaneously busy and by avoiding storage conflicts among different parallel subprograms. You do this *load balancing* by dividing your program into subtasks of approximately equal numerical intensity that will run simultaneously on different processors. The rule of thumb is to make the task with the largest granularity (workload) dominant by forcing it to execute first and to keep all the processors busy by having the number of tasks an integer multiple of the number of processors. This is not always possible.

The individual parallel threads can have *shared* or *local* data. The shared data may be used by all the machines, while the local data are private to only one thread. To avoid storage conflicts, design your program so that parallel subtasks use data that are independent of the data in the main task and in other parallel tasks. This means that these data should not be modified *or even examined* by different tasks simultaneously. In organizing these multiple tasks, reduce communication *overhead costs* by limiting communication and synchronization. These costs tend to be high for fine-grain programming where much coordination is necessary. However, *do not* eliminate communications that are necessary to ensure the scientific or mathematical validity of the results; bad science can do harm!

14.12 Practical Aspects of Message Passing for MIMD

It makes sense to run only the most numerically intensive codes on parallel machines. Frequently these are very large programs assembled over a number of years or decades by a number of people. It should come as no surprise, then, that the programming languages for parallel machines are primarily Fortran90, which has explicit structures for the compiler to parallelize, and C. (We have not attained good speedup with Java in parallel and therefore do not recommend it for parallel computing.) Effective parallel programming becomes more challenging as the number of processors increases. Computer scientists suggest that it is best *not* to attempt to modify a serial code but instead to rewrite it from scratch using algorithms and subroutine libraries best suited to parallel architecture. However, this may involve months or years of work, and surveys find that $\sim$70% of computational scientists revise existing codes [Pan 96].

Most parallel computations at present are done on a multiple-instruction, multiple-data computers via message passing. In Appendix D we give a tutorial on the use of MPI, the most common message-passing interface. Here we outline some practical concerns based on user experience [Dong 05, Pan 96].

Parallelism carries a price tag: There is a steep learning curve requiring intensive effort. Failures may occur for a variety of reasons, especially because parallel environments tend to change often and get "locked up" by a programming error. In addition, with multiple computers and multiple operating systems involved, the familiar techniques for debugging may not be effective.

Preconditions for parallelism: If your program is run thousands of times between changes, with execution time in days, and you must significantly increase the resolution of the output or study more complex systems, then parallelism is worth considering. Otherwise, and to the extent of the difference, parallelizing a code may not be worth the time investment.

The problem affects parallelism: You must analyze your problem in terms of how and when data are used, how much computation is required for each use, and the type of problem architecture:

- **Perfectly parallel:** The same application is run simultaneously on different data sets, with the calculation for each data set independent (e.g., running multiple versions of a Monte Carlo simulation, each with different seeds, or analyzing data from independent detectors). In this case it would be straightforward to parallelize with a respectable performance to be expected.
- **Fully synchronous:** The same operation applied in parallel to multiple parts of the same data set, with some waiting necessary (e.g., determining positions and velocities of particles simultaneously in a molecular dynamics simulation). Significant effort is required, and unless you balance the computational intensity, the speedup may not be worth the effort.
- **Loosely synchronous:** Different processors do small pieces of the computation but with intermittent data sharing (e.g., diffusion of groundwater from one location to another). In this case it would be difficult to parallelize and probably not worth the effort.
- **Pipeline parallel:** Data from earlier steps processed by later steps, with some overlapping of processing possible (e.g., processing data into images and then into animations). Much work may be involved, and unless you balance the computational intensity, the speedup may not be worth the effort.

14.12.1 High-Level View of Message Passing

Although it is true that parallel computing programs may become very complicated, the basic ideas are quite simple. All you need is a regular programming language like C or Fortran, plus four communication statements:[3]

send: One processor sends a message to the network. It is not necessary to indicate who will receive the message, but it must have a name.
receive: One processor receives a message from the network. This processor does not have to know who sent the message, but it has to know the message's name.
myid: An integer that uniquely identifies each processor.
numnodes: An integer giving the total number of nodes in the system.

Once you have made the decision to run your program on a computer cluster, you will have to learn the specifics of a message-passing system such as MPI (Appendix D). Here we give a broader view. When you write a message-passing program, you intersperse calls to the message-passing library with your regular Fortran or C program. The basic steps are

1. Submit your job from the command line or a job control system.

[3] Personal communication, Yuefan Deng.

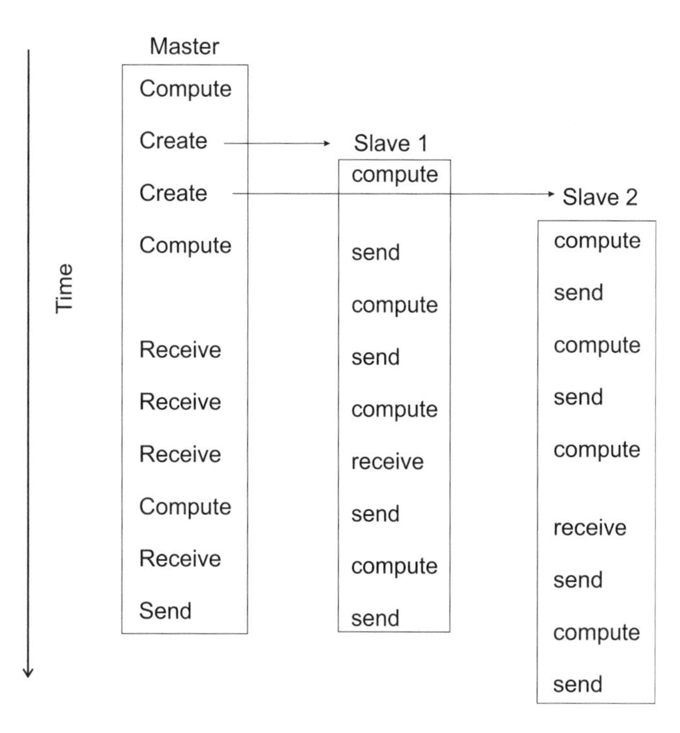

Figure 14.9 A master process and two slave processes passing messages. Notice how this program has more sends than receives and consequently may lead to results that depend on order of execution, or may even lock up.

2. Have your job start additional processes.
3. Have these processes exchange data and coordinate their activities.
4. Collect these data and have the processes stop themselves.

We show this graphically in Figure 14.9 where at the top we see a *master* process create two *slave* processes and then assign work for them to do (arrows). The processes then communicate with each other via message passing, output their data to files, and finally terminate.

What can go wrong: Figure 14.9 also illustrates some of the difficulties:

- The programmer is responsible for getting the processes to cooperate and for dividing the work correctly.
- The programmer is responsible for ensuring that the processes have the correct data to process and that the data are distributed equitably.
- The commands are at a lower level than those of a compiled language, and this introduces more details for you to worry about.
- Because multiple computers and multiple operating systems are involved, the user may not receive or understand the error messages produced.
- It is possible for messages to be sent or received not in the planned order.

- A *race condition* may occur in which the program results depend upon the specific ordering of messages. There is no guarantee that slave 1 will get its work done before slave 2, even though slave 1 may have started working earlier (Figure 14.9).
- Note in Figure 14.9 how different processors must wait for signals from other processors; this is clearly a waste of computer time and a potential for problems.
- Processes may *deadlock*, that is, wait for a message that never arrives.

14.13 Example of a Supercomputer: IBM Blue Gene/L

Whatever figures we give to describe the latest supercomputer will be obsolete by the time you read them. Nevertheless, for the sake of completeness and to set the scale for the present, we do it anyway. At the time of this writing, the fastest computer, in some aggregate sense, is the IBM Blue Gene series [Gara 05]. The name reflects its origin in a computer originally intended for gene research that is now sold as a general-purpose supercomputer (after approximately $600 million in development costs).

A building-block view of Blue Gene is given in Figure 14.10. In many ways this is a computer built by committee, with compromises made in order to balance cost, cooling, computing speed, use of existing technologies, communication speed, and so forth. As a case in point, the CPUs have dual cores, with one for computing and the other for communication. This reflects the importance of communication for distributed-memory computing (there are both on- and off-chip distributed memories). And while the CPU is fast at 5.6 GFLOPs, there are faster ones available, but they would generate so much heat that it would not be possible to obtain the extreme scalability up to 2^{16} = 65,536 dual-processor nodes. The next objective is to balance a low cost/performance ratio with a high performance/watt ratio.

Observe that at the lowest level Blue Gene contains two CPUs (dual cores) on a chip, with two chips on a card, with 16 cards on a board, with 32 boards in a cabinet, and up to 64 cabinets for a grand total of 65,536 CPUs (Figure 14.10). And if a way can be found to make all these chips work together for the common good on a single problem, they would turn out a peak performance of 360×10^{12} floating-point operations per second (360 tFLOPs). Each processor runs a Linux operating system (imagine what the cost in both time and money would be for Windows!) and utilizes the hardware by running a distributed memory MPI with C, C++, and Fortran90 compilers; this is not yet a good place for Java.

Blue Gene has three separate communication networks (Figure 14.11). At the heart of the memory system is a $64 \times 32 \times 32$ 3-D torus that connects all the memory blocks; Figure 14.11a shows the torus for $2 \times 2 \times 2$ memory blocks. The links are made by special link chips that also compute; they provide both direct neighbor–neighbor communications as well as cut through communication across the network. The result of this sophisticated network is that there is approximately the same effective bandwidth and latencies (response times) between all nodes,

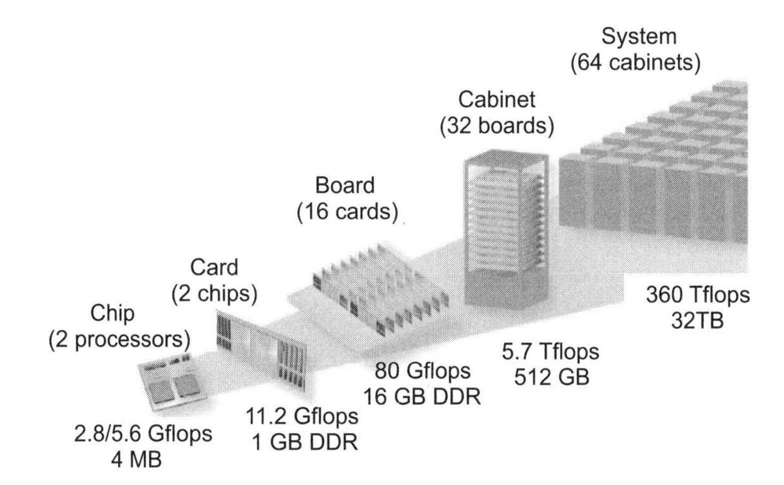

Figure 14.10 The building blocks of Blue Gene (adapted from (Gara 05)).

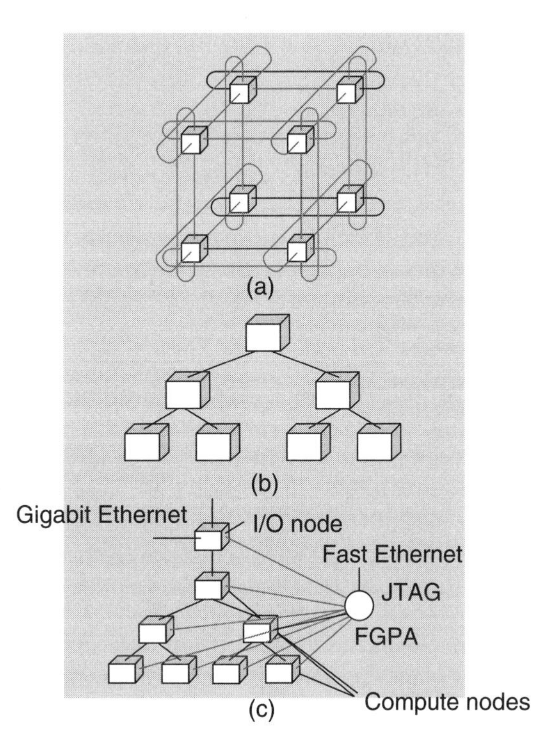

Figure 14.11 *(a)* A 3-D torus connecting $2 \times 2 \times 2$ memory blocks. *(b)* The global collective memory system. *(c)* The control and GB-Ethernet memory system (adapted from (Gara 05)).

yet in order to obtain high speed, it is necessary to keep communication local. For node-to-node communication a rate of 1.4 Gb/s $= 1/10^{-9}$ s $= 1$ ns is obtained.

The latency ranges from 100 ns for 1 hop, to 6.4 μs for 64 hops between processors. The collective network in Figure 14.11b is used to communicate with all the processors simultaneously, what is known as a *broadcast*, and does so at 4 b/cycle. Finally, the control and gigabit ethernet network (Figure 14.11c) is used for I/O to communicate with the switch (the hardware communication center) and with ethernet devices. Its total capacity is greater than 1 tb ($= 10^{12}$)/s.

The computing heart of Blue Gene is its integrated circuit and the associated memory system (Figure 14.12). This is essentially an entire computer system on a chip containing, among other things,

- two PowerPC 440s with attached floating-point units (for rapid processing of floating-point numbers); one CPU is for computing, and one is for I/O.
- a RISC architecture CPU with seven stages, three pipelines, and 32-b, 64-way associative cache lines,
- variable memory page size,
- embedded dynamic memory controllers,
- a gigabit ethernet adapter,
- a total of 512 MB/node (32 tB when summed over nodes),
- level-1 (L1) cache of 32 kB, L2 cache of 2 kB, and L3 cache of 4 MB.

14.14 Unit III. HPC Program Optimization

The type of optimization often associated with *high-performance* or *numerically intensive* computing is one in which sections of a program are rewritten and reorganized in order to increase the program's speed. The overall value of doing this, especially as computers have become so fast and so available, is often a subject of controversy between computer scientists and computational scientists. Both camps agree that using the optimization options of compilers is a good idea. However, computational scientists tend to run large codes with large amounts of data in order to solve real-world problems and often believe that you cannot rely on the compiler to do all the optimization, especially when you end up with time on your hands waiting for the computer to finish executing your program. Here are some entertaining, yet insightful, views [Har 96] reflecting the conclusions of those who have reflected upon this issue:

> More computing sins are committed in the name of efficiency (without necessarily achieving it) than for any other single reason—including blind stupidity.
>
> — W.A. Wulf

> We should forget about small efficiencies, say about 97% of the time: premature optimization is the root of all evil.
>
> — Donald Knuth

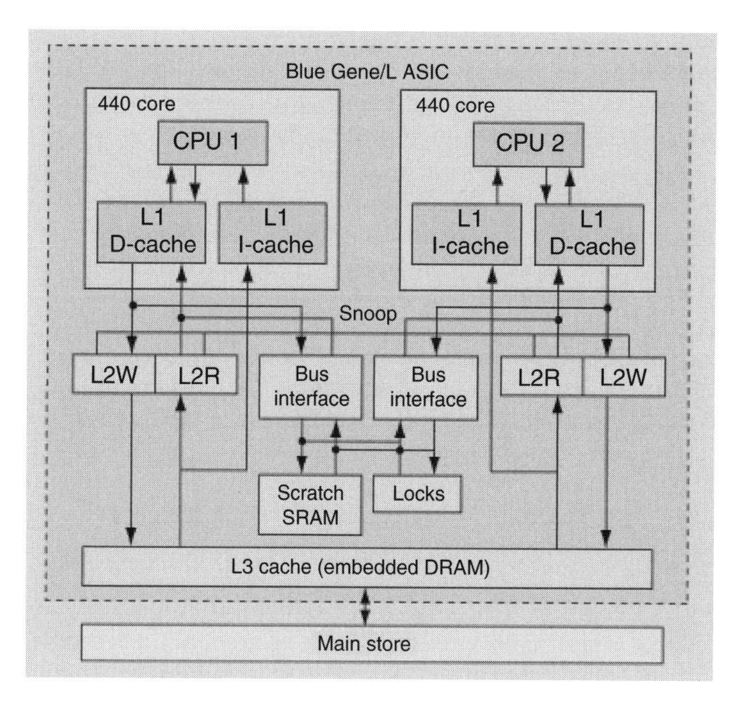

Figure 14.12 The single-node memory system (adapted from (Gara 05)).

The best is the enemy of the good.

— Voltaire

Rule 1: Do not do it.

Rule 2 (for experts only): Do not do it yet.

Do not optimize as you go: Write your program without regard to possible optimizations, concentrating instead on making sure that the code is clean, correct, and understandable. If it's too big or too slow when you've finished, then you can consider optimizing it.

Remember the 80/20 rule: In many fields you can get 80% of the result with 20% of the effort (also called the 90/10 rule—it depends on who you talk to). Whenever you're about to optimize code, use profiling to find out where that 80% of execution time is going, so you know where to concentrate your effort.

Always run "before" and "after" benchmarks: How else will you know that your optimizations actually made a difference? If your optimized code turns out to be only slightly faster or smaller than the original version, undo your changes and go back to the original, clear code.

Use the right algorithms and data structures: Do not use an $\mathcal{O}(n^2)$ DFT algorithm to do a Fourier transform of a thousand elements when there's an $\mathcal{O}(n \log n)$ FFT available. Similarly, do not store a thousand items in an array that requires an $\mathcal{O}(n)$ search when you could use an $\mathcal{O}(\log n)$ binary tree or an $\mathcal{O}(1)$ hash table.

14.14.1 Programming for Virtual Memory (Method)

While paging makes little appear big, you pay a price because your program's run time increases with each page fault. If your program does not fit into RAM all at once, it will run significantly slower. If virtual memory is shared among multiple programs that run simultaneously, they all can't have the entire RAM at once, and so there will be memory access *conflicts*, in which case the performance of all the programs will suffer. The basic rules for programming for virtual memory are as follows.

1. Do not waste your time worrying about reducing the amount of memory used (the *working set size*) unless your program is large. In that case, take a global view of your entire program and optimize those parts that contain the largest arrays.
2. Avoid page faults by organizing your programs to successively perform their calculations on subsets of data, each fitting completely into RAM.
3. Avoid simultaneous calculations in the same program to avoid competition for memory and consequent page faults. Complete each major calculation before starting another.
4. Group data elements close together in memory blocks if they are going to be used together in calculations.

14.14.2 Optimizing Programs; Java *versus* Fortran/C

Many of the optimization techniques developed for Fortran and C are also relevant for Java applications. Yet while Java is a good language for scientific programming and is the most universal and portable of all languages, at present Java code runs slower than Fortran or C code and does not work well if you use MPI for parallel computing (see Appendix D). In part, this is a consequence of the Fortran and C compilers having been around longer and thereby having been better refined to get the most out of a computer's hardware, and in part this is also a consequence of Java being designed for portability and not speed. Since modern computers are so fast, whether a program takes 1s or 3s usually does not matter much, especially in comparison to the hours or days of *your* time that it might take to modify a program for different computers. However, you may want to convert the code to C (whose command structure is very similar to that of Java) if you are running a computation that takes hours or days to complete and will be doing it many times.

Especially when asked to, Fortran and C compilers look at your entire code as a single entity and rewrite it for you so that it runs faster. (The rewriting is at a fairly basic level, so there's not much use in your studying the compiler's output as a way of improving your programming skills.) In particular, Fortran and C compilers are very careful in accessing arrays in memory. They also are careful to keep the cache lines full so as not to keep the CPU waiting with nothing to do. There is no fundamental reason why a program written in Java cannot be compiled to produce an highly efficient code, and indeed such compilers are being developed and becoming available. However, such code is optimized for a particular computer architecture and so is not portable. In contrast, the byte code (`.class` file) produced by Java is designed to be interpreted or recompiled by the *Java Virtual Machine* (just another program). When you change from Unix to Windows, for example, the Java Virtual Machine program changes, but the byte code is the same. This is the essence of Java's portability.

In order to improve the performance of Java, many computers and browsers run a *Just-in-Time* (JIT) Java compiler. If a JIT is present, the Java Virtual Machine feeds your byte code `Prog.class` to the JIT so that it can be recompiled into native code explicitly tailored to the machine you are using. Although there is an extra step involved here, the total time it takes to run your program is usually 10–30 times faster with a JIT as compared to line-by-line interpretation. Because the JIT is an integral part of the Java Virtual Machine on each operating system, this usually happens automatically.

In the experiments below you will investigate techniques to optimize both Fortran and Java programs and to compare the speeds of both languages for the same computation. If you run your Java code on a variety of machines (easy to do with Java), you should also be able to compare the speed of one computer to that of another. Note that a knowledge of Fortran is not required for these exercises.

14.14.2.1 GOOD AND BAD VIRTUAL MEMORY USE (EXPERIMENT)

To see the effect of using virtual memory, run these simple pseudocode examples on your computer (Listings 14.1–14.4). Use a command such as `time` to measure the time used for each example. These examples call functions `force12` and `force21`. You should write these functions and make them have significant memory requirements for both local and global variables.

```
for j = 1, n; { for i = 1, n; {
    f12(i,j) = force12(pion(i), pion(j))                    // Fill f12
    f21(i,j) = force21(pion(i), pion(j))                    // Fill f21
    ftot = f12(i,j) + f21(i,j) }}                          // Fill ftot
```

Listing 14.1 BAD program, too simultaneous.

You see (Listing 14.1) that each iteration of the for loop requires the data and code for all the functions as well as access to all the elements of the matrices and arrays. The working set size of this calculation is the sum of the sizes of the arrays f12(N,N), f21(N,N), and pion(N) plus the sums of the sizes of the functions force12 and force21.

A better way to perform the same calculation is to break it into separate components (Listing 14.2):

```
for j = 1, n; { for i = 1, n;    f12(i,j) = force12(pion(i), pion(j)) }
for j = 1, n; { for i = 1, n;    f21(i,j) = force21(pion(i), pion(j)) }
for j = 1, n; { for i = 1, n;    ftot = f12(i,j) + f21(i,j) }
```

Listing 14.2 GOOD program, separate loops.

Here the separate calculations are independent and the *working set size*, that is, the amount of memory used, is reduced. However, you do pay the additional overhead costs associated with creating extra for loops. Because the working set size of the first for loop is the sum of the sizes of the arrays f12(N, N) and pion(N), and of the function force12, we have approximately half the previous size. The size of the last for loop is the sum of the sizes for the two arrays. The working set size of the entire program is the larger of the working set sizes for the different for loops.

As an example of the need to group data elements close together in memory or common blocks if they are going to be used together in calculations, consider the following code (Listing 14.3):

```
Common zed, ylt(9), part(9), zpart1(50000), zpart2(50000), med2(9)
        for j = 1, n; ylt(j) = zed * part(j)/med2(9)        // Discontinuous variables
```

Listing 14.3 BAD Program, discontinuous memory.

Here the variables zed, ylt, and part are used in the same calculations and are adjacent in memory because the programmer grouped them together in Common (global variables). Later, when the programmer realized that the array med2 was needed, it was tacked onto the end of Common. All the data comprising the variables zed, ylt, and part fit onto one page, but the med2 variable is on a different page because the large array zpart2(50000) separates it from the other variables. In fact, the system may be forced to make the entire 4-kB page available in order to fetch the 72 B of data in med2. While it is difficult for a Fortran or C programmer to ensure the placement of variables within page boundaries, you will improve your chances by grouping data elements together (Listing 14.4):

```
Common zed, ylt(9), part(9), med2(9), zpart1(50000), zpart2(50000)
        for j = 1, n;          ylt(j) = zed*part(j)/med2(J)              // Continuous
```

Listing 14.4 GOOD program, continuous memory.

14.14.3 Experimental Effects of Hardware on Performance

In this section you conduct an experiment in which you run a complete program in several languages and on as many computers as are available. In this way you explore how a computer's architecture and software affect a program's performance.

Even if *you* do not know (or care) what is going on inside a program, some optimizing compilers are smart and caring enough to figure it out for you and then go about rewriting your program for improved performance. You control how completely the compiler does this when you add *optimization options* to the compile command:

```
> f90 –O tune.f90
```

Here –O turns on optimization (O is the capital letter "oh," not zero). The actual optimization that is turned on differs from compiler to compiler. Fortran and C compilers have a bevy of such options and directives that let you truly customize the resulting compiled code. Sometimes optimization options make the code run faster, sometimes not, and sometimes the faster-running code gives the wrong answers (but does so quickly).

Because computational scientists may spend a good fraction of their time running compiled codes, the compiler options tend to become quite specialized. As a case in point, most compilers provide a number of levels of optimization for the compiler to attempt (there are no guarantees with these things). Although the speedup obtained depends upon the details of the program, higher levels may give greater speedup, as well as a concordant greater risk of being wrong.

The **Forte/Sun** Fortran compiler options include

–O	Use the default optimization level (–O3)
–O1	Provide minimum statement-level optimizations
–O2	Enable basic block-level optimizations
–O3	Add loop unrolling and global optimizations
–O4	Add automatic inlining of routines from the same source file
–O5	Attempt aggressive optimizations (with profile feedback)

For the **Visual Fortran (Compaq, Intel)** compiler under windows, options are entered as /optimize and for optimization are

/optimize:0	Disable most optimizations
/optimize:1	Local optimizations in the source program unit
/optimize:2	Global optimization, including /optimize:1
/optimize:3	Additional global optimizations; speed at cost of code size: loop unrolling, instruction scheduling, branch code replication, padding arrays for cache

/optimize:4	Interprocedure analysis, inlining small procedures
/optimize:5	Activate loop transformation optimizations

The **gnu compilers** gcc, g77, g90 accept –O options as well as

–malign–double	Align doubles on 64-bit boundaries
–ffloat–store	For codes using IEEE-854 extended precision
–fforce–mem, –fforce–addr	Improves loop optimization
–fno–inline	Do not compile statement functions inline
–ffast–math	Try non-IEEE handling of floats
–funsafe–math–optimizations	Speeds up float operations; incorrect results possible
–fno–trapping–math	Assume no floating-point traps generated
–fstrength–reduce	Makes some loops faster
–frerun–cse–after–loop	
–fexpensive–optimizations	
–fdelayed–branch	
–fschedule–insns	
–fschedule–insns2	
–fcaller–saves	
–funroll–loops	Unrolls iterative DO loops
–funroll–all–loops	Unrolls DO WHILE loops

14.14.4 Java *versus* Fortran/C

The various versions of the program tune solve the matrix eigenvalue problem

$$\mathbf{H}\mathbf{c} = E\mathbf{c} \tag{14.11}$$

for the eigenvalues E and eigenvectors $\mathbf{c}$ of a Hamiltonian matrix $\mathbf{H}$. Here the individual Hamiltonian matrix elements are assigned the values

$$H_{i,j} = \begin{cases} i, & \text{for } i = j, \\ 0.3^{|i-j|}, & \text{for } i \neq j, \end{cases} = \begin{bmatrix} 1 & 0.3 & 0.14 & 0.027 & \ldots \\ 0.3 & 2 & 0.3 & 0.9 & \ldots \\ 0.14 & 0.3 & 3 & 0.3 & \ldots \\ \ddots & & & & \end{bmatrix}. \tag{14.12}$$

Because the Hamiltonian is almost diagonal, the eigenvalues should be close to the values of the diagonal elements and the eigenvectors should be close to N-dimensional unit vectors. For the present problem, the H matrix has dimension $N \times N \simeq 2000 \times 2000 = 4,000,000$, which means that matrix manipulations should take enough time for you to see the effects of optimization. If your computer has a large supply of central memory, you may need to make the matrix even larger to see what happens when a matrix does not all fit into RAM.

We find the solution to (14.11) via a variation of the *power* or *Davidson method.* We start with an arbitrary first guess for the eigenvector $\mathbf{c}$ and use it to calculate the energy corresponding to this eigenvector,[4]

$$\mathbf{c}_0 \simeq \begin{pmatrix} 1 \\ 0 \\ \vdots \\ 0 \end{pmatrix}, \qquad E \simeq \frac{\mathbf{c}_0^\dagger \mathbf{H} \mathbf{c}_0}{\mathbf{c}_0^\dagger \mathbf{c}_0}, \tag{14.13}$$

where $\mathbf{c}_0^\dagger$ is the row vector adjoint of $\mathbf{c}_0$. Because $\mathbf{H}$ is nearly diagonal with diagonal elements that increases as we move along the diagonal, this guess should be close to the eigenvector with the smallest eigenvalue. The heart of the algorithm is the guess that an improved eigenvector has the kth component

$$\mathbf{c}_1|_k \simeq \mathbf{c}_0|_k + \frac{[\mathbf{H} - E\mathbf{I}]\mathbf{c}_0|_k}{E - H_{k,k}}, \tag{14.14}$$

where k ranges over the length of the eigenvector. If repeated, this method converges to the eigenvector with the smallest eigenvalue. It will be the smallest eigenvalue since it gets the largest weight (smallest denominator) in (14.14) each time. For the present case, six places of precision in the eigenvalue are usually obtained after 11 iterations. Here are the steps to follow:

- Vary the variable `err` in `tune` that controls precision and note how it affects the number of iterations required.
- Try some variations on the initial guess for the eigenvector (14.14) and see if you can get the algorithm to converge to some of the other eigenvalues.
- Keep a table of your execution times *versus* technique.
- Compile and execute `tune.f90` and record the run time. On Unix systems, the compiled program will be placed in the file `a.out`. From a Unix shell, the compilation, timing, and execution can all be done with the commands

> **f90 tune.f90**	Fortran compilation
> **cc -lm tune.c**	C compilation, `gcc` also likely
> **time a.out**	Execution

  Here the compiled Fortran program is given the (default) name `a.out`, and the `time` command gives you the execution (user) time and `system` time in seconds to execute `a.out`.

[4] Note that the codes refer to the eigenvector c_0 as `coef`.

- As indicated in §14.14.3, you can ask the compiler to produce a version of your program optimized for speed by including the appropriate compiler option:

```
> f90 -O tune.f90
```

  Execute and time the optimized code, checking that it still gives the same answer, and note any speedup in your journal.
- Try out optimization options up to the highest levels and note the run time and accuracy obtained. Usually -O3 is pretty good, especially for as simple a program as tune with only a main method. With only one program unit we would not expect -O4 or -O5 to be an improvement over -O3. However, we do expect -O3, with its loop unrolling, to be an improvement over -O2.
- The program tune4 does some *loop unrolling* (we will explore that soon). To see the best we can do with Fortran, record the time for the most optimized version of tune4.f90.
- The program Tune.java in Listing 14.5 is the Java equivalent of the Fortran program tune.f90.
- To get an idea of what Tune.java does (and give you a feel for how hard life is for the poor computer), assume ldim =2 and work through one iteration of Tune *by hand*. Assume that the iteration loop has converged, follow the code to completion, and write down the values assigned to the variables.
- Compile and execute Tune.java. You do not have to issue the time command since we built a timer into the Java program (however, there is no harm in trying it). Check that you still get the same answer as you did with Fortran and note how much longer it takes with Java.
- Try the -O option with the Java compiler and note if the speed changes (since this just inlines methods, it should not affect our one-method program).
- You might be surprised how much slower Java is than Fortran and that the Java optimizer does not seem to do much good. To see what the actual Java byte code does, invoke the Java profiler with the command

```
> javap -c Tune
```

  This should produce a file, java.prof for you to look at with an editor. Look at it and see if you agree with us that scientists have better things to do with their time than trying to understand such files!
- We now want to perform a little experiment in which we see what happens to performance as we fill up the computer's memory. In order for this experiment to be reliable, it is best for you to *not* to be sharing the computer with any other users. On Unix systems, the who -a command shows you the other users (we leave it up to you to figure out how to negotiate with them).
- To get some idea of what aspect of our little program is making it so slow, compile and run Tune.java for the series of matrix sizes ldim = 10, 100, 250, 500, 750, 1025, 2500, and 3000. You may get an error message that Java is out of memory at 3000. This is because you have not turned on the use of virtual memory. In Java, the memory allocation pool for your program is called the

heap and it is controlled by the –Xms and –Xmx options to the Java interpreter java:

–Xms256m	Set initial heap size to 256 MB
–Xmx512m	Set maximum heap size to 512 MB

```java
// Tune.java: eigenvalue solution for performace tuning

public class Tune {

  public static void main(String[] argv) {
    final int Ldim = 2051;
    int i, j, iter = 0;
    double [][] ham = new double [Ldim] [Ldim]; double[] coef = new double [Ldim];
    double [] sigma = new double [Ldim]; double time, err, ener, ovlp, step = 0.;
    time = System.currentTimeMillis();                            // Initialize time
      for ( i = 1;  i <= Ldim-1;  i++ )  {                       // Init matrix & vector
        for ( j = 1;  j <= Ldim-1;  j++ )  {
          if (Math.abs(j-i) >10)  ham[j][i] = 0. ;
          else  ham[j][i] = Math.pow(0.3, Math.abs(j-i));
        }
        ham[i][i] = i ;     coef[i] = 0.;
      }
      coef[1] = 1.;     err = 1.;     iter = 0 ;                     // Start iteration
      while (iter  < 15 && err > 1.e-6)  {
        iter = iter + 1;   ener = 0.;    ovlp = 0.;
        for ( i= 1;  i <= Ldim-1;  i++ )  {                     // Compute E & normalize
          ovlp = ovlp + coef[i]*coef[i];    sigma[i] = 0.;
          for (j= 1; j <= Ldim-1;  j++)  sigma[i] = sigma[i]+coef[j]*ham[j][i];
          ener = ener + coef[i]*sigma[i] ;
        }
        ener = ener/ovlp;
        for (  i = 1;   i <= Ldim-1;   i++ )  { coef[i] = coef[i]/Math.sqrt(ovlp) ;
                                                sigma[i] = sigma[i]/Math.sqrt(ovlp); }
        err = 0.;
        for ( i = 2;  i <= Ldim-1;  i++ )  {                                 // Update
          step = (sigma[i] - ener*coef[i])/(ener-ham[i][i]);
          coef[i] = coef[i] + step; err = err +  step*step ;
        }
        err = Math.sqrt(err) ;
        System.out.println ("iter, ener, err " + iter + ",  " + ener + ",  " + err);
      }
      time = (System.currentTimeMillis() - time)/1000.;                // Elapsed time
      System.out.println("time = " + time + "s");
} }
```

Listing 14.5 Tune.java is meant to be numerically intensive enough to show the results of various types of optimizations. The program solves the eigenvalue problem iteratively for a nearly diagonal Hamiltonian matrix using a variation of the power method.

- Make a graph of run time *versus* matrix size. It should be similar to Figure 14.13, although if there is more than one user on your computer while you run, you may get erratic results. Note that as our matrix becomes larger than $\sim 1000 \times 1000$ in size, the curve sharply increases in slope with execution time, in our case increasing like the *third* power of the dimension. Since the number of elements to compute increases as the *second* power of the

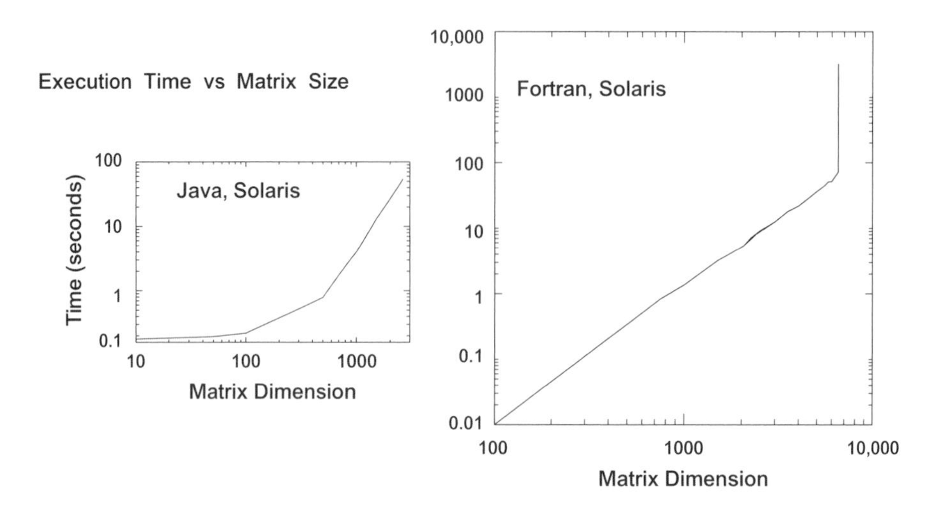

Figure 14.13 Running time *versus* dimension for an eigenvalue search using Tune.java and tune.f90.

dimension, something else is happening here. It is a good guess that the additional slowdown is due to page faults in accessing memory. In particular, accessing 2-D arrays, with their elements scattered all through memory, can be very slow.

- Repeat the previous experiment with tune.f90 that gauges the effect of increasing the ham matrix size, only now do it for ldim = 10, 100, 250, 500, 1025, 3000, 4000, 6000,.... You should get a graph like ours. Although our implementation of Fortran has automatic virtual memory, its use will be exceedingly slow, especially for this problem (possibly a 50-fold increase in time!). So if you submit your program and you get nothing on the screen (though you can hear the disk spin or see it flash busy), then you are probably in the virtual memory regime. If you can, let the program run for one or two iterations, kill it, and then scale your run time to the time it would have taken for a full computation.
- To test our hypothesis that the access of the elements in our 2-D array ham [i][j] is slowing down the program, we have modified Tune.java into Tune4.java in Listing 14.6.
- Look at Tune4.java and note where the nested for loop over i and j now takes step of $\Delta i = 2$ rather the unit steps in Tune.java. If things work as expected, the better memory access of Tune4.java should cut the run time nearly in half. Compile and execute Tune4.java. Record the answer in your table.

```
// Tune4.java: matrix algebra program, basic optimization

public class Tune4  {

  public static void main(String[] argv)  {
    final int Ldim = 2051;
    int i, j, iter = 0;
    double [][] ham = new double [Ldim] [Ldim]; double [] diag = new double [Ldim];
    double [] coef  = new double [Ldim]; double [] sigma = new double [Ldim];
    double err, ener, ovlp, ovlp1, ovlp2, step = 0., fact, time, t, t1, t2, u;
    time = System.currentTimeMillis();                          // Store initial time
    for ( i = 1;  i <= Ldim-1;  i++ )    {                      // Set up Hamiltonian
      for ( j = 1;  j <= Ldim-1;  j++ )  if (Math.abs(j-i) >10) ham[j][i] = 0. ;
        else  ham[j][i] = Math.pow(0.3, Math.abs(j-i));
    }                                                       // Iterate towards solution
    for (i=1;  i<Ldim-1;  i++) {ham[i][i] = i; coef[i] = 0.; diag[i] = ham [i][i];}
    coef[1] = 1.;   err = 1.;   iter = 0 ;
    while (iter  < 15 && err > 1.e-6)  {          // Compute current energy & normalize
      iter = iter + 1; ener = 0. ; ovlp1 = 0.;    ovlp2 = 0.;
      for ( i= 1;  i <= Ldim-2;  i = i + 2  ) {
        ovlp1 = ovlp1 + coef[i]*coef[i] ;
        ovlp2 = ovlp2 + coef[i+1]*coef[i+1] ;
        t1 =  t2 = 0.;
        for ( j=1;  j <= Ldim-1;  j++ )  { t1 = t1 + coef[j]*ham[j][i];
                                           t2 = t2 + coef[j]*ham[j][i+1];  }
        sigma[i] = t1;  sigma[i + 1] = t2;
        ener = ener + coef[i]*t1 + coef[i+1]*t2 ;
      }
      ovlp = ovlp1 + ovlp2 ;
      ener = ener/ovlp;
      fact = 1./Math.sqrt(ovlp);
      coef[1] = fact*coef[1];
      err = 0.;                                                // Update & error norm
      for (i = 2; i <= Ldim-1; i++) { t = fact*coef[i];  u = fact*sigma[i]-ener*t;
          step = u/(ener-diag[i]);  coef[i] = t + step;  err = err +  step*step; }
      err = Math.sqrt(err) ;
      System.out.println ("iter, ener, err "+iter+", " + ener + ", " + err);
    }
    time = (System.currentTimeMillis() - time)/1000;
    System.out.println("time = " + time + "s");                       // Elapsed time
} }
```

Listing 14.6 Tune4.java does some loop unrolling by explicitly writing out two steps of a for loop (steps of 2.) This results in better memory access and faster execution.

- In order to cut the number of calls to the 2-D array in half, we employed a technique know as *loop unrolling* in which we explicitly wrote out some of the lines of code that otherwise would be executed implicitly as the for loop went through all the values for its counters. This is not as clear a piece of code as before, but it evidently, permits the compiler to produce a faster executable. To check that Tune and Tune4 actually do the same thing, assume ldim =4 and run through one iteration of Tune4.java *by hand*. Hand in your manual trial.

14.15 Programming for the Data Cache (Method)

Data caches are small, very fast memory used as temporary storage between the ultrafast CPU registers and the fast main memory. They have grown in importance as high-performance computers have become more prevalent. For systems that use a data cache, this may well be the single most important programming consideration; continually referencing data that are not in the cache (*cache misses*) may lead to an order-of-magnitude increase in CPU time.

As indicated in Figures 14.2 and 14.14, the data cache holds a copy of some of the data in memory. The basics are the same for all caches, but the sizes are manufacturer-dependent. When the CPU tries to address a memory location, the *cache manager* checks to see if the data are in the cache. If they are not, the manager reads the data from memory into the cache, and then the CPU deals with the data directly in the cache. The cache manager's view of RAM is shown in Figure 14.14.

When considering how a matrix operation uses memory, it is important to consider the *stride* of that operation, that is, the number of array elements that are stepped through as the operation repeats. For instance, summing the diagonal elements of a matrix to form the trace

$$\mathrm{Tr}\, A = \sum_{i=1}^{N} a(i,i) \tag{14.15}$$

involves a large stride because the diagonal elements are stored far apart for large N. However, the sum

$$c(i) = x(i) + x(i+1) \tag{14.16}$$

has stride 1 because adjacent elements of x are involved. The basic rule in programming for a cache is

- Keep the stride low, preferably at 1, which in practice means.
- Vary the leftmost index first on Fortran arrays.
- Vary the rightmost index first on Java and C arrays.

14.15.1 Exercise 1: Cache Misses

We have said a number of times that your program will be slowed down if the data it needs are in virtual memory and not in RAM. Likewise, your program will also be slowed down if the data required by the CPU are not in the cache. For high-performance computing, you should write programs that keep as much of the data being processed as possible in the cache. To do this you should recall that Fortran matrices are stored in successive memory locations with the row index varying most rapidly (column-major order), while Java and C matrices are stored in successive memory locations with the column index varying most rapidly (row-major order). While it is difficult to isolate the effects of the cache from other elements of the computer's architecture, you should now estimate its importance by comparing the time it takes to step through the matrix elements row by row to the time it takes to step through the matrix elements column by column.

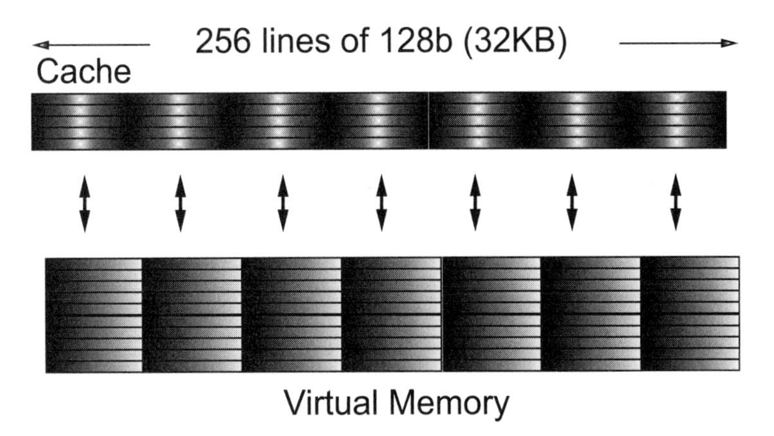

Figure 14.14 The cache manager's view of RAM. Each 128-B cache line is read into one of four lines in cache.

By actually running on machines available to you, check that the two simple codes in Listing 14.7 with the same number of arithmetic operations take significantly different times to run because one of them must make large jumps through memory with the memory locations addressed not yet read into the cache:

```
x(j) = m(1,j)                                   // Sequential column reference
```

```
for j = 1, 9999;
    x(j) = m(j,1)                                  // Sequential row reference
```

Listing 14.7 Sequential column and row references.

14.15.2 Exercise 2: Cache Flow

Test the importance of cache flow on your machine by comparing the time it takes to run the two simple programs in Listings 14.8 and 14.9. Run for increasing column size idim and compare the times for loop A *versus* those for loop B. A computer with very small caches may be most sensitive to stride.

```
Dimension Vec(idim,jdim)                                    // Stride 1 fetch (f90)
      for j = 1, jdim; { for i=1, idim;   Ans = Ans + Vec(i,j)*Vec(i,j) }
```

Listing 14.8 GOOD f90, BAD Java/C Program; minimum, maximum stride.

```
Dimension Vec(idim, jdim)                                    // Stride jdim fetch (f90)
      for i = 1, idim; { for j=1, jdim;   Ans = Ans + Vec(i,j)*Vec(i,j) }
```

Listing 14.9 BAD f90, GOOD Java/C Program; maximum, minimum stride.

Loop A steps through the matrix Vec in column order. Loop B steps through in row order. By changing the size of the columns (the rightmost Fortran index), we change the step size (*stride*) taken through memory. Both loops take us through all the elements of the matrix, but the stride is different. By increasing the stride in any language, we use fewer elements already present in the cache, require additional swapping and loading of the cache, and thereby slow down the whole process.

14.15.3 Exercise 3: Large-Matrix Multiplication

As you increase the dimensions of the arrays in your program, memory use increases geometrically, and at some point you should be concerned about efficient memory use. The penultimate example of memory usage is large-matrix multiplication:

$$[C] = [A] \times [B], \tag{14.17}$$

$$c_{ij} = \sum_{k=1}^{N} a_{ik} \times b_{kj}. \tag{14.18}$$

```
for i = 1, N; {                                              // Row
  for j = 1, N; {                                            // Column
    c(i,j) = 0.0                                             // Initialize
    for k = 1, N; {
      c(i,j) = c(i,j) + a(i,k)*b(k,j) }}}                    // Accumulate sum
```

Listing 14.10 BAD f90, GOOD Java/C Program; maximum, minimum stride.

This involves all the concerns with different kinds of memory. The natural way to code (14.17) follows from the definition of matrix multiplication (14.18), that is, as a sum over a row of A times a column of B. Try out the two code in Listings 14.10 and 14.11 on your computer. In Fortran, the first code has B with stride 1, but C with stride N. This is corrected in the second code by performing the initialization in another loop. In Java and C, the problems are reversed. On one of our machines, we found a factor of 100 difference in CPU times even though the number of operations is the same!

```
for j = 1, N; {                                           // Initialization
  for  i = 1, N; {
    c(i,j) = 0.0 }
    for k = 1, N; {
      for  i = 1, N; {c(i,j) = c(i,j) + a(i,k)*b(k,j) }}}
```

Listing 14.11 GOOD f90, BAD Java/C Program; minimum, maximum stride.

15

Thermodynamic Simulations & Feynman Quantum Path Integration

In Unit I of this chapter we describe how magnetic materials are simulated by using the Metropolis algorithm to solve the Ising model. This extends the techniques studied in Chapter 5, "Monte Carlo Simulations," to thermodynamics. Not only do thermodynamic simulations have important practical applications, but they also give us insight into what is "dynamic" in thermodynamics. In Unit II we describe a new Monte Carlo algorithm known as Wang–Landau sampling that in the last few years has shown itself to be far more efficient than the 50-year-old Metropolis algorithm. Wang–Landau sampling is an active subject in present research, and it is nice to see it fitting well into an elementary textbook. Unit III applies the Metropolis algorithm to Feynman's path integral formulation of quantum mechanics [F&H 65]. The theory, while the most advanced to be found in this book, forms the basis for field-theoretic calculations of quantum chromodynamics, some of the most fundamental and most time-consuming computations in existence. Basic discussions can be found in [Mann 83, MacK 85, M&N 87], with a recent review in [Potv 93].

15.1 Unit I. Magnets via the Metropolis Algorithm

Ferromagnets contain finite-size *domains* in which the spins of all the atoms point in the same direction. When an external magnetic field is applied to these materials, the different domains align and the materials become "magnetized." Yet as the temperature is raised, the total magnetism decreases, and at the Curie temperature the system goes through a *phase transition* beyond which all magnetization vanishes. Your **problem** is to explain the thermal behavior of ferromagnets.

15.2 An Ising Chain (Model)

As our model we consider N magnetic dipoles fixed in place on the links of a linear chain (Figure 15.1). (It is a straightforward generalization to handle 2-D and 3-D lattices.) Because the particles are fixed, their positions and momenta are not dynamic variables, and we need worry only about their spins. We assume that the particle at site i has spin s_i, which is either up or down:

$$s_i \equiv s_{z,i} = \pm\frac{1}{2}. \tag{15.1}$$

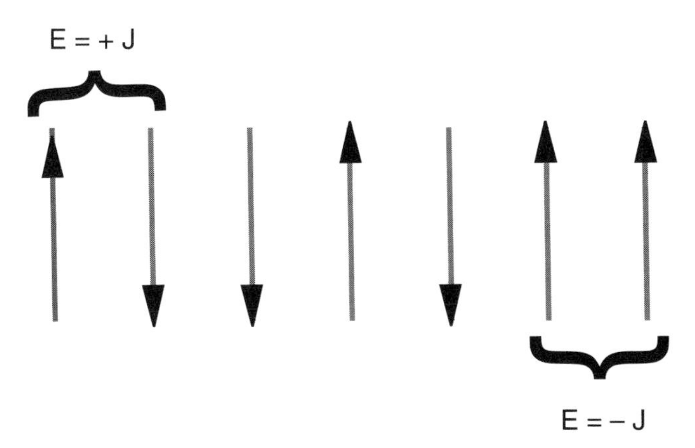

Figure 15.1 A 1-D lattice of N spins. The interaction energy $V = \pm J$ between nearest-neighbor pairs is shown for aligned and opposing spins.

Each configuration of the N particles is described by a quantum state vector

$$|\alpha_j\rangle = |s_1, s_2, \ldots, s_N\rangle = \left\{\pm\frac{1}{2}, \pm\frac{1}{2}, \ldots\right\}, \quad j = 1, \ldots, 2^N. \tag{15.2}$$

Because the spin of each particle can assume any one of *two* values, there are 2^N different possible states for the N particles in the system. Since fixed particles cannot be interchanged, we do not need to concern ourselves with the symmetry of the wave function.

The energy of the system arises from the interaction of the spins with each other and with the external magnetic field B. We know from quantum mechanics that an electron's spin and magnetic moment are proportional to each other, so a magnetic *dipole–dipole* interaction is equivalent to a *spin–spin* interaction. We assume that each dipole interacts with the external magnetic field and with its nearest neighbor through the potential:

$$V_i = -J\mathbf{s}_i \cdot \mathbf{s}_{i+1} - g\mu_b\,\mathbf{s}_i \cdot \mathbf{B}. \tag{15.3}$$

Here the constant J is called the *exchange energy* and is a measure of the strength of the spin–spin interaction. The constant g is the gyromagnetic ratio, that is, the proportionality constant between a particle's angular momentum and magnetic moment. The constant $\mu_b = e\hbar/(2m_e c)$ is the Bohr magneton, the basic measure for magnetic moments.

Even for small numbers of particles, the 2^N possible spin configurations gets to be very large ($2^{20} > 10^6$), and it is expensive for the computer to examine them all. Realistic samples with $\sim 10^{23}$ particles are beyond imagination. Consequently, statistical approaches are usually assumed, even for moderate values of N. Just how large N must be for this to be accurate is one of the things we want you to explore with your simulations.

The energy of this system in state α_k is the expectation value of the sum of the potential V over the spins of the particles:

$$E_{\alpha_k} = \left\langle \alpha_k \middle| \sum_i V_i \middle| \alpha_k \right\rangle = -J \sum_{i=1}^{N-1} s_i s_{i+1} - B\mu_b \sum_{i=1}^{N} s_i. \tag{15.4}$$

An apparent paradox in the Ising model occurs when we turn off the external magnetic field and thereby eliminate a preferred direction in space. This means that the average magnetization should vanish even though the lowest energy state would have all spins aligned. The answer to this paradox is that the system with $B = 0$ is unstable. Even if all the spins are aligned, there is nothing to stop their spontaneous reversal. Indeed, natural magnetic materials have multiple finite domains with all the spins aligned, but with the different domains pointing in different directions. The instabilities in which domains change direction are called *Bloch-wall transitions*. For simplicity we assume $B = 0$, which means that the spins interact just with each other. However, be cognizant of the fact that this means there is no preferred direction in space, and so you may have to be careful how you calculate observables. For example, you may need to take an absolute value of the total spin when calculating the magnetization, that is, to calculate $\langle |\sum_i s_i| \rangle$ rather than $\langle \sum_i s_i \rangle$.

The equilibrium alignment of the spins depends critically on the sign of the exchange energy J. If $J > 0$, the lowest energy state will tend to have neighboring spins aligned. If the temperature is low enough, the ground state will be a *ferromagnet* with all the spins aligned. If $J < 0$, the lowest energy state will tend to have neighbors with opposite spins. If the temperature is low enough, the ground state will be a *antiferromagnet* with alternating spins.

The simple 1-D Ising model has its limitations. Although the model is accurate in describing a system in thermal equilibrium, it is not accurate in describing the *approach* to thermal equilibrium (nonequilibrium thermodynamics is a difficult subject for which the theory is not complete). Second, as part of our algorithm we postulate that only one spin is flipped at a time, while real magnetic materials tend to flip many spins at a time. Other limitations are straightforward to improve, for example, the addition of longer-range interactions, the motion of the centers, higher-multiplicity spin states, and two and three dimensions.

A fascinating aspect of magnetic materials is the existence of a critical temperature, the *Curie temperature*, above which the gross magnetization essentially vanishes. Below the Curie temperature the quantum state of the material has long-range order extending over macroscopic dimensions; above the Curie temperature there is only short-range order extending over atomic dimensions. Even though the 1-D Ising model predicts realistic temperature dependences for the thermodynamic quantities, the model is too simple to support a phase transition. However, the 2-D and 3-D Ising models do support the Curie temperature phase transition.

15.3 Statistical Mechanics (Theory)

Statistical mechanics starts with the elementary interactions among a system's particles and constructs the macroscopic thermodynamic properties such as specific heats. The essential assumption is that all configurations of the system consistent with the constraints are possible. In some simulations, such as the molecular dynamics ones in Chapter 16, "Simulating Matter with Molecular Dynamics," the problem is set up such that the *energy* of the system is fixed. The states of this type of system are described by what is called a *microcanonical ensemble*. In contrast, for the thermodynamic simulations we study in this chapter, the temperature, volume, and number of particles remain fixed, and so we have what is called a *canonical ensemble*.

When we say that an object is *at* temperature T, we mean that the object's atoms are in thermodynamic equilibrium at temperature T such that each atom has an average energy proportional to T. Although this may be an equilibrium state, it is a dynamic one in which the object's energy fluctuates as it exchanges energy with its environment (it is thermo*dynamics* after all). Indeed, one of the most illuminating aspects of the simulation we shall develop is its visualization of the continual and random interchange of energy that occurs at equilibrium.

The energy E_{α_j} of state α_j in a canonical ensemble is not constant but is distributed with probabilities $P(\alpha_j)$ given by the Boltzmann distribution:

$$\mathcal{P}(E_{\alpha_j}, T) = \frac{e^{-E_{\alpha_j}/k_B T}}{Z(T)}, \qquad Z(T) = \sum_{\alpha_j} e^{-E_{\alpha_j}/k_B T}. \tag{15.5}$$

Here k is Boltzmann's constant, T is the temperature, and $Z(T)$ is the partition function, a weighted sum over states. Note that the sums in (15.5) are over the individual *states* or *configurations* of the system. Another formulation, such as the Wang–Landau algorithm in Unit II, sums over the *energies* of the states of the system and includes a density-of-states factor $g(E_i)$ to account for degenerate states with the same energy. While the present sum over states is a simpler way to express the problem (one less function), we shall see that the sum over energies is more efficient numerically. In fact, in this unit we even ignore the partition function $Z(T)$ because it cancels out when dealing with the *ratio* of probabilities.

15.3.1 Analytic Solutions

For very large numbers of particles, the thermodynamic properties of the 1-D Ising model can be solved analytically and determine [P&B 94]

$$U = \langle E \rangle \tag{15.6}$$

$$\frac{U}{J} = -N \tanh \frac{J}{k_B T} = -N \frac{e^{J/k_B T} - e^{-J/k_B T}}{e^{J/k_B T} + e^{-J/k_B T}} = \begin{cases} N, & k_B T \to 0, \\ 0, & k_B T \to \infty. \end{cases} \tag{15.7}$$

The analytic results for the specific heat per particle and the magnetization are

$$C(k_BT) = \frac{1}{N}\frac{dU}{dT} = \frac{(J/k_BT)^2}{\cosh^2(J/k_BT)} \tag{15.8}$$

$$M(k_BT) = \frac{Ne^{J/k_BT}\sinh(B/k_BT)}{\sqrt{e^{2J/k_BT}\sinh^2(B/k_BT) + e^{-2J/k_BT}}}. \tag{15.9}$$

The **2-D Ising model** has an analytic solution, but it was not easy to derive [Yang 52, Huang 87]. Whereas the internal energy and heat capacity are expressed in terms of elliptic integrals, the spontaneous magnetization per particle has the rather simple form

$$\mathcal{M}(T) = \begin{cases} 0, & T > T_c \\ \frac{(1+z^2)^{1/4}(1-6z^2+z^4)^{1/8}}{\sqrt{1-z^2}}, & T < T_c, \end{cases} \tag{15.10}$$

$$kT_c \simeq 2.269185J, \quad z = e^{-2J/k_BT}, \tag{15.11}$$

where the temperature is measured in units of the Curie temperature T_c.

15.4 Metropolis Algorithm

In trying to devise an algorithm that simulates thermal equilibrium, it is important to understand that the Boltzmann distribution (15.5) does not require a system to remain in the state of lowest energy but says that it is less likely for the system to be found in a higher-energy state than in a lower-energy one. Of course, as $T \to 0$, only the lowest energy state will be populated. For finite temperatures we expect the energy to fluctuate by approximately k_BT about the equilibrium value.

In their simulation of neutron transmission through matter, Metropolis, Rosenbluth, Teller, and Teller [Metp 53] invented an algorithm to improve the Monte Carlo calculation of averages. This *Metropolis algorithm* is now a cornerstone of computational physics. The sequence of configurations it produces (a *Markov chain*) accurately simulates the fluctuations that occur during thermal equilibrium. The algorithm randomly changes the individual spins such that, on the average, the probability of a configuration occurring follows a Boltzmann distribution. (We do not find the proof of this trivial or particularly illuminating.)

The Metropolis algorithm is a combination of the variance reduction technique discussed in §6.7.1 and the von Neumann rejection technique discussed in §6.7.3. There we showed how to make Monte Carlo integration more efficient by sampling random points predominantly where the integrand is large and how to generate random points with an arbitrary probability distribution. Now we would like to have spins flip randomly, have a system that can reach any energy in a finite number of steps (*ergodic* sampling), have a distribution of energies described by a Boltzmann distribution, yet have systems that equilibrate quickly enough to compute in reasonable times.

The Metropolis algorithm is implemented via a number of steps. We start with a fixed temperature and an initial spin configuration and apply the algorithm until a thermal equilibrium is reached (equilibration). Continued application of the algorithm generates the statistical fluctuations about equilibrium from which we deduce the thermodynamic quantities such as the magnetization $M(T)$. Then the temperature is changed, and the whole process is repeated in order to deduce the T dependence of the thermodynamic quantities. The accuracy of the deduced temperature dependences provides convincing evidence for the validity of the algorithm. Because the possible 2^N configurations of N particles can be a very large number, the amount of computer time needed can be very long. Typically, a small number of iterations $\simeq 10N$ is adequate for equilibration.

The explicit steps of the Metropolis algorithm are as follows.

1. Start with an arbitrary spin configuration $\alpha_k = \{s_1, s_2, \dots, s_N\}$.
2. Generate a trial configuration α_{k+1} by
   a. picking a particle i randomly and
   b. flipping its spin.[1]
3. Calculate the energy $E_{\alpha_{\text{tr}}}$ of the trial configuration.
4. If $E_{\alpha_{\text{tr}}} \le E_{\alpha_k}$, accept the trial by setting $\alpha_{k+1} = \alpha_{\text{tr}}$.
5. If $E_{\alpha_{\text{tr}}} > E_{\alpha_k}$, accept with relative probability $\mathcal{R} = \exp(-\Delta E/k_B T)$:
   a. Choose a uniform random number $0 \le r_i \le 1$.
   b. Set $\alpha_{k+1} = \begin{cases} \alpha_{\text{tr}}, & \text{if } \mathcal{R} \ge r_j \text{ (accept)}, \\ \alpha_k, & \text{if } \mathcal{R} < r_j \text{ (reject)}. \end{cases}$

The heart of this algorithm is its generation of a random spin configuration α_j (15.2) with probability

$$\mathcal{P}(E_{\alpha_j}, T) \propto e^{-E_{\alpha_j}/k_B T}. \tag{15.12}$$

The technique is a variation of von Neumann rejection (stone throwing in § 6.5) in which a random *trial* configuration is either accepted or rejected depending upon the value of the Boltzmann factor. Explicitly, the ratio of probabilities for a trial configuration of energy E_t to that of an initial configuration of energy E_i is

$$\mathcal{R} = \frac{\mathcal{P}_{\text{tr}}}{\mathcal{P}_i} = e^{-\Delta E/k_B T}, \quad \Delta E = E_{\alpha_{\text{tr}}} - E_{\alpha_i}. \tag{15.13}$$

[1] Large-scale, practical computations make a full sweep in which every spin is updated once and then use this as the new trial configuration. This is found to be more efficient and useful in removing some autocorrelations. (We thank G. Schneider for this observation.)

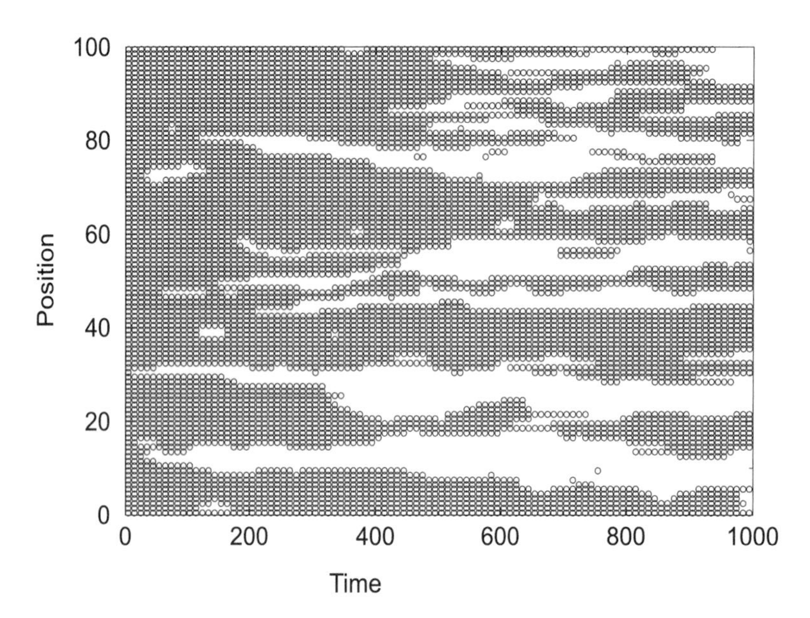

Figure 15.2 A 1-D lattice of 100 spins aligned along the ordinate. Up spins are indicated by circles, and down spins by blank spaces. The iteration number ("time") dependence of the spins is shown along the abscissa. Even though the system starts with all up spins (a "cold" start), the system is seen to form domains as it equilibrates.

If the trial configuration has a lower energy ($\Delta E \le 0$), the relative probability will be greater than 1 and we will accept the trial configuration as the new initial configuration without further ado. However, if the trial configuration has a higher energy ($\Delta E > 0$), we will not reject it out of hand but instead accept it with relative probability $\mathcal{R} = \exp(-\Delta E/k_B T) < 1$. To accept a configuration with a probability, we pick a uniform random number between 0 and 1, and if the probability is greater than this number, we accept the trial configuration; if the probability is smaller than the chosen random number, we reject it. (You can remember which way this goes by letting $E_{\alpha_{\text{tr}}} \to \infty$, in which case $\mathcal{P} \to 0$ and nothing is accepted.) When the trial configuration is rejected, the next configuration is identical to the preceding one.

How do you start? One possibility is to start with random values of the spins (a "hot" start). Another possibility (Figure 15.2) is a "cold" start in which you start with all spins parallel ($J > 0$) or antiparallel ($J < 0$). In general, one tries to remove the importance of the starting configuration by letting the calculation run a while ($\simeq 10N$ rearrangements) before calculating the equilibrium thermodynamic quantities. You should get similar results for hot, cold, or arbitrary starts, and by taking their average you remove some of the statistical fluctuations.

15.4.1 Metropolis Algorithm Implementation

1. Write a program that implements the Metropolis algorithm, that is, that produces a new configuration α_{k+1} from the present configuration α_k. (Alternatively, use the program `Ising.java` shown in Listing 15.1.)
2. Make the key data structure in your program an array `s[N]` containing the values of the spins s_i. For debugging, print out $+$ and $-$ to give the spin at each lattice point and examine the pattern for different trial numbers.
3. The value for the exchange energy J fixes the scale for energy. Keep it fixed at $J = 1$. (You may also wish to study antiferromagnets with $J = -1$, but first examine ferromagnets whose domains are easier to understand.)
4. The thermal energy k_BT is in units of J and is the independent variable. Use $k_BT = 1$ for debugging.
5. Use periodic boundary conditions on your chain to minimize end effects. This means that the chain is a circle with the first and last spins adjacent to each other.
6. Try $N \simeq 20$ for debugging, and larger values for production runs.
7. Use the printout to check that the system equilibrates for
   a. a totally ordered initial configuration (cold start); your simulation should resemble Figure 15.2.
   b. a random initial configuration (hot start). ▌

15.4.2 Equilibration, Thermodynamic Properties (Assessment)

1. Watch a chain of N atoms attain thermal equilibrium when in contact with a heat bath. At high temperatures, or for small numbers of atoms, you should see large fluctuations, while at lower temperatures you should see smaller fluctuations.
2. Look for evidence of instabilities in which there is a spontaneous flipping of a large number of spins. This becomes more likely for larger k_BT values.
3. Note how at thermal equilibrium the system is still quite dynamic, with spins flipping all the time. It is this energy exchange that determines the thermodynamic properties.
4. You may well find that simulations at small k_BT (say, $k_BT \simeq 0.1$ for $N = 200$) are slow to equilibrate. Higher k_BT values equilibrate faster yet have larger fluctuations.
5. Observe the formation of domains and the effect they have on the total energy. Regardless of the direction of spin within a domain, the atom–atom interactions are attractive and so contribute negative amounts to the energy of the system when aligned. However, the $\uparrow\downarrow$ or $\downarrow\uparrow$ interactions between domains contribute positive energy. Therefore you should expect a more negative energy at lower temperatures where there are larger and fewer domains.
6. Make a graph of average domain size *versus* temperature. ▌

```java
// Ising.java: 1-D Ising model with Metropolis algorithm
import java.io.*;                                              // Location of PrintWriter
import java.util.*;                                               // Location of Random

public class Ising {
  public static int N = 1000;                                        // Number of atoms
  public static double B = 1., mu = .33, J = .20,  k = 1., T = 100000000.;

  public static void main(String[] argv) throws IOException, FileNotFoundException {
    Random randnum = new Random(500767);                         // Seed random generator
    PrintWriter q   = new PrintWriter(new FileOutputStream ("ising.dat"), false);
    int i, j, M = 5000;                                         // Number of spin flips
    double[] state = new double[N]; double[] test = state;
    double ES = energy(state), p, ET;                             // State, test's energy
    for ( i=0 ;  i < N ;  i++ ) state[i] = -1.;                     // Set initial state
    for ( j=1 ;  j <= M ;  j++ ) {                               // Change state and test
      test = state;
      i = (int)(randnum.nextDouble()*(double)N);                       // Flip random atom
      test[i] *= -1.;
      ET = energy(test);
      p = Math.exp((ES-ET)/(k*T));
      if (p >= randnum.nextDouble())  { state = test; ES = ET;  }         // Test trial
      q.println(ES);                                           // Output energy to file
    }
    q.close();
  }

  public static double energy (double[] S)  {                    // Method to calc energy
    double FirstTerm = 0., SecondTerm = 0. ;
    int i;                                                            // Sum of energy
    for ( i=0 ;  i <= (N-2) ;  i++ )  FirstTerm += S[i]*S[i + 1];
    FirstTerm *= -J;
    for ( i=0 ;  i <= (N-1) ;  i++ )  SecondTerm += S[i];
    SecondTerm *= -B*mu;
    return (FirstTerm + SecondTerm);
} }
```

Listing 15.1 Ising.java implements the Metropolis algorithm for a 1-D Ising chain.

Thermodynamic Properties: For a given spin configuration α_j, the energy and magnetization are given by

$$E_{\alpha_j} = -J \sum_{i=1}^{N-1} s_i s_{i+1}, \quad \mathcal{M}_j = \sum_{i=1}^{N} s_i. \tag{15.14}$$

The internal energy $U(T)$ is just the average value of the energy,

$$U(T) = \langle E \rangle, \tag{15.15}$$

where the average is taken over a system in equilibrium. At high temperatures we expect a random assortment of spins and so a vanishing magnetization. At low temperatures when all the spins are aligned, we expect $\mathcal{M}$ to approach $N/2$. Although the specific heat can be computed from the elementary definition

$$C = \frac{1}{N}\frac{dU}{dT}, \tag{15.16}$$

the numerical differentiation may be inaccurate since U has statistical fluctuations. A better way to calculate the specific heat is to first calculate the fluctuations in

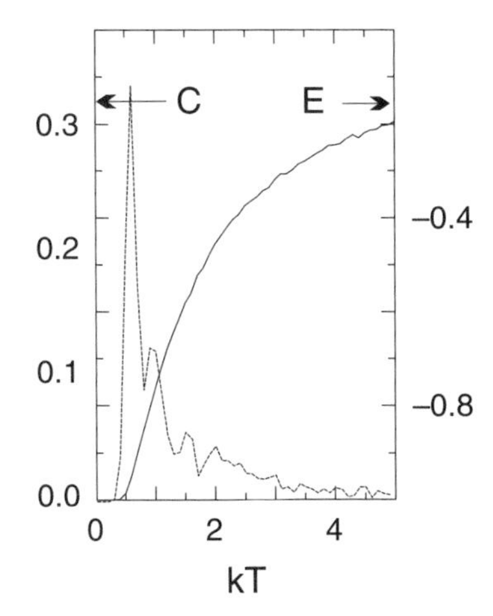

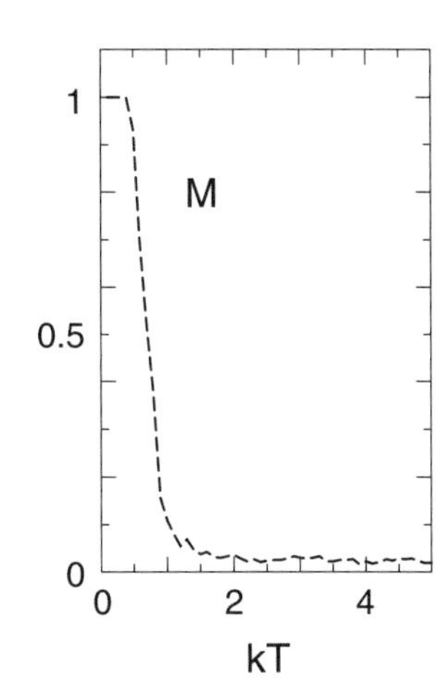

Figure 15.3 Simulation results for the energy, specific heat, and magnetization of a 1-D lattice of 100 spins as a function of temperature.

energy occurring during M trials and then determine the specific heat from the fluctuations:

$$U_2 = \frac{1}{M} \sum_{t=1}^{M} (E_t)^2, \tag{15.17}$$

$$C = \frac{1}{N^2} \frac{U_2 - (U)^2}{k_B T^2} = \frac{1}{N^2} \frac{\langle E^2 \rangle - \langle E \rangle^2}{k_B T^2}. \tag{15.18}$$

1. Extend your program to calculate the internal energy U and the magnetization $\mathcal{M}$ for the chain. Do not recalculate entire sums when only one spin changes.
2. Make sure to wait for your system to equilibrate before you calculate thermodynamic quantities. (You can check that U is fluctuating about its average.) Your results should resemble Figure 15.3.
3. Reduce statistical fluctuations by running the simulation a number of times with different seeds and taking the average of the results.
4. The simulations you run for small N may be realistic but may not agree with statistical mechanics, which assumes $N \simeq \infty$ (you may assume that $N \simeq 2000$ is close to infinity). Check that agreement with the analytic results for the thermodynamic limit is better for large N than small N.
5. Check that the simulated thermodynamic quantities are independent of initial conditions (within statistical uncertainties). In practice, your cold and hot start results should agree.
6. Make a plot of the internal energy U as a function of $k_B T$ and compare it to the analytic result (15.7).

7. Make a plot of the magnetization $\mathcal{M}$ as a function of k_BT and compare it to the analytic result. Does this agree with how you expect a heated magnet to behave?
8. Compute the energy fluctuations U_2 (15.17) and the specific heat C (15.18). Compare the simulated specific heat to the analytic result (15.8). ▌

15.4.3 Beyond Nearest Neighbors and 1-D (Exploration)

- Extend the model so that the spin–spin interaction (15.3) extends to next-nearest neighbors as well as nearest neighbors. For the ferromagnetic case, this should lead to more binding and less fluctuation because we have increased the couplings among spins and thus increased the thermal inertia.
- Extend the model so that the ferromagnetic spin–spin interaction (15.3) extends to nearest neighbors in two dimensions, and for the truly ambitious, three dimensions (the code `Ising3D.java` is available for instructors). Continue using periodic boundary conditions and keep the number of particles small, at least to start [G,T&C 06].

1. Form a square lattice and place $\sqrt{N}$ spins on each side.
2. Examine the mean energy and magnetization as the system equilibrates.
3. Is the temperature dependence of the average energy qualitatively different from that of the 1-D model?
4. Identify domains in the printout of spin configurations for small N.
5. Once your system appears to be behaving properly, calculate the heat capacity and magnetization of the 2-D Ising model with the same technique used for the 1-D model. Use a total number of particles of $100 \leq N \leq 2000$.
6. Look for a phase transition from ordered to unordered configurations by examining the heat capacity and magnetization as functions of temperature. The former should diverge, while the latter should vanish at the phase transition (Figure 15.4).

Exercise: Three fixed spin-$\frac{1}{2}$ particles interact with each other at temperature $T = 1/k_b$ such that the energy of the system is

$$E = -(s_1\, s_2 + s_2\, s_3).$$

The system starts in the configuration ↑↓↑. Do a simulation by hand that uses the Metropolis algorithm and the series of random numbers 0.5, 0.1, 0.9, 0.3 to determine the results of just two thermal fluctuations of these three spins. ▌

15.5 Unit II. Magnets via Wang–Landau Sampling ⊙

In Unit I we used a Boltzmann distribution to simulate the thermal properties of an Ising model. There we described the probabilities for explicit spin *states* α with

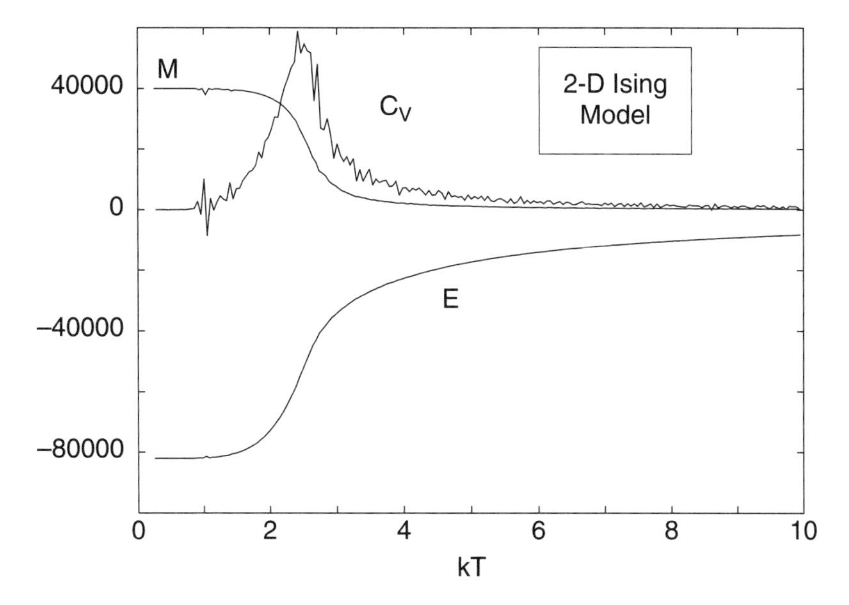

Figure 15.4 The energy, specific heat, and magnetization as a function of temperature for a 2-D lattice with 40,000 spins. The values of *C* and *E* have been scaled to fit on the same plot as *M*. (Courtesy of J. Wetzel.)

energy E_α for a system at temperature T and summed over various configurations. An equivalent formulation describes the probability that the system will have the explicit *energy* E at temperature T:

$$\mathcal{P}(E_i, T) = g(E_i)\,\frac{e^{-E_i/k_BT}}{Z(T)}, \qquad Z(T) = \sum_{E_i} g(E_i)\,e^{-E_i/k_BT}. \tag{15.19}$$

Here k_B is Boltzmann's constant, T is the temperature, $g(E_i)$ is the number of states of energy E_i $(i = 1, \ldots, M)$, $Z(T)$ is the partition function, and the sum is still over all M states of the system but now with states of the same energy entering just once owing to $g(E_i)$ accounting for their degeneracy. Because we again apply the theory to the Ising model with its discrete spin states, the energy also assumes only discrete values. If the physical system had an energy that varied continuously, then the number of states in the interval $E \to E + dE$ would be given by $g(E)\,dE$ and $g(E)$ would be called the *density of states*. As a matter of convenience, we call $g(E_i)$ the density of states even when dealing with discrete systems, although the term "degeneracy factor" may be more precise.

Even as the Metropolis algorithm has been providing excellent service for more than 50 years, recent literature shows increasing use of Wang–Landau sampling (WLS) [WL 04, Clark]. Because WLS determines the density of states and the associated partition function, it is not a direct substitute for the Metropolis algorithm and its simulation of thermal fluctuations. However, we will see that WLS provides an equivalent simulation for the Ising model.[2] Nevertheless, there are cases where the Metropolis algorithm can be used but WLS cannot, computation of the ground-state function of the harmonic oscillator (Unit III) being an example.

The advantages of WLS is that it requires much shorter simulation times than the Metropolis algorithm and provides a direct determination of $g(E_i)$. For these reasons it has shown itself to be particularly useful for first-order phase transitions where systems spend long times trapped in metastable states, as well as in areas as diverse as spin systems, fluids, liquid crystals, polymers, and proteins. The time required for a simulation becomes crucial when large systems are modeled. Even a spin lattice as small as 8×8 has $2^{64} \simeq 1.84 \times 10^{19}$ configurations, and it would be too expensive to visit them all.

In Unit I we ignored the partition function when employing the Metropolis algorithm. Now we focus on the partition function $Z(T)$ and the density-of-states function $g(E)$. Because $g(E)$ is a function of energy but not temperature, once it has been deduced, $Z(T)$ and all thermodynamic quantities can be calculated from it without having to repeat the simulation for each temperature. For example, the internal energy and the entropy are calculated directly as

$$U(T) \stackrel{\text{def}}{=} \langle E \rangle = \frac{\sum_{E_i} E_i\, g(E_i)\, e^{-E_i/k_BT}}{\sum_{E_i} g(E_i)\, e^{-E_i/k_BT}}, \tag{15.20}$$

$$S = k_BT \ln g(E_i). \tag{15.21}$$

The density of states $g(E_i)$ will be determined by taking the equivalent of a random walk in *energy* space. We flip a randomly chosen spin, record the energy of the new configuration, and then keep walking by flipping more spins to change the energy. The table $H(E_i)$ of the number of times each energy E_i is attained is called the energy *histogram* (an example of why it is called a histogram is given in Figure 15.5 on the right). If the walk were continued for a very long time, the histogram $H(E_i)$ would converge to the density of states $g(E_i)$. Yet with 10^{19}–10^{30} steps required even for small systems, this direct approach is unrealistically inefficient because the walk would rarely ever get away from the most probable energies.

Clever idea number 1 behind the Wang–Landau algorithm is to explore more of the energy space by increasing the likelihood of walking into less probable configurations. This is done by increasing the acceptance of less likely configurations while simultaneously decreasing the acceptance of more likely ones. In other words, we want to accept more states for which the density $g(E_i)$ is small and reject more states for which $g(E_i)$ is large (fret not these words, the equations are simple). To accomplish this trick, we accept a new energy E_i with a probability inversely

[2] We thank Oscar A. Restrepo of the Universidad de Antioquia for letting us use some of his material here.

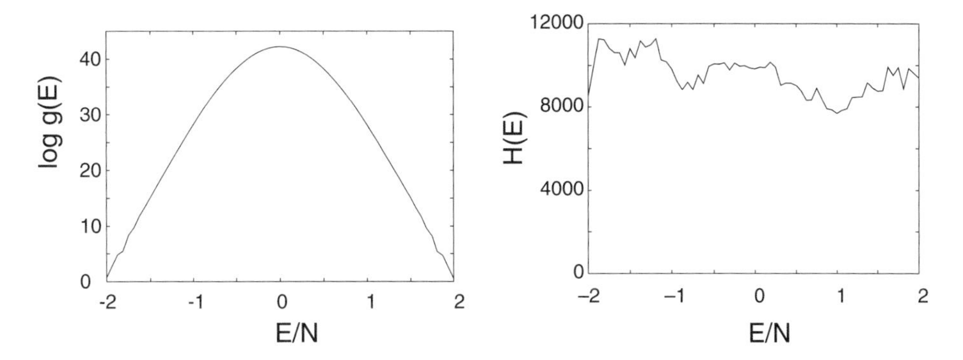

Figure 15.5 *Left:* Logarithm of the density of states log $g(E) \propto S$ *versus* the energy per particle for a 2-D Ising model on an 8×8 lattice. *Right:* The histogram $H(E)$ showing the number of states visited as a function of the energy per particle. The aim of WLS is to make this function flat.

proportional to the (initially unknown) density of states,

$$\mathcal{P}(E_i) = \frac{1}{g(E_i)}, \tag{15.22}$$

and then build up a histogram of visited states via a random walk.

The problem with clever idea number 1 is that $g(E_i)$ is unknown. WLS's clever idea 2 is to determine the unknown $g(E_i)$ simultaneously with the construction of the random walk. This is accomplished by improving the value of $g(E_i)$ via the multiplication $g(E_i) \rightarrow f\, g(E_i)$, where $f > 1$ is an empirical factor. When this works, the resulting histogram $H(E_i)$ becomes "flatter" because making the small $g(E_i)$ values larger makes it more likely to reach states with small $g(E_i)$ values. As the histogram gets flatter, we keep decreasing the multiplicative factor f until it is satisfactory close to 1. At that point we have a flat histogram and a determination of $g(E_i)$.

At this point you may be asking yourself, "Why does a flat histogram mean that we have determined $g(E_i)$?" Flat means that all energies are visited equally, in contrast to the peaked histogram that would be obtained normally without the $1/g(E_i)$ weighting factor. Thus, if by including this weighting factor we produce a flat histogram, then we have perfectly counteracted the actual peaking in $g(E_i)$, which means that we have arrived at the correct $g(E_i)$.

15.6 Wang-Landau Sampling

The steps in WLS are similar to those in the Metropolis algorithm, but now with use of the density-of-states function $g(E_i)$ rather than a Boltzmann factor:

1. Start with an arbitrary spin configuration $\alpha_k = \{s_1, s_2, \ldots, s_N\}$ and with arbitrary values for the density of states $g(E_i) = 1,\ i = 1, \ldots, M$, where $M = 2^N$ is the number of states of the system.
2. Generate a trial configuration α_{k+1} by
   a. picking a particle i randomly and
   b. flipping i's spin.
3. Calculate the energy $E_{\alpha_{\text{tr}}}$ of the trial configuration.
4. If $g(E_{\alpha_{\text{tr}}}) \leq g(E_{\alpha_k})$, accept the trial, that is, set $\alpha_{k+1} = \alpha_{\text{tr}}$.
5. If $g(E_{\alpha_{\text{tr}}}) > g(E_{\alpha_k})$, accept the trial with probability $\mathcal{P} = g(E_{\alpha_k})/(g(E_{\alpha_{\text{tr}}})$:
   a. choose a uniform random number $0 \leq r_i \leq 1$.
   b. set $\alpha_{k+1} = \begin{cases} \alpha_{\text{tr}}, & \text{if } \mathcal{P} \geq r_j \text{ (accept)}, \\ \alpha_k, & \text{if } \mathcal{P} < r_j \text{ (reject)}. \end{cases}$

   This acceptance rule can be expressed succinctly as

$$\mathcal{P}(E_{\alpha_k} \to E_{\alpha_{\text{tr}}}) = \min\left[1, \frac{g(E_{\alpha_k})}{g(E_{\alpha_{\text{tr}}})}\right], \tag{15.23}$$

   which manifestly always accepts low-density (improbable) states.
6. One we have a new state, we modify the current density of states $g(E_i)$ via the multiplicative factor f:

$$g(E_{\alpha_{k+1}}) \to f\, g(E_{\alpha_{k+1}}), \tag{15.24}$$

   and add 1 to the bin in the histogram corresponding to the new energy:

$$H(E_{\alpha_{k+1}}) \to H(E_{\alpha_{k+1}}) + 1. \tag{15.25}$$

7. The value of the multiplier f is empirical. We start with Euler's number $f = e = 2.71828$, which appears to strike a good balance between very large numbers of small steps (small f) and too rapid a set of jumps through energy space (large f). Because the entropy $S = k_B \ln g(E_i) \to k_B[\ln g(E_i) + \ln f]$, (15.24) corresponds to a uniform increase by k_B in entropy.
8. Even with reasonable values for f, the repeated multiplications in (15.24) lead to exponential growth in the magnitude of g. This may cause floating-point overflows and a concordant loss of information [in the end, the magnitude of $g(E_i)$ does not matter since the function is normalized]. These overflows are avoided by working with logarithms of the function values, in which case the update of the density of states (15.24) becomes

$$\ln g(E_i) \to \ln g(E_i) + \ln f. \tag{15.26}$$

9. The difficulty with storing $\ln g(E_i)$ is that we need the ratio of $g(E_i)$ values to calculate the probability in (15.23). This is circumvented by employing

the identity $x = \exp(\ln\, x)$ to express the ratio as

$$\frac{g(E_{\alpha_k})}{g(E_{\alpha_{\text{tr}}})} = \exp\left[\ln \frac{g(E_{\alpha_k})}{g(E_{\alpha_{\text{tr}}})}\right] = \exp\left[\ln\, g(E_{\alpha_k})\right] - \exp\left[\ln\, g(E_{\alpha_{\text{tr}}})\right]. \quad (15.27)$$

In turn, $g(E_k) = f \times g(E_k)$ is modified to $\ln\, g(E_k) \to \ln\, g(E_k) + \ln f$.

10. The random walk in E_i continues until a flat histogram of visited energy values is obtained. The flatness of the histogram is tested regularly (every 10,000 iterations), and the walk is terminated once the histogram is sufficiently flat. The value of f is then reduced so that the next walk provides a better approximation to $g(E_i)$. Flatness is measured by comparing the variance in $H(E_i)$ to its average. Although 90%–95% flatness can be achieved for small problems like ours, we demand only 80% (Figure 15.5):

$$\text{If}\quad \frac{H_{\text{max}} - H_{\text{min}}}{H_{\text{max}} + H_{\text{min}}} < 0.2, \text{stop, let}\ f \to \sqrt{f}\ (\ln\, f \to \ln\, f/2). \qquad (15.28)$$

11. Then keep the generated $g(E_i)$ and reset the histogram values $h(E_i)$ to zero.
12. The walks are terminated and new ones initiated until no significant correction to the density of states is obtained. This is measured by requiring the multiplicative factor $f \simeq 1$ within some level of tolerance; for example, $f \leq 1 + 10^{-8}$. If the algorithm is successful, the histogram should be flat (Figure 15.5) within the bounds set by (15.28).
13. The final step in the simulation is normalization of the deduced density of states $g(E_i)$. For the Ising model with N up or down spins, a normalization condition follows from knowledge of the total number of states [Clark]:

$$\sum_{E_i} g(E_i) = 2^N \quad \Rightarrow \quad g^{(\text{norm})}(E_i) = \frac{2^N}{\sum_{E_i} g(E_i)} g(E_i). \qquad (15.29)$$

Because the sum in (15.29) is most affected by those values of energy where $g(E_i)$ is large, it may not be precise for the low-E_i densities that contribute little to the sum. Accordingly, a more precise normalization, at least if your simulation has done a good job in occupying all energy states, is to require that there are just two grounds states with energies $E = -2N$ (one with all spins up and one with all spins down):

$$\sum_{E_i = -2N} g(E_i) = 2. \qquad (15.30)$$

In either case it is good practice to normalize $g(E_i)$ with one condition and then use the other as a check.

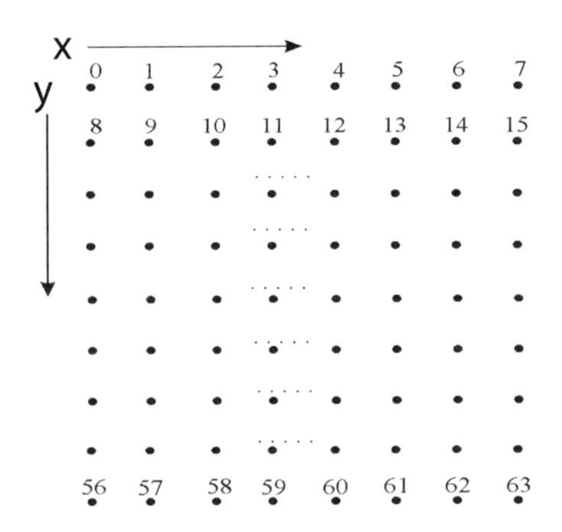

Figure 15.6 The numbering scheme used for our 8×8 2-D lattice of spins.

15.6.1 WLS Ising Model Implementation

We assume an Ising model with spin–spin interactions between nearest neighbors located in an $L \times L$ lattice (Figure 15.6). To keep the notation simple, we set $J = 1$ so that

$$E = -\sum_{i \leftrightarrow j}^{N} \sigma_i \sigma_j, \tag{15.31}$$

where $\leftrightarrow$ indicates nearest neighbors. Rather than recalculate the energy each time a spin is flipped, only the difference in energy is computed. For example, for eight spins in a 1-D array,

$$-E_k = \sigma_0\sigma_1 + \sigma_1\sigma_2 + \sigma_2\sigma_3 + \sigma_3\sigma_4 + \sigma_4\sigma_5 + \sigma_5\sigma_6 + \sigma_6\sigma_7 + \sigma_7\sigma_0, \tag{15.32}$$

where the 0–7 interaction arises because we assume periodic boundary conditions. If spin 5 is flipped, the new energy is

$$-E_{k+1} = \sigma_0\sigma_1 + \sigma_1\sigma_2 + \sigma_2\sigma_3 + \sigma_3\sigma_4 - \sigma_4\sigma_5 - \sigma_5\sigma_6 + \sigma_6\sigma_7 + \sigma_7\sigma_0, \tag{15.33}$$

and the difference in energy is

$$\Delta E = E_{k+1} - E_k = 2(\sigma_4 + \sigma_6)\sigma_5. \tag{15.34}$$

This is cheaper to compute than calculating and then subtracting two energies.

When we advance to two dimensions with the 8×8 lattice in Figure 15.6, the change in energy when spin $\sigma_{i,j}$ flips is

$$\Delta E = 2\sigma_{i,j}(\sigma_{i+1,j} + \sigma_{i-1,j} + \sigma_{i,j+1} + \sigma_{i,j-1}), \tag{15.35}$$

which can assume the values $-8, -4, 0, 4$, and 8. There are two states of minimum energy $-2N$ for a 2-D system with N spins, and they correspond to all spins pointing in the same direction (either up or down). The maximum energy is $+2N$, and it corresponds to alternating spin directions on neighboring sites. Each spin flip on the lattice changes the energy by four units between these limits, and so the values of the energies are

$$E_i = -2N, \quad -2N+4, \quad -2N+8, \ldots, 2N-8, \quad 2N-4, \quad 2N. \tag{15.36}$$

These energies can be stored in a uniform 1-D array via the simple mapping

$$E' = (E+2N)/4 \quad \Rightarrow \quad E' = 0,\ 1,\ 2,\ \ldots\ ,\ N. \tag{15.37}$$

Listing 15.2 displays our implementation of Wang–Landau sampling to calculate the density of states and internal energy $U(T)$ (15.20). We used it to obtain the entropy $S(T)$ and the energy histogram $H(E_i)$ illustrated in Figure 15.5. Other thermodynamic functions can be obtained by replacing the E in (15.20) with the appropriate variable. The results look like those in Figure 15.4. A problem that may be encountered when calculating these variables is that the sums in (15.20) can become large enough to cause overflows, even though the ratio would not. You work around that by factoring out a common large factor; for example,

$$\sum_{E_i} X(E_i)\, g(E_i)\, e^{-E_i/k_BT} = e^{\lambda} \sum_{E_i} X(E_i)\, e^{\ln g(E_i) - E_i/k_BT - \lambda}, \tag{15.38}$$

where λ is the largest value of $\ln g(E_i) - E_i/k_BT$ at each temperature. The factor e^{λ} does not actually need to be included in the calculation of the variable because it is present in both the numerator and denominator and so cancels out.

```java
// Wang Landau algorithm for 2-D spin system
// Author: Oscar A. Restrepo, Universidad de Antioquia, Medellin, Colombia

import java.io.*;

public class WangLandau {
  public static int L = 8, N =(L*L), sp[][] = new int[L][L];          // Grid size, spins
  public static int hist[] = new int[N+1], prhist[] = new int[N+1];       // Histograms
  public static double S[]         = new double[N+1];                // Entropy = log g(E)
  public static int iE(int e)     { return (e+2*N)/4; }

  public static void WL()   {
    double fac, Hinf=1.e10, Hsup=0., deltaS, tol=1.e-8;          // f, hist, entropy, tol
    double height, ave, percent;                           // Hist height, avg hist, % < 20%
    int i, xg, yg, j, iter, Eold, Enew;                         // Grid positions, energies
    int ip[] = new int[L], im[] = new int[L];                       // BC R or down, L or up
    height = Math.abs(Hsup-Hinf)/2.;                               // Initialize histogram
    ave = (Hsup+Hinf) / 2.;                                   // about average of histogram
    percent = height / ave;
    for (i=0; i<L; i++) for(j=0; j<L; j++) sp[i][j] = 1;                   // Initial spins
    for (i=0; i<L; i++)  { ip[i] = i+1;   im[i] = i-1; }                 // Case plus, mimus
    ip[L-1] = 0;  im[0] = L-1;                                                   // Borders
    Eold = -2*N;                                                        // Initialize energy
    for ( j = 0; j<=N; j++ ) S[j] = 0;                                // Entropy initialized
    iter = 0; fac = 1;                                                // Initial factor ln e
    while ( fac > tol )  {
      iter++;
      i = (int)(Math.random()*N);                                      // Select random spin
```

```
    xg = i%L;        yg = i/L;                                   // Localize grid point
    Enew = Eold + 2*( sp[ip[xg]][yg] + sp[im[xg]][yg] + sp[xg][ip[yg]]
           + sp[xg][im[yg]] ) * sp[xg][yg];                           // Change energy
    deltaS = S[iE(Enew)] - S[iE(Eold)];                              // Change entropy
    if ( deltaS <= 0 || Math.random() < Math.exp(-deltaS) ) { Eold = Enew;
                                                   sp[xg][yg] *= -1; }   // Flip spin
    hist[iE(Eold)]++;                                 // Change histogram, add 1, update
    S[iE(Eold)] += fac;                                              // Change entropy
    if( iter%10000 == 0) {                          // Check flatness every 10000 sweeps
      for(j = 0; j <= N; j++)  {
        if ( j == 0 )  { Hsup = 0;   Hinf = 1e10; }        // Initialize new histogram
        if ( hist[j] == 0 )  continue;                      // Energies never visited
        if ( hist[j] > Hsup ) Hsup = hist[j];
        if ( hist[j] < Hinf ) Hinf = hist[j];
      }
      height = Hsup-Hinf;
      ave = Hsup+Hinf;
      percent = height/ave;
      if( percent < 0.2 )  {                                         // Histogram flat?
        System.out.println(" iter "+iter +"   log(f) "+fac );
        for( j=0; j<=N; j++ )  {prhist[j] = hist[j];  hist[j] = 0;}      // Save hist
        fac *= 0.5;                                      // Equivalent to log(sqrt(f))
} } } }

  public static void IntEnergy()throws IOException, FileNotFoundException   {
    double T, maxL, Ener, U, sumdeno, sumnume, exponent = 0;        // Temp, max lambda
    int i;
    PrintWriter b  = new PrintWriter(new FileOutputStream ("IntEnergy.dat"), true);
    for( T = 0.2; T <= 8.0; T += 0.2 ) {                            // Select lambda max
      Ener = -2*N;
      maxL = 0.0;                                                          // Initialize
      for( i=0; i<N; i++ ) { if (S[i]!=0 && (S[i]-Ener/T)>maxL) maxL = S[i]-Ener/T;
                             Ener = Ener+4; }
      sumdeno = sumnume = 0;
      Ener = -2*N;
      for( i=0; i<N; i++ ){
        if( S[i] != 0) exponent = S[i]-Ener/T-maxL;
        sumnume += Ener*Math.exp(exponent);
        sumdeno += Math.exp(exponent);
        Ener = Ener+ 4.0;
      }
      U = sumnume/sumdeno/N;
      b.println("  "+T+"   "+U);
      System.out.println("Output data in IntEnergy.dat");
    }
  }

  public static void  main(String[] argv)throws IOException, FileNotFoundException  {
    PrintWriter q  = new PrintWriter( new FileOutputStream("wanglandau.dat"), true );
    int order, j ;
    double deltaS = 0.0;
    WL();                                                  // Call Wang Landau algorithm
    for( j=0; j <= N; j++ )  {
      order = j*4 - 2*N;
      deltaS = S[j] - S[0] + Math.log(2);
      if( S[j] != 0 ) q.println("   "+ 1.*order/N +"    " + deltaS+"   "+ prhist[j]);
    }
    IntEnergy();
    System.out.println("Output data in wanglandau.dat");
} }                                                                      // Main, class
```

Listing 15.2 WangLandau.java simulates the 2-D Ising model using Wang-Landau sampling to compute the density of states and from that the internal energy.

15.6.2 WLS Ising Model Assessment

Repeat the assessment conducted in §15.4.2 for the thermodynamic properties of the Ising model but use WLS in place of the Metropolis algorithm.

15.7 Unit III. Feynman Path Integrals ⊙

Problem: As is well known, a classical particle attached to a linear spring undergoes simple harmonic motion with a position in space as a function of time given by $x(t) = A\sin(\omega_0 t + \phi)$. Your **problem** is to take this classical space-time trajectory $x(t)$ and use it to generate the quantum wave function $\psi(x,t)$ for a particle bound in a harmonic oscillator potential.

15.8 Feynman's Space-Time Propagation (Theory)

Feynman was looking for a formulation of quantum mechanics that gave a more direct connection to classical mechanics than does Schrödinger theory and that made the statistical nature of quantum mechanics evident from the start. He followed a suggestion by Dirac that Hamilton's principle of least action, which can be used to derive classical mechanics, may be the $\hbar \to 0$ limit of a quantum least-action principle. Seeing that Hamilton's principle deals with the paths of particles through space-time, Feynman posultated that the quantum wave function describing the propagation of a free particle from the space-time point $a = (x_a{,}t_a)$ to the point $b = (x_b,\ t_b)$ can expressed as [F&H 65]

$$\psi(x_b, t_b) = \int dx_a\, G(x_b, t_b; x_a, t_a)\psi(x_a, t_a), \tag{15.39}$$

where G is the *Green's function* or *propagator*

$$G(x_b, t_b; x_a, t_a) \equiv G(b, a) = \sqrt{\frac{m}{2\pi i(t_b - t_a)}}\exp\left[i\frac{m(x_b - x_a)^2}{2(t_b - t_a)}\right]. \tag{15.40}$$

Equation (15.39) is a form of Huygens's wavelet principle in which each point on the wavefront $\psi(x_a, t_a)$ emits a spherical wavelet $G(b; a)$ that propagates forward in space and time. It states that a new wavefront $\psi(x_b, t_b)$ is created by summation over and interference with all the other wavelets.

Feynman imagined that another way of interpreting (15.39) is as a form of Hamilton's principle in which the probability amplitude (wave function ψ) for a particle to be at B is equal to the sum over all *paths* through space-time originating at time A and ending at B (Figure 15.7). This view incorporates the statistical nature of quantum mechanics by having different probabilities for travel along

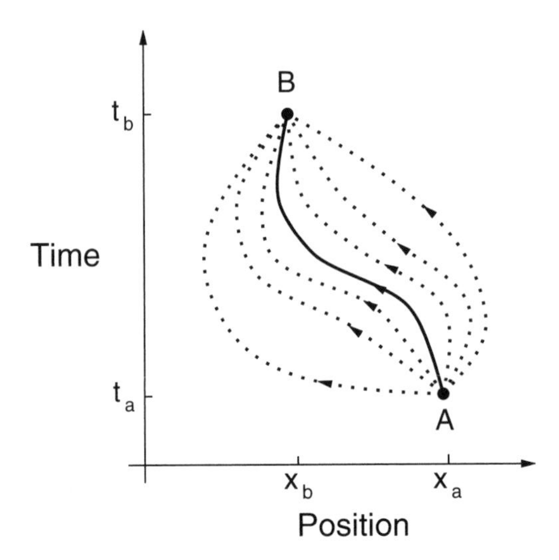

Figure 15.7 A collection of paths connecting the initial space-time point A to the final point B. The solid line is the trajectory followed by a classical particle, while the dashed lines are additional paths sampled by a quantum particle. A classical particle somehow "knows" ahead of time that travel along the classical trajectory minimizes the action S.

the different paths. All paths are possible, but some are more likely than others. (When you realize that Schrödinger theory solves for wave functions and considers paths a classical concept, you can appreciate how different it is from Feynman's view.) The values for the probabilities of the paths derive from *Hamilton's classical principle of least action*:

> The most general motion of a physical particle moving along the classical trajectory $\bar{x}(t)$ from time t_a to t_b is along a path such that the action $S[\bar{x}(t)]$ is an extremum:
>
> $$\delta S[\bar{x}(t)] = S[\bar{x}(t) + \delta x(t)] - S[\bar{x}(t)] = 0, \tag{15.41}$$
>
> with the paths constrained to pass through the endpoints:
>
> $$\delta(x_a) = \delta(x_b) = 0.$$

This formulation of classical mechanics, which is based on the calculus of variations, is equivalent to Newton's differential equations if the action S is taken as the line integral of the Lagrangian along the path:

$$S[\bar{x}(t)] = \int_{t_a}^{t_b} dt\, L\left[x(t), \dot{x}(t)\right], \quad L = T\left[x, \dot{x}\right] - V[x]. \tag{15.42}$$

Here T is the kinetic energy, V is the potential energy, $\dot{x} = dx/dt$, and square brackets indicate a *functional*[3] of the function $x(t)$ and $\dot{x}(t)$.

Feynman observed that the classical action for a free particle ($V = 0$),

$$S[b,a] = \frac{m}{2}\left(\dot{x}\right)^2 (t_b - t_a) = \frac{m}{2}\frac{(x_b - x_a)^2}{t_b - t_a}, \tag{15.43}$$

is related to the free-particle propagator (15.40) by

$$G(b,a) = \sqrt{\frac{m}{2\pi i(t_b - t_a)}}\, e^{iS[b,a]/\hbar}. \tag{15.44}$$

This is the much sought connection between quantum mechanics and Hamilton's principle. Feynman then postulated a reformulation of quantum mechanics that incorporated its statistical aspects by expressing $G(b,a)$ as the weighted sum over all *paths* connecting a to b,

$$\boxed{G(b,a) = \sum_{\text{paths}} e^{iS[b,a]/\hbar}} \quad \text{(path integral)}. \tag{15.45}$$

Here the classical action S (15.42) is evaluated along different paths (Figure 15.7), and the exponential of the action is summed over paths. The sum (15.45) is called a *path integral* because it sums over actions $S[b,a]$, each of which is an integral (on the computer an integral and a sum are the same anyway). The essential connection between classical and quantum mechanics is the realization that in units of $\hbar \simeq 10^{-34}$ Js, the action is a very large number, $S/\hbar \geq 10^{20}$, and so even though all paths enter into the sum (15.45), the main contributions come from those paths adjacent to the classical trajectory $\bar{x}$. In fact, because S is an extremum for the classical trajectory, it is a constant to first order in the variation of paths, and so nearby paths have phases that vary smoothly and relatively slowly. In contrast, paths far from the classical trajectory are weighted by a rapidly oscillating $\exp(iS/\hbar)$, and when many are included, they tend to cancel each other out. In the classical limit, $\hbar \to 0$, only the single classical trajectory contributes and (15.45) becomes Hamilton's principle of least action! In Figure 15.8 we show an example of a trajectory used in path-integral calculations.

[3] A *functional* is a number whose value depends on the complete behavior of some function and not just on its behavior at one point. For example, the derivative $f'(x)$ depends on the value of f at x, yet the integral $I[f] = \int_a^b dx\, f(x)$ depends on the entire function and is therefore a functional of f.

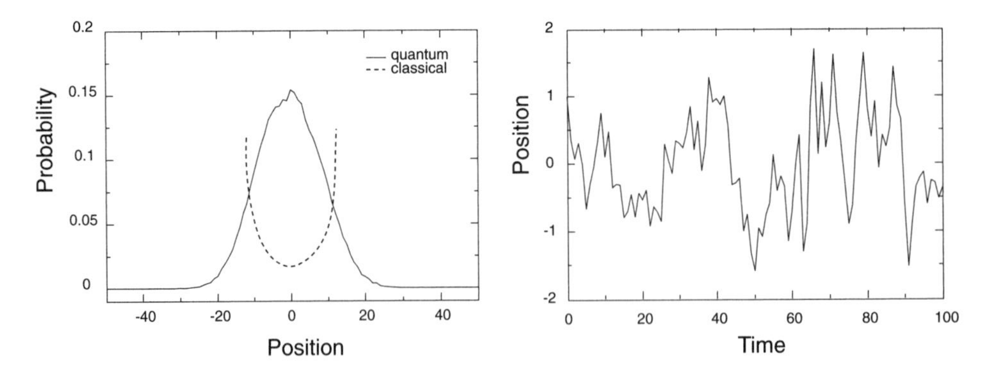

Figure 15.8 *Left:* The probability distribution for the harmonic oscillator ground state as determined with a path-integral calculation (the classical result has maxima at the two turning points). *Right:* A space-time trajectory used as a quantum path.

15.8.1 Bound-State Wave Function (Theory)

Although you may be thinking that you have already seen enough expressions for the Green's function, there is yet another one we need for our computation. Let us assume that the Hamiltonian $\tilde{H}$ supports a spectrum of eigenfunctions,

$$\tilde{H}\psi_n = E_n\psi_n,$$

each labeled by the index n. Because $\tilde{H}$ is Hermitian, the solutions form a complete orthonormal set in which we may expand a general solution:

$$\psi(x,t) = \sum_{n=0}^{\infty} c_n\, e^{-iE_n t}\,\psi_n(x), \quad c_n = \int_{-\infty}^{+\infty} dx\, \psi_n^*(x)\psi(x, t=0), \tag{15.46}$$

where the value for the expansion coefficients c_n follows from the orthonormality of the ψ_n's. If we substitute this c_n back into the wave function expansion (15.46), we obtain the identity

$$\psi(x,t) = \int_{-\infty}^{+\infty} dx_0 \sum_n \psi_n^*(x_0)\psi_n(x)e^{-iE_n t}\psi(x_0, t=0). \tag{15.47}$$

Comparison with (15.39) yields the eigenfunction expansion for G:

$$G(x,t;x_0,t_0=0) = \sum_n \psi_n^*(x_0)\psi_n(x)e^{-iE_n t}. \tag{15.48}$$

We relate this to the bound-state wave function (recall that our **problem** is to calculate that) by (1) requiring all paths to start and end at the space position $x_0 = x$, (2) by taking $t_0 = 0$, and (3) by making an analytic continuation of (15.48) to negative imaginary time (permissable for analytic functions):

$$G(x,-i\tau;x,0)=\sum_n |\psi_n(x)|^2 e^{-E_n\tau}=|\psi_0|^2 e^{-E_0\tau}+|\psi_1|^2 e^{-E_1\tau}+\cdots$$

$$\Rightarrow \quad \boxed{|\psi_0(x)|^2=\lim_{\tau\to\infty} e^{E_0\tau}G(x,-i\tau;x,0)}. \tag{15.49}$$

The limit here corresponds to long imaginary times τ, after which the parts of ψ with higher energies decay more quickly, leaving only the ground state ψ_0.

Equation (15.49) provides a closed-form solution for the ground-state wave function directly in terms of the propagator G. Although we will soon describe how to compute this equation, look now at Figure 15.8 showing some results of a computation. Although we start with a probability distribution that peaks near the classical turning points at the edges of the well, after a large number of iterations we end up with a distribution that resembles the expected Gaussian. On the right we see a trajectory that has been generated via statistical variations about the classical trajectory $x(t)=A\sin(\omega_0 t+\phi)$.

15.8.2 Lattice Path Integration (Algorithm)

Because both time and space are integrated over when evaluating a path integral, we set up a lattice of discrete points in space-time and visualize a particle's trajectory as a series of straight lines connecting one time to the next (Figure 15.9). We divide the time between points A and B into N equal steps of size ε and label them with the index j:

$$\varepsilon \stackrel{\text{def}}{=} \frac{t_b-t_a}{N} \quad \Rightarrow \quad t_j=t_a+j\varepsilon \quad (j=0,N). \tag{15.50}$$

Although it is more precise to use the actual positions $x(t_j)$ of the trajectory at the times t_j to determine the x_js (as in Figure 15.9), in practice we discretize space uniformly and have the links end at the nearest regular points. Once we have a lattice, it is easy to evaluate derivatives or integrals on a link[4]:

$$\frac{dx_j}{dt}\simeq\frac{x_j-x_{j-1}}{t_j-t_{j-1}}=\frac{x_j-x_{j-1}}{\varepsilon}, \tag{15.51}$$

$$S_j\simeq L_j\,\Delta t\simeq\frac{1}{2}m\frac{(x_j-x_{j-1})^2}{\varepsilon}-V(x_j)\varepsilon, \tag{15.52}$$

where we have assumed that the Lagrangian is constant over each link.

[4] Even though Euler's rule has a large error, it is often use in lattice calculations because of its simplicity. However, if the Lagrangian contains second derivatives, you should use the more precise central-difference method to avoid singularities.

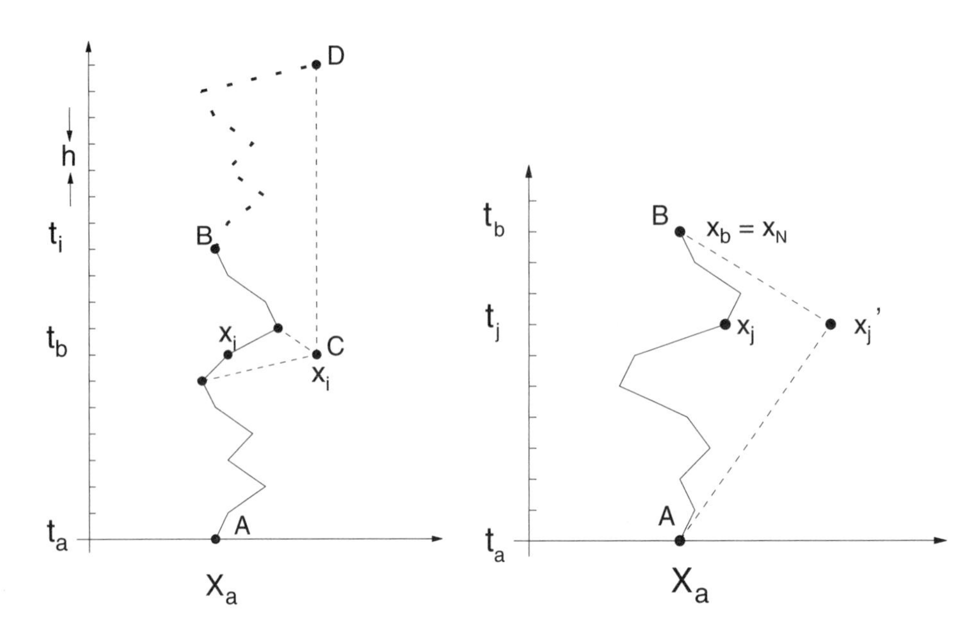

Figure 15.9 *Left:* A path through the space-time lattice that starts and ends at $x_a = x_b$. The action is an integral over this path, while the *path integral* is a sum of integrals over all paths. The dotted path *BD* is a transposed replica of path *AC*. *Right:* The dashed path joins the initial and final times in two equal time steps; the solid curve uses *N* steps each of size ε. The position of the curve at time t_j defines the position x_j.

Lattice path integration is based on the *composition theorem* for propagators:

$$G(b,a) = \int dx_j \, G(x_b, t_b; x_j, t_j) G(x_j, t_j; x_a, t_a) \quad (t_a < t_j,\ t_j < t_b). \tag{15.53}$$

For a free particle this yields

$$\begin{aligned} G(b,a) &= \sqrt{\frac{m}{2\pi i(t_b - t_j)}} \sqrt{\frac{m}{2\pi i(t_j - t_a)}} \int dx_j \, e^{i(S[b,j]+S[j,a])} \\ &= \sqrt{\frac{m}{2\pi i(t_b - t_a)}} \int dx_j \, e^{iS[b,a]}, \end{aligned} \tag{15.54}$$

where we have added the actions since line integrals combine as $S[b,j] + S[j,a] = S[b,a]$. For the N-linked path in Figure 15.9, equation (15.53) becomes

$$G(b,a) = \int dx_1 \cdots dx_{N-1} \, e^{iS[b,a]}, \quad S[b,a] = \sum_{j=1}^{N} S_j, \tag{15.55}$$

where S_j is the value of the action for link j. At this point the integral over the *single* path shown in Figure 15.9 has become an N-term sum that becomes an infinite sum as the time step ε approaches zero.

To summarize, Feynman's path-integral postulate (15.45) means that we sum over all paths connecting A to B to obtain Green's function $G(b, a)$. This means that we must sum not only over the links in one path but *also* over all the different paths in order to produce the variation in paths required by Hamilton's principle. The sum is constrained such that paths must pass through A and B and cannot double back on themselves (causality requires that particles move only forward in time). This is the essence of *path integration*. Because we are integrating over functions as well as along paths, the technique is also known as *functional integration*.

The propagator (15.45) is the sum over all paths connecting A to B, with each path weighted by the exponential of the action along that path, explicitly:

$$G(x,t;x_0,t_0) = \sum \!\!\!\!\!\! \int dx_1\, dx_2 \cdots dx_{N-1} e^{iS[x,x_0]}, \tag{15.56}$$

$$S[x,x_0] = \sum_{j=1}^{N-1} S[x_{j+1},x_j] \simeq \sum_{j=1}^{N-1} L\left(x_j,\dot{x}_j\right)\varepsilon, \tag{15.57}$$

where $L(x_j, \dot{x}_j)$ is the average value of the Lagrangian on link j at time $t = j\varepsilon$. The computation is made simpler by assuming that the potential $V(x)$ is independent of velocity and does not depend on other x values (local potential). Next we observe that G is evaluated with a negative imaginary time in the expression (15.49) for the ground-state wave function. Accordingly, we evaluate the Lagrangian with $t = -i\tau$:

$$L\left(x,\dot{x}\right) = T - V(x) = +\frac{1}{2}m\left(\frac{dx}{dt}\right)^2 - V(x), \tag{15.58}$$

$$\Rightarrow \quad L\left(x, \frac{dx}{-id\tau}\right) = -\frac{1}{2}m\left(\frac{dx}{d\tau}\right)^2 - V(x). \tag{15.59}$$

We see that the reversal of the sign of the kinetic energy in L means that L now equals the negative of the Hamiltonian evaluated at a real positive time $t = \tau$:

$$H\left(x, \frac{dx}{d\tau}\right) = \frac{1}{2}m\left(\frac{dx}{d\tau}\right)^2 + V(x) = E, \tag{15.60}$$

$$\Rightarrow \quad L\left(x, \frac{dx}{-id\tau}\right) = -H\left(x, \frac{dx}{d\tau}\right). \tag{15.61}$$

In this way we rewrite the t-path integral of L as a τ-path integral of H and so express the action and Green's function in terms of the Hamiltonian:

$$S[j+1,j] = \int_{t_j}^{t_{j+1}} L(x,t)\,dt = -i\int_{\tau_j}^{\tau_{j+1}} H(x,\tau)\,d\tau, \tag{15.62}$$

$$\Rightarrow \quad G(x,-i\tau;x_0,0) = \int dx_1 \ldots dx_{N-1}\, e^{-\int_0^\tau H(\tau')d\tau'}, \tag{15.63}$$

where the line integral of H is over an entire trajectory. Next we express the path integral in terms of the average energy of the particle on each link, $E_j = T_j + V_j$, and then sum over links[5] to obtain the summed energy $\mathcal{E}$:

$$\int H(\tau)\,d\tau \;\simeq\; \sum_j \varepsilon E_j = \varepsilon\mathcal{E}(\{x_j\}), \tag{15.64}$$

$$\mathcal{E}(\{x_j\}) \stackrel{\text{def}}{=} \sum_{j=1}^{N}\left[\frac{m}{2}\left(\frac{x_j - x_{j-1}}{\varepsilon}\right)^2 + V\left(\frac{x_j + x_{j-1}}{2}\right)\right]. \tag{15.65}$$

In (15.65) we have approximated each path link as a *straight line*, used Euler's derivative rule to obtain the velocity, and evaluated the potential at the midpoint of each link. We now substitute this G into our solution (15.49) for the ground-state wave function in which the initial and final points in space are the same:

$$\lim_{\tau\to\infty}\frac{G(x,-i\tau,x_0=x,0)}{\int dx\,G(x,-i\tau,x_0=x,0)} = \frac{\int dx_1\cdots dx_{N-1}\exp\left[-\int_0^\tau H\,d\tau'\right]}{\int dx\,dx_1\cdots dx_{N-1}\exp\left[-\int_0^\tau H\,d\tau'\right]}$$

$$\Rightarrow \quad |\psi_0(x)|^2 = \frac{1}{Z}\lim_{\tau\to\infty}\int dx_1\cdots dx_{N-1}\,e^{-\varepsilon\mathcal{E}}, \tag{15.66}$$

$$Z = \lim_{\tau\to\infty}\int dx\,dx_1\cdots dx_{N-1}e^{-\varepsilon\mathcal{E}}. \tag{15.67}$$

The similarity of these expressions to thermodynamics, even with a partition function Z, is no accident; by making the time parameter of quantum mechanics imaginary, we have converted the time-dependent Schrödinger equation to the heat diffusion equation:

$$i\frac{\partial\psi}{\partial(-i\tau)} = \frac{-\nabla^2}{2m}\psi \quad\Rightarrow\quad \frac{\partial\psi}{\partial\tau} = \frac{\nabla^2}{2m}\psi. \tag{15.68}$$

[5] In some cases, such as for an infinite square well, this can cause problems if the trial link causes the energy to be infinite. In that case, one can modify the algorithm to use the potential at the beginning of a link

It is not surprising then that the sum over paths in Green's function has each path weighted by the Boltzmann factor $\mathcal{P} = e^{-\varepsilon\mathcal{E}}$ usually associated with thermodynamics. We make the connection complete by identifying the temperature with the inverse time step:

$$\mathcal{P} = e^{-\varepsilon\mathcal{E}} = e^{-\mathcal{E}/k_B T} \quad \Rightarrow \quad k_B T = \frac{1}{\varepsilon} \equiv \frac{\hbar}{\varepsilon}. \tag{15.69}$$

Consequently, the $\varepsilon \to 0$ limit, which makes time continuous, is a "high-temperature" limit. The $\tau \to \infty$ limit, which is required to project the ground-state wave function, means that we must integrate over a path that is long in imaginary time, that is, long compared to a typical time $\hbar/\Delta E$. Just as our simulation of the Ising model in Unit I required us to wait a long time while the system equilibrated, so the present simulation requires us to wait a long time so that all but the ground-state wave function has decayed. Alas, this is the solution to our **problem** of finding the ground-state wave function.

To summarize, we have expressed the Green's function as a path integral that requires integration of the Hamiltonian along paths and a summation over all the paths (15.66). We evaluate this path integral as the sum over all the trajectories in a space-time lattice. Each trial path occurs with a probability based on its action, and we use the Metropolis algorithm to include statistical fluctuation in the links, as if they are in thermal equilibrium. This is similar to our work with the Ising model in Unit I, however now, rather than reject or accept a flip in spin based on the change in energy, we reject or accept a change in a link based on the change in energy. The more iterations we let the algorithm run for, the more time the deduced wave function has to equilibrate to the ground state.

In general, Monte Carlo Green's function techniques work best if we start with a good guess at the correct answer and have the algorithm calculate variations on our guess. For the present problem this means that if we start with a path in space-time close to the classical trajectory, the algorithm may be expected to do a good job at simulating the quantum fluctuations about the classical trajectory. However, it does not appear to be good at finding the classical trajectory from arbitrary locations in space-time. We suspect that the latter arises from $\delta S/\hbar$ being so large that the weighting factor $\exp(\delta S/\hbar)$ fluctuates wildly (essentially averaging out to zero) and so loses its sensitivity.

15.8.2.1 A TIME-SAVING TRICK

As we have formulated the computation, we pick a value of x and perform an expensive computation of line integrals over all space and time to obtain $|\psi_0(x)|^2$ at one x. To obtain the wave function at another x, the entire simulation must be repeated from scratch. Rather than go through all that trouble again and again, we will compute the entire x dependence of the wave function in one fell swoop. The trick is to insert a delta function into the probability integral (15.66), thereby fixing the initial position to be x_0, and then to integrate over all values for x_0:

$$|\psi_0(x)|^2 = \int dx_1 \cdots dx_N \, e^{-\varepsilon\mathcal{E}(x,x_1,\ldots)} \tag{15.70}$$

$$= \int dx_0 \cdots dx_N \delta(x - x_0) \, e^{-\varepsilon\mathcal{E}(x,x_1,\ldots)}. \tag{15.71}$$

This equation expresses the wave function as an average of a delta function over all paths, a procedure that might appear totally inappropriate for numerical computation because there is tremendous error in representing a singular function on a finite-word-length computer. Yet when we simulate the sum over all paths with (15.71), there will always be some x value for which the integral is nonzero, and we need to accumulate only the solution for various (discrete) x values to determine $|\psi_0(x)|^2$ for all x.

To understand how this works in practice, consider path AB in Figure 15.9 for which we have just calculated the summed energy. We form a new path by having one point on the chain jump to point C (which changes two links). If we replicate section AC and use it as the extension AD to form the top path, we see that the path CBD has the same summed energy (action) as path ACB and in this way can be used to determine $|\psi(x'_j)|^2$. That being the case, once the system is equilibrated, we determine new values of the wave function at new locations x'_j by flipping links to new values and calculating new actions. The more frequently some x_j is accepted, the greater the wave function at that point.

15.8.3 Lattice Implementation

The program QMC.java in Listing 15.3 evaluates the integral (15.45) by finding the average of the integrand $\delta(x_0 - x)$ with paths distributed according to the weighting function $\exp[-\varepsilon\mathcal{E}(x_0, x_1, \ldots, x_N)]$. The physics enters via (15.73), the calculation of the summed energy $\mathcal{E}(x_0, x_1, \ldots, x_N)$. We evaluate the action integral for the harmonic oscillator potential

$$V(x) = \frac{1}{2}x^2 \tag{15.72}$$

and for a particle of mass $m = 1$. Using a convenient set of natural units, we measure lengths in $\sqrt{1/m\omega} \equiv \sqrt{\hbar/m\omega} = 1$ and times in $1/\omega = 1$. Correspondingly, the oscillator has a period $T = 2\pi$. Figure 15.8 shows results from an application of the Metropolis algorithm. In this computation we started with an initial path close to the classical trajectory and then examined $\frac{1}{2}$ million variations about this path. All paths are constrained to begin and end at $x = 1$ (which turns out to be somewhat less than the maximum amplitude of the classical oscillation).

When the time difference $t_b - t_a$ equals a short time like $2T$, the system has not had enough time to equilibrate to its ground state and the wave function looks like the probability distribution of an excited state (nearly classical with the probability highest for the particle to be near its turning points where its velocity vanishes). However, when $t_b - t_a$ equals the longer time $20T$, the system has had enough time to decay to its ground state and the wave function looks like the expected Gaussian distribution. In either case (Figure 15.8 right), the trajectory through space-time fluctuates about the classical trajectory. This fluctuation is a consequence of the Metropolis algorithm occasionally going uphill in its search; if you modify the program so that searches go only downhill, the space-time trajectory will be a very smooth trigonometric function (the classical trajectory), but the wave function, which is a measure of the fluctuations about the classical trajectory, will vanish! The explicit steps of the calculation are

1. Construct a grid of N time steps of length ε (Figure 15.9). Start at $t = 0$ and extend to time $\tau = N\varepsilon$ [this means N time intervals and $(N+1)$ lattice points in time]. Note that time always increases monotonically along a path.
2. Construct a grid of M space points separated by steps of size δ. Use a range of x values several time larger than the characteristic size or range of the potential being used and start with $M \simeq N$.
3. When calculating the wave function, any x or t value falling between lattice points should be assigned to the closest lattice point.
4. Associate a position x_j with each time τ_j, subject to the boundary conditions that the initial and final positions always remain the same, $x_N = x_0 = x$.
5. Choose a path of straight-line links connecting the lattice points corresponding to the classical trajectory. Observe that the x values for the links of the path may have values that increase, decrease, or remain unchanged (in contrast to time, which always increases).
6. Evaluate the energy $\mathcal{E}$ by summing the kinetic and potential energies for each link of the path starting at $j = 0$:

$$\mathcal{E}(x_0, x_1, \ldots, x_N) \simeq \sum_{j=1}^{N} \left[\frac{m}{2} \left(\frac{x_j - x_{j-1}}{\varepsilon} \right)^2 + V\left(\frac{x_j + x_{j-1}}{2} \right) \right]. \tag{15.73}$$

7. Begin a sequence of repetitive steps in which a random position x_j associated with time t_j is changed to the position x'_j (point C in Figure 15.9). This changes *two* links in the path.
8. For the coordinate that is changed, use the Metropolis algorithm to weigh the change with the Boltzmann factor.
9. For each lattice point, establish a running sum that represents the value of the wave function squared at that point.
10. After each single-link change (or decision not to change), increase the running sum for the new x value by 1. After a sufficiently long running time, the sum divided by the number of steps is the simulated value for $|\psi(x_j)|^2$ at each lattice point x_j.
11. Repeat the entire link-changing simulation starting with a different seed. The average wave function from a number of intermediate-length runs is better than that from one very long run. ▌

```java
//QMC.java: Quantum MonteCarlo Feynman path integration
import java.io.*;                                              // Location of PrintWriter
import java.util.*;                                               // Location of Random
import java.lang.*;                                                 // Location of Math

public class QMC {
  public static void main(String[] argv) throws IOException, FileNotFoundException {
  PrintWriter q = new PrintWriter(new FileOutputStream("QMC.DAT"), true);
  int N = 100, M = 101, Trials = 25000, seedTrials = 200;              // t grid,x(odd)
  double path[] = new double[N], xscale = 10.;
  long prop[]   = new long[M],    seed = 10199435;
  for ( int count = 0; count < seedTrials*10; count += 10) {
    Random randnum = new Random(seed + count);
    double change = 0., newE = 0., oldE = 0. ;
    for ( int i=0; i < N; i++ )  path[i] = 0. ;                          // Initial path
    oldE = energy(path);                                               // Find E of path
    for ( int i=0; i < Trials; i++ )  {                             // Pick random element
      int element = randnum.nextInt(N);
      change = 1.8 * (0.5 - randnum.nextDouble());
      path[element] += change;                                             // Change path
      newE = energy(path);                                                  // Find new E
      if ( newE > oldE && Math.exp(-newE + oldE)
        <= randnum.nextDouble() )  path[element]-=change;                       // Reject
      for ( int j=0; j < N; j++ )  {                                  // Add probabilities
        element = (int)Math.round( (M-1)*(path[j]/xscale + .5) );
        if (element < M && element>=0)  prop[element]++ ;
      }
      oldE = newE;
    }                                                                           // t loop
  }                                                                          // Seed loop
  for ( int i=0; i < M; i++ )  q.println(xscale*(i-(M-1)/2)
          + " " + (double)prop[i]/((double)Trials*(double)seedTrials));
  System.out.println(" ");
  System.out.println("QMC Program Complete.");
  System.out.println("Data stored in QMC.DAT");
  System.out.println(" ");
  }

  public static double energy(double path[])  {
    int i = 0;
    double sum = 0. ;
    for ( i=0; i < path.length-2; i++ )
      {sum +=  (path[i+1] - path[i])*(path[i+1] - path[i]) ;}
    sum +=  path[i+1]*path[i+1];
    return sum;
} }                                                                          // End class
```

Listing 15.3 QMC.java solves for the ground-state probability distribution via a Feynman path integration using the Metropolis algorithm to simulate variations about the classical trajectory.

15.8.4 Assessment and Exploration

1. Plot some of the actual space-time paths used in the simulation along with the classical trajectory.
2. For a more continuous picture of the wave function, make the x lattice spacing smaller; for a more precise value of the wave function at any particular lattice site, sample more points (run longer) and use a smaller time step ε.

3. Because there are no sign changes in a ground-state wave function, you can ignore the phase, assume $\psi(x) = \sqrt{\psi^2(x)}$, and then estimate the energy via

$$E = \frac{\langle\psi|\, H \,|\psi\rangle}{\langle\psi|\psi\rangle} = \frac{\omega}{2\langle\psi|\psi\rangle} \int_{-\infty}^{+\infty} \psi^*(x) \left(-\frac{d^2}{dx^2} + x^2\right) \psi(x)\, dx,$$

   where the space derivative is evaluated numerically.
4. Explore the effect of making $\hbar$ larger and thus permitting greater fluctuations around the classical trajectory. Do this by decreasing the value of the exponent in the Boltzmann factor. Determine if this makes the calculation more or less robust in its ability to find the classical trajectory.
5. Test your ψ for the gravitational potential (see quantum bouncer below):

$$V(x) = mg|x|, \quad x(t) = x_0 + v_0 t + \tfrac{1}{2} g t^2.$$

15.9 Exploration: Quantum Bouncer's Paths ⊙

Another problem for which the classical trajectory is well known is that of a *quantum bouncer*. Here we have a particle dropped in a uniform gravitational field, hitting a hard floor, and then bouncing. When treated quantum mechanically, quantized levels for the particle result [Gibbs 75, Good 92, Whine 92, Bana 99, Vall 00]. In 2002 an experiment to discern this gravitational effect at the quantum level was performed by Nesvizhevsky et al. [Nes 02] and described in [Schw 02]. It consisted of dropping ultracold neutrons from a height of 14 μm unto a neutron mirror and watching them bounce. It found a neutron ground state at 1.4 peV.

We start by determining the analytic solution to this problem for stationary states and then generalize it to include time dependence.[6] The time-independent Schrödinger equation for a particle in a uniform gravitation potential is

$$-\frac{\hbar^2}{2m}\frac{d^2\psi(x)}{dx^2} + mxg\,\psi(x) = E\,\psi(x), \tag{15.74}$$

$$\psi(x \le 0) = 0, \quad \text{(boundary condition)}. \tag{15.75}$$

The boundary condition (15.75) is a consequence of the hard floor at $x = 0$. A change of variables converts (15.74) to a dimensionless form,

$$\frac{d^2\psi}{dz^2} - (z - z_E)\,\psi = 0, \tag{15.76}$$

$$z = x\left(\frac{2gm^2}{\hbar^2}\right)^{1/3}, \quad z_E = E\left(\frac{2}{\hbar^2 m g^2}\right)^{1/3}. \tag{15.77}$$

This equation has an analytic solution in terms of Airy functions Ai(z) [L 96]:

$$\psi(z) = N_n \,\text{Ai}(z - z_E), \tag{15.78}$$

[6] Oscar A. Restrepo assisted in the preparation of this section.

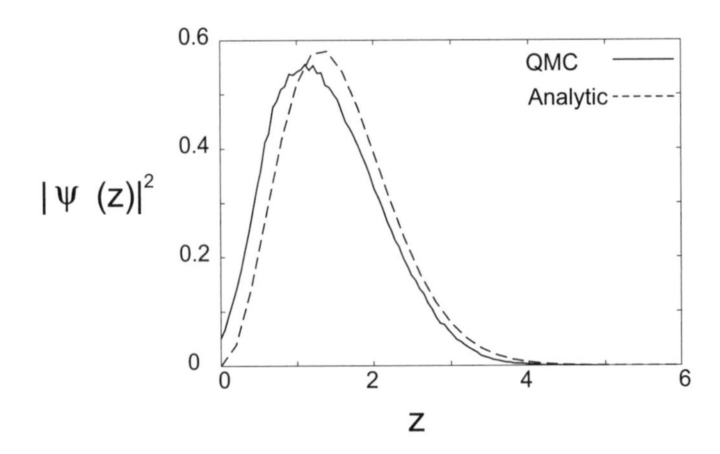

Figure 15.10 The Airy function squared (continuous line) and the Quantum Monte Carlo solution $|\psi_0(q)|^2$ (dashed line) after a million trajectories.

where N_n is a normalization constant and z_E is the scaled value of the energy. The boundary condition $\psi(0) = 0$ implies that

$$\psi(0) = N_E \text{ Ai}(-z_E) = 0, \tag{15.79}$$

which means that the allowed energies of the system are discrete and correspond to the zeros z_n of the Airy functions [Pres 00] at negative argument. To simplify the calculation, we take $\hbar = 1$, $g = 2$, and $m = \frac{1}{2}$, which leads to $z = x$ and $z_E = E$.

The time-dependent solution for the quantum bouncer is constructed by forming the infinite sum over all the discrete eigenstates, each with a time dependence appropriate to its energy:

$$\psi(z,t) = \sum_{n=1}^{\infty} C_n N_n \text{ Ai}(z - z_n) e^{-iE_n t/\hbar}, \tag{15.80}$$

where the C_n's are constants.

Figure 15.10 shows the results of solving for the quantum bouncer's ground-state probability $|\psi_0(z)|^2$ using Feynman's path integration. The time increment dt and the total time t were selected by trial and error in such a way as to make $|\psi(0)|^2 \simeq 0$ (the boundary condition). To account for the fact that the potential is infinite for negative x values, we selected trajectories that have positive x values over all their links. This incorporates the fact that the particle can never penetrate the floor. Our program is given in Listing 15.4, and it yields the results in Figure 15.10 after using 10^6 trajectories and a time step $\varepsilon = d\tau = 0.05$. Both results were normalized via a trapezoid integration. As can be seen, the agreement between the analytic result and the path integration is satisfactory.

```
// QMCbouncer.java: quantum bouncer wavefunction via path integration
//   Author: Oscar Restrepo, Universidad de Antioquia
import java.io.*;
import java.util.*;

public class QMCbouncer{
  static int max = 300000, N = 100;                          // Trajectories, array
  static double dt = 0.05, g  =  2.0;                        // Time step, gravity

  public static void main(String[] argv) throws IOException, FileNotFoundException  {
        PrintWriter q   = new PrintWriter(new FileOutputStream("bouncer.dat"), true);
    double change, newE, oldE, norm, h, firstlast, maxx, path[] = new double[101];
    int prop[] = new int[201], i, j, element, ele, maxel;           // Probabilities
    h=0.00;
    for ( j = 0; j <= N; j++ )  {path[j]=0.0; prop[j]=0;}              // Initialize
    oldE = energy(path);                                                 // Initial E
    maxel = 0;
    for (i = 0; i < max; i++ ) {
      element = (int)(Math.random()*N);
      if( element!=0 && element!= N )  {                             // Ends not allowed
        change  = ( (Math.random() -0.5)*20. )/10.;                           //-1 to 1
        if ( path[element] + change > 0.)  {                       // No negative paths
          path[element] += change;
          newE = energy(path);                                    // New trajectory E
          if (newE>oldE && Math.exp(-newE+oldE) <= Math.random() )
            { path[element] -= change; }                               // Link rejected
          ele = (int) (path[element]*1250./100.);                     // Scale changed
          if ( ele >= maxel ) maxel=ele;                          // Scale change 0 to N
          if ( element !=0 )  prop[ele]++;
          oldE = newE;                                                  // For next cicle
    } } }
    maxx=0.0;
    for ( i = 1; i < N; i++ ) if ( path[i] >= maxx) maxx=path[i];                 // Norm
    norm = 0.;
    h = maxx/maxel;
    firstlast = h*0.5* ( prop[0] + prop[maxel] );
    for ( i=1; i <= maxel; i++ ) norm = norm+prop[i];                         // Trap rule
    norm = norm*h + firstlast;
    for (i=0; i<=maxel; i++) q.println(" "+h*i+" "+(double)(prop[i])/norm);
    System.out.println("data stored in bouncer.dat \n");
  }                                                                                // main

  static double energy (double arr[]) {                                      // E of path
    int i;
    double sum=0.;
    for ( i = 0; i < N; i++ )                                                   // KE, PE
      {sum += 0.5*Math.pow( (arr[i+1]-arr[i])/dt, 2 )+ g * ( arr[i] + arr[i+1] )/2.;}
                           //  Linear V
    return (sum);
} }
```

Listing 15.4 QMCbouncer.java uses Feynman path integration to compute the path of a quantum particle in a gravitational field.

16

Simulating Matter with Molecular Dynamics

Problem: Determine whether a collection of argon molecules placed in a box will form an ordered structure at low temperature.

You may have seen in your introductory classes that the ideal gas law can be derived from first principles if gas molecules are treated as billiard balls bouncing off the walls but not interacting with each other. We want to extend this model so that we can solve for the motion of every molecule in a box interacting with every other molecule via a potential. We picked argon because it is an inert element with a closed shell of electrons and so can be modeled as almost-hard spheres.

16.1 Molecular Dynamics (Theory)

Molecular dynamics (MD) is a powerful simulation technique for studying the physical and chemical properties of solids, liquids, amorphous materials, and biological molecules. Even though we know that quantum mechanics is the proper theory for molecular interactions, MD uses Newton's laws as the basis of the technique and focuses on bulk properties, which do not depend much on small-r behaviors. In 1985 Car and Parrinello showed how MD can be extended to include quantum mechanics by applying density functional theory to calculate the force [C&P 85]. This technique, known as *quantum MD*, is an active area of research but is beyond the realm of the present chapter.[1] For those with more interest in the subject, there are full texts [A&T 87, Rap 95, Hock 88] on MD and good discussions [G,T&C 06, Thij 99, Fos 96], as well as primers [Erco] and codes, [NAMD, Mold, ALCMD] available on-line.

MD's solution of Newton's laws is conceptually simple, yet when applied to a very large number of particles becomes the "high school physics problem from hell." Some approximations must be made in order not to have to solve the 10^{23}–10^{25} equations of motion describing a realistic sample but instead to limit the problem to $\sim 10^6$ particles for protein simulations and $\sim 10^8$ particles for materials simulations. If we have some success, then it is a good bet that the model will improve if we incorporate more particles or more quantum mechanics, something that becomes easier as computing power continues to increase.

[1] We thank Satoru S. Kano for pointing this out to us.

In a number of ways, MD simulations are similar to the thermal Monte Carlo simulations we studied in Chapter 15, "Thermodynamic Simulations & Feynman Quantum Path Integration," Both typically involve a large number N of interacting particles that start out in some set configuration and then equilibrate into some dynamic state on the computer. However, in MD we have what statistical mechanics calls a *microcanonical ensemble* in which the energy E and volume V of the N particles are fixed. We then use Newton's laws to generate the dynamics of the system. In contrast, Monte Carlo simulations do not start with first principles but instead incorporate an element of chance and have the system in contact with a heat bath at a fixed temperature rather than keeping the energy E fixed. This is called a *canonical ensemble*.

Because a system of molecules is dynamic, the velocities and positions of the molecules change continuously, and so we will need to follow the motion of each molecule in time to determine its effect on the other molecules, which are also moving. After the simulation has run long enough to stabilize, we will compute time averages of the dynamic quantities in order to deduce the thermodynamic properties. We apply Newton's laws with the assumption that the net force on each molecule is the sum of the two-body forces with all other $(N-1)$ molecules:

$$m\frac{d^2\mathbf{r}_i}{dt^2} = \mathbf{F}_i(\mathbf{r}_0, \ldots, \mathbf{r}_{N-1}) \tag{16.1}$$

$$m\frac{d^2\mathbf{r}_i}{dt^2} = \sum_{i<j=0}^{N-1} \mathbf{f}_{ij}, \quad i = 0, \ldots, (N-1). \tag{16.2}$$

In writing these equations we have ignored the fact that the force between argon atoms really arises from the particle–particle interactions of the 18 electrons and the nucleus that constitute each atom (Figure 16.1). Although it may be possible to ignore this internal structure when deducing the long-range properties of inert elements, it matters for systems such as polyatomic molecules that display rotational, vibrational, and electronic degrees of freedom as the temperature is raised.[2]

We assume that the force on molecule i derives from a potential and that the potential is the sum of central molecule–molecule potentials:

$$\mathbf{F}_i(\mathbf{r}_0, \mathbf{r}_1, \ldots, \mathbf{r}_{N-1}) = -\nabla_{\mathbf{r}_i} U(\mathbf{r}_0, \mathbf{r}_1, \ldots, \mathbf{r}_{N-1}), \tag{16.3}$$

$$U(\mathbf{r}_0, \mathbf{r}_1, \ldots, \mathbf{r}_{N-1}) = \sum_{i<j} u(r_{ij}) = \sum_{i=0}^{N-2}\sum_{j=i+1}^{N-1} u(r_{ij}), \tag{16.4}$$

$$\Rightarrow \quad \mathbf{f}_{ij} = -\frac{du(r_{ij})}{dr_{ij}} \left(\frac{x_i - x_j}{r_{ij}}\hat{\mathbf{e}}_x + \frac{y_i - y_j}{r_{ij}}\hat{\mathbf{e}}_y + \frac{z_i - z_j}{r_{ij}}\hat{\mathbf{e}}_z \right). \tag{16.5}$$

Here $r_{ij} = |\mathbf{r}_i - \mathbf{r}_j| = r_{ji}$ is the distance between the centers of molecules i and j, and the limits on the sums are such that no interaction is counted twice. Because

[2] We thank Saturo Kano for clarifying this point.

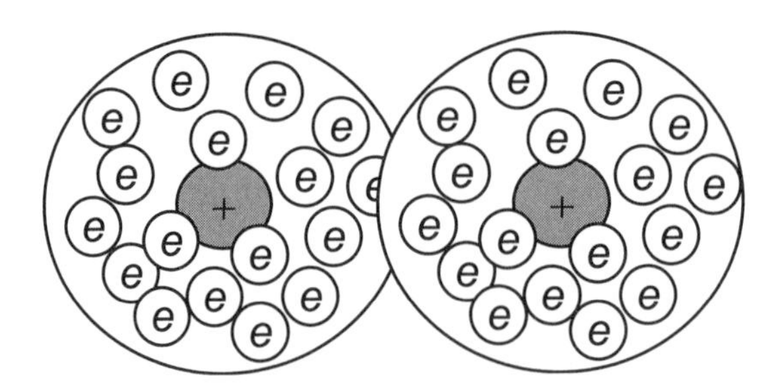

Figure 16.1 The molecule–molecule effective interaction arises from the many-body interaction of the electrons and nucleus in one electronic cloud with the electrons and nucleus in another electron cloud. (The size of the nucleus at the center is highly exaggerated relative to the size of the molecule, and the electrons are really just points.)

we have assumed a *conservative* potential, the total energy of the system, that is, the potential plus kinetic energies summed over all particles, should be conserved over time. Nonetheless, in a practical computation we "cut the potential off" [assume $u(r_{ij}) = 0$] when the molecules are far apart. Because the derivative of the potential produces an infinite force at this cutoff point, energy will no longer be precisely conserved. Yet because the cutoff radius is large, the cutoff occurs only when the forces are minuscule, and so the violation of energy conservation should be small relative to approximation and round-off errors.

In a first-principles calculation, the potential between any two argon atoms arises from the sum over approximately 1000 electron–electron and electron–nucleus Coulomb interactions. A more practical calculation would derive an effective potential based on a form of many-body theory, such as Hartree–Fock or density functional theory. Our approach is simpler yet. We use the Lennard–Jones potential,

$$u(r) = 4\epsilon\left[\left(\frac{\sigma}{r}\right)^{12} - \left(\frac{\sigma}{r}\right)^{6}\right], \tag{16.6}$$

$$\mathbf{f}(r) = -\frac{du}{dr} = \frac{48\epsilon}{r^2}\left[\left(\frac{\sigma}{r}\right)^{12} - \frac{1}{2}\left(\frac{\sigma}{r}\right)^{6}\right]\mathbf{r}. \tag{16.7}$$

TABLE 16.1
Parameter Values and Scales for the Lennard-Jones Potential

Quantity	Mass	Length	Energy	Time	Temperature
Unit	m	σ	ϵ	$\sqrt{m\sigma^2/\epsilon}$	ϵ/k_B
Value	6.7×10^{-26} kg	3.4×10^{-10} m	1.65×10^{-21} J	4.5×10^{-12} s	119 K

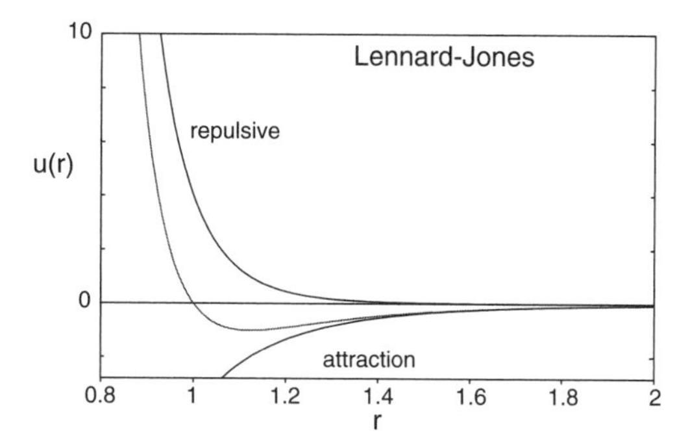

Figure 16.2 The Lennard-Jones potential. Note the sign change at $r = 1$ and the minimum at $r \simeq 1.1225$ (natural units). Note too that because the r axis does not extend to $r = 0$, the very high central repulsion is not shown.

Here the parameter ϵ governs the strength of the interaction, and σ determines the length scale. Both are deduced by fits to data, which is why this is called a "phenomenological" potential.

Some typical values for the parameters, and corresponding scales for the variables, are given in Table 16.1. In order to make the program simpler and to avoid under- and overflows, it is helpful to measure all variables in the natural units formed by these constants. The interparticle potential and force then take the forms

$$u(r) = 4\left[\frac{1}{r^{12}} - \frac{1}{r^6}\right], \qquad f(r) = \frac{48}{r}\left[\frac{1}{r^{12}} - \frac{1}{2r^6}\right]. \tag{16.8}$$

The Lennard-Jones potential is seen in Figure 16.2 to be the sum of a long-range attractive interaction $\propto 1/r^6$ and a short-range repulsive one $\propto 1/r^{12}$. The change from repulsion to attraction occurs at $r = \sigma$. The minimum of the potential occurs at $r = 2^{1/6}\sigma = 1.1225\sigma$, which would be the atom–atom spacing in a solid bound by this potential. The repulsive $1/r^{12}$ term in the Lennard-Jones potential (16.6) arises when the electron clouds from two atoms overlap, in which case the Coulomb interaction and the Pauli exclusion principle keep the electrons apart. The $1/r^{12}$ term dominates at short distances and makes atoms behave like hard spheres. The precise value of 12 is not of theoretical significance (although it's being large is) and was probably chosen because it is 2×6.

The $1/r^6$ term dominates at large distances and models the weak *van der Waals* induced dipole–dipole attraction between two molecules.[3] The attraction arises from fluctuations in which at some instant in time a molecule on the right tends to be more positive on the left side, like a dipole $\Leftarrow$. This in turn attracts the negative charge in a molecule on its left, thereby inducing a dipole $\Leftarrow$ in it. As long as

[3] There are also van der Waals forces that cause dispersion, but we are not considering those here.

the molecules stay close to each other, the polarities continue to fluctuate in synchronization $\Leftarrow\Leftarrow$ so that the attraction is maintained. The resultant dipole–dipole attraction behaves like $1/r^6$, and although much weaker than a Coulomb force, it is responsible for the binding of neutral, inert elements, such as argon for which the Coulomb force vanishes.

16.1.1 Connection to Thermodynamic Variables

We assume that the number of particles is large enough to use statistical mechanics to relate the results of our simulation to the thermodynamic quantities (the simulation is valid for any number of particles, but the use of statistics requires large numbers). The equipartition theorem tells us that for molecules in thermal equilibrium at temperature T, each molecular degree of freedom has an energy $k_BT/2$ on the average associated with it, where $k_B = 1.38 \times 10^{-23}$ J/K is Boltzmann's constant. A simulation provides the kinetic energy of translation[4]:

$$\text{KE} = \frac{1}{2}\left\langle \sum_{i=0}^{N-1} v_i^2 \right\rangle. \tag{16.9}$$

The time average of KE (three degrees of freedom) is related to temperature by

$$\langle \text{KE} \rangle = N\frac{3}{2}k_BT \;\Rightarrow\; T = \frac{2\langle \text{KE} \rangle}{3k_BN}. \tag{16.10}$$

The system's pressure P is determined by a version of the *Virial theorem*,

$$PV = Nk_BT + \frac{w}{3}, \qquad w = \left\langle \sum_{i<j}^{N-1} \mathbf{r}_{ij} \cdot \mathbf{f}_{ij} \right\rangle, \tag{16.11}$$

where the Virial w is a weighted average of the forces. Note that because ideal gases have no interaction forces, their Virial vanishes and we have the ideal gas law. The pressure is thus

$$P = \frac{\rho}{3N}\left(2\langle \text{KE} \rangle + w\right), \tag{16.12}$$

where $\rho = N/V$ is the density of the particles.

[4] Unless the temperature is very high, argon atoms, being inert spheres, have no rotational energy.

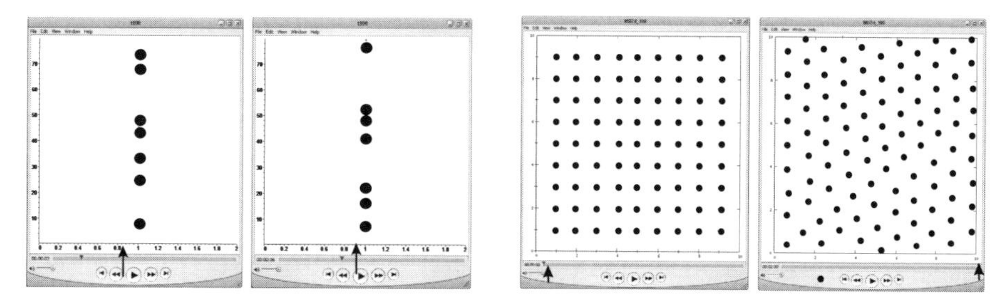

Figure 16.3 *Left:* Two frames from the animation of a 1-D simulation that starts with uniformly spaced atoms. Note how an image atom has moved in from the bottom after an atom leaves from the top. *Right:* Two frames from the animation of a 2-D simulation showing the initial and an equilibrated state. Note how the atoms start off in a simple cubic arrangement but then equilibrate to a face-centered-cubic lattice. In all cases, it is the interatomic forces that constrain the atoms to a lattice.

16.1.2 Setting Initial Velocity Distribution

Even though we start the system off with a velocity distribution characteristic of some temperature, since the system is not in equilibrium initially (some of the assigned KE goes into PE), this is not the true temperature of the system [Thij 99]. Note that this initial randomization is the only place where chance enters into our MD simulation, and it is there to speed the simulation along. Once started, the time evolution is determined by Newton's laws, in contrast to Monte Carlo simulations which are inherently stochastic. We produce a Gaussian (Maxwellian) velocity distribution with the methods discussed in Chapter 5, "Monte Carlo Simulations." In our sample code we take the average $\frac{1}{12}\sum_{i=1}^{12} r_i$ of uniform random numbers $0 \leq r_i \leq 1$ to produce a Gaussian distribution with mean $\langle r \rangle = 0.5$. We then subtract this mean value to obtain a distribution about 0.

16.1.3 Periodic Boundary Conditions and Potential Cutoff

It is easy to believe that a simulation of 10^{23} molecules should predict bulk properties well, but with typical MD simulations employing only 10^3–10^6 particles, one must be clever to make less seem like more. Furthermore, since computers are finite, the molecules in the simulation are constrained to lie within a finite box, which inevitably introduces artificial *surface effects* from the walls. Surface effects are particularly significant when the number of particles is small because then a large fraction of the molecules reside near the walls. For example, if 1000 particles are arranged in a $10 \times 10 \times 10 \times 10$ cube, there are 10^3–$8^3 = 488$ particles one unit

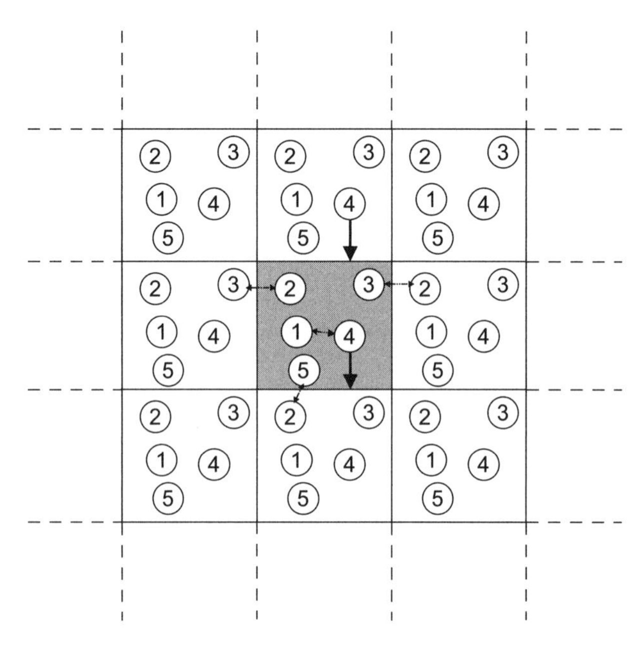

Figure 16.4 The infinite space generated by imposing periodic boundary conditions on the particles within the simulation volume (shaded box). The two-headed arrows indicate how a particle interacts with the nearest version of another particle, be that within the simulation volume or an image. The vertical arrows indicate how the image of particle 4 enters when particle 4 exits.

from the surface, that is, 49% of the molecules, while for 10^6 particles this fraction falls to 6%.

The imposition of *periodic boundary conditions* (PBCs) strives to minimize the shortcomings of both the small numbers of particles and of artificial boundaries. Even though we limit our simulation to an $L_x \times L_y \times L_z$ box, we imagine this box being replicated to infinity in all directions (Figure 16.4). Accordingly, after each time-integration step we examine the position of each particle and check if it has left the simulation region. If it has, then we bring an *image* of the particle back through the opposite boundary (Figure 16.4):

$$x \;\Rightarrow\; \begin{cases} x + L_x, & \text{if } x \leq 0, \\ x - L_x, & \text{if } x > L_x. \end{cases} \tag{16.13}$$

Consequently, each box looks the same and has continuous properties at the edges. As shown by the one-headed arrows in Figure 16.4, if a particle exits the simulation volume, its image enters from the other side, and so balance is maintained.

In principle a molecule interacts with all others molecules and their images, so even though there is a finite number of atoms in the interaction volume, there is an effective infinite number of interactions [Erco]. Nonetheless, because the Lennard–Jones potential falls off so rapidly for large r, $V(r = 3\sigma) \simeq V(1.13\sigma)/200$, far-off molecules do not contribute significantly to the motion of a molecule, and we pick a value $r_{\text{cut}} \simeq 2.5\sigma$ beyond which we ignore the effect of the potential:

$$u(r) = \begin{cases} 4\left(r^{-12} - r^{-6}\right), & \text{for } r < r_{\text{cut}}, \\ 0, & \text{for } r > r_{\text{cut}}. \end{cases} \tag{16.14}$$

Accordingly, if the simulation region is large enough for $u(r > L_i/2) \simeq 0$, an atom interacts with only the *nearest image* of another atom.

The only problem with the cutoff potential (16.14) is that since the derivative du/dr is singular at $r = r_{\text{cut}}$, the potential is no longer conservative and thus energy conservation is no longer ensured. However, since the forces are already very small at r_{cut}, the violation will also be very small.

16.2 Verlet and Velocity-Verlet Algorithms

A realistic MD simulation may require integration of the 3-D equations of motion for 10^{10} time steps for each of 10^3–10^6 particles. Although we could use our standard `rk4` ODE solver for this, time is saved by using a simple rule embedded in the program. The Verlet algorithm uses the central-difference approximation (Chapter 7, "Differentiation & Searching") for the second derivative to advance the solutions by a single time step h for all N particles simultaneously:

$$\mathbf{F}_i[\mathbf{r}(t), t] = \frac{d^2\mathbf{r}_i}{dt^2} \simeq \frac{\mathbf{r}_i(t+h) + \mathbf{r}_i(t-h) - 2\mathbf{r}_i(t)}{h^2}, \tag{16.15}$$

$$\Rightarrow \quad \mathbf{r}_i(t+h) \simeq 2\mathbf{r}_i(t) - \mathbf{r}_i(t-h) + h^2\mathbf{F}_i(t) + O(h^4), \tag{16.16}$$

where we have set $m = 1$. (Improved algorithms may vary the time step depending upon the speed of the particle.) Notice that even though the atom–atom force does not have an explicit time dependence, we include a t dependence in it as a way of indicating its dependence upon the atoms' positions at a particular time. Because this is really an implicit time dependence, energy remains conserved.

Part of the efficiency of the Verlet algorithm (16.16) is that it solves for the position of each particle without requiring a separate solution for the particle's velocity. However, once we have deduced the position for various times, we can use the central-difference approximation for the first derivative of $\mathbf{r}_i$ to obtain the velocity:

$$\mathbf{v}_i(t) = \frac{dr_i}{dt} \simeq \frac{\mathbf{r}_i(t+h) - \mathbf{r}_i(t-h)}{2h} + O(h^2). \tag{16.17}$$

Note, finally, that because the Verlet algorithm needs $\mathbf{r}$ from two previous steps, it is not self-starting and so we start it with the forward difference,

$$\mathbf{r}(t = -h) \simeq \mathbf{r}(0) - h\mathbf{v}(0) + \frac{h^2}{2}\mathbf{F}(0). \tag{16.18}$$

Velocity-Verlet Algorithm: Another version of the Verlet algorithm, which we recommend because of its increased stability, uses a forward-difference approximation for the derivative to advance *both* the position and velocity simultaneously:

$$\mathbf{r}_i(t+h) \simeq \mathbf{r}_i(t) + h\mathbf{v}_i(t) + \frac{h^2}{2}\mathbf{F}_i(t) + O(h^3), \tag{16.19}$$

$$\mathbf{v}_i(t+h) \simeq \mathbf{v}_i(t) + h\,\overline{\mathbf{a}(t)} + O(h^2) \tag{16.20}$$

$$\simeq \mathbf{v}_i(t) + h\left[\frac{\mathbf{F}_i(t+h) + \mathbf{F}_i(t)}{2}\right] + O(h^2). \tag{16.21}$$

Although this algorithm appears to be of lower order than (16.16), the use of updated positions when calculating velocities, and the subsequent use of these velocities, make both algorithms of similar precision.

Of interest is that (16.21) approximates the average force during a time step as $[\mathbf{F}_i(t+h) + \mathbf{F}_i(t)]/2$. Updating the velocity is a little tricky because we need the force at time $t+h$, which depends on the particle positions at $t+h$. Consequently, we must update all the particle positions and forces to $t+h$ before we update any velocities, while saving the forces at the earlier time for use in (16.21). As soon as the positions are updated, we impose periodic boundary conditions to ensure that we have not lost any particles, and then we calculate the forces.

16.3 1-D Implementation and Exercise

CD

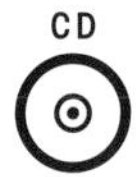

On the CD you will find a folder MDanimations that contains a number of 2-D animations (movies) of solutions to the MD equations. Some frames from these animations are shown in Figure 16.3. We recommend that you look at them in order to better visualize what the particles do during an MD simulation. In particular, these simulations use a potential and temperature that should lead to a solid or liquid system, and so you should see the particles binding together.

```
// MD.java, Molecular Dyanmics via Lennard-Jones potential, velocity Verlet algorithm
import java.io.*;
import java.util.*;

public class MD {
  static int L, Natom = 8, Nmax = 513;                                  // Class variables
  static double x[] = new double[Nmax], fx[][] = new double[Nmax][2];

  public static void main(String[] argv) throws IOException, FileNotFoundException {
    int t1, t2, i, Itemp, t, Nstep=5000, Nprint=100, Ndim=1;
    double h = 0.0004, hover2, PE, KE, T, Tinit = 10.0, vx[] = new double[Nmax];
    L = (int)Math.pow(1.*Natom, 1./Ndim);
    Natom = (int)Math.pow(L, Ndim);
    System.out.println("Natom = "+Natom+" L= "+L+"");
    i = -1;
    for ( int ix = 0; ix <= L-1; ix++ ) {                          // Set up lattice of side L
      i = i+1;
      x[i] = ix;                                                       // Initial velocities
      vx[i] =(Math.random()+Math.random()+Math.random()+Math.random()+Math.random()
             +Math.random()+Math.random()+Math.random()+Math.random()+
```

```
                    Math.random()+Math.random()+Math.random())/12.-0.5;
      vx[i] = vx[i]*Math.sqrt(Tinit);                          // Scale v with temperature
      System.out.println("init vx = "+vx[i]);
    }
    t1 = 0;     t2 = 1;                                                    // t, t+h indices
    hover2 = h/2.;
    t = 0;
    KE = 0.0;  PE = 0.0;                                              // initial KE & PE v
    PE = Forces(t1, PE);
    for ( i = 0; i <= Natom-1; i++ )      KE=KE+(vx[i]*vx[i])/2;
    System.out.println(t+" PE= "+PE+" KE = "+KE+" PE+KE = "+(PE+KE));
    for( t = 1; t < Nstep; t++ ) {
      for( i = 0; i <= Natom-1; i++ )  {                                      // Main loop
        PE = Forces(t1, PE);                                              // Velocity Verlet
        x[i] = x[i] + h*(vx[i] + hover2*fx[i][t1]);
        if (x[i] <= 0.) x[i] = x[i] + L;                                            // PBC
        if (x[i] >= L)  x[i] = x[i] - L;
      }
    PE = Forces(t2, PE);
    KE = 0.;
    for( i = 0; i <= Natom-1; i++)   {
      vx[i] = vx[i] + hover2*(fx[i][t1] + fx[i][t2]);
      KE = KE + (vx[i]*vx[i])/2;
    }
    T = 2.*KE / (3.*Natom);
    if (t%Nprint==0)System.out.println(t+" PE ="+PE+" KE = "+KE+" PE+KE = "+(PE+KE));
    Itemp = t1;                                                          // Time t and t+h
    t1 = t2;              t2 = Itemp;
    }
  }
                                                                  // Force = class variable
  public static double Forces(int t, double PE) {
    int i, j;
    double fijx, r2, invr2=0, dx, r2cut = 9.;
    PE = 0.;                                                                 // Initialize
    for (i=0; i<= Natom-1; i++)    {fx[i][t] = 0.; }
    for(i = 0; i<= Natom-2; i++)   {
      for(j = i+1; j<=Natom-1; j++) {
        dx = x[i]-x[j];
        if (Math.abs(dx) > 0.50*L) {dx = dx - sign(L,dx);}                          // PBC
        r2 =  dx*dx ;
        if (r2 < r2cut)  {                                                       // Cut off
          if ( r2 == 0.) r2 = 0.0001;
          invr2 = 1./r2;
          fijx =  48.*(Math.pow(invr2,3) -0.5)*Math.pow(invr2,3);
          fijx = fijx*invr2*dx;
          fx[i][t] = fx[i][t] + fijx;
          fx[j][t] = fx[j][t] - fijx;
          PE = PE + 4*Math.pow(invr2,3)*( Math.pow(invr2,3) - 1.);
        }
      }
    }
  return PE;
  }

  public static double  sign(double a,double b)
    {if (b >= 0.) return Math.abs(a);  else  return -Math.abs(a); }
}
```

Listing 16.1 MD.java performs a 1-D MD simulation with too small a number of large time steps for just a few particles. To be realistic the user should change the parameters and the number of random numbers added to form the Gaussian distribution.

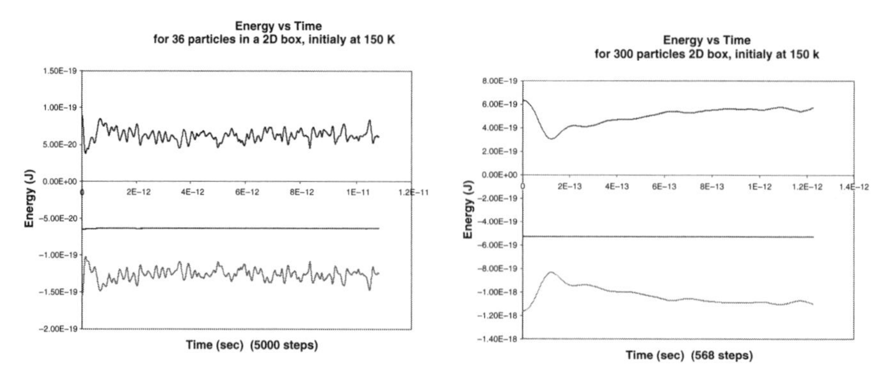

Figure 16.5 The kinetic, potential, and total energy for a 2-D MD simulation with 36 particles (*left*), and 300 particles (*right*), both with an initial temperature of 150 K. The potential energy is negative, the kinetic energy is positive, and the total energy is seen to be conserved (flat).

The program MD.java implements an MD simulation in 1-D using the velocity–Verlet algorithm. Use it as a model and do the following:

1. Ensure that you can run and visualize the 1-D simulation.
2. Place the particles initially at the sites of a face-centered-cubic (FCC) lattice, the equilibrium configuration for a Lennard-Jones system at low temperature. The particles will find their own ways from there. An FCC lattice has four quarters of a particle per unit cell, so an L^3 box with a lattice constant L/N contains (parts of) $4N^3$ = 32, 108, 256, ... particles.
3. To save computing time, assign initial particle velocities corresponding to a fixed-temperature Maxwellian distribution.
4. Print the code and indicate on it which integration algorithm is used, where the periodic boundary conditions are imposed, where the nearest image interaction is evaluated, and where the potential is cut off.
5. A typical time step is $\Delta t = 10^{-14}$ s, which in our natural units equals 0.004. You probably will need to make 10^4–10^5 such steps to equilibrate, which corresponds to a total time of only 10^{-9} s (a lot can happen to a speedy molecule in 10^{-9} s). Choose the *largest* time step that provides stability and gives results similar to Figure 16.5.
6. The PE and KE change with time as the system equilibrates. Even after that, there will be fluctuations since this is a dynamic system. Evaluate the time-averaged energies for an equilibrated system.
7. Compare the final temperature of your system to the initial temperature. Change the initial temperature and look for a simple relation between it and the final temperature (Figure 16.6).

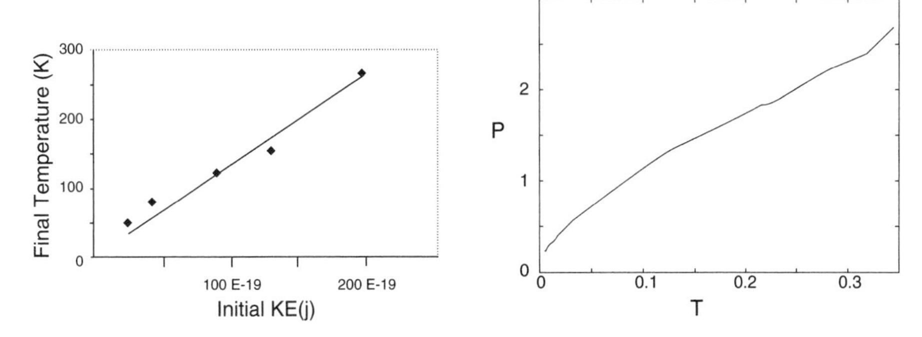

Figure 16.6 *Left:* The temperature after equilibration as a function of initial kinetic energy for a simulation with 36 particles in two dimensions. *Right:* The pressure *versus* temperature for a simulation with several hundred particles. (Courtesy of J. Wetzel.)

16.4 Trajectory Analysis

1. Modify your program so that it outputs the coordinates and velocities of some particles throughout the simulation. Note that you do not need as many time steps to follow a trajectory as you do to compute it and so you may want to use the *mod* operator %100 for output.
2. Start your assessment with a 1-D simulation at zero temperature. The particles should remain in place without vibration. Increase the temperature and note how the particles begin to move about and interact.
3. Try starting off all your particles at the minima in the Lennard-Jones potential. The particles should remain bound within the potential until you raise the temperature.
4. Repeat the simulations for a 2-D system. The trajectories should resemble billiard ball–like collisions.
5. Create an animation of the time-dependent locations of several particles.
6. Calculate and plot as a function of temperature the root-mean-square displacement of molecules:

$$R_{\text{rms}} = \sqrt{\left\langle |\mathbf{r}(t+\Delta t) - \mathbf{r}(t)|^2 \right\rangle}, \tag{16.22}$$

   where the average is over all the particles in the box. Verify that for a liquid R^2_{rms} grows linearly with time.
7. Test your system for time-reversal invariance. Stop it at a fixed time, reverse all the velocities, and see if the system retraces its trajectories back to the initial configuration.

CD

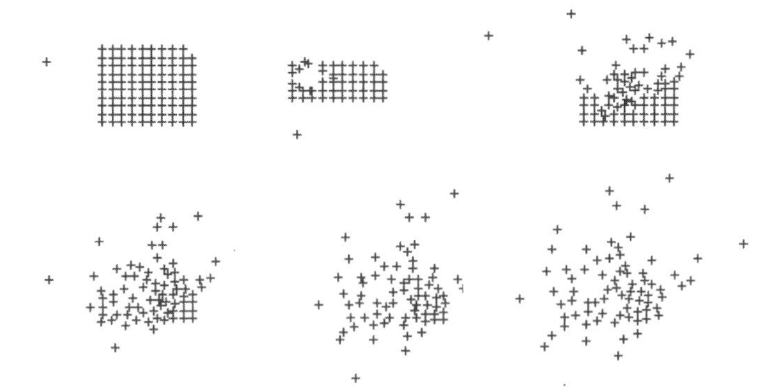

Figure 16.7 A simulation of a projectile shot into a group of particles. (Courtesy of J. Wetzel.)

16.5 Quiz

1. We wish to make an MD simulation *by hand* of the positions of particles 1 and 2 that are in a 1-D box of side 8. For an origin located at the center of the box, the particles are initially at rest and at locations $x_i(0) = -x_2(0) = 1$. The particles are subject to the force

$$F(x) = \begin{cases} 10, & \text{for } |x_1 - x_2| \le 1, \\ -1, & \text{for } 1 \le |x_1 - x_2| \le 3, \\ 0, & \text{otherwise.} \end{cases} \tag{16.23}$$

Use a simple algorithm to determine the positions of the particles up until the time they leave the box. Make sure to apply periodic boundary conditions. *Hint:* Since the configuration is symmetric, you know the location of particle 2 by symmetry and do not need to solve for it. We suggest the Verlet algorithm (no velocities) with a forward-difference algorithm to initialize it. To speed things along, use a time step of $h = 1/\sqrt{2}$. ▌

17

PDEs for Electrostatics & Heat Flow

17.1 PDE Generalities

Physical quantities such as temperature and pressure vary continuously in both space and time. Such being our world, the function or *field* $U(x, y, z, t)$ used to describe these quantities must contain independent space and time variations. As time evolves, the changes in $U(x, y, z, t)$ at any one position affect the field at neighboring points. This means that the dynamic equations describing the dependence of U on four independent variables must be written in terms of partial derivatives, and therefore the equations must be *partial differential equations* (PDEs), in contrast to ordinary differential equations (ODEs).

The most general form for a PDE with two independent variables is

$$A \frac{\partial^2 U}{\partial x^2} + 2B \frac{\partial^2 U}{\partial x \partial y} + C \frac{\partial^2 U}{\partial y^2} + D \frac{\partial U}{\partial x} + E \frac{\partial U}{\partial y} = F, \qquad (17.1)$$

where A, B, C, and F are arbitrary functions of the variables x and y. In the table below we define the classes of PDEs by the value of the discriminant d in the second row [A&W 01], with the next two rows being examples:

Elliptic	*Parabolic*	*Hyperbolic*
$d = AC - B^2 > 0$	$d = AC - B^2 = 0$	$d = AC - B^2 < 0$
$\nabla^2 U(x) = -4\pi\rho(x)$	$\nabla^2 U(\mathbf{x}, t) = a\, \partial U/\partial t$	$\nabla^2 U(\mathbf{x}, t) = c^{-2} \partial^2 U/\partial t^2$
Poisson's	Heat	Wave

We usually think of a parabolic equation as containing a first-order derivative in one variable and a second-order derivative in the other; a hyperbolic equation as containing second-order derivatives of all the variables, with opposite signs when placed on the same side of the equal sign; and an elliptic equation as containing second-order derivatives of all the variables, with all having the same sign when placed on the same side of the equal sign.

After solving enough problems, one often develops some physical intuition as to whether one has sufficient *boundary conditions* for there to exist a unique solution for a given physical situation (this, of course, is in addition to requisite *initial conditions*). For instance, a string tied at both ends and a heated bar placed in an

TABLE 17.1
The Relation Between Boundary Conditions and Uniqueness for PDEs

Boundary Condition	*Elliptic (Poisson Equation)*	*Hyperbolic (Wave Equation)*	*Parabolic (Heat Equation)*
Dirichlet open surface	Underspecified	Underspecified	*Unique and stable (1-D)*
Dirichlet closed surface	*Unique and stable*	Overspecified	Overspecified
Neumann open surface	Underspecified	Underspecified	*Unique and Stable (1-D)*
Neumann closed surface	*Unique and stable*	Overspecified	Overspecified
Cauchy open surface	Nonphysical	*Unique and stable*	Overspecified
Cauchy closed surface	Overspecified	Overspecified	Overspecified

infinite heat bath are physical situations for which the boundary conditions are adequate. If the boundary condition is the value of the solution on a surrounding closed surface, we have a *Dirichlet boundary condition*. If the boundary condition is the value of the normal derivative on the surrounding surface, we have a *Neumann boundary condition*. If the value of both the solution and its derivative are specified on a closed boundary, we have a *Cauchy boundary condition*. Although having an adequate boundary condition is necessary for a unique solution, having too many boundary conditions, for instance, both Neumann and Dirichlet, may be an overspecification for which no solution exists.[1]

Solving PDEs numerically differs from solving ODEs in a number of ways. First, because we are able to write all ODEs in a standard form,

$$\frac{d\mathbf{y}(t)}{dt} = \mathbf{f}(\mathbf{y}, t), \tag{17.2}$$

with t the single independent variable, we are able to use a standard algorithm, `rk4` in our case, to solve all such equations. Yet because PDEs have several independent variables, for example, $\rho(x, y, z, t)$, we would have to apply (17.2) simultaneously and independently to each variable, which would be very complicated. Second, since there are more equations to solve with PDEs than with ODEs, we need more information than just the two *initial conditions* $[x(0), \dot{x}(0)]$. In addition, because each PDE often has its own particular set of boundary conditions, we have to develop a special algorithm for each particular problem.

[1] Although conclusions drawn for exact PDEs may differ from those drawn for the finite-difference equations, they are usually the same; in fact, Morse and Feshbach [M&F 53] use the finite-difference form to derive the relations between boundary conditions and uniqueness for each type of equation shown in Table 17.1 [Jack 88].

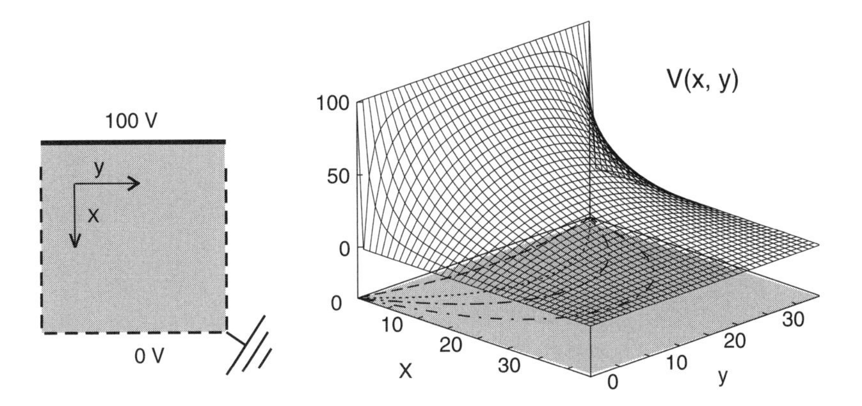

Figure 17.1 *Left:* The shaded region of space within a square in which we want to determine the electric potential. There is a wire at the top kept at a constant 100 V and a grounded wire (dashed) at the sides and bottom. *Right:* The electric potential as a function of x and y. The projections onto the shaded xy plane are equipotential surfaces or lines.

17.2 Unit I. Electrostatic Potentials

Your **problem** is to find the electric potential for all points *inside* the charge-free square shown in Figure 17.1. The bottom and sides of the region are made up of wires that are "grounded" (kept at 0 V). The top wire is connected to a battery that keeps it at a constant 100 V.

17.2.1 Laplace's Elliptic PDE (Theory)

We consider the entire square in Figure 17.1 as our boundary with the voltages prescribed upon it. If we imagine infinitesimal insulators placed at the top corners of the box, then we will have a closed boundary within which we will solve our problem. Since the values of the potential are given on all sides, we have Neumann conditions on the boundary and, according to Table 17.1, a unique and stable solution.

It is known from classical electrodynamics that the electric potential $U(\mathbf{x})$ arising from static charges satisfies Poisson's PDE [Jack 88]:

$$\nabla^2 U(\mathbf{x}) = -4\pi\rho(\mathbf{x}), \tag{17.3}$$

where $\rho(\mathbf{x})$ is the charge density. In charge-free regions of space, that is, regions where $\rho(\mathbf{x}) = 0$, the potential satisfies *Laplace's equation*:

$$\nabla^2 U(\mathbf{x}) = 0. \tag{17.4}$$

Both these equations are elliptic PDEs of a form that occurs in various applications. We solve them in 2-D rectangular coordinates:

$$\frac{\partial^2 U(x,y)}{\partial x^2} + \frac{\partial^2 U(x,y)}{\partial y^2} = \begin{cases} 0, & \text{Laplace's equation,} \\ -4\pi\rho(\mathbf{x}), & \text{Poisson's equation.} \end{cases} \tag{17.5}$$

In both cases we see that the potential depends simultaneously on x and y. For Laplace's equation, the charges, which are the source of the field, enter indirectly by specifying the potential values in some region of space; for Poisson's equation they enter directly.

17.3 Fourier Series Solution of a PDE

For the simple geometry of Figure 17.1 an analytic solution of Laplace's equation

$$\frac{\partial^2 U(x,y)}{\partial x^2} + \frac{\partial^2 U(x,y)}{\partial y^2} = 0 \tag{17.6}$$

exists in the form of an infinite series. If we assume that the solution is the product of independent functions of x and y and substitute the product into (17.6), we obtain

$$U(x,y) = X(x)Y(y) \;\Rightarrow\; \frac{d^2X(x)/dx^2}{X(x)} + \frac{d^2Y(y)/dy^2}{Y(y)} = 0. \tag{17.7}$$

Because $X(x)$ is a function of only x, and $Y(y)$ of only y, the derivatives in (17.7) are *ordinary* as opposed to *partial* derivatives. Since $X(x)$ and $Y(y)$ are assumed to be independent, the only way (17.7) can be valid for *all* values of x and y is for each term in (17.7) to be equal to a constant:

$$\frac{d^2Y(y)/dy^2}{Y(y)} = -\frac{d^2X(x)/dx^2}{X(x)} = k^2, \tag{17.8}$$

$$\Rightarrow \quad \frac{d^2X(x)}{dx^2} + k^2X(x) = 0, \qquad \frac{d^2Y(y)}{dy^2} - k^2Y(y) = 0. \tag{17.9}$$

We shall see that this choice of sign for the constant matches the boundary conditions and gives us periodic behavior in x. The other choice of sign would give periodic behavior in y, and that would not work with these boundary conditions.

The solutions for $X(x)$ are periodic, and those for $Y(y)$ are exponential:

$$X(x) = A\sin kx + B\cos kx, \quad Y(y) = Ce^{ky} + De^{-ky}. \tag{17.10}$$

The $x = 0$ boundary condition $U(x = 0, y) = 0$ can be met only if $B = 0$. The $x = L$ boundary condition $U(x = L, y) = 0$ can be met only for

$$kL = n\pi, \quad n = 1, 2, \ldots . \tag{17.11}$$

Such being the case, for each value of n there is the solution

$$X_n(x) = A_n \sin\left(\frac{n\pi}{L}x\right). \tag{17.12}$$

For each value of k_n that satisfies the x boundary conditions, $Y(y)$ must satisfy the y boundary condition $U(x, 0) = 0$, which requires $D = -C$:

$$Y_n(y) = C(e^{k_n y} - e^{-k_n y}) \equiv 2C \sinh\left(\frac{n\pi}{L}y\right). \tag{17.13}$$

Because we are solving linear equations, the principle of linear superposition holds, which means that the most general solution is the sum of the products:

$$U(x, y) = \sum_{n=1}^{\infty} E_n \sin\left(\frac{n\pi}{L}x\right) \sinh\left(\frac{n\pi}{L}y\right). \tag{17.14}$$

The E_n values are arbitrary constants and are fixed by requiring the solution to satisfy the remaining boundary condition at $y = L$, $U(x, y = L) = 100$ V:

$$\sum_{n=1}^{\infty} E_n \sin\frac{n\pi}{L}x \sinh n\pi = 100\,\text{V}. \tag{17.15}$$

We determine the constants E_n by projection: Multiply both sides of the equation by $\sin m\pi/Lx$, with m an integer, and integrate from 0 to L:

$$\sum_{n}^{\infty} E_n \sinh n\pi \int_0^L dx \sin\frac{n\pi}{L}x \sin\frac{m\pi}{L}x = \int_0^L dx\, 100 \sin\frac{m\pi}{L}x. \tag{17.16}$$

The integral on the LHS is nonzero only for $n = m$, which yields

$$E_n = \begin{cases} 0, & \text{for } n \text{ even}, \\ \frac{4(100)}{n\pi \sinh n\pi}, & \text{for } n \text{ odd}. \end{cases} \tag{17.17}$$

Finally, we obtain the potential at any point (x, y) as

$$U(x, y) = \sum_{n=1,3,5,\ldots}^{\infty} \frac{400}{n\pi} \sin\left(\frac{n\pi x}{L}\right) \frac{\sinh(n\pi y/L)}{\sinh(n\pi)}. \tag{17.18}$$

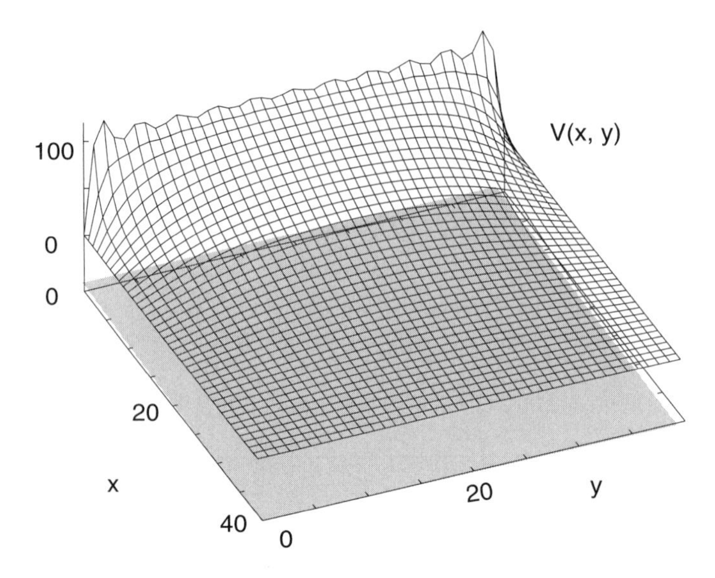

Figure 17.2 The analytic (Fourier series) solution of Laplace's equation showing Gibbs-overshoot oscillations near $x = 0$. The solution shown here uses 21 terms, yet the oscillations remain even if a large number of terms is summed.

17.3.1 Polynomial Expansion As an Algorithm

It is worth pointing out that even though a product of separate functions of x and y is an acceptable form for a solution to Laplace's equation, this does not mean that the solution to realistic problems will have this form. Indeed, a realistic solution can be expressed as an infinite sum of such products, but the sum is no longer separable. Worse than that, as an algorithm, we must stop the sum at some point, yet the series converges so painfully slowly that many terms are needed, and so round-off error may become a problem. In addition, the sinh functions in (17.18) overflow for large n, which can be avoided somewhat by expressing the quotient of the two sinh functions in terms of exponentials and then taking a large n limit:

$$\frac{\sinh(n\pi y/L)}{\sinh(n\pi)} = \frac{e^{n\pi(y/L-1)} - e^{-n\pi(y/L+1)}}{1 - e^{-2n\pi}} \rightarrow e^{n\pi(y/L-1)}. \tag{17.19}$$

A third problem with the "analytic" solution is that a Fourier series converges only in the *mean square* (Figure 17.2). This means that it converges to the *average* of the left- and right-hand limits in the regions where the solution is discontinuous [Krey 98], such as in the corners of the box. Explicitly, what you see in Figure 17.2 is a phenomenon known as the **Gibbs overshoot** that occurs when a Fourier series with a finite number of terms is used to represent a discontinuous function. Rather

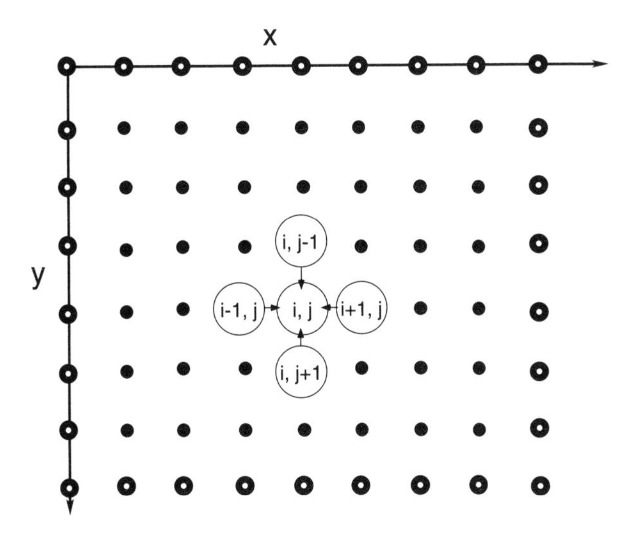

Figure 17.3 The algorithm for Laplace's equation in which the potential at the point $(x, y) = (i, j)\Delta$ equals the average of the potential values at the four nearest-neighbor points. The nodes with white centers correspond to fixed values of the potential along the boundaries.

than fall off abruptly, the series develops large oscillations that tend to overshoot the function at the corner. To obtain a smooth solution, we had to sum 40,000 terms, where, in contrast, the numerical solution required only hundreds of iterations.

17.4 Solution: Finite-Difference Method

To solve our 2-D PDE numerically, we divide space up into a lattice (Figure 17.3) and solve for U at each site on the lattice. Since we will express derivatives in terms of the finite differences in the values of U at the lattice sites, this is called a *finite-difference* method. A numerically more efficient, but also more complicated approach, is the *finite-element* method (Unit II), which solves the PDE for small geometric elements and then matches the elements.

To derive the finite-difference algorithm for the numeric solution of (17.5), we follow the same path taken in § 7.1 to derive the forward-difference algorithm for differentiation. We start by adding the two Taylor expansions of the potential to the right and left of (x, y) and above and below (x, y):

$$U(x+\Delta x, y) = U(x, y) + \frac{\partial U}{\partial x}\Delta x + \frac{1}{2}\frac{\partial^2 U}{\partial x^2}(\Delta x)^2 + \cdots, \tag{17.20}$$

$$U(x-\Delta x, y) = U(x, y) - \frac{\partial U}{\partial x}\Delta x + \frac{1}{2}\frac{\partial^2 U}{\partial x^2}(\Delta x)^2 - \cdots. \tag{17.21}$$

All odd terms cancel when we add these equations, and we obtain a central-difference approximation for the second partial derivative good to order Δ^4:

$$\frac{\partial^2 U(x,y)}{\partial x^2} \simeq \frac{U(x+\Delta x,y)+U(x-\Delta x,y)-2U(x,y)}{(\Delta x)^2}, \tag{17.22}$$

$$\frac{\partial^2 U(x,y)}{\partial y^2} \simeq \frac{U(x,y+\Delta y)+U(x,y-\Delta y)-2U(x,y)}{(\Delta y)^2}. \tag{17.23}$$

Substituting both these approximations in Poisson's equation (17.5) leads to a finite-difference form of the PDE:

$$\frac{U(x+\Delta x,y)+U(x-\Delta x,y)-2U(x,y)}{(\Delta x)^2} + \frac{U(x,y+\Delta y)+U(x,y-\Delta y)-2U(x,y)}{(\Delta y)^2} = -4\pi\rho.$$

We assume that the x and y grids are of equal spacings $\Delta x = \Delta y = \Delta$, and so the algorithm takes the simple form

$$U(x+\Delta,y)+U(x-\Delta,y)+U(x,y+\Delta)+U(x,y-\Delta)-4U(x,y) = -4\pi\rho. \tag{17.24}$$

The reader will notice that this equation shows a relation among the solutions at five points in space. When $U(x,y)$ is evaluated for the N_x x values on the lattice and for the N_y y values, we obtain a set of $N_x \times N_y$ simultaneous linear algebraic equations for U[i][j] to solve. One approach is to solve these equations explicitly as a (big) matrix problem. This is attractive, as it is a direct solution, but it requires a great deal of memory and accounting. The approach we follow here is based on the algebraic solution of (17.24) for $U(x,y)$:

$$U(x,y) \simeq \frac{1}{4}\left[U(x+\Delta,y)+U(x-\Delta,y)+U(x,y+\Delta)+U(x,y-\Delta)\right] + \pi\rho(x,y)\Delta^2, \tag{17.25}$$

where we would omit the $\rho(x)$ term for Laplace's equation. In terms of discrete locations on our lattice, the x and y variables are

$$x = x_0 + i\Delta, \quad y = y_0 + j\Delta, \quad i,j = 0,\ldots,N_{\text{max-1}}, \tag{17.26}$$

where we have placed our lattice in a square of side L. The finite-difference algorithm (17.25) becomes

$$\boxed{U_{i,j} = \frac{1}{4}\left[U_{i+1,j}+U_{i-1,j}+U_{i,j+1}+U_{i,j-1}\right] + \pi\rho(i\Delta, j\Delta)\Delta^2.} \tag{17.27}$$

This equation says that when we have a proper solution, it will be the average of the potential at the four nearest neighbors (Figure 17.3) plus a contribution from the local charge density. As an algorithm, (17.27) does not provide a direct solution to Poisson's equation but rather must be repeated many times to converge upon the solution. We start with an initial guess for the potential, improve it by sweeping through all space taking the average over nearest neighbors at each node, and keep repeating the process until the solution no longer changes to some level of precision or until failure is evident. When converged, the initial guess is said to have *relaxed* into the solution.

A reasonable question with this simple an approach is, "Does it always converge, and if so, does it converge fast enough to be useful?" In some sense the answer to the first question is not an issue; if the method does not converge, then we will know it; otherwise we have ended up with a solution and the path we followed to get there does not matter! The answer to the question of speed is that relaxation methods may converge slowly (although still faster than a Fourier series), yet we will show you two clever tricks to accelerate the convergence.

At this point it is important to remember that our algorithm arose from expressing the Laplacian ∇^2 in rectangular coordinates. While this does not restrict us from solving problems with circular symmetry, there may be geometries where it is better to develop an algorithm based on expressing the Laplacian in cylindrical or spherical coordinates in order to have grids that fit the geometry better.

17.4.1 Relaxation and Overrelaxation

There are a number of ways in which the algorithm (17.25) can be iterated so as to convert the boundary conditions to a solution. Its most basic form is the *Jacobi method* and is one in which the potential values are not changed until an entire sweep of applying (17.25) at each point is completed. This maintains the symmetry of the initial guess and boundary conditions. A rather obvious improvement on the Jacobi method employs the updated guesses for the potential in (17.25) as soon as they are available. As a case in point, if the sweep starts in the upper-left-hand corner of Figure 17.3, then the leftmost `(i -1, j)` and topmost `(i, j-1)` values of the potential used will be from the present generation of guesses, while the other two values of the potential will be from the previous generation:

$$U^{(\text{new})}_{i,j} = \frac{1}{4}\left[U^{(\text{old})}_{i+1,j} + U^{(\text{new})}_{i-1,j} + U^{(\text{old})}_{i,j+1} + U^{(\text{new})}_{i,j-1}\right] \quad \text{(Gauss–Seidel method)} \quad (17.28)$$

This technique, known as the *Gauss–Seidel* (GS) *method*, usually leads to accelerated convergence, which in turn leads to less round-off error. It also uses less memory as there is no need to store two generations of guesses. However, it does distort the symmetry of the boundary conditions, which one hopes is insignificant when convergence is reached.

A less obvious improvement in the relaxation technique, known as *successive overrelaxation* (SOR), starts by writing the algorithm (17.25) in a form that determines the new values of the potential $U^{(\text{new})}$ as the old values $U^{(\text{old})}$ plus a correction or residual r:

$$U^{(\text{new})}_{i,j} = U^{(\text{old})}_{i,j} + r_{i,j}. \tag{17.29}$$

While the Gauss–Seidel technique may still be used to incorporate the updated values in $U^{(\text{old})}$ to determine r, we rewrite the algorithm here in the general form:

$$\begin{aligned} r_{i,j} &\equiv U^{(\text{new})}_{i,j} - U^{(\text{old})}_{i,j} \\ &= \frac{1}{4}\left[U^{(\text{old})}_{i+1,j} + U^{(\text{new})}_{i-1,j} + U^{(\text{old})}_{i,j+1} + U^{(\text{new})}_{i,j-1}\right] - U^{(\text{old})}_{i,j}. \end{aligned} \tag{17.30}$$

The successive overrelaxation technique [Pres 94, Gar 00] proposes that if convergence is obtained by adding r to U, then even more rapid convergence might be obtained by adding more or less of r:

$$\boxed{U^{(\text{new})}_{i,j} = U^{(\text{old})}_{i,j} + \omega r_{i,j},} \quad \text{(SOR)}, \tag{17.31}$$

where ω is a parameter that amplifies or reduces the residual. The nonaccelerated relaxation algorithm (17.28) is obtained with $\omega = 1$, accelerated convergence (overrelaxation) is obtained with $\omega \geq 1$, and underrelaxation is obtained with $\omega < 1$. Values of $1 \leq \omega \leq 2$ often work well, yet $\omega > 2$ may lead to numerical instabilities. Although a detailed analysis of the algorithm is needed to predict the optimal value for ω, we suggest that you explore different values for ω to see which one works best for your particular problem.

17.4.2 Lattice PDE Implementation

In Listing 17.1 we present the code `LaplaceLine.java` that solves the square-wire problem (Figure 17.1). Here we have kept the code simple by setting the length of the box $L = N_{\text{max}}\Delta = 100$ and by taking $\Delta = 1$:

$$\begin{aligned} U(i, N_{\text{max}}) &= 99 \quad \text{(top)}, & U(1, j) &= 0 \quad \text{(left)}, \\ U(N_{\text{max}}, j) &= \ 0 \quad \text{(right)}, & U(i, 1) &= 0 \quad \text{(bottom)}, \end{aligned} \tag{17.32}$$

We run the algorithm (17.27) for a fixed 1000 iterations. A better code would vary Δ and the dimensions and would quit iterating once the solution converges to some tolerance. Study, compile, and execute the basic code.

```java
/* LaplaceLine.java:    Laplace eqn via finite difference mthd
                          wire in a grounded box, Output for 3D gnuplot */
import java.io.*;

public class LaplaceLine {
  static int Nmax = 100;                                        // Size of box

  public static void main(String[] argv) throws IOException, FileNotFoundException {
    double V[][] = new double[Nmax][Nmax];
    int i, j, iter;
    PrintWriter w = new PrintWriter(new FileOutputStream("LaplaceLine.dat"), true);
    for (i=0;  i<Nmax;  i++) for (j=0;  j<Nmax;  j++) V[i][j] = 0.;     // Initialize
    for ( i=0;  i < Nmax;  i++ ) V[i][0] = 100. ;                       // V[i][0] = 100 V
    for ( iter=0;  iter < 1000;  iter++ ) {                             // Iterations
      for ( i=1;  i < (Nmax-1);  i++ )  {                               // x-direction
        for ( j=1;  j < (Nmax-1);  j++ )                                // y-direction
          { V[i][j] = (V[i+1][j] + V[i-1][j] + V[i][j+1] + V[i][j-1])/4.; }
      }
    }
    for ( i=0;  i < Nmax ;  i=i + 2)  {                         // Data in gnuplot format
      for ( j=0;  j < Nmax;  j=j + 2)  {w.println("" + V[i][j] + "");}
      w.println("");                                   // Blank line separates rows for gnuplot
    }
    System.out.println("data stored in LaplaceLine.dat");
} }                                                            // End main, class
```

Listing 17.1 LaplaceLine.java solves Laplace's equation via relaxation. The various parameters need to be adjusted for an accurate solution.

17.5 Assessment via Surface Plot

After executing LaplaceLine.java, you should have a file with data in a format appropriate for a surface plot like Figure 17.1. Seeing that it is important to visualize your output to ensure the reasonableness of the solution, you should learn how to make such a plot before exploring more interesting problems. The 3-D surface plots we show in this chapter were made with both *gnuplot* and *OpenDX* (Appendix C). Below we repeat the commands used for Gnuplot with the produced output file:

> **gnuplot**	Start Gnuplot system from a shell
gnuplot> **set hidden3d**	Hide surface whose view is blocked
gnuplot> **set unhidden3d**	Show surface though hidden from view
gnuplot> **splot 'Laplace.dat' with lines**	Surface plot of Laplace.dat with lines
gnuplot> **set view 65,45**	Set x and y rotation viewing angles
gnuplot> **replot**	See effect of your change
gnuplot> **set contour**	Project contours onto xy plane
gnuplot> **set cntrparam levels 10**	10 contour levels
gnuplot> **set terminal PostScript**	Output in PostScript format for printing
gnuplot> **set output "Laplace.ps"**	Output to file Laplace.ps
gnuplot> **splot 'Laplace.dat' w l**	Plot again, output to file
gnuplot> **set terminal x11**	To see output on screen again

gnuplot> **set title 'Potential V(x,y) vs x,y'**	Title graph
gnuplot> **set xlabel 'x Position'**	Label x axis
gnuplot> **set ylabel 'y Position'**	Label y axis
gnuplot> **set zlabel 'V(x,y)'; replot**	Label z axis and replot
gnuplot> **help**	Tell me more
gnuplot> **set nosurface**	Do not draw surface; leave contours
gnuplot> **set view 0, 0, 1**	Look down directly onto base
gnuplot> **replot**	Draw plot again; may want to write to file
gnuplot> **quit**	Get out of Gnuplot

Here we have explicitly stated the viewing angle for the surface plot. Because Gnuplot 4 and later versions permit you to rotate surface plots interactively, we recommend that you do just that to find the best viewing angle. Changes made to a plot are seen when you redraw the plot using the `replot` command. For this sample session the default output for your graph is your terminal screen. To print a paper copy of your graph we recommend first saving it to a file as a *PostScript* document (suffix .ps) and then printing out that file to a PostScript printer. You create the PostScript file by changing the terminal type to `Postscript`, setting the name of the file, and then issuing the subcommand `splot` again. This plots the result out to a file. If you want to see plots on your screen again, set the terminal type back to `x11` again (for Unix's *X Windows System*) and then plot it again.

17.6 Alternate Capacitor Problems

We give you (or your instructor) a choice now. You can carry out the assessment using our wire-plus-grounded-box problem, or you can replace that problem with a more interesting one involving a realistic capacitor or nonplanar capacitors. We now describe the capacitor problem and then move on to the assessment and exploration.

Elementary textbooks solve the capacitor problem for the uniform field confined between two infinite plates. The field in a finite capacitor varies near the edges (edge effects) and extends beyond the edges of the capacitor (fringe fields). We model the realistic capacitor in a grounded box (Figure 17.4) as two plates (wires) of finite length. Write your simulation such that it is convenient to vary the grid spacing Δ and the geometry of the box and plate. We pose three versions of this problem, each displaying somewhat different physics. In each case the boundary condition $V = 0$ on the surrounding box must be imposed for all iterations in order to obtain a unique solution.

1. For the simplest version, assume that the plates are very thin sheets of conductors, with the top plate maintained at 100 V and the bottom at -100 V. Because the plates are conductors, they must be equipotential surfaces, and a battery can maintain them at constant voltages. Write or modify the

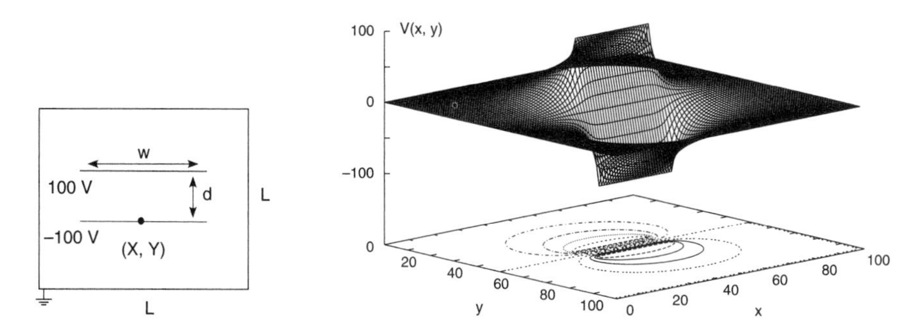

Figure 17.4 *Left:* A simple model of a parallel-plate capacitor within a box. A realistic model would have the plates close together, in order to condense the field, and the enclosing grounded box so large that it has no effect on the field near the capacitor. *Right:* A numerical solution for the electric potential for this geometry. The projection on the *xy* plane gives the equipotential lines.

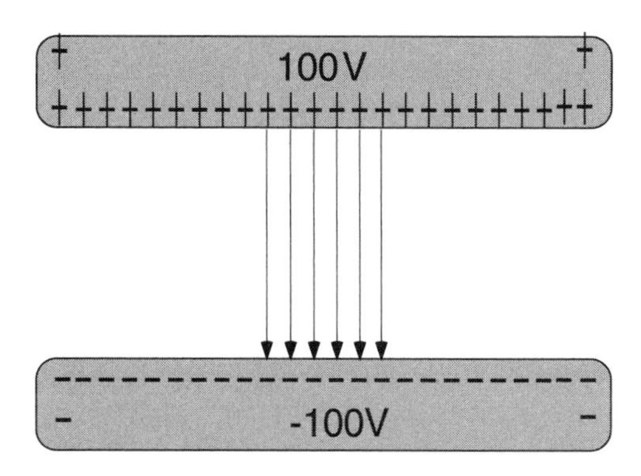

Figure 17.5 A guess as to how charge may rearrange itself on finite conducting plates.

given program to solve Laplace's equation such that the plates have fixed voltages.

2. For the next version of this problem, assume that the plates are composed of a line of dielectric material with uniform charge densities ρ on the top and $-\rho$ on the bottom. Solve Poisson's equation (17.3) in the region including the plates, and Laplace's equation elsewhere. Experiment until you find a numerical value for ρ that gives a potential similar to that shown in Figure 17.6 for plates with fixed voltages.
3. For the final version of this problem investigate how the charges on a capacitor with finite-thickness conducting plates (Figure 17.5) distribute themselves. Since the plates are conductors, they are still equipotential surfaces at 100 and $-100\,\text{V}$, only now they have a thickness of at least 2Δ (so we can see

the difference between the potential near the top and the bottom surfaces of the plates). Such being the case, we solve Laplace's equation (17.4) much as before to determine $U(x, y)$. Once we have $U(x, y)$, we substitute it into Poisson's equation (17.3) and determine how the charge density distributes itself along the top and bottom surfaces of the plates. *Hint:* Since the electric field is no longer uniform, we know that the charge distribution also will no longer be uniform. In addition, since the electric field now extends beyond the ends of the capacitor and since field lines begin and end on charge, some charge may end up on the edges and outer surfaces of the plates (Figure 17.4).

4. The numerical solution to our PDE can be applied to arbitrary boundary conditions. Two boundary conditions to explore are triangular and sinusoidal:

$$U(x) = \begin{cases} 200x/w, & \text{for } x \leq w/2, \\ 100(1 - x/w), & \text{for } x \geq w/2, \end{cases} \qquad U(x) = 100 \sin\left(\frac{2\pi x}{w}\right).$$

5. **Square conductors:** You have designed a piece of equipment consisting of a small metal box at 100 V within a larger grounded one (Figure 17.8). You find that sparking occurs between the boxes, which means that the electric field is too large. You need to determine where the field is greatest so that you can change the geometry and eliminate the sparking. Modify the program to satisfy these boundary conditions and to determine the field between the boxes. Gauss's law tells us that the field vanishes within the inner box because it contains no charge. Plot the potential and equipotential surfaces and sketch in the electric field lines. Deduce where the electric field is most intense and try redesigning the equipment to reduce the field.
6. **Cracked cylindrical capacitor:** You have designed the cylindrical capacitor containing a long outer cylinder surrounding a thin inner cylinder (Figure 17.8 right). The cylinders have a small crack in them in order to connect them to the battery that maintains the inner cylinder at -100 V and outer cylinder at 100 V. Determine how this small crack affects the field configuration. In order for a unique solution to exist for this problem, place both cylinders within a large grounded box. Note that since our algorithm is based on expansion of the Laplacian in rectangular coordinates, you cannot just convert it to a radial and angle grid.

17.7 Implementation and Assessment

1. Write or modify the CD program to find the electric potential for a capacitor within a grounded box. Use the labeling scheme on the left in Figure 17.4.
2. To start, have your program undertake 1000 iterations and then quit. During debugging, examine how the potential changes in some key locations as you iterate toward a solution.

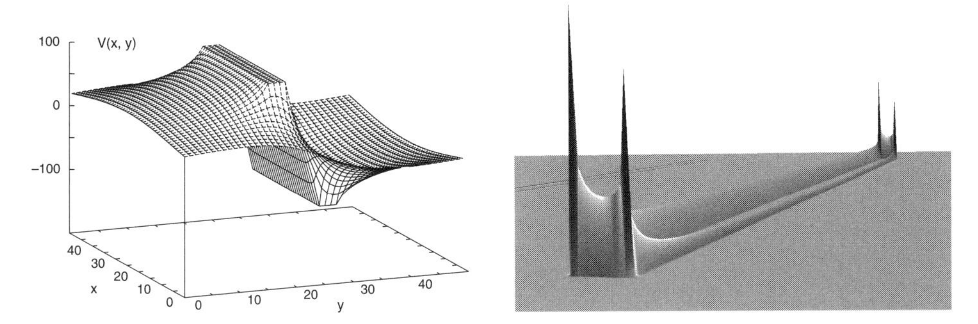

Figure 17.6 *Left:* A Gnuplot visualization of the computed electric potential for a capacitor with finite width plates. *Right:* An OpenDX visualization of the charge distribution along one plate determined by evaluating $\nabla^2 V(x, y)$ (courtesy of J. Wetzel). Note the "lightening rod" effect of charge accumulating at corners and points.

3. Repeat the process for different step sizes Δ and draw conclusions regarding the stability and accuracy of the solution.
4. Once your program produces reasonable solutions, modify it so that it stops iterating after convergence is reached, or if the number of iterations becomes too large. Rather than trying to discern small changes in highly compressed surface plots, use a numerical measure of precision, for example,

$$\texttt{trace} = \sum_i |\texttt{V[i][i]}|,$$

which samples the solution along the diagonal. Remember, this is a simple algorithm and so may require many iterations for high precision. You should be able to obtain changes in the trace that are less than 1 part in 10^4. (The `break` command or a `while` loop is useful for this type of test.)
5. Equation (17.31) expresses the **successive overrelaxation** technique in which convergence is accelerated by using a judicious choice of ω. Determine by trial and error the approximate best value of ω. You will be able to double the speed.
6. Now that the code is accurate, modify it to simulate a more realistic capacitor in which the plate separation is approximately $\frac{1}{10}$ of the plate length. You should find the field more condensed and more uniform between the plates.
7. If you are working with the wire-in-the-box problem, compare your numerical solution to the analytic one (17.18). Do not be surprised if you need to sum thousands of terms before the analytic solution converges! ∎

17.8 Electric Field Visualization (Exploration)

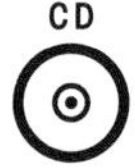
CD

Plot the equipotential surfaces on a separate 2-D plot. Start with a crude, hand-drawn sketch of the electric field by drawing curves orthogonal to the equipotential lines, beginning and ending on the boundaries (where the charges lie). The regions of high density are regions of high electric field. Physics tells us that the electric field **E** is the negative gradient of the potential:

$$\mathbf{E} = -\nabla U(x, y) = -\frac{\partial U(x, y)}{\partial x}\,\hat{\epsilon}_x - \frac{\partial U(x, y)}{\partial y}\,\hat{\epsilon}_y, \tag{17.33}$$

where $\hat{\epsilon}_i$ is a unit vector in the i direction. While at first it may seem that some work is involved in determining these derivatives, once you have a solution for $U(x, y)$ on a grid, it is simple to use the central-difference approximation for the derivative to determine the field, for example:

$$E_x \simeq \frac{U(x+\Delta, y) - U(x-\Delta, y)}{2\Delta} = \frac{U_{i+1,j} - U_{i-1,j}}{2\Delta}. \tag{17.34}$$

Once you have a data file representing such a vector field, it can be visualized by plotting arrows of varying lengths and directions, or with just lines (Figure 17.7). This is possible in Maple and Mathematica [L 05] or with `vectors` style in Gnuplot[2] where `N` is a normalization factor. ▮

17.9 Laplace Quiz

You are given a simple Laplace-type equation

$$\frac{\partial u}{\partial x} + \frac{\partial u}{\partial y} = -\rho(x, y),$$

where x and y are Cartesian spatial coordinates and $\rho(x, y)$ is the charge density in space.

1. Develop a simple algorithm that will permit you to solve for the potential u between two square conductors kept at fixed u, with a charge density ρ between them.
2. Make a simple sketch that shows with arrows how your algorithm works.
3. Make sure to specify how you start and terminate the algorithm.

[2] The Gnuplot command `plot "Laplace_field.dat" using 1:2:3:4 with Vectors` plots variable-length arrows at (x, y) with components `Dx` $\propto E_x$ and `Dy` $\propto E_y$. You determine empirically what scale factor gives you the best visualization (nonoverlapping arrows). Accordingly, you output data lines of the form `(x, y, Ex/N, Ey/N)`

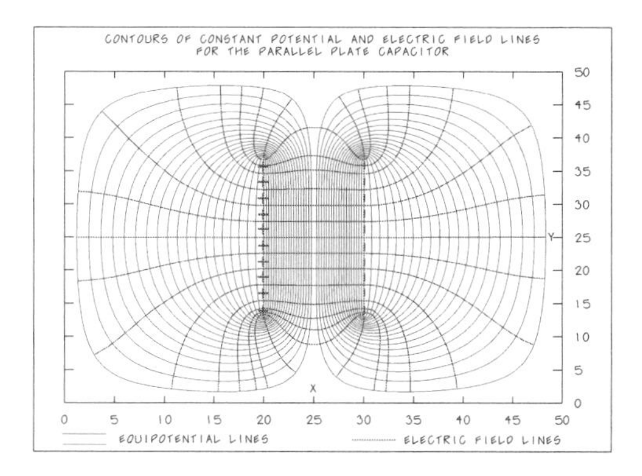

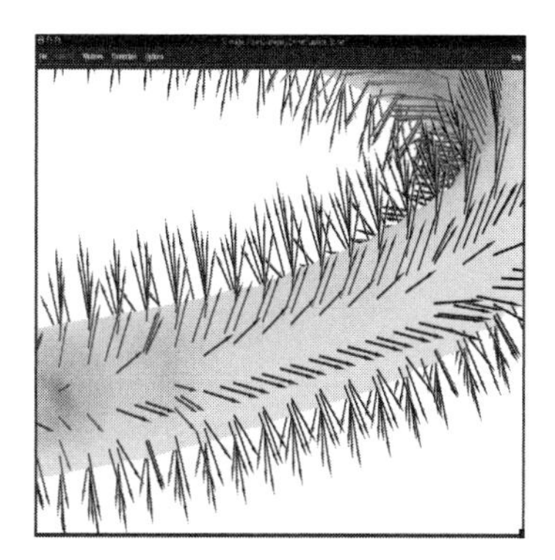

Figure 17.7 *Left:* Computed equipotential surfaces and electric field lines for a realistic capacitor. *Right:* Equipotential surfaces and electric field lines mapped onto the surface for a 3-D capacitor constructed from two tori (see OpenDX in Appendix C).

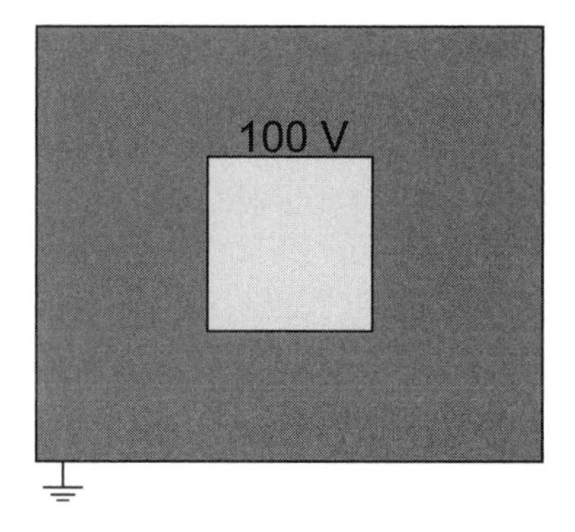

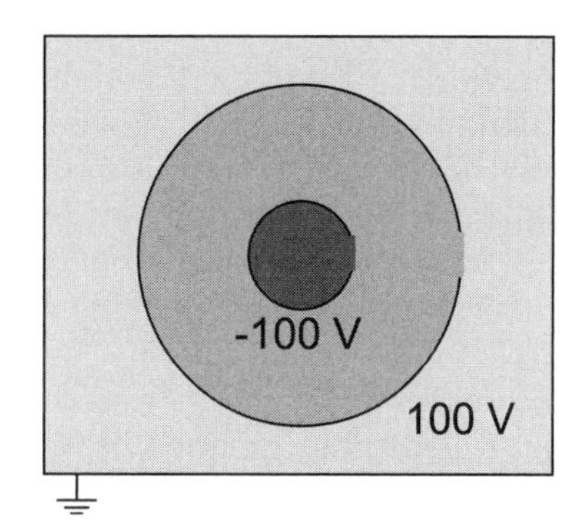

Figure 17.8 *Left:* The geometry of a capacitor formed by placing two long, square cylinders within each other. *Right:* The geometry of a capacitor formed by placing two long, circular cylinders within each other. The cylinders are cracked on the side so that wires can enter the region.

4. **Thinking outside the box**⊙: Find the electric potential for all points *outside* the charge-free square shown in Figure 17.1. Is your solution unique? ▌

17.10 Unit II. Finite-Element Method ⊙

In this unit we solve a simpler problem than the one in Unit I (1-D rather than 2-D), but we do it with a less simple algorithm (finite element). Our usual approach to PDEs in this text uses finite differences to approximate various derivatives in terms of the finite differences of a function evaluated upon a fixed grid. The finite-element method

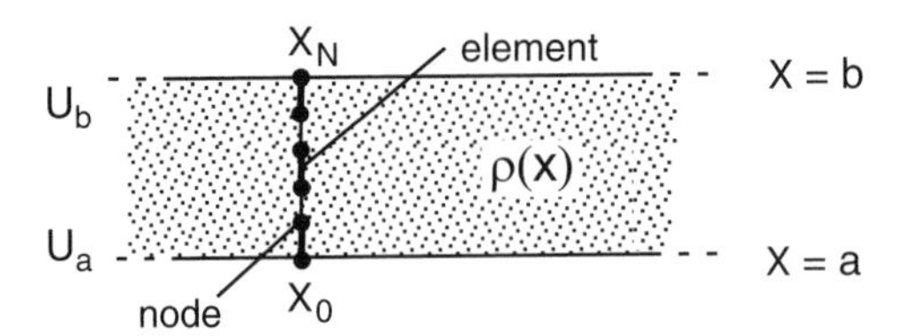

Figure 17.9 Two metal plates with a charge density between them. The dots are the nodes x_i, and the lines connecting the nodes are the finite elements.

(FEM), in contrast, breaks space up into multiple geometric objects (elements), determines an approximate form for the solution appropriate to each element, and then matches the solutions up at the domain edges.

The theory and practice of FEM as a numerical method for solving partial differential equations have been developed over the last 30 years and still provide an active field of research. One of the theoretical strengths of FEM is that its mathematical foundations allow for elegant proofs of the convergence of solutions to many delicate problems. One of the practical strengths of FEM is that it offers great flexibility for problems on irregular domains or highly varying coefficients or singularities. Although finite differences are simpler to implement than FEM, they are less robust mathematically and less efficient in terms of computer time for big problems. Finite elements in turn are more complicated to implement but more appropriate and precise for complicated equations and complicated geometries. In addition, the same basic finite-element technique can be applied to many problems with only minor modifications and yields solutions that may be evaluated for any value of x, not just those on a grid. In fact, the finite-elements method with various preprogrammed multigrid packages has very much become the standard for large-scale practical applications. Our discussion is based upon [Shaw 92, Li, Otto].

17.11 Electric Field from Charge Density (Problem)

You are given two conducting plates a distance $b-a$ apart, with the lower one kept at potential U_a, the upper plate at potential U_b, and a uniform charge density $\rho(x)$ placed between them (Figure 17.9). Your **problem** is to compute the electric potential between the plates.

17.12 Analytic Solution

The relation between charge density $\rho(\mathbf{x})$ and potential $U(\mathbf{x})$ is given by Poisson's equation (17.5). For our problem, the potential U changes only in the x direction, and so the PDE becomes the ODE:

$$\frac{d^2U(x)}{dx^2} = -4\pi\rho(x) = -1, \quad 0 < x < 1, \tag{17.35}$$

where we have set $\rho(x) = 1/4\pi$ to simplify the programming. The solution we want is subject to the Dirichlet boundary conditions:

$$U(x = a = 0) = 0, \quad U(x = b = 1) = 1, \tag{17.36}$$

$$\Rightarrow \quad U(x) = -\frac{x}{2}(x-3). \tag{17.37}$$

Although we know the analytic solution, we shall develop the finite-element method for solving the ODE as if it were a PDE (it would be in 2-D) and as if we did not know the solution. Although we will not demonstrate it, this method works equally well for any charge density $\rho(x)$.

17.13 Finite-Element (Not Difference) Methods

In a finite-element method, the domain in which the PDE is solved is split into finite subdomains, called *elements*, and a trial solution to the PDE in each subdomain is hypothesized. Then the parameters of the trial solution are adjusted to obtain a *best fit* (in the sense of Chapter 8, "Solving Systems of Equations with Matrices; Data Fitting") to the exact solution. The numerically intensive work is in finding the best values for these parameters and in matching the trial solutions for the different subdomains. A FEM solution follows six basic steps [Li]:

1. Derivation of a *weak form* of the PDE. This is equivalent to a least-squares minimization of the integral of the difference between the approximate and exact solutions.
2. Discretization of the computational domains.
3. Generation of interpolating or trial functions.
4. Assembly of the *stiffness matrix* and *load vector*.
5. Implementation of the boundary conditions.
6. Solution of the resulting linear system of equations.

17.13.1 Weak Form of PDE

Finite-difference methods look for an approximate solution of an approximate PDE. Finite-element methods strive to obtain the best possible global agreement of an approximate trial solution with the exact solution. We start with the differential equation

$$-\frac{d^2U(x)}{dx^2} = 4\pi\rho(x). \tag{17.38}$$

A measure of overall agreement must involve the solution $U(x)$ over some region of space, such as the integral of the trial solution. We can obtain such a measure by converting the differential equation (17.38) to its equivalent integral or *weak* form. We assume that we have an approximate or *trial solution* $\Phi(x)$ that vanishes at the endpoints, $\Phi(a) = \Phi(b) = 0$ (we satisfy the boundary conditions later). We next multiply both sides of the differential equation (17.38) by Φ:

$$-\frac{d^2U(x)}{dx^2}\Phi(x) = 4\pi\rho(x)\Phi(x). \tag{17.39}$$

Next we integrate (17.39) by parts from a to b:

$$-\int_a^b dx\, \frac{d^2U(x)}{dx^2}\Phi(x) = \int_a^b dx\, 4\pi\rho(x)\,\Phi(x), \tag{17.40}$$

$$-\frac{dU(x)}{dx}\Phi(x)\,|_a^b + \int_a^b dx\, \frac{dU(x)}{dx}\Phi'(x) = \int_a^b dx\, 4\pi\rho(x)\,\Phi(x),$$

$$\Rightarrow \boxed{\int_a^b dx\, \frac{dU(x)}{dx}\Phi'(x) = \int_a^b dx\, 4\pi\rho(x)\,\Phi(x).} \tag{17.41}$$

Equation (17.41) is the weak form of the PDE. The unknown exact solution $U(x)$ and the trial function Φ are still to be specified. Because the approximate and exact solutions are related by the integral of their difference over the entire domain, the solution provides a global best fit to the exact solution.

17.13.2 Galerkin Spectral Decomposition

The approximate solution to a weak PDE is found via a stepwise procedure. We split the full domain of the PDE into subdomains called *elements*, find approximate solutions within each element, and then match the elemental solutions onto each other. For our 1-D problem the subdomain elements are straight lines of equal length, while for a 2-D problem, the elements can be parallelograms or triangles (Figure 17.9). Although life is simpler if all the finite elements are the same size, this is not necessary. Indeed, higher precision and faster run times may be obtained by picking small domains in regions where the solution is known to vary rapidly, and picking large domains in regions of slow variation.

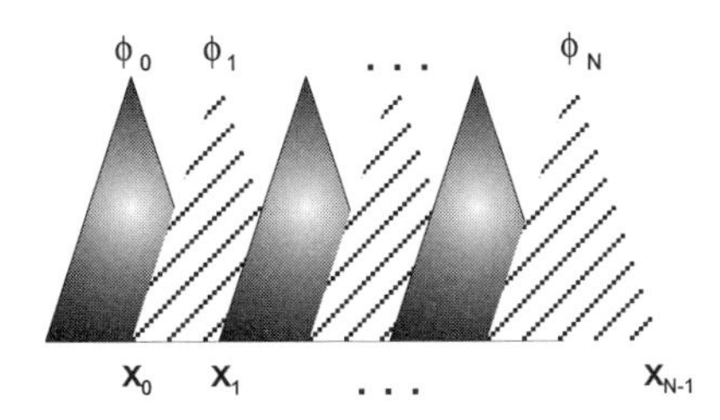

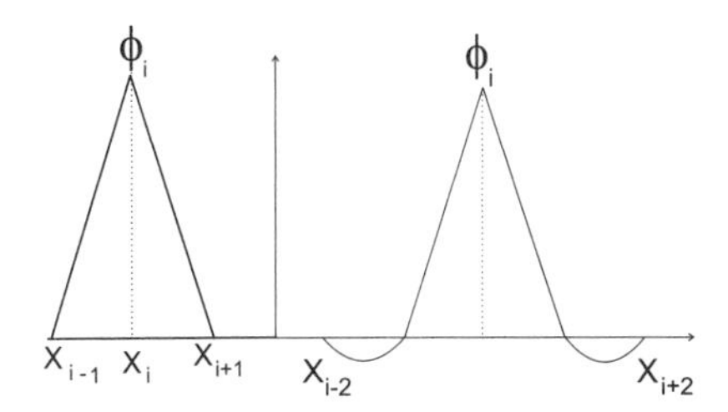

Figure 17.10 *Left:* A set of overlapping basis functions ϕ_i. Each function is a triangle from x_{i-1} to x_{i+1}. *Middle:* A Piecewise-linear function. *Right:* A piecewise-quadratic function.

The critical step in the finite-element method is expansion of the trial solution U in terms of a set of basis functions ϕ_i:

$$U(x) \simeq \sum_{j=0}^{N-1} \alpha_j \phi_j(x). \tag{17.42}$$

We chose ϕ_i's that are convenient to compute with and then determine the unknown expansion coefficients α_j. Even if the basis functions are not sines or cosines, this expansion is still called a *spectral* decomposition. In order to satisfy the boundary conditions, we will later add another term to the expansion. Considerable study has gone into the effectiveness of different basis functions. If the sizes of the finite elements are made sufficiently small, then good accuracy is obtained with simple piecewise-continuous basis functions, such as the triangles in Figure 17.10. Specifically, we use basis functions ϕ_i that form a triangle or "hat" between x_{i-1} and x_{i+1} and equal 1 at x_i:

$$\phi_i(x) = \begin{cases} 0, & \text{for } x < x_{i-1}, \text{ or } x > x_{i+1}, \\ \frac{x - x_{i-1}}{h_{i-1}}, & \text{for } x_{i-1} \le x \le x_i, \\ \frac{x_{i+1} - x}{h_i}, & \text{for } x_i \le x \le x_{i+1}, \end{cases} \qquad (h_i = x_{i+1} - x_i). \tag{17.43}$$

Because we have chosen $\phi_i(x_i) = 1$, the values of the expansion coefficients α_i equal the values of the (still-to-be-determined) solution at the nodes:

$$U(x_i) \simeq \sum_{i=0}^{N-1} \alpha_i \phi_i(x_i) = \alpha_i \phi_i(x_i) = \alpha_i, \tag{17.44}$$

$$\Rightarrow \quad U(x) \simeq \sum_{j=0}^{N-1} U(x_j) \phi_j(x). \tag{17.45}$$

Consequently, you can think of the hat functions as linear interpolations between the solution at the nodes.

17.13.2.1 SOLUTION VIA LINEAR EQUATIONS

Because the basis functions ϕ_i in (17.42) are known, solving for $U(x)$ involves determining the coefficients α_j, which are just the unknown values of $U(x)$ on the nodes. We determine those values by substituting the expansions for $U(x)$ and $\Phi(x)$ into the weak form of the PDE (17.41) and thereby convert them to a set of simultaneous linear equations (in the standard matrix form):

$$\mathbf{A}\mathbf{y} = \mathbf{b}. \tag{17.46}$$

We substitute the expansion $U(x) \simeq \sum_{j=0}^{N-1} \alpha_j \phi_j(x)$ into the weak form (17.41):

$$\int_a^b dx \frac{d}{dx}\left(\sum_{j=0}^{N-1} \alpha_j \phi_j(x)\right) \frac{d\Phi}{dx} = \int_a^b dx 4\pi\rho(x)\Phi(x).$$

By successively selecting $\Phi(x) = \phi_0, \phi_1, \ldots, \phi_{N-1}$, we obtain N simultaneous linear equations for the unknown α_j's:

$$\int_a^b dx \frac{d}{dx}\left(\sum_{j=0}^{N-1} \alpha_j \phi_j(x)\right) \frac{d\phi_i}{dx} = \int_a^b dx\, 4\pi\rho(x)\phi_i(x), \quad i = 0, \ldots, N-1. \tag{17.47}$$

We factor out the unknown α_j's and write out the equations explicitly:

$$\begin{aligned}
\alpha_0 \int_a^b \phi_0'\phi_0'\, dx + \alpha_1 \int_a^b \phi_0'\phi_1'\, dx + \cdots + \alpha_{N-1} \int_a^b \phi_0'\phi_{N-1}'\, dx &= \int_a^b 4\pi\rho\phi_0\, dx, \\
\alpha_0 \int_a^b \phi_1'\phi_0'\, dx + \alpha_1 \int_a^b \phi_1'\phi_1'\, dx + \cdots + \alpha_{N-1} \int_a^b \phi_1'\phi_{N-1}'\, dx &= \int_a^b 4\pi\rho\phi_1\, dx, \\
&\ddots \\
\alpha_0 \int_a^b \phi_{N-1}'\phi_0'\, dx + \alpha_1 \int \cdots + \alpha_{N-1} \int_a^b \phi_{N-1}'\phi_{N-1}'\, dx &= \int_a^b 4\pi\rho\phi_{N-1}\, dx.
\end{aligned}$$

Because we have chosen the ϕ_i's to be the simple hat functions, the derivatives are easy to evaluate analytically (otherwise they can be done numerically):

$$\frac{d\phi_{i,i+1}}{dx} = \begin{cases} 0, & x < x_{i-1}, \text{ or } x_{i+1} < x, \\ \frac{1}{h_{i-1}}, & x_{i-1} \le x \le x_i, \\ \frac{-1}{h_i}, & x_i \le x \le x_{i+1}, \end{cases} \quad \begin{cases} 0, & x < x_i, \text{ or } x_{i+2} < x \\ \frac{1}{h_i}, & x_i \le x \le x_{i+1}, \\ \frac{-1}{h_{i+1}}, & x_{i+1} \le x \le x_{i+2}. \end{cases} \tag{17.48}$$

The integrals to evaluate are

$$\int_{x_{i-1}}^{x_{i+1}} dx (\phi_i^{'})^2 = \int_{x_{i-1}}^{x_i} dx \frac{1}{(h_{i-1})^2} + \int_{x_i}^{x_{i+1}} dx \frac{1}{h_i^2} = \frac{1}{h_{i-1}} + \frac{1}{h_i},$$

$$\int_{x_{i-1}}^{x_{i+1}} dx\, \phi_i^{'} \phi_{i+1}^{'} = \int_{x_{i-1}}^{x_{i+1}} dx\, \phi_{i+1}^{'} \phi_i^{'} = \int_{x_i}^{x_{i+1}} dx \frac{-1}{h_i^2} = -\frac{1}{h_i},$$

$$\int_{x_{i-1}}^{x_{i+1}} dx\, (\phi_{i+1}^{'})^2 = \int_{x_i}^{x_{i+1}} dx\, (\phi_{i+1}^{'})^2 = \int_{x_i}^{x_{i+1}} dx\, \frac{+1}{h_i^2} = +\frac{1}{h_i}.$$

We rewrite these equations in the standard matrix form (17.46) with $\mathbf{y}$ constructed from the unknown α_j's, and the tridiagonal stiffness matrix $\mathbf{A}$ constructed from the integrals over the derivatives:

$$\mathbf{y} = \begin{pmatrix} \alpha_0 \\ \alpha_1 \\ \ddots \\ \alpha_{N-1} \end{pmatrix}, \quad \mathbf{b} = \begin{pmatrix} \int_{x_0}^{x_1} dx\, 4\pi\rho(x)\phi_0(x) \\ \int_{x_1}^{x_2} dx\, 4\pi\rho(x)\phi_1(x) \\ \ddots \\ \int_{x_{N-1}}^{x_N} dx 4\pi\rho(x)\phi_{N-1}(x) \end{pmatrix}, \tag{17.49}$$

$$\mathbf{A} = \begin{pmatrix} \frac{1}{h_0} + \frac{1}{h_1} & -\frac{1}{h_1} & -\frac{1}{h_0} & 0 & \cdots \\ -\frac{1}{h_1} & \frac{1}{h_1} + \frac{1}{h_2} & -\frac{1}{h_2} & 0 & \cdots \\ 0 & -\frac{1}{h_2} & \frac{1}{h_2} + \frac{1}{h_3} & -\frac{1}{h_3} & \cdots \\ \ddots & \ddots & -\frac{1}{h_{N-1}} & -\frac{1}{h_{N-2}} & \frac{1}{h_{N-2}} + \frac{1}{h_{N-1}} \end{pmatrix}. \tag{17.50}$$

The elements in $\mathbf{A}$ are just combinations of inverse step sizes and so do not change for different charge densities $\rho(x)$. The elements in $\mathbf{b}$ do change for different ρ's, but the required integrals can be performed analytically or with Gaussian quadrature (Chapter 6, "Integration"). Once $\mathbf{A}$ and $\mathbf{b}$ are computed, the matrix equations are solved for the expansion coefficients α_j contained in $\mathbf{y}$.

17.13.2.2 DIRICHLET BOUNDARY CONDITIONS

Because the basis functions vanish at the endpoints, a solution expanded in them also vanishes there. This will not do in general, and so we add the particular solution $U_a\phi_0(x)$, which satisfies the boundary conditions [Li]:

$$U(x) = \sum_{j=0}^{N-1} \alpha_j \phi_j(x) + U_a \phi_N(x) \qquad \text{(satisfies boundary conditions)}, \tag{17.51}$$

where $U_a = U(x_a)$. We substitute $U(x) - U_a\phi_0(x)$ into the weak form to obtain $(N+1)$ simultaneous equations, still of the form $\mathbf{Ay} = \mathbf{b}$ but now with

$$\mathbf{A} = \begin{pmatrix} A_{0,0} & \cdots & A_{0,N-1} & 0 \\ & \ddots & & \\ A_{N-1,0} & \cdots & A_{N-1,N-1} & 0 \\ 0 & 0 & \cdots & 1 \end{pmatrix}, \quad \mathbf{b}' = \begin{pmatrix} b_0 - A_{0,0}U_a \\ \ddots \\ b_{N-1} - A_{N-1,0}U_a \\ U_a \end{pmatrix}. \tag{17.52}$$

This is equivalent to adding a new element and changing the load vector:

$$b'_i = b_i - A_{i,0}U_a, \quad i = 1, \ldots, N-1, \quad b'_N = U_a. \tag{17.53}$$

To impose the boundary condition at $x = b$, we again add a term and substitute into the weak form to obtain

$$b'_i = b_i - A_{i,N-1}U_b, \quad i = 1, \ldots, N-1 \quad b'_N = U_b. \tag{17.54}$$

We now solve the linear equations $\mathbf{Ay} = \mathbf{b}'$. For 1-D problems, 100–1000 equations are common, while for 3-D problems there may be millions. Because the number of calculations varies approximately as N^2, it is important to employ an efficient and accurate algorithm because round-off error can easily accumulate after thousands of steps. We recommend one from a scientific subroutine library (see Chapter 8, "Solving Systems of Equations with Matrices; Data Fitting").

17.14 FEM Implementation and Exercises

In Listing 17.2 we give our program `LaplaceFEM.java` that determines the FEM solution, and in Figure 17.11 we show that solution. We see on the left that three elements do not provide good agreement with the analytic result, whereas $N = 11$ elements produces excellent agreement.

1. Examine the FEM solution for the choice of parameters

$$a = 0, \quad b = 1, \quad U_a = 0, \quad U_b = 1.$$

2. Generate your own triangulation by assigning explicit x values at the nodes over the interval $[0, 1]$.
3. Start with $N = 3$ and solve the equations for N values up to 1000.
4. Examine the stiffness matrix $\mathbf{A}$ and ensure that it is triangular.
5. Verify that the integrations used to compute the load vector $\mathbf{b}$ are accurate.
6. Verify that the solution of the linear equation $\mathbf{Ay} = \mathbf{b}$ is correct.
7. Plot the numerical solution for $U(x)$ for $N = 10$, 100, and 1000 and compare with the analytic solution.

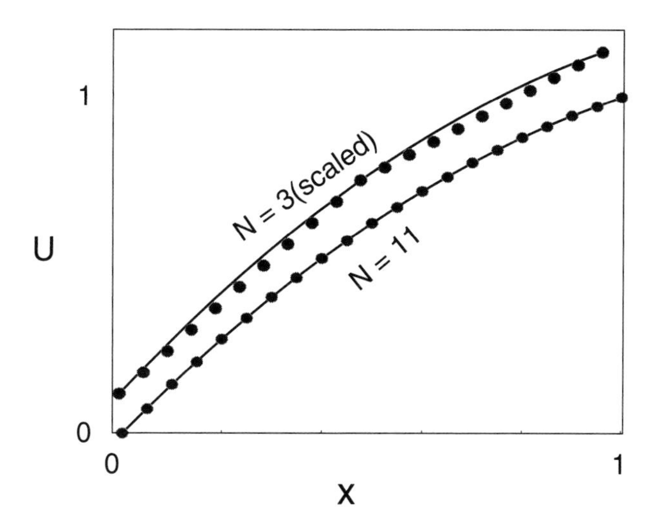

Figure 17.11 Exact (line) *versus* FEM solution (points) for the two-plate problem for $N = 3$ and $N = 11$ finite elements.

8. The log of the relative global error (number of significant figures) is

$$\mathcal{E} = \log_{10} \left| \frac{1}{b-a} \int_a^b dx \frac{U_{\text{FEM}}(x) - U_{\text{exact}}(x)}{U_{\text{exact}}(x)} \right| .$$

Plot the global error *versus* x for $N = 10$, 100, and 1000.

```
/*  LaplaceFEM3.java, solution of 1D Poisson equation
  using Finite Element Method with Galerkin approximation  */
import Jama.*;
import java.io.*;

public class LaplaceFEM3  {

  public static void main(String[] argv) throws IOException, FileNotFoundException {
    PrintWriter w =  new PrintWriter(new FileOutputStream("fem3.dat"), true);
    PrintWriter q =  new PrintWriter(new FileOutputStream("fem3t.dat"), true);
    PrintWriter t =  new PrintWriter(new FileOutputStream("fem3e.dat"), true);
    int i, j; int N = 11;
    double u[] = new double[N], A[][] = new double[N][N], b[][] = new double[N][1];
    double x2[] = new double[21], u_fem[] = new double[21], u_exact[] =new double
        [21];
    double error[] = new double[21], x[] = new double[N], h = 1./(N-1);
    for ( i=0; i <= N-1; i++ )  {
      x[i] = i*h;
      System.out.println("" + x[i] + "");
    }
    for ( i=0; i <= N-1; i++ ) {                                      // Initialize
      b[i][0] = 0. ;
      for ( j=0; j <= N-1; j++ )  A[i][j] = 0. ;
    }
```

```
    for ( i=1; i <= N-1; i++ )  {
      A[i-1][i-1] = A[i-1][i-1] + 1./h;
      A[i-1][i]   = A[i-1][i] - 1./h;
      A[i][i-1]   = A[i-1][i];
      A[i][i]     = A[i][i] + 1./h;
      b[i-1][0]   = b[i-1][0] + int2(x[i-1], x[i]);
      b[i][0]     = b[i][0] + int1(x[i-1], x[i]);
    }

    for ( i=1; i <= N-1; i++ ) {                      // Dirichlet BC @ left end
      b[i][0] = b[i][0]-0.*A[i][0];
      A[i][0] = 0. ;
      A[0][i] = 0. ;
    }
    A[0][0] = 1. ;
    b[0][0] = 0. ;
    for ( i=1; i <= N-1; i++ )  {                     // Dirichlet bc @ right end
      b[i][0] = b[i][0]-1.*A[i][N-1];
      A[i][N-1] = 0. ;
      A[N-1][i] = 0. ;
    }
    A[N-1][N-1] = 1. ;
    b[N-1][0] = 1. ;
    Matrix A1 = new Matrix(A);                                 // Jama matrix object
    Matrix b1 = new Matrix(b);
    A1.print (16, 14);                                             // Jama print A1
    b1.print (16, 14);                                             // Jama print b1
    Matrix sol = A1.solve(b1);                          // Jama solves linear system
    sol.print (16, 14);                                      // Jama print solution
    for ( i=0; i <= N-1; i++ ) u[i] += sol.get(i, 0);            // Get solution
    for ( i=0; i <= 20; i++ )  x2[i]=0.05*i;
    for ( i=0; i <= x2.length-1; i++ ) {
      u_fem[i] = numerical(x, u, x2[i]);
      u_exact[i] = exact(x2[i]);
      q.println(" " + 0.05*i + " " + u_exact[i] + " ");
      w.println(" " + 0.05*i + " " + u_fem[i] + " ");
      error[i] = u_fem[i]-u_exact[i];                             // Global error
      t.println(" " + 0.05*i + " " + error[i] + " ");
    }
}

  public static double int1 (double min, double max)  {                // Simpson
    int n, no = 1000;
    double interval, sum = 0., x;
    interval = ((max -min) /(no-1));
    for ( n=2;  n < no;  n += 2)  {                            // Loop odd points
      x = interval * (n-1);
      sum += 4 * f(x)*lin1(x, min, max);
    }
    for ( n=3;  n < no;  n += 2) {                            // Loop even points
      x = interval * (n-1);
      sum += 2 * f(x)*lin1(x, min, max);
    }
    sum += f(min)*lin1(min, min, max) + f(max)*lin1(max, min, max);
    sum *= interval/6.;
    return (sum);
  }

  public static double int2 (double min, double max)  {
    int n, no = 1000;
    double interval, sum = 0., x;
    interval = ((max -min) /(no-1));
    for ( n = 2;  n < no;  n += 2)     {                       // Loop odd points
      x = interval * (n-1);
      sum += 4 * f(x)*lin2(x, min, max);
    }
    for ( n=3;  n < no;  n += 2)  {                           // Loop even points
      x = interval * (n-1);
```

```
        sum += 2 * f(x)*lin2(x, min, max);
      }
      sum += f(min)*lin2(min, min, max) + f(max)*lin2(max, min, max);
      sum *= interval/6.;
      return (sum);
    }

    public static double int1f(double min, double max)  {                    // 2nd Simpson
      double xm, sum;
      xm = (min + max)*0.5;
      sum = (max-min)*(f(min)*lin1(min, min, max)
           + 4*f(xm)*lin1(xm, min, max) + f(max)*lin1(max, min, max) )/6;
      return sum;
    }

    public static double int2f(double min, double max)  {
      double xm, sum;
      xm = (min + max)*0.5;
      sum = (max-min)*(f(min)*lin2(min, min, max)
     + 4*f(xm)*lin2(xm, min, max) + f(max)*lin2(max, min, max))/6.;
    return sum;
   }

    public static double lin1(double x, double x1, double x2)                  // Hat funcs
         { return  (x-x1)/(x2-x1); }

    public static double lin2(double x, double x1, double x2)
         { return (x2-x)/(x2-x1); }

    public static double f(double x)                                   // RHS of the equation
     { return 1. ; }

    public static double numerical(double x[], double u[], double xp)   {
      int i, N = 11;                                      // interpolate numerical solution
      double y; y = 0. ;
      for ( i=0; i <= N-2; i++ )  {
        if ( xp >=x[i] && xp <= x[i + 1] )  y = lin2(xp, x[i],
                          x[i+1])*u[i] + lin1(xp, x[i], x[i + 1])*u[i + 1];
      }
    return y;
    }

    public static double exact(double x){                              // Analytic solution
      double u;
      u = -x*(x-3.)/2. ;
     return u;
    }
  }
```

Listing 17.2 LaplaceFEM.java provides a finite-element method solution of the 1-D Laplace equation via a Galerkin spectral decomposition. The resulting matrix equations are solved with Jama. Although the algorithm is more involved than the solution via relaxation (Listing 17.1), it is a direct solution with no iteration required.

17.15 Exploration

1. Modify your program to use piecewise-quadratic functions for interpolation and compare to the linear function results.

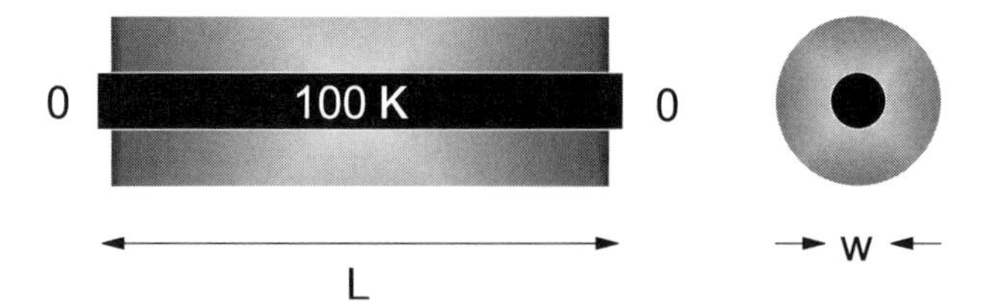

Figure 17.12 A metallic bar insulated along its length with its ends in contact with ice.

2. Explore the resulting electric field if the charge distribution between the plates has the explicit x dependences

$$\rho(x) = \frac{1}{4\pi} \begin{cases} \frac{1}{2} - x, \\ \sin x, \\ 1 \text{ at } x = 0, \quad -1 \text{ at } x = 1 \quad \text{(a capacitor).} \end{cases}$$

17.16 Unit III. Heat Flow via Time-Stepping (Leapfrogging)

Problem: You are given an aluminum bar of length $L = 1$ m and width w aligned along the x axis (Figure 17.12). It is insulated along its length but not at its ends. Initially the bar is at a uniform temperature of 100 K, and then both ends are placed in contact with ice water at 0 K. Heat flows out of the noninsulated ends only. Your **problem** is to determine how the temperature will vary as we move along the length of the bar at later times.

17.17 The Parabolic Heat Equation (Theory)

A basic fact of nature is that heat flows from hot to cold, that is, from regions of high temperature to regions of low temperature. We give these words mathematical expression by stating that the rate of heat flow $\mathbf{H}$ through a material is proportional to the gradient of the temperature T across the material:

$$\mathbf{H} = -K\nabla T(\mathbf{x}, t), \tag{17.55}$$

where K is the thermal conductivity of the material. The total amount of heat $Q(t)$ in the material at any one time is proportional to the integral of the temperature over the material's volume:

$$Q(t) = \int d\mathbf{x}\, C\rho(\mathbf{x})\, T(\mathbf{x}, t), \tag{17.56}$$

where C is the specific heat of the material and ρ is its density. Because energy is conserved, the rate of decrease in Q with time must equal the amount of heat flowing out of the material. After this energy balance is struck and the divergence theorem applied, the *heat equation* results:

$$\frac{\partial T(\mathbf{x},t)}{\partial t} = \frac{K}{C\rho}\nabla^2 T(\mathbf{x},t). \tag{17.57}$$

The heat equation (17.57) is a parabolic PDE with space and time as independent variables. The specification of this problem implies that there is no temperature variation in directions perpendicular to the bar (y and z), and so we have only one spatial coordinate in the Laplacian:

$$\frac{\partial T(x,t)}{\partial t} = \frac{K}{C\rho}\frac{\partial^2 T(x,t)}{\partial x^2}. \tag{17.58}$$

As given, the initial temperature of the bar and the boundary conditions are

$$T(x,t=0) = 100\,\mathrm{K}, \quad T(x=0,t) = T(x=L,t) = 0\,\mathrm{K}. \tag{17.59}$$

17.17.1 Solution: Analytic Expansion

Analogous to Laplace's equation, the analytic solution starts with the assumption that the solution separates into the product of functions of space and time:

$$T(x,t) = X(x)\mathcal{T}(t). \tag{17.60}$$

When (17.60) is substituted into the heat equation (17.58) and the resulting equation is divided by $X(x)\mathcal{T}(t)$, two noncoupled ODEs result:

$$\frac{d^2X(x)}{dx^2} + k^2X(x) = 0, \quad \frac{d\mathcal{T}(t)}{dt} + k^2\frac{C}{C\rho}\mathcal{T}(t) = 0, \tag{17.61}$$

where k is a constant still to be determined. The boundary condition that the temperature equals zero at $x = 0$ requires a sine function for X:

$$X(x) = A\sin kx. \tag{17.62}$$

The boundary condition that the temperature equals zero at $x = L$ requires the sine function to vanish there:

$$\sin kL = 0 \;\Rightarrow\; k = k_n = n\pi/L, \quad n = 1, 2, \ldots. \tag{17.63}$$

The time function is a decaying exponential with k in the exponent:

$$\mathcal{T}(t) = e^{-k_n^2 t/C\rho}, \;\Rightarrow\; T(x,t) = A_n \sin k_n x e^{-k_n^2 t/C\rho}, \tag{17.64}$$

where n can be any integer and A_n is an arbitrary constant. Since (17.58) is a linear equation, the most general solution is a linear superposition of all values of n:

$$T(x,t) = \sum_{n=1}^{\infty} A_n \sin k_n x \, e^{-k_n^2 t/C\rho}. \tag{17.65}$$

The coefficients A_n are determined by the initial condition that at time $t = 0$ the entire bar has temperature $T = 100$ K:

$$T(x, t=0) = 100 \;\Rightarrow\; \sum_{n=1}^{\infty} A_n \sin k_n x = 100. \tag{17.66}$$

Projecting the sine functions determines $A_n = 4T_0/n\pi$ for n odd, and so

$$T(x,t) = \sum_{n=1,3,\ldots}^{\infty} \frac{4T_0}{n\pi} \sin k_n x e^{-k_n^2 K t/(C\rho)}. \tag{17.67}$$

17.17.2 Solution: Time-Stepping

As we did with Laplace's equation, the numerical solution is based on converting the differential equation to a finite-difference ("difference") equation. We discretize space and time on a lattice (Figure 17.13) and solve for solutions on the lattice sites. The horizontal nodes with white centers correspond to the known values of the temperature for the initial time, while the vertical white nodes correspond to the fixed temperature along the boundaries. If we *also* knew the temperature for times along the bottom row, then we could use a relaxation algorithm as we did for Laplace's equation. However, with only the top row known, we shall end up with an algorithm that steps forward in time one row at a time, as in the children's game *leapfrog*.

As is often the case with PDEs, the algorithm is customized for the equation being solved and for the constraints imposed by the particular set of initial and boundary conditions. With only one row of times to start with, we use a forward-difference approximation for the time derivative of the temperature:

$$\frac{\partial T(x,t)}{\partial t} \simeq \frac{T(x, t+\Delta t) - T(x,t)}{\Delta t}. \tag{17.68}$$

Because we know the spatial variation of the temperature along the entire top row and the left and right sides, we are less constrained with the space derivative as with the time derivative. Consequently, as we did with the Laplace equation, we use the more accurate central-difference approximation for the space derivative:

$$\frac{\partial^2 T(x,t)}{\partial x^2} \simeq \frac{T(x+\Delta x, t) + T(x-\Delta x, t) - 2T(x,t)}{(\Delta x)^2}. \tag{17.69}$$

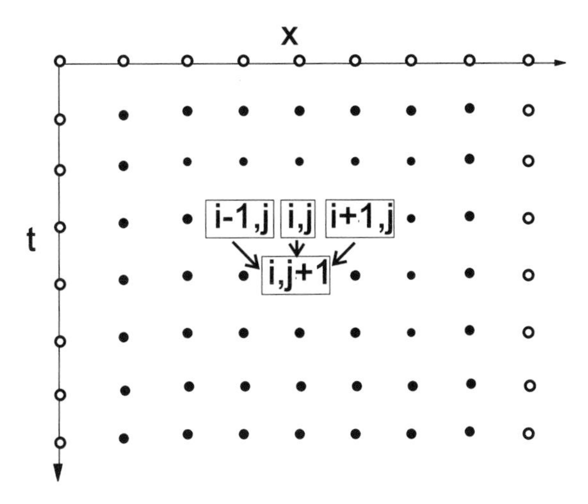

Figure 17.13 The algorithm for the heat equation in which the temperature at the location $x = i\Delta x$ and time $t = (j+1)\Delta t$ is computed from the temperature values at three points of an earlier time. The nodes with white centers correspond to known initial and boundary conditions. (The boundaries are placed artificially close for illustrative purposes.)

Substitution of these approximations into (17.58) yields the heat difference equation

$$\frac{T(x,t+\Delta t)-T(x,t)}{\Delta t} = \frac{K}{C\rho}\frac{T(x+\Delta x,t)+T(x-\Delta x,t)-2T(x,t)}{\Delta x^2}. \tag{17.70}$$

We reorder (17.70) into a form in which T can be stepped forward in t:

$$\boxed{T_{i,j+1} = T_{i,j} + \eta\left[T_{i+1,j} + T_{i-1,j} - 2T_{i,j}\right],} \qquad \eta = \frac{K\Delta t}{C\rho\Delta x^2}, \tag{17.71}$$

where $x = i\Delta x$ and $t = j\Delta t$. This algorithm is *explicit* because it provides a solution in terms of known values of the temperature. If we tried to solve for the temperature at all lattice sites in Figure. 17.13 simultaneously, then we would have an *implicit* algorithm that requires us to solve equations involving unknown values of the temperature. We see that the temperature at space-time point $(i, j+1)$ is computed from the three temperature values at an earlier time j and at adjacent space values $i \pm 1, i$. We start the solution at the top row, moving it forward in time for as long as we want and keeping the temperature along the ends fixed at 0 K (Figure 17.14).

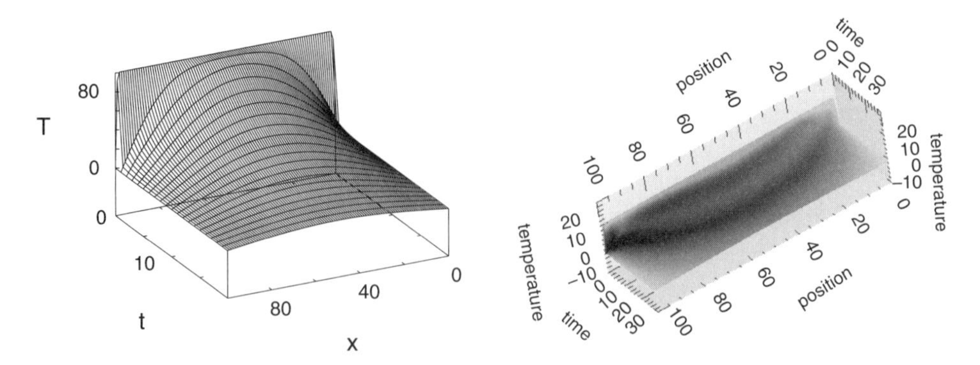

Figure 17.14 A numerical calculation of the temperature *versus* position and *versus* time, with isotherm contours projected onto the horizontal plane on the left and with a red-blue color scale used to indicate temperature on the right (the color is visible on the figures on the CD).

17.17.3 Von Neumann Stability Assessment

When we solve a PDE by converting it to a difference equation, we hope that the solution of the latter is a good approximation to the solution of the former. If the difference-equation solution diverges, then we know we have a bad approximation, but if it converges, then we may feel confident that we have a good approximation to the PDE. The *von Neumann stability analysis* is based on the assumption that eigenmodes of the difference equation can be written as

$$T_{m,j} = \xi(k)^j \, e^{ikm\Delta x}, \tag{17.72}$$

where $x = m\Delta x$ and $t = j\Delta t$, but $i = \sqrt{-1}$ is the imaginary number. The constant k in (17.72) is an unknown wave vector $(2\pi/\lambda)$, and $\xi(k)$ is an unknown complex function. View (17.72) as a basis function that oscillates in space (the exponential) with an amplitude or *amplification factor* $\xi(k)^j$ that increases by a power of ξ for each time step. If the general solution to the difference equation can be expanded in terms of these eigenmodes, then the general solution will be stable if the eigenmodes are stable. Clearly, for an eigenmode to be stable, the amplitude ξ cannot grow in time j, which means $|\xi(k)| < 1$ for all values of the parameter k [Pres 94, Anc 02].

Application of a stability analysis is more straightforward than it might appear. We just substitute the expression (17.72) into the difference equation (17.71):

$$\xi^{j+1} e^{ikm\Delta x} = \xi^{j+} e^{ikm\Delta x} + \eta \left[\xi^{j} e^{ik(m+1)\Delta x} + \xi^{j+} e^{ik(m-1)\Delta x} - 2\xi^{j+} e^{ikm\Delta x} \right].$$

After canceling some common factors, it is easy to solve for ξ:

$$\xi(k) = 1 + 2\eta[\cos(k\Delta x) - 1]. \tag{17.73}$$

In order for $|\xi(k)| < 1$ for all possible k values, we must have

$$\eta = \frac{K\,\Delta t}{C\rho\,\Delta x^2} < \frac{1}{2}. \tag{17.74}$$

This equation tells us that if we make the time step Δt smaller, we will always improve the stability, as we would expect. But if we decrease the space step Δx without a simultaneous quadratic *increase* in the time step, we will worsen the stability. The lack of space-time symmetry arises from our use of stepping in time but not in space.

In general, you should perform a stability analysis for every PDE you have to solve, although it can get complicated [Pres 94]. Yet even if you do not, the lesson here is that you may have to try different *combinations* of Δx and Δt variations until a stable, reasonable solution is obtained. You may expect, nonetheless, that there are choices for Δx and Δt for which the numerical solution fails and that simply decreasing an individual Δx or Δt, in the hope that this will increase precision, may not improve the solution.

```java
// EqHeat.java: Solve heat equation via finite differences
import java.io.*;                                                    // Import IO library

public class EqHeat {                                    // Class constants in MKS units
  public static final int Nx = 11, Nt = 300;                                // Grid sizes
  public static final double Dx = 0.01, Dt = 0.1;                           // Step sizes
  public static final double KAPPA = 210.;                        // Thermal conductivity
  public static final double SPH = 900.;                                 // Specific heat
  public static final double RHO = 2700.;                                   // Al density

  public static void main(String[] argv) throws IOException, FileNotFoundException {
    int ix, t;
    double T[][] = new double[Nx][2], cons;
    PrintWriter q = new PrintWriter (new FileOutputStream("EqHeat.dat"), true);
    for ( ix=1; ix < Nx-1; ix++ ) T[ix][0] = 100.;                          // Initialize
    T[0][0] = 0.; T[0][1] = 0.;                                        // Except the ends
    T[Nx-1][0] = 0.; T[Nx-1][1] = 0.;
    cons = KAPPA/(SPH*RHO)*Dt/(Dx*Dx);                              // Integration factor
    System.out.println("constant = " + cons);
    for ( t=1; t <= Nt; t++ ) {                                                 // t loop
      for (ix=1; ix < Nx-1; ix++)
        {T[ix][1] = T[ix][0] + cons*(T[ix + 1][0] + T[ix -1][0]-2.*T[ix][0]);}
      if ( t%10==0 || t==1 ) {                                       // Save every N steps
        for ( ix = 0; ix<Nx; ix++ ) q.println(T[ix][1]);
        q.println();                                               // Blank line ends row
      }
      for ( ix = 1; ix<Nx-1; ix++ ) T[ix][0] = T[ix][1];                    // New to old
    }                                                                       // End t loop
    System.out.println("data stored in EqHeat.dat");
  }                                                                           // End main
}                                                                            // End class
```

Listing 17.3 EqHeat.java solves the heat equation for a 1-D space and time by leapfrogging (time-stepping) the initial conditions forward in time. You will need to adjust the parameters to obtain a solution like those in the figures.

17.17.4 Heat Equation Implementation

Recollect that we want to solve for the temperature distribution within an aluminum bar of length $L = 1$ m subject to the boundary and initial conditions

$$T(x=0,t) = T(x=L,t) = 0\,\text{K}, \qquad T(x,t=0) = 100\,\text{K}. \tag{17.75}$$

The thermal conductivity, specific heat, and density for Al are

$$K = 237\,\text{W}/(\text{mK}), \quad C = 900\,\text{J}/(\text{kg K}), \quad \rho = 2700\,\text{kg}/\text{m}^3. \tag{17.76}$$

1. Write or modify `EqHeat.java` in Listing 17.3 to solve the heat equation.
2. Define a 2-D array `T[101][2]` for the temperature as a function of space and time. The first index is for the 100 space divisions of the bar, and the second index is for present and past times (because you may have to make thousands of time steps, you save memory by saving only two times).
3. For time $t = 0$ ($j = 1$), initialize `T` so that all points on the bar except the ends are at 100 K. Set the temperatures of the ends to 0 K.
4. Apply (17.68) to obtain the temperature at the next time step.
5. Assign the present-time values of the temperature to the past values:

```
T[i][1] = T[i][2], i = 1,... , 101.
```

6. Start with 50 time steps. Once you are confident the program is running properly, use thousands of steps to see the bar cool smoothly with time. For approximately every 500 time steps, print the time and temperature along the bar.

17.18 Assessment and Visualization

1. Check that your program gives a temperature distribution that varies smoothly along the bar and agrees with the boundary conditions, as in Figure 17.14.
2. Check that your program gives a temperature distribution that varies smoothly with time and attains equilibrium. You may have to vary the time and space steps to obtain well-behaved solutions.
3. Compare the analytic and numeric solutions (and the wall times needed to compute them). If the solutions differ, suspect the one that does not appear smooth and continuous.
4. Make surface plots of temperature *versus* position for several times.
5. Better yet, make a surface plot of temperature *versus* position *versus* time.
6. Plot the *isotherms* (contours of constant temperature).
7. **Stability test:** Check (17.74) that the temperature diverges in t if $\eta > \frac{1}{4}$.
8. **Material dependence:** Repeat the calculation for iron. Note that the stability condition requires you to change the size of the time step.

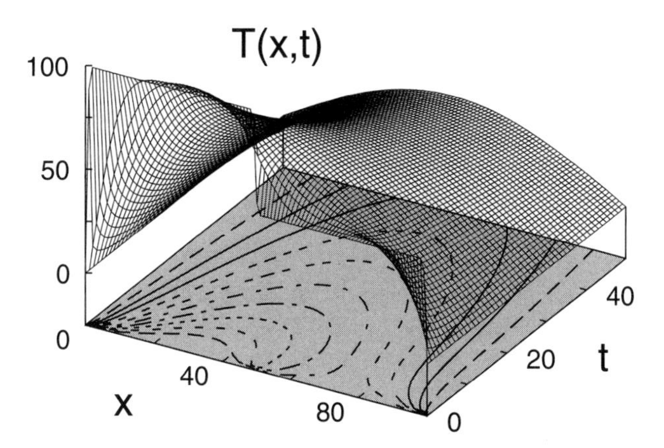

Figure 17.15 Temperature *versus* position and time when two bars at differing temperatures are placed in contact at $t = 0$. The projected contours show the isotherms.

9. **Initial sinusoidal distribution** $\sin(\pi x/L)$**:** Compare to the analytic solution,

$$T(x,t) = \sin(\pi x/L)e^{-\pi^2 K t/(L^2 C\rho)}.$$

10. **Two bars in contact:** Two identical bars 0.25 m long are placed in contact along one of their ends with their other ends kept at 0 K. One is kept in a heat bath at 100 K, and the other at 50 K. Determine how the temperature varies with time and location (Figure 17.15).
11. **Radiating bar (Newton's cooling):** Imagine now that instead of being insulated along its length, a bar is in contact with an environment at a temperature T_e. Newton's law of cooling (radiation) says that the rate of temperature change due to radiation is

$$\frac{\partial T}{\partial t} = -h(T - T_e), \tag{17.77}$$

where h is a positive constant. This leads to the modified heat equation

$$\frac{\partial T(x,t)}{\partial t} = \frac{K}{C\rho}\frac{\partial^2 T}{\partial^2 x} - hT(x,t). \tag{17.78}$$

Modify the algorithm to include Newton's cooling and compare the cooling of this bar with that of the insulated bar.

17.19 Improved Heat Flow: Crank–Nicolson Method

The Crank–Nicolson method [C&N 47] provides a higher degree of precision for the heat equation (17.57). This method calculates the time derivative with a central-difference approximation, in contrast to the forward-difference approximation used previously. In order to avoid introducing error for the initial time step, where only a single time value is known, the method uses a *split time step*,[3] so that time is advanced from time t to $t+\Delta t/2$:

$$\frac{\partial T}{\partial t}\left(x, t+\frac{\Delta t}{2}\right) \simeq \frac{T(x,t+\Delta t)-T(x,t)}{\Delta t} + O(\Delta t^2). \tag{17.79}$$

Yes, we know that this looks just like the forward-difference approximation for the derivative at time $t+\Delta t$, for which it would be a bad approximation; regardless, it is a better approximation for the derivative at time $t+\Delta t/2$, though it makes the computation more complicated. Likewise, in (17.68) we gave the central-difference approximation for the second space derivative for time t. For $t=t+\Delta t/2$, that becomes

$$\begin{aligned} &2(\Delta x)^2\frac{\partial^2 T}{\partial x^2}\left(x, t+\frac{\Delta t}{2}\right) \\ &\simeq [T(x-\Delta x,\, t+\Delta t) - 2T(x,\, t+\Delta t) + T(x+\Delta x,\, t+\Delta t)] \\ &+ [T(x-\Delta x,\, t) - 2T(x,\, t) + T(x+\Delta x,\, t)] + O(\Delta x^2). \end{aligned} \tag{17.80}$$

In terms of these expressions, the heat difference equation is

$$\begin{aligned} &T_{i,j+1} - T_{i,j} = \frac{\eta}{2}\left[T_{i-1,j+1} - 2T_{i,j+1} + T_{i+1,j+1} + T_{i-1,j} - 2T_{i,j} + T_{i+1,j}\right], \\ &x = i\Delta x, \quad t = j\Delta t, \quad \eta = \frac{K\Delta t}{C\rho\,\Delta x^2}. \end{aligned} \tag{17.81}$$

We group together terms involving the same temperature to obtain an equation with future times on the LHS and present times on the RHS:

$$\boxed{-T_{i-1,\,j+1} + \left(\frac{2}{\eta}+2\right)T_{i,\,j+1} - T_{i+1,\,j+1} = T_{i-1,\,j} + \left(\frac{2}{\eta}-2\right)T_{i,\,j} + T_{i+1,\,j}.} \tag{17.82}$$

[3] In §18.6.1 we develop another split-time algorithm for solution of the Schrödinger equation, where the real and imaginary parts of the wave function are computed at times that differ by $\Delta t/2$.

This equation represents an *implicit* scheme for the temperature $T_{i,j}$, where the word "implicit" means that we must solve simultaneous equations to obtain the full solution for all space. In contrast, an *explicit* scheme requires iteration to arrive at the solution. It is possible to solve (17.82) simultaneously for all unknown temperatures $(1 \le i \le N)$ at times j and $j+1$. We start with the initial temperature distribution throughout all of space, the boundary conditions at the ends of the bar for all times, and the approximate values from the first derivative:

$$\begin{array}{lll} T_{i,0}, \text{ known}, & T_{0,j}, \text{ known}, & T_{N,j}, \text{known}, \\ T_{0,j+1} = T_{0,j} = 0, & T_{N,j+1} = 0, & T_{N,j} = 0. \end{array}$$

We rearrange (17.82) so that we can use these known values of T to step the $j=0$ solution forward in time by expressing (17.82) as a set of simultaneous linear equations (in matrix form):

$$\begin{pmatrix} (\frac{2}{\eta}+2) & -1 & & & & \\ -1 & (\frac{2}{\eta}+2) & -1 & & & \\ & -1 & (\frac{2}{\eta}+2) & -1 & & \\ & & \ddots & \ddots & \ddots & \\ & & & -1 & (\frac{2}{\eta}+2) & -1 \\ & & & & -1 & (\frac{2}{\eta}+2) \end{pmatrix} \begin{pmatrix} T_{1,j+1} \\ T_{2,j+1} \\ T_{3,j+1)} \\ \vdots \\ T_{n-2,j+1} \\ T_{n-1,j+1} \end{pmatrix}$$

$$= \begin{pmatrix} T_{0,j+1} + T_{0,j} + (\frac{2}{\eta} - 2) T_{1,j} + T_{2,j} \\ T_{1,j} + (\frac{2}{\eta} - 2) T_{2,j} + T_{3,j} \\ T_{2,j} + (\frac{2}{\eta} - 2) T_{3,j} + T_{4,j} \\ \vdots \\ T_{n-3,j} + (\frac{2}{\eta} - 2) T_{n-2,j} + T_{n-1,j} \\ T_{n-2,j} + (\frac{2}{\eta} - 2) T_{n-1,j} + T_{n,j} + T_{n,j+1} \end{pmatrix}. \tag{17.83}$$

Observe that the T's on the RHS are all at the present time j for various positions, and at future time $j+1$ for the two ends (whose Ts are known for all times via the boundary conditions). We start the algorithm with the $T_{i,j=0}$ values of the initial conditions, then solve a matrix equation to obtain $T_{i,j=1}$. With that we know all the terms on the RHS of the equations ($j=1$ throughout the bar and $j=2$ at the ends) and so can repeat the solution of the matrix equations to obtain the temperature throughout the bar for $j=2$. So again we time-step forward, only now we solve matrix equations at each step. That gives us the spatial solution directly.

Not only is the Crank–Nicolson method more precise than the low-order time-stepping method of Unit III, but it also is stable for all values of Δt and Δx. To prove that, we apply the von Neumann stability analysis discussed in §17.17.3 to the Crank–Nicolson algorithm by substituting (17.71) into (17.82). This determines an amplification factor

$$\xi(k) = \frac{1-2\eta\sin^2(k\Delta x/2)}{1+2\eta\sin^2(k\Delta x/2)}. \tag{17.84}$$

Because $\sin^2()$ is positive-definite, this proves that $|\xi| \le 1$ for all Δt, Δx, and k.

17.19.1 Solution of Tridiagonal Matrix Equations ⊙

The Crank–Nicolson equations (17.83) are in the standard $[A]\mathbf{x} = \mathbf{b}$ form for linear equations, and so we can use our previous methods to solve them. Nonetheless, because the coefficient matrix $[A]$ is tridiagonal (zero elements except for the main diagonal and two diagonals on either side of it),

$$\begin{pmatrix} d_1 & c_1 & 0 & 0 & \cdots & \cdots & \cdots & 0 \\ a_2 & d_2 & c_2 & 0 & \cdots & \cdots & \cdots & 0 \\ 0 & a_3 & d_3 & c_3 & \cdots & \cdots & \cdots & 0 \\ \cdots & \cdots & \cdots & \cdots & \cdots & \cdots & \cdots & \cdots \\ 0 & 0 & 0 & 0 & \cdots & a_{N-1} & d_{N-1} & c_{N-1} \\ 0 & 0 & 0 & 0 & \cdots & 0 & a_N & d_N \end{pmatrix} \begin{pmatrix} x_1 \\ x_2 \\ x_3 \\ \ddots \\ x_{N-1} \\ x_N \end{pmatrix} = \begin{pmatrix} b_1 \\ b_2 \\ b_3 \\ \ddots \\ b_{N-1} \\ b_N \end{pmatrix},$$

a more robust and faster solution exists that makes this implicit method as fast as an explicit one. Because tridiagonal systems occur frequently, we now outline the specialized technique for solving them [Pres 94]. If we store the matrix elements $a_{i,j}$ using both subscripts, then we will need N^2 locations for elements and N^2 operations to access them. However, for a tridiagonal matrix, we need to store only the vectors $\{d_i\}_{i=1,N}$, $\{c_i\}_{i=1,N}$, and $\{a_i\}_{i=1,N}$, along, above, and below the diagonals. The single subscripts on a_i, d_i, and c_i reduce the processing from N^2 to $(3N-2)$ elements.

We solve the matrix equation by manipulating the individual equations until the coefficient matrix is *upper triangular* with all the elements of the main diagonal equal to 1. We start by dividing the first equation by d_1, then subtract a_2 times the first equation,

$$\begin{pmatrix} 1 & \frac{c_1}{d_1} & 0 & 0 & \cdots & \cdots & \cdots & 0 \\ 0 & d_2 - \frac{a_2 c_1}{d_1} & c_2 & 0 & \cdots & \cdots & \cdots & 0 \\ 0 & a_3 & d_3 & c_3 & \cdots & \cdots & \cdots & 0 \\ \cdots & \cdots & \cdots & \cdots & \cdots & \cdots & \cdots & \cdots \\ 0 & 0 & 0 & 0 & \cdots & a_{N-1} & d_{N-1} & c_{N-1} \\ 0 & 0 & 0 & 0 & \cdots & 0 & a_N & d_N \end{pmatrix} \begin{pmatrix} x_1 \\ x_2 \\ x_3 \\ \ddots \\ \cdot \\ x_N \end{pmatrix} = \begin{pmatrix} \frac{b_1}{d_1} \\ b_2 - \frac{a_2 b_1}{d_1} \\ b_3 \\ \ddots \\ \cdot \\ b_N \end{pmatrix},$$

and then dividing the second equation by the second diagonal element,

$$\begin{pmatrix} 1 & \frac{c_1}{d_1} & 0 & 0 & \cdots & \cdots & \cdots & 0 \\ 0 & 1 & \frac{c_2}{d_2 - a_2 \frac{c_1}{a_1}} & 0 & \cdots & & \cdots & 0 \\ 0 & a_3 & d_3 & c_3 & \cdots & \cdots & \cdots & 0 \\ \cdots & \cdots & \cdots & \cdots & \cdots & \cdots & \cdots & \cdots \\ 0 & 0 & 0 & 0 & & a_{N-1} & d_{N-1} & c_{N-1} \\ 0 & 0 & 0 & 0 & \cdots & 0 & a_N & d_N \end{pmatrix} \begin{pmatrix} x_1 \\ x_2 \\ x_3 \\ \ddots \\ \cdot \\ x_N \end{pmatrix} = \begin{pmatrix} \frac{b_1}{d_1} \\ \frac{b_2 - a_2 \frac{b_1}{d_1}}{d_2 - a_2 \frac{c_1}{d_1}} \\ b_3 \\ \ddots \\ \cdot \\ b_N \end{pmatrix}.$$

Assuming that we can repeat these steps without ever dividing by zero, the system of equations will be reduced to upper triangular form,

$$\begin{pmatrix} 1 & h_1 & 0 & 0 & \cdots & 0 \\ 0 & 1 & h_2 & 0 & \cdots & 0 \\ 0 & 0 & 1 & h_3 & \cdots & 0 \\ 0 & \cdots & \cdots & \ddots & \ddots & \cdots \\ 0 & 0 & 0 & 0 & \cdots & \cdots \\ 0 & 0 & 0 & \cdots & 0 & 1 \end{pmatrix} \begin{pmatrix} x_1 \\ x_2 \\ x_3 \\ \ddots \\ \cdot \\ x_N \end{pmatrix} = \begin{pmatrix} p_1 \\ p_2 \\ p_3 \\ \ddots \\ \cdot \\ p_N \end{pmatrix},$$

where $h_1 = c_1/d_1$ and $p_1 = b_1/d_1$. We then recur for the others elements:

$$h_i = \frac{c_i}{d_i - a_i h_{i-1}}, \quad p_i = \frac{b_i - a_i p_{i-1}}{d_i - a_i h_{i-1}}. \tag{17.85}$$

Finally, back substitution leads to the explicit solution for the unknowns:

$$x_i = p_i - h_i x_{i-1}; \quad i = n-1, n-2, \ldots, 1, \quad x_N = p_N. \tag{17.86}$$

In Listing 17.4 we give the program `HeatCNTridiag.java` that solves the heat equation using the Crank–Nicolson algorithm via a triadiagonal reduction.

```
// HeatCNTridiag.java: heat equation via Crank-Nicholson
// Output in gnuplot 3D grid format
import java.io.*;

public class HeatCNTridiag {

  public static void main(String[] argv)throws IOException, FileNotFoundException {
    PrintWriter w =  new PrintWriter(new FileOutputStream("HeatCNTriD.dat"), true);
    int Max =51, i, j, n=50, m=50;
    double Ta[] = new double[Max], Tb[] = new double[Max], Tc[] = new double[Max];
    double Td[] = new double[Max], a[]  = new double[Max], b[]  = new double[Max];
    double c[] = new double[Max], d[] =new double[Max], t[][]=new double [Max][Max];
    double width = 1.0, height = 0.1, ct = 1.0, k, r, h;              // Rectangle W & H
    double x[] = new double[Max], Pi = 3.1415926535;

    for ( i = 0; i < n; i++ ) t[i][0] = 0.0;                          // Initialize
    for ( i = 1; i < m; i++ ) t[0][i] = 0.0;
    h  = width / ( n - 1 );                        // Compute step sizes and constants
    k  = height / ( m - 1 );
    r  = ct * ct * k / ( h * h );
    for ( j = 1; j <= m; j++ ) { t[1][j] = 0.0; t[n][j] = 0.0; }                // BCs
    for ( i = 2; i <= n-1  ; i++ ) t[i][1] = Math.sin( Pi * h *i);              // ICs
    for ( i = 1; i <= n ; i++ )  Td[i] = 2. + 2./r;
    Td[1] = 1.; Td[n] = 1.;
    for ( i = 1; i <= n - 1 ; i++ ) {Ta[i] = -1.0;Tc[i] = -1.0;}       // Off diagonal
    Ta[n-1] = 0.0; Tc[1] = 0.0; Tb[1] = 0.0;Tb[n] = 0.0;
    for ( j = 2; j <= m; j++ )  {
      for (i = 2; i <= n-1; i++) Tb[i] = t[i-1][j-1]+t[i+1][j-1]+(2/r-2) * t[i][j-1];
      Tridiag(a,d,c,b,Ta,Td,Tc,Tb,x,n);                                // Solve system
      for ( i = 1; i <= n; i++ ) t[i][j] = x[i];
    }
    for (j=1;j<=m;j++) {
      for ( i = 1; i <= n; i++ ) w.println(""+t[i][j]+"");
      w.println();                                           // Empty line for gnuplot
    }
    System.out.println("data stored in HeatCNTridiag.dat");
  }

  public static void Tridiag(double a[],double d[],double c[],double b[],
                double Ta[],double Td[],double Tc[],double Tb[],double x[],int n) {
    int i, Max = 51;
    double h[] = new double[Max], p[] = new double[Max];
    for (i = 1; i <= n; i++) {a[i] = Ta[i]; b[i] = Tb[i]; c[i] = Tc[i]; d[i] =Td[i];}
    h[1] = c[1]/d[1];          p[1] = b[1]/d[1];
    for ( i = 2; i <= n; i++ ) {
      h[i]=c[i]/(d[i]-a[i]*h[i-1]);  p[i]=(b[i]-a[i]*p[i-1])/(d[i]-a[i]*h[i-1]); }
    x[n] = p[n];
    for ( i = n - 1; i >= 1; i-- ) x[i] = p[i] - h[i]*x[i+1];
  }
}
```

Listing 17.4 HeatCNTridiag.java is the complete program for solution of the heat equation in one space dimension and time via the Crank-Nicolson method. The resulting matrix equations are solved via a technique specialized to tridiagonal matrices.

17.19.2 Crank-Nicolson Method Implementation and Assessment

Use the Crank–Nicolson method to solve for the heat flow in the metal bar in §17.16.

1. Write a program using the Crank–Nicolson method to solve the heat equation for at least 100 time steps.

2. Solve the linear system of equations (17.83) using either JAMA or the special tridiagonal algorithm.
3. Check the stability of your solution by choosing different values for the time and space steps.
4. Construct a contoured surface plot of temperature *versus* position and versus time.
5. Compare the implicit and explicit algorithms used in this chapter for relative precision and speed. You may assume that a stable answer that uses very small time steps is accurate.

18

PDE Waves: String, Quantum Packet, and E&M

In this chapter we explore the numerical solution of a number of PDEs known as wave equations. *We have two purposes in mind. First, especially if you have skipped the discussion of the heat equation in Chapter 17,* "PDES for Electrostatics & Heat Flow," *we wish to give another example of how initial conditions in time are treated with a time-stepping or leapfrog algorithm. Second, we wish to demonstrate that once we have a working algorithm for solving a wave equation, we can include considerably more physics than is possible with analytic treatments. Unit I deals with a number of aspects of waves on a string. Unit II deals with quantum wave packets, which have their real and imaginary parts solved for at different (split) times. Unit III extends the treatment to electromagnetic waves that have the extra complication of being vector waves with interconnected E and H fields. Shallow-water waves, dispersion, and shock waves are studied in Chapter 19,* "Solitons and Computational Fluid Dynamics."

18.1 Unit I. Vibrating String

Problem: Recall the demonstration from elementary physics in which a string tied down at both ends is plucked "gently" at one location and a pulse is observed to travel along the string. Likewise, if the string has one end free and you shake it just right, a standing-wave pattern is set up in which the nodes remain in place and the antinodes move just up and down. Your **problem** is to develop an accurate model for wave propagation on a string and to see if you can set up traveling- and standing-wave patterns.[1]

18.2 The Hyperbolic Wave Equation (Theory)

Consider a string of length L tied down at both ends (Figure 18.1 left). The string has a constant density ρ per unit length, a constant tension T, is subject to no frictional forces, and the tension is high enough that we may ignore sagging due to gravity. We assume that the displacement of the string $y(x,t)$ from its rest position is in the vertical direction only and that it is a function of the horizontal location along the string x and the time t.

[1] Some similar but independent studies can also be found in [Raw 96].

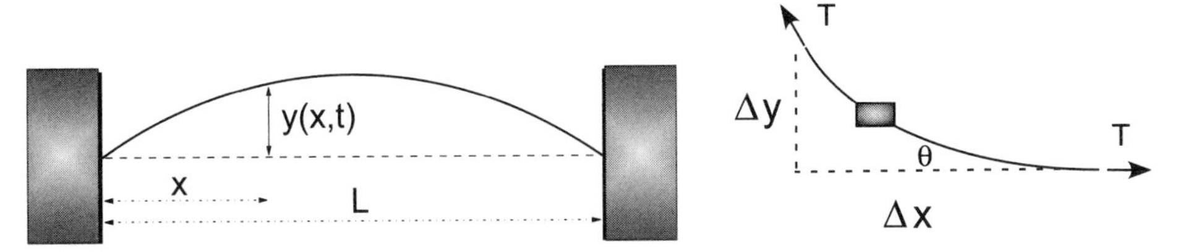

Figure 18.1 *Left:* A stretched string of length *L* tied down at both ends and under high enough tension to ignore gravity. The vertical disturbance of the string from its equilibrium position is $y(x, t)$. *Right:* A differential element of the string showing how the string's displacement leads to the restoring force.

To obtain a simple linear equation of motion (nonlinear wave equations are discussed in Chapter 19, "Solitons & Computational Fluid Dynamics"), we assume that the string's relative displacement $y(x,t)/L$ and slope $\partial y/\partial x$ are small. We isolate an infinitesimal section Δx of the string (Figure 18.1 right) and see that the difference in the vertical components of the tension at either end of the string produces the restoring force that accelerates this section of the string in the vertical direction. By applying Newton's laws to this section, we obtain the familiar wave equation:

$$\sum F_y = \rho\,\Delta x\,\frac{\partial^2 y}{\partial t^2}, \tag{18.1}$$

$$\sum F_y = T\sin\theta(x+\Delta x) - T\sin\theta(x) = T\left.\frac{\partial y}{\partial x}\right|_{x+\Delta x} - T\left.\frac{\partial y}{\partial x}\right|_{x} \simeq T\frac{\partial^2 y}{\partial x^2},$$

$$\Rightarrow \quad \frac{\partial^2 y(x,t)}{\partial x^2} = \frac{1}{c^2}\frac{\partial^2 y(x,t)}{\partial t^2}, \quad c = \sqrt{\frac{T}{\rho}}, \tag{18.2}$$

where we have assumed that θ is small enough for $\sin\theta \simeq \tan\theta = \partial y/\partial x$. The existence of two independent variables x and t makes this a PDE. The constant c is the velocity with which a disturbance travels along the wave and is seen to decrease for a heavier string and increase for a tighter one. Note that this signal velocity c is *not* the same as the velocity of a string element $\partial y/\partial t$.

The initial condition for our problem is that the string is plucked gently and released. We assume that the "pluck" places the string in a triangular shape with the center of triangle $\frac{8}{10}$ of the way down the string and with a height of 1:

$$y(x, t=0) = \begin{cases} 1.25x/L, & x \le 0.8L, \\ (5-5x/L), & x > 0.8L, \end{cases} \quad \text{(initial condition 1).} \tag{18.3}$$

Because (18.2) is second-order in time, a second initial condition (beyond initial displacement) is needed to determine the solution. We interpret the "gentleness" of the pluck to mean the string is released from rest:

$$\frac{\partial y}{\partial t}(x, t=0) = 0, \quad \text{(initial condition 2)}. \tag{18.4}$$

The boundary conditions have both ends of the string tied down for all times:

$$y(0,t) \equiv 0, \quad y(L,t) \equiv 0, \quad \text{(boundary conditions)}. \tag{18.5}$$

18.2.1 Solution via Normal-Mode Expansion

The analytic solution to (18.2) is obtained via the familiar separation-of-variables technique. We assume that the solution is the product of a function of space and a function of time:

$$y(x,t) = X(x)T(t). \tag{18.6}$$

We substitute (18.6) into (18.2), divide by $y(x,t)$, and are left with an equation that has a solution only if there are solutions to the two ODEs:

$$\frac{d^2T(t)}{dt^2} + \omega^2 T(t) = 0, \qquad \frac{d^2X(x)}{dt^2} + k^2 X(x) = 0, \qquad k \stackrel{\text{def}}{=} \frac{\omega}{c}. \tag{18.7}$$

The angular frequency ω and the wave vector k are determined by demanding that the solutions satisfy the boundary conditions. Specifically, the string being attached at both ends demands

$$X(x=0,t) = X(x=l,t) = 0 \tag{18.8}$$

$$\Rightarrow \quad X_n(x) = A_n \sin k_n x, \quad k_n = \frac{\pi(n+1)}{L}, \quad n = 0, 1, \ldots. \tag{18.9}$$

The time solution is

$$T_n(t) = C_n \sin \omega_n t + D_n \cos \omega_n t, \quad \omega_n = nck_0 = n\frac{2\pi c}{L}, \tag{18.10}$$

where the frequency of this nth *normal mode* is also fixed. In fact, it is the single frequency of oscillation that defines a normal mode. The *initial condition* (18.3) of zero velocity, $\partial y/\partial t(t=0) = 0$, requires the C_n values in (18.10) to be zero. Putting the pieces together, the normal-mode solutions are

$$y_n(x,t) = \sin k_n x \, \cos \omega_n t, \quad n = 0, 1, \ldots. \tag{18.11}$$

Since the wave equation (18.2) is linear in y, the principle of linear superposition holds and the most general solution for waves on a string with fixed ends can be written as the sum of normal modes:

$$y(x,t) = \sum_{n=0}^{\infty} B_n \sin k_n x \cos \omega_n t. \tag{18.12}$$

(Yet we will lose linear superposition once we include nonlinear terms in the wave equation.) The Fourier coefficient B_n is determined by the second initial condition (18.3), which describes how the wave is plucked:

$$y(x,t=0) = \sum_{n}^{\infty} B_n \sin nk_0 x. \tag{18.13}$$

Multiply both sides by $\sin mk_0x$, substitute the value of $y(x,0)$ from (18.3), and integrate from 0 to l to obtain

$$B_m = 6.25 \frac{\sin(0.8m\pi)}{m^2\pi^2}. \tag{18.14}$$

You will be asked to compare the Fourier series (18.12) to our numerical solution. While it is in the nature of the approximation that the precision of the numerical solution depends on the choice of step sizes, it is also revealing to realize that the precision of the analytic solution depends on summing an infinite number of terms, which can be done only approximately.

18.2.2 Algorithm: Time-Stepping

As with Laplace's equation and the heat equation, we look for a solution $y(x,t)$ only for discrete values of the independent variables x and t on a grid (Figure 18.2):

$$x = i\Delta x, \quad i = 1, \ldots, N_x, \quad t = j\Delta t, \quad j = 1, \ldots, N_t, \tag{18.15}$$

$$y(x,t) = y(i\Delta x,\, i\Delta t) \overset{\text{def}}{=} y_{i,j}. \tag{18.16}$$

In contrast to Laplace's equation where the grid was in two space dimensions, the grid in Figure 18.2 is in both space and time. That being the case, moving across a row corresponds to increasing x values along the string for a fixed time, while moving down a column corresponds to increasing time steps for a fixed position. Even though the grid in Figure 18.2 may be square, we cannot use a relaxation technique for the solution because we do not know the solution on all four sides. The boundary conditions determine the solution along the right and left sides, while the initial time condition determines the solution along the top.

As with the Laplace equation, we use the central-difference approximation to *discretize* the wave equation into a difference equation. First we express the second derivatives in terms of finite differences:

$$\frac{\partial^2 y}{\partial t^2} \simeq \frac{y_{i,j+1} + y_{i,j-1} - 2y_{i,j}}{(\Delta t)^2}, \quad \frac{\partial^2 y}{\partial x^2} \simeq \frac{y_{i+1,j} + y_{i-1,j} - 2y_{i,j}}{(\Delta x)^2}. \tag{18.17}$$

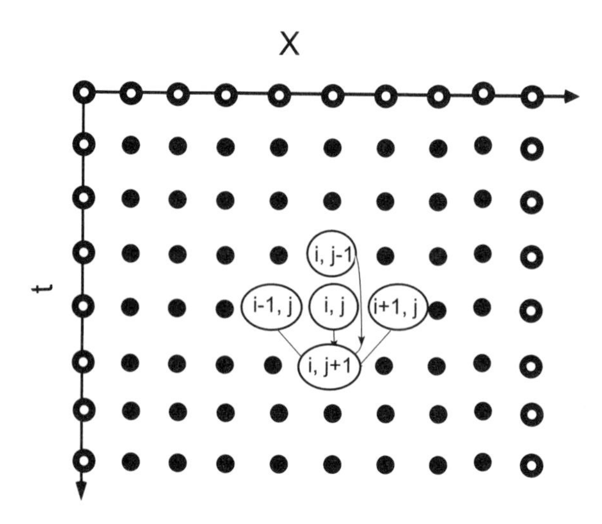

Figure 18.2 The solutions of the wave equation for four earlier space-time points are used to obtain the solution at the present time. The boundary and initial conditions are indicated by the white-centered dots.

Substituting (18.17) in the wave equation (18.2) yields the difference equation

$$\frac{y_{i,j+1} + y_{i,j-1} - 2y_{i,j}}{c^2(\Delta t)^2} = \frac{y_{i+1,j} + y_{i-1,j} - 2y_{i,j}}{(\Delta x)^2}. \tag{18.18}$$

Notice that this equation contains three time values: j+1 = the future, j = the present, and j−1 = the past. Consequently, we rearrange it into a form that permits us to predict the future solution from the present and past solutions:

$$\boxed{y_{i,j+1} = 2y_{i,j} - y_{i,j-1} + \frac{c^2}{c'^2}\left[y_{i+1,j} + y_{i-1,j} - 2y_{i,j}\right],} \qquad c' \stackrel{\text{def}}{=} \frac{\Delta x}{\Delta t}. \tag{18.19}$$

Here c' is a combination of numerical parameters with the dimension of velocity whose size relative to c determines the stability of the algorithm. The algorithm (18.19) propagates the wave from the two earlier times, j and $j-1$, and from three nearby positions, $i-1$, i, and $i+1$, to a later time $j+1$ and a single space position i (Figure 18.2).

As you have seen in our discussion of the heat equation, a leapfrog method is quite different from a relaxation technique. We start with the solution along the topmost row and then move down one step at a time. If we write the solution for present times to a file, then we need to store only three time values on the computer, which saves memory. In fact, because the time steps must be quite small to obtain high precision, you may want to store the solution only for every fifth or tenth time.

Initializing the recurrence relation is a bit tricky because it requires displacements from two earlier times, whereas the initial conditions are for only one time. Nonetheless, the rest condition (18.3) when combined with the *forward-difference* approximation lets us extrapolate to negative time:

$$\frac{\partial y}{\partial t}(x,0) \simeq \frac{y(x,0)-y(x,-\Delta t)}{\Delta t} = 0, \quad \Rightarrow \quad y_{i,0} = y_{i,1}. \tag{18.20}$$

Here we take the initial time as $j = 1$, and so $j = 0$ corresponds to $t = -\Delta t$. Substituting this relation into (18.19) yields for the initial step

$$y_{i,2} = y_{i,1} + \frac{c^2}{c'^2}\left[y_{i+1,1} + y_{i-1,1} - 2y_{i,1}\right] \quad (t = \Delta t \text{ only}). \tag{18.21}$$

Equation (18.21) uses the solution throughout all space at the initial time $t = 0$ to propagate (leapfrog) it forward to a time Δt. Subsequent time steps use (18.19) and are continued for as long as you like.

As is also true with the heat equation, the success of the numerical method depends on the relative sizes of the time and space steps. If we apply a von Neumann stability analysis to this problem by substituting $y_{m,j} = \xi^j \exp(ikm\,\Delta x)$, as we did in §17.17.3, a complicated equation results. Nonetheless, [Pres 94] shows that the difference-equation solution will be stable for the general class of transport equations if

$$c \le c' = \Delta x/\Delta t \quad \text{(Courant condition)}. \tag{18.22}$$

Equation (18.22) means that the solution gets better with smaller *time* steps but gets worse for smaller space steps (unless you simultaneously make the time step smaller). Having different sensitivities to the time and space steps may appear surprising because the wave equation (18.2) is symmetric in x and t, yet the symmetry is broken by the nonsymmetric initial and boundary conditions.

Exercise: Figure out a procedure for solving for the wave equation for all times in just one step. Estimate how much memory would be required.

Exercise: Try to figure out a procedure for solving for the wave motion with a relaxation technique. What would you take as your initial guess, and how would you know when the procedure has converged? ▮

18.2.3 Wave Equation Implementation

The program `EqString.java` in Listing 18.1 solves the wave equation for a string of length $L = 1$ m with its ends fixed and with the gently plucked initial conditions. Note that our use of $L = 1$ violates our assumption that $y/L \ll 1$ but makes it easy

```java
//  EqString.java: Leapfrog solution of wave equation, gnuplot output
import java.io.*;

public class EqString {
  final static double rho = 0.01, ten = 40., max = 100.;

  public static void main(String[] argv) throws IOException, FileNotFoundException {
    int i, k;
    double x[][] = new double[101][3], ratio, c, c1;
    PrintWriter w = new PrintWriter (new FileOutputStream("EqString.dat"), true);
    c = Math.sqrt(ten/rho);                                    // Propagation speed
    c1 = c;                                                    // CFL criteria
    ratio =  c*c/(c1*c1);
    for ( i=0;  i < 81;   i++ ) x[i][0] = 0.00125*i;           // Initial conds
    for ( i=81; i < 101;  i++ ) x[i][0] = 0.1-0.005*(i-80);
    for ( i=0;  i < 101;  i++ ) w.println("" + x[i][0] + "");  // First time step
    w.println("");
    for (i=1; i<100; i++) x[i][1] =x[i][0]+0.5*ratio*(x[i+1][0]+x[i-1][0]-2*x[i][0]);
      for ( k=1;  k < max;  k++ )  {                           // Later time steps
        for ( i=1;  i < 100;  i++ )  x[i][2] = 2.*x[i][1]
                           -x[i][0] + ratio*(x[i+1][1] + x[i-1][1] - 2*x[i][1]);
        for ( i=0;  i < 101;  i++ )  { x[i][0] = x[i][1]; x[i][1] = x[i][2]; }
        if ((k%5) == 0) {                                      // Print every 5th point
          for ( i=0;  i < 101;  i++ ) w.println("" + x[i][2] + "");  // Gnuplot 3D
          w.println("");                                       // Empty line for gnuplot
        }
      }
      System.out.println("data in EqString.dat, gnuplot format");
} }
```

Listing 18.1 EqString.java solves the wave equation via time stepping for a string of length $L = 1$ m with its ends fixed and with the gently plucked initial conditions. You will need to modify this code to include new physics.

to display the results; you should try $L = 1000$ to be realistic. The values of density and tension are entered as constants, $\rho = 0.01\,\mathrm{kg/m}$ and $T = 40\,\mathrm{N}$, with the space grid set at 101 points, corresponding to $\Delta = 0.01$ cm.

18.2.4 Assessment and Exploration

1. Solve the wave equation and make a surface plot of displacement *versus* time and position.
2. Explore a number of space and time step combinations. In particular, try steps that satisfy and that do not satisfy the Courant condition (18.22). Does your exploration conform with the stability condition?
3. Compare the analytic and numeric solutions, summing at least 200 terms in the analytic solution.
4. Use the plotted time dependence to estimate the peak's propagation velocity c. Compare the deduced c to (18.2).
5. Our solution of the wave equation for a plucked string leads to the formation of a wave packet that corresponds to the sum of multiple normal modes of the string. On the right in Figure 18.3 we show the motion resulting from the string initially placed in a single normal mode (standing wave),

$$y(x, t=0) = 0.001\,\sin 2\pi x, \qquad \frac{\partial y}{\partial t}(x, t=0) = 0.$$

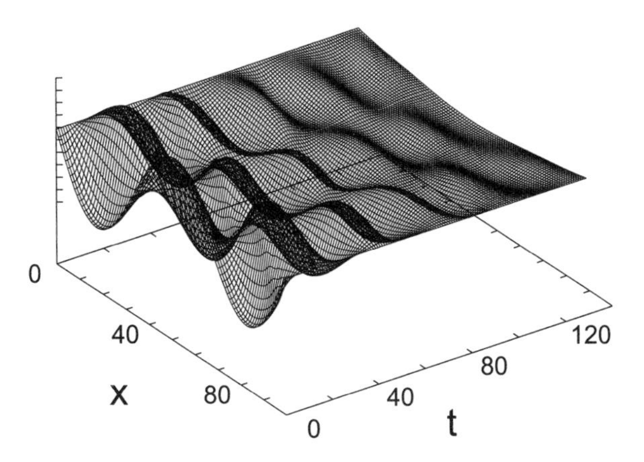

Figure 18.3 The vertical displacement as a function of position *x* and time *t*. A string initially placed in a standing wave on a string with friction. Notice how the standing wave moves up and down with time. (Courtesy of J. Wiren.)

Modify the program to incorporate this initial condition and see if a normal mode results.

6. Observe the motion of the wave for initial conditions corresponding to the sum of two adjacent normal modes. Does beating occur?
7. When a string is plucked near its end, a pulse reflects off the ends and bounces back and forth. Change the initial conditions of the model program to one corresponding to a string plucked exactly in its middle and see if a traveling or a standing wave results.
8. ⊙ Figure 18.4 shows the wave packets that result as a function of time for initial conditions corresponding to the double pluck indicated on the left in the figure. Verify that initial conditions of the form

$$\frac{y(x,t=0)}{0.005} = \begin{cases} 0, & 0.0 \le x \le 0.1, \\ 10x-1, & 0.1 \le x \le 0.2, \\ -10x+3, & 0.2 \le x \le 0.3, \\ 0, & 0.3 \le x \le 0.7, \\ 10x-7, & 0.7 \le x \le 0.8, \\ -10x+9, & 0.8 \le x \le 0.9, \\ 0, & 0.9 \le x \le 1.0 \end{cases}$$

lead to this type of a repeating pattern. In particular, observe whether the pulses move or just oscillate up and down. ▮

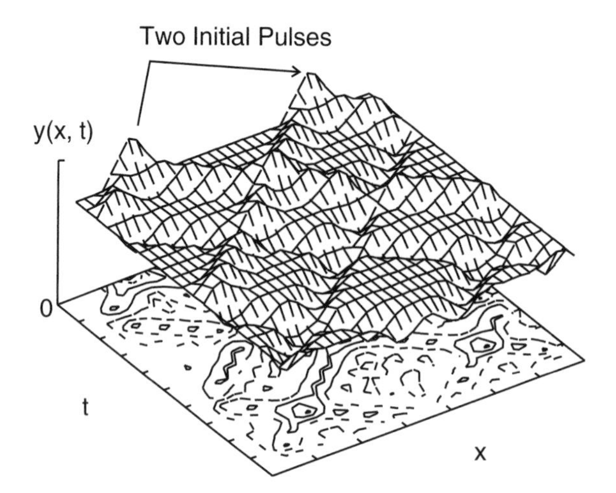

Figure 18.4 The vertical displacement as a function of position and time of a string initially plucked simultaneously at two points, as shown by arrows. Note that the initial peaks break up into waves traveling to the right and to the left. The traveling waves invert on reflection from the fixed ends. As a consequence of these inversions, the $t \simeq 12$ wave is an inverted $t = 0$ wave.

18.3 Waves with Friction (Extension)

The string problem we have investigated so far can be handled by either a numerical or an analytic technique. We now wish to extend the theory to include some more realistic physics. These extensions have only numerical solutions.

Real plucked strings do not vibrate forever because the real world contains friction. Consider again the element of a string between x and $x + dx$ (Figure 18.1 right) but now imagine that this element is moving in a viscous fluid such as air. An approximate model has the frictional force pointing in a direction opposite the (vertical) velocity of the string and proportional to that velocity, as well as proportional to the length of the string element:

$$F_f \simeq -2\kappa\, \Delta x\, \frac{\partial y}{\partial t}, \tag{18.23}$$

where κ is a constant that is proportional to the viscosity of the medium in which the string is vibrating. Including this force in the equation of motion changes the wave equation to

$$\frac{\partial^2 y}{\partial t^2} = c^2 \frac{\partial^2 y}{\partial x^2} - \frac{2\kappa}{\rho} \frac{\partial y}{\partial t}. \tag{18.24}$$

In Figure 18.3 we show the resulting motion of a string plucked in the middle when friction is included. Observe how the initial pluck breaks up into waves traveling to the right and to the left that are reflected and inverted by the fixed ends. Because those parts of the wave with the higher velocity experience greater friction, the peak tends to be smoothed out the most as time progresses.

Exercise: Generalize the algorithm used to solve the wave equation to now include friction and check if the wave's behavior seems physical (damps in time). Start with $T = 40\,\mathrm{N}$ and $\rho = 10\,\mathrm{kg/m}$ and pick a value of κ large enough to cause a noticeable effect but not so large as to stop the oscillations. As a check, reverse the sign of κ and see if the wave grows in time (which would eventually violate our assumption of small oscillations). ▌

18.4 Waves for Variable Tension and Density (Extension)

We have derived the propagation velocity for waves on a string as $c = \sqrt{T/\rho}$. This says that waves move slower in regions of high density and faster in regions of high tension. If the density of the string varies, for instance, by having the ends thicker in order to support the weight of the middle, then c will no longer be a constant and our wave equation will need to be extended. In addition, if the density increases, then so will the tension because it takes greater tension to accelerate a greater mass. If gravity acts, then we will also expect the tension at the ends of the string to be higher than in the middle because the ends must support the entire weight of the string.

To derive the equation for wave motion with variable density and tension, consider again the element of a string (Figure 18.1 right) used in our derivation of the wave equation. If we do not assume the tension T is constant, then Newton's second law gives

$$F = ma \tag{18.25}$$

$$\Rightarrow \quad \frac{\partial}{\partial x}\left[T(x)\,\frac{\partial y(x,t)}{\partial x}\right]\Delta x = \rho(x)\Delta x\,\frac{\partial^2 u(x,t)}{\partial t^2} \tag{18.26}$$

$$\Rightarrow \quad \boxed{\frac{\partial T(x)}{\partial x}\,\frac{\partial y(x,t)}{\partial x} + T(x)\,\frac{\partial^2 y(x,t)}{\partial x^2} = \rho(x)\,\frac{\partial^2 y(x,t)}{\partial t^2}.} \tag{18.27}$$

If $\rho(x)$ and $T(x)$ are known functions, then these equations can be solved with just a small modification of our algorithm.

In §18.4.1 we will solve for the tension in a string due to gravity. Readers interested in an **alternate easier problem** that still shows the new physics may assume that the density and tension are proportional:

$$\rho(x) = \rho_0 e^{\alpha x}, \quad T(x) = T_0 e^{\alpha x}. \tag{18.28}$$

While we would expect the tension to be greater in regions of higher density (more mass to move and support), being proportional is clearly just an approximation. Substitution of these relations into (18.27) yields the new wave equation:

$$\frac{\partial^2 y(x,t)}{\partial x^2} + \alpha \frac{\partial y(x,t)}{\partial x} = \frac{1}{c^2}\frac{\partial^2 y(x,t)}{\partial t^2}, \quad c^2 = \frac{T_0}{\rho_0}. \tag{18.29}$$

Here c is a constant that would be the wave velocity if $\alpha = 0$. This equation is similar to the wave equation with friction, only now the first derivative is with respect to x and not t. The corresponding difference equation follows from using central-difference approximations for the derivatives:

$$y_{i,j+1} = 2y_{i,j} - y_{i,j-1} + \frac{\alpha c^2 (\Delta t)^2}{2\Delta x}[y_{i+1,j} - y_{i,j}] + \frac{c^2}{c'^2}[y_{i+1,j} + y_{i-1,j} - 2y_{i,j}],$$

$$y_{i,2} = y_{i,1} + \frac{c^2}{c'^2}[y_{i+1,1} + y_{i-1,1} - 2y_{i,1}] + \frac{\alpha c^2 (\Delta t)^2}{2\Delta x}[y_{i+1,1} - y_{i,1}]. \tag{18.30}$$

18.4.1 Waves on a Catenary

Up until this point we have been ignoring the effect of gravity upon our string's shape and tension. This is a good approximation if there is very little sag in the string, as might happen if the tension is very high and the string is light. Even if there is some sag, our solution for $y(x,t)$ could be used as the disturbance about the equilibrium shape. However, if the string is massive, say, like a chain or heavy cable, then the sag in the middle caused by gravity could be quite large (Figure 18.5), and the resulting variation in shape and tension needs to be incorporated into the wave equation. Because the tension is no longer uniform, waves travel faster near the ends of the string, which are under greater tension since they must support the entire weight of the string.

18.4.2 Derivation of a Catenary Shape

Consider a string of uniform density ρ acted upon by gravity. To avoid confusion with our use of $y(x)$ to describe a disturbance on a string, we call $u(x)$ the equilibrium shape of the string (Figure 18.5). The statics problem we need to solve is to determine the shape $u(x)$ and the tension $T(x)$. The inset in Figure 18.5 is a free-body diagram of the midpoint of the string and shows that the weight W of this section of arc length s is balanced by the vertical component of the tension T. The

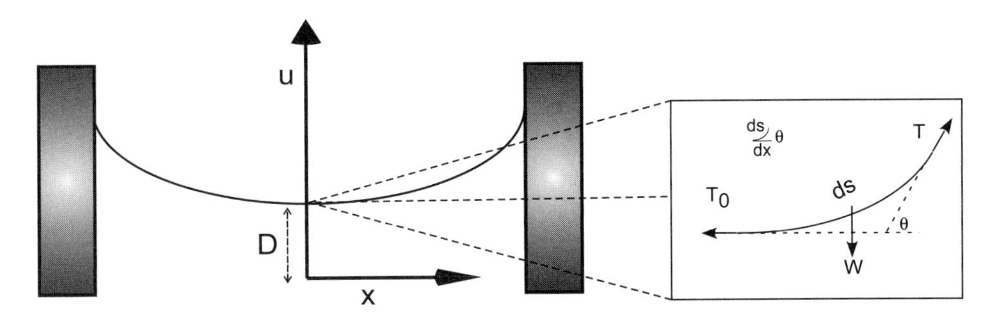

Figure 18.5 *Left:* A uniform string suspended from its ends in a gravitational field assumes a catenary shape. *Right:* A force diagram of a section of the catenary at its lowest point. The tension now varies along the string.

horizonal tension T_0 is balanced by the horizontal component of T:

$$T(x)\sin\theta = W = \rho g s, \quad T(x)\cos\theta = T_0, \tag{18.31}$$

$$\Rightarrow \quad \tan\theta = \rho g s / T_0. \tag{18.32}$$

The trick is to convert (18.32) to a differential equation that we can solve. We do that by replacing the slope $\tan\theta$ by the derivative du/dx and taking the derivative with respect to x:

$$\frac{du}{dx} = \frac{\rho g}{T_0} s, \quad \Rightarrow \quad \frac{d^2u}{dx^2} = \frac{\rho g}{T_0}\frac{ds}{dx}. \tag{18.33}$$

Yet since $ds = \sqrt{dx^2 + du^2}$, we have our differential equation

$$\frac{d^2u}{dx^2} = \frac{1}{D}\frac{\sqrt{dx^2+du^2}}{dx} = \frac{1}{D}\sqrt{1+\left(\frac{du}{dx}\right)^2}, \tag{18.34}$$

$$D = T_0/\rho g, \tag{18.35}$$

where D is a combination of constants with the dimension of length. Equation (18.34) is the equation for the *catenary* and has the solution [Becker 54]

$$\boxed{u(x) = D\cosh\frac{x}{D}.} \tag{18.36}$$

Here we have chosen the x axis to lie a distance D below the bottom of the catenary (Figure 18.5) so that $x = 0$ is at the center of the string where $y = D$ and $T = T_0$. Equation (18.33) tells us the arc length $s = Ddu/dx$, so we can solve for $s(x)$ and, via (18.31), for the tension $T(x)$:

$$s(x) = D\sinh\frac{x}{D}, \quad \Rightarrow \quad T(x) = T_0\frac{ds}{dx} = \rho g u(x) = T_0\cosh\frac{x}{D}. \tag{18.37}$$

It is this variation in tension that causes the wave velocity to change for different positions on the string.

18.4.3 Catenary and Frictional Wave Exercises

We have given you the program `EqString.java` (Listing 18.1) that solves the wave equation. Modify it to produce waves on a catenary including friction or for the assumed density and tension given by (18.28) with $\alpha = 0.5$, $T_0 = 40$ N, and $\rho_0 = 0.01$ kg/m. (The instructor's CD contains the programs `CatFriction.java` and `CatString.java` that do this.)

1. Look for some interesting cases and create surface plots of the results.
2. Explain in words how the waves dampen and how a wave's velocity appears to change. The behavior you obtain may look something like that shown in Figure 18.6.
3. **Normal modes:** Search for normal-mode solutions of the variable-tension wave equation, that is, solutions that vary as

$$u(x,t) = A\cos(\omega t)\sin(\gamma x).$$

   Try using this form to start your program and see if you can find standing waves. Use large values for ω.
4. When conducting physics demonstrations, we set up standing-wave patterns by driving one end of the string periodically. Try doing the same with your program; that is, build into your code the condition that for all times

$$y(x=0,t) = A\sin\omega t.$$

   Try to vary A and ω until a normal mode (standing wave) is obtained.
5. (For the exponential density case.) If you were able to find standing waves, then verify that this string acts like a high-frequency filter, that is, that there is a frequency below which no waves occur.
6. For the catenary problem, plot your results showing *both* the disturbance $u(x,t)$ about the catenary and the actual height $y(x,t)$ above the horizontal for a plucked string initial condition.

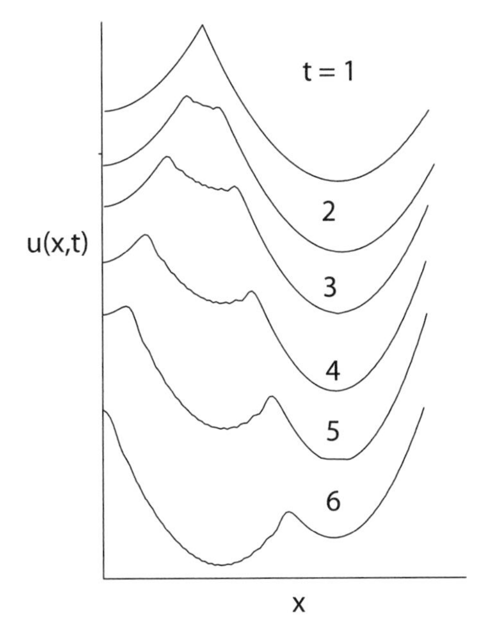

Figure 18.6 The wave motion of a plucked catenary with friction. (Courtesy of Juan Vanegas.)

7. Try the first two normal modes for a uniform string as the initial conditions for the catenary. These should be close to, but not exactly, normal modes.
8. We derived the normal modes for a uniform string after assuming that $k(x) = \omega/c(x)$ is a constant. For a catenary without too much x variation in the tension, we should be able to make the approximation

$$c(x)^2 \simeq \frac{T(x)}{\rho} = \frac{T_0 \cosh(x/d)}{\rho}.$$

See if you get a better representation of the first two normal modes if you include some x dependence in k.

18.5 Unit II. Quantum Wave Packets

Problem: An experiment places an electron with a definite momentum and position in a 1-D region of space the size of an atom. It is confined to that region by some kind of attractive potential. Your **problem** is to determine the resultant electron behavior in time and space.

18.6 Time-Dependent Schrödinger Equation (Theory)

Because the region of confinement is the size of an atom, we must solve this problem quantum mechanically. Nevertheless, it is different from the problem of a particle confined to a box considered in Chapter 9, "Differential Equation Applications," because now we are starting with a particle of definite momentum and position. In Chapter 9 we had a time-independent situation in which we had to solve the eigenvalue problem. Now the definite momentum and position of the electron imply that the solution is a wave packet, which is not an eigenstate with a uniform time dependence of $\exp(-i\omega t)$. Consequently, we must now solve the time-dependent Schrödinger equation.

We model an electron initially localized in space at $x=5$ with momentum k_0 ($\hbar=1$ in our units) by a wave function that is a wave packet consisting of a Gaussian multiplying a plane wave:

$$\psi(x,t=0) = \exp\left[-\frac{1}{2}\left(\frac{x-5}{\sigma_0}\right)^2\right] e^{ik_0 x}. \tag{18.38}$$

To solve the **problem** we must determine the wave function for all later times. If (18.38) were an eigenstate of the Hamiltonian, its $\exp(-i\omega t)$ time dependence can be factored out of the Schrödinger equation (as is usually done in textbooks). However, $\tilde{H}\psi \neq E\psi$ for this ψ, and so we must solve the full time-dependent Schrödinger equation. To show you where we are going, the resulting wave packet behavior is shown in Figures 18.7 and 18.8.

The time and space evolution of a quantum particle is described by the 1-D time-dependent Schrödinger equation,

$$i\,\frac{\partial\psi(x,t)}{\partial t} = \tilde{H}\psi(x,t) \tag{18.39}$$

$$i\,\frac{\partial\psi(x,t)}{\partial t} = -\frac{1}{2m}\frac{\partial^2\psi(x,t)}{\partial x^2} + V(x)\psi(x,t), \tag{18.40}$$

where we have set $2m=1$ to keep the equations simple. Because the initial wave function is complex (in order to have a definite momentum associated with it), the wave function will be complex for all times. Accordingly, we decompose the wave function into its real and imaginary parts:

$$\psi(x,t) = R(x,t) + i\,I(x,t), \tag{18.41}$$

$$\Rightarrow \quad \frac{\partial R(x,t)}{\partial t} = -\frac{1}{2m}\frac{\partial^2 I(x,t)}{\partial x^2} + V(x)I(x,t), \tag{18.42}$$

$$\frac{\partial I(x,t)}{\partial t} = +\frac{1}{2m}\frac{\partial^2 R(x,t)}{\partial x^2} - V(x)R(x,t), \tag{18.43}$$

where $V(x)$ is the potential acting on the particle.

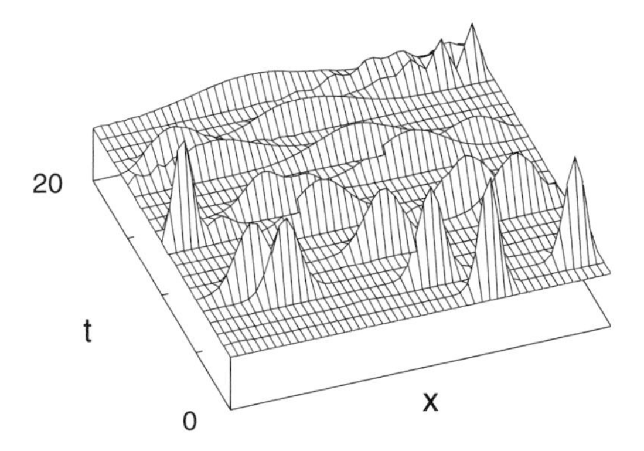

Figure 18.7 The position as a function of time of a localized electron confined to a square well (computed with the code `SqWell.java` available on the instructor's CD). The electron is initially on the right with a Gaussian wave packet. In time, the wave packet spreads out and collides with the walls.

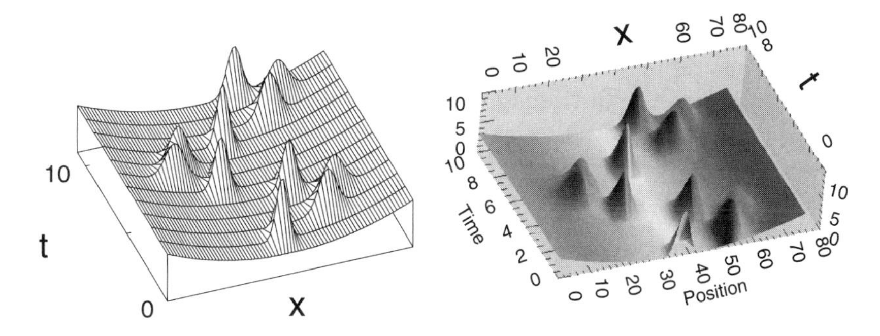

Figure 18.8 The probability density as a function of time for an electron confined to a 1-D harmonic oscillator potential well. On the left is a conventional surface plot from Gnuplot, while on the right is a color visualization from OpenDX.

18.6.1 Finite-Difference Algorithm

The time-dependent Schrödinger equation can be solved with both implicit (large-matrix) and explicit (leapfrog) methods. The extra challenge with the Schrödinger equation is to ensure that the integral of the probability density $\int_{-\infty}^{+\infty} dx\ \rho(x,t)$ remains constant (conserved) to a high level of precision for all time. For our project we use an *explicit* method that improves the numerical conservation of probability by solving for the real and imaginary parts of the wave function at slightly different or "staggered" times [Ask 77, Viss 91, MLP 00]. Explicitly, the real part R is determined at times 0, $\Delta t, \ldots$, and the imaginary part I at $\frac{1}{2}\Delta t$, $\frac{3}{2}\Delta t, \ldots$. The algorithm is based on (what else?) the Taylor expansions of R and I:

$$R\left(x, t+\frac{1}{2}\Delta t\right) = R\left(x, t-\frac{1}{2}\Delta t\right) + [4\alpha + V(x)\,\Delta t]I(x,t)$$

$$-2\alpha[I(x+\Delta x, t) + I(x-\Delta x, t)], \tag{18.44}$$

where $\alpha = \Delta t/2(\Delta x)^2$. In discrete form with $R_{x=i\Delta x}^{t=n\Delta t}$, we have

$$R_i^{n+1} = R_i^n - 2\left\{\alpha\left[I_{i+1}^n + I_{i-1}^n\right] - 2\left[\alpha + V_i\,\Delta t\right] I_i^n\right\}, \tag{18.45}$$

$$I_i^{n+1} = I_i^n + 2\left\{\alpha\left[R_{i+1}^n + R_{i-1}^n\right] - 2\left[\alpha + V_i\,\Delta t\right] R_i^n\right\}, \tag{18.46}$$

where the superscript n indicates the time and the subscript i the position.

The probability density ρ is defined in terms of the wave function evaluated at three different times:

$$\rho(t) = \begin{cases} R^2(t) + I\left(t+\frac{\Delta t}{2}\right)I\left(t-\frac{\Delta t}{2}\right), & \text{for integer } t, \\ I^2(t) + R\left(t+\frac{\Delta t}{2}\right)R\left(t-\frac{\Delta t}{2}\right), & \text{for half-integer } t. \end{cases} \tag{18.47}$$

Although probability is not conserved exactly with this algorithm, the error is two orders higher than that in the wave function, and this is usually quite satisfactory. If it is not, then we need to use smaller steps. While this definition of ρ may seem strange, it reduces to the usual one for $\Delta t \to 0$ and so can be viewed as part of the art of numerical analysis. You will investigate just how well probability is conserved. We refer the reader to [Koon 86, Viss 91] for details on the stability of the algorithm.

18.6.2 Wave Packet Implementation and Animation

In Listing 18.2 you will find the program `Harmos.java` that solves for the motion of the wave packet (18.38) inside a harmonic oscillator potential. The program `Slit.java` on the instructor's CD solves for the motion of a Gaussian wave packet as it passes through a slit (Figure 18.10). You should solve for a wave packet confined to the square well:

$$V(x) = \begin{cases} \infty, & x < 0, \text{ or } x > 15, \\ 0, & 0 \leq x \leq 15. \end{cases}$$

1. Define arrays `psr[751][2]` and `psi[751][2]` for the real and imaginary parts of ψ, and `Rho[751]` for the probability. The first subscript refers to the x position on the grid, and the second to the present and future times.

2. Use the values $\sigma_0 = 0.5$, $\Delta x = 0.02$, $k_0 = 17\pi$, and $\Delta t = \frac{1}{2}\Delta x^2$.
3. Use equation (18.38) for the initial wave packet to define psr[j][1] for all j at $t = 0$ and to define psi[j][1] at $t = \frac{1}{2}\Delta t$.
4. Set Rho[1] = Rho[751] = 0.0 because the wave function must vanish at the infinitely high well walls.
5. Increment time by $\frac{1}{2}\Delta t$. Use (18.45) to compute psr[j][2] in terms of psr[j][1], and (18.46) to compute psi[j][2] in terms of psi[j][1].
6. Repeat the steps through all of space, that is, for $i = 2$–750.
7. Throughout all of space, replace the present wave packet (second index equal to 1) by the future wave packet (second index 2).
8. After you are sure that the program is running properly, repeat the time-stepping for ~5000 steps.

```
// Harmos.java: t-dependent Schro eqn, for wavepacket in harmonic oscillator V
import java.io.*;

public class Harmos {

  public static void main(String[] argv) throws IOException, FileNotFoundException {
    PrintWriter w = new PrintWriter(new FileOutputStream("Harmos.dat"), true);
    double psr[][] = new double[751][2], psi[][] = new double[751][2];
    double p2[] = new double[751], v[] = new double[751], dx=0.02, k0, dt, x, pi;
    int i, n, max = 750;
    pi  = 3.14159265358979323846;  k0 = 3.0*pi;  dt = dx*dx/4.0;   x = -7.5;// i.c.
    for( i=0; i < max; i++) {
      psr[i][0] = Math.exp(-0.5*(Math.pow((x/0.5), 2.))) * Math.cos(k0*x);  // RePsi
      psi[i][0] = Math.exp(-0.5*(Math.pow((x/0.5), 2.))) * Math.sin(k0*x);  // ImPsi
      v[i] = 5.0*x*x;                                                        // Potential
      x    = x + dx;
    }
    for ( n=0; n < 20000; n++)  {                                  // Propagate in time
      for( i=1; i < max-1; i++ )  {                                           // RePsi
        psr[i][1] = psr[i][0] - dt*(psi[i+1][0] + psi[i-1][0]
                  -2.*psi[i][0])/(dx*dx)+dt*v[i]*psi[i][0];
        p2[i] = psr[i][0]*psr[i][1]+psi[i][0]*psi[i][0];
      }
      for ( i=1; i < max-1; i++ )  { psi[i][1] = psi[i][0] + dt*(psr[i+1][1]
          + psr[i-1][1] -2.*psr[i][1])/(dx*dx)-dt*v[i]*psr[i][1]; }       // ImPsi
      if ( (n == 0) || (n%2000 == 0) ) {                       // Output every 2000 steps
        for( i=0; i<max; i=i+10 ) w.println(""+(p2[i]+0.0015*v[i])+"");
        w.println("");
      }
      for ( i=0; i<max; i++ )  {psi[i][0] = psi[i][1]; psr[i][0] = psr[i][1];}
    }
    System.out.println("data saved in Harmos.dat");
} }
```

Listing 18.2 Harmos.java solves the time-dependent Schrödinger equation for a particle described by a Gaussian wave packet moving within a harmonic oscillator potential.

1. **Animation:** Output the probability density after every 200 steps for use in animation.
2. Make a surface plot of probability *versus* position *versus* time. This should look like Figure 18.7 or 18.8.
3. Make an animation showing the wave function as a function of time.

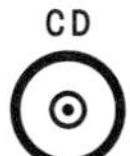

4. Check how well the probability is conserved for early and late times by determining the integral of the probability over all of space, $\int_{-\infty}^{+\infty} dx\, \rho(x)$, and seeing by how much it changes in time (its specific value doesn't matter because that's just normalization).
5. What might be a good explanation of why collisions with the walls cause the wave packet to broaden and break up? (*Hint:* The collisions do not appear so disruptive when a Gaussian wave packet is confined within a harmonic oscillator potential well.)

18.7 Wave Packets in Other Wells (Exploration)

1-D Well: Now confine the electron to lie within the harmonic oscillator potential:

$$V(x) = \frac{1}{2}x^2 \quad (-\infty \le x \le \infty).$$

Take the momentum $k_0 = 3\pi$, the space step $\Delta x = 0.02$, and the time step $\Delta t = \frac{1}{4}\Delta x^2$. Note that the wave packet broadens yet returns to its initial shape!

2-D Well ⊙: Now confine the electron to lie within a 2-D parabolic tube (Figure 18.9):

$$V(x,y) = 0.9x^2, \quad -9.0 \le x \le 9.0, \quad 0 \le y \le 18.0.$$

The extra degree of freedom means that we must solve the 2-D PDE:

$$i\frac{\partial \psi(x,y,t)}{\partial t} = -\left(\frac{\partial^2 \psi}{\partial x^2} + \frac{\partial^2 \psi}{\partial y^2}\right) + V(x,y)\psi. \tag{18.48}$$

Assume that the electron's initial localization is described by the 2-D Gaussian wave packet:

$$\psi(x,y,t=0) = e^{ik_{0x}x}\, e^{ik_{0y}y} \exp\left[-\frac{(x-x_0)^2}{2\sigma_0^2}\right] \exp\left[-\frac{(y-y_0)^2}{2\sigma_0^2}\right]. \tag{18.49}$$

Note that you can solve the 2-D equation by extending the method we just used in 1-D or you can look at the next section where we develop a special algorithm.

18.8 Algorithm for the 2-D Schrödinger Equation

One way to develop an algorithm for solving the time-dependent Schrödinger equation in 2-D is to extend the 1-D algorithm to another dimension. Rather than do that, we apply quantum theory directly to obtain a more powerful algorithm

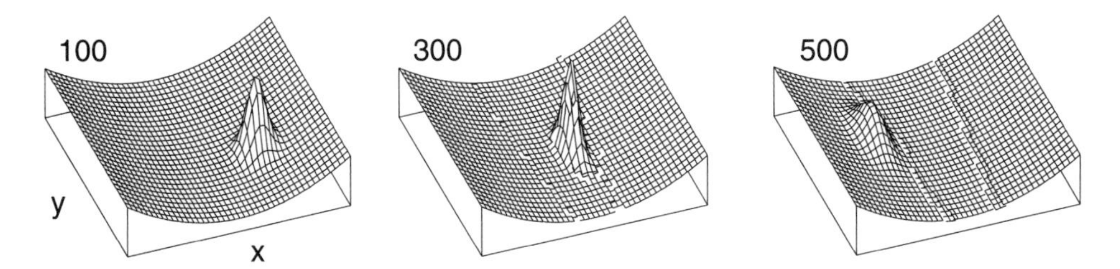

Figure 18.9 The probability density as a function of *x* and *y* of an electron confined to a 2-D parabolic tube. The electron's initial localization is described by a Gaussian wave packet in both the *x* and *y* directions. The times are 100, 300, and 500 steps.

[MLP 00]. First we note that equation (18.48) can be integrated in a formal sense [L&L,M 76] to obtain the operator solution:

$$\psi(x,y,t) = U(t)\psi(x,y,t=0) \tag{18.50}$$

$$U(t) = e^{-i\tilde{H}t}, \quad \tilde{H} = -\left(\frac{\partial^2}{\partial x^2} + \frac{\partial^2}{\partial y^2}\right) + V(x,y),$$

where $U(t)$ is an operator that translates a wave function by an amount of time t and $\tilde{H}$ is the Hamiltonian operator. From this formal solution we deduce that a wave packet can be translated ahead by time Δt via

$$\psi_{i,j}^{n+1} = U(\Delta t)\psi_{i,j}^{n}, \tag{18.51}$$

where the superscripts denote time $t = n\,\Delta t$ and the subscripts denote the two spatial variables $x = i\Delta x$ and $y = j\Delta y$. Likewise, the inverse of the time evolution operator moves the solution back one time step:

$$\psi^{n-1} = U^{-1}(\Delta t)\psi^{n} = e^{+i\tilde{H}\Delta t}\psi^{n}. \tag{18.52}$$

While it would be nice to have an algorithm based on a direct application of (18.52), the references show that the resulting algorithm is not stable. That being so, we base our algorithm on an indirect application [Ask 77], namely, the relation between the difference in ψ^{n+1} and ψ^{n-1}:

$$\psi^{n+1} = \psi^{n-1} + [e^{-i\tilde{H}\Delta t} - e^{+i\tilde{H}\Delta t}]\psi^{n}, \tag{18.53}$$

where the difference in sign of the exponents is to be noted. The algorithm derives from combining the $O(\Delta x^2)$ expression for the second derivative obtained from the Taylor expansion,

$$\frac{\partial^2 \psi}{\partial x^2} \simeq -\frac{1}{2}\left[\psi^n_{i+1,j} + \psi^n_{i-1,j} - 2\psi^n_{i,j}\right], \tag{18.54}$$

with the corresponding-order expansion of the evolution equation (18.53). Substituting the resulting expression for the second derivative into the 2-D time-dependent Schrödinger equation results in[2]

$$\psi^{n+1}_{i,j} = \psi^{n-1}_{i,j} - 2i\left[\left(4\alpha + \tfrac{1}{2}\Delta t V_{i,j}\right)\psi^n_{i,j} - \alpha\left(\psi^n_{i+1,j} + \psi^n_{i-1,j} + \psi^n_{i,j+1} + \psi^n_{i,j-1}\right)\right],$$

where $\alpha = \Delta t/2(\Delta x)^2$. We convert this complex equations to coupled real equations by substituting in the wave function $\psi = R + iI$,

$$R^{n+1}_{i,j} = R^{n-1}_{i,j} + 2\left[\left(4\alpha + \tfrac{1}{2}\Delta t V_{i,j}\right)I^n_{i,j} - \alpha\left(I^n_{i+1,j} + I^n_{i-1,j} + I^n_{i,j+1} + I^n_{i,j-1}\right)\right],$$

$$I^{n+1}_{i,j} = I^{n-1}_{i,j} - 2\left[\left(4\alpha + \tfrac{1}{2}\Delta t V_{i,j}\right)R^n_{i,j} + \alpha\left(R^n_{i+1,j} + R^n_{i-1,j} + R^n_{i,j+1} + R^n_{i,j-1}\right)\right].$$

This is the algorithm we use to integrate the 2-D Schrödinger equation. To determine the probability, we use the same expression (18.47) used in 1-D.

18.8.0.1 EXPLORATION: A BOUND AND DIFFRACTED 2-D PACKET

1. Determine the motion of a 2-D Gaussian wave packet within a 2-D harmonic oscillator potential:

$$V(x,y) = 0.3(x^2 + y^2), \quad -9.0 \le x \le 9.0, \quad -9.0 \le y \le 9.0. \tag{18.55}$$

2. Center the initial wave packet at $(x, y) = (3.0, -3)$ and give it momentum $(k_{0x}, k_{0y}) = (3.0, 1.5)$.
3. Young's single-slit experiment has a wave passing through a small slit with the transmitted wave showing interference effects. In quantum mechanics, where we represent a particle by a wave packet, this means that an interference pattern should be formed when a particle passes through a small slit. Pass a Gaussian wave packet of width 3 through a slit of width 5 (Figure 18.10) and look for the resultant quantum interference.

[2] For reference sake, note that the constants in the equation change as the dimension of the equation changes; that is, there will be different constants for the 3-D equation, and therefore our constants are different from the references!

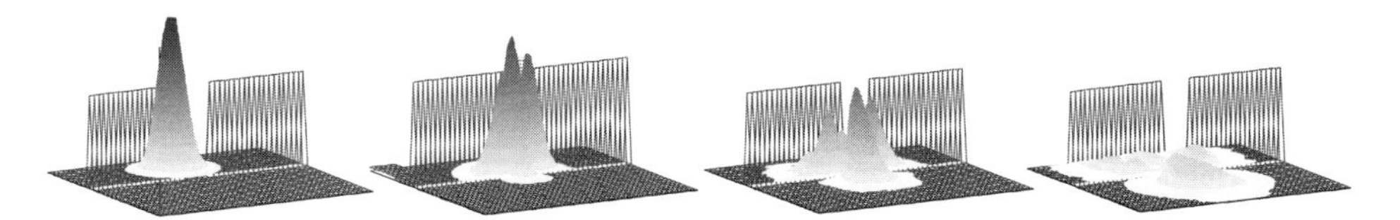

Figure 18.10 The probability density as a function of position and time for an electron incident upon and passing through a slit.

18.9 Unit III. E&M Waves via Finite-Difference Time Domain ⊙

Problem: You are given a 1-D resonant cavity with perfectly conducting walls. An initial electric pulse with the shape

$$E_x(z=3) = \exp\left[\frac{1}{2}\left(\frac{40-t}{12}\right)^2\right]\cos(2\pi f t), \quad 0 \le t \le T, \tag{18.56}$$

is placed in this cavity. Determine the motion of this pulse at all later times for $T = 10^{-8}$ s and $f = 700$ MHz.

Simulations of electromagnetic waves are of tremendous practical importance. Indeed, the fields of nanotechnology and spintronics rely heavily upon such simulations. The basic techniques used to solve for electromagnetic waves are essentially the same as those we used in Units I and II for string and quantum waves: Set up a grid in space and time and then step the initial solution forward in time one step at a time. For E&M simulations, this technique is known as the finite difference time domain *(FDTD) method. What is new for E&M waves is that they are vector fields, with the variations of one generating the other, so that the components of* **E** *and* **B** *are coupled to each other. Our treatment of FDTD does not do justice to the wealth of physics that can occur, and we recommend [Sull 00] for a more complete treatment and [Ward 04] (and their Web site) for modern applications.*

18.10 Maxwell's Equations

The description of electromagnetic (EM) waves via Maxwell's equations is given in many textbooks. For propagation in just one dimension (z) and for free space with no sinks or sources, four coupled PDEs result:

$$\vec{\nabla}\cdot\mathbf{E} = 0 \;\Rightarrow\; \frac{\partial E_x(z,t)}{\partial x} = 0 \tag{18.57}$$

$$\vec{\nabla}\cdot\mathbf{H} = 0 \;\Rightarrow\; \frac{\partial H_y(z,t)}{\partial y} = 0, \tag{18.58}$$

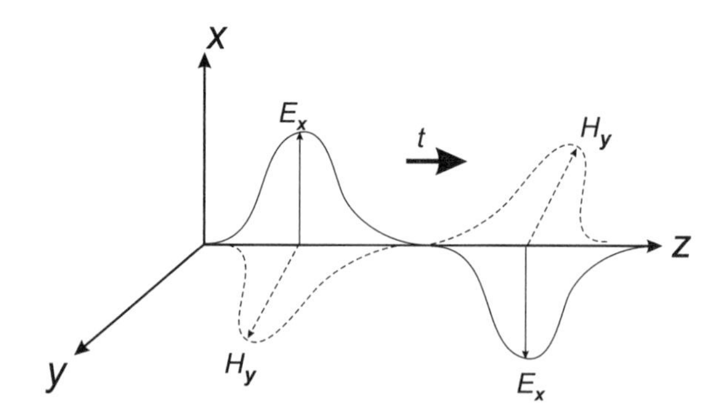

Figure 18.11 A single electromagnetic pulse traveling along the z axis. The two pulses correspond to two different times, and the coupled electric and magnetic fields are indicated by solid and dashed curves, respectively.

$$\frac{\partial \mathbf{E}}{\partial t} = +\frac{1}{\epsilon_0}\vec{\nabla} \times \mathbf{H} \Rightarrow \frac{\partial E_x}{\partial t} = -\frac{1}{\epsilon_0}\frac{\partial H_y(z,t)}{\partial z}, \tag{18.59}$$

$$\frac{\partial \mathbf{H}}{\partial t} = -\frac{1}{\mu_0}\vec{\nabla} \times \mathbf{E} \Rightarrow \frac{\partial H_y}{\partial t} = -\frac{1}{\mu_0}\frac{\partial E_x(z,t)}{\partial z}. \tag{18.60}$$

As indicated in Figure 18.11, we have chosen the electric field $\mathbf{E}(z,t)$ to oscillate (be polarized) in the x direction and the magnetic field $\mathbf{H}(z,t)$ to be polarized in the y direction. As indicated by the bold arrow in Figure 18.11, the direction of power flow for the assumed transverse electromagnetic (TEM) wave is given by the right-hand rule for $\mathbf{E} \times \mathbf{H}$. Note that although we have set the initial conditions such that the EM wave is traveling in only one dimension (z), its electric field oscillates in a perpendicular direction (x) and its magnetic field oscillates in yet a third direction (y); so while some may call this a 1-D wave, the vector nature of the fields means that the wave occupies all three dimensions.

18.11 FDTD Algorithm

We need to solve the two coupled PDEs (18.59) and (18.60) appropriate for our problem. As is usual for PDEs, we approximate the derivatives via the central-difference approximation, here in both time and space. For example,

$$\frac{\partial E(z,t)}{\partial t} \simeq \frac{E(z,t+\frac{\Delta t}{2}) - E(z,t-\frac{\Delta t}{2})}{\Delta t}, \tag{18.61}$$

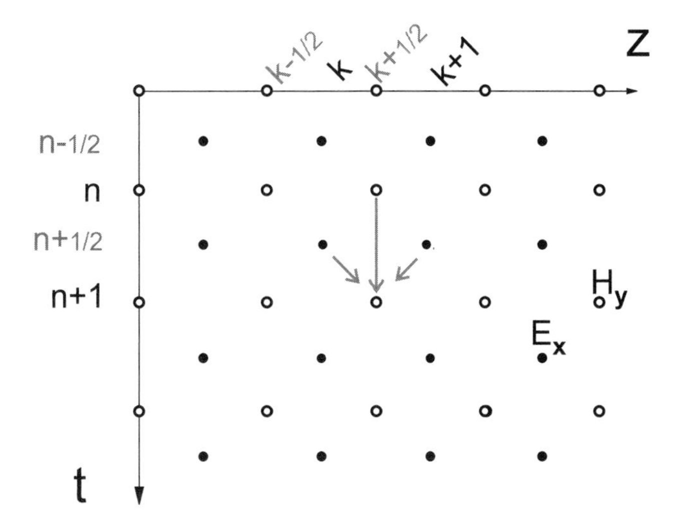

Figure 18.12 The scheme for using the known values of E_x and H_y at three earlier times and three different space positions to obtain the solution at the present time. Note that the values of E_x are determined on the lattice of filled circles, corresponding to integer space indices and half-integer time indices. In contrast, the values of H_y are determined on the lattice of open circles, corresponding to half-integer space indices and integer time indices.

$$\frac{\partial E(z,t)}{\partial z} \simeq \frac{E(z+\frac{\Delta z}{2},t) - E(z-\frac{\Delta z}{2},t)}{\Delta z}. \tag{18.62}$$

We next substitute the approximations into Maxwell's equations and rearrange the equations into the form of an algorithm that advances the solution through time. Because only first derivatives occur in Maxwell's equations, the equations are simple, although the electric and magnetic fields are intermixed.

As introduced by Yee [Yee 66], we set up a space-time lattice (Figure 18.12) in which there are half-integer time steps as well as half-integer space steps. The magnetic field will be determined at integer time sites and half-integer space sites (open circles), and the electric field will be determined at half-integer time sites and integer space sites (filled circles). Because the fields already have subscripts indicating their vector nature, we indicate the lattice position as superscripts, for example,

$$E_x(z,t) \to E_x(k\Delta z, n\Delta t) \to E_x^{k,n}. \tag{18.63}$$

Maxwell's equations (18.59) and (18.60) now become the discrete equations

$$\frac{E_x^{k,n+1/2} - E_x^{k,n-1/2}}{\Delta t} = -\frac{H_y^{k+1/2,n} - H_y^{k-1/2,n}}{\epsilon_0 \Delta z},$$

$$\frac{H_y^{k+1/2,n+1} - H_y^{k+1/2,n}}{\Delta t} = -\frac{E_x^{k+1,n+1/2} - E_x^{k,n+1/2}}{\mu_0 \Delta z}.$$

To repeat, this formulation solves for the electric field at integer space steps (k) but half-integer time steps (n), while the magnetic field is solved for at half-integer space steps but integer time steps.

We convert these equations into two simultaneous algorithms by solving for E_x at time $n+\frac{1}{2}$, and H_y at time n:

$$E_x^{k,n+1/2} = E_x^{k,n-1/2} - \frac{\Delta t}{\epsilon_0 \, \Delta z}\left(H_y^{k+1/2,n} - H_y^{k-1/2,n}\right), \tag{18.64}$$

$$H_y^{k+1/2,n+1} = H_y^{k+1/2,n} - \frac{\Delta t}{\mu_0 \Delta z}\left(E_x^{k+1,n+1/2} - E_x^{k,n+1/2}\right). \tag{18.65}$$

The algorithms must be applied simultaneously because the space variation of H_y determines the time derivative of E_x, while the space variation of E_x determines the time derivative of H_y (Figure 18.12). This algorithm is more involved than our usual time-stepping ones in that the electric fields (filled circles in Figure 18.12) at future times $t=n+\frac{1}{2}$ are determined from the electric fields at one time step earlier $t=n-\frac{1}{2}$, and the magnetic fields at half a time step earlier $t=n$. Likewise, the magnetic fields (open circles in Figure 18.12) at future times $t=n+1$ are determined from the magnetic fields at one time step earlier $t=n$, and the electric field at half a time step earlier $t=n+\frac{1}{2}$. In other words, it is as if we have two interleaved lattices, with the electric fields determined for half-integer times on lattice 1 and the magnetic fields at integer times on lattice 2.

Although these half-integer times appear to be the norm for FDTD methods [Taf 89, Sull 00], it may be easier for some readers to understand the algorithm by doubling the index values and referring to even and odd times:

$$E_x^{k,n} = E_x^{k,n-2} - \frac{\Delta t}{\epsilon_0 \Delta z}\left(H_y^{k+1,n-1} - H_y^{k-1,n-1}\right), \quad k \text{ even}, n \text{ odd}, \tag{18.66}$$

$$H_y^{k,n} = H_y^{k,n-2} - \frac{\Delta t}{\mu_0 \Delta z}\left(E_x^{k+1,n-1} - E_x^{k-1,n-1}\right), \quad k \text{ odd}, n \text{ even}. \tag{18.67}$$

This makes it clear that E is determined for even space indices and odd times, while H is determined for odd space indices and even times.

We simplify the algorithm and make its stability analysis simpler by normalizing the electric fields to have the same dimensions as the magnetic fields,

$$\tilde{E} = \sqrt{\frac{\mu_0}{\epsilon_0}}\, E. \tag{18.68}$$

The algorithm (18.64) and (18.65) now becomes

$$\boxed{\tilde{E}_x^{k,n+1/2} = \tilde{E}_x^{k,n-1/2} + \beta\left(H_y^{k-1/2,n} - H_y^{k+1/2,n}\right),} \tag{18.69}$$

$$\boxed{H_y^{k+1/2,n+1} = H_y^{k+1/2,n} + \beta\left(\tilde{E}_x^{k,n+1/2} - \tilde{E}_x^{k+1,n+1/2}\right),} \tag{18.70}$$

$$\beta = \frac{c}{\Delta z/\Delta t}, \qquad c = \frac{1}{\sqrt{\epsilon_0 \mu_0}}. \tag{18.71}$$

Here c is the speed of light in a vacuum and β is the ratio of the speed of light to grid velocity $\Delta z/\Delta t$.

The space step Δz and the time step Δt must be chosen so that the algorithm is stable. The scales of the space and time dimensions are set by the wavelength and frequency, respectively, of the propagating wave. As a minimum, we want at least 10 grid points to fall within a wavelength:

$$\Delta z \le \frac{\lambda}{10}. \tag{18.72}$$

The time step is then determined by the Courant stability condition [Taf 89, Sull 00] to be

$$\beta = \frac{c}{\Delta z/\Delta t} \le \frac{1}{2}. \tag{18.73}$$

As we have seen before, (18.73) implies that making the time step smaller improves precision and maintains stability, but making the space step smaller must be accompanied by a simultaneous decrease in the time step in order to maintain stability (you should check this).

18.11.1 Implementation

In Listing 18.3 we provide a simple implementation of the FDTD algorithm for a z lattice of 200 sites. The initial condition corresponds to a Gaussian pulse in time for the E field, located at the midpoint of the z lattice (in Δt and Δz units):

$$E_x(z=100, t) = \exp\left[\frac{1}{2}\left(\frac{40-t}{12}\right)^2\right]. \tag{18.74}$$

```
// FDTD.java  FDTD solution of Maxwell's equations in 1-D
import java.io.*;
import java.util.*;

public class FDTD {
  public static void main(String[] argv) throws IOException, FileNotFoundException  {
    double dx, dt, beta, c0;
    int time = 100, max = 200, i, n, j, k;
    beta = 0.5;                                  // beta = c/(dz/dt) < 0.5 for  stability
    double Ex[] = new double[max+1]; double Hy[] = new double[max+1];        // Ex & Hy
                 //    Gaussian Pulse Variables
    int m = max/2;                                          // Pulse in center of space
    double t = 40, width = 12.;                           // Center, width of the pulse
    PrintWriter w =new PrintWriter( new FileOutputStream("E.dat"), true );
    PrintWriter q =new PrintWriter( new FileOutputStream("H.dat"), true );
      // Intial conditions
    for( k = 1; k < max; k++ ) Ex[k] = Hy[k]=0.;

    for ( n = 0; n < time; n++ ) {
      for( k = 1; k < max; k++ ) Ex[k]  = Ex[k] + beta * ( Hy[k-1]-Hy[k] );   //Eq 1
      Ex[m] =  Ex[m] + Math.exp(  -0.5*((t-n)/width)*((t-n)/width) );          // Pulse
      for( j = 0; j < max-1; j++) Hy[j] = Hy[j] + beta * ( Ex[j]-Ex[j+1] );   // Eq 2
    }
    for ( k = 0; k < max; k++ ) {
      w.println(""+k+" "+Ex[k]+" " +Hy[k]+" ");
      q.println(""+k+" "+Hy[k]+" ");
    }
} }
```

Listing 18.3 FDTD.java solves Maxwell's equations via FDTD time stepping (finite-difference time domain) for linearly polarized wave propagation in the z direction in free space.

The algorithm then steps out in time for as long you the user desires (although it makes no sense to have the pulse extend beyond the integration region). Note that the initial condition (18.74) is somewhat unusual in that it is imposed over the time period during which the E and H fields are being stepped out in time. However, the initial pulse dies out quickly, and so, for example, is not present in the simulation results seen at $t = 100$ in Figure 18.13.

Our implementation is also unusual in that we have not imposed boundary conditions on the solution. Of course a unique solution to a PDE requires proper boundary conditions, so it must be that our finite lattice size is imposing effective boundary conditions (something to be explored in the assessment).

18.11.2 Assessment

1. Impose boundary conditions such that all fields vanish on the boundaries. Compare the solutions so obtained to those without explicit conditions for times less than and greater than those at which the pulses hit the walls.
2. Examine the stability of the solution for different values of Δz and Δt and thereby test the Courant condition (18.73).
3. Extend the algorithm to include the effect of entering, propagating through, and exiting a dielectric material placed within the z integration region.
   a. Ensure that you see both transmission and reflection at the boundaries.

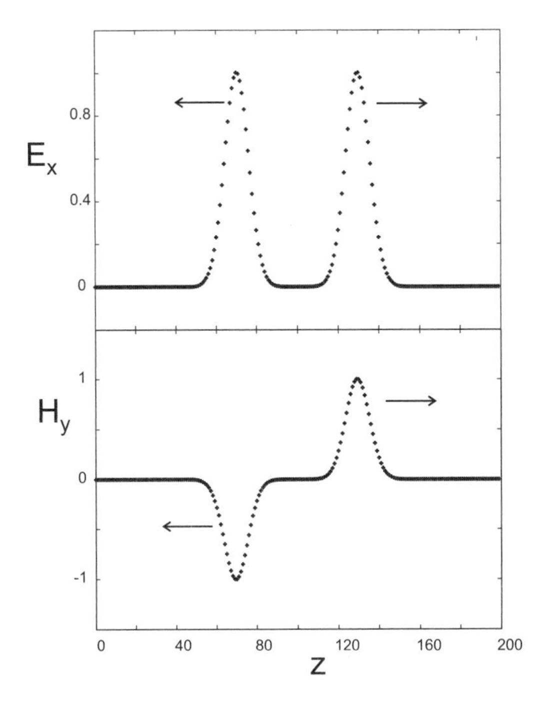

Figure 18.13 The E and H fields at time 100 propagating outward in free space from an initial electrical pulse at $z = 50$. The arrows indicate the directions of propagation with time.

b. Investigate the effect of a dielectric with an index of refraction less than 1.

4. The direction of propagation of the pulse is given $\mathbf{E} \times \mathbf{H}$, which depends on the relative phase between the E and H fields. (With no initial $\mathbf{H}$ field, we obtain pulses both to the right and the left.)
   a. Modify the algorithm such that there is an initial H pulse as well as an initial E pulse, both with a Gaussian times a sinusoidal shape.
   b. Verify that the direction of pulse propagation changes if the E and H fields have relative phases of 0 or π.
5. Investigate the resonator modes of a wave guide by picking the initial conditions corresponding to plane waves with nodes at the boundaries.
6. Determine what happens when you try to set up standing waves with wavelengths longer than the size of the integration region.
7. Simulate unbounded propagation by building in periodic boundary conditions into the algorithm.
8. ⊙ Place a medium with periodic permittivity in the integration volume. This should act as a frequency-dependent filter, which does not propagate certain frequencies at all.

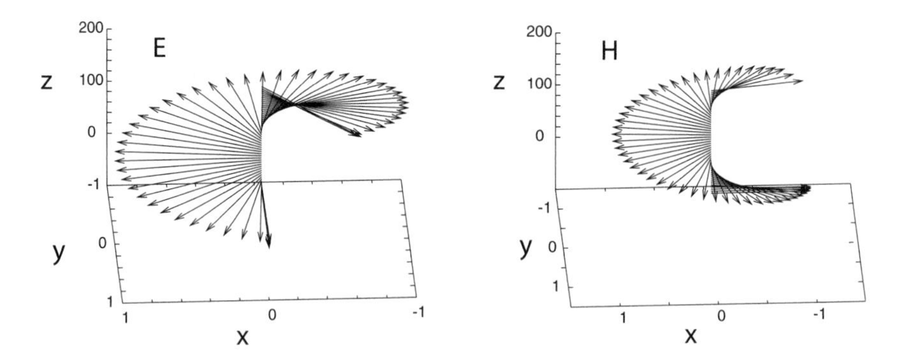

Figure 18.14 The E and H fields at time 100 for a circularly polarized EM wave in free space.

18.11.3 Extension: Circularly Polarized EM Waves

We now extend our treatment to EM waves in which the **E** and **H** fields, while still transverse and propagating in the z direction, are not restricted to linear polarizations along just one axis. Accordingly, we add to (18.59) and (18.60):

$$\frac{\partial H_x}{\partial t} = \frac{1}{\mu_0}\frac{\partial E_y}{\partial z}, \tag{18.75}$$

$$\frac{\partial E_y}{\partial t} = \frac{1}{\epsilon_0}\frac{\partial H_x}{\partial z}. \tag{18.76}$$

When discretized in the same way as (18.64) and (18.65), we obtain

$$H_x^{k+1/2,n+1} = H_x^{k+1/2,n} + \frac{\Delta t}{\mu_0\,\Delta z}\left(E_y^{k+1,n+1/2} - E_y^{k,n+1/2}\right), \tag{18.77}$$

$$E_y^{k,n+1/2} = E_y^{k,n-1/2} + \frac{\Delta t}{\epsilon_0\,\Delta z}\left(H_y^{k+1/2,n} - H_y^{k-1/2,n}\right). \tag{18.78}$$

To produce a circularly polarized traveling wave, we set the initial conditions:

$$E_x = \cos\left(t - \frac{z}{c} + \phi_y\right), \quad H_x = \sqrt{\frac{\epsilon_0}{\mu_0}}\cos\left(t - \frac{z}{c} + \phi_y\right), \tag{18.79}$$

$$E_y = \cos\left(t - \frac{z}{c} + \phi_x\right), \quad H_y = \sqrt{\frac{\epsilon_0}{\mu_0}}\cos\left(t - \frac{z}{c} + \phi_x + \pi\right). \tag{18.80}$$

We take the phases to be $\phi_x = \pi/2$ and $\phi_y = 0$, so that their difference $\phi_x - \phi_y = \pi/2$, which leads to circular polarization. We include the initial conditions in the same manner as we did the Gaussian pulse, only now with these cosine functions.

Listing 18.4 gives our implementation EMcirc.java for waves with transverse two-component **E** and **H** fields. Some results of the simulation are shown in Figure 18.14, where you will note the difference in phase between **E** and **H**.

```
// EMcirc.java: FDTD Propgation of circularly polarized EM wave
import java.io.*;
import java.util.*;

public class EMcirc {

  public static double Exini( int tim, double x, double phx)                 // Initial Ex
    { return Math.cos(tim-2*Math.PI*x/200. +phx ); }
  public static double Eyini( int tim, double x, double phy)                 // Initial Ey
    { return Math.cos(tim-2*Math.PI*x/200. +phy ); }
 public static double Hxini( int tim, double x, double phy)                   // Inital Hx
   { return Math.cos(tim-2*Math.PI*x/200. +phy+Math.PI); }
 public static double Hyini( int tim, double x, double phx)                  // Initial Hy
   { return Math.cos(tim-2*Math.PI*x/200.0 +phx); }

  public static void main(String[] argv) throws IOException, FileNotFoundException {
    double dx, dt, beta, c0, phx, phy;
    int time = 100, max = 200, i, n, j, k;
    phx = 0.5*Math.PI; phy = 0.0;                            // Phase, difference is pi/2
    beta = 0.1;                                    // beta = c/(dz/dt) < 0.5 for stability
    double Ex[] = new double[max+1]; double Hy[] = new double[max+1];          // Ex, Hy
    double Ey[] = new double[max+1]; double Hx[] = new double[max+1];          // Ey, Hx
    for (i=0; i < max; i++) {Ex[i]=0;  Ey[i]=0;  Hx[i]=0;  Hy[i]=0;}     // Initialize
    for(k = 0; k < max; k++)                                              // Initialize
      { Ex[k]=Exini(0,(double)k,phx);  Ey[k]=Eyini(0,(double)k,phy);
        Hx[k]=Hxini(0,(double)k,phy);  Hy[k]=Hyini(0,(double)k,phx);       }
    PrintWriter w = new PrintWriter( new FileOutputStream( "Efield.dat" ), true);
    PrintWriter q = new PrintWriter( new FileOutputStream( "Hfield.dat" ), true);
    for (n = 0; n < time; n++ ) {
      for(k = 1; k < max; k++ ) {                                         // New Ex, Ey
       Ex[k] = Ex[k] + beta * (Hy[k-1] - Hy[k]);
       Ey[k] = Ey[k] + beta * (Hx[k]   - Hx[k-1]);
      }
      for (j = 0; j < max-1; j++ ) {                                      // New Hx, Hy
        Hy[j] = Hy[j] + beta * (Ex[j]   - Ex[j+1]);
        Hx[j] = Hx[j] + beta * (Ey[j+1] - Ey[j]);
      }
    }                                                                         // Time
    for ( k = 0; k < max; k = k+4 ) {                          // Plot every 4 points
      w.println(""+0.0+" "+0.0+" "+k+"  "+Ex[k]+" "+Ey[k]+"  "+0.0);
      q.println(""+0.0+" "+0.0+" "+k+"  "+Hx[k]+" "+Hy[k]+"  "+0.0);
} } }
```

Listing 18.4 EMcirc.java solves Maxwell's equations via FDTD time-stepping for circularly polarized wave propagation in the *z* direction in free space.

19

Solitons & Computational Fluid Dynamics

In Unit I of this chapter we discuss shallow-water soliton waves. This extends the discussion of waves in Chapter 18, "PDE Waves: String, Quantum Packet, and E&M," *by progressively including nonlinearities, dispersion, and hydrodynamic effects. In Unit II we confront the more general equations of computational fluid dynamics (CFD) and their solutions.*[1] *The mathematical description of the motion of fluids, though not a new subject, remains a challenging one. The equations are complicated and nonlinear, there are many degrees of freedom, the nonlinearities may lead to instabilities, analytic solutions are rare, and the boundary conditions for realistic geometries (like airplanes) are not intuitive. These difficulties may explain why fluid dynamics is often absent from undergraduate and even graduate physics curricula. Nonetheless, as an essential element of the real world that also has tremendous practical importance, we encourage its study. We recommend [F&W 80, L&L,F 87] for those interested in the derivations, and [Shaw 92] for more details about the computations.*

19.1 Unit I. Advection, Shocks, and Russell's Soliton

In 1834, J. Scott Russell observed on the Edinburgh-Glasgow canal [Russ 44]:

> I was observing the motion of a boat which was rapidly drawn along a narrow channel by a pair of horses, when the boat suddenly stopped—not so the mass of water in the channel which it had put in motion; it accumulated round the prow of the vessel in a state of violent agitation, then suddenly leaving it behind, rolled forward with great velocity, assuming the form of a large solitary elevation, a rounded, smooth and well-defined heap of water, which continued its course along the channel apparently without change of form or diminution of speed. I followed it on horseback, and overtook it still rolling on at a rate of some eight or nine miles an hour, preserving its original figure some thirty feet long and a foot to a foot and a half in height. Its height gradually diminished, and after a chase of one or two miles I lost it in the windings of the channel. Such, in the month of August 1834, was my first chance interview with that singular and beautiful phenomenon....

Russell also noticed that an initial arbitrary waveform set in motion in the channel evolves into two or more waves that move at different velocities and progressively move apart until they form individual solitary waves. In Figure 19.2 we see a single steplike wave breaking up into approximately eight of these solitary waves (also

[1] We acknowledge some helpful reading of Unit I by Satoru S. Kano.

called *solitons*). These eight solitons occur so frequently that some consider them the normal modes for this nonlinear system. Russell went on to produce these solitary waves in a laboratory and empirically deduced that their speed c is related to the depth h of the water in the canal and to the amplitude A of the wave by

$$c^2 = g(h+A), \tag{19.1}$$

where g is the acceleration due to gravity. Equation (19.1) implies an effect not found for linear systems, namely, that waves with greater amplitudes A travel faster than those with smaller amplitudes. Observe that this is similar to the formation of shock waves but different from *dispersion* in which waves of different wavelengths have different velocities. The dependence of c on amplitude A is illustrated in Figure 19.3, where we see a taller soliton catching up with and passing through a shorter one.

Problem: Explain Russell's observations and see if they relate to the formation of *tsunamis*. The latter are ocean waves that form from sudden changes in the level of the ocean floor and then travel over long distances without dispersion or attenuation until they wreak havoc on a distant shore.

19.2 Theory: Continuity and Advection Equations

The motion of a fluid is described by the continuity equation and the Navier–Stokes equation [L&L,M 76]. We will discuss the former here and the latter in §19.7 of Unit II. The continuity equation describes conservation of mass:

$$\boxed{\frac{\partial \rho(\mathbf{x},t)}{\partial t} + \vec{\nabla}\cdot\mathbf{j} = 0,} \qquad \mathbf{j} \stackrel{\text{def}}{=} \rho\mathbf{v}(\mathbf{x},t). \tag{19.2}$$

Here $\rho(\mathbf{x},t)$ is the mass density, $\mathbf{v}(\mathbf{x},t)$ is the velocity of the fluid, and the product $\mathbf{j} = \rho\mathbf{v}$ is the mass current. As its name implies, the divergence $\vec{\nabla}\cdot\mathbf{j}$ describes the spreading of the current in a region of space, as might occur if there were a current source. Physically, the continuity equation (19.2) states that changes in the density of the fluid within some region of space arise from the flow of current in and out of that region.

For 1-D flow in the x direction and for a fluid that is moving with a constant velocity $v = c$, the continuity equation (19.2) takes the simple form

$$\frac{\partial \rho}{\partial t} + \frac{\partial (c\rho)}{\partial x} = 0, \tag{19.3}$$

$$\boxed{\frac{\partial \rho}{\partial t} + c\frac{\partial \rho}{\partial x} = 0.} \tag{19.4}$$

This equation is known as the *advection equation*, where the term "advection" is used to describe the horizontal transport of a conserved quantity from one region of space to another due to a velocity field. For instance, advection describes dissolved salt transported in water.

The advection equation looks like a first-derivative form of the wave equation, and indeed, the two are related. A simple substitution proves that any function with the form of a traveling wave,

$$u(x,t) = f(x - ct), \tag{19.5}$$

will be a solution of the advection equation. If we consider a surfer riding along the crest of a traveling wave, that is, remaining at the same position relative to the wave's shape as time changes, then the surfer does not see the shape of the wave change in time, which implies that

$$x - ct = \text{constant} \quad \Rightarrow \quad x = ct + \text{constant}. \tag{19.6}$$

The speed of the surfer is therefore $dx/dt = c$, which is a constant. Any function $f(x - ct)$ is clearly a traveling wave solution in which an arbitrary pulse is carried along by the fluid at velocity c without changing shape.

19.2.1 Advection Implementation

Although the advection equation is simple, trying to solve it by a simple differencing scheme (the leapfrog method) may lead to unstable numerical solutions. As we shall see when we look at the nonlinear version of this equation, there are better ways to solve it. Listing 19.1 presents our code for solving the advection equation using the Lax–Wendroff method (a better method).

19.3 Theory: Shock Waves via Burgers' Equation

In a later section we will examine use of the KdeV equation to describe Russell's solitary waves. In order to understand the physics contained in that equation, we study some terms in it one at a time. To start, consider Burgers' equation [Burg 74]:

$$\frac{\partial u}{\partial t} + \epsilon u \frac{\partial u}{\partial x} = 0, \tag{19.7}$$

$$\frac{\partial u}{\partial t} + \epsilon \frac{\partial (u^2/2)}{\partial x} = 0, \tag{19.8}$$

where the second equation is the *conservative form*. This equation can be viewed as a variation on the advection equation (19.4) in which the wave speed $c = \epsilon u$ is proportional to the amplitude of the wave, as Russell found for his waves. The second nonlinear term in Burgers' equation leads to some unusual behaviors. Indeed, John von Neumann studied this equation as a simple model for turbulence [F&S].

```
// AdvecLax.java Lax-Wendroff solution of advection eq
import java.io.*;

public class AdvecLax {
  public static void main(String[] argv) throws IOException, FileNotFoundException {
    PrintWriter w = new PrintWriter(new FileOutputStream("numerical.dat"), true);
    PrintWriter q = new PrintWriter(new FileOutputStream("initial.dat"), true); // ic
    PrintWriter r = new PrintWriter(new FileOutputStream("exact.dat"), true);
    int m = 100, i, j, n;;
    double u[] = new double[m + 1], u0[] = new double[m + 1];
    double uf[] = new double[m + 1],    c = 1. ;                    // Wave speed
    double x, dx, dt, T_final, beta = 0.8;                       // beta = c*dt/dx
    dx = 1./m;        dt = beta*dx/c;          T_final = 0.5;
    n = (int)(T_final/dt); System.out.println("" + n + "");
    for ( i=0; i < m-1; i++ ) {
      x = i*dx;
      u0[i] = Math.exp(-300.* (x-0.12)*(x-0.12) );          // Gaussian initial data
      q.println("" + 0.01*i + " " + u0[i] + " ");
      uf[i] = Math.exp(-300.*(x-0.12-c*T_final)*(x-0.12-c*T_final) );          // Exact
    r.println("" + 0.01*i + " " + uf[i] + " ");
    }
    for ( j=1; j < n; j++ ) {
      for ( i=0; i < m-1; i++ ) {                                  // Lax-Wendroff scheme
        u[i + 1] = (1.-beta*beta)*u0[i + 1] -
        (0.5*beta)*(1.-beta)*u0[i+2] + (0.5*beta)*(1.+beta)*u0[i];
        u[0] = 0. ;  u[m-1] = 0. ;
        u0[i] = u[i];
  } }
    for ( j=0; j < m-1; j++ ){w.println(" "+0.01*j +" "+u[j]+" "); }
    System.out.println("Solution saved in numerical.dat");
} }
```

Listing 19.1 AdvecLax.java solves the advection equation via the Lax-Wendroff scheme.

In the advection equation (19.4), all points on the wave move at the same speed c, and so the shape of the wave remains unchanged in time. In Burgers' equation (19.7), the points on the wave move ("advect") themselves such that the local speed depends on the local wave's amplitude, with the high parts of the wave moving progressively faster than the low parts. This changes the shape of the wave in time; if we start with a wave packet that has a smooth variation in height, the high parts will speed up and push their way to the front of the packet, thereby forming a sharp leading edge known as a *shock wave* [Tab 89]. A shock wave solution to Burgers' equation with $\epsilon = 1$ is shown in Figure 19.1.

19.3.1 Algorithm: The Lax-Wendroff Method for Burgers' Equation

We first solve Burgers' equation (19.4) via the usual approach in which we express the derivatives as central differences. This leads to a leapfrog scheme for the future

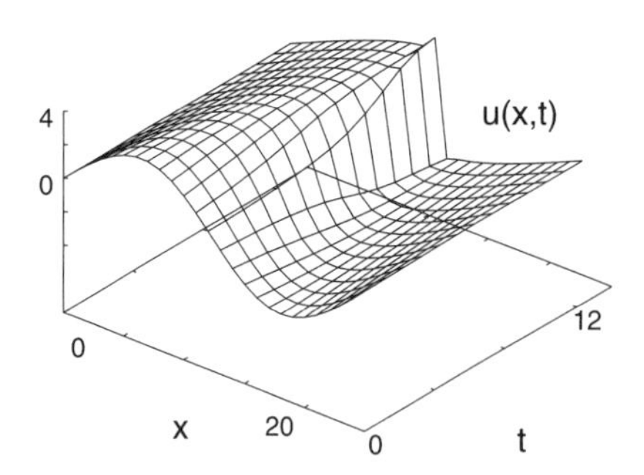

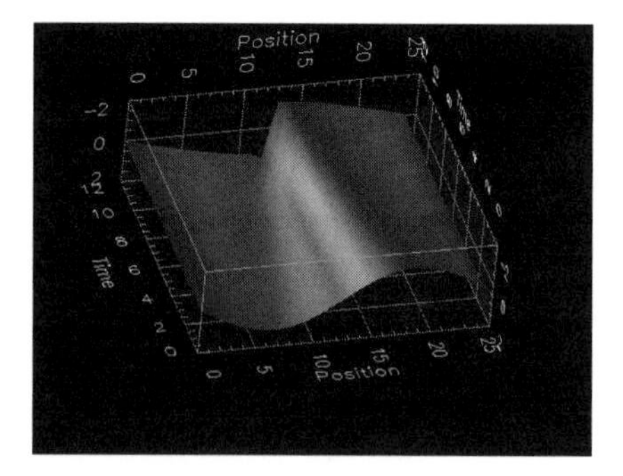

Figure 19.1 Formation of a shock wave. Wave height *versus* position for increasing times as visualized with Gnuplot (left) and with OpenDX (right).

solution in terms of present and past ones:

$$u(x, t+\Delta t) = u(x, t-\Delta t) - \beta\left[\frac{u^2(x+\Delta x, t) - u^2(x-\Delta x, t)}{2}\right],$$

$$u_{i,j+1} = u_{i,j-1} - \beta\left[\frac{u^2_{i+1,j} - u^2_{i-1,j}}{2}\right], \beta = \frac{\epsilon}{\Delta x/\Delta t}. \tag{19.9}$$

Here u^2 is the square of u and is not its second derivative, and β is a ratio of constants known as the *Courant–Friedrichs–Lewy* (CFL) *number*. As you should prove for yourself, $\beta < 1$ is required for stability.

While we have used a leapfrog method with success in the past, its low-order approximation for the derivative becomes inaccurate when the gradients can get large, as happens with shock waves, and the algorithm may become unstable [Pres 94]. The *Lax–Wendroff method* attains better stability and accuracy by retaining second-order differences for the time derivative:

$$u(x, t+\Delta t) \simeq u(x, t) + \frac{\partial u}{\partial t}\Delta t + \frac{1}{2}\frac{\partial^2 u}{\partial t^2}\Delta t^2. \tag{19.10}$$

To covert (19.10) to an algorithm, we use Burgers' equation $\partial u/\partial t = -\epsilon\partial(u^2/2)/\partial x$ for the first-order time derivative. Likewise, we use this equation to express the second-order time derivative in terms of space derivatives:

$$\frac{\partial^2 u}{\partial t^2} = \frac{\partial}{\partial t}\left[-\epsilon\frac{\partial}{\partial x}\left(\frac{u^2}{2}\right)\right] = -\epsilon\frac{\partial}{\partial x}\frac{\partial}{\partial t}\left(\frac{u^2}{2}\right) \tag{19.11}$$

$$= -\epsilon\frac{\partial}{\partial x}\left(u\frac{\partial u}{\partial t}\right) = \epsilon^2\frac{\partial}{\partial x}\left[u\frac{\partial}{\partial x}\left(\frac{u^2}{2}\right)\right].$$

We next substitute these derivatives into the Taylor expansion (19.10) to obtain

$$u(x,t+\Delta t) = u(x,t) - \Delta t\epsilon \frac{\partial}{\partial x}\left(\frac{u^2}{2}\right) + \frac{(\Delta t)^2}{2}\epsilon^2 \frac{\partial}{\partial x}\left[u\frac{\partial}{\partial x}\left(\frac{u^2}{2}\right)\right].$$

We now replace the outer x derivatives by central differences of spacing $\Delta x/2$:

$$u(x,t+\Delta t) = u(x,t) - \frac{\Delta t\epsilon}{2}\frac{u^2(x+\Delta x,t) - u^2(x-\Delta x,t)}{2\Delta x} + \frac{(\Delta t)^2 \epsilon^2}{2}$$
$$\times \frac{1}{2\Delta x}\left[u\left(x+\frac{\Delta x}{2},t\right)\frac{\partial}{\partial x}u^2\left(x+\frac{\Delta x}{2},t\right) - u\left(x-\frac{\Delta x}{2},t\right)\right.$$
$$\left.\frac{\partial}{\partial x}u^2\left(x-\frac{\Delta x}{2},t\right)\right].$$

Next we approximate $u(x \pm \Delta x/2, t)$ by the average of adjacent grid points,

$$u(x \pm \frac{\Delta x}{2}, t) \simeq \frac{u(x,t) + u(x\pm\Delta x, t)}{2},$$

and apply a central-difference approximation to the second derivatives:

$$\frac{\partial u^2(x\pm\Delta x/2,t)}{\partial x} = \frac{u^2(x\pm\Delta x,t) - u^2(x,t)}{\pm\Delta x}.$$

Finally, putting all these derivatives together yields the discrete form

$$u_{i,j+1} = u_{i,j} - \frac{\beta}{4}\left(u_{i+1,j}^2 - u_{i-1,j}^2\right) + \frac{\beta^2}{8}\left[\left(u_{i+1,j}+u_{i,j}\right)\left(u_{i+1,j}^2 - u_{i,j}^2\right)\right.$$
$$\left. -\left(u_{i,j}+u_{i-1,j}\right)\left(u_{i,j}^2 - u_{i-1,j}^2\right)\right], \qquad (19.12)$$

where we have substituted the CFL number β. This Lax–Wendroff scheme is explicit, centered upon the grid points, and stable for $\beta < 1$ (small nonlinearities).

19.3.2 Implementation and Assessment of Burgers' Shock Equation

1. Write a program to solve Burgers' equation via the leapfrog method.
2. Define arrays u0[100] and u[100] for the initial data and the solution.
3. Take the initial wave to be sinusoidal, u0[i]= $3\sin(3.2x)$, with speed $c = 1$.
4. Incorporate the boundary conditions u[0]=0 and u[100]=0.
5. Keep the CFL number $\beta < 1$ for stability.

6. Now modify your program to solve Burgers' shock equation (19.8) using the Lax–Wendroff method (19.12).
7. Save the initial data and the solutions for a number of times in separate files for plotting.
8. Plot the initial wave and the solution for several time values on the same graph in order to see the formation of a shock wave (like Figure 19.1).
9. Run the code for several increasingly large CFL numbers. Is the stability condition $\beta < 1$ correct for this nonlinear problem?
10. Compare the leapfrog and Lax–Wendroff methods. With the leapfrog method you should see shock waves forming but breaking up into ripples as the square edge develops. The ripples are numerical artifacts. The Lax–Wendroff method should give a a better shock wave (square edge), although some ripples may still occur. ▌

The listing below presents the essentials of the Lax–Wendroff method.

```
public class Shock {
    public static void main(String[] argv)
        throws IOException, FileNotFoundException {
        PrintWriter w = new PrintWriter(new FileOutputStream("Shock.dat"), true);
        PrintWriter q = new PrintWriter(new FileOutputStream("Initial.dat"), true);
        int m = 100, i,j,n;                                          // Number grid points
        double u[] = new double[m+1], u0[] = new double[m+1];             // solutions
        double epsilon = 1.0, beta = 0.1;                            // Speed, CFL number
        double x, dx, dt, T_final = 0.15;
        dx =2./m;        dt = beta*dx/epsilon;                       // Space, time steps
        n = (int)(T_final/dt);     System.out.println(""+n+"");
        for( i=0; i < m-1; i++ ) {
          x = i*dx;
          u0[i] = 3.0 * Math.sin( 3.2*x );
          q.println("" + 0.01*i + " " + u0[i] + " ");
        }
        for ( j = 1; j < n; j++) {                                   // Lax-Wendroff scheme
          for ( i = 0; i < m-1; i++ ) {
            u[i+1] = u0[i+1]- (Math.pow(u0[i+2],2)-Math.pow(u0[i],2))*(0.25*beta)+
                    (((u0[i+2] + u0[i+1])/2.)*(Math.pow(u0[i+2], 2)
                    - Math.pow(u0[i+1], 2)) - ((u0[i+1] + u0[i])/2.)
                    *(Math.pow(u0[i+1], 2) - Math.pow(u0[i], 2)))*0.25*beta*beta;
            u[0] = 0.; u[m-1]=0.;
            u0[i] = u[i];                                            // Shift new to old
          }
        }
        for ( j=0; j<m-1; j++ )  w.println(" "+0.01*j+" "+u[j]+" ");
        System.out.println("Solution in Shock.dat");                 // Gnu 3-D format
} }                                                                  //End main, End class
```

19.4 Including Dispersion

We have just seen that Burgers' equation can turn an initially smooth wave into a square-edged shock wave. An inverse wave phenomenon is *dispersion*, in which a waveform disperses or broadens as it travel through a medium. Dispersion does not cause waves to lose energy and attenuate but rather to lose information with time. Physically, dispersion may arise when the propagating medium has structures with a spatial regularity equal to some fraction of a wavelength. Mathematically, dispersion may arise from terms in the wave equation that contain higher-order space derivatives. For example, consider the waveform

$$u(x,t) = e^{\pm i(kx-\omega t)} \tag{19.13}$$

corresponding to a plane wave traveling to the right ("traveling" because the phase $kx - \omega t$ remains unchanged if you increase x with time). When this $u(x,t)$ is substituted into the advection equation (19.4), we obtain

$$\omega = ck. \tag{19.14}$$

This equation is an example of a *dispersion relation*, that is, a relation between frequency ω and wave vector k. Because the *group velocity* of a wave

$$v_g = \frac{\partial \omega}{\partial k}, \tag{19.15}$$

the linear dispersion relation (19.14) leads to all frequencies having the same group velocity c and thus *dispersionless* propagation.

Let us now imagine that a wave is propagating with a small amount of *dispersion*, that is, with a frequency that has somewhat less than a linear increase with the wave number k:

$$\omega \simeq ck - \beta k^3. \tag{19.16}$$

Note that we skip the even powers in (19.16), so that the group velocity,

$$v_g = \frac{d\omega}{dk} \simeq c - 3\beta k^2, \tag{19.17}$$

is the same for waves traveling to the left the or the right. If plane-wave solutions like (19.13) were to arise from a wave equation, then (as verified by substitution) the ω term of the dispersion relation (19.16) would arise from a first-order time derivative, the ck term from a first-order space derivative, and the k^3 term from a third-order space derivative:

$$\frac{\partial u(x,t)}{\partial t} + c\frac{\partial u(x,t)}{\partial x} + \beta\frac{\partial^3 u(x,t)}{\partial x^3} = 0. \tag{19.18}$$

We leave it as an exercise to show that solutions to this equation do indeed have waveforms that disperse in time.

19.5 Shallow-Water Solitons, the KdeV Equation

CD

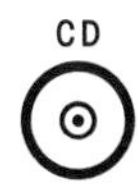

In this section we look at shallow-water soliton waves. Though including some complications, this subject is fascinating and is one for which the computer has been absolutely essential for discovery and understanding. In addition, we recommend that you look at some of the soliton animation we have placed in the animations folder on the CD.

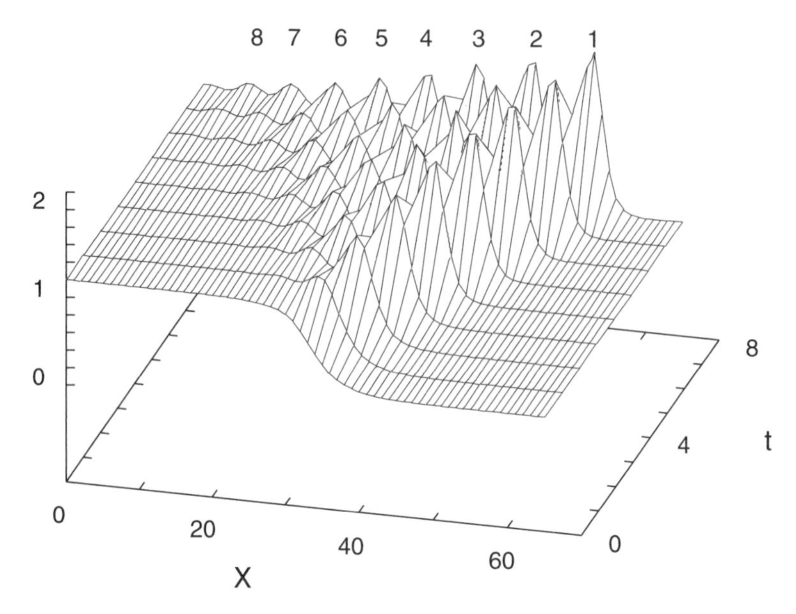

Figure 19.2 A single two-level waveform at time zero progressively breaks up into eight solitons (labeled) as time increases. The tallest soliton (1) is narrower and faster in its motion to the right.

We want to understand the unusual water waves that occur in shallow, narrow channels such as canals [Abar 93, Tab 89]. The analytic description of this "heap of water" was given by Korteweg and deVries (KdeV) [KdeV 95] with the partial differential equation

$$\frac{\partial u(x,t)}{\partial t} + \varepsilon u(x,t)\,\frac{\partial u(x,t)}{\partial x} + \mu\,\frac{\partial^3 u(x,t)}{\partial x^3} = 0. \tag{19.19}$$

As we discussed in §19.1 in our study of Burgers' equation, the nonlinear term $\varepsilon u\,\partial u/\partial t$ leads to a sharpening of the wave and ultimately a *shock* wave. In contrast, as we discussed in our study of dispersion, the $\partial^3 u/\partial x^3$ term produces broadening. These together with the $\partial u/\partial t$ term produce traveling waves. For the proper parameters and initial conditions, the dispersive broadening exactly balances the nonlinear narrowing, and a stable traveling wave is formed.

KdeV solved (19.19) analytically and proved that the speed (19.1) given by Russell is in fact correct. Seventy years after its discovery, the KdeV equation was rediscovered by Zabusky and Kruskal [Z&K 65], who solved it numerically and found that a $\cos(x/L)$ initial condition broke up into eight solitary waves (Figure 19.2). They also found that the parts of the wave with larger amplitudes moved faster than those with smaller amplitudes, which is why the higher peaks tend to be on the right in Figure 19.2. As if wonders never cease,

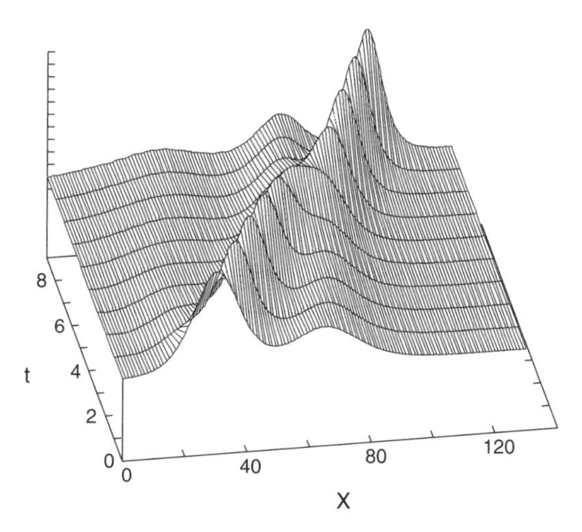

Figure 19.3 Two shallow-water solitary waves crossing each other (computed with the code `SolCross.java` on the instructor's CD). The taller soliton on the left catches up with and overtakes the shorter one at $t \simeq 5$.

Zabusky and Kruskal, who coined the name *soliton* for the solitary wave, also observed that a faster peak actually passed through a slower one unscathed (Figure 19.3).

19.5.1 Analytic Soliton Solution

The trick in analytic approaches to these types of nonlinear equations is to substitute a guessed solution that has the form of a traveling wave,

$$u(x,t) = u(\xi = x - ct). \tag{19.20}$$

This form means that if we move with a constant speed c, we will see a constant wave form (but now the speed will depend on the magnitude of u). There is no guarantee that this form of a solution exists, but it is a lucky guess because substitution into the KdeV equation produces a solvable ODE and its solution:

$$-c\,\frac{\partial u}{\partial \xi} + \epsilon\, u\,\frac{\partial u}{\partial \xi} + \mu\,\frac{d^3 u}{d\xi^3} = 0, \tag{19.21}$$

$$u(x,t) = \frac{-c}{2}\operatorname{sech}^2\left[\frac{1}{2}\sqrt{c}(x - ct - \xi_0)\right], \tag{19.22}$$

where ξ_0 is the initial phase. We see in (19.22) an amplitude that is proportional to the wave speed c, and a sech^2 function that gives a single lumplike wave. This is a typical analytic form for a soliton.

19.5.2 Algorithm for KdeV Solitons

The KdeV equation is solved numerically using a finite-difference scheme with the time and space derivatives given by central-difference approximations:

$$\frac{\partial u}{\partial t} \simeq \frac{u_{i,j+1} - u_{i,j-1}}{2\Delta t}, \qquad \frac{\partial u}{\partial x} \simeq \frac{u_{i+1,j} - u_{i-1,j}}{2\Delta x}. \tag{19.23}$$

To approximate $\partial^3 u(x,t)/\partial x^3$, we expand $u(x,t)$ to $\mathcal{O}(\Delta t)^3$ about the four points $u(x \pm 2\Delta x, t)$ and $u(x \pm \Delta x, t)$,

$$u(x \pm \Delta x, t) \simeq u(x,t) \pm (\Delta x)\frac{\partial u}{\partial x} + \frac{(\Delta x)^2}{2!}\frac{\partial^2 u}{\partial^2 x} \pm \frac{(\Delta x)^3}{3!}\frac{\partial^3 u}{\partial x^3}, \tag{19.24}$$

which we solve for $\partial^3 u(x,t)/\partial x^3$. Finally, the factor $u(x,t)$ in the second term of (19.19) is taken as the average of three x values all with the same t:

$$u(x,t) \simeq \frac{u_{i+1,j} + u_{i,j} + u_{i-1,j}}{3}. \tag{19.25}$$

These substitutions yield the algorithm for the KdeV equation:

$$\begin{aligned} u_{i,j+1} \simeq u_{i,j-1} &- \frac{\epsilon}{3}\frac{\Delta t}{\Delta x}\left[u_{i+1,j} + u_{i,j} + u_{i-1,j}\right]\left[u_{i+1,j} - u_{i-1,j}\right] \\ &- \mu \frac{\Delta t}{(\Delta x)^3}\left[u_{i+2,j} + 2u_{i-1,j} - 2u_{i+1,j} - u_{i-2,j}\right]. \end{aligned} \tag{19.26}$$

To apply this algorithm to predict future times, we need to know $u(x,t)$ at present and past times. The initial-time solution $u_{i,1}$ is known for all positions i via the initial condition. To find $u_{i,2}$, we use a forward-difference scheme in which we expand $u(x,t)$, keeping only two terms for the time derivative:

$$\begin{aligned} u_{i,2} \simeq u_{i,1} &- \frac{\epsilon\,\Delta t}{6\,\Delta x}\left[u_{i+1,1} + u_{i,1} + u_{i-1,1}\right]\left[u_{i+1,1} - u_{i-1,1}\right] \\ &- \frac{\mu}{2}\frac{\Delta t}{(\Delta x)^3}\left[u_{i+2,1} + 2u_{i-1,1} - 2u_{i+1,1} - u_{i-2,1}\right]. \end{aligned} \tag{19.27}$$

The keen observer will note that there are still some undefined columns of points, namely, $u_{1,j}$, $u_{2,j}$, $u_{N_{\max}-1,j}$, and $u_{N_{\max},j}$, where $N_{\max}$ is the total number of grid points. A simple technique for determining their values is to assume that $u_{1,2} = 1$ and $u_{N_{\max},2} = 0$. To obtain $u_{2,2}$ and $u_{N_{\max}-1,2}$, assume that $u_{i+2,2} = u_{i+1,2}$ and $u_{i-2,2} = u_{i-1,2}$ (avoid $u_{i+2,2}$ for $i = N_{\max} - 1$, and $u_{i-2,2}$ for $i = 2$). To carry out these steps, approximate (19.27) so that

$$u_{i+2,2} + 2u_{i-1,2} - 2u_{i+1,2} - u_{i-2,2} \rightarrow u_{i-1,2} - u_{i+1,2}.$$

The truncation error and stability condition for our algorithm are related:

$$\mathcal{E}(u) = \mathcal{O}[(\Delta t)^3] + \mathcal{O}[\Delta t(\Delta x)^2], \tag{19.28}$$

$$\frac{1}{(\Delta x/\Delta t)}\left[\epsilon|u| + 4\frac{\mu}{(\Delta x)^2}\right] \leq 1. \tag{19.29}$$

The first equation shows that smaller time and space steps lead to a smaller approximation error, yet because round-off error increases with the number of steps, the total error does not necessarily decrease (Chapter 2, "Errors & Uncertainties in Computations). Yet we are also limited in how small the steps can be made by the stability condition (19.29), which indicates that making Δx too small always leads to instability. Care and experimentation are required.

19.5.3 Implementation: KdeV Solitons

Modify or run the program Soliton.java in Listing 19.2 that solves the KdeV equation (19.19) for the initial condition

$$u(x, t = 0) = \frac{1}{2}\left[1 - \tanh\left(\frac{x - 25}{5}\right)\right],$$

with parameters $\epsilon = 0.2$ and $\mu = 0.1$. Start with $\Delta x = 0.4$ and $\Delta t = 0.1$. These constants are chosen to satisfy (19.28) with $|u| = 1$.

1. Define a 2-D array u[131][3] with the first index corresponding to the position x and the second to the time t. With our choice of parameters, the maximum value for x is $130 \times 0.4 = 52$.
2. Initialize the time to $t = 0$ and assign values to u[i][1].
3. Assign values to u[i][2], $i = 3, 4, \ldots, 129$, corresponding to the next time interval. Use (19.27) to advance the time but note that you cannot start at $i = 1$ or end at $i = 131$ because (19.27) would include u[132][2] and u[–1][1], which are beyond the limits of the array.
4. Increment the time and assume that u[1][2] $= 1$ and u[131][2] $= 0$. To obtain u[2][2] and u[130][2], assume that u[i+2][2] = u[i+1][2] and u[i–2][2] = u[i–1][2]. Avoid u[i+2][2] for i = 130, and u[i–2][2] for i = 2. To do this, approximate (19.27) so that (19.28) is satisfied.
5. Increment time and compute u[i][j] for j = 3 and for i = 3, 4, . . . , 129, using equation (19.26). Again follow the same procedures to obtain the missing array elements u[2][j] and u[130][j] (set u[1][j] = 1. and u[131][j] = 0). As you print out the numbers during the iterations, you will be convinced that it was a good choice.
6. Set u[i][1] = u[i][2] and u[i][2] = u[i][3] for all i. In this way you are ready to find the next u[i][j] in terms of the previous two rows.

7. Repeat the previous two steps about 2000 times. Write your solution to a file after approximately every 250 iterations.
8. Use your favorite graphics tool (we used Gnuplot) to plot your results as a 3-D graph of disturbance u *versus* position *and versus* time.
9. Observe the wave profile as a function of time and try to confirm Russell's observation that a taller soliton travels faster than a smaller one. ▮

```
// Soliton.java: Solves Kortewg-deVries Equation
import java.io.*;

public class Soliton {
  static double ds = 0.4;                                          // Delta x
  static double dt = 0.1;                                          // Delta t
  static int max = 2000;                                           // Time steps
  static double mu = 0.1;                              // Mu from KdeV equation
  static double eps = 0.2;                            // Epsilon from KdeV eq

  public static void main(String[] argv) throws IOException, FileNotFoundException {
    int i, j, k;
    double a1, a2, a3, fac, time, u[][] = new double[131][3];
    PrintWriter w = new PrintWriter(new FileOutputStream("soliton.dat"), true);
    for ( i=0;  i < 131;  i++ ){ u[i][0] = 0.5*(1.-((Math.exp(2*(0.2*ds*i - 5.))-1)
                        /(Math.exp(2*(0.2*ds*i - 5.)) + 1))); }    // Initial wave
    u[0][1] = 1.; u[0][2] = 1.; u[130][1] = 0.;  u[130][2] = 0.;       // End points
    fac = mu*dt/(ds*ds*ds);
    time = dt;
    for ( i=1;  i < 130;  i++ ) {                                  // First time step
      a1 = eps*dt*(u[i + 1][0] + u[i][0] + u[i-1][0]) / (ds*6.);
      if (i>1 && i < 129) a2 = u[i+2][0] + 2.*u[i-1][0] - 2.*u[i+1][0] - u[i-2][0];
      else a2 = u[i-1][0] - u[i + 1][0];
      a3 = u[i + 1][0]-u[i-1][0];
      u[i][1] = u[i][0] - a1*a3 - fac*a2/3.;
    }
    for ( j=1;  j < max;  j++ )  {                              // Other time steps
      time += dt;
      for ( i=1;  i < 130;  i++ )  {
         a1 = eps*dt*(u[i + 1][1] + u[i][1] + u[i-1][1]) / (3.*ds);
         if (i>1 && i < 129) a2 = u[i+2][1] + 2.*u[i-1][1] - 2.*u[i+1][1] -u[i-2][1];
         else a2 = u[i-1][1] - u[i+1][1];
         a3       = u[i+1][1] - u[i-1][1];
         u[i][2] = u[i][0] - a1*a3 - 2.*fac*a2/3.;
      }
      for (k=0; k < 131; k++) {u[k][0] = u[k][1]; u[k][1] = u[k][2];}
      if (j%200 == 0) { for ( k=0;  k < 131;  k += 2)  w.println("" + u[k][2] + "");
                        w.println( "");  }          // For gnuplot, every 200th step

    }
    System.out.println("data stored in soliton.dat");
} }
```

Listing 19.2 Soliton.java solves the KdeV equation for 1-D solitons corresponding to a "bore" initial conditions.

19.5.4 Exploration: Solitons in Phase Space and Crossing

1. Explore what happens when a tall soliton collides with a short one.
   a. Start by placing a tall soliton of height 0.8 at $x = 12$ and a smaller soliton in front of it at $x = 26$:

$$u(x,t=0) = 0.8\left[1-\tanh^2\left(\frac{3x}{12}-3\right)\right] + 0.3\left[1-\tanh^2\left(\frac{4.5x}{26}-4.5\right)\right].$$

b. Do they reflect from each other? Do they go through each other? Do they interfere? Does the tall soliton still move faster than the short one after the collision (Figure 19.3)?

2. Construct phase–space plots [$\dot{u}(t)$ *versus* $u(t)$] of the KdeV equation for various parameter values. Note that only very specific sets of parameters produce solitons. In particular, by correlating the behavior of the solutions with your phase–space plots, show that the soliton solutions correspond to the *separatrix* solutions to the KdeV equation. In other words, the stability in time for solitons is analogous to the infinite period for a pendulum balanced straight upward. ∎

19.6 Unit II. River Hydrodynamics

Problem: In order to give migrating salmon a place to rest during their arduous upstream journey, the Oregon Department of Environment is thinking of placing objects in several deep, wide, fast-flowing streams. One such object is a long beam of rectangular cross section (Figure 19.4 left), and another is a set of plates (Figure 19.4 right). The objects are far enough below the surface so as not to disturb the surface flow, and far enough from the bottom of the stream so as not to disturb the flow there either.

Your **problem** is to determine the spatial dependence of the stream's velocity and, specifically, whether the wake of the object will be large enough to provide a resting place for a meter-long salmon.

19.7 Hydrodynamics, the Navier–Stokes Equation (Theory)

We continue with the assumption made in Unit I that water is *incompressible* and thus that its density ρ is constant. We also simplify the theory by looking only at steady-state situations, that is, ones in which the velocity is not a function of time. However, to understand how water flows around objects, like our beam, it is essential to include the complication of frictional forces (*viscosity*).

For the sake of completeness, we repeat here the first equation of hydrodynamics, the continuity equation (19.2):

$$\boxed{\frac{\partial\rho(\mathbf{x},t)}{\partial t} + \vec{\nabla}\cdot\mathbf{j} = 0,} \qquad \mathbf{j} \stackrel{\text{def}}{=} \rho\mathbf{v}(\mathbf{x},t). \tag{19.30}$$

Before proceeding to the second equation, we introduce a special time derivative, the *hydrodynamic* derivative $D\mathbf{v}/Dt$, which is appropriate for a quantity contained

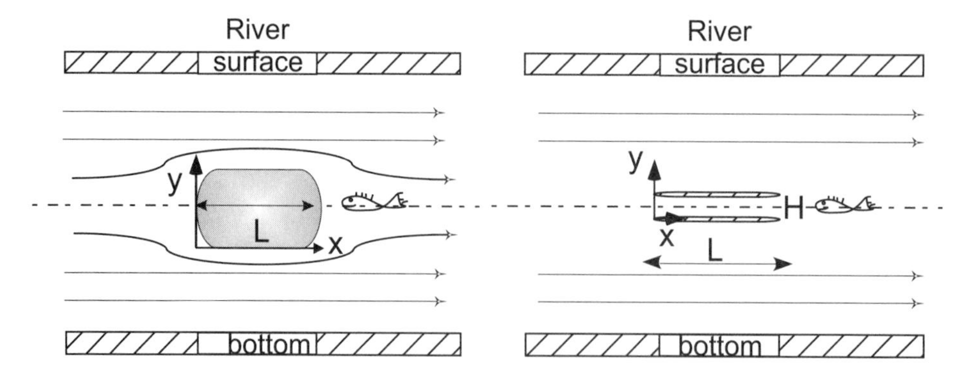

Figure 19.4 Cross-sectional view of the flow of a stream around a submerged beam (*left*) and two parallel plates (*right*). Both beam and plates have length L along the direction of flow. The flow is seen to be symmetric about the centerline and to be unaffected at the bottom and surface by the submerged object.

in a moving fluid [F&W 80]:

$$\frac{D\mathbf{v}}{Dt} \stackrel{\text{def}}{=} (\mathbf{v} \cdot \vec{\nabla})\mathbf{v} + \frac{\partial \mathbf{v}}{\partial t}. \tag{19.31}$$

This derivative gives the rate of change, as viewed from a stationary frame, of the velocity of material in *an element of fluid* and so incorporates changes due to the motion of the fluid (first term) as well as any explicit time dependence of the velocity. Of particular interest is that $D\mathbf{v}/Dt$ is second order in the velocity, and so its occurrence reflects nonlinearities into the theory. You may think of these nonlinearities as related to the fictitious (inertial) forces that would occur if we tried to describe the motion in the fluid's rest frame (an accelerating frame).

The material derivative is the leading term in the *Navier–Stokes equation,*

$$\boxed{\frac{D\mathbf{v}}{Dt} = \nu \nabla^2 \mathbf{v} - \frac{1}{\rho} \vec{\nabla} P(\rho, T, x),} \tag{19.32}$$

$$\frac{\partial v_x}{\partial t} + \sum_{j=x}^{z} v_j \frac{\partial v_x}{\partial x_j} = \nu \sum_{j=x}^{z} \frac{\partial^2 v_x}{\partial x_j^2} - \frac{1}{\rho}\frac{\partial P}{\partial x},$$

$$\frac{\partial v_y}{\partial t} + \sum_{j=x}^{z} v_j \frac{\partial v_y}{\partial x_j} = \nu \sum_{j=x}^{z} \frac{\partial^2 v_y}{\partial x_j^2} - \frac{1}{\rho}\frac{\partial P}{\partial y},$$

$$\frac{\partial v_z}{\partial t} + \sum_{j=x}^{z} v_j \frac{\partial v_z}{\partial x_j} = \nu \sum_{j=x}^{z} \frac{\partial^2 v_z}{\partial x_j^2} - \frac{1}{\rho}\frac{\partial P}{\partial z}. \tag{19.33}$$

Here ν is the kinematic viscosity, P is the pressure, and (19.33) shows the derivatives in Cartesian coordinates. This equation describes transfer of the momentum of the fluid within some region of space as a result of forces and flow (think $d\mathbf{p}/dt = \mathbf{F}$), there being a simultaneous equation for each of the three velocity components. The $\mathbf{v}\cdot\nabla\mathbf{v}$ term in $D\mathbf{v}/Dt$ describes transport of momentum in some region of space resulting from the fluid's flow and is often called the *convection* or *advection* term.[2] The $\vec{\nabla}P$ term describes the velocity change as a result of pressure changes, and the $\nu\nabla^2\mathbf{v}$ term describes the velocity change resulting from viscous forces (which tend to dampen the flow).

The explicit functional dependence of the pressure on the fluid's density and temperature $P(\rho, T, x)$ is known as the *equation of state of the fluid* and would have to be known before trying to solve the Navier–Stokes equation. To keep our problem simple we assume that the pressure is independent of density and temperature, which leaves the four simultaneous partial differential equations (19.30) and (19.32) to solve. Because we are interested in *steady-state* flow around an object, we assume that all time derivatives of the velocity vanish. Because we assume that the fluid is incompressible, the time derivative of the density also vanishes, and (19.30) and (19.32) become

$$\vec{\nabla}\cdot\mathbf{v} \equiv \sum_i \frac{\partial v_i}{\partial x_i} = 0, \tag{19.34}$$

$$(\mathbf{v}\cdot\vec{\nabla})\mathbf{v} = \nu\nabla^2\mathbf{v} - \frac{1}{\rho}\vec{\nabla}P. \tag{19.35}$$

The first equation expresses the equality of inflow and outflow and is known as the *condition of incompressibility*. In as much as the stream in our problem is much wider than the width (z dimension) of the beam and because we are staying away from the banks, we will ignore the z dependence of the velocity. The explicit PDEs we need to solve are then:

$$\frac{\partial v_x}{\partial x} + \frac{\partial v_y}{\partial y} = 0, \tag{19.36}$$

$$\boxed{\nu\left(\frac{\partial^2 v_x}{\partial x^2} + \frac{\partial^2 v_x}{\partial y^2}\right) = v_x\frac{\partial v_x}{\partial x} + v_y\frac{\partial v_x}{\partial y} + \frac{1}{\rho}\frac{\partial P}{\partial x},} \tag{19.37}$$

$$\boxed{\nu\left(\frac{\partial^2 v_y}{\partial x^2} + \frac{\partial^2 v_y}{\partial y^2}\right) = v_x\frac{\partial v_y}{\partial x} + v_y\frac{\partial v_y}{\partial y} + \frac{1}{\rho}\frac{\partial P}{\partial y}.} \tag{19.38}$$

[2] We discuss pure advection in §19.1. In oceanology or meteorology, convection implies the transfer of mass in the vertical direction where it overcomes gravity, whereas advection refers to transfer in the horizontal direction.

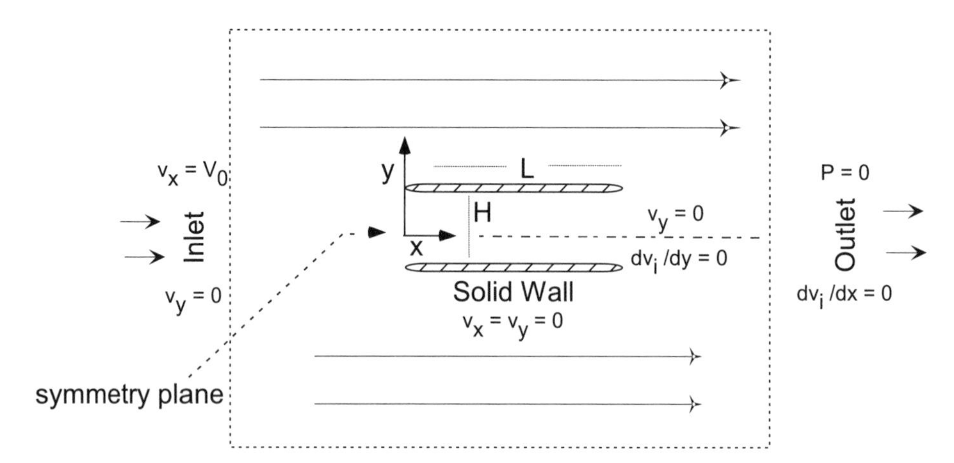

Figure 19.5 The boundary conditions for two thin submerged plates. The surrounding box is the integration volume within which we solve the PDEs and upon whose surface we impose the boundary conditions. In practice the box is much larger than L and H.

19.7.1 Boundary Conditions for Parallel Plates

The plate problem is relatively easy to solve analytically, and so we will do it! This will give us some experience with the equations as well as a check for our numerical solution. To find a unique solution to the PDEs (19.36)–(19.38), we need to specify boundary conditions. As far as we can tell, picking boundary conditions is somewhat of an acquired skill, and it becomes easier with experience (similar to what happens after solving hundreds of electrostatics problems). Some of the boundary conditions apply to the flow at the surfaces of submerged objects, while others apply to the "box" or tank that surrounds the fluid. As we shall see, sometimes these boundary conditions relate to the velocities directly, while at other times they relate to the derivatives of the velocities.

We assume that the submerged parallel plates are placed in a stream that is flowing with a constant velocity V_0 in the horizontal direction (Figure 19.4 right). If the velocity V_0 is not too high or the kinematic viscosity ν is sufficiently large, then the flow should be smooth and without turbulence. We call such flow *laminar*. Typically, a fluid undergoing laminar flow moves in smooth paths that do not close on themselves, like the smooth flow of water from a faucet. If we imagine attaching a vector to each element of the fluid, then the path swept out by that vector is called a *streamline* or *line of motion* of the fluid. These streamlines can be visualized experimentally by adding a colored dye to the stream. We assume that the plates are so thin that the flow remains laminar as it passes around and through them.

If the plates are thin, then the flow upstream of them is not affected, and we can limit our solution space to the rectangular region in Figure 19.5. We assume that the length L and separation H of the plates are small compared to the size of the stream, so the flow returns to uniform as we get far downstream from the plates. As seen in Figure 19.5, there are boundary conditions at the *inlet* where the fluid enters our solution space, at the *outlet* where it leaves, and at the stationary plates. In addition, since the plates are far from the stream's bottom and surface, we may assume that the dotted-dashed centerline is a plane of symmetry, with identical flow above and below the plane. We thus have four different types of boundary conditions to impose on our solution:

Solid plates: Since there is friction (viscosity) between the fluid and the plate surface, the only way to have laminar flow is to have the fluid's velocity equal to the plate's velocity, which means both are zero:

$$v_x = v_y = 0.$$

Such being the case, we have smooth flow in which the negligibly thin plates lie along streamlines of the fluid (like a "streamlined" vehicle).

Inlet: The fluid enters the integration domain at the inlet with a horizontal velocity V_0. Since the inlet is far upstream from the plates, we assume that the fluid velocity at the inlet is unchanged:

$$v_x = V_0, \quad v_y = 0.$$

Outlet: Fluid leaves the integration domain at the outlet. While it is totally reasonable to assume that the fluid returns to its unperturbed state there, we are not told what that is. So, instead, we assume that there is a physical outlet at the end with the water just shooting out of it. Consequently, we assume that the water pressure equals zero at the outlet (as at the end of a garden hose) and that the velocity does not change in a direction normal to the outlet:

$$P = 0, \quad \frac{\partial v_x}{\partial x} = \frac{\partial v_y}{\partial x} = 0.$$

Symmetry plane: If the flow is symmetric about the $y = 0$ plane, then there cannot be flow through the plane and the spatial derivatives of the velocity components normal to the plane must vanish:

$$v_y = 0, \quad \frac{\partial v_y}{\partial y} = 0.$$

This condition follows from the assumption that the plates are along streamlines and that they are negligibly thin. It means that all the streamlines are parallel to the plates as well as to the water surface, and so it must be that $v_y = 0$ everywhere. The fluid enters in the horizontal direction, the plates do not change the vertical y component of the velocity, and the flow remains symmetric about the centerline. There is a retardation of the flow around the plates due to the viscous nature of the flow and due to the $\mathbf{v} = 0$ boundary layers formed on the plates, but there are no actual v_y components.

19.7.2 Analytic Solution for Parallel Plates

For steady flow around and through the parallel plates, with the boundary conditions imposed and $v_y \equiv 0$, the continuity equation (19.36) reduces to

$$\frac{\partial v_x}{\partial x} = 0. \tag{19.39}$$

This tells us that v_x does not vary with x. With these conditions, the Navier–Stokes equations (19.38) in x and y reduce to the linear PDEs

$$\frac{\partial P}{\partial x} = \rho\nu \frac{\partial^2 v_x}{\partial y^2}, \qquad \frac{\partial P}{\partial y} = 0. \tag{19.40}$$

(Observe that if the effect of gravity were also included in the problem, then the pressure would increase with the depth y.) Since the LHS of the first equation describes the x variation, and the RHS the y variation, the only way for the equation to be satisfied in general is if both sides are constant:

$$\frac{\partial P}{\partial x} = C, \quad \rho\nu \frac{\partial^2 v_x}{\partial y^2} = C. \tag{19.41}$$

Double integration of the second equation with respect to y and replacement of the constant by $\partial P/\partial x$ yields

$$\rho\nu \frac{\partial v_x}{\partial y} = \frac{\partial P}{\partial x} y + C_1, \quad \Rightarrow \quad \rho\nu v_x = \frac{\partial P}{\partial x} \frac{y^2}{2} + C_1 y + C_2,$$

where C_1 and C_2 are constants. The values of C_1 and C_2 are determined by requiring the fluid to stop at the plate, $v_x(0) = v_x(H) = 0$, where H is the distance between plates. This yields

$$\rho\nu\, v_x(y) = \frac{1}{2} \frac{\partial P}{\partial x} \left(y^2 - yH\right). \tag{19.42}$$

Because $\partial P/\partial y = 0$, the pressure does not vary with y. The continuity and smoothness of P over the region,

$$\frac{\partial^2 P}{\partial x \partial y} = \frac{\partial^2 P}{\partial y \partial x} = 0, \tag{19.43}$$

are a consequence of laminar flow. Such being the case, we may assume that $\partial P/\partial x$ has no y dependence, and so (19.42) describes a velocity profile varying as y^2.

A check on our numerical CFD simulation ensures that it also gives a parabolic velocity profile for two parallel plates. To be even more precise, we can determine

$\partial P/\partial x$ for this problem and thereby produce a purely numerical answer for comparison. To do that we examine a volume of current that at the inlet starts out with $0 \le y \le H$. Because there is no vertical component to the flow, this volume ends up flowing between the plates. If the volume has a unit z width, then the mass flow (mass/unit time) at the inlet is

$$Q(\text{mass/time}) = \rho \times 1 \times H \times \frac{dx}{dt} = \rho H v_x = \rho H V_0. \tag{19.44}$$

When the fluid is between the plates, the velocity has a parabolic variation in height y. Consequently, we integrate over the area of the plates without changing the net mass flow between the plates:

$$Q = \int \rho v_x dA = \rho \int_0^H v_x(y)\, dy = \frac{1}{2\nu}\frac{\partial P}{\partial x}\left(\frac{H^3}{3} - \frac{H^3}{2}\right). \tag{19.45}$$

Yet we know that $Q = \rho H V_0$, and substitution gives us an expression for how the pressure gradient depends upon the plate separation:

$$\frac{\partial P}{\partial x} = -12\frac{\rho \nu V_0}{H^2}. \tag{19.46}$$

We see that there is a pressure drop as the fluid flows through the plates and that the drop increases as the plates are brought closer together (the Bernoulli effect). To program the equations, we assign the values $V_0 = 1\,\text{m/s}$ ($\simeq 2.24\,\text{mi/h}$), $\rho = 1\,\text{kg/m}^3$ ($\ge$air), $\nu = 1\,\text{m}^2/\text{s}$ (somewhat less viscous than glycerin), and $H = 1\,\text{m}$ (typical boulder size):

$$\frac{\partial P}{\partial x} = -12 \quad \Rightarrow \quad v_x = 6y(1-y). \tag{19.47}$$

19.7.3 Finite-Difference Algorithm and Overrelaxation

Now we develop an algorithm for solution of the Navier–Stokes and continuity PDEs using successive overrelaxation. This is a variation of the method used in Chapter 17, "PDEs for Electrostatics & Heat Flow," to solve Poisson's equation. We divide space into a rectangular grid with the spacing h in both the x and y directions:

$$x = ih, \quad i = 0, \ldots, N_x; \qquad y = jh, \quad j = 0, \ldots, N_y.$$

We next express the derivatives in (19.36)–(19.38) as finite differences of the values of the velocities at the grid points using central-difference approximations.

For $\nu = 1\,\mathrm{m}^2/\mathrm{s}$ and $\rho = 1\,\mathrm{kg/m}^3$, this yields

$$\begin{aligned}
&v^x_{i+1,j} - v^x_{i-1,j} + v^y_{i,j+1} - v^y_{i,j-1} = 0,\\
&v^x_{i+1,j} + v^x_{i-1,j} + v^x_{i,j+1} + v^x_{i,j-1} - 4v^x_{i,j}\\
&\quad = \frac{h}{2} v^x_{i,j}\left[v^x_{i+1,j} - v^x_{i-1,j}\right] + \frac{h}{2} v^y_{i,j}\left[v^x_{i,j+1} - v^x_{i,j-1}\right] + \frac{h}{2}\left[P_{i+1,j} - P_{i-1,j}\right],\\
&v^y_{i+1,j} + v^y_{i-1,j} + v^y_{i,j+1} + v^y_{i,j-1} - 4v^y_{i,j}\\
&\quad = \frac{h}{2} v^x_{i,j}\left[v^y_{i+1,j} - v^y_{i-1,j}\right] + \frac{h}{2} v^y_{i,j}\left[v^y_{i,j+1} - v^y_{i,j-1}\right] + \frac{h}{2}\left[P_{i,j+1} - P_{i,j-1}\right].
\end{aligned}$$

Since $v^y \equiv 0$ for this problem, we rearrange terms to obtain for v^x:

$$\begin{aligned}
4v^x_{i,j} = {} & v^x_{i+1,j} + v^x_{i-1,j} + v^x_{i,j+1} + v^x_{i,j-1} - \frac{h}{2} v^x_{i,j}\left[v^x_{i+1,j} - v^x_{i-1,j}\right]\\
& -\frac{h}{2} v^y_{i,j}\left[v^x_{i,j+1} - v^x_{i,j-1}\right] - \frac{h}{2}\left[P_{i+1,j} - P_{i-1,j}\right].
\end{aligned} \tag{19.48}$$

We recognize in (19.48) an algorithm similar to the one we used in solving Laplace's equation by relaxation. Indeed, as we did there, we can accelerate the convergence by writing the algorithm with the new value of v^x given as the old value plus a correction (residual):

$$v^x_{i,j} = v^x_{i,j} + r_{i,j}, \qquad r \overset{\text{def}}{=} v^{x(\text{new})}_{i,j} - v^{x(\text{old})}_{i,j} \tag{19.49}$$

$$\begin{aligned}
\Rightarrow \quad r = {} & \frac{1}{4}\left\{ v^x_{i+1,j} + v^x_{i-1,j} + v^x_{i,j+1} + v^x_{i,j-1} - \frac{h}{2} v^x_{i,j}\left[v^x_{i+1,j} - v^x_{i-1,j}\right]\right.\\
& \left. - \frac{h}{2} v^y_{i,j}\left[v^x_{i,j+1} - v^x_{i,j-1}\right] - \frac{h}{2}\left[P_{i+1,j} - P_{i-1,j}\right]\right\} - v^x_{i,j}.
\end{aligned} \tag{19.50}$$

As done with the Poisson equation algorithm, successive iterations sweep the interior of the grid, continuously adding in the residual (19.49) until the change becomes smaller than some set level of tolerance, $|r_{i,j}| < \varepsilon$.

A variation of this method, *successive overrelaxation*, increases the speed at which the residuals approach zero via an amplifying factor ω:

$$v^x_{i,j} = v^x_{i,j} + \omega\, r_{i,j} \quad \text{(SOR)}. \tag{19.51}$$

The standard relaxation algorithm (19.49) is obtained with $\omega = 1$, an accelerated convergence (*overrelaxation*) is obtained with $\omega \geq 1$, and *underrelaxation* occurs for $\omega < 1$. Values $\omega > 2$ lead to numerical instabilities and so are not recommended. Although a detailed analysis of the algorithm is necessary to predict the optimal value for ω, we suggest that you test different values for ω to see which one provides the fastest convergence for your problem.

19.7.4 Successive Overrelaxation Implementation

Here is a program fragment that implements SOR for v^x and v^y:

```
omega = 0.55;     iter=0;     err=1.;                    // SOR param,
iteration no, error while ( (err > 0.1) && (iter <=100) ) {
   err=0.0;
   for(i=1;i<Nx; i++) {                                          // x sweep
      for(j=1;j<Ny; j++) {                                       // y sweep
   vx[i+1][j]=vx[i-1][j]+vy[i][j-1]-vy[i][j+1];
   r1=omega*(vx[i+1][j]+vx[i-1][j]+vx[i][j+1]+vx[i][j-1]
   - 0.5*vx[i][j]*(vx[i+1][j]-vx[i-1][j])-
   0.5*vy[i][j]*(vx[i][j+1]-vx[i][j-1])
   - 0.5*(P[i+1][j]-P[i-1][j]))/4.0-vx[i][j];
   vx[i][j] += r1;
   r2=omega*(vy[i+1][j]+vy[i-1][j]+vy[i][j+1]+vy[i][j-1]
    -0.5*vx[i][j]*(vy[i+1][j]-vy[i-1][j])
    -0.5*vy[i][j]*(vy[i][j+1]-vy[i][j-1])
    -0.5*(P[i][j+1]-P[i][j-1]))/4.0-vy[i][j];
   vy[i][j] += r2;
   if( Math.abs(r1) > err ) err = Math.abs(r1);
   if( Math.abs(r2) > err ) err = Math.abs(r2);
      }
   }
      iter++;
}
```

1. Use the program twoplates.java on the disk, or write your own, to solve the Navier–Stokes equation for the velocity of a fluid in 2-D flow. Represent the x and y components of the velocity by the arrays vx[Nx][Ny] and vy[Nx][Ny].
2. Specialize your solution to the rectangular domain and boundary conditions indicated in Figure 19.5.
3. Use of the following parameter values,

$$\nu = 1\,\mathrm{m}^2/\mathrm{s}, \quad \rho = 1\,\mathrm{kg/m}^3, \quad \text{(flow parameters)},$$

$$N_x = 400, \quad N_y = 40, \quad h = 1, \quad \text{(grid parameters)},$$

   leads to the equations

$$\frac{\partial P}{\partial x} = -12, \quad \frac{\partial P}{\partial y} = 0, \quad v^x = \frac{3j}{20}\left(1 - \frac{j}{40}\right), \quad v^y = 0.$$

4. For the relaxation method, output the iteration number and the computed v^x and then compare the analytic and numeric results.
5. Repeat the calculation and see if SOR speeds up the convergence.

19.8 2-D Flow over a Beam

Now that the comparison with an analytic solution has shown that our CFD simulation works, we return to determining if the beam in Figure 19.4 might produce a good resting place for salmon. While we have no analytic solution with which to compare, our canoeing and fishing adventures have taught us that *standing waves* with fish in them are often formed behind rocks in streams, and so we expect there will be a standing wave formed behind the beam.

19.9 Theory: Vorticity Form of the Navier–Stokes Equation

We have seen how to solve numerically the hydrodynamics equations

$$\vec{\nabla}\cdot\mathbf{v}=0, \quad (\mathbf{v}\cdot\vec{\nabla})\mathbf{v}=-\frac{1}{\rho}\vec{\nabla}P+\nu\nabla^2\mathbf{v}. \tag{19.52}$$

These equations determine the components of a fluid's velocity, pressure, and density as functions of position. In analogy to electrostatics, where one usually solves for the simpler scalar potential and then takes its gradient to determine the more complicated vector field, we now recast the hydrodynamic equations into forms that permit us to solve two simpler equations for simpler functions, from which the velocity is obtained via a gradient operation.[3]

We introduce the *stream function* $\mathbf{u}(\mathbf{x})$ from which the velocity is determined by the curl operator:

$$\mathbf{v} \stackrel{\text{def}}{=} \vec{\nabla}\times\mathbf{u}(\mathbf{x})=\hat{\epsilon}_x\left(\frac{\partial u_z}{\partial y}-\frac{\partial u_y}{\partial z}\right)+\hat{\epsilon}_y\left(\frac{\partial u_x}{\partial z}-\frac{\partial u_z}{\partial x}\right). \tag{19.53}$$

Note the absence of the z component of velocity $\mathbf{v}$ for our problem. Since $\vec{\nabla}\cdot(\vec{\nabla}\times\mathbf{u})\equiv 0$, we see that any $\mathbf{v}$ that can be written as the curl of $\mathbf{u}$ automatically satisfies the continuity equation $\vec{\nabla}\cdot\mathbf{v}=0$. Further, since $\mathbf{v}$ for our problem has only x and y components, $\mathbf{u}(\mathbf{x})$ needs to have only a z component:

$$u_z\equiv u \quad \Rightarrow \quad v_x=\frac{\partial u}{\partial y}, \quad v_y=-\frac{\partial u}{\partial x}. \tag{19.54}$$

(Even though the vorticity has just one component, it is a pseudoscalar and not a scalar because it reverses sign upon reflection.) It is worth noting that in 2-D flows, the contour lines $u=$ constant are the *streamlines*.

[3] If we had to solve only the simpler problem of *irrotational flow* (no turbulence), then we would be able to use a scalar velocity potential, in close analogy to electrostatics [Lamb 93]. For the more general *rotational flow*, two vector potentials are required.

The second simplifying function is the *vorticity* field $\mathbf{w}(\mathbf{x})$, which is related physically and alphabetically to the angular velocity ω of the fluid. Vorticity is defined as the curl of the velocity (sometimes with a – sign):

$$\mathbf{w} \stackrel{\text{def}}{=} \vec{\nabla} \times \mathbf{v}(\mathbf{x}). \tag{19.55}$$

Because the velocity in our problem does not change in the z direction, we have

$$w_z = \left(\frac{\partial v_y}{\partial x} - \frac{\partial v_x}{\partial y} \right). \tag{19.56}$$

Physically, we see that the vorticity is a measure of how much the fluid's velocity curls or rotates, with the direction of the vorticity determined by the right-hand rule for rotations. In fact, if we could pluck a small element of the fluid into space (so it would not feel the internal strain of the fluid), we would find that it is rotating like a solid with angular velocity $\omega \propto \mathbf{w}$ [Lamb 93]. That being the case, it is useful to think of the vorticity as giving the local value of the fluid's angular velocity vector. If $\mathbf{w} = 0$, we have *irrotational* flow.

The field lines of w are continuous and move as if attached to the particles of the fluid. A uniformly flowing fluid has vanishing curl, while a nonzero vorticity indicates that the current curls back on itself or rotates. From the definition of the stream function (19.53), we see that the vorticity $\mathbf{w}$ is related to it by

$$\mathbf{w} = \vec{\nabla} \times \mathbf{v} = \vec{\nabla} \times (\vec{\nabla} \times \mathbf{u}) = \vec{\nabla}(\vec{\nabla} \cdot \mathbf{u}) - \nabla^2 \mathbf{u}, \tag{19.57}$$

where we have used a vector identity for $\vec{\nabla} \times (\vec{\nabla} \times \mathbf{u})$. Yet the divergence $\vec{\nabla} \cdot \mathbf{u} = 0$ since $\mathbf{u}$ has only a z component that does not vary with z (or because there is no source for $\mathbf{u}$). We have now obtained the basic relation between the stream function $\mathbf{u}$ and the vorticity $\mathbf{w}$:

$$\boxed{\vec{\nabla}^2 \mathbf{u} = -\mathbf{w}.} \tag{19.58}$$

Equation (19.58) is analogous to Poisson's equation of electrostatics, $\nabla^2 \phi = -4\pi\rho$, only now each component of vorticity $\mathbf{w}$ is a source for the corresponding component of the stream function $\mathbf{u}$. If the flow is irrotational, that is, if $\mathbf{w} = 0$, then we need only solve Laplace's equation for each component of u. Rotational flow, with its coupled nonlinearities equations, leads to more interesting behavior.

As is to be expected from the definition of $\mathbf{w}$, the vorticity form of the Navier–Stokes equation is obtained by taking the curl of the velocity form, that is, by operating on both sides with $\vec{\nabla} \times$. After significant manipulations we obtain

$$\nu \nabla^2 \mathbf{w} = [(\vec{\nabla} \times \mathbf{u}) \cdot \vec{\nabla}]\mathbf{w}. \tag{19.59}$$

This and (19.58) are the two simultaneous PDEs that we need to solve. In 2-D, with $\mathbf{u}$ and $\mathbf{w}$ having only z components, they are

$$\boxed{\frac{\partial^2 u}{\partial x^2} + \frac{\partial^2 u}{\partial y^2} = -w,} \tag{19.60}$$

$$\boxed{\nu\left(\frac{\partial^2 w}{\partial x^2} + \frac{\partial^2 w}{\partial y^2}\right) = \frac{\partial u}{\partial y}\frac{\partial w}{\partial x} - \frac{\partial u}{\partial x}\frac{\partial w}{\partial y}.} \tag{19.61}$$

So after all that work, we end up with two simultaneous, nonlinear, elliptic PDEs for the functions $w(x,y)$ and $u(x,y)$ that look like a mixture of Poisson's equation with the frictional and variable-density terms of the wave equation. The equation for u is Poisson's equation with source w and must be solved simultaneously with the second. It is this second equation that contains mixed products of the derivatives of u and w and thus introduces nonlinearity.

19.9.1 Finite Differences and the SOR Algorithm

We solve (19.60) and (19.61) on an $N_x \times N_y$ grid of uniform spacing h with

$$x = i\Delta x = ih, \quad i = 0, \ldots, N_x, \quad y = j\Delta y = jh, \quad j = 0, \ldots, N_y.$$

Since the beam is symmetric about its centerline (Figure 19.4 left), we need the solution only in the upper half-plane. We apply the now familiar central-difference approximation to the Laplacians of u and w to obtain the difference equation

$$\frac{\partial^2 u}{\partial x^2} + \frac{\partial^2 u}{\partial y^2} \simeq \frac{u_{i+1,j} + u_{i-1,j} + u_{i,j+1} + u_{i,j-1} - 4u_{i,j}}{h^2}. \tag{19.62}$$

Likewise, for the first derivatives,

$$\frac{\partial u}{\partial y}\frac{\partial w}{\partial x} \simeq \frac{u_{i,j+1} - u_{i,j-1}}{2h}\,\frac{w_{i+1,j} - w_{i-1,j}}{2h}. \tag{19.63}$$

The difference form of the vorticity Navier–Stokes equation (19.60) becomes

$$u_{i,j} = \frac{1}{4}\left(u_{i+1,j} + u_{i-1,j} + u_{i,j+1} + u_{i,j-1} + h^2 w_{i,j}\right), \tag{19.64}$$

$$w_{i,j} = \frac{1}{4}(w_{i+1,j} + w_{i-1,j} + w_{i,j+1} + w_{i,j-1}) - \frac{R}{16}\{[u_{i,j+1} - u_{i,j-1}]$$

$$\times [w_{i+1,j} - w_{i-1,j}] - [u_{i+1,j} - u_{i-1,j}][w_{i,j+1} - w_{i,j-1}]\}, \tag{19.65}$$

$$R = \frac{1}{\nu} = \frac{V_0 h}{\nu} \quad \text{(in normal units)}. \tag{19.66}$$

Note that we have placed $u_{i,j}$ and $w_{i,j}$ on the LHS of the equations in order to obtain an algorithm appropriate to solution by relaxation.

The parameter R in (19.66) is related to the *Reynolds number*. When we solve the problem in natural units, we measure distances in units of grid spacing h, velocities in units of initial velocity V_0, stream functions in units of $V_0 h$, and vorticity in units of V_0/h. The second form is in regular units and is dimensionless. This R is known as the *grid Reynolds number* and differs from the physical R, which has a pipe diameter in place of the grid spacing h.

The grid Reynolds number is a measure of the strength of the coupling of the nonlinear terms in the equation. When R is small, the viscosity acts as a frictional force that damps out fluctuations and keeps the flow smooth. When the physical R is large ($R \simeq 2000$), physical fluids undergo phase transitions from laminar to turbulent flow in which turbulence occurs at a cascading set of smaller and smaller space scales [Rey 83]. However, simulations that produce the onset of turbulence have been a research problem since Reynolds first experiments in 1883 [Rey 83, F&S], possibly because laminar flow simulations are stable against small perturbations and some large-scale "kick" appears necessary to change laminar to turbulent flow. Recent research along these lines have been able to find unstable, traveling-wave solutions to the Navier–Stokes equations, and the hope is that these may lead to a turbulent transition [Fitz 04].

As discussed in §19.7.3, the finite-difference algorithm can have its convergence accelerated by the use of successive overrelaxation (19.64):

$$u_{i,j} = u_{i,j} + \omega\, r^{(1)}_{i,j}, \qquad w_{i,j} = w_{i,j} + \omega\, r^{(2)}_{i,j} \quad \text{(SOR)}. \tag{19.67}$$

Here ω is the overrelaxation parameter and should lie in the range $0 < \omega < 2$ for stability. The residuals are just the changes in a single step, $r^{(1)} = u^{\text{new}} - u^{\text{old}}$ and $r^{(2)} = w^{\text{new}} - w^{\text{old}}$:

$$r^{(1)}_{i,j} = \frac{1}{4}(u_{i+1,j} + u_{i-1,j} + u_{i,j+1} + u_{i,j-1} + w_{i,j}) - u_{i,j}, \tag{19.68}$$

$$r^{(2)}_{i,j} = \frac{1}{4}\Bigg(w_{i+1,j} + w_{i-1,j} + w_{i,j+1} + w_{i,j-1} - \frac{R}{4}\{[u_{i,j+1} - u_{i,j-1}]$$

$$\times [w_{i+1,j} - w_{i-1,j}] - [u_{i+1,j} - u_{i-1,j}][w_{i,j+1} - w_{i,j-1}]\}\Bigg) - w_{i,j}.$$

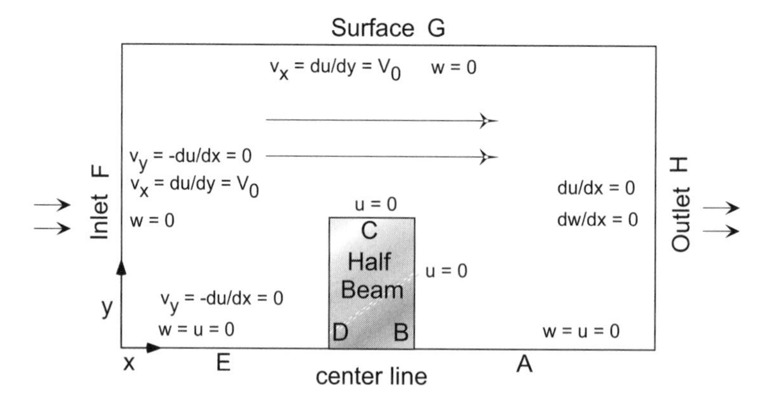

Figure 19.6 Boundary conditions for flow around the beam in Figure 19.4. The flow is symmetric about the centerline, and the beam has length L in the x direction (along flow).

19.9.2 Boundary Conditions for a Beam

A well-defined solution of these elliptic PDEs requires a combination of (less than obvious) boundary conditions on u and w. Consider Figure 19.6, based on the analysis of [Koon 86]. We assume that the inlet, outlet, and surface are far from the beam, which may not be evident from the not-to-scale figure.

Freeflow: If there were no beam in the stream, then we would have free flow with the entire fluid possessing the inlet velocity:

$$v_x \equiv V_0, \quad v_y = 0, \quad \Rightarrow \quad u = V_0 y, \quad w = 0. \tag{19.69}$$

(Recollect that we can think of $w = 0$ as indicating no fluid rotation.) The centerline divides the system along a symmetry plane with identical flow above and below it. If the velocity is symmetric about the centerline, then its y component must vanish there:

$$v_y = 0, \quad \Rightarrow \quad \frac{\partial u}{\partial x} = 0 \quad \text{(centerline AE)}. \tag{19.70}$$

Centerline: The centerline is a streamline with $u =$ constant because there is no velocity component perpendicular to it. We set $u = 0$ according to (19.69). Because there cannot be any fluid flowing into or out of the beam, the normal component of velocity must vanish along the beam surfaces. Consequently, the streamline $u = 0$ is the entire lower part of Figure 19.6, that is, the centerline and the beam surfaces. Likewise, the symmetry of the problem permits us to set the vorticity $w = 0$ along the centerline.

Inlet: At the inlet, the fluid flow is horizontal with uniform x component V_0 at all heights and with no rotation:

$$v_y = -\frac{\partial u}{\partial x} = 0, \quad w = 0 \quad \text{(inlet F)}, \quad v_x = \frac{\partial u}{\partial y} = V_0. \tag{19.71}$$

Surface: We are told that the beam is sufficiently submerged so as not to disturb the flow on the surface of the stream. Accordingly, we have free-flow conditions on the surface:

$$v_x = \frac{\partial u}{\partial y} = V_0, \quad w = 0 \quad \text{(surface G)}. \tag{19.72}$$

Outlet: Unless something truly drastic is happening, the conditions on the far downstream outlet have little effect on the far upstream flow. A convenient choice is to require the stream function and vorticity to be constant:

$$\frac{\partial u}{\partial x} = \frac{\partial w}{\partial x} = 0 \quad \text{(outlet H)}. \tag{19.73}$$

Beamsides: We have already noted that the normal component of velocity v_x and stream function u vanish along the beam surfaces. In addition, because the flow is viscous, it is also true that the fluid "sticks" to the beam somewhat and so the tangential velocity also vanishes along the beam's surfaces. While these may all be true conclusions regarding the flow, specifying them as boundary conditions would overrestrict the solution (see Table 17.1 for elliptic equations) to the point where no solution may exist. Accordingly, we simply impose the *no-slip* boundary condition on the vorticity w. Consider a grid point (x, y) on the upper surface of the beam. The stream function u at a point $(x, y+h)$ above it can be related via a Taylor series in y:

$$u(x, y+h) = u(x, y) + \frac{\partial u}{\partial y}(x, y)h + \frac{\partial^2 u}{\partial y^2}(x, y)\frac{h^2}{2} + \cdots. \tag{19.74}$$

Because **w** has only a z component, it has a simple relation to $\nabla \times \mathbf{v}$:

$$w \equiv w_z = \frac{\partial v_y}{\partial x} - \frac{\partial v_x}{\partial y}. \tag{19.75}$$

Because of the fluid's viscosity, the velocity is stationary along the beam top:

$$v_x = \frac{\partial u}{\partial y} = 0 \quad \text{(beam top)}. \tag{19.76}$$

Because the current flows smoothly along the top of the beam, v_y must also vanish. In addition, since there is no x variation, we have

$$\frac{\partial v_y}{\partial x} = 0 \Rightarrow w = -\frac{\partial v_x}{\partial y} = -\frac{\partial^2 u}{\partial y^2}. \tag{19.77}$$

After substituting these relations into the Taylor series (19.74), we can solve for w and obtain the finite-difference version of the top boundary condition:

$$w \simeq -2\frac{u(x, y+h) - u(x, y)}{h^2} \Rightarrow w_{i,j} = -2\frac{u_{i,j+1} - u_{i,j}}{h^2} \quad \text{(top)}. \tag{19.78}$$

Similar treatments applied to other surfaces yield the following boundary conditions.

$u = 0;$	$w = 0$	Centerline EA
$u = 0,$	$w_{i,j} = -2(u_{i+1,j} - u_{i,j})/h^2$	Beam back B
$u = 0,$	$w_{i,j} = -2(u_{i,j+1} - u_{i,j})/h^2$	Beam top C
$u = 0,$	$w_{i,j} = -2(u_{i-1,j} - u_{i,j})/h^2$	Beam front D
$\partial u/\partial x = 0,$	$w = 0$	Inlet F
$\partial u/\partial y = V_0,$	$w = 0$	Surface G
$\partial u/\partial x = 0,$	$\partial w/\partial x = 0$	Outlet H

19.9.3 SOR on a Grid Implementation

Beam.java in Listing 19.3 is our solution of the vorticity form of the Navier–Stokes equation. You will notice that while the relaxation algorithm is rather simple, some care is needed in implementing the many boundary conditions. Relaxation of the stream function and of the vorticity is done by separate methods, and the file output format is that for a Gnuplot surface plot.

```
// Beam.java solves Navier-Stokes equations for flow over plate
import java.io.*;
import java.util.*;

public class Beam {
  static int Nxmax = 70;  static int Nymax = 20;                    // Grid parameters
  static double u[][] = new double [Nxmax + 1][Nymax + 1];                   // Stream
  static double w[][] = new double [Nxmax + 1][Nymax + 1];                // Vorticity
  static double V0 = 1.;  static double R;                   // Initial v, Reynold's #
  static double omega = 0.1;                                // Relaxation parameter
  static int IL = 10; static int H = 8; static int T = 8;                  // Geometry
  static double h = 1.;  static double nu = 1. ;                       // h, viscosity

  public static void main(String[] argv) throws IOException, FileNotFoundException {
    PrintWriter q = new PrintWriter(new FileOutputStream("flow1.dat.dat"), true);
    int i, j, iter;
    iter = 0;                                                // Number of iterations
```

```
    R = V0*h/nu;                                      // Reynold number in normal units
    borders();
    while (iter <= 400){ iter++ ;  relax(); }
      for ( i=0; i <= Nxmax; i++ ) {                           // Output for gnuplot 3D
        for ( j=0; j <= Nymax; j++ ) {
          q.println(" " + u[i][j]/(V0*h) + " ");                      // Stream in V0h
        }
        q.println("");                                        // Empty line for gnuplot
      }
      System.out.println("data stored in flow1.dat");
  }                                                                         // End main

  public static void borders() {                 // Initialize stream,vorticity, sets BC
    int i, j;
    for ( i=0; i <= Nxmax; i++ ) {                        // Initialize stream function
        for ( j=0; j <= Nymax; j++ )                                    // And vorticity
          { w[i][j] = 0.;  u[i][j] = j*V0; }
    }
    for ( i=0; i <= Nxmax; i++ ) {                                      // Fluid surface
      u[i][Nymax] = u[i][Nymax-1] + V0*h;
      w[i][Nymax-1] = 0. ;
    }
    for ( j=0; j <= Nymax; j++ ) { u[1][j] = u[0][j];   w[0][j] = 0. ; }       // Inlet
    for ( i=0; i <= Nxmax; i++ ){                                          // Centerline
      if ((i <= IL)||(i>=IL + T)) {  u[i][0] = 0.;  w[i][0] = 0. ; }
    }
    for ( j=1; j <= Nymax-1; j++ ) {                                          // Outlet
        w[Nxmax][j] = w[Nxmax-1][j];
        u[Nxmax][j] = u[Nxmax-1][j];
  } }                                                                       // Borders

  public static void beam() {                                           // BC for beam
    int i, j;
    for ( j=0; j <= H; j++ ) {                                          // Beam sides
      w[IL][j] = -2*u[IL-1][j]/(h*h);                                   // Front side
      w[IL + T][j] = -2*u[IL + T + 1][j]/(h*h);                          // Back side
    }
    for ( i=IL; i <= IL + T; i++ ) w[i][H-1] = -2*u[i][H]/(h*h);                // Top
    for ( i=IL; i <= IL + T; i++ ) {
      for ( j=0; j <= H; j++ ) {
        u[IL][j] = 0. ;                                                       // Front
        u[IL + T][j] = 0. ;                                                    // Back
        u[i][H] = 0;                                                            // Top
  } } }

  public static void  relax() {                              // Method to relax stream
    int i, j;
    double r1, r2;
    beam();                                                // Reset conditions at beam
    for ( i=1; i <= Nxmax-1; i++ )  {                         // Relax stream function
      for ( j=1; j <= Nymax-1; j++ ) {
        r1 = omega*((u[i + 1][j] + u[i-1][j] + u[i][j + 1] + u[i][j-1]
            + h*h*w[i][j])*(1./4.)-u[i][j]);
        u[i][j] += r1;
      }
    }
    for ( i=1; i < Nxmax-1; i++ )  {                                 // Relax vorticity
      for ( j=1; j < Nymax-1; j++ ) {
        r2 = omega*((w[i + 1][j] + w[i-1][j] + w[i][j + 1] + w[i][j-1]-
            (R/4.)*((u[i][j + 1]-u[i][j-1])*(w[i + 1][j]-w[i-1][j])-
            (u[i + 1][j]-u[i-1][j])*(w[i][j+1]-w[i][j-1])))/4. -w[i][j]);
        w[i][j]  += r2;
} } } }
```

Listing 19.3 Beam.java solves the Navier-Stokes equation for the flow over a plate.

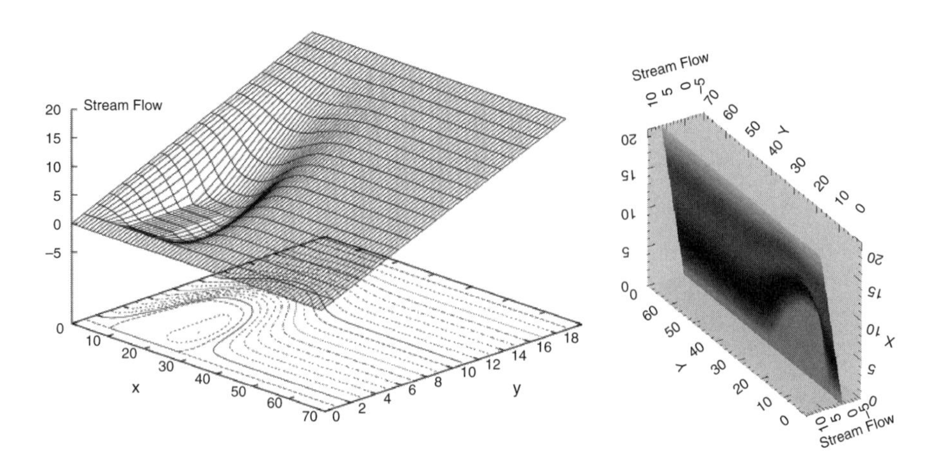

Figure 19.7 *Left:* Gnuplot surface plot with contours of the stream function u for $R = 5$. *Right:* Contours of stream function for $R = 1$ visualized with colors by OpenDX.

19.9.4 Assessment

1. Use `Beam.java` as a basis for your solution for the stream function u and the vorticity w using the finite-differences algorithm (19.64).
2. A good place to start your simulation is with a beam of size $L = 8h$, $H = h$, Reynolds number $R = 0.1$, and intake velocity $V_0 = 1$. Keep your grid small during debugging, say, $N_x = 24$ and $N_y = 70$.
3. Explore the **convergence** of the algorithm.
   a. Print out the iteration number and u values upstream from, above, and downstream from the beam.
   b. Determine the number of iterations necessary to obtain three-place convergence for successive relaxation ($\omega = 0$).
   c. Determine the number of iterations necessary to obtain three-place convergence for successive overrelaxation ($\omega \simeq 0.3$). Use this number for future calculations.
4. Change the beam's horizontal placement so that you can see the undisturbed current entering on the left and then developing into a standing wave. Note that you may need to increase the size of your simulation volume to see the effect of all the boundary conditions.
5. Make surface plots including contours of the stream function u and the vorticity w. Explain the behavior seen.
6. Is there a region where a big fish can rest behind the beam?
7. The results of the simulation (Figure 19.7) are for the one-component stream function u. Make several visualizations showing the fluid velocity throughout the simulation region. Note that the velocity is a vector with two components

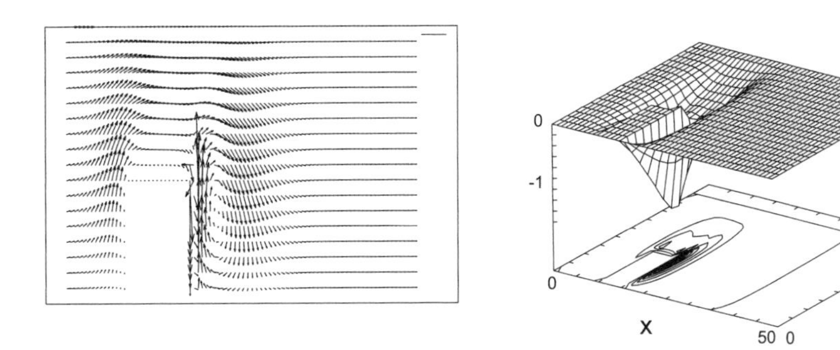

Figure 19.8 *Left:* The velocity field around the beam as represented by vectors. *Right:* The vorticity as a function of x and y. Rotation is seen to be largest behind the beam.

(or a magnitude and direction), and both degrees of freedom are interesting to visualize. A plot of vectors would work well here (Gnuplot and OpenDX make *vector* plots for this purpose, Mathematica has *plotfield*, and Maple has *fieldplot*, although the latter two require some work for numerical data).

8. Explore how increasing the Reynolds number R changes the flow pattern. Start at $R = 0$ and gradually increase R while watching for numeric instabilities. To overcome the instabilities, reduce the size of the relaxation parameter ω and continue to larger R values.
9. Verify that the flow around the beam is smooth for small R values, but that it separates from the back edge for large R, at which point a small vortex develops.

19.9.5 Exploration

1. Determine the flow behind a circular rock in the stream.
2. The boundary condition at an outlet far downstream should not have much effect on the simulation. Explore the use of other boundary conditions there.
3. Determine the pressure variation around the beam.

20

Integral Equations in Quantum Mechanics

The power and accessibility of high-speed computers have changed the view as to what kind of equations are soluble. In Chapter 9, "Differential Equation Applications," and Chapter 12, "Discrete & Continuous Nonlinear Dynamics," we saw how even nonlinear differential equations can be solved easily and can give new insight into the physical world. In this chapter we examine how the integral equations of quantum mechanics can be solved for both bound and scattering states. In Unit I we extend our treatment of the eigenvalue problem, solved as a coordinate-space differential equation in Chapter 9, to the equivalent integral-equation problem in momentum space. In Unit II we treat the singular integral equations for scattering, a more difficult problem. After studying this chapter, we hope that the reader will view both integral and differential equations as soluble.

20.1 Unit I. Bound States of Nonlocal Potentials

Problem: A particle undergoes a many-body interaction with a medium (Figure 20.1) that results in the particle experiencing an effective potential at $\mathbf{r}$ that depends on the wave function at the $\mathbf{r}'$ values of the other particles [L 96]:

$$V(r)\psi(r) \rightarrow \int dr'\, V(r,r')\psi(r'). \tag{20.1}$$

This type of interaction is called *nonlocal* and leads to a Schrödinger equation that is a combined integral and differential ("integrodifferential") equation:

$$-\frac{1}{2\mu}\frac{d^2\psi(r)}{dr^2} + \int dr'\, V(r,r')\psi(r') = E\psi(r). \tag{20.2}$$

Your **problem** is to figure out how to find the bound-state energies E and wave functions ψ for the integral equation in (20.2).[1]

[1] We use natural units in which $\hbar \equiv 1$ and omit the traditional bound-state subscript n on E and ψ in order to keep the notation simpler.

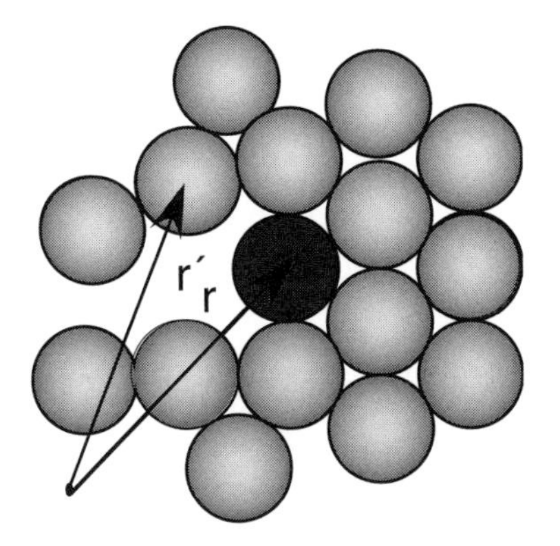

Figure 20.1 A dark particle moving in a dense multiparticle medium. The nonlocality of the potential felt by the dark particle at **r** arises from the particle interactions at all **r**′.

20.2 Momentum-Space Schrödinger Equation (Theory)

One way of dealing with equation (20.2) is by going to momentum space where it becomes the integral equation [L 96]

$$\frac{k^2}{2\mu}\psi(k) + \frac{2}{\pi}\int_0^\infty dp p^2 V(k,p)\psi(p) = E\psi(k), \tag{20.3}$$

where we restrict our solution to $l = 0$ partial waves. In (20.3), $V(k,p)$ is the momentum-space representation (double Fourier transform) of the potential,

$$V(k,p) = \frac{1}{kp}\int_0^\infty dr \sin(kr) V(r) \sin(pr), \tag{20.4}$$

and $\psi(k)$ is the (unnormalized) momentum-space wave function (the probability amplitude for finding the particle with momentum k),

$$\psi(k) = \int_0^\infty dr kr\psi(r) \sin(kr). \tag{20.5}$$

Equation (20.3) is an integral equation for $\psi(k)$, in contrast to an integral representation of $\psi(k)$, because the integral in it cannot be evaluated until $\psi(p)$ is known. Although this may seem like an insurmountable barrier, we will transform this equation into a matrix equation that can be solved with the matrix techniques discussed in Chapter 8, "Solving Systems of Equations with Matrices; Data Fitting."

k_1 k_2 k_3 k_N

Figure 20.2 The grid of momentum values on which the integral equation is solved.

20.2.1 Integral to Linear Equations (Method)

We approximate the integral over the potential as a weighted sum over N integration points (usually Gauss quadrature[2]) for $p = k_j,\ j = 1, N$:

$$\int_0^\infty dp p^2 V(k,p)\psi(p) \simeq \sum_{j=1}^{N} w_j k_j^2 V(k,k_j)\psi(k_j). \tag{20.6}$$

This converts the integral equation (20.3) to the algebraic equation

$$\frac{k^2}{2\mu}\psi(k) + \frac{2}{\pi}\sum_{j=1}^{N} w_j k_j^2 V(k,k_j)\psi(k_j) = E. \tag{20.7}$$

Equation (20.7) contains the N unknowns $\psi(k_j)$, the single unknown E, and the unknown function $\psi(k)$. We eliminate the need to know the entire function $\psi(k)$ by restricting the solution to the same values of k_i as used to approximate the integral. This leads to a set of N coupled linear equations in $(N+1)$ unknowns:

$$\frac{k_i^2}{2\mu}\psi(k_i) + \frac{2}{\pi}\sum_{j=1}^{N} w_j k_j^2\, V(k_i,k_j)\psi(k_j) = E\psi(k_i), \quad i = 1, N. \tag{20.8}$$

As a case in point, if $N = 2$, we would have the two simultaneous linear equations

$$\frac{k_1^2}{2\mu}\psi(k_1) + \frac{2}{\pi}w_1 k_1^2\, V(k_1,k_1)\psi(k_1) + w_2 k_2^2\, V(k_1,k_2) = E\psi(k_1),$$

$$\frac{k_2^2}{2\mu}\psi(k_2) + \frac{2}{\pi}w_1 k_1^2\, V(k_2,k_1)\psi(k_1) + w_2 k_2^2\, V(k_2,k_2)\psi(k_2) = E\psi(k_2).$$

We write our coupled dynamic equations in matrix form as

$$\boxed{[H][\psi] = E[\psi]} \tag{20.9}$$

[2] See Chapter 6, "Integration," for a discussion of numerical integration.

or as explicit matrices

$$\begin{pmatrix} \frac{k_1^2}{2\mu}+\frac{2}{\pi}V(k_1,k_1)k_1^2w_1 & \frac{2}{\pi}V(k_1,k_2)k_2^2w_2 & \cdots & \frac{2}{\pi}V(k_1,k_N)k_N^2w_N \\ \frac{2}{\pi}V(k_2,k_1)k_1^2w_1 & \frac{2}{\pi}V(k_2,k_2)k_2^2w_2+\frac{k_2^2}{2\mu} & \cdots & \\ \ddots & & & \\ \cdots & \cdots & \cdots & \frac{k_N^2}{2\mu}+\frac{2}{\pi}V(k_N,k_N)k_N^2w_N \end{pmatrix}$$

$$\times \begin{pmatrix} \psi(k_1) \\ \psi(k_2) \\ \ddots \\ \psi(k_N) \end{pmatrix} = E \begin{pmatrix} \psi(k_1) \\ \psi(k_2) \\ \ddots \\ \psi(k_N) \end{pmatrix}. \tag{20.10}$$

Equation (20.9) is the matrix representation of the Schrödinger equation (20.3). The wave function $\psi(k)$ on the grid is the $N \times 1$ vector

$$[\psi(k_i)] = \begin{pmatrix} \psi(k_1) \\ \psi(k_2) \\ \ddots \\ \psi(k_N) \end{pmatrix}. \tag{20.11}$$

The astute reader may be questioning the possibility of solving N equations for $(N+1)$ unknowns. That reader is wise; only sometimes, and only for certain values of E (eigenvalues) will the computer be able to find solutions. To see how this arises, we try to apply the matrix inversion technique (which we will use successfully for scattering in Unit II). We rewrite (20.9) as

$$[H - EI][\psi] = [0] \tag{20.12}$$

and multiply both sides by the inverse of $[H - EI]$ to obtain the formal solution

$$[\psi] = [H - EI]^{-1}[0]. \tag{20.13}$$

This equation tells us that (1) if the inverse exists, then we have the *trivial* solution $\psi \equiv 0$, which is not very interesting, and (2) for a nontrivial solution to exist, our assumption that the inverse exists must be incorrect. Yet we know from the theory of linear equations that the inverse fails to exist when the determinant vanishes:

$$\det[H - EI] = 0 \qquad \text{(bound-state condition)}. \tag{20.14}$$

Equation (20.14) is the additional equation needed to find unique solutions to the eigenvalue problem. Nevertheless, there is no guarantee that solutions of (20.14) can always be found, but if they are found, they are the desired *eigenvalues* of (20.9).

20.2.2 Delta-Shell Potential (Model)

To keep things simple and to have an analytic answer to compare with, we consider the local delta-shell potential:

$$V(r) = \frac{\lambda}{2\mu}\delta(r-b). \tag{20.15}$$

This would be a good model for an interaction that occurs in 3-D when two particles are predominantly a fixed distance b apart. We use (20.4) to determine its momentum-space representation:

$$V(k',k) = \frac{1}{k'k}\int_0^\infty \sin(k'r')\frac{\lambda}{2\mu}\delta(r-b)\sin(kr)\,dr = \frac{\lambda}{2\mu}\frac{\sin(k'b)\sin(kb)}{k'k}. \tag{20.16}$$

Beware: We have chosen this potential because it is easy to evaluate the momentum-space matrix element of the potential. However, its singular nature in r space leads to a very slow falloff in k space, and this causes the integrals to converge so slowly that numerics are not as precise as we would like.

If the energy is parameterized in terms of a wave vector κ by $E = -\kappa^2/2\mu$, then for this potential there is, at most, one bound state and it satisfies the transcendental equation [Gott 66]

$$e^{-2\kappa b} - 1 = \frac{2\kappa}{\lambda}. \tag{20.17}$$

Note that bound states occur only for attractive potentials and only if the attraction is strong enough. For the present case this means that we must have $\lambda < 0$.

Exercise: Pick some values of b and λ and use them to verify with a hand calculation that (20.17) can be solved for κ.

20.2.3 Binding Energies Implementation

An actual computation may follow two paths. The first calls subroutines to evaluate the determinant of the $[H - EI]$ matrix in (20.14) and then to *search* for those values of energy for which the computed determinant vanishes. This provides E, but not wave functions. The other approach calls an eigenproblem solver that may give some or all eigenvalues and eigenfunctions. In both cases the solution is obtained iteratively, and you may be required to guess starting values for both the eigenvalues and eigenvectors. In Listing 20.1 we present our solution of the integral equation for bound states of the delta-shell potential using the JAMA matrix library and the gauss method for Gaussian quadrature points and weights.

```
// Bound.java: Bound states in momentum space for delta shell potential
//                         uses JAMA and includes Gaussian integration
import Jama.*;

public class Bound {
  static double min =0., max =200., u =0.5, b =10. ;                // Class variables

  public static void main(String[] args) {
    System.out.println("M, lambda, eigenvalue");
    for ( int M = 16; M <= 128; M += 8) {
      for ( int lambda = -1024; lambda < 0; lambda /= 2) {
        double A[][] = new double[M][M],                             // Hamiltonian
        WR[] = new double[M], VR,                          // RE eigenvalues, potential
        k[] = new double[M], w[] = new double[M];                         // Pts & wts
        gauss(M, min, max, k, w);                          // Call gauss integration
        for ( int i=0; i < M; i++ )                               // Set Hamiltonian
          for ( int j=0; j < M; j++ ) {
           VR = lambda/2/u*Math.sin(k[i]*b)/k[i]*Math.sin(k[j]*b)/k[j];
           A[i][j] = 2/Math.PI*VR*k[j]*k[j]*w[j];
           if (i == j) A[i][j] += k[i]*k[i]/2/u;
          }
        EigenvalueDecomposition E = new EigenvalueDecomposition(new Matrix(A));
        WR = E.getRealEigenvalues();                                // RE eigenvalues
     // Matrix V = E.getV();                                         // Eigenvectors
        for ( int j=0; j < M; j++ ) if (WR[j] < 0) {
          System.out.println(M + " " + lambda + " " + WR[j]);
          break;
  } } } }

  // Method gauss: pts & wts for Gauss quadrature, uniform [a, b]
  private static void gauss(int npts, double a, double b, double[] x, double[] w) {
    int m = (npts + 1)/2;
    double t, t1, pp = 0, p1, p2, p3, eps = 3.e-10;                 // eps = accuracy
    for ( int i=1; i <= m; i++ ) {
      t = Math.cos(Math.PI*(i-0.25)/(npts + 0.5)); t1 = 1;
      while((Math.abs(t-t1))>=eps) {
        p1 = 1. ; p2 = 0. ;
        for ( int j=1; j <= npts; j++ )
          {p3 = p2; p2 = p1; p1 = ((2*j-1)*t*p2-(j-1)*p3)/j;}
        pp = npts*(t*p1-p2)/(t*t-1);
        t1 = t; t = t1 - p1/pp;
      }
      x[i-1] = -t; x[npts-i] = t;
      w[i-1] = 2./((1-t*t)*pp*pp); w[npts-i] = w[i-1];
    }
    for ( int i=0; i < npts ; i++ ) {
      x[i] = x[i]*(b-a)/2. + (b + a)/2. ;
      w[i] = w[i]*(b-a)/2. ;
} } }
```

Listing 20.1 Bound.java solves the Lippmann–Schwinger integral equation for bound states within a delta-shell potential. The integral equations are converted to matrix equations using Gaussian grid points, and they are solved with JAMA.

1. Write your own program, or modify the code on the CD, to solve the integral equation (20.9) for the delta-shell potential (20.16). Either evaluate the determinant of $[H - EI]$ and then find the E for which the determinant vanishes *or* find the eigenvalues and eigenvectors for this H.
2. Set the scale by setting $2\mu = 1$ and $b = 10$.

3. Set up the potential and Hamiltonian matrices $V(i,j)$ and $H(i,j)$ for Gaussian quadrature integration with at least $N = 16$ grid points.
4. Adjust the value and sign of λ for bound states. A good approach is to start with a large negative value for λ and then make it less negative. You should find that the eigenvalue moves up in energy.
5. Try increasing the number of grid points in steps of 8, for example, 16, 24, 32, 64, . . ., and see how the energy changes.
6. *Note:* Your eigenenergy solver may return several eigenenergies. The true bound state will be at negative energy and will be stable as the number of grid points changes. The others are numerical artifacts.
7. Extract the best value for the bound-state energy and estimate its precision by seeing how it changes with the number of grid points.
8. Check your solution by comparing the RHS and LHS in the matrix multiplication $[H][\psi] = E[\psi]$.
9. Verify that, regardless of the potential's strength, there is only a single bound state and that it gets deeper as the magnitude of λ increases.

20.2.4 Wave Function (Exploration)

1. Determine the momentum-space wave function $\psi(k)$. Does it fall off at $k \to \infty$? Does it oscillate? Is it well behaved at the origin?
2. Determine the coordinate-space wave function via the Bessel transform

$$\psi(r) = \int_0^{\infty} dk\psi(k)\frac{\sin(kr)}{kr}k^2. \tag{20.18}$$

Does $\psi_0(r)$ fall off as you would expect for a bound state? Does it oscillate? Is it well behaved at the origin?
3. Compare the r dependence of this $\psi_0(r)$ to the analytic wave function:

$$\psi_0(r) \propto \begin{cases} e^{-\kappa r} - e^{\kappa r}, & \text{for } r < b, \\ e^{-\kappa r}, & \text{for } r > b. \end{cases} \tag{20.19}$$

20.3 Unit II. Nonlocal Potential Scattering ⊙

Problem: Again we have a particle interacting with the nonlocal potential discussed in Unit I (Figure 20.3 left), only now the particle has sufficiently high energy that it scatters rather than binds with the medium. Your **problem** is to determine the scattering cross section for scattering from a nonlocal potential.

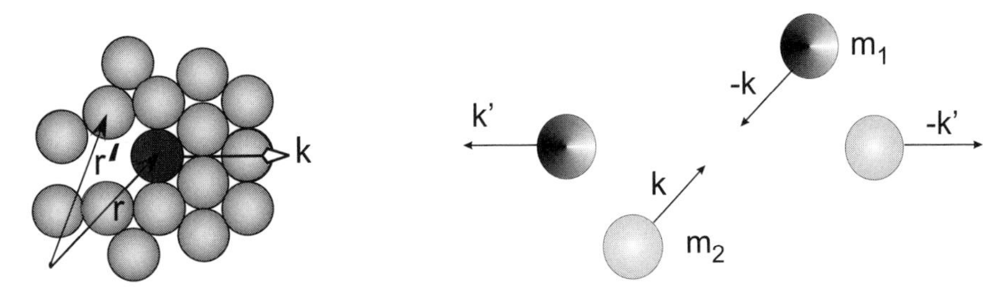

Figure 20.3 *Left:* A projectile of momentum k (dark particle at r) scattering from a dense medium. *Right:* The same process viewed in the CM system, where the k's are CM momenta.

20.4 Lippmann-Schwinger Equation (Theory)

Because experiments measure scattering amplitudes and not wave functions, it is more direct to have a theory dealing with amplitudes rather than wave functions. An integral form of the Schrödinger equation dealing with the reaction amplitude or R matrix is the *Lippmann–Schwinger equation*:

$$R(k',k) = V(k',k) + \frac{2}{\pi}\mathcal{P}\int_0^\infty dp\,\frac{p^2 V(k',p)R(p,k)}{(k_0^2 - p^2)/2\mu}. \tag{20.20}$$

As in Unit I, the equations are for partial wave $l = 0$ and $\hbar = 1$. In (20.20) the momentum k_0 is related to the energy E and the reduced mass μ by

$$E = \frac{k_0^2}{2\mu}, \qquad \mu = \frac{m_1 m_2}{m_1 + m_2}, \tag{20.21}$$

and the initial and final COM momenta k and k' are the momentum-space variables. The experimental observable that results from a solution of (20.20) is the diagonal matrix element $R(k_0, k_0)$, which is related to the scattering phase shift δ_0 and thus the cross section:

$$R(k_0, k_0) = -\frac{\tan\delta_l}{\rho}, \qquad \rho = 2\mu k_0. \tag{20.22}$$

Note that (20.20) is not just the evaluation of an integral; it is an integral equation in which $R(p,k)$ is integrated over all p, yet since $R(p,k)$ is unknown, the integral cannot be evaluated until after the equation is solved! The symbol $\mathcal{P}$ in (20.20) indicates the Cauchy principal-value prescription for avoiding the singularity arising from the zero of the denominator (we discuss how to do that next).

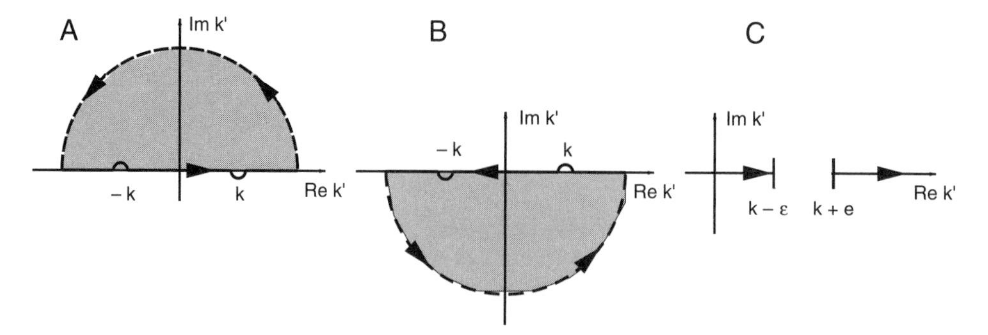

Figure 20.4 Three different paths in the complex k' plane used to evaluate line integrals when there are singularities. Here the singularities are at k and $-k$, and the integration variable is k'.

20.4.1 Singular Integrals (Math)

A *singular* integral

$$\mathcal{G} = \int_a^b g(k)\, dk, \tag{20.23}$$

is one in which the integrand $g(k)$ is singular at a point k_0 within the interval $[a, b]$ and yet the integral $\mathcal{G}$ is still finite. (If the integral itself were infinite, we could not compute it.) Unfortunately, computers are notoriously bad at dealing with infinite numbers, and if an integration point gets too near the singularity, overwhelming subtractive cancellation or overflow may occur. Consequently, we apply some results from complex analysis before evaluating singular integrals numerically.[3]

In Figure 20.4 we show three ways in which the singularity of an integrand can be avoided. The paths in Figures 20.4A and 20.4B move the singularity slightly off the real k axis by giving the singularity a small imaginary part $\pm i\epsilon$. The Cauchy principal-value prescription $\mathcal{P}$ (Figure 20.4C) is seen to follow a path that "pinches" both sides of the singularity at k_0 but does not to pass through it:

$$\mathcal{P}\int_{-\infty}^{+\infty} f(k)\, dk = \lim_{\epsilon\to 0}\left[\int_{-\infty}^{k_0-\epsilon} f(k)\, dk + \int_{k_0+\epsilon}^{+\infty} f(k)\, dk\right]. \tag{20.24}$$

The preceding three prescriptions are related by the identity

$$\int_{-\infty}^{+\infty} \frac{f(k)\, dk}{k - k_0 \pm i\epsilon} = \mathcal{P}\int_{-\infty}^{+\infty} \frac{f(k)\, dk'}{k - k_0} \mp i\pi f(k_0), \tag{20.25}$$

which follows from Cauchy's residue theorem.

[3] [S&T 93] describe a different approach using Maple and Mathematica.

20.4.2 Numerical Principal Values

A numerical evaluation of the principal value limit (20.24) is awkward because computers have limited precision. A better algorithm follows from the theorem

$$\mathcal{P}\int_{-\infty}^{+\infty}\frac{dk}{k-k_0}=0. \tag{20.26}$$

This equation says that the curve of $1/(k-k_0)$ as a function of k has equal and opposite areas on both sides of the singular point k_0. If we break the integral up into one over positive k and one over negative k, a change of variable $k \to -k$ permits us to rewrite (20.26) as

$$\mathcal{P}\int_{0}^{+\infty}\frac{dk}{k^2-k_0^2}=0. \tag{20.27}$$

We observe that the principal-value exclusion of the singular point's contribution to the integral is equivalent to a simple subtraction of the zero integral (20.27):

$$\mathcal{P}\int_{0}^{+\infty}\frac{f(k)\,dk}{k^2-k_0^2}=\int_{0}^{+\infty}\frac{[f(k)-f(k_0)]\,dk}{k^2-k_0^2}. \tag{20.28}$$

Notice that there is no $\mathcal{P}$ on the RHS of (20.28) because the integrand is no longer singular at $k=k_0$ (it is proportional to the df/dk) and can therefore be evaluated numerically using the usual rules. The integral (20.28) is called the *Hilbert transform* of f and also arises in inverse problems.

20.4.3 Reducing Integral Equations to Matrix Equations (Algorithm)

Now that we can handle singular integrals, we can go about reducing the integral equation (20.20) to a set of linear equations that can be solved with matrix methods. We start by rewriting the principal-value prescription as a definite integral [H&T 70]:

$$R(k',k)=V(k',k)+\frac{2}{\pi}\int_0^{\infty}dp\,\frac{p^2V(k',p)R(p,k)-k_0^2V(k',k_0)R(k_0,k)}{(k_0^2-p^2)/2\mu}. \tag{20.29}$$

We convert this integral equation to linear equations by approximating the integral as a sum over N integration points (usually Gaussian) k_j with weights w_j:

$$R(k,k_0)\simeq V(k,k_0)+\frac{2}{\pi}\sum_{j=1}^{N}\frac{k_j^2V(k,k_j)R(k_j,k_0)w_j}{(k_0^2-k_j^2)/2\mu}$$

$$-\frac{2}{\pi}k_0^2V(k,k_0)R(k_0,k_0)\sum_{m=1}^{N}\frac{w_m}{(k_0^2-k_m^2)/2\mu}. \tag{20.30}$$

We note that the last term in (20.30) implements the principal-value prescription and cancels the singular behavior of the previous term. Equation (20.30) contains the $(N+1)$ unknowns $R(k_j, k_0)$ for $j = 0, N$. We turn it into $(N+1)$ simultaneous equations by evaluating it for $(N+1)$ k values on a grid (Figure 20.2) consisting of the observable momentum k_0 and the integration points:

$$k = k_i = \begin{cases} k_j, & j = 1, N \quad \text{(quadrature points)}, \\ k_0, & i = 0 \quad\quad\ \text{(observable point)}. \end{cases} \tag{20.31}$$

There are now $(N+1)$ linear equations for $(N+1)$ unknowns $R_i \equiv R(k_i, k_0)$:

$$R_i = V_i + \frac{2}{\pi}\sum_{j=1}^{N} \frac{k_j^2 V_{ij} R_j w_j}{(k_0^2 - k_j^2)/2\mu} - \frac{2}{\pi} k_0^2 V_{i0} R_0 \sum_{m=1}^{N} \frac{w_m}{(k_0^2 - k_m^2)/2\mu}. \tag{20.32}$$

We express these equations in matrix form by combining the denominators and weights into a single denominator vector D:

$$D_i = \begin{cases} +\frac{2}{\pi}\frac{w_i k_i^2}{(k_0^2 - k_i^2)/2\mu}, & \text{for } i = 1, N, \\ -\frac{2}{\pi}\sum_{j=1}^{N} \frac{w_j k_0^2}{(k_0^2 - k_j^2)/2\mu}, & \text{for } i = 0. \end{cases} \tag{20.33}$$

The linear equations (20.32) now assume that the matrix form

$$R - DVR = [1 - DV]\,R = V, \tag{20.34}$$

where R and V are *column vectors* of length $N+1$:

$$[R] = \begin{pmatrix} R_{0,0} \\ R_{1,0} \\ \ddots \\ R_{N,0} \end{pmatrix}, \quad [V] = \begin{pmatrix} V_{0,0} \\ V_{1,0} \\ \ddots \\ V_{N,0} \end{pmatrix}. \tag{20.35}$$

We call the matrix $[1 - DV]$ in (20.34) the wave matrix F and write the integral equation as the matrix equation

$$[F][R] = [V], \qquad F_{ij} = \delta_{ij} - D_j V_{ij}. \tag{20.36}$$

With R the unknown vector, (20.36) is in the standard form $AX = B$, which can be solved by the mathematical subroutine libraries discussed in Chapter 8, "Solving Systems of Equations with Matrices; Data Fitting."

20.4.4 Solution via Inversion or Elimination

An elegant (but alas not efficient) solution to (20.36) is by matrix inversion:

$$[R] = [F]^{-1}[V]. \tag{20.37}$$

```
// Scatt.java: Soln of Lippmann Schwinger in p space for scattering

import Jama.*;
import java.io.*;
import java.util.*;

public class Scatt {
  public static void main(String[] argv) throws IOException, FileNotFoundException {
    PrintWriter q = new PrintWriter(new FileOutputStream("sin2.dat"), true);
    int n, i, j, m, Row, Column, M = 300;
    double pot, lambda, scale, ko, Temp, shift, shiftan, sin2, k2;
    double pi = 3.1415926535897932384626, b = 10., RN1, potlast=0.0;
    double[][] F = new double[M][M]; double[] k = new double[M];
    double[] w = new double[M]; double[]D = new double[M]; double[] r =new double[M];
    double[] V = new double[M]; double[][]P = new double[M][M];
    double[][]L = new double[M][M]; double[][]U = new double[M][M];
    n = 26;            scale = n/2;          pot = 0. ;
    shiftan = 0.;      lambda =1.5;                          // Set up Gauss points
    Gauss.gauss(n, 2, 0., scale, k, w);
    ko = 0.02;
    for ( m=1;m<901;m++)  {                                  // Set up D matrix
      k[n] = ko;
      for ( i=0; i<= n-1; i++ ){
      D[i]=2/pi*w[i]*k[i]*k[i]/(k[i]*k[i]-ko*ko);
      }
      D[n] = 0. ;
      for (j=0; j <= n-1;j++) D[n]=D[n]+w[j]*ko*ko/(k[j]*k[j]-ko*ko);
      D[n] = D[n]*(-2./pi);
      for ( i=0; i <= n; i++ ) {                       // Set up F matrix and V vector
        for ( j=0; j <= n; j++ )  {
          pot = -b*b * lambda * Math.sin(b*k[i])* Math.sin(b*k[j])/(k[i]*b*k[j]*b);
          F[i][j] = pot*D[j];
          if (i==j) F[i][j] = F[i][j] + 1.;
        }
        V[i] = pot;
      }                                                // Change arrays into matrices
      Matrix Fmat = new Matrix(F, n+1, n+1);
      Matrix Vvec = new Matrix( n+1, 1);
      Matrix Finv = Fmat.inverse();
      for ( i=0; i <= n; i++ )  Vvec.set(i, 0, V[i]);
      Matrix R = Finv.times(Vvec);                               // Invert matrix
      RN1 = R.get(n, 0);                                    // Get last value of R
                                                             // Define phase shift
      shift = Math.atan(-RN1*ko);
      sin2 = Math.sin(shift)*Math.sin(shift);
      q.println(ko*b + "   " + sin2);
      ko=ko+0.2*3.141592/1000.0;
    }
    System.out.println("Output in sin2.dat");
  }
}
```

Listing 20.2 Scatt.java solves the Lippmann–Schwinger integral equation for scattering from a delta-shell potential. The singular integral equations are regularized by a subtraction, converted to matrix equations using Gaussian grid points, and then solved with JAMA.

Because the inversion of even complex matrices is a standard routine in mathematical libraries, (20.37) is a *direct solution* for the R matrix. Unless you need the inverse for other purposes (like calculating wave functions), a more efficient approach is to use Gaussian *elimination* to find an $[R]$ that solves $[F][R] = [V]$ without computing the inverse.

20.4.5 Scattering Implementation

For the scattering problem, we use the same delta-shell potential (20.16) discussed in §20.2.2 for bound states:

$$V(k', k) = \frac{-|\lambda|}{2\mu k' k} \sin(k'b) \sin(kb). \tag{20.38}$$

This is one of the few potentials for which the Lippmann–Schwinger equation (20.20) has an analytic solution [Gott 66] with which to check:

$$\tan \delta_0 = \frac{\lambda b \sin^2(kb)}{kb - \lambda b \sin(kb) \cos(kb)}. \tag{20.39}$$

Our results were obtained with $2\mu = 1$, $\lambda b = 15$, and $b = 10$, the same as in [Gott 66]. In Figure 20.5 we give a plot of $\sin^2 \delta_0$ *versus* kb, which is proportional to the scattering cross section arising from the $l = 0$ phase shift. It is seen to reach its maximum values at energies corresponding to resonances. In Listing 20.2 we present our program for solving the scattering integral equation using the JAMA matrix library and the `gauss` method for quadrature points. For your implementation:

1. Set up the matrices `V[]`, `D[]`, and `F[][]`. Use at least $N = 16$ Gaussian quadrature points for your grid.
2. Calculate the matrix F^{-1} using a library subroutine.
3. Calculate the vector R by matrix multiplication $R = F^{-1}V$.
4. Deduce the phase shift δ from the $i = 0$ element of R:

$$R(k_0, k_0) = R_{0,0} = -\frac{\tan \delta}{\rho}, \quad \rho = 2\mu k_0. \tag{20.40}$$

5. Estimate the precision of your solution by increasing the number of grid point in steps of two (we found the best answer for $N = 26$). If your phase shift changes in the second or third decimal place, you probably have that much precision.
6. Plot $\sin^2 \delta$ *versus* energy $E = k_0^2/2\mu$ starting at zero energy and ending at energies where the phase shift is again small. Your results should be similar to those in Figure 20.5. Note that a *resonance* occurs when δ_l increases rapidly through $\pi/2$, that is, when $\sin^2 \delta_0 = 1$.
7. Check your answer against the analytic results (20.39).

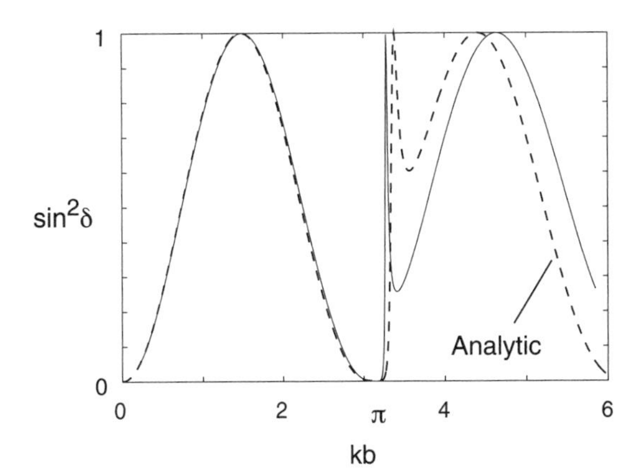

Figure 20.5 The energy dependence of the cross section for $l = 0$ scattering from an attractive delta-shell potential with $\lambda b = 15$. The dashed curve is the analytic solution (20.39), and the solid curve results from numerically solving the integral Schrödinger equation, either via direct matrix inversion or via LU decomposition.

20.4.6 Scattering Wave Function (Exploration)

1. The F^{-1} matrix that occurred in our solution to the integral equation

$$R = F^{-1}V = (1 - VG)^{-1}V \tag{20.41}$$

is actually quite useful. In scattering theory it is known as the *wave matrix* because it is used in expansion of the wave function:

$$u(r) = N_0 \sum_{i=1}^{N} \frac{\sin(k_i r)}{k_i r} F(k_i, k_0)^{-1}. \tag{20.42}$$

Here N_0 is a normalization constant and the R matrix gives standing-wave boundary conditions. Plot $u(r)$ and compare it to a free wave. ❚

Appendix A: Glossary

absolute value — The value of a quantity expressed as a positive number, for example, $|f(x)|$.
accuracy — The degree of exactness provided by a description or theory. *Accuracy* usually refers to an absolute quality, while *precision* usually refers to the number of digits used to represent a number.
address — The numerical designation of a location in memory. An identifier, such as a label, that points to an address in memory or a data source.
algorithm — A set of rules for solving a problem in a finite number of steps. Usually independent of the software or hardware.
allocate — To assign a resource for use, often memory.
alphanumeric — The combination of alphabetic letters, numerical digits, and special characters, such as %, $, and /.
analog — The mapping of a continuous physical observable to numbers, for example, a car's speed to the numbers on its speedometer.
animation — A process in which motion is simulated by presenting a series of slightly different pictures (frames) in succession.
append — To add on, especially at the end of an object or word.
application — A self-contained, executable program containing tasks to be performed by a computer, usually for a practical purpose.
architecture — The overall design of a computer in terms of its major components: memory, processor, I/O, and communication.
archive — To copy programs and data to an auxiliary medium or file system for long-term, compact storage.
argument — A parameter passed from one program part to another or to a command.
arithmetic unit — The part of the central processing unit that performs arithmetic.
array (matrix) — A group of numbers stored together in rows and columns that may be referenced by one or more subscripts. Each number in an array is an array element.
assignment statement — A command that sets a value to a variable or symbol.
B — The abbreviation for byte (8 bits).
b — The abbreviation for bit (binary digit).
background — (1) A technique of having a programming run at low priority ("in the background") while a higher-priority program runs "in the foreground." (2) The part of a video display not containing windows.
base — The radix of a number system. (For example, 10 is the radix of the decimal system.)
basic machine language — Instructions telling the hardware to do basic a operation such as store or add binary numbers.
batch — The running of programs without user interaction, often in the background.

baud — The number of signal elements per unit time, often 1 bit per second.
binary — Related to the number system with base 2.
BIOS — Basic input/output system.
bit — Contraction of "binary digit"; digit 0 or 1 in binary representation.
Boolean algebra — A branch of symbolic logic dealing with logical relations as opposed to numerical values.
boot — To "bootstrap"; to start a computer by loading the operating system.
branch — To pick a path within a program based on the values of variables.
bug — A mistake in a computer program or operating system; a malfunction.
bus — A communication channel (a bunch of wires) used for transmitting information quickly among computer parts.
byte — Eight bits of storage. Java uses two bytes to store a single character in extended unicode.
byte code — Compiled code read by all computer systems but still needing to be interpreted (or recompiled); contained in a class file.
cache — Small, very fast memory used as temporary storage between very fast CPU registers and main memory or between disk and RAM.
calling sequence — The data and setup needed to call a method or subprogram.
central processing unit (CPU) — The part of a computer that accepts and acts on instructions; where calculations are done and communications controlled.
checkpoint — A statement within a program that stops normal execution and provides output to assist in debugging.
checksum — The summation of digits or bits used to check the integrity of data.
child — An object created by a parent object.
class — (1) A group of objects or methods having a common characteristic. (2) A collection of data types and associated methods. (3) An instance of an object. (4) The byte code version of a Java program.
clock — Electronics that generate periodic signals to control execution.
code — A program or the writing of a program (often compiled).
column — The vertical line of numbers in an array.
column-major order — The method used by Fortran to store matrices in which the leftmost subscript attains its maximum value before the subscript to the right is incremented. (Java and C use row-major order.)
command — A computer instruction; a control signal.
command key — A keyboard key, or combination of keys, that performs a predefined function.
compilation — The translation of a program written in a high-level language to (more) basic language.
compiler — A program that translates source code from a high-level computer language to more basic machine language.
concatenate — To join together two or more strings head to tail.
concurrent processing — The same as parallel processing; the simultaneous execution of several related instructions.
conditional statement — A statement executed only under certain conditions.

control character — A character that modifies or controls the running of a program (e.g., control + *c*).
control statement — A statement within a program that transfers control to another section of the program.
copy — To transfer data *without* removing the original.
CPU — See *central processing unit.*
crash — The abnormal termination of a program or a piece of hardware.
cycle time (clock speed) — The time needed for a CPU to execute a simple instruction.
data — Information stored in numerical form; plural of datum.
data dependence — Occurs when two statements are addressing identical storage locations.
dependence — Relation among program statements in which the results depend on the order in which the statements are executed.
data type — Definitions that permit proper interpretation of a character string.
debug — To detect, locate, and remove mistakes in software or hardware.
default — The assumption made when no specific directive is given.
delete — To remove and leave no record.
DFT — Discrete Fourier transform.
digital — The representation of quantities in discrete form; contrast analog.
dimension of an array The number of elements that may be referenced by an array index. The *logical dimension* is the largest value actually used by the program.
directory — A collection of files given their own name.
discrete — Related to distinct elements.
disk, disc — A circular magnetic medium used for storage.
double precision — The use of two memory words to store a number.
download — To transfer data *from* a remote computer *to* a local computer.
DRAM — See dynamic RAM. Contrast with *SRAM.*
driver — A set of instructions needed to transmit data to or from an external device.
dump — Data resulting from the listing of all information in memory.
dynamic RAM — Computer memory needing frequent refreshment.
E — A symbol for "exponent." To illustrate, `1.97E2` $= 1.97 \times 10^2$.
element — An item of data within an array; a component of a language.
enable — To make a computer part operative.
ethernet — A high-speed local area network (LAN) composed of specific cable technology and communication protocols.
executable program — A set of instructions that can be loaded into a computer's memory and executed.
executable statement — A statement that causes a certain computational action, such as assigning a value to a variable.
fetch — To locate and retrieve information from storage.
FFT — Fast Fourier transform.
flash memory — Memory that does not require power to retain its contents.
floating point — The finite storage of numbers in scientific notation.
FLOP — Floating-point operations per second.
foreground — Running high-priority programs before lower-priority programs.

Fortran — Acronym for **for**mula **tran**slation; a classic language.
fragmentation — File storage in many small, dispersed pieces.
garbage — Meaningless numbers, usually the result of error or improper definition. Obsolete data in memory waiting to be removed ("collected").
giga, G — Prefix indicating 1 billion, 10^9, of something (US).
GUI — Graphical user interface; a windows environment.
hard disk — A circular, spinning, storage device using magnetic memory.
hardware — The physical components of a computer system.
hashing — A transformation that converts keystrokes to data values.
heuristic — A trial-and-error approach to problem solving.
hexadecimal — Base 16; {0,1,2,3,4,5,6,7,8,9,A,B,C,D,E,F}.
hidden line surface — The part of a graphics object normally hidden from view.
high-level language — A programming language similar to normal language.
host computer — A central or remote computer serving other computers.
HPC — High-performance computing.
HTML — Hypertext markup language.
icon — A small on-screen symbol that activates an application.
increment — The amount added to a variable, especially an array index.
index — The symbol used to locate a variable in an array; the subscript.
infinite loop — The endless repetition of a set of instructions.
input — The introduction of data from an external device into main storage.
instructions — Commands to the hardware to do basic things.
instruction stack — The ordered group of instructions currently in use.
interpolation — Finding values between known values.
interpreter — A language translator that sequentially converts each line of source code to machine code and immediately executes each line.
interrupt — A command that stops the execution of a program when an abnormal condition is encountered.
iterate — To repeat a series of steps automatically.
jump — A departure from the linear processing of code; branch, transfer.
just-in-time compiler — A program that recompiles a Java class file into more efficient machine code.
kernel — The inner or central part of a large program or of an operating system that is not modified (much) when run on different computers.
kill — To delete or stop a process.
kilo, k — Prefix indicating 1 thousand, 10^3.
LAN — Local area network.
LAPACK — A linear algebra package (a subroutine library).
language — Rules, representations, and conventions used to communicate information.
LHS — Left-hand side.
library (lib) — A collection of programs or methods usually on a related topic.
linking — Connecting separate pieces of code to form an executable program.
literal — A symbol that defines itself, such as the letter *A*.
load — To read information into a computer's memory.
load module — A program that is loaded into memory and run immediately.

log in (on) — To sign onto a computer; to begin a session.
loop — A set of instructions executed repeatedly as long as some condition is met.
low-level language — Machine-related commands not for humans.
machine language — Commands understood by computer hardware.
machine precision — The maximum positive number that, when added to the number stored as 1, does not change it.
macro — A single higher-level statement resulting in several lower-level ones.
main method — The section of an application program where execution begins.
main storage — The fast electronic memory; physical memory.
mantissa — Significant digits in a floating-point number; for example, 1.2 in 1.2E3.
mega, M — A prefix denoting a million, or $1,048,576 = 2^{20}$.
method — A subroutine used to calculate a function or manipulate data.
MIMD — Multiple-instruction, multiple-data computer.
modular programming — The technique of writing programs with many reusable independent parts.
modulo (mod) — A function that yields a remainder after the division of numbers.
multiprocessors — Computers with more than one processor.
multitasking — The system by which several jobs reside in a computer's memory simultaneously; they may run in parallel or sequentially.
NAN — Not a number; a computer error message.
nesting — Embedding a group of statements within another group.
object — A software component with multiple parts or properties.
object-oriented programming — A modular programming style focused on classes of data objects and associated methods to interact with the objects.
object program (code) — A program in basic machine language produced by compiling a high-level language.
octal — Base 8; easy to convert to or from binary.
ODE — Ordinary differential equation.
1-D — One-dimensional.
operating system (OS) — The program that controls a computer and runs applications, processes I/O, and shells.
OOP — Object-oriented programming.
optimization — The modification of a program to make it run more quickly.
overflow — The result of trying to store too large a number.
package — A collection of related programs or classes.
page — A segment of memory that is read as a single block.
parallel (concurrent) processing — Simultaneous or independent processing in different CPUs.
parallelization — Rewriting an existing program to run in parallel.
partition — The section of memory assigned to a program during its execution.
PC — Personal computer.
PDE — Partial differential equation.
physical memory — The fast electronic memory of a computer; main memory; contrast with *virtual memory*.
physical record — The physical unit of data for input or output that may contain a number of logical records.

pipeline (segmented) arithmetic units — An assembly-line approach to central processing; the CPU simultaneously gathers, stores, and processes data.
pixel — A picture element; a dot on the screen. See also *voxel*.
Portable Document Format, .pdf — A document format developed by Adobe that is of high quality and readable by an extended browser.
PostScript, .ps — A language developed by Adobe for printing high-quality text and graphics.
precision — The degree of exactness with which a quantity is presented. High-precision numbers are not necessarily *accurate*.
program — A set of instructions that a computer interprets and executes.
protocol — A set of rules or conventions.
pseudocode — A mixture of normal language and coding that provides a symbolic guide to a program.
queue — An ordered group of items waiting to be acted upon in turn.
radix — The base number in a number system that is raised to powers.
RAM — Random-access (central) memory that is reached directly.
random access — Reading or writing memory independent of storage order.
record — A collection of data items treated as a unit.
recurrence/recursion — Repetition producing new values from previous ones.
registers — Very high-speed memory used by the central processing unit.
reserved words — Words that cannot be used in an application program.
RHS — Right-hand side.
RISC — A CPU design for a reduced instruction set computer.
row-major order — The method used by Java to store matrices in which the rightmost subscript varies most rapidly and attains its maximum value before the left subscript is incremented.
run — To execute a program.
scalar — A data value or number, for example, π.
serial/scalar processing — Calculations in which numbers are processed in sequence. Contrast with *vector processing* and *parallel processing*.
shell — A command-line interpreter; the part of the operating system where the user enters commands.
SIMD — A single instruction, multiple-data computer.
simulation — The modeling of a real system by a computer program.
single precision — The use of one computer word to store a variable.
SISD — A single-instruction, single-data computer.
software — Programs or instructions.
source code — A program in a high-level language needing compilation to run.
SRAM — See static RAM.
Static RAM. — Memory that retains its contents as long as power is applied. Contrast with *DRAM*.
stochastic — A process in which there is an element of chance.
stride — The number of array elements stepped through as an operation repeats.
string — A connected sequence of characters treated as a single object.
structure — The organization or arrangement of a program or a computer.
subprogram — Part of a program invoked by another program unit; a *subroutine*.

supercomputer — The class of fastest and most powerful computers available.
superscalar — A later-generation RISC computer.
syntax — The rules governing the structure of a language.
TCP/IP — Transmission control protocol/internet protocol.
telnet — Protocols for computer–computer communications.
tera, T — Prefix indicating 10^{12}.
top-down programming — Designing a program from the most general view of the problem down to the specific subroutines.
unary — An operation that uses only one operand; monadic.
underflow — The result of trying to store too small a number.
unit — A device having a special function.
upload — Data transfer *from* a local *to* a remote computer; the opposite of *download*.
URL — Universal resource locator; web address.
utility programs — Programs to enhance other programs or do chores.
vector — A group of N numbers in memory arranged in 1-D order.
vector processing — Calculations in which an entire vector of numbers is processed with one operation.
virtual memory — Memory on the slow, hard disk and not in fast RAM.
visualization — Conversion of numbers to 2-D and 3-D pictures or graphs.
volume — A physical unit of a storage medium, such as a disk.
voxel — A volume element on a regular 3-D grid.
word — A unit of main storage, usually 1, 2, 4, 6, or 8 bytes.
word length — The amount of memory used to store a computer word.
WWW — World wide web.

Appendix B: Installing PtPlot & Java Developer's Kit Packages

The first step in setting up PtPlot is to download the latest version and then uncompress it.[1] The plotting package we call PtPlot is part of a larger Java project called *Ptolemy* [PtPlot], all free for downloading. After you have properly unzipped or untarred the PtPlot package, a directory such as ptplot5.6 should be created. The number 5.6 here is the version number of the PtPlot package that we are now using; there may be a newer version when you do your download. For our examples we have renamed the directory in which PtPlot resides as simply ptplot, with no version number attached. On Unix, we assume that your ptplot directory will be ~/java_packages/ptplot, where the ~ indicates your home directory. On Windows, we assume that your ptplot directory will be C:\ptplot.[2] Advanced users may prefer to keep the version number in the directory name or use a different organizational system. However, if this is the first time that you have installed a Java package, we recommend that you use the same directory names that we have.

Now that we have placed the PtPlot package in its own directory, we need to tell Java where to find it. As a matter of convention, the Java compiler javac and the Java interpreter java assume that the value of a variable named CLASSPATH contains the information on where packages such as PtPlot are stored. This type of variable that controls the environment in which programs run is called an *environment variable*. Because the programs in the packages having already been compiled into class files, the variable that directs Java to the classes is called CLASSPATH. To get PtPlot to work under Java, you need to modify the CLASSPATH variable to include the location where PtPlot is stored.

Windows 95: Open the autoexec.bat file (usually C:\autoexec.bat) in a text editor such as *Notepad*. At the end of the file, add the line

```
SET CLASSPATH=%CLASSPATH%;C:\ptplot
```

Save the autoexec.bat file and restart your computer.

Windows 98/ME: Click Start and then Run under that. Key in the command name msconfig and press Enter. The System Configuration Utility should appear. Select the tab Autoexec.bat and look for a line that says SET CLASSPATH. If you cannot find that line, click New and enter SET CLASSPATH = C:\ptplot. If the

[1] If you have never done this before, you may want to do it with a friend who has. In Windows, you use *WinZip* to unzip files. In Unix, try double-clicking on an icon of the file, or decompress it from within a shell with gunzip and tar -xvf.

[2] If you use the Windows automatic installer and you want to follow our examples verbatim, you should install to C:\ptplot rather than C:\ptolemy\ptplot5.6.

SET CLASSPATH line already exists, select it, choose Edit, add a semicolon ; to the end of the line, and then add C:\ptplot after the semicolon. The semicolon is a separator that tells Java that it should look in more than one location for class files. Make sure that the checkbox next to the CLASSPATH line is checked and then click OK. Answer Yes when Windows asks you whether you want to restart the computer.

Windows NT/2000/XP: Open the Control Panel (Start, Settings, Control Panel in NT/2000, or Start, Control Panel in XP). Open the System icon (you may need to switch to the Classic View in Windows XP). Under the System Properties window, select the Advanced tab and choose Environment Variables. Two lists should be shown. One contains environment variables just for you, and one contains environment variables for all users on the system. On your personal environment variable list, look for CLASSPATH. If you cannot find the CLASSPATH variable, click New and enter CLASSPATH for the variable name and C:\ptplot for the value. If the CLASSPATH variable already exists, select it, choose Edit, add a semicolon ; to the end of the current value, and then add C:\ptplot after the semicolon. The semicolon is a separator that tells Java that it should look in more than one location for class files. Click OK until you get back to the Control Panel and then restart your machine.

Unix: We assume that you do not have system authority for your computer and so will install PtPlot in your home directory. We suggest that you make a subdirectory called java_packages in your home directory and install PtPlot there:

```
> cd                                    Change to my home directory
> mkdir java_packages                   Create subdirectory here
> mkdir java_packages/ptplot            Create subdirectory in java_packages
```

If the ~ is used to represent your home directory, this assumes that you will be installing the PtPlot package in the ~/java_packages/ptplot/ directory.

Your CLASSPATH variable that needs updating is contained in the initiation (init) file .cshrc in your home directory. Because this file name begins with a . it is usually hidden from view. *Beware*: It is easy to mess up your .cshrc file and end up in a quagmire where you cannot enter commands. For this reason we suggest that you first create a backup copy of your .cshrc file in case anything goes wrong:

```
> cp .cshrc .cshrc_bk                   Make a backup, just in case
```

Next, open the .cshrc file in a text editor. Because this file may contain some very long lines that must be kept intact, if your text editor has an automatic word wrap feature, make sure it is turned off. Next look for a line that starts with setenv CLASSPATH. If you cannot find that line, add setenv CLASSPATH ~/java_packages/ptplot on its own line at the end of your .cshrc file. If the setenv CLASSPATH line already exists, add a colon : to the end of the existing line and then add ~/java_packages/ptplot after the colon. The colon is a separator that tells

Java that it should look in more than one location for class files. Save your .cshrc file and then close and reopen all the shells that you have open (or log off and then back on).

Once you have the CLASSPATH variable set, you should make sure that it is working. To check, go to a command prompt in a shell and enter

```
> echo %CLASSPATH%                                    Windows check
> echo $CLASSPATH                                     Unix check
```

The complete value for the CLASSPATH variable should be printed to the screen, for example:

```
/home/jan/java/classes:/home/jan:/home/jan/mpi:/usr/local/mpiJava/
lib/classes:/home/jan/java_packages:/home/jab/java_packages/ptplot:
```

If your changes do not take, carefully follow the directions again, and if you still have problems, ask for help.

At this point you are ready to try out PtPlot. Get the file EasyPtPlot.java containing a sample plot program from the CD and enter

```
> javac EasyPtPlot.java                               Compile sample plot program
> java EasyPtPlot                                     Run sample plot program
```

If the PtPlot package was installed correctly, you should get a nice graph on your screen. If this does not work, ask for help.

B.1 Installing Java Developer's Kit

Sun Microsystem's Java Web site [SunJ] contains the latest (free) version of Java Developer's Kit (JDK) (Java SE or Java 2, presently JDK6.3). Though most operating systems have all that is needed to *run* Java programs, you need the developer's kit to compile Java programs from a source file. The actual JDK tools occupy about 7 MB, with its (optional) documentation occupying more than 10 times as much space (the documentation is in HTML). This means that you may want to install only the tools if space is an issue.

On Windows computers, JDK is usually placed in C:\Program Files\Java\jdk1.5\bin, while on Unix it is placed in /usr/local/jdk1.5. Once installed, you will need to update the PATH and the CLASSPATH environment variables so that the shell knows where to find the Java compiler and classes you make. This is essentially the same procedure we used to include the PtPlot classes in CLASSPATH. Under Windows go to the System Control Panel and then select the Environment tab. Then add ;C:\jdk1.5\bin (or whatever file JDK was installed in) to the PATH variable. Likewise, CLASSPATH needs to be added, for example,

```
> set CLASSPATH=C:\...
```

Finally, reboot Windows for the changes to take effect, or log off and on under Unix.

TABLE B.1
Some of Java's Packages and the Classes They Contain

Java Package	*Classes for*
java.lang	Basic elements of the Java language
java.util	Utilities; random-number generators, date, time, and so on
java.awt	Abstract Windowing Toolkit; creating graphical interfaces
java.applet	Creating applets and interacting with browsers
java.beans	Creating reusable software components
java.io	Data input and output

B.2 Using Classes and Packages

Most of the programs we give as examples contain a single class in a file with a .java or .class extension. That class contains a main method and possibly some other methods. Just as we encourage you to modify old programs rather than write all programs from scratch, so we encourage you to include in your new programs methods that you and others have already written and debugged. Java contains a large number of *libraries* consisting of collections of methods for various purposes, and you should think of these libraries as *toolboxes* from which you can extract individual tools needed for your programs. In fact, the *object-oriented* approach of Java is specifically designed to make the reuse of components easier and safer.

B.2.1 Including Packages

A collection of related methods is called a class, and a broader collection, a library (such as the Math library). In strict Java naming convention, each method would be in a class file, and the libraries would be called *packages*. In general, there are two types of packages: the standard Java packages that constitute the Java language, and user-defined packages that extend standard Java. The PtPlot package is an example of a user-defined package. Some of the standard Java packages are given in Table B.1.

Because these Java packages contain hundreds or thousands of classes, some organization is necessary to keep track of what each method does. Java does this with a hierarchical directory structure in which there are parent packages containing subpackages, with the subpackages containing more subpackages or various classes. To make the name of each class unique, it is preceded by the names of the package and subpackages that contain this class. As an example, consider the command

System.out.println

that we use to print a line of output on the screen. Here System is the name of the class (classes begin with capital letters) containing many of Java's methods. When we give the combination System.out, we are referring to an object representing the standard *out*put stream. The final println that is affixed to System.out is the name of the method that prints lines.

To summarize, the convention is to have the names of classes capitalized, while the names of packages and methods are lowercase. This is relevant because the System.out.println command is in the java.lang package, and so the proper full name of the command is actually

java.lang.System.out.println

which contains the package name as well. Because java.lang is the most basic package, the Java compiler automatically looks there to find the methods we invoke. This means we can, fortunately, omit the java.lang prefix. Technically, java is the main package and lang is a subpackage, but it is easier to just say java.lang is the package. We must admit that we sometimes find Java's naming conventions overwhelming. Nevertheless, we will use them when importing packages and using methods from other classes, so some familiarity is helpful.

The classes from a package are included with the import command. It may be given in one of two forms:

```
import <packageName>.<specific classes>          // Import specific classes
import <packageName>.*                 // Import all classes from packageName
```

The import command tells the compiler to look in the package packageName for methods that it might not find otherwise. However, for the importation to work, the compiler must know *where* the package of classes and their methods are stored on your particular computer. In other words, the compiler needs to know the path to follow through the local disk memory to get to the directory or folder where the classes are stored. For this purpose each computer system, be it Windows, Unix, or Mac OS, has an environmental variable named CLASSPATH that contains the explicit path to where the classes are stored on that particular computer. As we show in Appendix B on installing PtPlot, you need to modify this variable before your program can import the package.

Even though what follows is more advanced programming than we do in this book, for completeness we indicate how you can create your own packages. This is done by placing several classes in the same .java file and then including a package command at the beginning of the file:

```
package < mypackage_name >;
public class < myclass1_name >
       { < normal class structure , multiple methods OK > }
        public class <myclass2_name>
       {                < normal class structure , multiple methods OK > }
```

Your package may be a collection of methods *without* any main method, for instance, mathematical subroutines that are called from all the programs you write. However, there must be one main method someplace if the program is to run since execution always begins in a main method. Likewise, the main method must be a public class.

Appendix C: OpenDX: Industrial-Strength Data Visualization

Most of the visualizations we use in this book are 2-D, $y(x)$ plots, or 3-D, $z(x, y)$ (surface) plots. Some of the applications, especially the applets, use animations, which may also be 2-D or 3-D. Samples are found on the CD. We use and recommend *Grace* (2-D) and *Gnuplot* (2-D, 3-D) for stand-alone visualizations, and *PtPlot* for visualizations callable from Java programs [L 05]. All have the power and flexibility for scientific work, as well as being free or open source.

An industrial-strength tool for data visualization, which we also recommend, is *OpenDX*, or *DX* for short.[1] It was originally developed as *IBM Data Explorer* but has now joined the ranks of open-source software [DX1, DX2] and is roughly equivalent to the commercial package AVS. DX works under Linux, Unix, or a Linux emulator (Cygwin) on a PC. The design goals of DX were to

- Run under many platforms with various data formats.
- Be user-friendly via visual programming, a modern technique in which programs are written by connecting lines between graphical objects, as opposed to issuing textual commands (Figure C.3 right).
- Handle large multidimensional data sets via volume rendering of $f(x, y, z)$ as well as *slicing* and *dicing* of higher-dimensional data sets.
- Create attractive graphics without expensive hardware.
- Have a graphical user interface that avoids the necessity of learning many commands.

The price to pay for all this power is the additional time spent learning how to do new things. Nevertheless, it is important for students of computational physics to get some experience with state-of-the-art visualization.

This appendix is meant as an introduction to DX. We use it to visualize some commonly encountered data such as scalar and vector fields and 3-D probability densities. DX can do much more than that, and, indeed, is a standard tool in visualization laboratories around the world. However, the visualizations in this text are just in gray, while DX uses color as a key element. Consequently, we recommend that you examine the DX visualizations on the CD to appreciate their beauty and effectiveness.

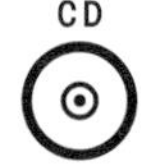

Analogous to the philosophy behind the Unix operating system, DX is a toolbox containing tools that permit you to create visualizations customized to your data. These visualizations may be created by command-line programming (for experienced users) or by visual programming. Typically, six separate steps are

[1] Juan Manuel Vanegas Moller and Guillermo Avendaño assisted in the preparation of this appendix.

involved:

1. Store the data in a file using a standard format.
2. Use the Data Prompter to describe the data's organization and to place that description in a .general file.
3. Import the data into DX.
4. Chose a visualization method appropriate for the data type.
5. Create a visual program with the Visual Program Editor by networking modules via drawn lines.
6. Create and manipulate an image.

C.1 Getting DX and Unix Running (for Windows)

If you do not have DX running on your computer, you can download it free from the DX home page [DX1]. Once running, just issue the dx or DX command and the system starts up. DX uses the Unix X-Windows system. If you are running under MS Windows, then you will first need to start an X-server program and then start DX at the resulting X-Windows prompt. We discussed this in Chapter 3, where we recommended the free Unix shell emulator *Cygwin* [CYG].

In order to run DX with Cygwin, you must start your X server *prior* to launching DX. You may also have to tell the X server where to place the graphical display:

```
> startxwin.sh                              Starts X server under Cygwin
> set environment DISPLAY localhost:0       Sets X11 display
> dx                                        Starts DX; or start menu
```

Here we issued the dx command from an X11 window. This works if your PATH variable includes the location of DX. We have had problems getting the dx command to work on some of our Cygwin installations, and in those cases we followed the alternative approach of loading DX as a regular MS Windows application. To do that, we *first* start the X11 server from a Cygwin bash shell with the command startxwin.sh and *then* start DX from the Start menu. While both approaches work fine, DX run from a Unix shell looks for your files within the Unix file system at /home/userid, while DX run under Windows looks within the MS Windows file system at C:\Documents and Settings\userid\My Documents\. Another approach, which is easier but costs money, is to install a commercial X11 server such as *X-Win 32* [XWIN32] coupled with your DX. Opening DX through MS Windows then automatically opens the X11 server with no effort on your part.

C.2 Test Drive of DX Visual Programming

Here we lead you through a test drive that creates a color surface plot with DX. This is deliberately cookbook-style in order to get you on the road and running. We then go on to provide a more systematic discussion and exploration of DX features, but without all the details.

If you want to see some of the capabilities of DX without doing any work, look at some of the built-in Samples accessible from the main menu (we recommend `ThunderStreamlines.net` and `RubberTube.net`). Once the pretty picture appears, go to Options/ViewControl/Mode and try out Rotate and Navigate. If you want to see how the graphical program created the visualization, go to Windows/Open Visual Program Editor and rearrange the program elements.

1. **Prepare input data:** Run the program `ShockLax.java` that describes shock wave formation and produces the file `Shock.dat` we wish to visualize; alternatively, copy the file `ShockLax.dat` from the CD (Codes/JavaCodes). If you are running DX under Cygwin, you will need to copy this file to the appropriate folder in `/home/userid`, where `userid` is your name on the computer.
2. **Examine input data:** Take note of the structure of the data in `Shock.dat`. It is in a file format designed for creating a surface plot of $z(x, y)$ with Gnuplot. It contains a column of 25 data values, a blank line, and then another 25 values followed by another blank lines, and so forth, for a total of 13 columns each with 25 data values:

```
                0.0
0.6950843693971483
 1.355305208363503
1.9461146066793003
               ...
-1.0605832625347442
-0.380140746321537
      (blank line)
               0.0
0.6403868757235301
1.2556172093991282
               ...
2.3059977070286473
 2.685151549102467
      (blank line)
2.9987593603912095
               ...
```

   Recall that the data are the z values, the locations in the column are the x values, and the column number gives the y values. The blank lines were put in to tell Gnuplot that we are starting a new column.
3. **Edit data:** While you can leave the blank lines in the file and still produce a DX visualization, they are not necessary because we explicitly told DX that we have one column of 25×13 data elements. Try both and see!
4. **Start DX:** Either key in `dx` at the Unix prompt or start an X-Windows server and then start DX. A Data Explorer main window (Figure C.1 left) should appear.

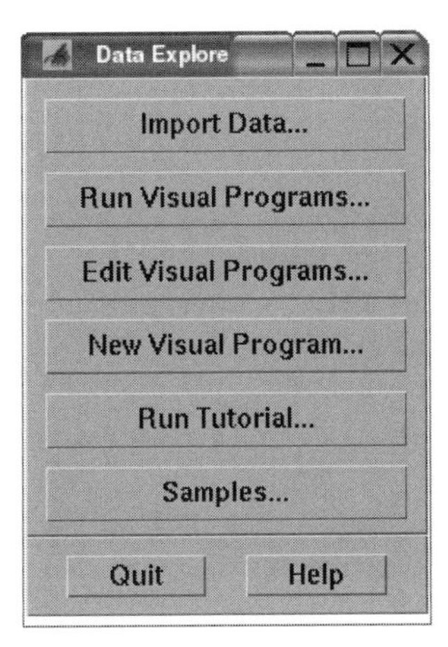

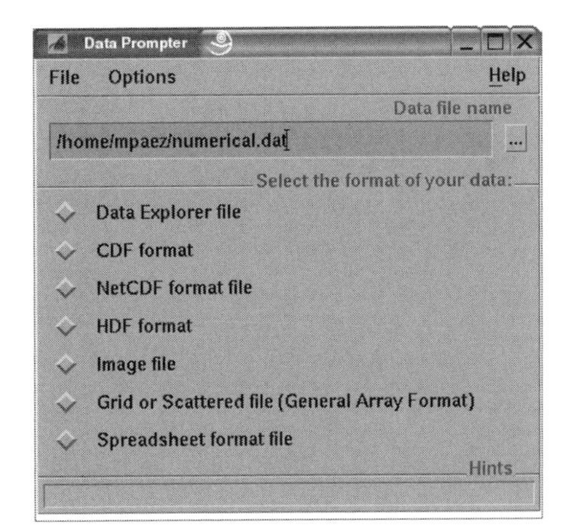

Figure C.1 *Left:* The main Data Explorer window. *Right:* The Data Prompter window that opens when you select Import Data.

5. **Import data:** Press the Import Data button in the main Data Explorer window. The Data Prompter window (Figure C.1 right) appears after a flash and a bang. Either key in the full path name of the data file (`/home/mpaez/numerical.dat` in our case) or click on Select Data File from the File menu. We recommend that after entering the file name, you depress the Browse Data button in the Data Prompter. This gives you a window (Figure C.2 right) showing you what DX thinks is in your data file. If you do not agree with DX, then you need to find some way to settle the dispute.
6. **Describe data:** Press the Grid or Scattered file button in the Data Prompter window. The window now expands (Figure C.2 left) to one that lets you pick out a description of the data.
   a. Select the leftmost Grid type, that is, the one with the regular array of small squares.
   b. For these data, select Single time step, which is appropriate because our data contain only a single column with no explicit (x, y) values.
   c. Because we have only the one z component to plot (a scalar field), set the slider at 1.
   d. Press the Describe Data button. This opens another window that tells DX the structure of the data file. Enter 25 as the first entry for the Grid size, and 13 as the second entry. This describes the 13 groups of 25 data elements in a single long column. Press the button Column.
   e. Under the File menu, select Save as and enter a name with a .general extension, for example, `Shock.general`. The file will contain

```
file = /home/mpaez/Shock.dat grid = 25 x 13 format = ascii
interleaving = field majority = column field = field0 structure =
```

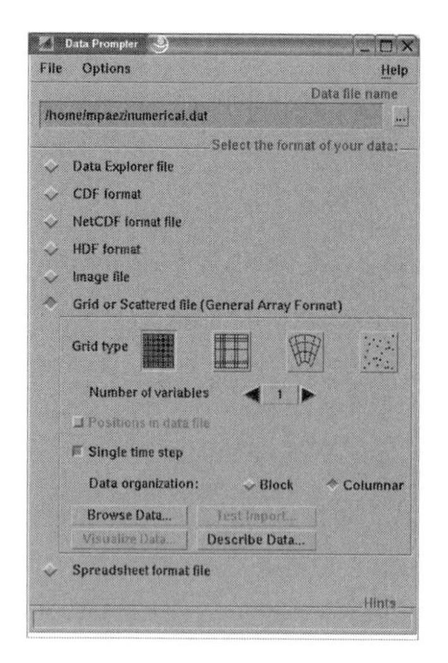

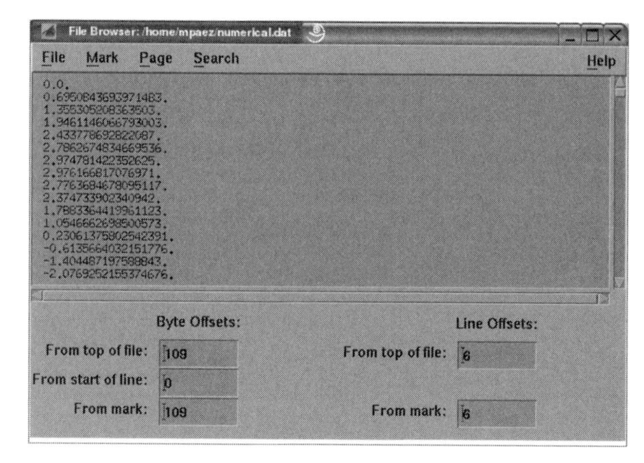

Figure C.2 *Left:* The expanded Data Prompter window obtained by selecting Grid or Scattered File. *Right:* The File Browser window used to examine a data file.

```
scalar type = float dependency = positions positions = regular,
regular, 0, 1, 0, 1 end
```

Now close all the Data Prompter windows.

7. **Program visualization with a visual editor:** Next design the program for your visualization by drawing lines to connect building blocks (modules). This is an example of *visual programming* in which drawing a flowchart replaces writing lines of code.
   a. Press the button New Visual Program from the Data Explorer main window. The Visual Program Editor window appears (Figure C.3 left). There is a blank space to the right known as the *canvas*, upon which you will draw, and a set of Tools on the left. (We list the names of all the available tools in §C.3, with their descriptions available from DX's built-in user's manual.)
   b. Click on the [+] to the left of Import and Export. The box changes to [-] and opens up a list of categories. Select Import so that it remains highlighted and then click near the top of the canvas. This should place the Import module on the canvas (Figure C.3 left).
   c. Close Import and Export and open the Transformation box. Select AutoColor so that it remains highlighted and then click below the Import module. The AutoColor image that appears on the canvas will be used to color the surface based on the z values.

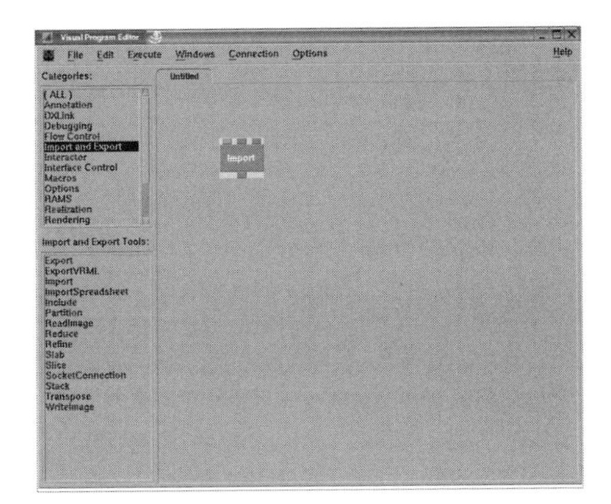

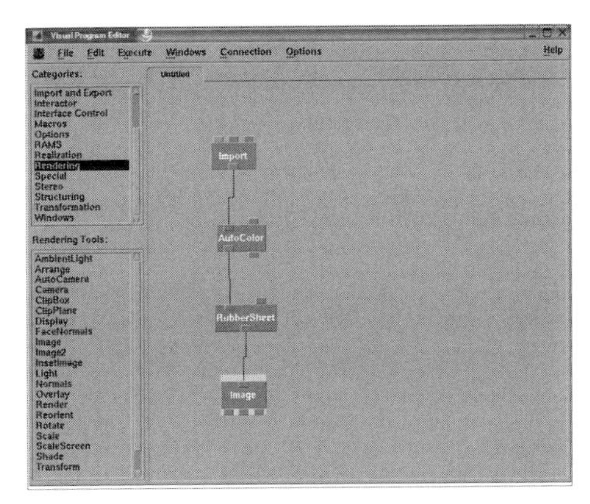

Figure C.3 The Visual Program Editor with Tools or Categories on the left and the canvas on the right. In the left window a single module has been selected from Categories, while in the right window several modules have been selected and networked together.

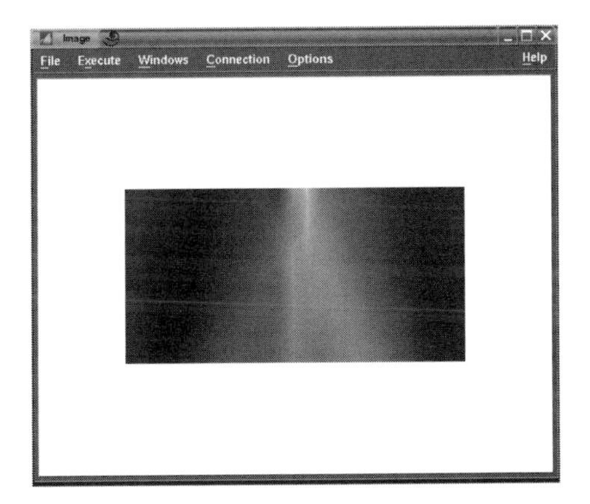

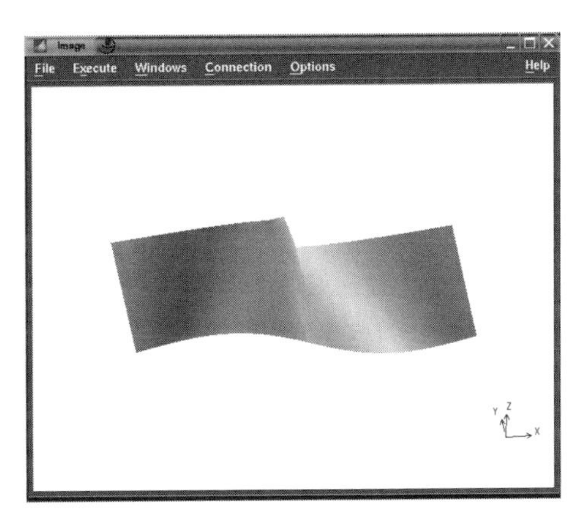

Figure C.4 *Left:* The image resulting from selecting Execute Once in the Visual Program Editor. *Right:* The same image after being rotated with the mouse reveals its rubber sheet nature.

d. Connect the output of Import to the left input of Autocolor by dragging your mouse, keeping the left button pressed until the two are connected.

e. Close the Transformation box and look under Realization. Select RubberSheet (a descriptive name for a surface plot), place it on the canvas below Autocolor, and connect the output of Autocolor to the left input tag of RubberSheet.

f. Close the Realization box and look under the Rendering tool. Select Image and connect its input to the output of RubberSheet. You should now have a canvas (Figure C.3 right).

g. A quick double-click on the Import block should bring up the Import window. Push the Name button so that it turns green and ensure that the

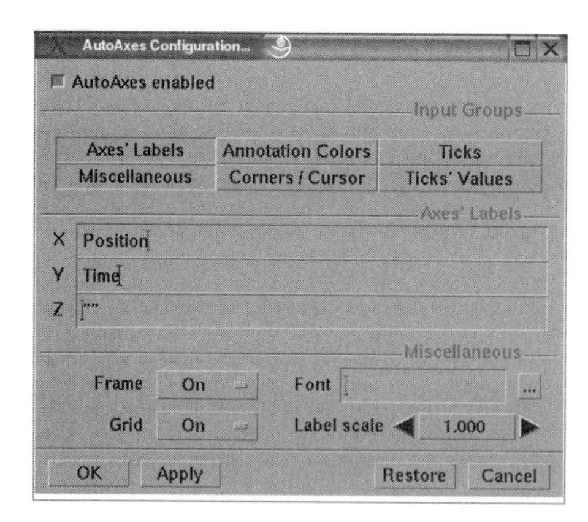

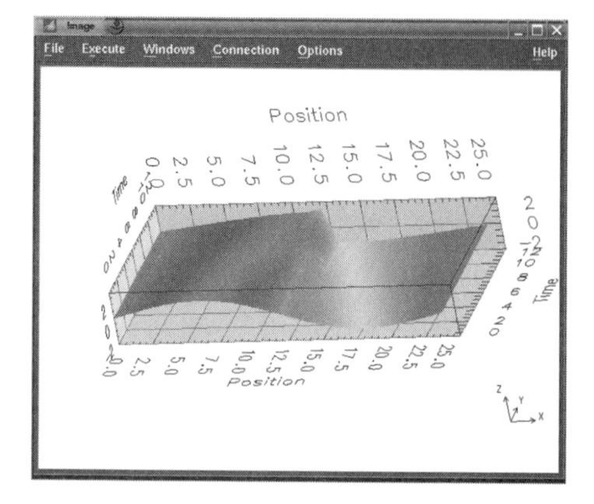

Figure C.5 *Left:* The AutoAxes Configuration window brought up as an Option from the Visual Program Editor. *Right:* The final image with axes and labels.

.general file that you created previously appears under Value. If not, enter it by hand. In our case it has the name "/home/mpaez/numer.general". Next press the Apply button and then OK to close the window.

h. Return to the Visual Program Editor. On the menu across the top, press Execute/Execute Once, and an image (Figure C.4 left) should appear.

8. **Manipulating the image:** Try grabbing and rotating the image with the mouse. If that does not work, you need to tell DX what you want:
   a. From the menu at the top of the Image window select Options; in the drop-down window select Mode; in the next drop-down window select Rotate. For instance, the rather flat-looking image on the left in Figure C.4 acquires depth when rotated to the image on the right.
   b. Again select Options from the Image window's menu (Figure C.5 left), but this time select AutoAxes from the drop-down window. In the new window key in Position in the X space and Time in the Y space. Press Apply/OK, and an image (Figure C.5 right) should appear.
   c. Now that you have the image you desire, you need to save it. From the File menu of the Image window, select Save Image. A pop-up window appears, from which you select Format, then the format you desire for the image, then the Output file name, and finally Apply.
   d. You want to ensure that the visual program is saved (in addition to the data file). In the Image window select File/Save Program As (we selected Shock2.net).
9. **Improved scale:** Often the effectiveness of a surface plot like Figure C.5 can be improved by emphasizing the height variation of the surface. The height is changed in DX by changing the *scale*:
   a. Go to the Visual Program Editor, select Rendering/scale (Figure C.6 right). Place the Scale icon between the RubberSheet and Image icons and connect.
   b. Double-click the Image icon, and a Scale window opens (Figure C.6 right).

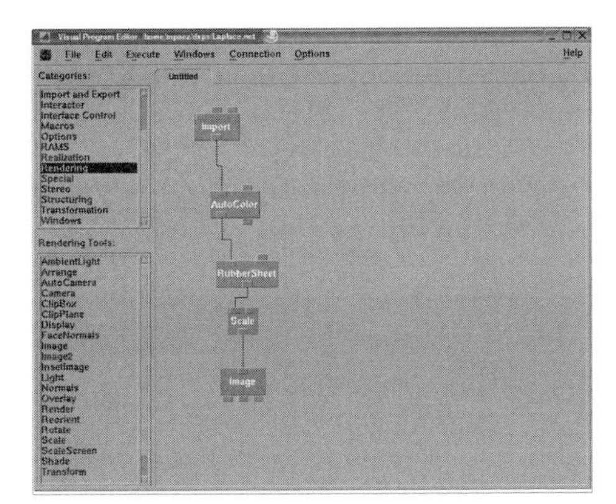

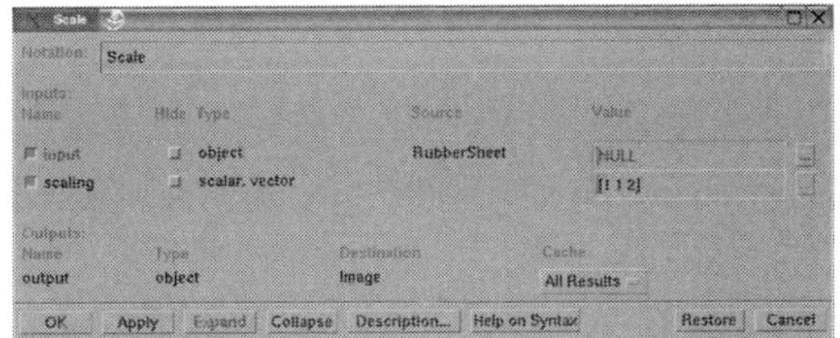

Figure C.6 On the left the Scale option is selected under Rendering and placed on the canvas. Double-clicking on the Image icon produces the Scale window on the right. By changing the second line under Value to [112], the z scale is doubled.

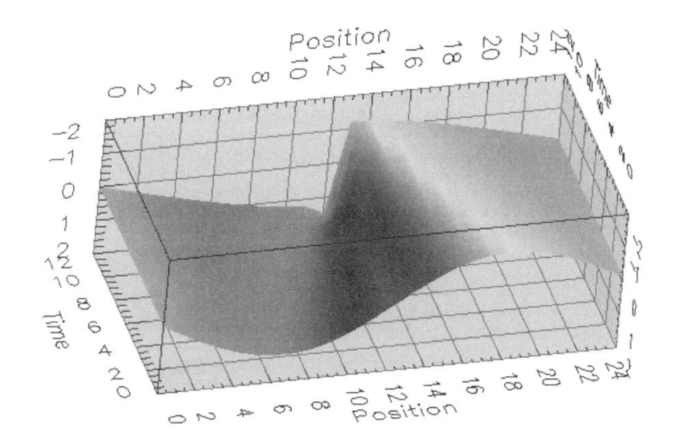

Figure C.7 The graph with a better scale for the z axis.

c. Change the second line under Value from [111] to [112] to double the z scale. Press Apply/OK.

d. Select Execute, and then under Option, select Execute once. The improved image in Figure C.7 appears.

C.3 DX Tools Summary

DX has built-in documentation that includes a *QuickStart Guide*, a *User's Guide*, a *User's Reference*, a *Programmer's Reference*, and an *Installation and Configuration Guide*. Here we give a short summary of some of the available tools. Although you can find all this information under the Tools categories in the Visual Program Editor, it helps to know what you are looking for! The DX on-line help provides more information about each tool. In what follows we list the tools in each category.

Categories

Annotation	DXLink	Debugging	Flow control	Import and export
Interactor	Interface control	Realization	Rendering	Special
Structuring	Transformation	Windows	All	

Annotation: adds various information to visualization

Annotation	DXLink	Debugging	Flow control	Import	Export
AutoAxes	AutoGlyph	Caption	ColorBar	Format	Glyph
Legend	Parse	Plot	Ribbon	Text	Tube

DXLink: DX control from other programs

DXLInput	DXLInputNamed	DXLOutput

Debugging: analyzes program's execution

Message	Echo	Print	Describe	System	Trace
Usage	Verify	VisualObject			

Flow Control: execution flow in a visual program

Done	Execute	First	ForEachMember	ForEachN	GetGlobal
GetLocal	Route	SetGlobal	SetLocal	Switch	

Import and Export: data flow and processing in a visual program

Export	Import	ImportSpreadsheet	Include	Partition	ReadImage
Reduce	Refine	Slab	Slice	Stack	Transpose
WriteImage					

Interactor: interactive control of input to modules via an interface

FileSelector	Integer	IntegerList	Reset	Scalar	ScalarList
Selector	String	SelectorList	StringList	Toggle	Value
ValueList	VectorList	Vector			

Interface Control: control DX tools from within a visual program		
ManageColormapEditor	ManageControlPanel	ManageImageWindow
ManageSequencer		

Realization: create structures for rendering and display					
AutoGrid	ShowConnections	Connect	Construct	Enumerate	Grid
Isolate	ShowBoundary	MapToPlane	Regrid	RubberSheet	Sample
Isosurface	ShowPositions	Band	ShowBox	Streakline	Streamline

Rendering: create/modify an image					
AmbientLight	Arrange	AutoCamera	Camera	ClipBox	ClipPlane
Display	FaceNormals	Image	Light	Normals	Overlay
Render	Reorient	Rotate	Scale	ScaleScreen	Shade
Transform	Translate	UpdateCamera			

Special: miscellaneous				
Colormap	Input	Output	Pick	Probe
ProbeList	Receiver	Sequencer	Transmitter	

Structuring: manipulate DX data structures					
Append	Attribute	ChangeGroupMember	CopyContainer	Collect	Rename
Extract	Replace	ChangeGroupType	CollectNamed	Inquire	List
Mark	Remove	CollectMultiGrid	CollectSeries	Select	Unmark

Transformation: modify or add to an input Field					
AutoColor	AutoGrayScale	Categorize	CategoryStatistics	Color	Compute
Compute2	Convert	DFT	Direction	DivCurl	Equalize
FFT	Filter	Gradient	Histogram	Lookup	Map
Measure	Morph	Post	QuantizeImage	Sort	Statistics

Windows: create or supervise image windows		
ReadImageWindow	SuperviseState	SuperviseWindow

C.4 DX Data Structure and Storage

Good organization or structuring of data is important for the efficient extraction (*mining*) of signals from data. Data that are collected in *fixed* steps for the independent variables, for example, $y(x = 1), y(x = 2), \ldots, y(x = 100)$, fit a *regular grid.*

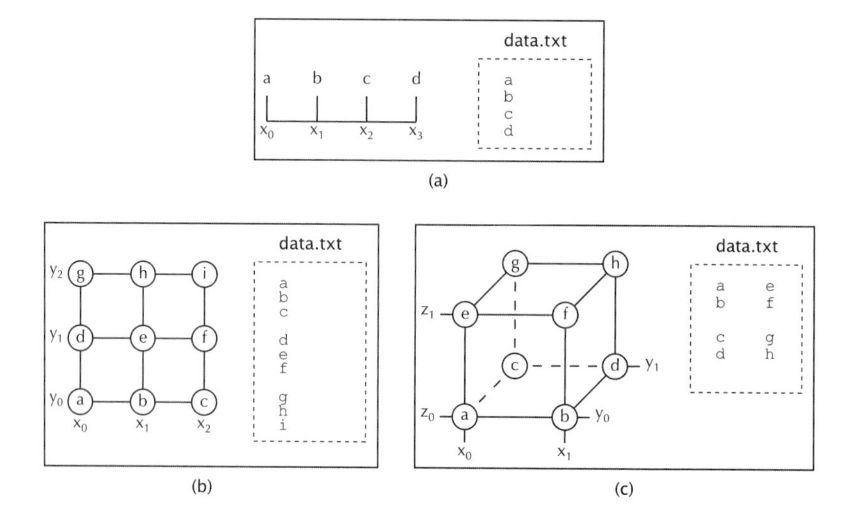

Figure C.8 Geometric and file representations of 1-D, 2-D, and 3-D data sets. The letters a, b, c, d, ... represent numerical values that may be scalars or vectors.

The corresponding data are called *regular data*. The coordinates of a single datum within a regular set can be generated from three parameters:

1. The origin of the data, for instance, the first datum at $(x = 0,\ y = 0)$.
2. The size of the steps in each dimension for the independent variables, for example, $(\Delta x,\ \Delta y)$.
3. The total number of data in the set, for example, N.

The data structures used by spreadsheets store dependent variables with their associated independent variables, for instance, $\{x, y, z(x, y)\}$. However, as already discussed for the Gnuplot surface plot (Figure 3.8 right), if the data are regular, then storing the actual values of x and y is superfluous for visualization since only the fact that they are evenly spaced matters. That being the case, regular data need only contain the values of the dependent variable $f(x, y)$ and the three parameters discussed previously. In addition, in order to ensure the general usefulness of a data set, a description of the data and its structure should be placed at the beginning of the set or in a separate file.

Figure C.8 gives a geometric representation of 1-D, 2-D, and 3-D data structures. Here each row of data is a group of lines, where each line ends with a *return* character. The columns correspond to the different positions in the row. The different rows are separated by blank lines. For the 3-D structure, we separate values for the third dimension by blank spaces from other elements in the same row.

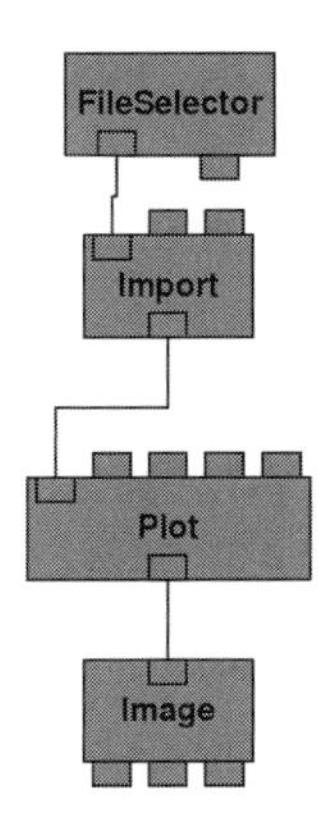

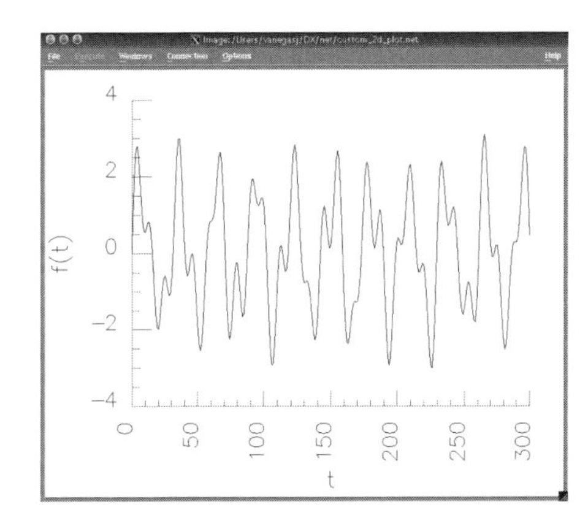

Figure C.9 The visual program Simple_linear_plot.net on the left imports a simple linear data file and produces the simple $x+y$ plot shown on the right.

C.5 Sample Visual Programs

We have included on the CD a number of DX programs we have found useful for some of the problems in this book and which produce the visualizations we are about to show. Because color is such an important part of these visualizations, we suggest that you look at these visualizations on the CD as well. Although you are free to try these programs, please note that you will first have to edit them so that the file paths point to your directory, for example, /home/userid/DXdata, rather than the named directories in the files.

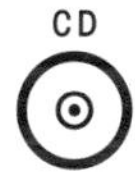

C.5.1 Sample 1: Linear Plot

The visual program simple_linear_plot.net (Figure C.9 left) reads data from the file simple_linear_data.dat and produces the simple linear plot shown on the right. These data are the sums of four sine functions,

$$f(t) = 1.8\sin(25t) + 0.22\sin(38t) + 0.5\sin(47t) + 0.95\sin(66t), \tag{3.1}$$

stored at *regular* values of t.

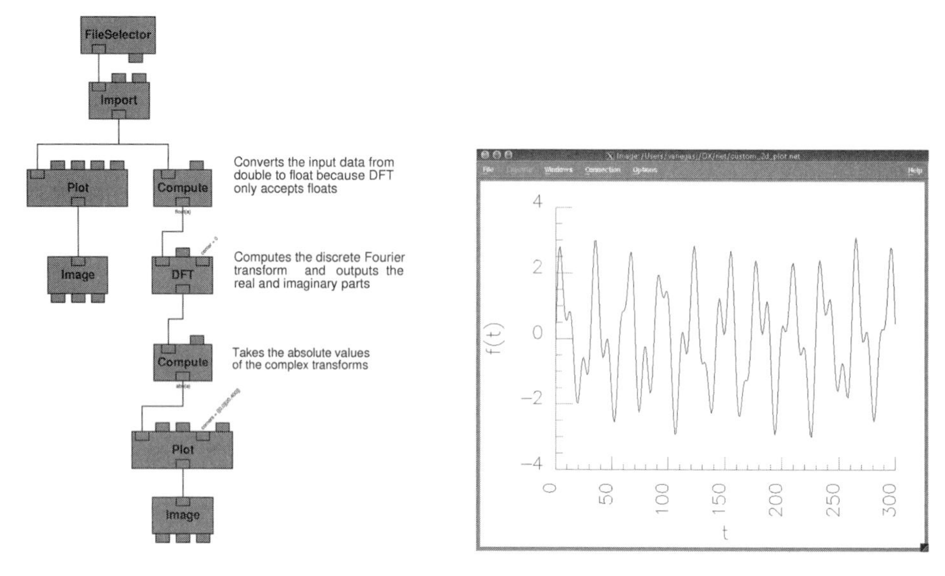

Figure C.10 The visual program on the left computes the discrete Fourier transform of the function (3.1), read in as data, resulting in the plot on the right.

C.5.2 Sample 2: Fourier Transform

As discussed in Chapter 10,"Fourier Analysis; Signals and Filters," the discrete Fourier transform (DFT) and the fast Fourier transform (FFT) are standard tools used to analyze oscillatory signals. DX contains modules for both tools. On the left in Figure C.10 you see how we have extended our visual program to calculate the DFT. The right side of Figure C.10 shows the resulting discrete Fourier transform with the expected four peaks at $\omega = 25$, 38, 47, and 66 rad/s, as expected from the function (3.1). Note how in the visual program we passed the data through a Compute module to convert it from doubles to floats. This is necessary because DX's DFT accepts only floats. After computing the DFT, we pass the data through another Compute module to take the absolute value of the complex transform, and then we plot them.

C.5.3 Sample 3: Potential of a 2-D Capacitor

While a vector quantity such as the electric field may be more directly related to experiment than a scalar potential, the potential is simpler to visualize. This is what we did in Chapter 17, "PDEs for Electrostatics and Heat Flow," with the program `Lapace.java` when we solved Laplace's equation numerically. Some output

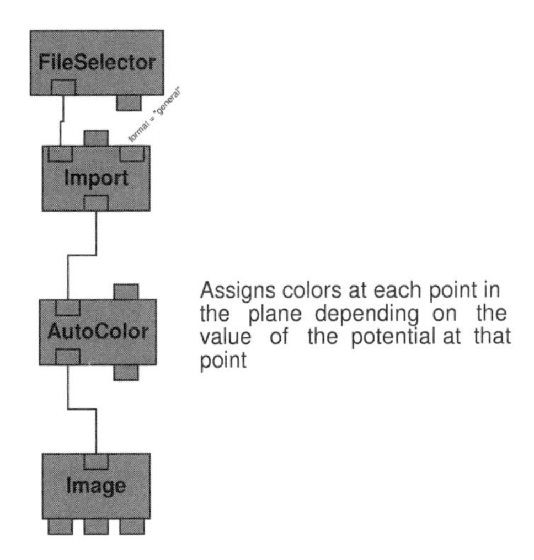

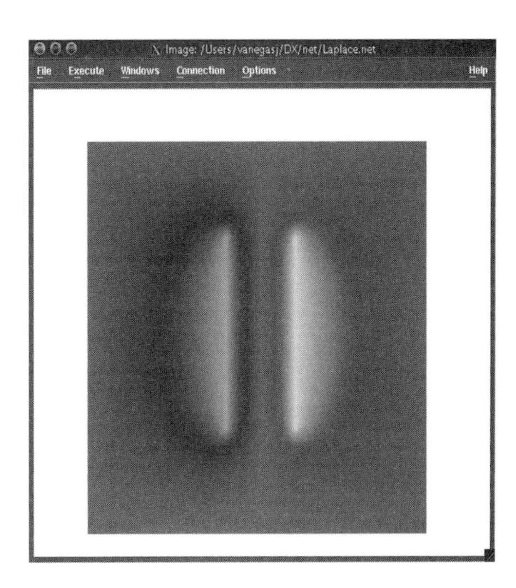

Figure C.11 The visual program on the left produces visualization of the electric potential on a 2-D plane containing a capacitor. Colors represent different potential values.

from that simulation is used in the DX visualization of the electric potential in a plane containing a finite parallel-plate capacitor (Figure C.11). The visual program `Laplace.net1` shown on the left creates the visualization, with the module AutoColor coloring each point in the plane according to its potential value.

Another way to visualize the same potential field is to plot the potential as the height of a 3-D surface, with the addition of the surface's color being determined by the value of the potential. The visual program `Laplace.net_2` on the left in Figure C.12 produces the visualization on the right. The module RubberSheet creates the 3-D surface from potential data. The range of colors can be changed, as well as the *opacity* of the surface, through the use of the Colormap and Color modules. The ColorBar module is also included to show how the value of the potential is mapped to each color in the plot.

C.5.4 Sample 4: Vector Field Plots

Visualization of the scalar field $V(x, y)$ required us to display one number at each point in a plane. A visualization of the vector electric field $\mathbf{E}(x, y) = -\vec{\nabla} V$ requires us to display three numbers at each point. In addition to it being more work to visualize a vector field, the increased complexity of the visualization may make the physics less clear even if the mathematics is more interesting. This is for you, and your mind's eye, to decide.

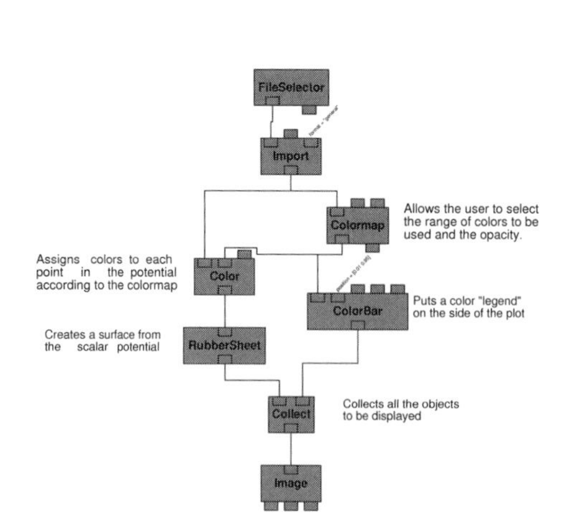

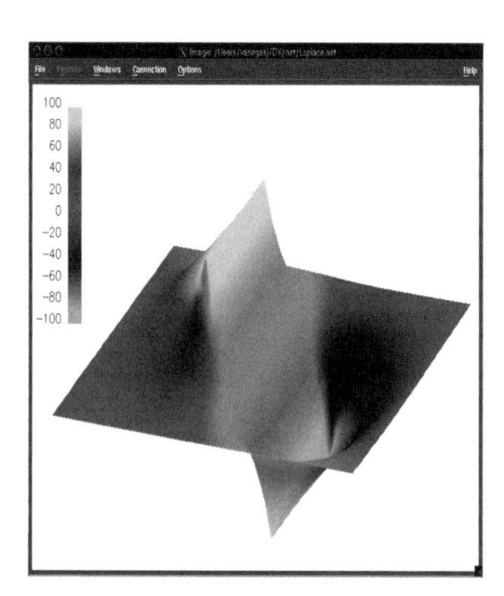

Figure C.12 The visual program on the left produces the surface plot of the electric potential on a 2-D plane containing a capacitor. The surface height and color vary in accord with the values of the potential.

DX makes the plotting of vector fields easy by providing the module Gradient to compute the gradient of a scalar field. The visual program Laplace.net_3 on the left in Figure C.13 produces the visualization of the capacitor's electric field shown on the right. Here the AutoGlyph module is used to visualize the vector nature of the field as small vectors (*glyphs*), and the Isosurface module is used to plot the *equipotential* lines. If the 3-D surface of the potential were used in place of the lines, much of the electric field would be obstructed.

C.5.5 Sample 5: 3-D Scalar Potentials

We leave it as an exercise for you to solve Laplace's equation for a 3-D capacitor composed of two concentric tori. Although the extension of the simulation from 2-D to 3-D is straightforward, the extension of the visualization is not. To illustrate, instead of equipotential lines, there are now equipotential surfaces $V(x, y, z)$, each of which is a solid figure with other surfaces hidden within. Likewise, while we can again use the Gradient module to compute the electric field, a display of arrows at all points in space is messy. Such being the case, one approach is to *map* the electric field onto a surface, with a display of only those vectors that are parallel or perpendicular to the surface. Typically, the surface might be an equipotential surface or the xy or yz plane. Because of the symmetry of the tori, the planes appear to provide the best visualization.

The visual program Laplace-3d.net_1 on the left in (Figure C.14) plots an electric field mapped onto an equipotential surface. The visualization on the right shows

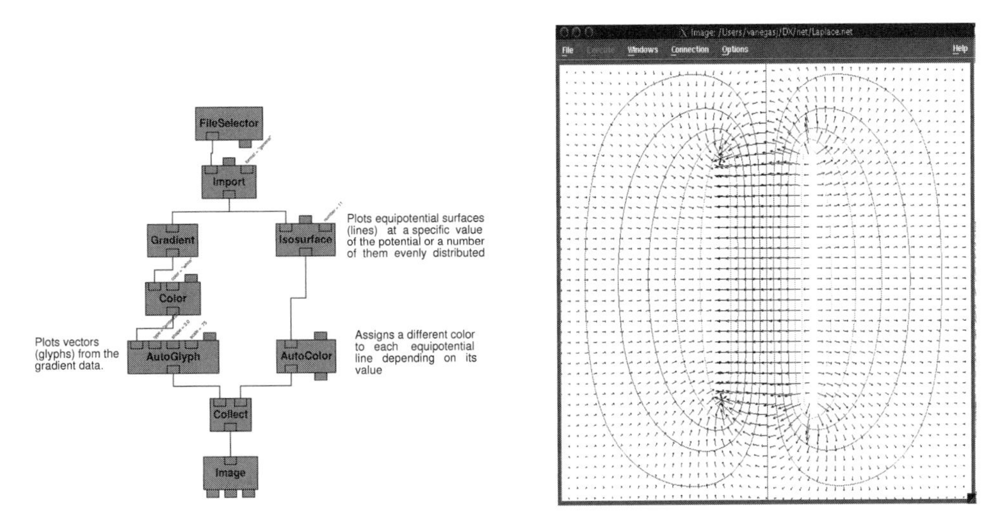

Figure C.13 The visual program on the left produces the electric field visualization on the right.

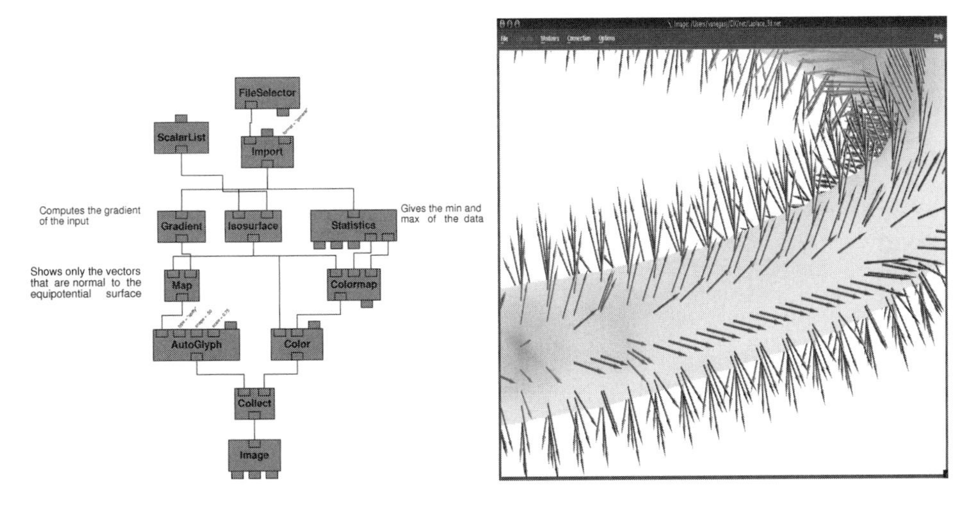

Figure C.14 The visual program on the left plots a torus as a surface and maps the electric field onto this surface. The visualization on the right results.

that the electric field is perpendicular to the surface but does not provide information about the behavior of the field in space. The program Laplace-3d.net_2 (Figure C.15) plots the electric field on the $y + z$ plane going through the tori. The

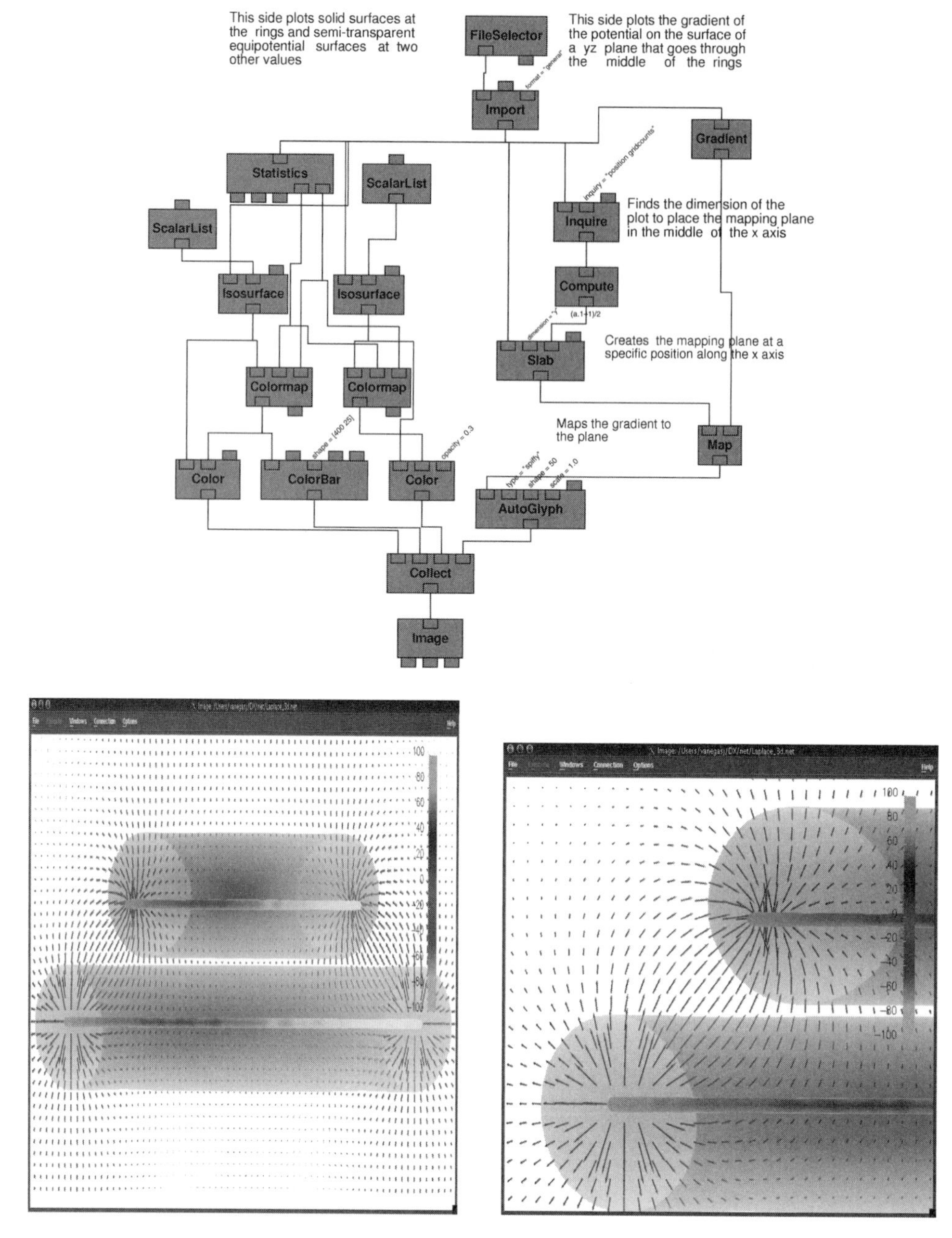

Figure C.15 The visual program at the top plots two equipotential surfaces in yellow and green, as well as the electric field on a $y+z$ plane going through the tori. The blue and red surfaces are the charged tori. The image on the far right is a close-up of the middle image.

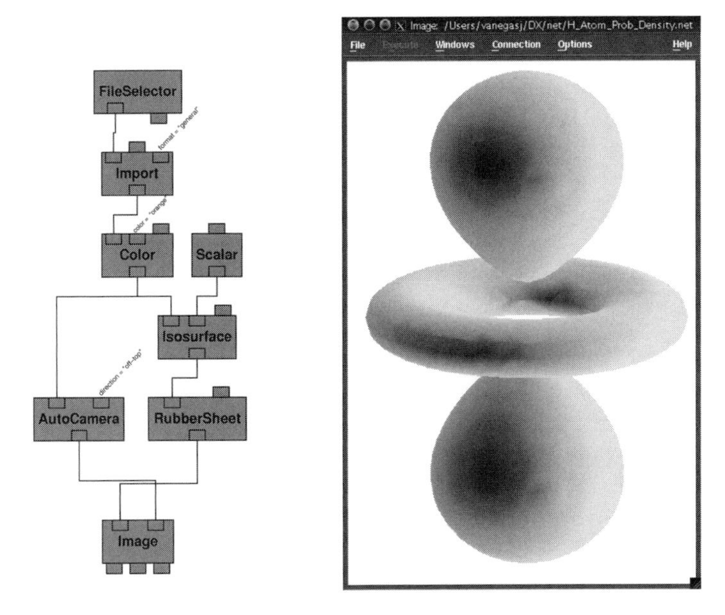

Figure C.16 The visual program on the left produces the surface plot of constant probability density for the 3-D state on the right.

plane is created by the module Slab. In addition, an equipotential surface surrounding each torus is plotted, with the surfaces made semitransparent in order to view their relation to the electric field. The bending of the arrows is evident in the closeup on the far right.

C.5.6 Sample 6: 3-D Functions, the Hydrogen Atom

The electron probability density $\rho_{nlm}(r, \theta, \phi)$ (calculated by H_atom_wf.java) is a scalar function that depends on three spatial variables [Libb 03]. Yet because the physical interpretation of a density is different from that of a potential, different visualization techniques are required. One way of visualizing a density is to draw 3-D surfaces of constant density. Because the density has a single fixed value, the surfaces indicates regions of space that have equal likelihood of having an electron present. The visual program H_atom_prob_density.net_1 produced the visualization in Figure C.16. This visualization does an excellent job of conveying the shape and rotational symmetry of the state but does not provide information about the variation of the density throughout space.

A modern technique for visualizing densities is *volume rendering*. This method, which is DX's default for 3-D scalar functions, represents the density as a translucent

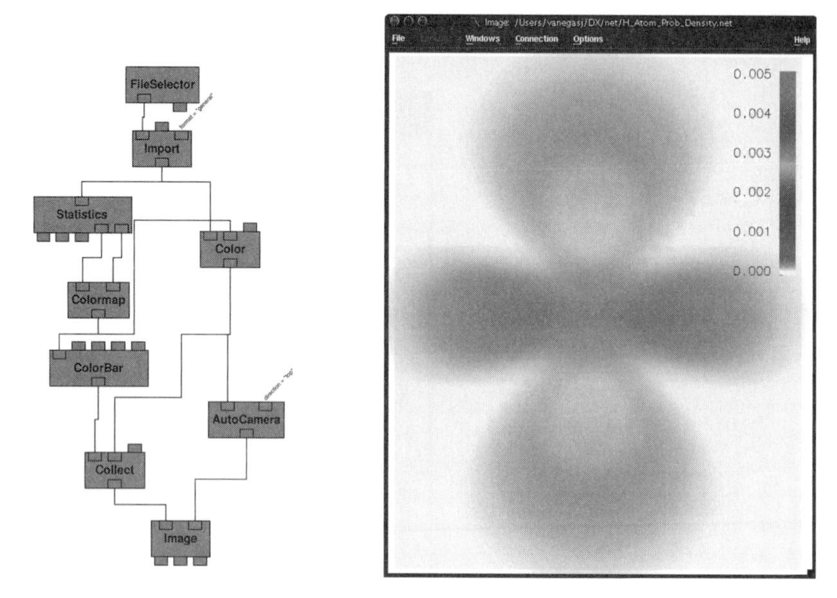

Figure C.17 The visual program on the left produces the 3-D cloud density on the right.

cloud or gel with light shining through it. The program H_atom_prob_density.net_2 produces the electron cloud shown in Figure C.17. We see a fuzziness and a color change indicative of the variation of the density, but no abrupt edges. However, in the process we lose the impression of how the state fills space and of its rotational symmetry. In addition, our eye equates the fuzziness with a low-quality image, both because it is less sharp and because the colors are less intense. We now alleviate that shortcoming.

As we have seen with the electric field, visualizing the variation of a field along two planes is a good way to convey its 3-D nature. Because the electron cloud is rotationally symmetric about the z axis, we examine the variation along the xy and yz planes (the xz plane is equivalent to the yz one). Figure C.18 shows the visualization obtained with the visual program H_atom_prob_density.net_3. Here Slab modules are used to create the 2-D plots of the density on the two planes, and the planes are made semitransparent for ease of interpretation. This visualization is clearly an improvement over those in Figure C.17, both in sharpness and in conveying the 3-D nature of the density.

C.6 Animations with OpenDX

An *animation* is a collection of images called *frames* that when viewed in sequence convey the sensation of continuous motion. It is an excellent way to visualize the behavior in time of a simulation or of a function $f(\mathbf{x}, t)$, as might occur in wave

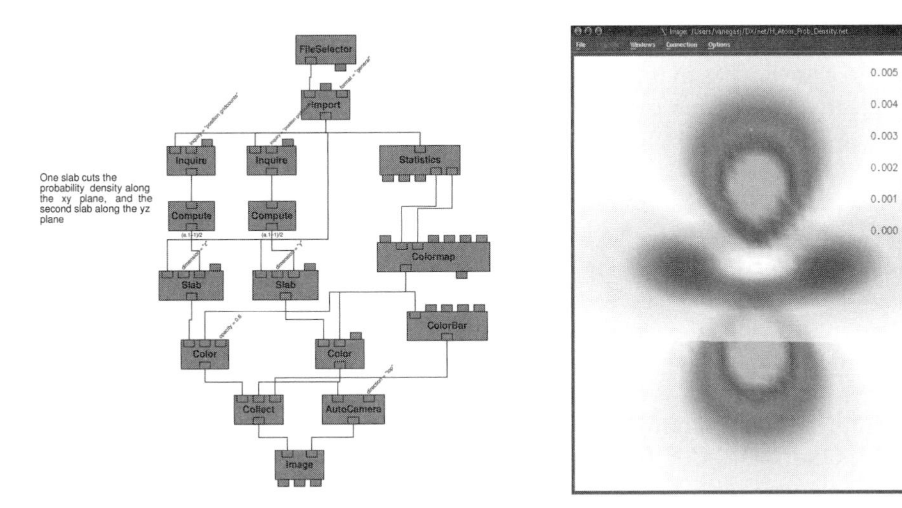

Figure C.18 The visual program on the left produces the 3-D probability density on the right. The density is plotted only on the surface of the *xy* and *yz* planes, with both planes made semitransparent.

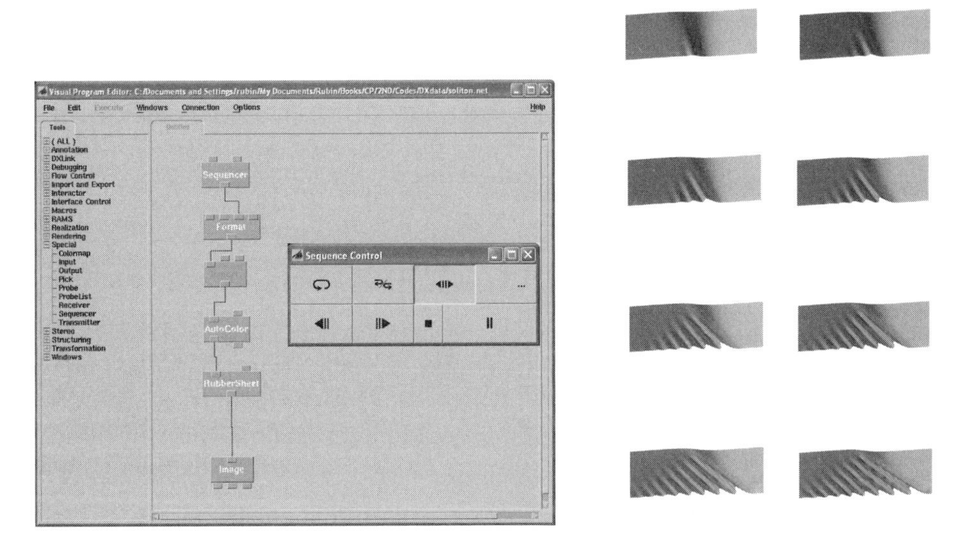

Figure C.19 *Left:* The visual DX program used to create an animation of the formation of a soliton wave. The arrows on the Sequence Control are used to move between the frames. *Right:* Nine frames from the DX animation.

motion or heat flow. In the Codes section of the CD, we give several sample animations of the figures in this book and we recommend that you try them out to see how effective animation is as a tool.

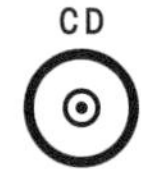

The easiest animation format to view is an *animated gif*, which can be viewed with any Web browser. (The file ends with the familiar .gif extender but has a series of images that load in sequence.) Otherwise, animations tend to be in multimedia/video file formats such as MPEG, DivX, mp3, yuv, ogg, and avi. Not all popular multimedia players such as *RealPlayer*, *Windows Media Player*, and *QuickTime player* play all formats, so you may need to experiment. We have found that the free multimedia players *VLC* and *ImageMagick* work well for various audio and video formats, as well as for DVDs, VCDs, and *streaming* protocols (the player starts before all the media are received).

A simple visual program to create animations of the formation of solitons (Chapter 19, "Solitons and Computational Fluid Dynamics") with OpenDX is shown on the left in Figure C.19. The new item here is the Sequencer module, which provides motion control (it is found under Tools/Special). As its name implies, Sequencer produces a sequence of integers corresponding to the frame numbers in the animation (the user sets the minimum, maximum, increment, and starting frame number). It is connected to the Format module, which is used to create file names given the three input values: (1) soliton, (2) the output of the Sequencer, and (3) a .general for the suffix. The Import module is used to read in the series of data files as the frame number increases.

On the right in Figure C.19 we show a series of eight frames that are merged to form the animation (a rather short animation). They are plots of the data files soliton001.dat, soliton002.dat, ... soliton008.dat, with one file for each frame and with a specified Field in each file imported at each time step. The Sequencer outputs the integer 2, and the output string from the Format becomes soliton002.general. This string is then input to Import as the name of the .general file to read in. In this way we import and image a whole series of files.

C.6.1 Scripted Animation with OpenDX

By employing the scripting ability of OpenDX, it is possible to generate multiple images with the Visual Editor being invoked only for the first image. The multiple images can then be merged into a movie. Here we describe the steps followed to create an .mpg movie:

1. Assemble a sequence of data files, one for each frame of the movie.
2. Use OpenDX to create an image file from the first data file. Also create .general and .net files.
3. Employ OpenDX with the –script option invoked to create a script. Read in the data file and the .net file and create a .jpg image. (Other image types are also possible, but this format produces good quality without requiring excessive disk space.)

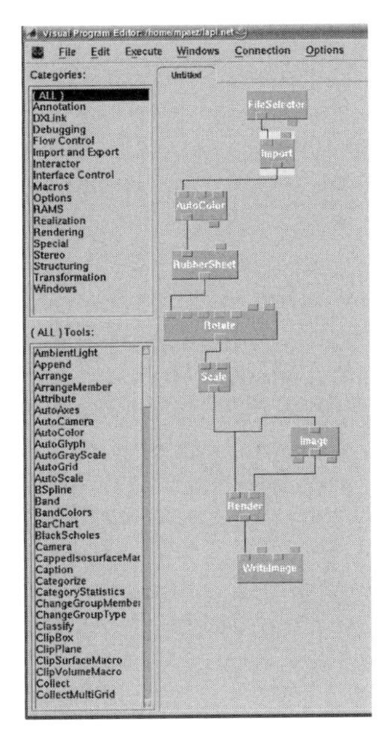

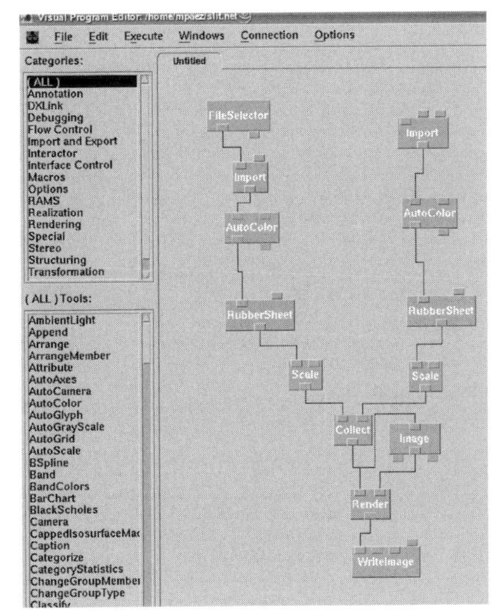

Figure C.20 *Left:* Use of the Visual Program Editor to produce the image of a wave packet that passes through a slit. *Right:* The visual program used to produce frames for an OpenDX animation.

4. Merge the .jpg image files into a movie by using a program such as Mplayer in Linux. We suggest producing the movie in the .avi or .mpg format.

To illustrate these steps, assume that you have a solution to Laplace's equation for a two-plate capacitor, with the voltage on the plates having a sinusoidal time dependence. Data files lapl000.dat, lapl001.dat, ..., lapl120.dat are produced in the Gnuplot 3-D format, each corresponding to a different time. To produce the first OpenDX image, lapl000.dat is copied to data.dat, which is a generic name used for importing data into OpenDX, and the file lapl.general describing the data is saved:

```
file = /home/mpaez/data.dat grid = 50 x 50 format = ascii
interleaving = field majority = column field = field0 structure =
scalar type = float dependency = positions positions = regular,
regular, 0, 1, 0, 1 end
```

Next, the Visual Program Editor is used to create the visual program (Figure C.20 right), and the program is saved as the file lapl.net.

To manipulate the first image so that the rest of them have the same appearance, we assembled the visual program in Figure C.20.

1. In the Control panel, double-click FileSelector and enter the path and name (lapl.general) of the .general file saved in the previous step.
2. The images should be plotted to the same scale for the movie to look continuous. This means you do *not* want autoscale on. Select AutoColor and expand the box. For min select −100 and for max select 100. Do the same in RubberSheet. This sets the scale for all the figures.
3. From the rotate icon, select axis, enter x, and in rotation enter −85. For the next axis button enter y, and in the next rotation button enter 10.0. These angles ensure a good view of the figure.
4. To obtain a better figure, on the Scale icon change the default scale from [1 1 1] to [1 4 3].
5. Next the internal DX script dat2image.sh is written by using the Script option.
6. Now we write an image to the disk. Select WriteImage and in the second box enter image as the name of a generic image. Then for format select tiff to create image.tiff.
7. You now have a script that can be used to convert other data files to images.

We have scripted the basic DX procedure for producing an image from data. To repeat the process for all the frames that will constitute a movie, we have written a separate shell script called dat2image.sh (it is on the CD) that

1. Individually copies each data file to a generic file data.dat, which is what DX uses as input with lapl.general.
2. Then OpenDX is called via
   dx -script lapl.net
   to generate images on the disk.
3. The *tiff* image image.tiff is transformed to the *jpg* format with names such as image.001. jpg, image.002. jpg, and so on (on Linux this is done with convert).
4. The Linux program MPlayer comes with an encoder called mencoder that takes the . jpg files and concatenates them into an .avi movie clip that you can visualize with Linux programs such as Xine, Kaffeine, and MPlayer. The last-mentioned program can be downloaded for Windows, but it works only in the System Symbol window or shell:

   > **mplayer -loop 0 output.avi**

   You will see the film running continuously. This file is called dat2image.sh and is given in Listing C.1. To run it in Linux,

   > **sh dat2image.sh**

```
do
if test $i -lt 10
    then
    cp ${prefix}00$i.dat data.dat
    dx -script lapl.net
    convert image.tiff image.00$i.jpg
```

```
    elif test $i -lt 100
        then
        cp ${prefix}0$i.dat data.dat
        dx -script lapl.net
        convert image.tiff image.0$i.jpg
    elif test $i -lt 1000
        then
        cp ${prefix}$i.dat data.dat
        dx -script lapl.net
        convert image.tiff image.$i.jpg
    fi
    i='expr $i + 1'
done
mencoder"mf:// image.*.jpg" -mf fps=25 -o output.avi -ovc lavc
                                        -lavcopts vcodec=mpeg4
```

Listing C.1 The shell script **dat2image.sh** creates an animation. The script is executed by entering its name at a shell prompt.

In each step of this shell script, the script copies a data file, for example, lapl014.dat in the fifteenth iteration, to the generic file data.dat. It then calls dx in a script option with the generic file lapl.net. An image image.14.tiff is produced and written to disk. This file is then transformed to image.14.jpg using the Linux convert utility. Once you have 120 .jpg files, Mplayer is employed to make the .avi animation, which can be visualized with the same Mplayer software or with any other that accepts .avi format, for example, Kaffeine, Xine, and MPlayer in Linux. In Windows you have to download MPlayer for Windows, which works only in the DOS shell (system symbol window). To use our program, you have to edit lapl.net to use your file names. Ensure that lapl.general and file data.dat are in the same directory.

C.6.2 Wave Packet and Slit Animation

On the left in Figure C.20 we show frames from an OpenDX movie that visualizes the solution of the Schrödinger equation for a Gaussian wave packet passing through a narrow slit. There are 140 frames (slit001.dat– slit140.dat), with each generated after five time iterations, with 45 $\times$ 30 data in a column. A separate program generates the data file SlitBarrier.dat that has only the potential (the barrier with the slit). The images produced by these programs are combined into one showing the wave packet and the potential. On the left side of the visual program:

- Double-click FileSelector and enter /home/mpaez/slit.general, the path, and the name of the file in the Control Panel under Import name.
- Double-click AutoColor and enter 0.0 for min; for max enter 0.8.
- Double-click RubberSheet and repeat the last step for min and max.
- Double-click the left Scale icon and change [111] to [113].
- Double-click WriteImage and write the name of the generic image slitimage and format = tiff.

On the right side of the visual program:

- Double-click Import and enter the path and the .general file name /home/mpaez/SlitBarrier.general.
- Double-click AutoColor and select 0.0 for min (use expand) and 0.8 for max.
- Double-click the right-hand Scale and enter [1 1 2]. The scale for the barrier is smaller than the one for the wave packet so that the barrier does not hide too much of the wave packet.

In this scheme we use the Image icon to manipulate the image to obtain the best view. This needs to be done just once.

Appendix D: An MPI Tutorial

In this appendix we present a tutorial on the use of MPI on a small Beowulf cluster composed of Unix or Linux computers.[1] *This follows our philosophy of "learning while doing." Our presentation is meant to help the user from the ground up, something that might not be needed if you were working at a central computing center with a reasonable level of support. Although your* ***problem*** *is still to take the program you have written to generate the bifurcation plot for bug populations and run different ranges of μ values simultaneously on several CPUs, in a more immediate sense your task is to get the experience of running MPI, to understand some of the MPI commands within the programs, and then to run a timing experiment. In §D.9 at the end of the appendix we give a listing and a brief description of the MPI commands and data types. General information about MPI is given in [MPI], detailed information about the syntax of MPI commands appears in [MPI2], and other useful material can be found in [MPImis]. The standard reference on the C language is [K&R 88], although we prefer [OR]. MPI is very much the standard software protocol for parallel computing and is at a higher level than its predecessor PVM [PVM] (which has its own tutorial on the CD).*

WHILE IN THE PAST we have run Java programs with a version of MPI, the difference in communication protocols used by MPI and Java have led to poor performance or to additional complications needed to improve performance [Fox 03]. In addition, you usually would not bother parallelizing a program unless it requires very large amounts of computing time, and those types of programs are usually written in Fortran or C (both for historical reasons and because Java is slower). So it makes sense for us to use Fortran or C for our MPI examples. We will use C because it is similar to Java.

D.1 Running on a Beowulf

A Beowulf cluster is a collection of independent computers each with its own memory and operating system that are connected to each other by a fast communication network over which messages are exchanged among processors. *MPI* is a library of commands that make communication between programs running on the different computers possible. The messages are sent as data contained in arrays. Because different processors do not directly access the memory on some other computer, when a variable is changed on one computer, it is *not* changed automatically in

[1] This material was developed with the help of Kristopher Wieland, Kevin Kyle, Dona Hertel, and Phil Carter. Some of the other materials derive from class notes from the Ohio Super Computer Center, which were written in part by Steve Gordon.

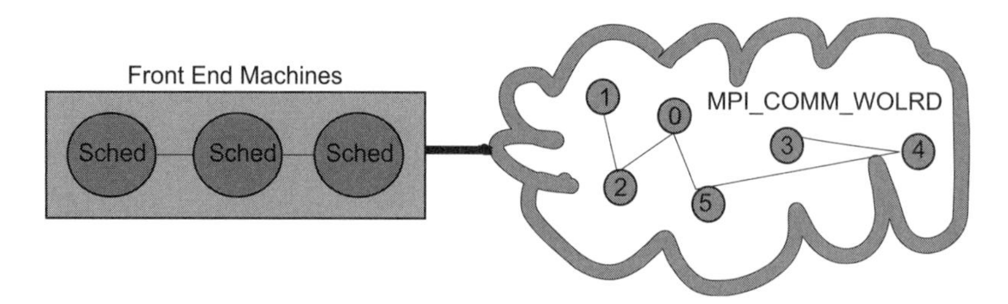

Figure D.1 A schematic view of a cluster (cloud) connected to front-end machines (box).

the other copies of the program running on other processors. This is an example of where MPI comes into play.

In Figure D.1 we show a typical, but not universal, configuration for a Beowulf cluster. Almost all clusters have the common feature of using MPI for communication among computers and Unix/Linux for the operating system. The cluster in Figure D.1 is shown within a cloud. The cloud symbolizes the grouping and connection of what are still independent computers communicating via MPI (the lines). The `MPI_COMM_WORLD` within the cloud is an MPI data type containing all the processors that are allowed to communicate with each other (in this case six). The box in Figure D.1 represents the *front end* or *submit hosts*. These are the computers from which users submit their jobs to the Beowulf and later work with the output from the Beowulf. We have placed the front-end computers outside the Beowulf cloud, although they could be within. This type of configuration frees the Beowulf cluster from administrative chores so that it can concentrate on number crunching, and is useful when there are multiple users on the Beowulf.

Finally, note that we have placed the letters "Sched" within the front-end machines. This represents a configuration in which these computers are also running some type of a *scheduler*, *grid engine*, or queueing system that oversees the running of jobs submitted to MPI by a number of users. For instance, if we have a cluster of 20 computers and user A requests 10 machines and user B requests 8 machines, then the grid engine will permit both users to run simultaneously and assign their jobs to different computers. However, if user A has requested 16 machines and user B 8, then the grid engine will make one of the users wait until the other finishes their work.

Some setup is required before you can run MPI on several computers at once. If someone has already done this for you, then you may skip the rest of this section and move on to § D.3. Our instructions have been run on a cluster of Sun computers running Solaris Unix (in a later section we discuss how to do this using the Torque scheduler on a Linux system). You will have to change the computer names and such for your purposes, but the steps should remain the same.

- First you need to have an active account on each of the computers in the Beowulf cluster.

- Open a shell on one of the machines designated for users to sign onto (the *front end*). You do not have to be sitting at the front end but instead can use ssh or *telnet*. Make the directory mpi in your home directory:

> **cd ~**	Change to home directory
> **mkdir output**	Screen output gets stored here first
> **mkdir output/error**	Place to store error messages
> **mkdir mpi**	A place to store your mpi stuff

- You need to have your Beowulf account configured so that Unix can find the MPI commands that you issue from the command line or from your programs. When you log onto the computer, the operating system reads a configuration file .cshrc residing in your home directory. It contains the places where the operating system looks for commands. (We are assuming here that you are using either cshell or tcshell, if not, then modify your .login, which should work regardless of the shell.) When a file name begins with a dot, it is usually hidden from view when you list the files, but it can be seen with the command ls -la. The list of places where Unix looks for commands is an *environmental variable* called PATH, and it should include the current version of the mpich-n.m/bin directory where the scripts for MPI reside. For us this is

  /usr/local/cluster/mpich-1.2.6/bin — This should be in your PATH

  Here the directory name cluster and 1.2.6 may need to be replaced by the name and number on your local system.
- Because the .cshrc file controls your environment, having an error in this file can lead to a nonfunctional computer. And since the format is rather detailed and unforgiving, it is easy to make mistakes. So before you work on your existing .cshrc file, make a backup copy of it:

  > **cp .cshrc .cshrc_bk**

  You can use this backup file as a reference or copy it back to .cshrc if things get to be too much of a mess. If you have really messed things up, your system administrator may have to copy the file back for you.
- Edit your .cshrc file so that it contains a line in which setenv PATH includes /usr/local/cluster/mpich-1.2.6/bin. If you do not have a .cshrc file, just create one. Find a line containing setenv PATH and add this in after one of the colons, making sure to separate the path names with colons. As an example, the .cshrc file for user rubin is

```
# @(#)cshrc 1.11 89/11/29 SMI umask 022
setenv PATH /usr/local/bin:/opt/SUNWspro/bin:/opt/SUNWrtvc/bin:
/opt/SUNWste/bin:/usr/bin/X11:/usr/openwin/bin:/usr/dt/bin:/usr/ucb/:
/usr/ccs/bin/:/usr/bin:/bin:/usr/sbin/:/sbin:
/usr/local/cluster/mpich-1.2.6/bin: setenv PAGER less setenv
CLASSPATH /home/rubin:/home/rubin/dev/java/chapmanjava/classes/:
/home/rubin/dev/565/javacode/:
/home/rubin/dev/565/currproj/:/home/rubin:/home/rubin/mpiJava:
/usr/local/mpiJava/lib/classes:
set prompt="%~::%m> "
```

- If you are editing your .login file, enter as the last line in the file:

```
set path = $path /usr/local/cluster/mpich-1.2.6/bin
```

- Because dot files are read by the system when you first log on, you will have to log off and back on for your changes to take effect. (Alternatively, you can use the source command to avoid logging off and on.) Once you have logged back on, check the values of the PATH environmental variable:

> **echo $PATH**	From Unix shell, tells you what Unix thinks

- Let us now take a look at what is done to the computers to have them run as a Beowulf cluster. On Unix systems the "slash" directory / is the root or top directory. Change the directory to /

> **cd /**	Change to root directory

  You should see files there, such as the kernel and the devices, that are part of the operating system. You should not modify these files, as that could cause real problems (it is the sort of thing that hackers and system administrators do).
- MPI is a *local* addition to the operating system. We have MPI and the *Sun Grid Engine* (SGE) in the /usr/local/cluster directory. Here the first / indicates the root directory and usr is the directory name under the root. Change the directory to /usr/local/cluster, or wherever MPI is kept on your system, and notice the directories scripts and mpich-1.2.6 (or maybe just a link to mpich). Feel free to explore these directories. The directory scripts contains various scripts designed to make running your MPI programs easier. (Scripts are small programs containing shell commands that are executed in order when the file is run.)
- In the mpich-1.2.6 directory you will notice that there are examples in C, C++, and Fortran. Feel free to copy these to your home directory and try them:

```
> cp -r examples /home/userid/mpi
```

  where userid is your name. We encourage you to try out the examples, although some may need modification to work on your local system.
- Further documentation can be found in

/usr/local/cluster/mpich-1.2.6/doc/mpichman-chp4.pdf	MPI documentation
/usr/local/cluster/sge/doc/SGE53AdminUserDoc.pdf	SGE documentation
/usr/local/cluster/sge/doc/SGE53Ref.pdf	SGE reference
man qstat	Manual page on qstat

- Copy the script run_mpi.sh from the Codes/MPIcodes directory on the CD to your personal mpi directory. This script contains the commands needed to run a program on the cluster.
- Copy the file /usr/local/cluster/mpich/share/machines.solaris to your home directory and examine it. (The solaris extender is there because we are using the *Solaris* version of the Unix operating system on our Beowulf; you may need to change this for your local system.) This file contains a list

of all the computers that are on the Beowulf cluster and available to MPI for use (though there is no guarantee that all the machines are operative):

```
# Change this file to contain the machines that you want to use
# to run MPI jobs on. Format: 1 host per line ,  either hostname
# or hostname:n, where n is the number of processors.
# hostname should be the same as        output from "hostname" command
paul
rose
tomek
manuel
```

D.2 Running MPI

If you are the only one working on a Beowulf cluster, then it may make sense to submit your jobs directly to MPI. However, if there is the possibility that a number of people may be using the cluster, or that you may be submitting a number of jobs to the cluster, then it is a good idea to use some kind of a queue management system to look after your jobs. This can avoid the inefficiency of having different jobs compete with each other for time and memory or having the entire cluster "hang" because a job has requested a processor that is not available. In this section we describe the use of the *Sun Grid Engine* [SGE]. In a later section we describe the use of the *Torque/Portable Batch System* (PBS) scheduler on a Linux system; the two are similar in purpose and commands, work under many operating systems, and are free.

On the left in Figure D.2 is a schematic view of how a C program containing MPI commands is executed. On the right in this figure is a schematic view of how a scheduling system takes an executable program and runs it under MPI on several systems. When a program is submitted to a cluster via a management system, the system installs a copy of the same program on each computer assigned to run the program.

There are a number of scripts that interpret the MPI commands you give within your programs (the commands are not part of the standard Fortran or C language), and then call the standard compilers. These scripts are called *wrappers* because they surround the standard compilers as a way of extending them to include MPI commands:

mpicc	C compiler	**mpicxx**	C++ compiler
mpif77	Fortran 77 compiler	**mpif90**	Fortran 90 compiler
mpiifort	Intel Fortran compilers	**mpiicc**	Intel C compiler

Typically you compile your programs on the front end of the Beowulf, or the master machines, but not on the execution nodes. You use these commands just as you use regular compiler commands, only now you may include MPI commands in your source program:

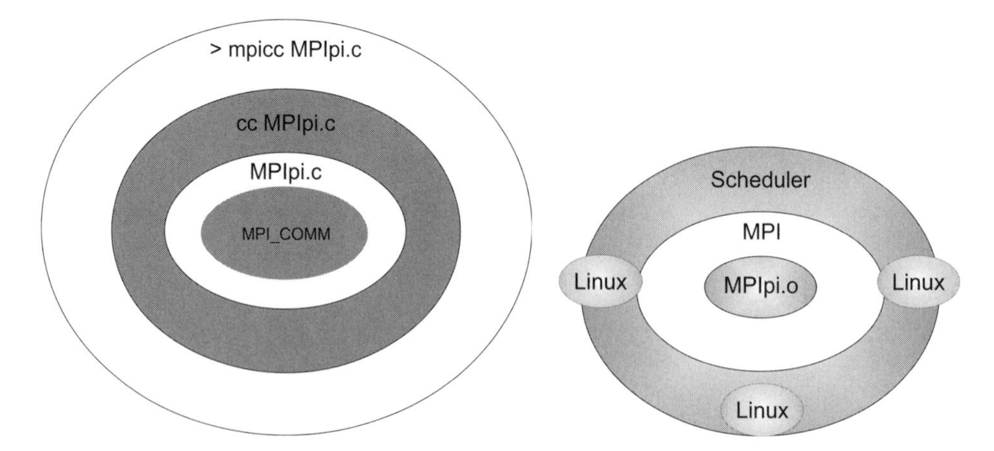

Figure D.2 *Left:* A schematic view of the MPI command `MPI_COMM` contained within the C program `MPI.c`. On the outer wrapper, the program is compiled with the shell command `mpicc`, which expands the MPI commands and invokes the C compiler `cc`. *Right:* A schematic view of how a scheduler runs the executable program `MPIpi.o` under MPI and on several Linux CPUs.

> **mpicc –o name name.c**	Compile `name.c` with MPI wrapper script
> **mpif77 –o name name.f**	Compile `name.f` with MPI wrapper script

D.2.1 MPI under the SGE Queueing System

Table D.1 lists some of the key number of Sun grid engine commands used to execute compiled programs. Other queueing systems have similar commands. The usual method of executing a program at the prompt runs only on the local computer. In order to run on a number of machines, the program needs to be submitted to the queue management system. We show this in Listing D.1, which uses the `run_mpi.sh` script and the `qsub` command to submit jobs to run in *batch* mode:

> **qsub run_mpi.sh name**	Submit `name` to run on cluster

This command returns a job ID number, which you should record to keep track of your program. Note that in order for this script to work, both `runMPI.sh` and the program `name` must be in the current directory. If you need to pass parameters to your program, place the program name as well as the parameters in quotes:

> **qsub run_mpi.sh "name -r 10"**	Parameters and program name in quotes

TABLE D.1
Some Common SGE Commands

Command	Action
qsub myscript	Submit batch script or job **myscript**
qhost	Show job/host status
qalter <job_id>	Change parameters for job in queue
qdel job_id	Remove job_id
qstat	Display status of batch jobs
qstat -f	Full listing for qstat
qstat -u <username>	User only for qstat
qmon	X-Window front end (integrated functionality)

```
# You may want to modify the parameters for
# "-N" (job queue name), "-pe" (queue type and number of requested CPUs),
# "myjob" (your compiled executable).


# You can compile you code, for example myjob.c (*.f), with GNU mpicc or
#  mpif77 compilers as follows:
# "mpicc -o myjob myjob.c" or "mpif77 -o myjob myjob.f"

# You can monitor your jobs with command
# "qstat -u your_username" or "qstat -f" to see all queues.
# To remove your job, run "qdel job_id"
# To kill running job, use "qdel -f job_id"

# ------Attention: #$ is a special CODINE symbol, not a comment -----
#
#   The name, which will identify your job in the queue system
#$ -N MPI_job
#
#   Queue request, mpich. You can specify the number of requested CPUs,
#   for example, from 2 to 3
#$ -pe class_mpi 4-6
#
# ---------------------------
#$ -cwd
#$ -o  $HOME/output/$JOB_NAME-$JOB_ID
#$ -e  $HOME/output/error/$JOB_NAME-$JOB_ID.error
#$ -v  MPIR_HOME=/usr/local/cluster/mpich-1.2.6
# ---------------------------

echo "Got $NSLOTS slots."


#  Don't modify the line below if you don't know what it is
$MPIR_HOME/bin/mpirun -np $NSLOTS $1
```

Listing D.1 The script **runMPI.sh** used to run an MPI program.

D.2.1.1 STATUS OF SUBMITTED PROGRAMS

After your program is successfully submitted, SGE places it in a queue where it waits for the requested number of processors to become available. SGE then executes the program on the cluster and *directs the output to a file* in the output subdirectory within your home directory. The program itself uses MPI and C/Fortran commands. In order to check the status of your submitted program, use qstat along with your job ID number:

```
> qstat 1263                                              Tell me the status of Job 1263
job-ID prior name user state submit/start at queue master ja-task-ID
1263 0 Test_MPI_J dhertel qw 07/20/2005 12:13:51
```

This is a typical qstat output. The qw in the state column indicates that the program is in the **q**ueue and **w**aiting to be executed.

```
> qstat 1263                                              Same as above, but at later time
job-ID prior name user state submit/start at queue master ja-task-ID
1263 0 Test_MPI_J dhertel t 07/20/2005 12:14:06 eigen11.q MASTER
1263 0 Test_MPI_J dhertel t 07/20/2005 12:14:06 eigen11.q SLAVE
1263 0 Test_MPI_J dhertel t 07/20/2005 12:14:06 eigen12.q SLAVE
1263 0 Test_MPI_J dhertel t 07/20/2005 12:14:06 eigen3.q SLAVE
1263 0 Test_MPI_J dhertel t 07/20/2005 12:14:06 eigen5.q SLAVE
1263 0 Test_MPI_J dhertel t 07/20/2005 12:14:06 eigen8.q SLAVE
```

Here the program has been assigned a set of nodes (eigenN is the name of the computers), with the last column indicating whether that node is a master, host, slave, or guest (to be discussed further in §D.3.1). At this point the state column will have either a t, indicating transfer, or an r, indicating running.

The **output** from your run is sent to the file Test_MPI.<jobID>.out in the output subdirectory within your home directory. Error messages are sent to a corresponding file in the error subdirectory. Of course you can still output to a file in the current working directory, as well as **input** from a file.

D.2.2 MPI Under the Torque/PBS Queueing System

Most Beowulf clusters use Linux, a version of the Unix operating system that runs on PC architecture. A popular, commercially supported version of Linux that runs well for CPUs using 64-bit words is *SUSE* [SUSE]. We have used this setup with MPI libraries, Intel compilers, and the cluster edition of the Math Kernel Library [Intel]. Although we could run the SGE scheduler and resource manager on this system, the compilers come with the Torque open source resource manager [Torque], which works quite well. Torque is based on and uses the same commands as Portable Batch System [PBS]. In this section we give a tutorial on the use of Torque for submitting jobs to MPI.[2] As we shall see, the steps and commands are very similar to those for SGE and follow the system outlined in Figure D.2.

D.2.2.1 RUNNING SERIAL JOBS WITH TORQUE

Sometimes you may have a serial job that runs too long for you to wait for its completion, or maybe you want to submit a number of long jobs and not have them compete with each other for resources. Either case can be handled by using the queueing system usually used for parallel processing, only now with multiple jobs on just one computer. In this case there are no MPI commands to deal with, but just three torque commands and a shell script that initiates your program:

qsub	Submit jobs to queue via Torque
qstat	Check status of jobs in queue
qdel	Delete a submitted job
script	Shell script that initiates program

Note that you cannot give Torque a complied binary program to run but rather just a shell script[3] that calls your executable. This is probably for security and reliability. For example:

> **icc SerProg.c –o SerProg**	Intel C Compiler, out to SerProg
> **qsub SerProg**	Submit SerProg to queue
qsub: file must be an ascii script	Torque's output
> **qsub script1**	Submit a script that calls SerProg
JobID = 12; output from SerProg	Successful submission

Here is a simple script1 for initiating the serial job in the file SerProg (you should copy this into a file so you can use it):

```
#!/bin/bash
cd $PBS_O_WORKDIR
./SerProg
```

Observe the #!/bin/bash statement at the beginning of the script. A statement of this form is required to tell the operating system which shell (command line interpreter) to use. This line is for the bash shell, with other choices including tcsh and ksh. The next command cd $PBS_O_WORKDIR (which you should *not* modify) sets a PBS environmental variable so that the script looks in the current working directory.

[2] We thank Justin Elser for setting up this system and for preparing the original form of this section.

[3] Recall that a shell script is just a file containing commands that would otherwise be entered at a shell's prompt. When the file name is given execution permission and entered as a command, all the commands within the file are executed.

The last command, ./SerProg, contains the name of the file you wish to execute and is the only line you should modify. (The ./ means the current directory, but you can also use ../SerProg for a file one directory up.)

As we indicated before, once you have submitted a job to Torque, you can log off (to get a mug of your favorite beverage) and the job will remain in the compute queue or continue to run. Alternatively, you may want to submit the job to a cluster machine different from the one on which you are working, log off from that one, and then continue working on your present machine. The job will run remotely with other users still able to run on that remote machine. (Other than a possible slowdown in speed, they may not even realize that they are sharing it with you.) Before you log off, it is probably a good idea to determine the status of your jobs in the queue. This is done with the qstat command:

```
> qstat                                                        What is queue's status?
JobID      Name          User      Time Use    S    Queue      Reply
170.phy    ScriptName    Justin       0        R    batch
> qdel 170                                                     Delete (kill) my job
```

This output indicates that Justin's job has id 170 (the .phy indicates the server phy), that it was submitted via the script ScriptName, that it is in the state R for running (Q if queued, E if executing), and that it is in the batch queue (default). You can delete *your* job by issuing the qdel command with your JobID.

If your program normally outputs to a file, then it will still output to that file under Torque. However, if your program outputs to stdout (the screen), the output will be redirected to the file ScriptName.oJobID and any errors will be redirected to the file ScriptName.eJobID. Here ScriptName is the shell script you used to submit the job, and JobID is the ID given to it by Torque. For example, here is the output from the long list command ll:

```
> ll                                                     Unix long list command
total 32
-rwxr-xr-x 1 q users 17702 2006-11-15 16:38 SerProg
-rw-r–r– 1 q users 1400 2006-11-15 16:38 SerProg.c
-rw-r–r– 1 q users 585 2006-11-15 16:39 ScriptName
-rw—— 1 q users 0 2006-11-15 16:39 ScriptName.e164
-rw—— 1 q users 327 2006-11-15 16:39 ScriptName.o164
```

D.2.3 Running Parallel Jobs with Torque

The basic Torque steps for submitting a script and ascertaining its status are the same for parallel jobs as for serial jobs, only now the script must have more commands in it and must call MPI. The first thing that must be done is to create a Multiprocessors Daemon (MPD) *secretword* (*not* your password) so that Torque can keep multiple jobs separate from all the others:

> **cd $HOME** Dotfiles stored in your home directory
> **echo "MPD_SECRETWORD=MySecretword" » .mpd.conf** Replace MySecretword
> **chmod 600 .mpd.conf** Change permission for file .mpd.conf

Note that you do not really need to remember your secretword because Torque uses it "behind the scenes"; you just need to have it stored in the file .mpd.conf in your home directory. (This file is not normally visible because it begins with a dot.)

```
#!/bin/bash
#
# All lines starting with "#PBS" are PBS commands
#
# Request 2 nodes with 2 processor per node (ppn) (= 4 processors)
# ppn can either be 1 or 2
#
#PBS -l nodes=2:ppn=2
#
# Set wall clock max time to 0 hours, 15 minutes and 0 seconds
#PBS -l walltime=00:15:00
#
# cd to working directory
cd $PBS_O_WORKDIR
# name of executable
myprog=MPIpi
#
# Number of processors is $NP
NP=4
#
# Run MYPROG with appropriate mpirun script
mpirun -r ssh -n $NP $myprog
#
# make sure to exit the script, else job won't finish properly
exit 0
```

Listing D.2 The script **TorqueScript.sh** used to submit an MPI program.

In Listing D.2 and on the CD we give the script TorqueScript.sh used to submit MPI jobs to the Torque scheduler. The script is submitted to Torque from a shell via the qsub command:

> **qsub TorqueScript.sh**

Observe again that the script must start with the line #!/bin/bash, to indicate which shell it should run under, and that the lines beginning with #PBS are commands to the Torque scheduler, not comments! The lines

myprog=MPIpi
mpirun -r ssh -n $NP $myprog

in the script run the compiled version of the program MPIpi.c, which is also on the CD. Here NP is the total number of processors you want to run on, and its value is written into the script. You will need to change MPIpi to the name of the compiled program you wish to run. (As long as the line $PBS_O_WORKDIR precedes this one,

Torque knows that your program is in the working directory.) The line

#PBS -l nodes=2:ppn=2

tells Torque to *reserve* two computers (nodes) with two processors per node (ppn) for a total of four processors. This value of ppn is appropriate for *dual-core* computers with two CPUs on their chips. If you have computers with four cores on each chip, then you should set ppn=4. If you want to use only one processor per machine, for example, to gauge the speedup from multiple cores, then set ppn=1. Even though we have reserved nodes*ppn processors for Torque to use, the actual number of processors used by MPI is given by the variable NP in the call

mpirun -r ssh -n NP **myprog**

Accordingly, we must set the value for NP as $\leq$nodes*ppn within the script. The maximum wall clock time that your job can run is set to 15 min via

#PBS -l walltime=00:15:00

This is actual run time, not the total time in the queue, and so this clock does not start ticking until the job starts executing. In general it is a good idea to use a walltime command with about twice the time you expect your job to run just in case something goes wrong. (Not only is this a nice thing to do for others, but it can also keep you from wasting a finite resource or waiting around forever.) Next observe that the mpirun command, which starts MPI, has the argument -r ssh. This is required on our installation for the machines to be able to communicate with each other using ssh and scp rather than the default rsh and rcp. The latter are less secure. Finally, the script ends with exit 0. This gives the script exit status 0 and thus provides a graceful ending. Graceless endings may not give the operating system enough time to clear your output files from buffers, which means that you may not see all your output!

D.3 Your First MPI Program: MPIhello.c

Listing D.3 gives the simple MPI program MPIhello.c. It has each of the processors print Hello World, followed by the processor's rank. Compile MPIhello.c using:

> **mpicc MPIhello.c -o hello**	Compilation via compiler wrapper

After successful compilation, an executable file hello should be in the directory in which you did the compilation. The program is executed via the script run_mpi.sh either directly or by use of the management command qsub:

> **run_mpi.sh hello**	Run directly under MPI
> **qsub run_mpi.sh hello**	Run under management system

This script sets up the running of the program on the 10 processors, with processor 0 the host and processors 1–9 the guests.

```
//  MPIhello.c          has each processor prints hello to screen
#include "mpi.h"
#include <stdio.h>

int main( int argc, char *argv[] ) {
  int myrank;
  MPI_Init( &argc, &argv );                                    // Initialize MPI
  MPI_Comm_rank( MPI_COMM_WORLD, &myrank );                    // Get CPU's rank
  printf( "Hello World from processor %d\n", myrank );
  MPI_Finalize( );                                             // Finalize MPI
  return 0;
}
```

Listing D.3 MPIhello.c gets each processor to say hello via MPI.

D.3.1 MPIhello.c Explained

Here is what is contained in MPIhello.c:

- The inclusion of MPI headers via the #include "mpi.h" statement on lines 2–3. These are short files that assist the C compiler by telling it the type of arguments that MPI functions use for input and output without giving any details about the functions. (In Fortran we used include "/usr/local/cluster/mpich-2.1.6/include/mpif.h" after the program line.)
- The main method is declared with an int main(int argc, char *argv[]) statement, where argc is a pointer to the number of arguments and argv is a pointer to the argument vector passed to main when you run the program from a shell. (*Pointers* are variable types that give the locations in memory where the values of the variables reside rather than the variables' actual values.) These arguments are passed to MPI to tell it how many processors you desire.
- The int myrank statement declares the variable myrank, which stands for the *rank* of the computer. Each processor running the program is assigned a unique number called its *rank* by MPI. This is how you tell the difference among identical programs running on different CPUs.
- The processor that executes the highest level of the program is called the *host* or *master*, and all other machines are called *guests* or *slaves*. The host always has myrank $= 0$, while all the other processors, based on who responds first, have their processor numbers assigned to myrank. This means that myrank $= 1$ for the first guest to respond, 2 for the second, and so on. Giving each processor a unique value for myrank is a critical element in parallel processing.
- The MPI_init() and MPI_Finalize() commands in MPIhello.c initialize and terminate MPI, respectively. All MPI programs must have these lines, with the MPI commands always placed between them. The MPI_Init(&argv, &argc) function call takes two arguments, both beginning with a & that indicates a pointer. These arguments are used for communication between the operating system and MPI.

- The MPI_Comm_rank(MPI_COMM_WORLD, &myrank) call returns a different value for rank for each processor running the program. The first argument is a predefined constant telling MPI which grouping of processors to communicate with. Unless you have set up groups of processors, just use the default *communicator* MPI_COMM_WORLD. The second argument is an integer that is returned with the rank of the individual program.

When MPIhello.c is executed, each processor prints its rank to the screen. Notice that it does not print the ranks in order and that the order will probably be different each time you run the program. Take a look at the output (in the file output/MPI_job-xxxx). It should look something like this:

```
"Hello, world!" from node 3 of 4 on eigen3.science.oregonstate.local
"Hello, world!" from node 2 of 4 on eigen2.science.oregonstate.local
Node 2 reporting
"Hello, world!" from node 1 of 4 on eigen1.science.oregonstate.local
"Hello, world!" from node 0 of 4 on eigen11.science.oregonstate.local
```

If the processing order matters for proper execution, call MPI_Barrier(MPI_COMM_WORLD) to synchronize the processors. It is similar to inserting a starting line at a relay race; a processor stops and waits at this line until all the other processors reach it, and then they all set off at the same time. However, modern programming practice suggests that you try to design programs so that the processors do not have to synchronize often. Having a processor stop and wait obviously slows down the number crunching and consequently removes some of the advantage of parallel computing. However, as a scientist it is more important to have correct results than fast ones, and so do not hesitate to insert barriers if needed.

Exercise: Modify MPIhello.c so that only the guest processors say hello. *Hint:* What do the guest processors all have in common? ▮

D.3.2 Send/Receive Messages: MPImessage2.c

Sending and receiving data constitute the heart of parallel computing. Guest processors need to transmit the data they have processed back to the host, and the host has to assemble the data and then assign new work to the guests. An important aspect of MPI communication is that if one processor sends data, another processor must receive those data. Otherwise, the sending processor may wait indefinitely for a signal that its data have been received or the receiving processor may wait indefinitely until it receives the data it is programmed to expect.

Argument Name	Use in **MPI_Send** and **MPI_Recv**
msg	Pointer (& in front) to array to be sent/received
msg_size	Size of array sent; may be bigger than actual size
MPI_TYPE	Predefined constant indicating variable type within array, other possible constants: MPI_INTEGER, MPI_DOUBLE
dest	Rank of processor receiving message
tag	Number that uniquely identifies a message
comm	A *communicator*, for example, predefined constant MPI_COMM_WORLD
source	Rank of processor sending message; if receiving messages from any source, use predefined constant MPI_ANY_SOURCE
status	Pointer to variable type MPI_Status containing status info

```
//  MPImessage2.c:  source node sends message to dest
#include "mpi.h"
#include <stdio.h>

int main(int argc,char *argv[])  {
  int rank, msg_size = 6, tag = 10, source = 0, dest = 1;

  MPI_Status status;
  MPI_Init( &argc, &argv );                                     // Initialize MPI
  MPI_Comm_rank( MPI_COMM_WORLD, &rank );                       // Get CPU's rank
  if ( rank == source ) {
    char *msg = "Hello";
    printf("Host about to send message: %s\n",msg);  // Send, may block till recieved
    MPI_Send( msg, msg_size, MPI_CHAR, dest, tag, MPI_COMM_WORLD );
  }
    else if ( rank == dest ) {
      char buffer[msg_size+1];                                          // Receive
      MPI_Recv( &buffer, msg_size, MPI_CHAR, source, tag, MPI_COMM_WORLD, &status );
      printf("Message recieved by %d: %s\n", rank, buffer);
    }
      printf("NODE %d done.\n", rank);                         //  All nodes print
      MPI_Finalize();                                               // Finalize MPI
      return 0;
}
```

Listing D.4 MPImessage2.c uses MPI commands to both send and receive messages. Note the possibility of blocking, in which the program waits for a message.

There is a basic MPI command MPI_Send to send a message from a *source* node, and another basic command MPI_Recv is needed for a *destination* node to receive it. The message itself must be an array even if there is only one element in the array. We see these commands in use in MPImessage2.c in Listing D.4. This program accomplishes the same thing as MPIhello.c but with send and receive commands.

The host sends the message and prints out a message, while the guests print out when they receive a message. The forms of the commands are

MPI_Send(msg, msg_size, MPI_TYPE, dest, tag, MPI_COMM_WORLD); Send
MPI_Recv(msg, msg_size, MPI_TYPE, source, tag, comm, status); Receive

The arguments and their descriptions are given in §D.3.2. The criteria for successfully sending and receiving a message are

1. The sender must specify a valid destination rank, and the processor with that rank must call MPI_recv.
2. The receiver must specify a valid source rank or MPI_ANY_SOURCE.
3. The send and receive *communicators* must be the same.
4. The tags must match.
5. The receiver's message array must be large enough to hold the array.

Exercise: Modify MPImessage2.c so that all processors say hello.

D.3.3 Receive More Messages: MPImessage3.c

```
// MPImessage3.c: guests send rank to the host, who prints them
#include "mpi.h"
#include <stdio.h>

int main(int argc,char *argv[])  {
  int rank, size, msg_size = 6, tag = 10, host = 0, n[1], r[1], i;      // 1-D Arrays

  MPI_Status status;
  MPI_Init( &argc, &argv );                                            // Initialize MPI
  MPI_Comm_rank( MPI_COMM_WORLD, &rank );                              // Get CPU's rank
  MPI_Comm_size( MPI_COMM_WORLD, &size );                          // Get number of CPUs
  if ( rank != host ) {
    n[0] = rank;
    printf("node %d about to send message\n", rank);
    MPI_Send( &n, 1, MPI_INTEGER, host, tag, MPI_COMM_WORLD );
  }
  else  {
      for ( i = 1; i < size; i++ )  {
      MPI_Recv( &r, 1, MPI_INTEGER, MPI_ANY_SOURCE, tag, MPI_COMM_WORLD, &status );
      printf("Message recieved: %d\n", r[0]);  }
  }
  MPI_Finalize();                                                  // Finalize MPI
  return 0;
}
```

Listing D.5 MPImessage3.c contains MPI commands that have each guest processor send a message to the host processor who then prints out the rank of that guest.

A bit more advanced use of message passing is shown by MPImessage3.c in Listing D.5. Here each guest sends a message to the host who then prints out the rank of the guest that sent the message. The host loops through all the guests since otherwise it would stop looking for more messages after the first one arrives. The host calls MPI_Comm_size to determine the number of processors.

D.3.4 Broadcast Messages

If we used the same technique to send a message from one node to several other nodes, we would have to loop over calls to MPI_Send. In MPIpi.c in Listing D.6, we see an easy way to send a message to all the other nodes.

```
// MPIpi.c computes pi in parallel by stone throwing
#include "mpi.h" #include <stdio.h> #include <math.h>

double f(double);

int main(int argc,char *argv[]) {
  int n, myid, numprocs, i, namelen;
  double PI25DT = 3.141592653589793238462643;
  double mypi, pi, h, sum, x, startwtime=0., endwtime;
  char   processor_name[ MPI_MAX_PROCESSOR_NAME ];

  MPI_Init( &argc, &argv);
  MPI_Comm_size( MPI_COMM_WORLD, &numprocs);
  MPI_Comm_rank( MPI_COMM_WORLD, &myid);
  MPI_Get_processor_name( processor_name, &namelen);
  fprintf(stdout,"Process %d of %d is on %s\n", myid, numprocs, processor_name);
  fflush( stdout );
  n = 10000;                                          // Default # of rectangles
  if ( myid == 0 ) startwtime = MPI_Wtime();
  MPI_Bcast( &n, 1, MPI_INT, 0, MPI_COMM_WORLD );
  h   = 1./(double) n;
  sum = 0.;
  for ( i = myid + 1; i <= n; i += numprocs ) {       // Better if worked back
    x = h * ((double)i - 0.5);
    sum += f(x);
  }
  mypi = h * sum;
  MPI_Reduce( &mypi, &pi, 1, MPI_DOUBLE, MPI_SUM, 0, MPI_COMM_WORLD );
  if (myid == 0) {
    endwtime = MPI_Wtime();
    printf("pi is approximately %.16f, Error is %.16f\n", pi, fabs(pi - PI25DT));
    printf("wall clock time = %f\n", endwtime-startwtime);
    fflush(stdout);
  }
  MPI_Finalize();
  return 0;
}

double f(double a) { return (4./(1. + a*a));}        // Function f(a)
```

Listing D.6 MPIpi.c uses a number of processors to compute π by a Monte Carlo rejection (stone throwing).

This simple program computes π in parallel using the Monte Carlo "stone throwing" technique discussed in Chapter 5, "Monte Carlo Simulation." Notice the new MPI commands:

- **MPI_Wtime** is used to return the wall time in seconds (the time as given by a clock on the wall). This is useful when computing speedup curves (Figure D.3).

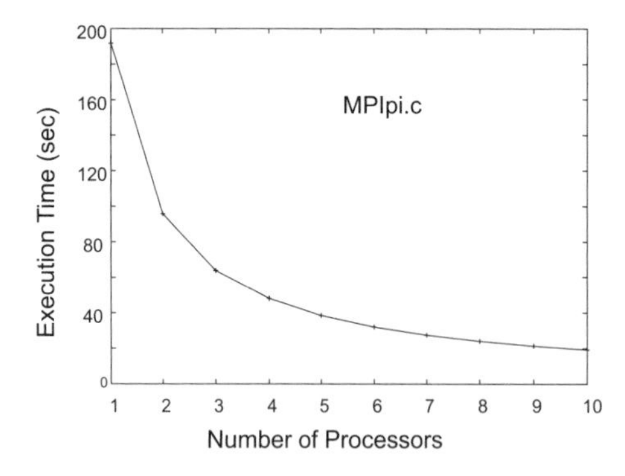

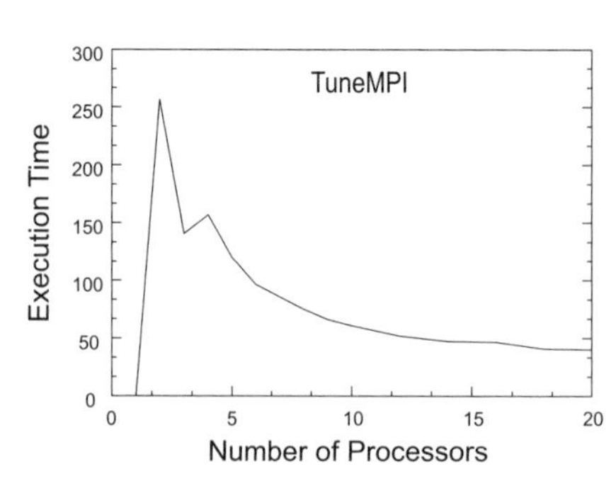

Figure D.3 Execution time *versus* number of processors. *Left:* For the calculation of π with MPIpi.c. *Right:* For the solution of an eigenvalue problem with TuneMPI.c. Note that the single-processor result here does *not* include the overhead for running MPI.

- **MPI_Bcast** sends out data from one processor to all the others. In our case the host broadcasts the number of iterations to the guests, which in turn replace their current values of n with the one received from the host.
- **MPI_Allreduce** is a glorified broadcast command. It collects the values of the variable mypi from each of the processors, performs an operation on them with MPI_SUM, and then broadcasts the result via the variable pi.

D.3.5 Exercise

On the left in Figure D.3 we show our results for the speedup obtained by calculating π in parallel with MPIpi.c. This exercise leads you through the steps required to obtain your own speedup curve:

1. Two versions of a parallel program are possible. In the *active host* version the host acts just like a guest and does some work. In the *lazy host* version the host does no work but instead just controls the action. Does MPIpi.c contain an active or a lazy host? Change MPIpi.c to the other version and record the difference in execution times.
2. Make a plot of the time *versus* the number of processors required for the calculation of π.
3. Make a *speedup* plot, that is, a graph of the computation time divided by the time for one processor *versus* the number of processors.
4. Record how long each of your runs takes and how accurate the answers are. Does round-off error enter in? What could you do to get a more accurate value for π?

D.4 Parallel Tuning

Recall the Tune program with which we experimented in Chapter 14, "High-Performance Computing Hardware, Tuning, and Parallel Computing," to determine how memory access for a large matrix affects the running time of programs. You may also recall that as the size of the matrix was made larger, the execution time increased more rapidly than the number of operations the program had to perform, with the increase coming from the time it took to transfer the needed matrix elements in and out of central memory.

Because parallel programming on a multiprocessor also involves a good deal of data transfer, the Tune program is also a good teaching tool for seeing how communication costs affect parallel computations. Listing D.7 gives the program TuneMPI.c, which is a modified version of the Tune program in which each row of the large-matrix multiplication is performed on a different processor using MPI:

$$[H]_{N\times N} \times [\Psi]_{N\times 1} = \begin{bmatrix} \Rightarrow & \text{rank 1} & \Rightarrow \\ \Rightarrow & \text{rank 2} & \Rightarrow \\ \Rightarrow & \text{rank 3} & \Rightarrow \\ \Rightarrow & \text{rank 1} & \Rightarrow \\ \Rightarrow & \text{rank 2} & \Rightarrow \\ \Rightarrow & \text{rank 3} & \Rightarrow \\ \Rightarrow & \text{rank 1} & \Rightarrow \\ & \ddots & \end{bmatrix}_{N\times N} \times \begin{bmatrix} \psi_1 & \Downarrow \\ \psi_2 & \Downarrow \\ \psi_3 & \Downarrow \\ \psi_4 & \Downarrow \\ \psi_5 & \Downarrow \\ \psi_6 & \Downarrow \\ \psi_7 & \Downarrow \\ \ddots & \end{bmatrix}_{N\times 1} . \tag{D.4.1}$$

Here the arrows indicate how each row of H is multiplied by the single column of Ψ, with the multiplication of each row performed on a different processor (rank). The assignment of rows to processors continues until we run out of processors, and then it starts all over again. Since this multiplication is repeated for a number of iterations, this is the most computationally intensive part of the program, and so it makes sense to parallelize it.

On the right in Figure D.3 is the speedup curve we obtained by running TuneMPI.c. However, even if the matrix is large, the Tune program is not computationally intensive enough to overcome the cost of communication among nodes inherent in parallel computing. Consequently, to increase computing time we have inserted an inner for loop over k that takes up time but accomplishes nothing (we've all had days like that). Slowing down the program should help make the speedup curve more realistic.

```
/* TuneMPI.c: a matrix algebra program to be tuned for performace
 N X N Matrix speed tests using MPI  */

#include "mpi.h"
#include <stdio.h>
#include <time.h>
#include <math.h>
```

```
int main(int argc,char *argv[]) {
  MPI_Status status;
  time_t systime_i, systime_f;
  int N = 200, MAX = 15, h = 1, myrank, nmach, i, j, k, iter = 0;
  long difftime = 0l;
  double ERR = 1.0e-6, dummy = 2., timempi[2], ham[N][N], coef[N], sigma[N];
  double ener[1], err[1], ovlp[1], mycoef[1], mysigma[1], myener[1], myerr[1];
  double myovlp[1], step = 0.0;
                                                      // MPI Initialization
  MPI_Init(&argc, &argv);
  MPI_Comm_rank( MPI_COMM_WORLD, &myrank );
  MPI_Comm_size( MPI_COMM_WORLD, &nmach );
  MPI_Barrier( MPI_COMM_WORLD );
  if ( myrank == 0 ) {
    timempi[0] = MPI_Wtime();                          // Store initial time
    systime_i = time(NULL);
  }
  printf("\n\t Processor %d checking in...\n", myrank);
  fflush(stdout);
  for ( i = 1; i < N; i++ ) {        // Set up Hamiltonian and starting vector
    for ( j = 1; j < N; j++ ) {
      if ( abs(j-i) > 10 ) ham[j][i] = 0.0;
        else ham[j][i] = pow(0.3, abs(j-i));
    }
    ham[i][i] = i;
    coef[i] = 0.0;
  }
  coef[1] = 1.0  ;
  err[0] = 1.0;
  iter = 0 ;
  if ( myrank == 0 )  {          //  Start iterating towards the solution
    printf( "\nIteration #\tEnergy\t\tERR\t\tTotal Time\n" );
    fflush(stdout);
  }
  while ( iter < MAX && err[0] > ERR ) {                // Start while loop
    iter = iter + 1;
    mycoef[0]=0.0;
    ener[0] = 0.0; myener[0] = 0.0 ;
    ovlp[0] = 0.0; myovlp[0] = 0.0 ;
    err[0] = 0.0 ; myerr[0] = 0.0;
    for (i= 1; i < N; i++) {
      h = (int)(i)%(nmach-1)+1 ;
      if (myrank == h)  {
        myovlp[0] = myovlp[0]+coef[i]*coef[i];
        mysigma[0] = 0.0;
        for ( j= 1; j < N; j++ ) mysigma[0] =  mysigma[0] + coef[j]*ham[j][i];
        myener[0] =  myener[0]+coef[i]*mysigma[0] ;
        MPI_Send( &mysigma, 1, MPI_DOUBLE, 0, h, MPI_COMM_WORLD );
      }
    if ( myrank == 0 )  {
      MPI_Recv( &mysigma, 1, MPI_DOUBLE, h, h, MPI_COMM_WORLD, &status );
      sigma[i]=mysigma[0];
    }
  }                                                        // End of for(i...
  MPI_Allreduce( &myener, &ener, 1, MPI_DOUBLE, MPI_SUM, MPI_COMM_WORLD );
  MPI_Allreduce( &myovlp, &ovlp, 1, MPI_DOUBLE, MPI_SUM, MPI_COMM_WORLD );
  MPI_Bcast( &sigma, N-1, MPI_DOUBLE, 0, MPI_COMM_WORLD );
  ener[0] = ener[0]/(ovlp[0]);
  for (  i = 1; i< N; i++) {
    h = (int)(i)%(nmach-1)+1 ;
    if (myrank == h) {
      mycoef[0] = coef[i]/sqrt(ovlp[0]);
      mysigma[0] = sigma[i]/sqrt(ovlp[0]);
      MPI_Send( &mycoef,  1, MPI_DOUBLE, 0, nmach+h+1,   MPI_COMM_WORLD );
      MPI_Send( &mysigma, 1, MPI_DOUBLE, 0, 2*nmach+h+1, MPI_COMM_WORLD );
    }
    if (myrank == 0) {
```

```
      MPI_Recv( &mycoef,   1, MPI_DOUBLE, h, nmach+h+1,    MPI_COMM_WORLD, &status );
      MPI_Recv( &mysigma,  1, MPI_DOUBLE, h, 2*nmach+h+1, MPI_COMM_WORLD, &status );
      coef[i]=mycoef[0];
      sigma[i]=mysigma[0];
    }
  }                                                             // End of for(i...
    MPI_Bcast( &sigma, N-1, MPI_DOUBLE, 0, MPI_COMM_WORLD );
    MPI_Bcast( &coef,  N-1, MPI_DOUBLE, 0, MPI_COMM_WORLD );
    for ( i = 2; i < N ; i++ )  {
      h = (int)(i)%(nmach-1)+1;
      if ( myrank == h )  {
        step = (sigma[i] - ener[0]*coef[i])/(ener[0]-ham[i][i]);
        mycoef[0] = coef[i] + step;
        myerr[0] =  myerr[0]+ pow(step,2);
        for ( k= 0; k <= N*N; k++ )                             // Slowdown loop
          { dummy = pow(dummy,dummy);          dummy = pow(dummy,1.0/dummy); }
        MPI_Send( &mycoef, 1, MPI_DOUBLE, 0, 3*nmach+h+1, MPI_COMM_WORLD );
      } // end of if(myrank..
      if (myrank == 0)   {
        MPI_Recv(&mycoef, 1, MPI_DOUBLE, h,3*nmach+h+1, MPI_COMM_WORLD, &status);
        coef[i]=mycoef[0];
      }
    }                                                           // End of for(i...
    MPI_Bcast(      &coef, N-1,      MPI_DOUBLE, 0,        MPI_COMM_WORLD );
    MPI_Allreduce( &myerr, &err , 1, MPI_DOUBLE, MPI_SUM, MPI_COMM_WORLD );
    err[0] = sqrt(err[0]);
    if ( myrank==0 ) { printf("\t#%d\t%g\t%g\n", iter, ener[0], err[0]);
                 fflush(stdout); }
  }                                                             // End while
  if (myrank == 0) {
    systime_f = time(NULL);                                     // Output elapsed time
    difftime = ((long) systime_f) - ((long) systime_i);
    printf( "\n\tTotal wall time = %d s\n", difftime );
    fflush(stdout);
    timempi[1] = MPI_Wtime();
    printf("\n\tMPItime= %g s\n", (timempi[1]-timempi[0]) );
    fflush(stdout);
  }
  MPI_Finalize();
}
```

Listing D.7 The C program **TuneMPI.c** is a parallel version of Tune.java, which we used to test the effects of various optimization modifications.

D.4.0.1 TUNEMPI.C EXERCISE

1. Compile TuneMPI.c:

   > **mpicc TuneMPI.c -lm -o TuneMPI** Compilation

   Here -lm loads the math library and -o places the object in TuneMPI. This is the base program. It will use one processor as the host and another one to do the work.
2. To determine the speedup with multiple processors, you need to change the run_mpi.sh script. Open it with an editor and find a line of the form

   #$ -pe class_mpi 1-4 A line in run_mpi.sh script

The last number on this line tells the cluster the maximum number of processors to use. Change this to the number of processors you want to use. Use a number from 2 to 10; starting with one processor leads to an error message, as that leaves no processor to do the work. After changing run_mpi.sh, run the program on the cluster. With the SEG management system this is done via

```
> qsub run_mpi.sh TuneMPI                    Submit to queue via SGE
```

3. You are already familiar with the scalar version of the Tune program. Find the scalar version of Tune.c (and add the extra lines to slow the program down) or modify the present one so that it runs on only one processor. Run the scalar version of TuneMPI and record the time it takes. Because there is overhead associated with running MPI, we expect the scalar program to be faster than an MPI program running on a single processor.
4. Open another window and watch the processing of your MPI jobs on the host computer. Check that all the temporary files are removed.
5. You now need to collect data for a plot of running time *versus* number of machines. Make sure your matrix size is large, say, with N=200 and up. Run TuneMPI on a variable number of machines, starting at 2, until you find no appreciable speedup (or an actual slowdown) with an increasing number of machines.
6. *Warning:* While you will do no harm running on the Beowulf when others are also running on it, in order to get meaningful, repeatable speedup graphs, you should have the cluster all to yourself. Otherwise, the time it takes to switch jobs around and to set up and drop communications may slow down your runs significantly. A management system should help with this. If you are permitted to log in directly to the Beowulf machines, you can check what is happening via who:

```
> rsh rose who                    Who is running on rose?
> rsh rubin who                   Who is running on rubin?
> rsh emma who                    Who is running on emma?
```

7. Increase the matrix size in steps and record how this affects the speedups. Remember, once the code is communications-bound, distributing it over many processors probably will make it run slower, not faster!

D.5 A String Vibrating in Parallel

The program MPIstring.c given in Listing D.8 is a parallel version of the solution of the wave equation (eqstring.c) discussed in Chapter 18, "PDE Waves: String, Wave Packet, and Electromagnetic." The algorithm calculates the future ([2]) displacement of a given string element from the present ([1]) displacements immediately to the left and right of that section, as well as the present and past ([0]) displacements of that element. The program is parallelized by assigning different sections of the string to different nodes.

```
// Code listing for MPIstring.c
#include <stdio.h>
#include <math.h>
#include "mpi.h"
#define maxt 10000                              // Number of time steps to take
#define L 10000                             // Number of divisions of the string
#define rho 0.01                                    // Density per length (kg/m)
#define ten 40.0                                                 // Tension (N)
#define deltat 1.0e-4                                            // Delta t (s)
#define deltax .01                                               // Delta x (m)
#define skip 50                        // Number of time steps to skip before printing
/* Need sqrt(ten/rho) <= deltax/deltat for a stable solution
            Decrease deltat for more accuracy, c' = deltax/deltat  */

main(int argc, char *argv[]) {
  const double scale = pow( deltat/deltax, 2) * ten/rho;
  int i, j, k, myrank, numprocs, start, stop, avgwidth, maxwidth, len;
  double left, right, startwtime, init\_string(int index);
  FILE *out;

  MPI_Init( &argc, &argv );
  MPI_Comm_rank( MPI_COMM_WORLD, &myrank );                       // Get my rank
  MPI_Comm_size( MPI_COMM_WORLD, &numprocs );               // Number of processors
  MPI_Status status;
    if (myrank == 0)  {
      startwtime = MPI_Wtime();
      out = fopen("eqstringmpi.dat","w");
    }
   // assign string to each node 1st and last points (0 and L-1) -must =0, else error
              // Thus L-2 segments for numprocs processors
    avgwidth = (L-2)/numprocs;
    start = avgwidth*myrank+1;
    if (myrank < numprocs - 1)   stop = avgwidth*(myrank+1);
    else stop = L-2;
    if (myrank == 0)    maxwidth = L-2 - avgwidth*(numprocs-1);
    else maxwidth = 0;
    double results[maxwidth];                              // Holds print for master
    len = stop - start;                                    // Length of the array - 1
    double x[3][len+1];
    for (i = start; i <= stop; i++)    x[0][i-start] = init_string(i);
    x[1][0] = x[0][0]+0.5*scale*(x[0][1]+init_string(start-1)-2.*x[0][0]); //1st step
    x[1][len] = x[0][len] +0.5*scale*(init_string(stop+1)+x[0][len-1]-2.0*x[0][len]);
    for (i = 1; i < len; i++)
        { x[1][i] = x[0][i] + 0.5*scale*(x[0][i+1] + x[0][i-1] - 2.0*x[0][i]); }
    for(k=1; k<maxt; k++) {                                      //  Later time steps
      if (myrank == 0) { MPI_Send( &x[1][len], 1, MPI_DOUBLE, 1, 1, MPI_COMM_WORLD);
                                             left = 0.0; } // Send to R, get from L
      else if (myrank < numprocs - 1) MPI_Sendrecv(&x[1][len],1,
MPI_DOUBLE, myrank+1, 1, &left, 1, MPI_DOUBLE, myrank-1, 1, MPI_COMM_WORLD, &status);
      else MPI_Recv( &left, 1, MPI_DOUBLE, myrank-1, 1, MPI_COMM_WORLD, &status );
      if (myrank == numprocs - 1) {                          // Send to L & get from R
        MPI_Send( &x[1][0], 1, MPI_DOUBLE, myrank-1, 2, MPI_COMM_WORLD );
        right = 0.0;
      }
      else if (myrank > 0) MPI_Sendrecv( &x[1][0], 1, MPI_DOUBLE, myrank-1, 2,
                &right, 1, MPI_DOUBLE, myrank+1, 2, MPI_COMM_WORLD, &status);
      else MPI_Recv( &right, 1, MPI_DOUBLE, 1, 2, MPI_COMM_WORLD, &status );
      x[2][0] = 2.*x[1][0] - x[0][0] + scale * (x[1][1] + left - 2.*x[1][0]);
      for (i = 1; i < len; i++)
        { x[2][i] = 2.*x[1][i] - x[0][i] + scale *(x[1][i+1]+x[1][i-1]-2.*x[1][i]); }
      x[2][len] = 2.*x[1][len] - x[0][len] + scale*(right+x[1][len-1] -2.*x[1][len]);
      for (i = 0; i <= len; i++)  { x[0][i] = x[1][i];  x[1][i] = x[2][i]; }
      if( k%skip == 0) {                            // Print using gnuplot 3D grid format
        if (myrank != 0) MPI_Send(&x[2][0], len+1, MPI_DOUBLE, 0, 3, MPI_COMM_WORLD);
        else {
          fprintf(out,"%f\n",0.0);                           // Left edge of (always 0)
```

```
          for ( i=0; i < avgwidth; i++ ) fprintf(out,"%f\n",x[2][i]);
          for ( i=1; i < numprocs-1; i++ ) {
            MPI_Recv( results , avgwidth, MPI_DOUBLE, i, 3, MPI_COMM_WORLD, &status );
            for (j = 0; j < avgwidth; j++) fprintf(out, "%f\n",results[j]);
          }
    MPI_Recv( results , maxwidth, MPI_DOUBLE, numprocs-1, 3, MPI_COMM_WORLD, &status );
          for ( j=0; j < maxwidth; j++ ) fprintf(out,"%f\n",results[j]);
          fprintf(out,"%f\n",0.0);                                        // R edge
          fprintf(out,"\n");                                   // Empty line for gnuplot
        }
      }
    }
    if (myrank == 0)
      printf("Data stored in eqstringmpi.dat\nComputation time: %f s\n",
             MPI_Wtime()-startwtime);
      MPI_Finalize();
      exit(0);
  }

  double init_string(int index) {
    if (index < (L-1)*4/5) return 1.0*index/((L-1)*4/5);
    return 1.0*(L-1-index)/((L-1)-(L-1)*4/5);
                // Half of a sine wave
  }
```

Listing D.8 MPIstring.c solves the wave equation for a string using several processors via MPI commands.

Notice how the MPI_Recv() and MPI_Send() commands require a pointer as their first argument, or an array element. When sending more than one element of an array to MPI_Send(), send a pointer to the first element of the array as the first argument, and then the number of elements as the second argument. Observe near the end of the program how the MPI_Send() call is used to send len + 1 elements of the 2-D array x[][], starting with the element x[2][0]. MPI sends these elements in the order in which they are stored in memory. In C, arrays are stored in row-major order with the first element of the second row immediately following the last element of the first row, and so on. Accordingly, this MPI_Send() call sends len + 1 elements of row 2, starting with column 0, which means all of row 2. If we had specified len + 5 elements instead, MPI would have sent all of row 2 plus the first four elements of row 3.

In MPIstring.c, the calculated future position of the string is stored in row 2 of x[3][len + 1], with different sections of the string stored in different columns. Row 1 stores the present positions, and row 0 stores the past positions. This is different from the column-based algorithm used in the serial program eqstring.c, following the original Fortran program, which was column-, rather than row-based. This was changed because MPI reads data by rows. The initial displacement of the string is given by the user-supplied function init_string(). Because the first time step requires only the initial displacement, no message passing is necessary. For later times, each node sends the displacement of its rightmost point to the node with the next highest rank. This means that the rightmost node (rank = numprocs

−1) does not send data and that the master (rank = 0) does not receive data. Communication is established via the MPI_Sendrecv() command, with the different sends and receives using tags to ensure proper delivery of the messages.

Next in the program, the nodes (representing string segments) send to and receive data from the segment to their right. All these sends and receives have a tag of 2. After every 50 iterations, the master collects the displacement of each segment of the string and outputs it. Each slave sends the data for the future time with a tag of 3. The master first outputs its own data and then calls MPI_Recv() for each node, one at a time, printing the data it receives.

D.5.1 MPIstring.c Exercise

1. Ensure that the input data (maxt, L, scale, skip) in MPIstring.c are the same as those in eqstring.c.
2. Ensure that init_string() returns the initial configuration used in eqstring.c.
3. Compile and run eqstring.c via

```
> gcc eqstring.c -o eqstring -lm          Compile C code
> ./eqstring                              Run in present directory
```

4. Run both programs to ensure that they produce the same output. (In Unix this is easy to check with the diff command.)

```
// deadlock-fixed.c: MPI deadlock.c without deadlock by Phil Carter
#include <stdio.h>
#include "mpi.h"
#define MAXLEN 100

main(int argc, char *argv[])  {
  int myrank, numprocs, torank, i;
  char tosend[MAXLEN], received[MAXLEN];
  MPI_Status status;
  MPI_Init( &argc, &argv );
  MPI_Comm_rank( MPI_COMM_WORLD, &myrank );
  MPI_Comm_size( MPI_COMM_WORLD, &numprocs );
  if ( myrank == numprocs - 1 ) torank = 0;
    else torank = myrank + 1;                    // Save string to send in tosend:
  sprintf( tosend, "Message sent from node %d to node %d\n", myrank, torank );
  for ( i = 0; i < numprocs; i++ )  {
    if ( myrank == i ) MPI_Send(tosend, MAXLEN, MPI_CHAR, torank, i, MPI_COMM_WORLD);
         else if ( myrank == i+1 || (i == numprocs - 1 && myrank == 0) )
              MPI_Recv( received, MAXLEN, MPI_CHAR, i, i, MPI_COMM_WORLD, &status );
    }
    printf("%s", received);                     // Print string after successful receive
    MPI_Finalize();
    exit(0);
}
```

Listing D.9 The MPI program **MPIdeadlock.c** illustrates deadlock (waiting to receive). The code MPIdeadlock-fixed.c in Listing D.6 removes the block.

D.6 Deadlock

It is important to avoid *deadlock* when using the MPI_Send() and MPI_Recv() commands. Deadlock occurs when one or more nodes wait for a nonoccurring event to take place. This can arise if each node waits to receive a message that is not sent. Compile and execute deadlock.c:

> **mpicc deadlock.c -o deadlock** Compile
> **qsub run_mpi.sh deadlock** Execute

Take note of the job ID returned, which we will call xxxx. Wait a few seconds and then look at the output of the program:

> **cat output/MPI_job-xxxx** Examine output

The output should list how many nodes (slots) were assigned to the job. Because all these nodes are now deadlocked, we have to cancel this job:

> **qdel xxxx** Cancel deadlocked job
> **qdel all** Alternate cancel

There are a number of ways to avoid deadlock. The program MPIstring.c used the function MPI_Sendrecv() to handle much of the message passing, and this does not cause deadlock. It is possible to use MPI_Send() and MPI_Recv(), but you should be careful to avoid deadlock, as we do in MPIdeadlock-fixed.c in Listing D.6.

```
// deadlock-fixed.c: MPI deadlock.c without deadlock by Phil Carter
#include <stdio.h>
#include "mpi.h"
#define MAXLEN 100

main(int argc, char *argv[]) {
  int myrank, numprocs, torank, i;
  char tosend[MAXLEN], received[MAXLEN];
  MPI_Status status;
  MPI_Init( &argc, &argv );
  MPI_Comm_rank( MPI_COMM_WORLD, &myrank );
  MPI_Comm_size( MPI_COMM_WORLD, &numprocs );
  if ( myrank == numprocs - 1 ) torank = 0;
    else torank = myrank + 1;                    // Save string to send in tosend:
  sprintf( tosend, "Message sent from node %d to node %d\n", myrank, torank );
  for ( i = 0; i < numprocs; i++ ) {
    if ( myrank == i ) MPI_Send(tosend, MAXLEN, MPI_CHAR, torank, i, MPI_COMM_WORLD);
        else if ( myrank == i+1 || (i == numprocs - 1 && myrank == 0) )
            MPI_Recv( received, MAXLEN, MPI_CHAR, i, i, MPI_COMM_WORLD, &status );
    }
    printf("%s", received);                    // Print string after successful receive
    MPI_Finalize();
    exit(0);
}
```

D.6.1 Nonblocking Communication

MPI_Send() and MPI_Recv(), as we have said, are susceptible to deadlock because they block the program from continuing until the send or receive is finished. This method of message passing is called *blocking communication*. One way to avoid deadlock is to use nonblocking communication such as MPI_Isend(), which returns before the send is complete and thus frees up the node. Likewise, a node can call MPI_Irecv(), which does not wait for the receive to be completed. Note that a node can receive a message with MPI_Recv() even if it was sent using MPI_Isend(), and similarly, receive a message using MPI_Irecv() even if it was sent with MPI_Send().

There are two ways to determine whether a nonblocking send or receive is finished. One is to call MPI_Test(). It is your choice as to whether you want to wait for the communication to be completed (e.g., to ensure that all string segments are at current time and not past time). To wait, call MPI_Wait(), which blocks execution until communication is complete. When you start a nonblocking send or receive, you get a request handle of data type MPI_Request to identify the communication you may need to wait for. A disadvantage of using nonblocking communication is that you have to ensure that you do not use the data being communicated until the communication has been completed. You can check for this using MPI_Test() or wait for completion using MPI_Wait().

Exercise: Rewrite MPIdeadlock.c so that it avoids deadlock by using nonblocking communication. *Hint:* Replace MPI_Recv() by MPI_Irecv().

D.6.2 Collective Communication

MPI contains several commands that automatically and simultaneously exchange messages among multiple nodes. This is called *collective communication*, in contrast to point-to-point communication between two nodes. The program MPIpi.c has already introduced the MPI_Reduce() command. It receives data from multiple nodes, performs an operation on the data, and stores the result on one node. The program tuneMPI.c used a similar function MPI_Allreduce() that does the same thing but stores the result on every node. The latter program also used MPI_Bcast(), which allows a node to send the same message to multiple nodes.

Collective commands communicate among a group of nodes specified by a communicator, such as MPI_COMM_WORLD. For example, in MPIpi.c we called MPI_Reduce() to receive results from every node, including itself. Other collective communication functions include MPI_Scatter(), which has one node send messages to every other node. This is similar to MPI_Bcast(), but the former sends a different message to each node by breaking up an array into pieces of specified lengths and sending the pieces to nodes. Likewise, `MPI_Gather()` gathers data from every node (including the root node) and places it in an array, with data from node 0 placed first, followed by node 1, and so on. A similar function, `MPI_Allgather()`, stores the data on every node rather than just the root node.

D.7 Bootable Cluster CD ⊙

One of the difficulties in learning how to parallel compute is the need for a parallel computer. Even though there may be many computers around that you may be able to use, knitting them all together into a parallel machine takes time and effort. However, if your interest is in learning about and experiencing distributed parallel computing, and not in setting up one of the fastest research machines in the world, then there is an easy way. It is called a *bootable cluster CD* (BCCD) and is a file on a CD. When you start your computer with the CD in place, you are given the option of having the computer ignore your regular operating system and instead boot from the CD into a preconfigured distributed computing environment. The new system does not change your system but rather is a nondestructive overlay on top of the existing hardware that runs a full-fledged parallel computing environment on just about any workstation-class system, including Macs. You boot up every machine you wish to have on your cluster this way, and if needed, set up a domain name system (DNS) and dynamic host configuration protocol (DHCP) servers, which are also included. Did we mention that the system is free? [BCCD]

D.8 Parallel Computing Exercises

1. **Bifurcation plot:** If you have not yet done so, take the program you wrote to generate the bifurcation plot for bug populations and run different ranges of μ values simultaneously on several CPUs.
2. **Processor ring:** Write a program in which
   a. a set of processors are arranged in a ring.
   b. each processor stores its rank in `MPI_COMM_WORLD` in an integer.
   c. each processor passes this rank on to its neighbor on the right.
   d. each processor keeps passing until it receives its rank back.
3. **Ping pong:** Write a program in which two processors repeatedly pass a message back and forth. Insert timing calls to measure the time taken for one message and determine how the time taken varies with the size of the message.
4. **Broadcast:** Have processor 1 send the same message to all the other processors and then receive messages of the same length from all the other processors. How does the time taken vary with the size of the messages and the number of processors?

D.9 List of MPI Commands

MPI Data Types and Operators

MPI defines some of its own data types. The following are data types used as arguments to MPI commands.

- **MPI_Comm:** A communicator, used to specify group of nodes, most commonly `MPI_COMM_WORLD` for all the nodes.
- **MPI_Status:** A variable holding status information returned by functions such as `MPI_Recv()`.
- **MPI_Datatype:** A predefined constant indicating the type of data being passed in a function such as `MPI_Send()` (see below).
- **MPI_O:** A predefined constant indicating the operation to be performed on data in functions such as `MPI_Reduce()` (see below).
- **MPI_Request:** A request handle to identify a nonblocking send or receive, for example, when using `MPI_Wait()` or `MPI_Test()`.

Predefined Constants: MPI_Op and MPI_Datatype

For a more complete list of constants used in MPI, see
`http://www-unix.mcs.anl.gov/mpi/www/www3/Constants.html`.

MPI_OP	*Description*	*MPI_Datatype*	*C Data Type*
MPI_MAX	Maximum	**MPI_CHAR**	char
MPI_MIN	Minimum	**MPI_SHORT**	short
MPI_SUM	Sum	**MPI_INT**	int
MPI_PROD	Product	**MPI_LONG**	long
MPI_LAND	Logical and	**MPI_FLOAT**	float
MPI_BAND	Bitwise and	**MPI_DOUBLE**	double
MPI_LOR	Logical or	**MPI_UNSIGNED_CHAR**	unsigned char
MPI_BOR	Bitwise or	**MPI_UNSIGNED_SHORT**	unsigned short
MPI_LXOR	Logical exclusive or	**MPI_UNSIGNED**	unsigned int
MPI_BXOR	Bitwise exclusive or	**MPI_UNSIGNED_LONG**	unsigned long
MPI_MINLOC	Find node's min and rank		
MPI_MAXLOC	Find node's max and rank		

MPI Commands

Below we list and identify the MPI commands used in this appendix. For the syntax of each command, along with many more MPI commands, see `http://www-unix.mcs.anl.gov/mpi/www/`, where each command is a hyperlink to a full description.

Basic MPI Commands

- **MPI_Send:** Sends a message.
- **MPI_Recv:** Receives a message.
- **MPI_Sendrecv:** Sends and receives a message.
- **MPI_Init:** Starts MPI at the beginning of the program.
- **MPI_Finalize:** Stops MPI at the end of the program.
- **MPI_Comm_rank:** Determines a node's rank.
- **MPI_Comm_size:** Determines the number of nodes in the communicator.
- **MPI_Get_processor_name:** Determines the name of the processor.
- **MPI_Wtime:** Returns the wall time in seconds since an arbitrary time in the past.
- **MPI_Barrier:** Blocks until all the nodes have called this function.

Collective Communication

- **MPI_Reduce:** Performs an operation on all copies of a variable and stores the result on a single node.
- **MPI_Allreduce:** Like `MPI_Reduce`, but stores the result on all the nodes.
- **MPI_Gather:** Gathers data from a group of nodes and stores them on one node.
- **MPI_Allgather:** Like `MPI_Gather` but stores the result on all the nodes.
- **MPI_Scatter:** Sends different data to all the other nodes (opposite of `MPI_Gather`).
- **MPI_Bcast:** Sends the same message to all the other processors.

Nonblocking Communication

- **MPI_Isend:** Begins a nonblocking send.
- **MPI_Irecv:** Begins a nonblocking receive.
- **MPI_Wait:** Waits for an MPI send or receive to be completed.
- **MPI_Test:** Tests for the completion of a send or receive.

Appendix E: Calling LAPACK from C

Calling a Fortran-based matrix library from Fortran is less trouble than calling it from some other language. However, if you refuse to be a Fortran programmer, then there are often C and Java implementations of Fortran libraries available, such as JAMA, JLAPACK, LAPACK++, and TNT, for use from these languages.

Some of our research projects have required us to have Fortran and C programs calling each other, and, after some experimentation, we have had success doing it [L&F 93]. Care is needed in accounting for the somewhat different ways compilers store subroutine names, for the quite different ways they store arrays with more than one subscript, and for the different data types available in the two languages.

The first thing you must do is ensure that the data types of the variables in the two languages are matched. The matching data types are given in Table E.1. Note that if the data are stored in arrays, the C calling program must convert to the storage scheme used in the Fortran subroutine before you can call the Fortran subroutine (and then convert back to the C scheme after the subroutine does its job if you have overwritten the original array and intend to use it again).

When a function is called in the C language, usually the actual value of the argument is passed to the function. In contrast, Fortran passes the address in memory where the value of the argument is to be found (a *reference pass*). If you do not ensure that both languages have the same passing protocol, your program will process the numerical value of the address of a variable as if it were the actual value of the variable (we are willing to place a bet on the correctness of the result). Here are some procedures for **calling Fortran from C**:

1. Use pointers for all the arguments in your C program. Generally this is done with the address operator &.
2. Do not have your program make calls such as `sub(1, N)` where the actual value of the constant `1` is fed to the subroutine. Instead, assign the value 1 to a variable and feed that variable (actually a pointer to it) to the subroutine. For example:

```
one = 1.                                    // Assign value to variable
sub(one)                  // All Fortran calls are reference calls
sub(&one)                            // In C, make pointer explicit
```

   This is important because the value `1` in the Fortran subroutine call is actually the address where the value 1 is stored. If the subroutine modifies that variable, it will modify the value of `1` every place in your program!
3. Depending on the particular operating system you are using, you may have to append an underscore _ to the called Fortran subprogram names; for example, `sub(one, N)` → `sub_(one, N)`. Generally, the Fortran compiler appends an underscore automatically to the names of its subprograms, while the C compiler does not (but you'll need to experiment).

TABLE E.1
Matching Data Types in C and Fortran

C	*Fortran*	C	*Fortran*
Char	Character	Unsigned int	Logical*4
Signed char	Integer*1	Float	Real (Real*4)
Unsigned char	Logical*1	Structure of two floats	Complex
Short signed int	Integer*2	Double	Real*8
Short unsigned int	Logical*2	Structure of two doubles	Complex*16
Signed int (long int)	Integer*4	Char[n]	Character*n

4. Use lowercase letters for the names of external functions. The exception is when the Fortran subprogram being called was compiled with a -U option, or the equivalent, for retaining uppercase letters.

E.1 Calling LAPACK Fortran from C

```
// Calling LAPACK from C          to solve AX=B
#include <stdio.h>}                                           // I/O
headers #define size 100                     // Dimension of Hilbert matrix

  main() {
    int i, j , c1, c2, pivot[size], ok;
    double matrix[size][size], help[size*size], result[size];
    c1 = size;                                  // Pointers for function call
    c2 = 1;                                        // Numbers as variables
    for ( i = 0; i < c1; i++ )  {                  // Create Hilbert matrix
      for ( j = 0; j < c1; j++ )  matrix[i][j] = 1.0/(i+j+1);
      result[i] = 1. / (i+1);                      // Create solution vector
    }
    for (i=0; i<size; i++) { for(j=0; j<size; j++) help[j+size*i] = matrix[j][i]; }
    dgesv_(&c1, &c2, help, &c1, pivot, result, &c1, &ok);
    for ( j=0; j < size; j++ ) printf("%e\n", result[j]);
  }
```

Notice here that the call to the Fortran subroutine dgesv is made as

```
dgesv_(&c1, &c2, help, &c1, pivot, result, &c1, &ok)
```

That is, lowercase letters are used and an underscore is added to the subroutine name. In addition, we convert the matrix A,

```
for ( i=0; i<size; i++ )                              // Matrix transformation
        for ( j=0; j<size; j++ )
                help[j+size*i] = matrix[j][i];
```

This changes C's row-major order to Fortran's column-major order.

E.2 Compiling C Programs with Fortran Calls

Multilanguage programs are actually created when the compiler links the object files together. The tricky part is that while Fortran automatically includes its math library, if your final linking is done with the C compiler, you may have to explicitly include the Fortran library as well as others. Here we give some examples that have worked at one time or another on one machine or another; you probably will need to read the user's guide and experiment to get this to work with your system:

> **cc -O call.c f77sub.o -lm -ldxml**	DEC extended mathlibe
> **cc -O call.c f77sub.o -L/usr/lang/SC0.0 -lF77**	Link, SunOS
> **cc -O c_fort c_fort.o area_f.o -lxlf**	Link, AIX
> **gcc -Wall call.c /usr/lib/libm.a -o call**	gcc explicit
> **gcc -Wall call.c -lm -o call**	gcc shorthand version of above

Appendix F: Software on the CD

JavaCodes Contents by Section

Section	*Program*	*Description*	*Section*	*Program*	*Description*
1.4	AreaScanner	Java 1.5 scanner class	6.2.1	Trap	Trapezoid rule integration
1.4.2	Area	First program	6.2.5	IntegGauss	Gaussian quadrature
1.5.4	Limits	Machine precision	7.5	Diff	Numerical differentiation
1.5.4	ByteLimit	Precise machine precision	7.9	Bisection	Bisection root finding
2.2.2	Bessel	Bessel function recurrence	7.10	Newton_cd	Newton–Raphson roots
3.2	EasyPtPlot	Simple use of PtPlot	7.10	Newton_fd	Newton–Raphson roots
3.2	TwoPlotExample	Two plots, one graph	8.3.3	JamaEigen	JAMA eigenvalues
4.4	Complex	Complex number objects	8.3.3	JamaFit	JAMA least-squares fit
4.4.4	ComplexDyn	Dynamic complex objects	8.5.5	SplineAppl	Cubic spline fit
4.6.1	Beats	Procedural beats	8.7.1	Fit	Linear least-squares fit
4.6.1	OOPBeats	Objective beats	8.2.2	Newton_Jama2	Newton–Raphson search
4.6.2	Moon	Procedural moon orbits	14.14.4	Tune, Tune4	Optimization testing
4.6.2	OOPMoon	Objective moon orbits	9.5.2	rk2, rk4	2nd, 4th O Runge-Kutta
4.6.2	OOPlanet	Objective moon orbits	9.5.2	rk45	Adaptive step rk solver
5.2.2	RandNum	Java random utility	9.5.2	ABM	Predictor-corrector solver
5.4.2	Walk	Random-walk simulation	9.11	Numerov	Quantum bound states
5.6	Decay	Spontaneous decay	9.11	QuantumEigen	rk quantum bound states
6.6.3	Int10d	10-D Monte Carlo integral			

JavaCodes Contents by Section *Continued*

Section	*Program*	*Description*	*Section*	*Program*	*Description*
10.4.1	DFT	Discrete Fourier transform	17.17.4	EqHeat	Heat equation via leapfrog
10.6	Filter	Autocorrelation filtering	17.19.1	HeatCNTridiag	Crank–Nicolson heat equation
10.7	Noise	Filtering	18.2.3	EqString	Waves on a string
11.4.2	CWT	Continuous wavelet TF	18.4.3	CatString/Friction	Catenary waves
			CPI	TwoDsol	2-D, Sine–Gordon solitons
11.5.2	Scale, Daub4	Generate Daub4 wavelets	18.6.2	SqWell, Harmos	Quantum wave packets
G	DaubCompress	Wavelet compression			
G	DaubMariana	Wavelet image compression	18.6.2	Slit	Wave packet through slit
12.5	Bugs	Bifurcation diagram	19.9.3	Beam	2-D Navier–Stokes fluid
12.8	LyapLog	Lyapunov exponents	19.1	AdvecLax	Advection
			19.3.1	Shock	Shock waves
12.8.1	Entropy	Shannon logistic entropy	19.5	SolCross	Solitary waves crossing
12.19	PredatorPrey	Lotka–Volterra model	19.5.3	Soliton	KdeV solitons
13.4.1	Film, FilmDim	Random deposition	20.2.3	Bound	Bound integral equation
13.9	CellAut, OutofEight	Cellular automata	20.4.5	Scatt	Scattering integral equation
13.9	LifeGame	Game of Life			
16.3	MD	1-D MD simulation	15.2	Ising, Ising3D	Ising model
17.4.2	LaplaceLine	Finite differential Laplace equation	15.8.3	QMC	Feynman path integration
17.14	LaplaceFEM	Finite element Laplace equation	15.9	QMCbouncer	Quantum bouncer

JavaCodes Contents by Name

Program	*Section*	*Description*	*Program*	*Section*	*Description*
ABM	9.5.2	Predictor-corrector solver	Entropy	12.8.1	Shannon logistic entropy
AdvecLax	19.1	Advection	EqHeat	17.17.4	Heat equation via leapfrog
Area	1.4.2	First program	EqString	18.2.3	Waves on a string
AreaScanner	1.4	Java 1.5 scanner class	Film, FilmDim	13.4.1	Random deposition
Beam	19.9.3	2-D Navier–Stokes fluid	Filter	10.6	Autocorrelation filtering
Beats	4.6.1	Procedural beats	Fit	8.7.1	Linear least–squares fit
Bessel	2.2.2	Bessel function recurrence	HeatCNTridiag	17.19.1	Crank–Nicolson heat equation
Bisection	7.9	Bisection root finding	IntegGauss	6.2.5	Gaussian quadrature
Bound	20.2.3	Bound integral equation	Int10d	6.6.3	10-D Monte Carlo integral
Bugs	12.5	Bifurcation diagram	Ising, Ising3D	15.2	Ising model
ByteLimit	1.5.4	Precise machine precision	JamaEigen	8.3.3	JAMA eigenvalues
CatString/Friction	18.4.3	Catenary waves	JamaFit	8.3.3	JAMA Least-square fit
CellAut	13.9	Cellular automata	LaplaceFEM	17.14	Finite element Laplace equation
Complex	4.4	Complex number objects	LaplaceLine	17.4.2	Finite differential Laplace equation
ComplexDyn	4.4.4	Dynamic complex objects	LifeGame	13.9	Game of Life
CWT	11.4.2	Continuous wavelet TF	Limits	1.5.4	Machine precision
DaubCompress	G	Wavelet compression	LyapLog	12.8	Lyapunov exponents
DaubMariana	G	Wavelet image compression	MD	16.3	1-D MD simulation
Daub4	11.5.2	Generate Daub4 wavelets	Moon	4.6.2	Procedural moon orbits
Decay	5.6	Spontaneous decay	Newton_cd	7.10	Newton–Raphson roots
DFT	10.4.1	Discrete Fourier transform	Newton_fd	7.10	Newton–Raphson roots
Diff	7.5	Numerical differentiation	Newton_Jama2	8.2.2	Newton–Raphson search
EasyPtPlot	3.2	Simple use of PtPlot	Noise	10.7	Filtering
			Numerov	9.11	Quantum bound states

JavaCodes Contents by Name *Continued*

Program	*Section*	*Description*	*Program*	*Section*	*Description*
OOPBeats	4.6.1	Objective beats	Scatt	20.4.5	Scattering integral equation
OOPMoon	4.6.2	Objective moon orbits	Shock	19.3.1	Shock waves
OOPlanet	4.6.2	Objective moon orbits	Slit	18.6.2	Wave packet through slit
OutofEight	13.9	Cellular automata objects	SolCross	19.5	Solitary waves crossing
RandNum	5.2.2	Java random utility	Soliton	19.5.3	KdeV solitons
PredatorPrey	12.19	Lotka–Volterra model	SplineAppl	8.5.5	Cubic spline fit
QMC	15.8.3	Feynman path integration	SqWell, Harmos	18.6.2	Quantum wave packets
QMCbouncer	15.9	Quantum bouncer	Trap	6.2.1	Trapezoid rule integration
QuantumEigen	9.11	rk quantum bound states	Tune, Tune4	14.14.4	Optimization testing
rk2, rk4	9.5.2	2nd- and 4th-order Runge–Kutta	TwoDsol	CPI	2-D, Sine-Gordon solitons
rk45	9.5.2	Adaptive step rk solver	TwoPlotExample	3.2	More involved PtPlot
Scale	11.5.2	Generate Daub4 wavelets	Walk	5.4.2	Random walk simulation

Animations Contents
(requires player VLC or QuickTime for mpeg, avi; browser for gifs)

Directory	*Chapter*	*Directory*	*Chapter*
DoublePendulum (also two pendulums; see also applets)	12	Fractals (see also applets)	13
MapleWaveMovie (need Maple; includes source files)	18	Laplace (DX movie)	17
MD	16	TwoSlits (includes DX source files)	18
2-Dsoliton (includes DX source files)	18, 19	Utilities (scripts, colormaps)	3
Waves (animated gifs need browser)	18		

Ccodes **Contents by Name**

Program	*Section*	*Description*	*Program*	*Section*	*Description*
accm2d.cpp	4.11	OOP example	int_10d.c	6.6.3	10-D Monte Carlo integral
AdvecLax.c	19.1	Advection	invfour.c	10.4.1	Inverse Fourier transform
area.c	1.4.2	First program	ising.c	15.2	Ising model
Beam.c	19.9.3	2-D Navier–Stokes fluid	laplace.c	17.4.2	Finite differential Laplace equation
bessel.c	2.2.2	Bessel function recurrence	laplaceAnal.c	17.4.2	Analytic solution for Laplace equation
bound.c	20.2.3	Bound integral equation	LaplaceSOR.c	17.4.2	Finite differential Laplace equation
bugs.c	12.5	Bifurcation diagram	limit.c	1.5.4	Machine precision
call.c	5.2.2	Calls random numbers	lineq.c	8.3.3	Linear equations as matrices
column.c	13.4.1	Fractal deposition	MD1D.c	16.3	1-D MD simulation
CWT3.c	11.4.2	Continuous wavelet TF	numerov.c	9.11	Quantum bound states
Daub4.c	11.4.2	Discrete wavelet TF	over.c	1.5.4	Overflow limits
Daub4Compress.c	11.4.2	Discrete wavelet TF	pond.c	6.5	Monte Carlo integration
decay.c	5.6	Spontaneous decay	qmc.c	15.8.3	Feynman path integration
diff.c	7.5	Numerical differentiation	random.c	5.2.2	Java random utility
dla.c	13.7.1	Diffusion–limited agregation	rk4.c	9.5.2	2nd- and 4th-order Runge–Kutta
eqheat.c	17.17.4	Heat equation via leapfrog	rk45.c	9.5.2	Adaptive step rk solver
eqstring.c	18.2.3	Waves on a string	shms.cpp	4.11	OOP example
exp-bad.c	1.6	Summing series	Shock.c	19.3.1	Shock waves
exp-good.c	1.6	Summing series	sierpin.c	13.2	Sierpiński gasket
fern.c	13.3.2	Fractal fern	soliton.c	19.5.3	KdeV solitons
film.c	13.4.1	Random deposition	SplineAppl.c	8.5.5	Cubic spline fit
fit.c	8.7.1	Linear least-squares fit	tree.c	13.3.3	Fractal tree
fourier.c	10.4.1	Discrete Fourier transform	twodsol.c	CPI	2-D, Sine–Gordon solitons
gauss.c	6.2.5	Gaussian quadrature	unim1d.cpp	4.11	OOP example
HeatCNTridiag.c	17.19.1	Crank–Nicolson heat equation	unimot2d.cpp	4.11	OOP example
integ.c	6.2.5	Integration	walk.c	5.4.2	Random-walk simulation

Applets Directory Contents
(see index.html; requires browser or Java appletviewer)

Applet	*Chapter*	*Applet*	*Chapter*
The chaotic pendulum	12	Four-centers chaotic scattering	12
Planetary orbits	9	Waves on a string, normal mode	18
Cellular automata for Sierpiński	13	Solitons	18
Spline interpolation	8	Relativistic scattering	—
Lagrange interpolation	8	Young's two slit interference	18
Wavelet compression	11	Starbrite (H-R diagram)	—
HearData: a sound player for data	—	Photoelectric effect	—
Visualizing physics with Sound	—	Create Lissajous figures	—
Wavepacket-wavepacket collision movies	18	Heat equation	17
Coping with Unix	—	Wave function (SqWell), (HarmOs)	18
Wave function (Asym), (PotBarrier)	18	Fractals (Sierpiński) (fern) (tree)	13
Fractals (film) (column) (dla)	13	Feynman path integrals	15

Fortran95codes **Contents by Name**

Program	*Section*	*Description*	*Program*	*Section*	*Description*
AdvecLax.f95	19.1	Advection	exp-good.f95	1.6	Series summation
BeamFlow.f95	19.9.3	2-D Navier–Stokes fluid	fit.f95	8.7.1	Linear least–squares fit
bessel.f95	2.2.2	Bessel function recurrence	harmos.f95	18.6.2	Quantum wave packets
CWTRHL.f95	11.4.2	Continuous wavelet TF	HeatCNTridiag.f95	19.1	Crank–Nicolson heat equation
DAUB4.f95	G	Wavelet compression	int10d.f95	6.6.3	10-D Monte Carlo integral
decay.f95	5.6	Spontaneous decay	integ.f95	6.2.5	Gaussian quadrature
diff.f95	7.5	Numerical differentiation	Ising3D.f95	15.2	Ising model
eqheat.f95	17.17.4	Heat equation via leapfrog	lagrange.f95	8.5.1	Lagrange interpolation
eqstring.f95	18.2.3	Waves on a string	LaplaceSOR.f95	17.4.2	Finite differential Laplace equation
exp-bad.f95	1.6	Series summation	limit.f95	1.5.4	Machine precision

(continued)

Fortran95codes Contents by Name *(continued)*

Program	*Section*	*Description*	*Program*	*Section*	*Description*
MDrhl.f95	16.3	1-D MD simulation	scatt.f95	20.4.5	Scattering integral equation
Newton_cd.f95	7.10	Newton–Raphson roots	shock.f95	19.3.1	Shock waves
Newton_fd.f95	7.10	Newton–Raphson roots	slit.f95	18.6.2	Wave packet through slit
overflow.f95	1.5.4	Machine precision	soliton.f95	19.5.3	KdeV solitons
pond.f95	6.5	Monte Carlo integration	Spline.f95	8.5.5	Cubic spline fit
qmc.f95	15.8.3	Feynman path integration	sqwell.f95	18.6.2	Quantum wave packets
random.f95	5.2.2	Java random utility	tune.f95	14.14.4	Optimization testing
rk4.f95	9.5.2	4th-order Runge–Kutta	twodsol.f95	CPI	2-D Sine–Gordon solitons
rk45.f95	9.5.2	Adaptive step rk solver	twoplates.f95	19.9.3	2-D Navier–Stokes fluid
			walk.f95	5.4.2	Random-walk simulation

MPIcodes Contents

Program	*Description*	*Program*	*Description*
MPIhello.c	First MPI program	MPIdeadlock.c	Deadlock examples
MPIdeadlock-fixed.c	Fixed deadlocks	MPImessage2.c	Send and receive messages
MPImessage3.c	More messages	MPIpi.c	Parallel computation of π
MPIstring.c	Parallel string equation	run_mpi.sh	Template script for grid engine
TuneMPI.c	Optimization tuning for MPI		

Fortran77codes Contents by Name

Program	*Section*	*Description*	*Program*	*Section*	*Description*
area.f	1.4.2	First program	decay.f	5.6	Spontaneous decay
bessel.f	2.2.2	Bessel function recurrence	diff.f	7.5	Numerical differentiation
bound.f	20.2.3	Bound integral equation	eqheat.f	17.17.4	Heat equation via leapfrog
bugs.f	12.5	Bifurcation diagram	eqstring.f	18.2.3	Waves on a string
complex.f	4.4	Complex number objects	exp-bad.f	1.6	Series summation

(continued)

Fortran77codes **Contents by Name** *(continued)*

Program	*Section*	*Description*	*Program*	*Section*	*Description*
exp-good.f	1.6	Series summation	qmc.f	15.8.3	Feynman path integration
fit.f	8.7.1	Linear least–squares fit	random.f	5.2.2	Java random utility
fourier.f	10.4.1	Discrete Fourier transform	rk4.f	9.5.2	4th-order Runge–Kutta
gauss.f	6.2.5	Gaussian quadrature	scatt.f	20.4.5	Scattering integral equation
harmos.f	18.6.2	Quantum wave packets			
int10d.f	6.6.3	10-D Monte Carlo integral	slit.f	18.6.2	Wave packet through slit
integ.f	6.2.5	Gaussian quadrature	soliton.f	19.5.3	KdeV solitons
ising.f	15.2	Ising model	spline.f	8.5.5	Cubic spline fit
lagrange.f	8.5.1	Lagrange interpolation	sqwell.f	18.6.2	Quantum wave packets
laplace.f	17.4.2	Finite differential Laplace equation	twodsol.f	CPI	2-D Sine–Gordon solitons
limit.f	1.5.4	Machine precision	walk.f	5.4.2	Random-walk simulation
over.f	1.5.4	Machine precision			
pond.f	6.5	Monte Carlo integration			

PVMcodes **Contents**

Program	*Description*	*Program*	*Description*
bugs.f	Serial bifurcation mapping	input	Simple input file
Makefile	Makefile for master/ worker program	PVMbugsMaster.c	Bifurcation master (C)
PVMbugsMaster.f	Bifurcation master (Fortran)	PVMbugsSlave.c	Bifurcation worker (C)
PVMbugsSlave.f	Bifurcation worker (Fortran)	PVMcommunMaster.c	Communications master
PVMcommunSlave.c	Communications worker	PVMmonteMaster.c	Monte Carlo integration master
PVMmonteSlave1.c	Monte Carlo integration worker 1	PVMmonteSlave2.c	Monte Carlo integration worker 2
PVMmonteSlave3.c	Monte Carlo integration worker 4	PVMmonteSlave4.c	Monte Carlo integration worker 4

OpenDX Contents

Color DX Figures			
DXdata			
.DS_Store	custom_linear_plot.cfg	custom_linear_plot.net	custom_linear_plot2.cfg
custom_linear_plot2.net	H_atom_prob_density-3D0.dat	laplace-colormap.cm	laplace_3d.cfg
H_atom_prob_density-3P1.general	H_atom_prob_density.cfg	H_atom_prob_density.cm	H_atom_prob_density.net
H_atom_prob_density-3D0.general	laplace.cfg	laplace.net	laplace3d-colormap.cm
laplace_3d-rings.dat	laplace_3d-rings.general	shock.dat	laplace_3d.dat
laplace_3d.general	laplace_3d.net	laplace_dx.dat	laplace_dx.general
shock.cfg	H_atom_prob_density-3P1.dat	shock.general	shock.net
simple_linear_data.dat	simple_linear_data.general	simple_linear_plot.cfg	simple_linear_plot.net
DXjava			
.DS_store	H_atom_wf.java	laplace.java	laplace_3d.java
laplace_3d_v1.java	laplace_circular.java	simple_2d_data.java	simple_linear_data.java

Appendix G: Compression via DWT with Thresholding

An important application of discrete wavelet transformation is image compression. Anyone who has stored high-resolution digital images or videos knows that such files can be very large. DWT can reduce image size significantly with just a minimal loss of quality by storing only a small number of smooth components and only as many detailed components as needed for a fixed resolution. To compress an image, we compute and store the DWT of each row of the image, setting all wavelet coefficients smaller than a certain threshold to zero (*thresholding*). When the image is needed, the inverse wavelet transform is used to reconstruct each row of the original image. For example, Table G.1 shows file sizes resulting from using DWT for image compression. We note (columns 1 and 2) that there is a factor-of-2 reduction in size arising from storing the DWT rather than the data themselves, and that this is independent of the threshold criteria (it still stores the zeros). After the compression program *WinZip* removes the zeros (columns 3 and 4), we note that the DWT file is a factor of 3 smaller than the compressed original, and a factor of 11 smaller than the noncompressed original.

A usual first step in dealing with digital images is to convert from one picture format to another.[1] We started with a 3073×2304 *pixel* (picture element) jpg (Joint Photographic Experts Group) color image. For our example (Figure G.1 left) we reduced the image size to 512×512 (a power of 2) and converted it to gray scale. You can do this with the free program IrfanView [irfanview] (Windows), the GNU image editor The Gimp (Unix/Linux) [Gimp], or *Netpbm* [Netpbm]. We used Netpbm's utility jpegtopnm to convert the .jpg to the portable gray map image in Netpbm format:

> **jpegtopnm marianabw.jpg > mariana.pnm** Convert jpeg to pnm

If you open mariana.pnm (available on the CD), you will find strings such as

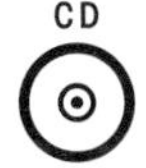


```
P5
512 512
255
yvttsojgrvv|yngcawbRQ] ...
```

This format is not easy to work with, so we converted it to mariana,

> **pnmtopnm -plain mariana.pnm>mariana** Convert pnm format to integers

Except for the first three lines that contain information about the internal code (width, height, 0 for black, and 255 for white), we now have integers:

[1] We thank Guillermo Avendanño-Franco for help with image programs.

TABLE G.1
Compression of a Data File Using DWT with a 20% Cutoff Threshold

Original File	DWT Reconstructed	WinZipped Original	WinZipped DWT
25,749	11,373	7,181	2,447

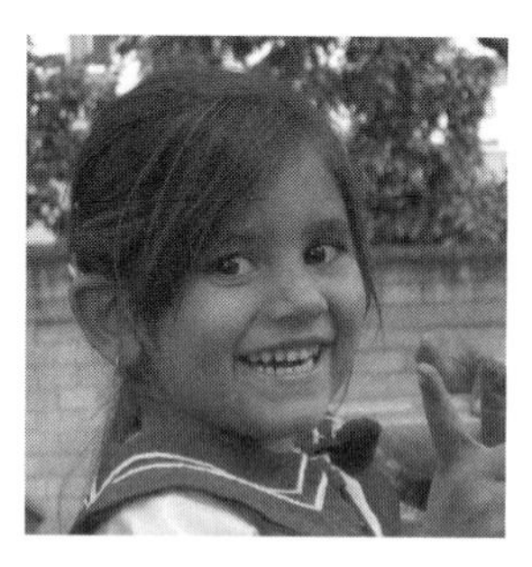

Figure G.1 *Left:* The original picture of Mariana. *Middle:* The reconstituted picture for a compression ratio of 8 ($\epsilon = 9$). *Right:* The reconstituted picture for a compression ratio of 46 ($\epsilon = 50$). The black dots in the images are called "salt and pepper" and can be eliminated.

```
P2
512 512
255
143 174 209 235 249 250 250 255 255 255 255 255 255 255 255 255 255
255 255 255 255 254 253 253 254 254 255 255 255 255 255 255 255 255
. . .
```

Because Java is not very flexible with I/O formats, we wrote a small C program to convert mariana to the one-column format mariana.dat:

```
#include <string.h>
#include <stdlib.h>
// Reads mariana.dat in decimal ascii, form 1-column mariana.dat
main()  {
  int i, j, k;
  char cd[10], c;
  FILE *pf, *pout;
  pf=fopen("mariana", "r");
  pout=fopen("mariana.dat","w");
  fscanf(pf,"%s",&cd);          printf("%s\n",cd);
  fscanf(pf,"%s",&cd);          printf("%s\n",cd);
  fscanf(pf,"%s",&cd);          printf("%s\n",cd);
  fscanf(pf,"%s",&cd);          printf("%s\n",cd);
  for ( k = 0; k < 512; k++ )  {
    for ( j = 0; j < 512; j++ ) fscanf(pf,"%d",&i); fprintf(pout,"%d\n",i);
  }
  fclose(pf);  fclose(pout);
}
```

Now that we have an accessible file format, we compress and expand it:

1. Read the one-column file and form the array fg[512][512] containing all the information about the image.
2. Use the program DaubMariana.java on the CD to apply the Daubechies 4-wavelet transform to each row of fg. This forms a new 2-D array containing the transformed matrix.
3. Use DaubMariana.java again, now transforming each column of the transformed matrix and saving the result in a different array.
4. Compress the image by selecting the tolerance level eps=9 and eliminating all signals that are smaller than eps:
   if abs(fg[i][j]) < eps, fg[i][j]=0
5. Apply the inverse wavelet transform to the modified array, column by column and then row by row.
6. Save the reconstituted image to Marianarec via output redirection:
   > **java DaubMariana > Marianarec**
7. The program also creates the file comp-info.dat containing information about the compression:

```
Number of nonzero coefs before compression: 262136
If abs(coef)<9coef=0
Number of nonzero coefficients after compression: 32753
Compression ratio:8.003419534088481
```

If the program is compiled and run again, the resulting output will contain the compression image in a file with many zeros:

```
P2
512 512
255
143 1543 2296 −405 594 −311 50 68 444 −375 0 20 0 274 −138 423
 371 −19 −51 −100 24 0 118 120 62 −59 404 −111 61 −82 306 −255
204 0 81 −60 0 −29 83 −214 55 −12 −34 24 −32 46 −37 29 −14 45 47
. . .
0 0 0 0 0 0 16 −24 0 0 0 −13 0 −32 0 0 0 0 0 9 −29 28 −21 13
0 −11 0 18 9 −18 0 −11 11 0 0 0 0 0 10 −20 0 0 0 0 −11 0 0
0 0 0 0 0 27 −29 0 −18 0 21 24 −54 15 0 0 0 0 0 0 0 0 0 0 0
−12 9 −9 9 0 19 0 0 0 0 0 0 −9 0 0 0 0 0 0 0 0 72 −53 0 0 0 0
0 0 44 35 0 0 0 0 0 0 0 0 0 0 0 0 0 0 0 0 0 0 0 0 0 0 −12 0 0
0 0 0 0 0 0 0 0 0 0 0 0 0 0 0 0 0 0 0 0 0 0 0 0 −19 0 11 0 −18
. . .
```

Although this is an image file, it is not meant for viewing as an image (it will be mainly black). ∎

G.1 More on Thresholding

Often in practical applications, a good number of the wavelet coefficients are nearly equal to zero. When these coefficients are set to zero via some thresholding scheme

[D&J 94], DWT files containing long strings of zeros result. Through a type of compression known as *entropy coding*, the amount of memory needed to store files of this sort can be greatly reduced.

Before we go on to compression algorithms, it is important to note that there are different types of thresholding. In *hard thresholding*, a rather arbitrary tolerance is selected, and any wavelet coefficient whose absolute value is below this tolerance is set to zero. Presumably, these small numbers have only small effects on the image. In *soft thresholding*, a tolerance h is selected, and, as before, any wavelet coefficient whose absolute value is below this tolerance is set to zero. However, all other entries d are replaced with $sign(d) \mid |d| - h|$. Soft thresholding can be thought of as a translation of the signal toward zero by the amount h. Finally, in *quantile thresholding*, a percentage p of entries are selected, and p percent of those entries with the smallest absolute values are set to zero.

DWT with thresholding is useful in analyzing signals and for compressing signals so that less memory is needed to store the transforms than to store the original signals. However, we have yet to take advantage of the frequent occurrence of zeros as wavelet coefficients. *Huffman entropy coding* is well suited for compressing data that contain many zeros. With this method, an integer sequence q is changed to a shorter sequence e that is stored as 8-bit integers. Strings of zeros are coded by the numbers 1–100, 105, and 106, while nonzero integers in q are coded by 101–104 and 107–254. The idea is to use two or three numbers for coding, with the first being a signal that a large number or a long zero sequence is coming. Entropy coding is designed so that the numbers that are expected to appear the most often in q need the least amount of space in e.

A step in compression, known as *quantization*, converts a sequence w of floating-point numbers to a sequence q of integers. The simplest technique is to round the floats to the nearest integer. Another option is to multiply each number in w by a constant k and then round to the nearest integer. Quantization is called *lossy* because information is lost when a float is converted to an integer. In Table G.1 we showed the effect of compression using the *WinZip* data compression algorithm. This is a hybrid of LZ77 and Huffman coding also known as *Deflate*.

G.2 Wavelet Implementation and Assessment

1. Write a program to plot *Daub4* wavelets. (Our sample program is `Daub4.java`.) Observe the behavior of the wavelet functions for different values of the coefficients. In order to do this, place a 1 in the coefficient vector for the wavelet structure you want and place 0's in all other locations. Then perform the inverse transform to produce the physical domain representation of the wavelet.
2. Run the code `Daub4.java` for different threshold values.
3. Run the code `DaubCompress.java` that uses other functions to give input data.
4. Write a Java program or applet that compresses a 512×512 image using *Daub4* wavelets. To do this, extend method `wt1` so that it performs a 2-D wavelet transform. Note that you may need some methods from the `java.awt.image`

package to plot the images. First, create an Image object using Image img. The MemoryImageSource class is used to create an image from an array of pixel values using the constructor MemoryImageSource(int w, int h, int pix[], int depls, int scan). Here w and h are the dimensions of the image, pix[] is an array containing the pixels values, depls is the deployment of data in pix[], and scan is the length of a row in pix[]. Finally, draw the image using drawImage(Image img, int x, int y, this), where x and y are the coordinates of the left corner of the image. Alternatively, you can use a program such as Matlab or Maple to display images from an array of integers.

5. Modify the program DaubCompress.java so that it outputs the DWT to a file.
6. Pick a different function to analyze, the more complicated the better.
7. Plot the resulting DWT data in a meaningful way.
8. Show in your plots the effects of increasing the threshold parameter in order to cut out more of the smaller transform values.
9. Examine the reconstituted signal for various threshold values including zero and note the effect of the cutoff on the image quality. ∎

BIBLIOGRAPHY

[Abar 93] Abarbanel, H. D. I., M. I. Rabinovich, and M. M. Sushchik (1993). *Introduction to Nonlinear Dynamics for Physicists,* World Scientific, Singapore.

[A&S 72] Abramowitz, M., and I. A. Stegun (1972). *Handbook of Mathematical Functions,* 10th Ed., U.S. Government Prnting Office, Washington, DC.

[Add 02] Addison, P. S. (2002). *The Illustrated Wavelet Transform Handbook,* Institute of Physics Publishing, Bristol and Philadelphia, PA.

[ALCMD] Morris, J., D. Turner, and K.-M. Ho. *AL_CMD, Ames Laboratory Classical Molecular Dynamics,* `cmp.ameslab.gov/cmp/CMP_Theory/cmd/alcmd_source.html`.

[A&T 87] Allan, M. P., and J. P. Tildesley (1987). *Computer Simulations of Liquids,* Oxford Science Publications, Oxford, UK.

[Amd 67] Amdahl, G. (1967). *Validity of the single-processor approach to achieving large-scale computing capabilities,* Proc. AFIPS., **30**, 483.

[Anc 02] Ancona, M. G. (2002). *Computational Methods for Applied Science and Engineering,* Rinton Press, Princeton, NJ.

[A&W 01] Arfken, G. B., and H. J. Weber (2001). *Mathematical Methods for Physicists,* Harcourt/Academic Press, San Diego.

[Argy 91] Argyris, J., M. Haase, and J. C. Heinrich (1991). *Finite element approximation to two-dimensional Sine–Gordon solitons,* Comput. Methods Appl. Mech. Eng. **86**, 1.

[Arm 91] Armin, B., and H. Shlomo, Eds. (1991). *Fractals and Disordered Systems,* Springer-Verlag, Berlin.

[Ask 77] Askar, A., and A. S. Cakmak (1977). *Explicit integration method for the time-dependent Schrödinger equation for collision problems,* J. Chem. Phys. **68**, 2794.

[Bai 05] Bailey, M. *OSU ChromaDepth Scientific Visualization Gallery,* `web.engr.oregonstate.edu/~mjb/chromadepth/`.

[Bana 99] Banacloche, J. G. (1999). *A quantum bouncing ball,* Am. J. Phys. **67**, 776.

[Barns 93] Barnsley, M. F., and L. P. Hurd (1993). *Fractal Image Compression,* A. K. Peters, Wellesley, MA.

[Becker 54] Becker, R. A. (1954). *Introduction to Theoretical Mechanics,* McGraw-Hill, New York.

[Berry] Berryman, A. A. *Predator-Prey Dynamics,* `classes.entom.wsu.edu/543/`.

[B&R 02] Bevington, P. R., and D. K. Robinson (2002). *Data Reduction and Error Analysis for the Physical Sciences,* 3rd Ed., McGraw-Hill, New York.

[Bin 01] Binder, K., and D. W. Heermann (2001). *Monte Carlo Methods,* Springer-Verlag, Berlin.

[Bleh 90] Bleher, S., C. Grebogi, and E. Ott (1990). *Bifurcations in chaotic scattering,* Phys. D, **46**, 87.

[Burg 74] Burgers, J. M. (1974). *The Non-Linear Diffusion Equation: Asymptotic Solutions and Statistical Problems,* Reidel, Boston.

[DX2] Braun, J. R. Ford, and D. Thompson (2001). *OpenDX: Paths to Visualization,* Visualization and Imagery Solutions, Missoula, MT.

[B&H 95] Briggs, W. L., and V. E. Henson (1995). *The DFT: An Owner's Manual*, SIAM, Philadelphia.

[C&P 85] R. Car, and M. Parrinello (1985). *Unified approach for molecular dynamics and density-functional theory*, Phys. Rev. Lett. **55,** 2471.

[C&P 88] Carrier, G. F., and C. E. Pearson (1988). *Partial Differential Equations*, Academic Press, San Diego.

[C&L 81] Christiansen, P. L., and P. S. Lomdahl (1981). *Numerical solutions of 2 + 1 dimensional Sine–Gordon solitons* Phys. **2D,** 482.

[CPUG] *CPUG*, Computational physics degree program for undergraduates, Oregon State University, www.physics.oregonstate.edu/CPUG.

[C&N 47] Crank, J., and P. Nicolson (1946). *A practical method for numerical evaluation of solutions of partial differential equations of the heat conduction type*, Proc. Cambridge Phil. Soc. **43,** 50.

[C&O 78] Christiansen, P. L., and O. H. Olsen (1978). *Ring-shaped quasi-soliton solutions to the two- and three-dimensional Sine–Gordon equation*, Phys. Lett. **68A,** 185; (1979) Phys. Scr. **20**, 531.

[Chrom] ChromaDepth Technologies, www.chromatek.com/.

[Clark] Clark University, *Statistical and Thermal Physics Curriculum Development Project*, stp.clarku.edu/; *Density of States of the 2D Ising Model*, stp.clarku.edu/simulations/ising/wanglandau.html.

[Co 65] Cooley, J. W., and J. W. Tukey, (1965). *An algorithm for the machine calculation of complex Fourier series*, Math. Comput. **19**, 297.

[Cour 28] Courant, R., K. Friedrichs, and H. Lewy (1928). *Über die partiellen Differenzengleichungen der mathematischen Physik*, Math. Ann. **100**, 32.

[Cre 81] Creutz, M., and B. Freedman (1981). *A statistical approach to quantum mechanics*, Ann. Phys. (N.Y.) **132**, 427.

[CYG] Cygwin, a Linux-like environment for Windows, x.cygwin.com/.

[Da 42] Danielson, G. C., and C. Lanczos (1942). *Some improvements in practical Fourier analysis and their application to X-ray scattering from liquids*, J. Franklin Inst. **233**, 365.

[Daub 95] Daubechies, I. (1995). *Wavelets and other phase domain localization methods*, Proc. Int. Cong. Math. Basel, **1, 2** 56, Birkhäuser.

[DeJ 92] De Jong, M. L. (1992). *Chaos and the simple pendulum*, Phys. Teacher **30**, 115.

[DeV 95] DeVries, P. L. (1996). *Resource letter CP-1: Computational Physics*, Am. J. Phys. **64**, 364.

[Dong 05] Dongarra, J., T. Sterling, H. Simon, and E. Strohmaier (2005). *High-performance computing*, IEEE/AIP Comput. Sci. Eng. **7**, 51.

[Donn 05] Donnelly, D., and B. Rust (2005). *The fast Fourier transform for experimentalists*, IEEE/AIP Comput. Sci. Eng. **7**, 71.

[D&J 94] Donoho, D. L., and I. M. Johnstone (1994). *Ideal denoising in an orthonormal basis chosen from a library of bases*, Compt. Rend. Acad. Sci. Paris Ser. A, **319**, 1317.

[DX1] Dx, Open DX, The open source software project based on IBM's Visualization Data Explorer, www.opendx.org/.

[Jtut] Eck, D., (2002), *Introduction to Programming Using Java, Version 4* (a free textbook), math.hws.edu/javanotes/.

[Eclipse] Eclipse, an open development platform, www.eclipse.org/.

[Erco] Ercolessi, F., *A Molecular Dynamics Primer*, www.ud.infn.it/~ercolessi/md/.

[E&P 88] Eugene, S. H., and M. Paul (1988). *Multifractal phenomena in physics and chemistry*, Nature **335**, 405.

[F&S] Falkovich, G., and K. R. Sreenivasan (2006). *Lesson from hydrodynamic turbulence*, Phys. Today **59**, 43.

[Fam 85] Family, F., and T. Vicsek (1985). *Scaling of the active zone in the Eden process on percolation networks and the ballistic deposition model*, J. Phys. A **18**, L75.

[Feig 79] Feigenbaum, M. J. (1979). *The universal metric properties of nonlinear transformations*, J. Stat. Phys. **21**, 669.

[F&W 80] Fetter, A. L., and J. D. Walecka (1980). *Theoretical Mechanics of Particles and Continua*, McGraw-Hill, New York.

[F&H 65] Feynman, R. P., and A. R. Hibbs (1965). *Quantum Mechanics and Path Integrals*, McGraw-Hill, New York.

[Fitz 04] Fitzgerald, R. (2004). *New experiments set the scale for the onset of turbulence in pipe flow*, Phys. Today **57**, 21.

[Fos 96] Fosdick L. D, E. R. Jessup, C. J. C. Schauble, and G. Domik (1996). *An Introduction to High Performance Scientific Computing*, MIT Press, Cambridge, MA.

[Fox 03] Fox, G. (2003). *HPJava: A data parallel programming alternative*, IEEE/AIP Comput. Sci, Eng. **5**, 60.

[Fox 94] Fox, G. (1994). *Parallel Computing Works!* Morgan Kaufmann, San Diego.

[Gara 05] Gara, A., M. A. Blumrich, D. Chen, G. L.-T. Chiu, P. Coteus, M. E. Giampapa, R. A. Haring, P. Heidelberger, D. Hoenicke, G. V. Kopcsay, T. A. Liebsch, M. Ohmacht, B. D. Steinmacher-Burow, T. Takken, and P. Vranas, *Overview of the Blue Gene/L system architecure* (2005). IBM J. Res. Dev. **49**, 195.

[Gar 00] Garcia, A. L. (2000). *Numerical Methods for Physics*, 2nd Ed., Prentice Hall, Upper Saddle River, NJ.

[Gibbs 75] Gibbs, R. L. (1975). *The quantum bouncer*, Am. J. Phys. **43**, 25–28.

[Good 92] Goodings, D. A., and T. Szeredi (1992), *The quantum bouncer by the path integral method*, Am. J. Phys. **59**, 924–930.

[Gimp] GIMP, the GNU image manipulation program, www.gimp.org/.

[GNU] Gnuplot, a portable command-line driven interactive data and function plotting utility, www.gnuplot.info/.

[Gold 67] Goldberg, A., H. M. Schey, and J. L. Schwartz (1967). *Computer-generated motion pictures of one-dimensional quantum-mechanical transmission and reflection phenomena*, Am. J. Phys. **35**, 177.

[Gos 99] Goswani, J. C., and A. K. Chan (1999). *Fundamentals of Wavelets*, John Wiley, New York.

[Gott 66] Gottfried, K. (1966), *Quantum Mechanics*, Benjamin, New York.

[G,T&C 06] Gould, H., J. Tobochnik, and W. Christian (2006). *An Introduction to Computer Simulation Methods*, 3rd Ed., Addison-Wesley, Reading, MA.

[Grace] Grace; A WYSIWYG 2D plotting tool for the X Window System (descendant of ACE/gr, Xmgr), plasma-gate.weizmann.ac.il/Grace/.

[Graps 95] Graps, A. (1995). *An introduction to wavelets*, IEEE/AIP Comput. Sci. Eng. **2**, 50.

[BCCD] Gray, P., and T. Murphy (2006). *Something wonderful this way comes*, Comput. Sci. Eng. **8**, 82; bccd.cs.uni.edu/.

[Gurney] Gurney, W. S. C., and R. M. Nisbet (1998). *Ecological Dynamics*, Oxford University Press, Oxford, UK.

[H&T 70] Haftel, M. I., and F. Tabakin (1970). *Off-shell effects in nuclear matter*, Nucl. Phys. **158**, 1.

[Har 96] Hardwich, J. *Rules for Optimization*, www.cs.cmu.edu/~jch/java.

[Hart 98] Hartmann, W. M. (1998). *Signals, Sound, and Sensation*, AIP Press, Springer-Verlag, New York.

[Hi,76] Higgins, R. J. (1976). *Fast Fourier transform: An introduction with some minicomputer experiments*, Am. J. Phys. **44**, 766.

[Hock 88] Hockney, R.W., and J. W. Eastwood (1988). *Computer Simulation Using Particles*, Adam Hilger, Bristol, UK.

[Huang 87] Hunag, K. (1987). *Statistical Mechanics*, John Wiley, New York.

[Intel] *Intel Cluster Tools*, www3.intel.com/cd/software/products/asmo-na/eng/cluster/244171.htm; *Intel Compilers*, www3.intel.com/cd/software/products/asmo-na/eng/compilers/284264.htm.

[irfanview] irfanview, www.irfanview.com/.

[Jack 88] Jackson, J. D. (1988). *Classical Electrodynamics*, 3rd Ed., John Wiley, New York.

[Jama] JAMA, a Java matrix package; Java Numerics, math.nist.gov/javanumerics/jama/.

[J&S 98] José, J. V, and E. J. Salatan (1988). *Classical Dynamics*, Cambridge University Press, Cambridge, UK.

[jEdit] jEdit, a mature programmer's text editor, www.jedit.org/.

[K&R 88] Kernighan, B., and D. Ritchie (1988). *The C Programming Language*, 2nd Ed., Prentice Hall, Englewood Cliffs, NJ.

[Koon 86] Koonin, S. E. (1986). *Computational Physics*, Benjamin, Menlo Park, CA.

[KdeV 95] Korteweg, D. J., and G. deVries (1895). *On the change of form of long waves advancing in a rectangular canal, and on a new type of long stationary waves*, Phil. Mag. **39**, 4.

[Krey 98] Kreyszig, E. (1998). *Advanced Engineering Mathematics*, 8th Ed., John Wiley, New York.

[Kutz] Kutz N., *Scientific Computing*, www.amath.washington.edu/courses/581-autumn-2003/.

[Lamb 93] Lamb, H. (1993). *Hydrodynamics*, 6th Ed., Cambridge University Press, Cambridge, UK.

[L&L,F 87] Landau, L. D., and E. M. Lifshitz (1987). *Fluid Mechanics*, 2nd Ed., Butterworth-Heinemann, Oxford, UK.

[L&L,M 76] Landau, L. D., and E. M. Lifshitz (1976). *Quantum Mechanics*, Pergamon, Oxford, UK.

[L&L,M 77] Landau, L. D., and E. M. Lifshitz (1976). *Mechanics*, 3rd Ed., Butterworth-Heinemann, Oxford, UK.

[L 05] Landau, R. H. (2005), *A First Course in Scientific Computing*, Princeton University Press, Princeton, NJ.

[L 96] Landau, R. H. (1996). *Quantum Mechanics II: A Second Course in Quantum Theory*, 2nd Ed., John Wiley, New York.

[L&F 93] Landau, R. H., and P. J. Fink (1993). *A Scientist's and Engineer's Guide to Workstations and Supercomputers*, John Wiley, New York.

[Lang] Lang, W. C., and K. Forinash (1998). *Time-frequency analysis with the continuous wavelet transform*, Am. J. Phys. **66**, 794.

[LAP 00] Anderson, E., Z. Bai, C. Bischof, J. Demmel, J. Dongarra, J. Du Croz, A. Greenbaum, S. Hammarling, A. McKenney, S. Ostrouchov, and D. Sorensen (2000). *LAPACK User's Guide*, 3rd Ed., SIAM, Philadelphia, netlib.org.

[Li] Li, Z., *Numerical Methods for Partial Differential Equations—Finite Element Method*, www4.ncsu.edu/~zhilin/TEACHING/MA587/.

[Libb 03] Liboff, R. L. (2003). *Introductory Quantum Mechanics*, Addison Wesley, Reading, MA.

[Lot 25] Lotka, A. J. (1925). *Elements of Physical Biology*, Williams & Wilkins, Baltimore.

[MacK 85] MacKeown, P. K. (1985). *Evaluation of Feynman path integrals by Monte Carlo methods*, Am. J. Phys. **53**, 880.

[Lusk 99] Lusk, W. E., and A. Skjellum (1999). *Using MPI: Portable Parallel Programming with the Message-Passing Interface*, 2nd Ed., MIT Press, Cambridge, MA.

[M&N 87] MacKeown, P. K., and D. J. Newman (1987). *Computational Techniques in Physics*, Adam Hilger, Bristol, UK.

[MLP 00] Maestri, J. J. V., R. H. Landau, and M. J. Paez (2000). *Two-particle Schrödinger equation animations of wave packet–wave packet scattering*, Am. J. Phys. **68**, 1113.

[Mallat 89] Mallat, P. G. (1982). *A theory for multiresolution signal decomposition: The wavelet representation*, IEEE Transa. Pattern Anal. Machine Intelligence, **11**, 674.

[Mand 67] Mandelbrot, B. (1967). *How long is the coast of Britain?* Science, **156**, 638.

[Mand 82] Mandelbrot, B. (1982). *The Fractal Geometry of Nature*, Freeman, San Francisco, p.29.

[Mann 90] Manneville, P. (1990). *Dissipative Structures and Weak Turbulence*, Academic Press, San Diego.

[Mann 83] Mannheim, P. D. (1983). *The physics behind path integrals in quantum mechanics*, Am. J. Phys. **51**, 328.

[M&T 03] Marion, J. B., and S. T. Thornton (2003). *Classical Dynamics of Particles and Systems*, 5th Ed., Harcourt Brace Jovanovich, Orlando, FL.

[Math 02] Mathews, J. (2002). *Numerical Methods for Mathematics, Science, and Engineering*, Prentice Hall, Upper Saddle River, NJ.

[Math 92] Mathews, J. (1992). *Numerical Methods for Mathematics, Science, and Engineering*, Prentice Hall, Englewood Cliffs, NJ.

[M&W 65] Mathews, J., and R. L. Walker (1965). *Mathematical Methods of Physics*, Benjamin, Reading, MA.

[MW] Mathworks, Matlab Wavelet Toolbox, www.mathworks.com/.

[Metp 53] Metropolis, M., A. W. Rosenbluth, M. N. Rosenbluth, A. H. Teller, and E. Teller (1953). *Equation of state calculations by fast computing machines*, J. Chem. Phys. **21**, 1087.

[Mold] Refson, K. *Moldy, A General-Purpose Molecular Dynamics Simulation Program*, www.earth.ox.ac.uk/~keithr/moldy.html.

[M&L 85] Moon, F. C., and G.-X. Li (1985). *Fractal basin boundaries and homoclinic orbits for periodic motion in a two-well potential*, Phys. Rev. Lett. **55**, 1439.

[M&F 53] Morse, P. M., and H. Feshbach (1953). *Methods of Theoretical Physics*, McGraw-Hill, New York.

[MPI] Math. and Computer Science Division, Argonne National Laboratory (2006). *The Message Passing Interface (MPI) Standard* (updated May 9, 2006), www-unix.mcs.anl.gov/mpi/.

[MPI2] Mathematics and Computer Science Division, Argonne National Laboratory (2004). *Web Pages for MPI and MPE* (updated August 4, 2004), www-unix.mcs.anl.gov/mpi/www.

[MPImis] Academic Computing and Communications Center, University of Illinois at Chicago (2004). *Argo Beowulf Cluster: MPI Commands and Examples* (updated December 3, 2004), www.uic.edu/depts/accc/hardware/argo/mpi_routines.html.

[NAMD] Nelson, M., W. Humphrey, A. Gursoy, A. Dalke, L. Kale, R. D. Skeel, and K. Schulten (1996). *NAMD—Scalable Molecular Dynamics*, J. Supercomput. Appl. High Performance Comput., www.ks.uiuc.edu/Research/namd/.

[NSF] Nation Science Foundation Supercomputer Centers: Cornell Theory Center, www.tc.cornell.edu; National Center for Supercomputing Applications, www.ncsa.uiuc.edu; Pittsburgh Supercomputing Center, www.psc.edu; San Diego Supercomputing Center, www.sdsc.edu; National Center for Atmospheric Research, www.ucar.edu.

[Nes 02] Nesvizhevsky, V. V., H. G. Borner, A. K. Petukhov, H. Abele, S. Baessler, F. J. Ruess, T. Stoferle, A. Westphal, A. M. Gagarski, G. A. Petrov, and A. V. Strelkov, (2002). *Quantum states of neutrons in the Earth's gravitational field*, Nature **415**, 297.

[Netpbm] Netpbm, a package of graphics programs and programming library, netpbm.sourceforge.net/doc/.

[Ott 02] Ott, E. (2002). *Chaos in Dynamical Systems*, Cambridge University Press, Cambridge, UK.

[Otto] Otto A., *Numerical Simulations of Fluids and Plasmas*, what.gi.alaska.edu/ao/sim/chapters/chap6.pdf.

[OR] Oualline, S. (1997). *Practical C Programming*, O'Reilly and Associates, Sebastopol.

[Pach 97] Pacheco, P. S. (1997). *Parallel Programming with MPI*, Morgan Kaufmann, San Diego.

[Pan 96] Pancake, C. M. (1996). *Is Parallelism for You?* IEEE Comput. Sci. Eng. **3**, 18.

[PBS] *Portable Batch System*, www.openpbs.org/.

[P&D 81] Pedersen, N. F., and A. Davidson (1981). *Chaos and noise rise in Josephson junctions*, Appl. Phys. Lett. **39**, 830.

[Peit 94] Peitgen, H.-O., H. Jürgens, and D. Saupe (1992). *Chaos and Fractals*, Springer-Verlag, New York.

[Penn 94] Penna, T. J. P. (1994). *Fitting curves by simulated annealing*, Comput. Phys. **9**, 341.

[Perlin] Perlin, K., NYU Media Research Laboratory, mrl.nyu.edu/~perlin.

[P&R 95] Phatak, S. C., and S. S. Rao (1995). *Logistic map: A possible random-number generator*, Phys. Rev. E **51**, 3670.

[PhT 88] Physics Today, Special issue on chaos, December 1988.

[P&W 91] Pinson, L. J., and R. S. Wiener (1991). *Objective-C Object-Oriented Programming Techniques*, Addison-Wesley, Reading, MA.

[P&B 94] Plischke, M., and B. Bergersen (1994). *Equilibrium Statistical Physics*, 2nd Ed., World Scientific, Singapore.

[Polikar] Polikar, R., *The Wavelet Tutorial*, users.rowan.edu/~polikar/WAVELETS/WTtutorial.html.

[Potv 93] Potvin, J. (1993). *Computational quantum field theory*, Comput. Phys. **7**, 149.

[Pov-Ray] *Persistence of Vision Raytracer*, www.povray.org.

[Pres 94] Press, W. H., B. P. Flannery, S. A. Teukolsky, and W. T. Vetterling (1994). *Numerical Recipes*, Cambridge University Press, Cambridge, UK.

[Pres 00] Press, W. H., B. P. Flannery, S. A. Teukolsky, and W. T. Vetterling (2000). *Numerical Recipes in C++*, 2nd Ed., Cambridge University Press, Cambridge, UK.

[PtPlot] PtPlot, a 2-D data plotter and histogram tool implemented in Java, ptolemy.eecs.berkeley.edu/java/ptplot/.

[PVM] A. Geist, A. Beguelin, J. Dongarra, W. Jiang, R. Manchek, and V. Sunderam (1994). *PVM: Parallel Virtual Machine—A User's Guide and Tutorial for Networked Parallel Computing*, Oak Ridge National Laboratory, Oak Ridge, TN.

[Quinn 04] Quinn, M. J. (2004). *Parallel Programming in C with MPI and OpenMP*, McGraw Hill, New York.

[Ram 00] Ramasubramanian, K., and M. S. Sriram (2000). *A comparative study of computation of Lyapunov spectra with different algorithms*, Physica D **139**, 72.

[Rap 95] Rapaport, D.C (1995). *The Art of Molecular Dynamics Simulation*, Cambridge University Press, Cambridge, UK.

[Rash 90] Rasband, S. N. (1990). *Chaotic Dynamics of Nonlinear Systems*, John Wiley, New York.

[Raw 96] Rawitscher, G., I. Koltracht, H. Dai, and C. Ribetti (1996). *The vibrating string: A fertile topic for teaching scientific computing*, Comput. Phys. **10**, 335.

[R&M93] Reitz, J. R., F. J. Milford, and Christy, R. W. (1993). *Foundations of Electromagnetic Theory*, 4th Ed., Addison-Wesley, Reading, MA.

[Rey 83] Reynolds, O. (1883). *An experimental investigation of the circumstances which determine whether the motion of water in parallel channels shall be direct or sinuous and of the law of resistance in parallel channels*, Proc. R. Soc. Lond. **35**, 84.

[Rhei 74] Rheinbold, W. C. (1974). *Methods for Solving Systems of Nonlinear Equations*, SIAM, Philadelphia.

[Rich 61] Richardson. L. F. (1961). *Problem of contiguity: An appendix of statistics of deadly quarrels*, Gen. Systems Yearbook, **6**, 139.

[Riz] Riznichenko G. Y., *Mathematical Models in Biophysics*, www.biophysics.org/education/galina.pdf.

[Rowe 95] Rowe, A. C. H., and P. C. Abbott (1995). *Daubechies Wavelets and Mathematica*, Comput. Phys. **9**, 635–548.

[Russ 44] Russell, J. S. (1844), *Report of the 14th Meeting of the British Association for the Advancement of Science*, John Murray, London.

[Sand 94] Sander, E., L. M. Sander, and R. M. Ziff (1994). *Fractals and fractal correlations*, Comput. Phys. **8**, 420.

[Schk 94] Scheck, F. (1994). *Mechanics, from Newton's Laws to Deterministic Chaos*, 2nd Ed., Springer-Verlag, New York.

[Schd 00] Schmid, E. W., G. Spitz, and W. Lösch (2000). *Theoretical Physics on the Personal Computer*, 2nd Ed., Springer-Verlag, Berlin.

[Shannon 48] Shannon, C. E. (1948). *A mathematical theory of communication, Bell System Tech. J.* **27,** 379.

[Shar] Sharov, A., *Quantitative Population Ecology*, www.gypsymoth.ento.vt.edu/~sharov/PopEcol/.

[Shaw 92] Shaw C. T. (1992), *Using Computational Fluid Dynamics*, Prentice Hall, Englewood Cliffs. NJ.

[S&T 93] Singh, P. P., and W. J. Thompson (1993), *Exploring the complex plane: Green's functions, Hilbert transforms, analytic continuation*, Comput. Phys. **7**, 388.

[Sipp 96] Sipper., M. (1997). *Evolution of Parallel Cellular Machines* Springer-Verlag, Heidelberg; www.cs.bgu.ac.il/~sipper/ca.html; *Cellular Automata*, cell-auto.com/.

[Smi 91] Smith, D. N. (1991). *Concepts of Object-Oriented Programming*, McGraw-Hill, New York.

[Smi 99] Smith, S. W. (1999). *The Scientist and Engineer's Guide to Digital Signal Processing*, California Technical Publishing, San Diego.

[Sterl 99] Sterling, T., J. Salmon, D. Becker, and D. Savarese (1999), *How to Build a Beowulf*, MIT Press, Cambridge, MA.

[Stez 73] Stetz, A., J. Carroll, N. Chirapatpimol, M. Dixit, G. Igo, M. Nasser, D. Ortendahl, and V. Perez-Mendez (1973). *Determination of the axial vector form factor in the radiative decay of the pion*, LBL 1707. Paper presented at the Symposium of the Division of Nuclear Physics, Washington, DC, April 1973.

[Sull 00] Sullivan, D. (2000). *Electromagnetic Simulations Using the FDTD Methods*, IEEE Press, New York.

[SunJ] Sun Java Developer's site, java.sun.com/.

[SGE] Sun N1 Grid Engine, www.sun.com/software/gridware/.

[SUSE] The openSUSE Project, en.opensuse.org/Welcome_to_openSUSE.org.

[Tab 89] Tabor, M. (1989). *Chaos and Integrability in Nonlinear Dynamics*, John Wiley, New York.

[Taf 89] Taflove, A., and S. Hagness. (2000). *Computational Electrodynamics: The Finite Difference Time Domain Method*, 2nd Ed., Artech House, Boston.

[Tait 90] Tait, R. N., T. Smy, and M. J. Brett (1990). *A ballistic deposition model for films evaporated over topography*, Thin Solid Films **187**, 375.

[Thij 99] Thijssen J. M. (1999). *Computational Physics*, Cambridge University Press, Cambridge, UK.

[Thom 92] Thompson, W. J. (1992), *Computing for Scientists and Engineers*, John Wiley, New York.

[Tick 04] Tickner, J. (2004), *Simulating nuclear particle transport in stochastic media using Perlin noise functions*, Nuclear Instrum. Methods B, **203**, 124.

[Torque] *TORQUE Resource Manager*, www.clusterresources.com/pages/products/torque-resource-manager.php.

[UCES] Undergraduate Computational Engineering and Science, www.krellinst.org/UCES/.

[Vall 00] Vallée, O. (2000). *Comment on a quantum bouncing ball by Julio Gea Banacloche*, Am. J. Phys. **68**, 672.

[VdeV 94] van de Velde, E. F. (1994). *Concurrent Scientific Computing*, Springer-Verlag, New York.

[VdB 99] van den Berg, J. C., Ed. (1999). *Wavelets in Physics*, Cambridge University Press, Cambridge. UK.

[Vida 99] Vidakovic, B. (1999). *Statistical Modeling by Wavelets*, John Wiley, New York.

[Viss 91] Visscher, P. B. (1991). *A fast explicit algorithm for the time-dependent Schrödinger equation*, Comput. Phys. **5**, 596.

[Vold 59] Vold, M. J. (1959), *Microscopic and macroscopic compaction of cohesive powders*, J. Colloid. Sci. **14**, 168.

[Volt 26] Volterra, V. (1926), *Variazioni e fluttuazioni del numero d'individui in specie animali conviventi*, Mem. R. Accad. Naz. dei Lincei. Ser. VI, **2**.

[Ward 04] Ward, D. W, and K. A. Nelson (2004). *Finite Difference Time Domain (FDTD) Simulations of Electromagnetic Wave Propagation Using a Spreadsheet*, ArXiv Phys. 0402091, 1–8.

[WL 04] Landau, D. P, S.-H. Tsai, and M. Exler (2004). *A new approach to Monte Carlo simulations in statistical physics: Wang–Landau sampling*, Am. J. Phys. **72**, 1294. Landau, D. P, and F. Wang (2001). *Determining the density of states for classical statistical models: A random walk algorithm to produce a flat histogram*, Phys. Rev. E **64**, 056101.

[WW 04] Warburton, R. D. H., and J. Wang (2004). *Analysis of asymptotic projectile motion with air resistance using the Lambert W function*, Am. J. Phys. **72**, 1404.

[Whine 92] Whineray, J. (1992). *An energy representation approach to the quantum bouncer*, Am. J. Phys. **60**, 948–950.

[Wiki] Wikipedia, the free encyclopedia, en.wikipedia.org/.

[Will 97] Williams, G. P. (1997). *Chaos Theory Tamed*, Joseph Henry Press, Washington, DC.

[W&S 83] Witten, T. A., and L. M. Sander (1981). *Diffusion-limited aggregation, a kinetic critical phenomenon*, Phys. Rev. Lett. **47**, 1400; (1983); *Diffusion-limited aggregation in three dimensions*, Phys. Rev. B **27**, 5686.

[Wolf 85] Wolf, A., J. B. Swift, H. L. Swinney, and J. A. Vastano, (1985). *Determining Lyapunov exponents from a time series*, Physica D, **16**, 285.

[Wolf 83] Wolfram S. (1983). *Statistical mechanics of cellular automata*, Rev. Mod. Phys. **55**, 601.

[XWIN32] X-Win32, a focused PC X server, www.starnet.com/products/xwin32/.

[Yang 52] Yang, C. N. (1952). *The Spontaneous Magnetization of a Two-Dimensional Ising Model*, Phys. Rev. **85**, 809.

[Yee 66] Yee, K. (1966). *Numerical solution of initial value problems involving Maxwell's equations in isotropic media*, IEEE Trans. Antennas Propagation **AP-14**, 302.

[Z&K 65] Zabusky, N. J., and M. D. Kruskal (1965). *Interaction of "solitons" in a collisionless plasma and the recurrence of initial states*, Phys. Rev. Lett. **15**, 240.

[Zucker] Zucker, M., *The Perlin noise FAQ*; www.cs.cmu.edu/~mzucker/code/perlin-noise-math-faq.html; see also Jönsson, A., *Generating Perlin Noise*, www.angelcode.com/dev/perlin/perlin.asp.

INDEX